IGMacdonald

June 1981

Differential Geometry,
Lie Groups,
and
Symmetric Spaces

This is a volume in
PURE AND APPLIED MATHEMATICS

A Series of Monographs and Textbooks

Editors: Samuel Eilenberg and Hyman Bass

A list of recent titles in this series appears at the end of this volume.

Differential Geometry, Lie Groups, *and* Symmetric Spaces

Sigurdur Helgason

Department of Mathematics
Massachusetts Institute of Technology
Cambridge, Massachusetts

ACADEMIC PRESS New York San Francisco London 1978
A Subsidiary of Harcourt Brace Jovanovich, Publishers

ACADEMIC PRESS, INC.
111 Fifth Avenue, New York, New York 10003

United Kingdom Edition published by
ACADEMIC PRESS, INC. (LONDON) LTD.
24/28 Oval Road, London NW1 7DX

Library of Congress Cataloging in Publication Data

Helgason, Sigurdur, Date
Differential geometry, Lie groups, and symmetric spaces.

(Pure and applied mathematics, a series of monographs and textbooks ;)
Includes bibliographies and index.
1. Geometry, Differential. 2. Lie groups. 3. Symmetric spaces. I. Title. II. Series.
QA3.P8 [QA641] 510'.8s [516'.36] 78–17916
ISBN 0–12–338460–5

PRINTED IN THE UNITED STATES OF AMERICA

CONTENTS

CHAPTER III

Structure of Semisimple Lie Algebras

CHAPTER IV

Symmetric Spaces

CHAPTER V

Decomposition of Symmetric Spaces

CHAPTER VI

Symmetric Spaces of the Noncompact Type

CHAPTER VII

Symmetric Spaces of the Compact Type

CHAPTER VIII

Hermitian Symmetric Spaces

CHAPTER IX

Structure of Semisimple Lie Groups

CHAPTER X

The Classification of Simple Lie Algebras and of Symmetric Spaces

PREFACE

The present book is intended as a textbook and a reference work on the three topics in the title. Together with a volume in progress on "Groups and Geometric Analysis" it supersedes my "Differential Geometry and Symmetric Spaces," published in 1962. Since that time several branches of the subject, particularly the function theory on symmetric spaces, have developed substantially. I felt that an expanded treatment might now be useful.

This first volume is an extensive revision of a part of "Differential Geometry and Symmetric Spaces." Apart from numerous minor changes the following material has been added:

> Chapter I, §15; Chapter II, §7-§8; Chapter III, §8; Chapter VII, §§7, 10, 11 and most of §2 and of §8; Chapter VIII, part of §7; all of Chapter IX and most of Chapter X. Many new exercises have been added, and solutions to the old and new exercises are now included and placed toward the end of the book.

The book begins with a general self-contained exposition of *differential* and *Riemannian geometry*, discussing affine connections, exponential mapping, geodesics, and curvature. Chapter II develops the basic theory of *Lie groups* and *Lie algebras*, homogeneous spaces, the adjoint group, etc. The Lie groups that are locally isomorphic to products of simple groups are called *semisimple*. These Lie groups have an extremely rich structure theory which at an early stage led to their complete classification, and which presumably accounts for their pervasive influence on present-day mathematics. Chapter III deals with their preliminary structure theory with emphasis on compact real forms.

Chapter IV is an introductory geometric study of *symmetric spaces*. According to its original definition, a symmetric space is a Riemannian manifold whose curvature tensor is invariant under all parallel translations. The theory of symmetric spaces was initiated by É. Cartan in 1926 and was vigorously developed by him in the late 1920s. By their definition, symmetric spaces form a special topic in Riemannian geometry; their theory, however, has merged with the theory of semisimple Lie groups. This circumstance is the source of very detailed and extensive information about these spaces. They can therefore often serve as examples on the basis of which general conjectures in differential geometry can be made and tested.

The definition above does not immediately suggest the special nature of symmetric spaces (especially if one recalls that all Riemannian manifolds and all Kähler manifolds possess tensor fields invariant under parallelism). However, the theory leads to the remarkable fact that symmetric spaces are locally just the Riemannian manifolds of the form $\boldsymbol{R}^n \times G/K$ where $\boldsymbol{R}^n$ is a Euclidean n-space, G is a semisimple Lie group that has an involutive automorphism whose fixed point set is the (essentially) compact group K, and G/K is provided with a G-invariant Riemannian structure. É. Cartan's classification of all real simple Lie algebras now led him quickly to an explicit classification of symmetric spaces in terms of the classical and exceptional simple Lie groups. On the other hand, the semisimple Lie group G (or rather the local isomorphism class of G) above is completely arbitrary; in this way valuable geometric tools become available to the theory of semisimple Lie groups. In addition, the theory of symmetric spaces helps to unify and explain in a general way various phenomena in classical geometries. Thus the isomorphisms that occur among the classical groups of low dimensions are geometrically interpreted by means of isometries; the analogy between the spherical geometries and the hyperbolic geometries is a special case of a general duality for symmetric spaces.

In Chapter V we give the local decomposition of a symmetric space into $\boldsymbol{R}^n$ and the two main types of symmetric spaces, the compact type and the noncompact type. These dual types are already distinguished by the sign of their sectional curvature. In Chapter VI we study the symmetric spaces of noncompact type. Since these spaces are completely determined by their isometry group, this chapter is primarily a global study of noncompact semisimple Lie groups. In Chapter IX this study is carried quite a bit further in the form of Cartan, Iwasawa, Bruhat, and Jordan decompositions.

In Chapter VII we derive topological and differential geometric properties of the compact symmetric space U/K by studying the isotropy action of K on U/K and on its tangent space at the origin. Chapter VIII deals with Hermitian symmetric spaces; we are primarily concerned with the noncompact ones and the Cartan-Harish-Chandra representation of these as bounded domains.

The book concludes with a classification of symmetric spaces by means of the Killing-Cartan classification of simple Lie algebras over $\boldsymbol{C}$

and Cartan's classification of simple Lie algebras over ***R***. The latter is carried out by means of Kač's classification of finite-order automorphisms of simple Lie algebras over ***C***.

Each chapter begins with a short summary and ends with references to source material. Given the enormity of the subject, I am aware that the result is at best an approximation as regards completeness and accuracy. Nevertheless, I hope that the notes will help the serious student gain a historical perspective, particularly as regards Cartan's magnificent papers on Lie groups and symmetric spaces, which are found in the two first volumes of his collected works. For example, he can witness Cartan's rather informal arguments in his climactic paper [10], written at the age of 58, leading him to a global classification of symmetric spaces; an example of a different kind is Cartan's paper in *Leipziger Berichte* (1893) where he indicates models of the exceptional groups as contact transformations or as invariance groups of Pfaffian equations and which, to my knowledge, have never been verified in print. Being the first such models, they have distinct historical interest although simpler models are now known.

In addition to papers and books utilized in the text, the bibliography lists many items on topics that are at best only briefly discussed in the text, but are nevertheless closely related, for example, pseudo-Riemannian symmetric spaces, trisymmetric spaces, reflexion spaces, homogeneous domains, discrete isometry groups, cohomology and Betti numbers of locally symmetric spaces. This part of the bibliography is selective, and no completeness is intended; in particular, papers on analysis and representation theory related to the topics of "Groups and Geometric Analysis" are not listed unless used in the present volume.

This book grew out of lectures given at the University of Chicago in 1958, at Columbia University in 1959-1960, and at various times at MIT since then. At Columbia I had the privilege of many long and informative discussions with Harish-Chandra, to whom I am deeply grateful. I am also indebted to A. Korányi, K. de Leeuw, E. Luft, H. Federer, I. Namioka, and M. Flensted-Jensen, who read parts of the manuscript and suggested several improvements. I want also to thank H.-C. Wang for putting the material in Exercise A.9, Chapter VI at my disposal. Finally, I am most grateful to my friend and colleague Victor Kač who provided me with an account of his method for classifying automorphisms of finite order (§5, Chapter X) of which only a short sketch was available in print.

With the sequel to this book in mind I will be grateful to readers who take the trouble of bringing errors in the text to my attention.

SUGGESTIONS TO THE READER

Since this book is intended for readers with varied backgrounds, we give some suggestions for its use.

Introductory Differential Geometry. Chapter I, Chapter IV, §1, and Chapter VIII, §1-§3 can be read independently of the rest of the book. These 120-odd pages, including the offered exercises, have on occasion served as the text for a one-term course on differential geometry, with only advanced calculus and some point set topology as prerequisites.

Introduction to Lie Groups. Chapter I, §1-§6, Chapter II, and Chapter III could similarly be used for a one-term course on Lie groups, assuming some familiarity with topological groups.

The chapters are rather independent after the fourth one and could for the most part be read in any order.

Exercises. Each chapter ends with a few exercises. Some of these furnish examples illuminating the theory developed in the text, while others produce extensions and ramifications of the theory. With a few possible exceptions (indicated with a star) the exercises can be worked out with methods from the text. Since the exercises present additional material and since some exercise groups furnish suitable topics for student seminars, I felt that leaving out the solutions would be counterproductive and might turn the exercises into unnecessary obstacles. Accordingly, solutions are provided at the end of the book which each reader can use to the extent he wishes.

S. Helgason

TENTATIVE CONTENTS OF THE SEQUEL

Groups and Geometric Analysis

CHAPTER I

ELEMENTARY DIFFERENTIAL GEOMETRY

This introductory chapter divides in a natural way into three parts: §1-§3 which deal with tensor fields on manifolds, §4-§8 which treat general properties of affine connections, and §9-§14 which give an introduction to Riemannian geometry with some emphasis on topics needed for the later treatment of symmetric spaces.

§1-§3. When a Euclidean space is stripped of its vector space structure and only its differentiable structure retained, there are many ways of piecing together domains of it in a smooth manner, thereby obtaining a so-called differentiable manifold. Local concepts like a differentiable function and a tangent vector can still be given a meaning whereby the manifold can be viewed "tangentially," that is, through its family of tangent spaces as a curve in the plane is, roughly speaking, determined by its family of tangents. This viewpoint leads to the study of tensor fields, which are important tools in local and global differential geometry. They form an algebra $\mathfrak{D}(M)$, the mixed tensor algebra over the manifold M. The alternate covariant tensor fields (the differential forms) form a submodule $\mathfrak{A}(M)$ of $\mathfrak{D}(M)$ which inherits a multiplication from $\mathfrak{D}(M)$, the exterior multiplication. The resulting algebra is called the Grassmann algebra of M. Through the work of É. Cartan the Grassmann algebra with the exterior differentiation d has become an indispensable tool for dealing with submanifolds, these being analytically described by the zeros of differential forms. Moreover, the pair $(\mathfrak{A}(M), d)$ determines the cohomology of M via de Rham's theorem, which however will not be dealt with here.

§4-§8. The concept of an affine connection was first defined by Levi-Civita for Riemannian manifolds, generalizing significantly the notion of parallelism for Euclidean spaces. On a manifold with a countable basis an affine connection always exists (see the exercises following this chapter). Given an affine connection on a manifold M there is to each curve $\gamma(t)$ in M associated an isomorphism between any two tangent spaces $M_{\gamma(t_1)}$ and $M_{\gamma(t_2)}$. Thus, an affine connection makes it possible to relate tangent spaces at distant points of the manifold. If the tangent vectors of the curve $\gamma(t)$ all correspond under these isomorphisms we have the analog of a straight line, the so-called geodesic. The theory of affine connections mainly amounts to a study of the mappings $\text{Exp}_p : M_p \rightarrow M$ under which straight lines (or segments of them) through the origin in the tangent space M_p correspond to geodesics through p in M. Each mapping Exp_p is a diffeomorphism of a neighborhood of 0 in M_p into M, giving the so-called normal coordinates at p. Some other local properties of Exp_p are given in §6, the existence of convex neighborhoods and a formula for the differential of Exp_p.

An affine connection gives rise to two important tensor fields, the curvature tensor field and the torsion tensor field which in turn describe the affine connection through É. Cartan's structural equations [(6) and (7), §8)].

§9-§14. A particularly interesting tensor field on a manifold is the so-called Riemannian structure. This gives rise to a metric on the manifold in a canonical fashion. It also determines an affine connection on the manifold, the Riemannian connection; this affine connection has the property that the geodesic forms the shortest curve between any two (not too distant) points. The relation between the metric and geodesics is further developed in §9-§10. The treatment is mainly based on the structural equations of É. Cartan and is independent of the Calculus of Variations.

The higher-dimensional analog of the Gaussian curvature of a surface was discovered by Riemann. Riemann introduced a tensor field which for any pair of tangent vectors at a point measures the corresponding sectional curvature, that is, the Gaussian curvature of the surface generated by the geodesics tangent to the plane spanned by the two vectors. Of particular interest are Riemannian manifolds for which the sectional curvature always has the same sign. The irreducible symmetric spaces are of this type. Riemannian manifolds of negative curvature are considered in §13 owing to their importance in the theory of symmetric spaces. Much progress has been made recently in the study of Riemannian manifolds whose sectional curvature is bounded from below by a constant > 0. However, no discussion of these is given since it is not needed in later chapters. The next section deals with totally geodesic submanifolds which are characterized by the condition that a geodesic tangent to the submanifold at a point lies entirely in it. In contrast to the situation for general Riemannian manifolds, totally geodesic submanifolds are a common occurrence for symmetric spaces.

§1. Manifolds

Let $\boldsymbol{R}^m$ and $\boldsymbol{R}^n$ denote two Euclidean spaces of m and n dimensions, respectively. Let O and O' be open subsets, $O \subset \boldsymbol{R}^m$, $O' \subset \boldsymbol{R}^n$ and suppose φ is a mapping of O into O'. The mapping φ is called *differentiable* if the coordinates $y_j(\varphi(p))$ of $\varphi(p)$ are differentiable (that is, indefinitely differentiable) functions of the coordinates $x_i(p)$, $p \in O$. The mapping φ is called *analytic* if for each point $p \in O$ there exists a neighborhood U of p and n power series P_j $(1 \leqslant j \leqslant n)$ in m variables such that $y_j(\varphi(q)) = P_j(x_1(q) - x_1(p), ..., x_m(q) - x_m(p))$ $(1 \leqslant j \leqslant n)$ for $q \in U$. A differentiable mapping $\varphi : O \rightarrow O'$ is called *a diffeomorphism* of O onto O' if $\varphi(O) = O'$, φ is one-to-one, and the inverse mapping φ^{-1} is differentiable. In the case when $n = 1$ it is customary to replace the term "mapping" by the term "function."

An analytic function on $\boldsymbol{R}^m$ which vanishes on an open set is identically 0. For differentiable functions the situation is completely different. In fact, if A and B are disjoint subsets of $\boldsymbol{R}^m$, A compact and B closed, then there exists a differentiable function φ which is identically 1 on A and identically 0 on B. The standard procedure for constructing such a function φ is as follows:

Let $0 < a < b$ and consider the function f on $\boldsymbol{R}$ defined by

$$f(x) = \begin{cases} \exp\left(\dfrac{1}{x-b} - \dfrac{1}{x-a}\right) & \text{if } a < x < b, \\ 0 & \text{otherwise.} \end{cases}$$

Then f is differentiable and the same holds for the function

$$F(x) = \int_x^b f(t)\, dt \Big/ \int_a^b f(t)\, dt,$$

which has value 1 for $x \leqslant a$ and 0 for $x \geqslant b$. The function ψ on $\boldsymbol{R}^m$ given by

$$\psi(x_1, ..., x_m) = F(x_1^2 + ... + x_m^2)$$

is differentiable and has values 1 for $x_1^2 + ... + x_m^2 \leqslant a$ and 0 for $x_1^2 + ... + x_m^2 \geqslant b$. Let S and S' be two concentric spheres in $\boldsymbol{R}^m$, S' lying inside S. Starting from ψ we can by means of a linear transformation of $\boldsymbol{R}^m$ construct a differentiable function on $\boldsymbol{R}^m$ with value 1 in the interior of S' and value 0 outside S. Turning now to the sets A and B we can, owing to the compactness of A, find finitely many spheres S_i $(1 \leqslant i \leqslant n)$, such that the corresponding open balls B_i $(1 \leqslant i \leqslant n)$, form a covering of A (that is, $A \subset \bigcup_{i=1}^n B_i$) and such that the closed balls $\bar{B}_i$ $(1 \leqslant i \leqslant n)$ do not intersect B. Each sphere S_i can be shrunk to a concentric sphere S_i' such that the corresponding open balls B_i' still form a covering of A. Let ψ_i be a differentiable function on $\boldsymbol{R}^m$ which is identically 1 on B_i' and identically 0 in the complement of B_i. Then the function

$$\varphi = 1 - (1 - \psi_1)(1 - \psi_2) ... (1 - \psi_n)$$

is a differentiable function on $\boldsymbol{R}^m$ which is identically 1 on A and identically 0 on B.

Let M be a topological space. We assume that M satisfies the Hausdorff separation axiom which states that any two different points in M can be separated by disjoint open sets. An *open chart* on M is a pair (U, φ) where U is an open subset of M and φ is a homeomorphism of U onto an open subset of $\boldsymbol{R}^m$.

Definition. Let M be a Hausdorff space. A *differentiable structure* on M of dimension m is a collection of open charts $(U_\alpha, \varphi_\alpha)_{\alpha \in A}$ on M where $\varphi_\alpha(U_\alpha)$ is an open subset of $\boldsymbol{R}^m$ such that the following conditions are satisfied:

(M_1) $M = \bigcup_{\alpha \in A} U_\alpha$.

(M_2) For each pair $\alpha, \beta \in A$ the mapping $\varphi_\beta \circ \varphi_\alpha^{-1}$ is a differentiable mapping of $\varphi_\alpha(U_\alpha \cap U_\beta)$ onto $\varphi_\beta(U_\alpha \cap U_\beta)$.

(M_3) The collection $(U_\alpha, \varphi_\alpha)_{\alpha \in A}$ is a maximal family of open charts for which (M_1) and (M_2) hold.

A *differentiable manifold* (or C^∞ *manifold* or simply *manifold*) of dimension m is a Hausdorff space with a differentiable structure of dimension m. If M is a manifold, a *local chart* on M (or a *local coordinate system* on M) is by definition a pair $(U_\alpha, \varphi_\alpha)$ where $\alpha \in A$. If $p \in U_\alpha$ and $\varphi_\alpha(p) = (x_1(p), \ldots, x_m(p))$, the set U_α is called a *coordinate neighborhood* of p and the numbers $x_i(p)$ are called *local coordinates* of p. The mapping $\varphi_\alpha : q \to (x_1(q), \ldots, x_m(q))$, $q \in U_\alpha$, is often denoted $\{x_1, \ldots, x_m\}$.

Remark 1. Condition (M_3) will often be cumbersome to check in specific instances. It is therefore important to note that the condition (M_3) is not essential in the definition of a manifold. In fact, if only (M_1) and (M_2) are satisfied, the family $(U_\alpha, \varphi_\alpha)_{\alpha \in A}$ can be extended in a unique way to a larger family $\mathfrak{M}$ of open charts such that (M_1), (M_2), and (M_3) are all fulfilled. This is easily seen by defining $\mathfrak{M}$ as the set of all open charts (V, φ) on M satisfying: (1) $\varphi(V)$ is an open set in $\boldsymbol{R}^m$; (2) for each $\alpha \in A$, $\varphi_\alpha \circ \varphi^{-1}$ is a diffeomorphism of $\varphi(V \cap U_\alpha)$ onto $\varphi_\alpha(V \cap U_\alpha)$.

Remark 2. If we let $\boldsymbol{R}^m$ mean a single point for $m = 0$, the preceding definition applies. The manifolds of dimension 0 are then the discrete topological spaces.

Remark 3. A manifold is connected if and only if it is pathwise connected. The proof is left to the reader.

An *analytic structure* of dimension m is defined in a similar fashion. In (M_2) we just replace "differentiable" by "analytic." In this case M is called an *analytic manifold.*

In order to define a *complex manifold* of dimension m we replace $\boldsymbol{R}^m$ in the definition of differentiable manifold by the m-dimensional complex space $\boldsymbol{C}^m$. The condition (M_2) is replaced by the condition that the m coordinates of $\varphi_\beta \circ \varphi_\alpha^{-1}(p)$ should be holomorphic functions of the coordinates of p. Here a function $f(z_1, \ldots, z_m)$ of m complex variables is called *holomorphic* if at each point $(z_1^0, \ldots, z_m^0)$ there exists a power series

$$\sum a_{n_1 \ldots n_m} (z_1 - z_1^0)^{n_1} \ldots (z_m - z_m^0)^{n_m},$$

which converges absolutely to $f(z_1, \ldots, z_m)$ in a neighborhood of the point.

The manifolds dealt with in the later chapters of this book (mostly

Lie groups and their coset spaces) are analytic manifolds. From Remark 1 it is clear that we can always regard an analytic manifold as a differentiable manifold. It is often convenient to do so because, as pointed out before for $\boldsymbol{R}^m$, the class of differentiable functions is much richer than the class of analytic functions.

Let f be a real-valued function on a C^∞ manifold M. The function f is called *differentiable* at a point $p \in M$ if there exists a local chart $(U_\alpha, \varphi_\alpha)$ with $p \in U_\alpha$ such that the composite function $f \circ \varphi_\alpha^{-1}$ is a differentiable function on $\varphi_\alpha(U_\alpha)$. The function f is called *differentiable* if it is differentiable at each point $p \in M$. If M is analytic, the function f is said to be *analytic* at $p \in M$ if there exists a local chart $(U_\alpha, \varphi_\alpha)$ with $p \in U_\alpha$ such that $f \circ \varphi_\alpha^{-1}$ is an analytic function on the set $\varphi_\alpha(U_\alpha)$.

Let M be a differentiable manifold of dimension m and let $\mathfrak{F}$ denote the set of all differentiable functions on M. The set $\mathfrak{F}$ has the following properties:

($\mathfrak{F}_1$) Let $\varphi_1, \ldots, \varphi_r \in \mathfrak{F}$ and let u be a differentiable function on $\boldsymbol{R}^r$. Then $u(\varphi_1, \ldots, \varphi_r) \in \mathfrak{F}$.

($\mathfrak{F}_2$) Let f be a real function on M such that for each $p \in M$ there exists a function $g \in \mathfrak{F}$ which coincides with f in some neighborhood of p. Then $f \in \mathfrak{F}$.

($\mathfrak{F}_3$) For each $p \in M$ there exist m functions $\varphi_1, \ldots, \varphi_m \in \mathfrak{F}$ and an open neighborhood U of p such that the mapping $q \to (\varphi_1(q), \ldots, \varphi_m(q))$ $(q \in U)$ is a homeomorphism of U onto an open subset of $\boldsymbol{R}^m$. The set U and the functions $\varphi_1, \ldots, \varphi_m$ can be chosen in such a way that each $f \in \mathfrak{F}$ coincides on U with a function of the form $u(\varphi_1, \ldots, \varphi_m)$ where u is a differentiable function on $\boldsymbol{R}^m$.

The properties ($\mathfrak{F}_1$) and ($\mathfrak{F}_2$) are obvious. To establish ($\mathfrak{F}_3$) we pick a local chart $(U_\alpha, \varphi_\alpha)$ such that $p \in U_\alpha$ and write $\varphi_\alpha(q) = (x_1(q), \ldots, x_m(q)) \in \boldsymbol{R}^m$ for $q \in U_\alpha$. Let S be a compact neighborhood of $\varphi_\alpha(p)$ in $\boldsymbol{R}^m$ such that S is contained in the open set $\varphi_\alpha(U_\alpha)$. Then as shown earlier, there exists a differentiable function ψ on $\boldsymbol{R}^m$ such that ψ has compact support† contained in $\varphi_\alpha(U_\alpha)$ and such that $\psi(s) = 1$ for all $s \in S$. Let $U = \varphi_\alpha^{-1}(\mathring{S})$ where $\mathring{S}$ is the interior of S and define the function φ_i $(1 \leqslant i \leqslant m)$ on M by

$$\varphi_i(q) = \begin{cases} 0 & \text{if } q \notin U_\alpha, \\ x_i(q)\, \psi(\varphi_\alpha(q)) & \text{if } q \in U_\alpha. \end{cases}$$

Then the set U and the functions $\varphi_1, \ldots, \varphi_m$ have the property stated in ($\mathfrak{F}_3$). In fact, if $f \in \mathfrak{F}$, then the function $f \circ \varphi_\alpha^{-1}$ is differentiable on the set $\varphi_\alpha(U_\alpha)$.

† The *support* of a function is the closure of the set where the function is different from 0.

Proposition 1.1. *Suppose M is a Hausdorff space and m an integer > 0. Assume $\mathfrak{F}$ is a collection of real-valued functions on M with the properties $\mathfrak{F}_1$, $\mathfrak{F}_2$, and $\mathfrak{F}_3$. Then there exists a unique collection $(U_\alpha, \varphi_\alpha)_{\alpha \in A}$ of open charts on M such that (M_1), (M_2), and (M_3) are satisfied and such that the differentiable functions on the resulting manifold are precisely the members of $\mathfrak{F}$.*

For the proof we select for each $p \in M$ the functions $\varphi_1, \ldots, \varphi_m$ and the neighborhood U of p given by $\mathfrak{F}_3$. Putting $U_\alpha = U$ and $\varphi_\alpha(q) = (\varphi_1(q), \ldots, \varphi_m(q))$ $(q \in U)$ we obtain a collection $(U_\alpha, \varphi_\alpha)_{\alpha \in A}$ of open charts on M satisfying (M_1). The condition (M_2) is also satisfied in virtue of $\mathfrak{F}_3$. As remarked earlier, the collection $(U_\alpha, \varphi_\alpha)_{\alpha \in A}$ can then be extended to a collection $(U_\alpha, \varphi_\alpha)_{\alpha \in A^*}$ which satisfies (M_1), (M_2), and (M_3). This induces a differentiable structure on M and each $g \in \mathfrak{F}$ is obviously a differentiable function. On the other hand, suppose that f is a differentiable function on the constructed manifold. If $p \in M$, there exists a local chart $(U_\alpha, \varphi_\alpha)$ where $\alpha \in A^*$ such that $p \in U_\alpha$ and such that $f \circ \varphi_\alpha^{-1}$ is a differentiable function on an open neighborhood of $\varphi_\alpha(p)$. Owing to (M_2) we may assume that $\alpha \in A$. There exists a differentiable function u on $\boldsymbol{R}^m$ such that $f \circ \varphi_\alpha^{-1}(x) = u(x_1, \ldots, x_m)$ for all points $x = (x_1, \ldots, x_m)$ in some open neighborhood of $\varphi_\alpha(p)$. This means (in terms of the φ_i above) that

$$f = u(\varphi_1, \ldots, \varphi_m)$$

in some neighborhood of p. Since $p \in M$ is arbitrary we conclude from $\mathfrak{F}_2$ and $\mathfrak{F}_1$ that $f \in \mathfrak{F}$. Finally, let $(V_\beta, \psi_\beta)_{\beta \in B}$ be another collection of open charts satisfying (M_1), (M_2), and (M_3) and giving rise to the same $\mathfrak{F}$. Writing $f \circ \varphi_\alpha^{-1} = f \circ \psi_\beta^{-1} \circ (\psi_\beta \circ \varphi_\alpha^{-1})$ for $f \in \mathfrak{F}$ we see that $\psi_\beta \circ \varphi_\alpha^{-1}$ is differentiable on $\varphi_\alpha(U_\alpha \cap V_\beta)$, so by the maximality (M_3), $(U_\alpha, \varphi_\alpha) \in (V_\beta, \psi_\beta)_{\beta \in B}$ and the uniqueness follows.

We shall often write $C^\infty(M)$ instead of $\mathfrak{F}$ and will sometimes denote by $C^\infty(p)$ the set of functions on M which are differentiable at p. The set $C^\infty(M)$ is an algebra over $\boldsymbol{R}$, the operations being

$$(\lambda f)(p) = \lambda f(p),$$

$$(f + g)(p) = f(p) + g(p),$$

$$(fg)(p) = f(p)\, g(p)$$

for $\lambda \in \boldsymbol{R}$, $p \in M$, $f, g \in C^\infty(M)$.

Lemma 1.2. *Let C be a compact subset of a manifold M and let V be an open subset of M containing C. Then there exists a function $\psi \in C^\infty(M)$ which is identically 1 on C, identically 0 outside V.*

This lemma has already been established in the case $M = \boldsymbol{R}^m$. We shall now show that the general case presents no additional difficulties.

Let $(U_\alpha, \varphi_\alpha)$ be a local chart on M and S a compact subset of U_α. There exists a differentiable function f on $\varphi_\alpha(U_\alpha)$ such that f is identically 1 on $\varphi_\alpha(S)$ and has compact support contained in $\varphi_\alpha(U_\alpha)$. The function F on M given by

$$F(q) = \begin{cases} f(\varphi_\alpha(q)) & \text{if } q \in U_\alpha, \\ 0 & \text{otherwise} \end{cases}$$

is a differentiable function on M which is identically 1 on S and identically 0 outside U_α. Since C is compact and V open, there exist finitely many coordinate neighborhoods $U_1, ..., U_n$ and compact sets $S_1, ..., S_n$ such that

$$C \subset \bigcup_1^n S_i, \qquad S_i \subset U_i$$

$$\left(\bigcup_1^n U_i\right) \subset V.$$

As shown previously, there exists a function $F_i \in C^\infty(M)$ which is identically 1 on S_i and identically 0 outside U_i. The function

$$\psi = 1 - (1 - F_1)(1 - F_2) ... (1 - F_n)$$

belongs to $C^\infty(M)$, is identically 1 on C and identically 0 outside V.

Let M be a C^∞ manifold and $(U_\alpha, \varphi_\alpha)_{\alpha \in A}$ a collection satisfying (M_1), (M_2), and (M_3). If U is an open subset of M, U can be given a differentiable structure by means of the open charts $(V_\alpha, \psi_\alpha)_{\alpha \in A}$ where $V_\alpha = U \cap U_\alpha$ and ψ_α is the restriction of φ_α to V_α. With this structure, U is called an *open submanifold* of M. In particular, since M is locally connected, each connected component of M is an open submanifold of M.

Let M and N be two manifolds of dimension m and n, respectively. Let $(U_\alpha, \varphi_\alpha)_{\alpha \in A}$ and $(V_\beta, \psi_\beta)_{\beta \in B}$ be collections of open charts on M and N, respectively, such that the conditions (M_1), (M_2), and (M_3) are satisfied. For $\alpha \in A$, $\beta \in B$, let $\varphi_\alpha \times \psi_\beta$ denote the mapping $(p, q) \to (\varphi_\alpha(p), \psi_\beta(q))$ of the product set $U_\alpha \times V_\beta$ into $\boldsymbol{R}^{m+n}$. Then the collection $(U_\alpha \times V_\beta, \varphi_\alpha \times \psi_\beta)_{\alpha \in A, \beta \in B}$ of open charts on the product space $M \times N$ satisfies (M_1) and (M_2) so by Remark 1, $M \times N$ can be turned into a manifold the *product* of M and N.

An immediate consequence of Lemma 1.2 is the following fact which will often be used: Let V be an open submanifold of M, f a function in $C^\infty(V)$, and p a point in V. Then there exists a function $\tilde{f} \in C^\infty(M)$ and an open neighborhood N, $p \in N \subset V$ such that f and $\tilde{f}$ agree on N.

Definition. Let M be a topological space and $V \subset M$. A *covering* of V is a collection of open subsets of M whose union contains V. A covering $\{U_\alpha\}_{\alpha \in A}$ of M is said to be *locally finite* if each $p \in M$ has a neighborhood which intersects only finitely many of the sets U_α.

Definition. A Hausdorff space M is called *paracompact* if for each covering $\{U_\alpha\}_{\alpha \in A}$ of M there exists a locally finite covering $\{V_\beta\}_{\beta \in B}$ which is a refinement of $\{U_\alpha\}_{\alpha \in A}$ (that is, each V_β is contained in some U_α).

Definition. A topological space is called *normal* if for any two disjoint closed subsets A and B there exist disjoint open subsets U and V such that $A \subset U$, $B \subset V$.

It is known that a locally compact Hausdorff space which has a countable base is paracompact and that every paracompact space is normal (see Propositions 15.1 and 15.2).

Theorem 1.3. (partition of unity). *Let M be a normal manifold and $\{U_\alpha\}_{\alpha \in A}$ a locally finite covering of M. Assume that each $\bar{U}_\alpha$ is compact. Then there exists a system $\{\varphi_\alpha\}_{\alpha \in A}$ of differentiable functions on M such that*

(i) *Each φ_α has compact support contained in U_α.*

(ii) $\varphi_\alpha \geqslant 0$, $\Sigma_{\alpha \in A}\, \varphi_\alpha = 1$.

We shall make use of the following fact (see Proposition 15.3):

Let $\{U_\alpha\}_{\alpha \in A}$ be a locally finite covering of a normal space M. Then each set U_α can be shrunk to a set V_α, such that $\bar{V}_\alpha \subset U_\alpha$ and $\{V_\alpha\}_{\alpha \in A}$ is still a covering of M.

To prove Theorem 1.3 we first shrink the U_α as indicated and thus get a new covering $\{V_\alpha\}_{\alpha \in A}$. Owing to Lemma 1.2 there exists a function $\psi_\alpha \in C^\infty(M)$ of compact support contained in U_α such that ψ_α is identically 1 on V_α and $\psi_\alpha \geqslant 0$ on M. Owing to the local finiteness the sum $\Sigma_{\alpha \in A}\, \psi_\alpha = \psi$ exists. Moreover, $\psi \in C^\infty(M)$ and $\psi(p) > 0$ for each $p \in M$. The functions $\varphi_\alpha = \psi_\alpha/\psi$ have the desired properties (i) and (ii).

The system $\{\varphi_\alpha\}_{\alpha \in A}$ is called a partition of unity *subordinate to the covering* $\{U_\alpha\}_{\alpha \in A}$. For a variation of Theorem 1.3 see Exercise A.6.

§ 2. Tensor Fields

1. Vector Fields and 1-Forms

Let A be an algebra over a field K. A *derivation* of A is a mapping $D: A \to A$ such that

(i) $D(\alpha f + \beta g) = \alpha Df + \beta Dg$ for $\alpha, \beta \in K$, $f, g \in A$;

(ii) $D(fg) = f(Dg) + (Df)g$ for $f, g \in A$.

Definition. A *vector field* X on a C^∞ manifold is a derivation of the algebra $C^\infty(M)$.

Let $\mathfrak{D}^1$ (or $\mathfrak{D}^1(M)$) denote the set of all vector fields on M. If $f \in C^\infty(M)$ and $X, Y \in \mathfrak{D}^1(M)$, then fX and $X + Y$ denote the vector fields

$$fX : g \to f(Xg), \qquad g \in C^\infty(M),$$

$$X + Y : g \to Xg + Yg, \qquad g \in C^\infty(M).$$

This turns $\mathfrak{D}^1(M)$ into a module over the ring $\mathfrak{F} = C^\infty(M)$. If $X, Y \in \mathfrak{D}^1(M)$, then $XY - YX$ is also a derivation of $C^\infty(M)$ and is denoted by the bracket $[X, Y]$. As is customary we shall often write $\theta(X)\,Y = [X, Y]$. The operator $\theta(X)$ is called the *Lie derivative* with respect to X. The bracket satisfies the *Jacobi identity* $[X, [Y, Z]] + [Y, [Z, X]] + [Z, [X, Y]] = 0$ or, otherwise written $\theta(X)\,([Y, Z]) = [\theta(X)\,Y, Z] + [Y, \theta(X)\,Z]$.

It is immediate from (ii) that if f is constant and $X \in \mathfrak{D}^1$, then $Xf = 0$. Suppose now that a function $g \in C^\infty(M)$ vanishes on an open subset $V \subset M$. Let p be an arbitrary point in V. According to Lemma 1.2 there exists a function $f \in C^\infty(M)$ such that $f(p) = 0$, and $f = 1$ outside V. Then $g = fg$ so

$$Xg = f(Xg) + g(Xf),$$

which shows that Xg vanishes at p. Since p was arbitrary, $Xg = 0$ on V. We can now define Xf on V for every function $f \in C^\infty(V)$. If $p \in V$, select $\tilde{f} \in C^\infty(M)$ such that f and $\tilde{f}$ coincide in a neighborhood of p and put $(Xf)\,(p) = (X\tilde{f})\,(p)$. The consideration above shows that this is a valid definition, that is, independent of the choice of $\tilde{f}$. This shows that a vector field on a manifold induces a vector field on any open submanifold.

On the other hand, let Z be a vector field on an open submanifold $V \subset M$ and p a point in V. Then there exists a vector field $\tilde{Z}$ on M and an open neighborhood N, $p \in N \subset V$ such that $\tilde{Z}$ and Z induce the same vector field on N. In fact, let C be any compact neighborhood of p contained in V and let N be the interior of C. Choose $\psi \in C^\infty(M)$ of compact support contained in V such that $\psi = 1$ on C. For any $g \in C^\infty(M)$, let g_V denote its restriction to V and define $\tilde{Z}g$ by

$$\tilde{Z}(g)\,(q) = \begin{cases} \psi(q)\,(Zg_V)\,(q) & \text{for } q \in V, \\ 0 & \text{if } \ q \notin V. \end{cases}$$

Then $g \to \tilde{Z}g$ is the desired vector field on M.

Now, let (U, φ) be a local chart on M, X a vector field on U, and let p be an arbitrary point in U. We put $\varphi(q) = (x_1(q), ..., x_m(q))$ $(q \in U)$, and $f^* = f \circ \varphi^{-1}$ for $f \in C^\infty(M)$. Let V be an open subset of U such that $\varphi(V)$ is an open ball in $\boldsymbol{R}^m$ with center $\varphi(p) = (a_1, \ldots, a_m)$. If $(x_1, \ldots, x_m) \in \varphi(V)$, we have

$$\begin{aligned} f^*(x_1, \ldots, x_m) &\\ = f^*(a_1, \ldots, a_m) &+ \int_0^1 \frac{\partial}{\partial t} f^*(a_1 + t(x_1 - a_1), \ldots, a_m + t(x_m - a_m))\, dt \\ = f^*(a_1, \ldots, a_m) &+ \sum_{j=1}^m (x_j - a_j) \int_0^1 f_j^*(a_1 + t(x_1 - a_1), \ldots, a_m + t(x_m - a_m))dt. \end{aligned}$$

(Here f_j^* denotes the partial derivative of f^* with respect to the jth argument.) Transferring this relation back to M we obtain

$$f(q) = f(p) + \sum_{i=1}^m (x_i(q) - x_i(p))\, g_i(q) \qquad (q \in V), \tag{1}$$

where $g_i \in C^\infty(V)$ $(1 \leqslant i \leqslant m)$, and

$$g_i(p) = \left(\frac{\partial f^*}{\partial x_i}\right)_{\varphi(p)}.$$

It follows that

$$(Xf)(p) = \sum_{i=1}^m \left(\frac{\partial f^*}{\partial x_i}\right)_{\varphi(p)} (Xx_i)(p) \qquad \text{for } p \in U. \tag{2}$$

The mapping $f \to (\partial f^*/\partial x_i) \circ \varphi$ $(f \in C^\infty(U))$ is a vector field on U and is denoted $\partial/\partial x_i$. We write $\partial f/\partial x_i$ instead of $\partial/\partial x_i(f)$. Now, by (2)

$$X = \sum_{i=1}^m (Xx_i) \frac{\partial}{\partial x_i} \qquad \text{on } U. \tag{3}$$

Thus, $\partial/\partial x_i$ $(1 \leqslant i \leqslant m)$ is a basis of the module $\mathfrak{D}^1(U)$.

For $p \in M$ and $X \in \mathfrak{D}^1$, let X_p denote the linear mapping X_p: $f \to (Xf)(p)$ of $C^\infty(p)$ into $\boldsymbol{R}$. The set $\{X_p : X \in \mathfrak{D}^1(M)\}$ is called the *tangent space* to M at p; it will be denoted by $\mathfrak{D}^1(p)$ or M_p and its elements are called the *tangent vectors* to M at p. Relation (2) shows that M_p is a vector space over $\boldsymbol{R}$ spanned by the m linearly independent vectors

$$e_i : f \to \left(\frac{\partial f^*}{\partial x_i}\right)_{\varphi(p)}, \qquad f \in C^\infty(M).$$

This tangent vector e_i will often be denoted by $(\partial/\partial x_i)_p$. A linear mapping $L : C^\infty(p) \to \boldsymbol{R}$ is a tangent vector to M at p if and only if the condition

$L(fg) = f(p)\, L(g) + g(p)\, L(f)$ is satisfied for all $f, g \in C^\infty(p)$. In fact, the necessity of the condition is obvious and the sufficiency is a simple consequence of (1). Thus, a vector field X on M can be identified with a collection $X_p (p \in M)$ of tangent vectors to M with the property that for each $f \in C^\infty(M)$ the function $p \to X_p f$ is differentiable.

Suppose the manifold M is analytic. The vector field X on M is then called *analytic* at p if Xf is analytic at p whenever f is analytic at p.

Remark. Let V be a finite-dimensional vector space over $\boldsymbol{R}$. If $X_1, \ldots, X_n$ is any basis of V, the mapping $\sum_{i=1}^n x_i X_i \to (x_1, \ldots, x_n)$ is an open chart valid on the entire V. The resulting differentiable structure is independent of the choice of basis. If $X \in V$, the tangent space V_X is identified with V itself by the formula

$$(Yf)(X) = \left\{ \frac{d}{dt} f(X + tY) \right\}_{t=0}, \qquad f \in C^\infty(V),$$

which to each $Y \in V$ assigns a tangent vector to V at X.

Let A be a commutative ring with identity element, E a module over A. Let E^* denote the set of all A-linear mappings of E into A. Then E^* is an A-module in an obvious fashion. It is called the *dual* of E.

Definition. Let M be a C^∞ manifold and put $\mathfrak{F} = C^\infty(M)$. Let $\mathfrak{D}_1(M)$ denote the dual of the $\mathfrak{F}$-module $\mathfrak{D}^1(M)$. The elements of $\mathfrak{D}_1(M)$ are called the *differential 1-forms* on M (or just 1-forms on M).

Let $X \in \mathfrak{D}^1(M)$, $\omega \in \mathfrak{D}_1(M)$. Suppose that X vanishes on an open set V. Then the function $\omega(X)$ vanishes on V. In fact, if $p \in V$, there exists a function $f \in C^\infty(M)$ such that $f = 0$ in a compact neighborhood of p and $f = 1$ outside V. Then $fX = X$ and since ω is $\mathfrak{F}$-linear, $\omega(X) = f\omega(X)$. Hence $(\omega(X))(p) = 0$. This shows also that a 1-form on M induces a 1-form on any open submanifold of M. Using (3) we obtain the following lemma.

Lemma 2.1. *Let $X \in \mathfrak{D}^1(M)$ and $\omega \in \mathfrak{D}_1(M)$. If $X_p = 0$ for some $p \in M$, then the function $\omega(X)$ vanishes at p.*

This lemma shows that given $\omega \in \mathfrak{D}_1(M)$, we can define the linear function ω_p on M_p by putting $\omega_p(X_p) = (\omega(X))(p)$ for $X \in \mathfrak{D}^1(M)$. The set $\mathfrak{D}_1(p) = \{\omega_p : \omega \in \mathfrak{D}_1(M)\}$ is a vector space over $\boldsymbol{R}$.

We have seen that a 1-form on M induces a 1-form on any open submanifold. On the other hand, suppose θ is a 1-form on an open submanifold V of M and p a point in V. Then there exists a 1-form $\tilde{\theta}$ on M, and an open neighborhood N of p, $p \in N \subset V$, such that θ and $\tilde{\theta}$ induce the same 1-form on N. In fact, let C be a compact neighborhood of p contained in V and let N be the interior of C. Select $\psi \in C^\infty(M)$ of compact support contained in V such that $\psi = 1$ on C. Then a 1-form $\tilde{\theta}$ with the desired property can be defined by

$$\tilde{\theta}(X) = \psi\theta(X_V) \text{ on } V, \qquad \tilde{\theta}(X) = 0 \text{ outside } V,$$

where $X \in \mathfrak{D}^1(M)$ and X_V denotes the vector field on V induced by X.

Lemma 2.2. *The space $\mathfrak{D}_1(p)$ coincides with M_p^*, the dual of M_p.*

We already know that $\mathfrak{D}_1(p) \subset M_p^*$. Now let $\{x_1, ..., x_m\}$ be a system of coordinates valid on an open neighborhood U of p. Owing to (3), there exist 1-forms ω^i on U such that† $\omega^i(\partial/\partial x_j) = \delta_{ij}$ $(1 \leqslant i, j \leqslant m)$. Let $L \in M_p^*$, $l_i = L((\partial/\partial x_i)_p)$ and $\theta = \sum_{i=1}^m l_i\omega^i$. Then there exists a 1-form $\tilde{\theta}$ on M and an open neighborhood N of p $(N \subset U)$ such that $\tilde{\theta}$ and θ induce the same form on N. Then $(\tilde{\theta})_p = L$ and the lemma is proved.

Each $X \in \mathfrak{D}^1(M)$ induces an $\mathfrak{F}$-linear mapping $\omega \rightarrow \omega(X)$ of $\mathfrak{D}_1(M)$ into $\mathfrak{F}$. If $X_1 \neq X_2$, the induced mappings are different (due to Lemma 2.2). Thus, $\mathfrak{D}^1(M)$ can be regarded as a subset of $(\mathfrak{D}_1(M))^*$.

Lemma 2.3. *The module $\mathfrak{D}^1(M)$ coincides with the dual of the module $\mathfrak{D}_1(M)$.*

Proof. Let $F \in \mathfrak{D}_1(M)^*$. Then $F(f\omega) = fF(\omega)$ for all $f \in C^\infty(M)$ and all $\omega \in \mathfrak{D}_1(M)$. This shows that if ω vanishes on an open set V, $F(\omega)$ also vanishes on V. Let $p \in M$ and $\{x_1, ..., x_m\}$ a system of local coordinates valid on an open neighborhood U of p. Each 1-form on U can be written $\sum_{i=1}^m f_i\omega^i$ where $f_i \in C^\infty(U)$ and ω^i has the same meaning as above. It follows easily that $F(\omega)$ vanishes at p whenever $\omega_p = 0$; consequently, the mapping $\omega_p \rightarrow (F(\omega))(p)$ is a well-defined linear function on $\mathfrak{D}_1(p)$. By Lemma 2.2 there exists a unique vector $X_p \in M_p$ such that $(F(\omega))(p) = \omega_p(X_p)$ for all $\omega \in \mathfrak{D}_1(M)$. Thus, F gives rise to a family X_p $(p \in M)$ of tangent vectors to M. For each $q \in U$ we can write

$$X_q = \sum_{i=1}^m a_i(q) \left(\frac{\partial}{\partial x_i}\right)_q,$$

where $a_i(q) \in \boldsymbol{R}$. For each i $(1 \leqslant i \leqslant m)$ there exists a 1-form $\tilde{\omega}^i$ on M which coincides with ω^i in an open neighborhood N_p of p, $(N_p \subset U)$. Then $(F(\tilde{\omega}^i))(q) = \tilde{\omega}_q^i(X_q) = a_i(q)$ for $q \in N_p$. This shows that the functions a_i are differentiable. If $f \in C^\infty(M)$ and we denote the function $q \rightarrow X_qf$ $(q \in M)$ by Xf, then the mapping $f \rightarrow Xf$ is a vector field on M which satisfies $\omega(X) = F(\omega)$ for all $\omega \in \mathfrak{D}_1(M)$. This proves the lemma.

† As usual, $\delta_{ij} = 0$ if $i \neq j$, $\delta_{ij} = 1$ if $i = j$.

2. The Tensor Algebra

Let A be a commutative ring with identity element. Let I be a set and suppose that for each $i \in I$ there is given an A-module E_i. The product set $\Pi_{i \in I} E_i$ can be turned into an A-module as follows: If $e = \{e_i\}$, $e' = \{e'_i\}$ are any two elements in ΠE_i (where $e_i, e'_i \in E_i$), and $a \in A$, then $e + e'$ and ae are given by

$$(e + e')_i = e_i + e'_i, \quad (ae)_i = ae_i \qquad \text{for } i \in I.$$

The module ΠE_i is called the *direct product* of the modules E_i. The *direct sum* $\Sigma_{i \in I} E_i$ is defined as the submodule of ΠE_i consisting of those elements $e = \{e_i\}$ for which all $e_i = 0$ except for finitely many i.

Suppose now the set I is finite, say $I = \{1, ..., s\}$. A mapping $f : E_1 \times ... \times E_s \to F$ where F is an A-module is said to be *A-multilinear* if it is A-linear in each argument. The set of all A-multilinear mappings of $E_1 \times ... \times E_s$ into F is again an A-module as follows:

$$(f + f')(e_1, ..., e_s) = f(e_1, ..., e_s) + f'(e_1, ..., e_s),$$
$$(af)(e_1, ..., e_s) = a(f(e_1, ..., e_s)).$$

Suppose that all the factors E_i coincide. The A-multilinear mapping f is called *alternate* if $f(X_1, ..., X_s) = 0$ whenever at least two X_i coincide.

Now, let M be a C^∞ manifold and as usual we put $\mathfrak{F} = C^\infty(M)$. If s is an integer, $s \geqslant 1$, we consider the $\mathfrak{F}$-module

$$\mathfrak{D}^1 \times \mathfrak{D}^1 \times ... \times \mathfrak{D}^1 \qquad (s \text{ times})$$

and let $\mathfrak{D}_s$ denote the $\mathfrak{F}$-module of all $\mathfrak{F}$-multilinear mappings of $\mathfrak{D}^1 \times ... \times \mathfrak{D}^1$ into $\mathfrak{F}$. Similarly $\mathfrak{D}^r$ denotes the $\mathfrak{F}$-module of all $\mathfrak{F}$-multilinear mappings of

$$\mathfrak{D}_1 \times \mathfrak{D}_1 \times ... \times \mathfrak{D}_1 \qquad (r \text{ times})$$

into $\mathfrak{F}$. This notation is permissible since we have seen that the modules $\mathfrak{D}^1$ and $\mathfrak{D}_1$ are duals of each other. More generally, let $\mathfrak{D}^r_s$ denote the $\mathfrak{F}$-module of all $\mathfrak{F}$-multilinear mappings of

$$\mathfrak{D}_1 \times ... \times \mathfrak{D}_1 \times \mathfrak{D}^1 \times ... \times \mathfrak{D}^1 \qquad (\mathfrak{D}_1 \ r \text{ times}, \ \mathfrak{D}^1 \ s \text{ times})$$

into $\mathfrak{F}$. We often write $\mathfrak{D}^r_s(M)$ instead of $\mathfrak{D}^r_s$. We have $\mathfrak{D}^r_0 = \mathfrak{D}^r$, $\mathfrak{D}^0_s = \mathfrak{D}_s$ and we put $\mathfrak{D}^0_0 = \mathfrak{F}$.

A *tensor field* T on M of type (r, s) is by definition an element of $\mathfrak{D}^r_s(M)$. This tensor field T is said to be *contravariant* of degree r, *covariant* of degree s. In particular, the tensor fields of type (0, 0), (1, 0), and (0, 1) on M are just the differentiable functions on M, the vector fields on M and the 1-forms on M, respectively.

If p is a point in M, we define $\mathfrak{D}^r_s(p)$ as the set of all $\boldsymbol{R}$-multilinear mappings of

$$M_p^* \times \dots \times M_p^* \times M_p \times \dots \times M_p \qquad (M_p^* \ \ r \text{ times}, M_p \ \ s \text{ times})$$

into $\boldsymbol{R}$. The set $\mathfrak{D}^r_s(p)$ is a vector space over $\boldsymbol{R}$ and is nothing but the tensor product

$$M_p \otimes \dots \otimes M_p \otimes M_p^* \otimes \dots \otimes M_p^* \qquad (M_p \ \ r \text{ times}, M_p^* \ \ s \text{ times})$$

or otherwise written

$$\mathfrak{D}^r_s(p) = \otimes^r M_p \otimes {}^s M_p^*.$$

We also put $\mathfrak{D}^0_0(p) = \boldsymbol{R}$. Consider now an element $T \in \mathfrak{D}^r_s(M)$. We have

$$T(g_1\theta_1, \dots, g_r\theta_r, f_1Z_1, \dots, f_sZ_s) = g_1 \dots g_r f_1 \dots f_s T(\theta_1, \dots, \theta_r, Z_1, \dots, Z_s)$$

for $f_i, g_j \in C^\infty(M)$, $Z_i \in \mathfrak{D}^1(M)$, $\theta_j \in \mathfrak{D}_1(M)$. It follows from Lemma 1.2 that if some θ_j or some Z_i vanishes on an open set V, then the function $T(\theta_1, \dots, \theta_r, Z_1, \dots, Z_s)$ vanishes on V. Let $\{x_1, \dots, x_m\}$ be a system of coordinates valid on an open neighborhood U of p. Then there exist vector fields X_i $(1 \leqslant i \leqslant m)$ and 1-forms ω_j $(1 \leqslant j \leqslant m)$ on M and an open neighborhood N of p, $p \in N \subset U$ such that on N

$$X_i = \frac{\partial}{\partial x_i}, \qquad \omega_j(X_i) = \delta_{ij} \qquad (1 \leqslant i, j \leqslant m).$$

On N, Z_i and θ_j can be written

$$Z_i = \sum_{k=1}^m f_{ik} X_k, \qquad \theta_j = \sum_{l=1}^m g_{jl}\omega_l,$$

where $f_{ik}, g_{jl} \in C^\infty(N)$, and by the remark above we have for $q \in N$,

$$T(\theta_1, \dots, \theta_r, Z_1, \dots, Z_s)(q)$$

$$= \sum_{l_j=1, k_i=1}^m g_{1l_1} \dots g_{rl_r} f_{1k_1} \dots f_{sk_s} T(\omega_{l_1}, \dots, \omega_{l_r}, X_{k_1}, \dots, X_{k_s})(q).$$

This shows that $T(\theta_1, \dots, \theta_r, Z_1, \dots, Z_s)(p) = 0$ if some θ_j or some Z_i vanishes at p. We can therefore define an element $T_p \in \mathfrak{D}^r_s(p)$ by the condition

$$T_p((\theta_1)_p, \dots, (\theta_r)_p, (Z_1)_p, \dots, (Z_s)_p) = T(\theta_1, \dots, \theta_r, Z_1, \dots, Z_s)(p).$$

The tensor field T thus gives rise to a family $T_p, p \in M$, where $T_p \in \mathfrak{D}^r_s(p)$. It is clear that if $T_p = 0$ for all p, then $T = 0$. The element $T_p \in \mathfrak{D}^r_s(p)$ depends differentiably on p in the sense that if N is a coordinate neighborhood of p and T_q (for $q \in N$) is expressed as above in terms of bases for $\mathfrak{D}_1(N)$ and $\mathfrak{D}^1(N)$, then the coefficients are differentiable functions on N. On the other hand, if there is a rule $p \to T(p)$ which to each $p \in M$ assigns a member $T(p)$ of $\mathfrak{D}^r_s(p)$ in a differentiable manner (as described above), then there exists a tensor field T of type (r, s) such that $T_p = T(p)$ for all $p \in M$. In the case when M is analytic it is clear how to define analyticity of a tensor field T, generalizing the notion of an analytic vector field.

The vector spaces $\mathfrak{D}^r_s(p)$ and $\mathfrak{D}^s_r(p)$ are dual to each other under the nondegenerate bilinear form $\langle\ ,\ \rangle$ on $\mathfrak{D}^r_s(p) \times \mathfrak{D}^s_r(p)$ defined by the formula

$$\langle e_1 \otimes \cdots \otimes e_r \otimes f_1 \otimes \cdots \otimes f_s, e'_1 \otimes \cdots \otimes e'_s \otimes f'_1 \otimes \cdots \otimes f'_r \rangle = \prod_{i,j} f_j(e'_j) f'_i(e_i),$$

where e_i, e'_j are members of a basis of M_p, f_j, f'_i are members of a dual basis of M_p^*. It is then obvious that the formula holds if e_i, e'_j are arbitrary elements of M_p and f_j, f'_i are arbitrary elements of M_p^*. In particular, the form $\langle\ ,\ \rangle$ is independent of the choice of basis used in the definition.

Each $T \in \mathfrak{D}^r_s(M)$ induces an $\mathfrak{F}$-linear mapping of $\mathfrak{D}^s_r(M)$ into $\mathfrak{F}$ given by the formula

$$(T(S))(p) = \langle T_p, S_p \rangle \qquad \text{for } S \in \mathfrak{D}^s_r(M).$$

If $T(S) = 0$ for all $S \in \mathfrak{D}^s_r(M)$, then $T_p = 0$ for all $p \in M$, so $T = 0$. Consequently, $\mathfrak{D}^r_s(M)$ can be regarded as a subset of $(\mathfrak{D}^s_r(M))^*$. We have now the following generalization of Lemma 2.3.

Lemma 2.3′. *The module $\mathfrak{D}^r_s(M)$ is the dual of $\mathfrak{D}^s_r(M)$ $(r, s \geqslant 0)$.*

Except for a change in notation the proof is the same as that of Lemma 2.3. To emphasize the duality we sometimes write $\langle T, S \rangle$ instead of $T(S)$, $(T \in \mathfrak{D}^r_s, S \in \mathfrak{D}^s_r)$.

Let $\mathfrak{D}$ (or $\mathfrak{D}(M)$) denote the direct sum of the $\mathfrak{F}$-modules $\mathfrak{D}^r_s(M)$,

$$\mathfrak{D} = \sum_{r,s=0}^{\infty} \mathfrak{D}^r_s.$$

Similarly, if $p \in M$ we consider the direct sum

$$\mathfrak{D}(p) = \sum_{r,s=0}^{\infty} \mathfrak{D}^r_s(p).$$

The vector space $\mathfrak{D}(p)$ can be turned into an associative algebra over $\boldsymbol{R}$ as follows: Let $a = e_1 \otimes \dots \otimes e_r \otimes f_1 \otimes \dots \otimes f_s$, $b = e'_1 \otimes \dots \otimes e'_\rho \otimes f'_1 \dots \otimes f'_\sigma$, where e_i, e'_i are members of a basis for M_p, f_j, f'_j are members of a dual basis for M_p^*. Then $a \otimes b$ is defined by the formula

$$a \otimes b = e_1 \otimes \dots \otimes e_r \otimes e'_1 \otimes \dots \otimes e'_\rho \otimes f_1 \otimes \dots \otimes f_s \otimes f'_1 \otimes \dots \otimes f'_\sigma.$$

We put $a \otimes 1 = a$, $1 \otimes b = b$ and extend the operation $(a, b) \to a \otimes b$ to a bilinear mapping of $\mathfrak{D}(p) \times \mathfrak{D}(p)$ into $\mathfrak{D}(p)$. Then $\mathfrak{D}(p)$ is an associative algebra over $\boldsymbol{R}$. The formula for $a \otimes b$ now holds for arbitrary elements e_i, $e'_i \in M_p$ and f_j, $f'_j \in M_p^*$. Consequently, the multiplication in $\mathfrak{D}(p)$ is independent of the choice of basis.

The tensor product $\otimes$ in $\mathfrak{D}$ is now defined as the $\mathfrak{F}$-bilinear mapping $(S, T) \to S \otimes T$ of $\mathfrak{D} \times \mathfrak{D}$ into $\mathfrak{D}$ such that

$$(S \otimes T)_p = S_p \otimes T_p, \qquad S \in \mathfrak{D}^r_s,\ T \in \mathfrak{D}^\rho_\sigma,\ p \in M.$$

This turns the $\mathfrak{F}$-module $\mathfrak{D}$ into a ring satisfying

$$f(S \otimes T) = fS \otimes T = S \otimes fT$$

for $f \in \mathfrak{F}$, S, $T \in \mathfrak{D}$. In other words, $\mathfrak{D}$ is an associative algebra over the ring $\mathfrak{F}$. The algebras $\mathfrak{D}$ and $\mathfrak{D}(p)$ are called the *mixed tensor algebras* over M and M_p, respectively. The submodules

$$\mathfrak{D}^* = \sum_{r=0}^{\infty} \mathfrak{D}^r, \qquad \mathfrak{D}_* = \sum_{s=0}^{\infty} \mathfrak{D}_s$$

are subalgebras of $\mathfrak{D}$ (also denoted $\mathfrak{D}^*(M)$ and $\mathfrak{D}_*(M)$) and the subspaces

$$\mathfrak{D}^*(p) = \sum_{r=0}^{\infty} \mathfrak{D}^r(p), \qquad \mathfrak{D}_*(p) = \sum_{s=0}^{\infty} \mathfrak{D}_s(p)$$

are subalgebras of $\mathfrak{D}(p)$.

Now let r, s be two integers $\geqslant 1$, and let i, j be integers such that $1 \leqslant i \leqslant r$, $1 \leqslant j \leqslant s$. Consider the $\boldsymbol{R}$-linear mapping $C^i{}_j$: $\mathfrak{D}^r_s(p) \to \mathfrak{D}^{r-1}_{s-1}(p)$ defined by

$$C^i{}_j(e_1 \otimes \dots \otimes e_r \otimes f_1 \otimes \dots \otimes f_s) = \langle e_i, f_j \rangle (e_1 \otimes \dots \hat{e}_i \dots \otimes e_r \otimes f_1 \otimes \dots \hat{f}_j \dots \otimes f_s),$$

where $e_1, \dots, e_r$ are members of a basis of M_p, $f_1, \dots, f_s$ are members of the dual basis of M_p^*. (The symbol ^ over a letter means that the letter is missing.) Now that the existence of $C^i{}_j$ is established, we note that the formula for $C^i{}_j$ holds for arbitrary elements $e_1, \ldots, e_r \in M_p$, $f_1, \ldots, f_s \in M_p^*$. In particular, $C^i{}_j$ is independent of the choice of basis.

There exists now a unique $\mathfrak{F}$-linear mapping $C^i{}_j$: $\mathfrak{D}^r_s(M) \to \mathfrak{D}^{r-1}_{s-1}(M)$ such that

$$(C^i{}_j(T))_p = C^i{}_j(T_p)$$

for all $T \in \mathfrak{D}^r_s(M)$ and all $p \in M$. This mapping satisfies the relation

$$C^i{}_j(X_1 \otimes \ldots \otimes X_r \otimes \omega_1 \otimes \ldots \otimes \omega_s)$$
$$= \langle X_i, \omega_j \rangle (X_1 \otimes \ldots \hat{X}_i \ldots \otimes X_r \otimes \omega_1 \otimes \ldots \hat{\omega}_j \ldots \otimes \omega_s)$$

for all $X_1, \ldots, X_r \in \mathfrak{D}^1$, $\omega_1, \ldots, \omega_s \in \mathfrak{D}_1$. The mapping $C^i{}_j$ is called the *contraction* of the ith contravariant index and the jth covariant index.

3. The Grassmann Algebra

As before, M denotes a C^∞ manifold and $\mathfrak{F} = C^\infty(M)$. If s is an integer $\geqslant 1$, let $\mathfrak{A}_s$ (or $\mathfrak{A}_s(M)$) denote the set of alternate $\mathfrak{F}$-multilinear mappings of $\mathfrak{D}^1 \times \ldots \times \mathfrak{D}^1$ (s times) into $\mathfrak{F}$. Then $\mathfrak{A}_s$ is a submodule of $\mathfrak{D}_s$. We put $\mathfrak{A}_0 = \mathfrak{F}$ and let $\mathfrak{A}$ (or $\mathfrak{A}(M)$) denote the direct sum $\mathfrak{A} = \sum_{s=0}^\infty \mathfrak{A}_s$ of the $\mathfrak{F}$-modules $\mathfrak{A}_s$. The elements of $\mathfrak{A}(M)$ are called *exterior differential forms* on M. The elements of $\mathfrak{A}_s$ are called differential s-forms (or just s-forms).

Let $\mathfrak{S}_s$ denote the group of permutations of the set $\{1, 2, \ldots, s\}$. Each $\sigma \in \mathfrak{S}_s$ induces an $\mathfrak{F}$-linear mapping of $\mathfrak{D}^1 \times \ldots \times \mathfrak{D}^1$ onto itself given by

$$(X_1, \ldots, X_s) \to (X_{\sigma^{-1}(1)}, \ldots, X_{\sigma^{-1}(s)}) \qquad (X_i \in \mathfrak{D}^1).$$

This mapping will also be denoted by σ. Since each $d \in \mathfrak{D}_s$ is a multilinear map of $\mathfrak{D}^1 \times \ldots \times \mathfrak{D}^1$ into $\mathfrak{F}$, the mapping $d \circ \sigma^{-1}$ is well defined. Moreover, the mapping $d \to d \circ \sigma^{-1}$ is a one-to-one $\mathfrak{F}$-linear mapping of $\mathfrak{D}_s$ onto itself. If we write $\sigma \cdot d = d \circ \sigma^{-1}$ we have $\sigma\tau \cdot d = \sigma \cdot (\tau \cdot d)$. Let $\epsilon(\sigma) = 1$ or -1 according to whether σ is an even or an odd permutation. Consider the linear transformation A_s: $\mathfrak{D}_s \to \mathfrak{D}_s$ given by

$$A_s(d_s) = \frac{1}{s!} \sum_{\sigma \in \mathfrak{S}_s} \epsilon(\sigma)\, \sigma \cdot d_s, \qquad d_s \in \mathfrak{D}_s.$$

If $s = 0$, we put $A_s(d_s) = d_s$. We extend A_s to an $\mathfrak{F}$-linear mapping A: $\mathfrak{D}_* \to \mathfrak{D}_*$ by putting $A(d) = \sum_{s=0}^\infty A_s(d_s)$ if $d = \sum_{s=0}^\infty d_s$, $d_s \in \mathfrak{D}_s$.

If $\tau \in \mathfrak{S}_s$, we have

$$\tau A_s(d_s) = \frac{1}{s!} \sum_{\sigma \in \mathfrak{S}_s} \epsilon(\sigma)\, \tau \cdot (\sigma \cdot d_s) = \frac{1}{s!} \sum_{\sigma \in \mathfrak{S}_s} \epsilon(\sigma)\, (\tau\sigma) \cdot d_s$$

$$= \epsilon(\tau) \frac{1}{s!} \sum_{\sigma \in \mathfrak{S}_s} \epsilon(\sigma)\, \sigma \cdot d_s.$$

Hence, $\tau \cdot (A_s(d_s)) = \epsilon(\tau)\, A_s(d_s)$. This shows that $A_s(\mathfrak{D}_s) \subset \mathfrak{A}_s$ and $A(\mathfrak{D}_*) \subset \mathfrak{A}$. On the other hand, if $d_s \in \mathfrak{A}_s$, then $\sigma \cdot d_s = \epsilon(\sigma)\, d_s$ for each $\sigma \in \mathfrak{S}_s$. Since $\epsilon(\sigma)^2 = 1$, we find that

$$A_s(d_s) = d_s \qquad \text{if } d_s \in \mathfrak{A}_s.$$

It follows that $A^2 = A$ and $A(\mathfrak{D}_*) = \mathfrak{A}$; in other words, A is a projection of $\mathfrak{D}_*$ onto $\mathfrak{A}$. The mapping A is called *alternation.*

Let N denote the kernel of A. Obviously N is a submodule of $\mathfrak{D}_*$.

Lemma 2.4.† *The module N is a two-sided ideal in $\mathfrak{D}_*$.*

It suffices to show that if $n_r \in N \cap \mathfrak{D}_r$, $d_s \in \mathfrak{D}_s$, then $A_{r+s}(n_r \otimes d_s) = A_{s+r}(d_s \otimes n_r) = 0$. Let $b_{r+s} = A_{r+s}(n_r \otimes d_s)$; then

$$(r+s)!\, b_{r+s} = \sum_{\sigma \in \mathfrak{S}_{r+s}} \epsilon(\sigma)\, \sigma \cdot (n_r \otimes d_s),$$

where

$$\sigma \cdot (n_r \otimes d_s)\,(X_1, \ldots, X_{r+s}) = n_r(X_{\sigma(1)}, \ldots, X_{\sigma(r)})\, d_s(X_{\sigma(r+1)}, \ldots, X_{\sigma(r+s)}).$$

The elements in $\mathfrak{S}_{r+s}$ which leave each number $r+1, \ldots, r+s$ fixed constitute a subgroup G of $\mathfrak{S}_{r+s}$, isomorphic to $\mathfrak{S}_r$. Let S be a subset of $\mathfrak{S}_{r+s}$ containing exactly one element from each left coset $\sigma_0 G$ of $\mathfrak{S}_{r+s}$. Then, since $\epsilon(\sigma_1\sigma_2) = \epsilon(\sigma_1)\,\epsilon(\sigma_2)$,

$$\sum_{\sigma \in \mathfrak{S}_{r+s}} \epsilon(\sigma)\, \sigma \cdot (n_r \otimes d_s) = \sum_{\sigma_0 \in S} \epsilon(\sigma_0) \sum_{\tau \in G} \epsilon(\tau)\, (\sigma_0\tau) \cdot (n_r \otimes d_s).$$

Let $X_i \in \mathfrak{D}^1$ $(1 \leqslant i \leqslant r+s)$, $(Y_1, \ldots, Y_{r+s}) = \sigma_0^{-1}(X_1, \ldots, X_{r+s})$. Then

$$\sum_{\tau \in G} \epsilon(\tau)\, ((\sigma_0\tau) \cdot (n_r \otimes d_s))\,(X_1, \ldots, X_{r+s})$$
$$= d_s(Y_{r+1}, \ldots, Y_{r+s}) \sum_{\tau \in \mathfrak{S}_r} \epsilon(\tau)\, (\tau \cdot n_r)\,(Y_1, \ldots, Y_r) = 0.$$

This shows that $b_{r+s} = 0$. Similarly one proves $A_{s+r}(d_s \otimes n_r) = 0$.

† Chevalley [2], p. 142.

For any two $\theta, \omega \in \mathfrak{A}$ we can now define the *exterior product*

$$\theta \wedge \omega = A(\theta \otimes \omega).$$

This turns the $\mathfrak{F}$-module $\mathfrak{A}$ into an associative algebra, isomorphic to $\mathfrak{D}_*/N$. The module $\mathfrak{A}(M)$ of alternate $\mathfrak{F}$-multilinear functions with the exterior multiplication is called the *Grassmann algebra* of the manifold M.

We can also for each $p \in M$ define the Grassmann algebra $\mathfrak{A}(p)$ of the tangent space M_p. The elements of $\mathfrak{A}(p)$ are the alternate, $\boldsymbol{R}$-multilinear, real-valued functions on M_p and the product (also denoted $\wedge$) satisfies

$$\theta_p \wedge \omega_p = (\theta \wedge \omega)_p, \qquad \theta, \omega \in \mathfrak{A}.$$

This turns $\mathfrak{A}(p)$ into an associative algebra containing the dual space M_p^*. If $\theta, \omega \in M_p^*$, we have $\theta \wedge \omega = -\omega \wedge \theta$; as a consequence one derives easily the following rule:

Let $\theta^1, ..., \theta^l \in M_p^*$ and let $\omega^i = \sum_{j=1}^l a_{ij}\theta^j$, $1 \leqslant i, j \leqslant l$, $(a_{ij} \in \boldsymbol{R})$. Then

$$\omega^1 \wedge ... \wedge \omega^l = \det (a_{ij})\, \theta^1 \wedge ... \wedge \theta^l.$$

For convenience we write down the exterior multiplication explicitly. Let $f, g \in C^\infty(M)$, $\theta \in \mathfrak{A}_r$, $\omega \in \mathfrak{A}_s$, $X_i \in \mathfrak{D}^1$. Then

$$f \wedge g = fg,$$

$$(f \wedge \theta)(X_1, ..., X_r) = f\,\theta(X_1, ..., X_r),$$

$$(\omega \wedge g)(X_1, ..., X_s) = g\,\omega(X_1, ..., X_s),$$

$$(\theta \wedge \omega)(X_1, ..., X_{r+s}) \tag{4}$$

$$= \frac{1}{(r+s)!} \sum_{\sigma \in \mathfrak{S}_{r+s}} \epsilon(\sigma)\, \theta(X_{\sigma(1)}, ..., X_{\sigma(r)})\, \omega(X_{\sigma(r+1)}, ..., X_{\sigma(r+s)}).$$

We also have the relation

$$\theta \wedge \omega = (-1)^{rs}\, \omega \wedge \theta. \tag{5}$$

4. Exterior Differentiation

Let M be a C^∞ manifold, $\mathfrak{A}(M)$ the Grassmann algebra over M. The operator d, the *exterior differentiation*, is described in the following theorem.

Theorem 2.5. *There exists a unique* $\boldsymbol{R}$*-linear mapping* $d: \mathfrak{A}(M) \to \mathfrak{A}(M)$ *with the following properties:*

(i) $d\mathfrak{A}_s \subset \mathfrak{A}_{s+1}$ *for each* $s \geqslant 0$.

(ii) *If* $f \in \mathfrak{A}_0$ $(= C^\infty(M))$, *then* df *is the* 1*-form given by* $df(X) = Xf$, $X \in \mathfrak{D}^1(M)$.

(iii) $d \circ d = 0$.

(iv) $d(\omega_1 \wedge \omega_2) = d\omega_1 \wedge \omega_2 + (-1)^r \omega_1 \wedge d\omega_2$ *if* $\omega_1 \in \mathfrak{A}_r$, $\omega_2 \in \mathfrak{A}(M)$.

Proof. Assuming the existence of d for M as well as for open submanifolds of M, we first prove a formula for d ((9) below) which then has the uniqueness as a corollary. Let $p \in M$ and $\{x_1, ..., x_m\}$ a coordinate system valid on an open neighborhood U of p. Let V be an open subset of U such that $\bar{V}$ is compact and $p \in V$, $\bar{V} \subset U$. From (ii) we see that the forms dx_i $(1 \leqslant i \leqslant m)$ on U satisfy $dx_i(\partial/\partial x_j) = \delta_{ij}$ on U. Hence dx_i $(1 \leqslant i \leqslant m)$ is a basis of the $C^\infty(U)$-module $\mathfrak{D}_1(U)$; thus each element in $\mathfrak{D}_*(U)$ can be expressed in the form

$$\sum F_{i_1 \ldots i_r} dx_{i_1} \otimes \ldots \otimes dx_{i_r}, \qquad F_{i_1 \ldots i_r} \in C^\infty(U).$$

It follows that if $\theta \in \mathfrak{A}(M)$ and if θ_U denotes the form induced by θ on U, then θ_U can be written

$$\theta_U = \sum f_{i_1 \ldots i_r} dx_{i_1} \wedge \ldots \wedge dx_{i_r}, \qquad f_{i_1 \ldots i_r} \in C^\infty(U). \tag{6}$$

This is called an *expression* of θ_U on U. We shall prove the formula

$$d(\theta_V) = (d\theta)_V.$$

Owing to Lemma 1.2 there exist functions $\psi_{i_1 \ldots i_r} \in C^\infty(M)$, $\varphi_i \in C^\infty(M)$ $(1 \leqslant i \leqslant m)$ such that

$$\psi_{i_1 \ldots i_r} = f_{i_1 \ldots i_r}, \qquad \varphi_1 = x_1, ..., \varphi_m = x_m \text{ on } V.$$

We consider the form

$$\omega = \sum \psi_{i_1 \ldots i_r} d\varphi_{i_1} \wedge \ldots \wedge d\varphi_{i_r}$$

on M. We have obviously $\omega_V = \theta_V$. Moreover, since $d(f(\theta - \omega)) = df \wedge (\theta - \omega) + fd(\theta - \omega)$ for each $f \in C^\infty(M)$, we can, choosing f identically 0 outside V, identically 1 on an open subset of V, deduce that $(d\theta)_V = (d\omega)_V$.

Since

$$d\omega = \sum d\psi_{i_1 \ldots i_r} \wedge d\varphi_{i_1} \wedge \ldots \wedge d\varphi_{i_r}$$

owing to (iii) and (iv), and since $d(f_V) = (df)_V$ for each $f \in C^\infty(M)$, we conclude that

$$(d\omega)_V = \sum df_{i_1 \ldots i_r} \wedge dx_{i_1} \wedge \ldots \wedge dx_{i_r}. \tag{7}$$

This proves the relation

$$(d\theta)_V = d(\theta_V) = \sum df_{i_1 \ldots i_r} \wedge dx_{i_1} \wedge \ldots \wedge dx_{i_r}. \tag{8}$$

On M itself we have the formula

$$\begin{aligned}(p+1)\, d\omega(X_1, \ldots, X_{p+1}) &= \sum_{i=1}^{p+1} (-1)^{i+1} X_i \cdot \omega(X_1, \ldots, \hat{X}_i, \ldots, X_{p+1}) \\ &\quad + \sum_{i<j} (-1)^{i+j} \omega([X_i, X_j], X_1, \ldots, \hat{X}_i, \ldots, \hat{X}_j, \ldots, X_{p+1}) \end{aligned} \tag{9}$$

for $\omega \in \mathfrak{A}_p(M)$ $(p \geqslant 1)$, $X_i \in \mathfrak{D}^1(M)$. In fact, it suffices to prove it in a coordinate neighborhood of each point; in that case it is a simple consequence of (8). The uniqueness of d is now obvious.

On the other hand, to prove the existence of d, we *define* d by (9) and (ii). Using the relation $[X, fY] = f[X, Y] + (Xf)\, Y$ $(f \in \mathfrak{F};\ X, Y \in \mathfrak{D}^1)$, it follows quickly that the right-hand side of (9) is $\mathfrak{F}$-linear in each variable X_i and vanishes whenever two variables coincide. Hence $d\omega \in \mathfrak{A}_{p+1}$ if $\omega \in \mathfrak{A}_p$. If $X \in \mathfrak{D}^1$, let X_V denote the vector field induced on V. Then $[X, Y]_V = [X_V, Y_V]$ and therefore the relation $(d\theta)_V = d(\theta_V)$ follows from (9). Next we observe that (8) follows from (9) and (ii). Also

$$d(fg) = f dg + g df \tag{10}$$

as a consequence of (ii). To show that (iii) and (iv) hold, it suffices to show that they hold in a coordinate neighborhood of each point of M. But on V, (iv) is a simple consequence of (10) and (8). Moreover, (8) and (ii) imply $d(dx_i) = 0$; consequently (using (iv)),

$$d(df) = d\left(\sum_j \frac{\partial f}{\partial x_j} dx_j\right) = \sum_{i,j} \frac{\partial^2 f}{\partial x_i \partial x_j} dx_i \wedge dx_j = 0$$

for each $f \in C^\infty(U)$. The relation (iii) now follows from (8) and (iv).

§ 3. Mappings

1. The Interpretation of the Jacobian

Let M and N be C^∞ manifolds and Φ a mapping of M into N. Let $p \in M$. The mapping Φ is called *differentiable at* p if $g \circ \Phi \in C^\infty(p)$ for each $g \in C^\infty(\Phi(p))$. The mapping Φ is called *differentiable* if it is differentiable at each $p \in M$. Similarly *analytic* mappings are defined. Let $\psi: q \to (x_1(q), ..., x_m(q))$ be a system of coordinates on a neighborhood U of $p \in M$ and $\psi': r \to (y_1(r), ..., y_n(r))$ a system of coordinates on a neighborhood U' of $\Phi(p)$ in N. Assume $\Phi(U) \subset U'$. The mapping $\psi' \circ \Phi \circ \psi^{-1}$ of $\psi(U)$ into $\psi'(U')$ is given by a system of n functions

$$y_j = \varphi_j(x_1, ..., x_m) \qquad (1 \leqslant j \leqslant n), \tag{1}$$

which we call the *expression* of Φ in coordinates. The mapping Φ is differentiable at p if and only if the functions φ_i have partial derivatives of all orders in some fixed neighborhood of $(x_1(p), ..., x_m(p))$.

The mapping Φ is called a *diffeomorphism* of M onto N if Φ is a one-to-one differentiable mapping of M onto N and Φ^{-1} is differentiable. If in addition M, N, Φ, and Φ^{-1} are analytic, Φ is called an *analytic diffeomorphism*.

If Φ is differentiable at $p \in M$ and $A \in M_p$, then the linear mapping $B : C^\infty(\Phi(p)) \to \boldsymbol{R}$ given by $B(g) = A(g \circ \Phi)$ for $g \in C^\infty(\Phi(p))$ is a tangent vector to N at $\Phi(p)$. The mapping $A \to B$ of M_p into $N_{\Phi(p)}$ is denoted $d\Phi_p$ (or just Φ_p) and is called the *differential* of Φ at p. We have seen that the vectors

$$e_i : f \to \left(\frac{\partial f^*}{\partial x_i}\right)_{\psi(p)} \qquad (1 \leqslant i \leqslant m), \qquad f^* = f \circ \psi^{-1},$$

$$\bar{e}_j : g \to \left(\frac{\partial g^*}{\partial y_j}\right)_{\psi'(\Phi(p))} \qquad (1 \leqslant j \leqslant n), \qquad g^* = g \circ (\psi')^{-1}$$

form a basis of M_p and $N_{\Phi(p)}$, respectively. Then

$$d\Phi_p(e_i)\, g = e_i(g \circ \Phi) = \left(\frac{\partial (g \circ \Phi)^*}{\partial x_i}\right)_{\psi(p)}.$$

But

$$(g \circ \Phi)^*(x_1, ..., x_m) = g^*(y_1, ..., y_n),$$

where $y_j = \varphi_j(x_1, ..., x_m)$ $(1 \leqslant j \leqslant n)$. Hence

$$d\Phi_p(e_i) = \sum_{j=1}^{n} \left(\frac{\partial \varphi_j}{\partial x_i}\right)_{\psi(p)} \bar{e}_j. \tag{2}$$

This shows that if we use the bases e_i $(1 \leqslant i \leqslant m)$, $\bar{e}_j$ $(1 \leqslant j \leqslant n)$ to express the linear transformation $d\Phi_p$ in matrix form, then the matrix we obtain is just the Jacobian of the system (1). From a standard theorem on the Jacobian (the inverse function theorem), we can conclude:

Proposition 3.1. *If $d\Phi_p$ is an isomorphism of M_p onto $N_{\Phi(p)}$, then there exist open submanifolds $U \subset M$ and $V \subset N$ such that $p \in U$ and Φ is a diffeomorphism of U onto V.*

Remark. If $N = \boldsymbol{R}$, $N_{\Phi(p)}$ is identified with $\boldsymbol{R}$ (Remark, §2) and thus $d\Phi_p$ becomes a linear function on M_p. This is the same linear function as we obtain by considering $d\Phi$ as a differential form on M. In fact, if $X \in M_p$, the tangent vector $d\Phi_p(X)$ and the tangent vector

$$f \rightarrow \left\{\frac{d}{dt} f(\Phi(p) + tX\Phi)\right\}_{t=0}, \qquad f \in C^\infty(\boldsymbol{R}),$$

both assign to f the number $f'(\Phi(p))\,(X\Phi)$.

Definition.

Let M and N be differentiable (or analytic) manifolds.

(a) A mapping $\Phi : M \rightarrow N$ is called *regular* at $p \in M$ if Φ is differentiable (analytic) at $p \in M$ and $d\Phi_p$ is a one-to-one mapping of M_p into $N_{\Phi(p)}$.

(b) M is called a *submanifold* of N if (1) $M \subset N$ (set theoretically); (2) the identity mapping I of M into N is regular at each point of M.

For example, the sphere $x_1^2 + x_2^2 + x_3^2 = 1$ is a submanifold of $\boldsymbol{R}^3$ and a topological subspace as well. However, a submanifold M of a manifold N is not necessarily a topological subspace of N. For example, let N be a torus and let M be a curve on N without double points, dense in N (Chapter II, §2). Proposition 3.1 shows that a submanifold M of a manifold N is an open submanifold of N if and only if $\dim M = \dim N$.

Proposition 3.2. *Let M be a submanifold of a manifold N and let $p \in M$. Then there exists a coordinate system $\{x_1, \ldots, x_n\}$ valid on an open neighborhood V of p in N such that $x_1(p) = \ldots = x_n(p) = 0$ and such that the set*

$$U = \{q \in V : x_j(q) = 0 \text{ for } m + 1 \leqslant j \leqslant n\}$$

together with the restrictions of $(x_1, \ldots, x_m)$ to U form a local chart on M containing p.

Proof. Let $\{y_1, \ldots, y_m\}$ and $\{z_1, \ldots, z_n\}$ be coordinate systems valid on open neighborhoods of p in M and N, respectively, such that $y_i(p) = z_j(p) = 0$, $(1 \leqslant i \leqslant m,\ 1 \leqslant j \leqslant n)$. The expression of the identity mapping $I: M \to N$ is (near p) given by a system of functions $z_j = \varphi_j(y_1, ..., y_m)$, $1 \leqslant j \leqslant n$. The Jacobian matrix $(\partial \varphi_j / \partial y_i)$ of this system has rank m at p since I is regular at p. Without loss of generality we may assume that the square matrix $(\partial \varphi_j / \partial y_i)_{1 \leqslant i,j \leqslant m}$ has determinant $\neq 0$ at p. In a neighborhood of $(0, ..., 0)$ we have therefore $y_i = \psi_i(z_1, ..., z_m)$, $1 \leqslant i \leqslant m$, where each ψ_i is a differentiable function. If we now put

$$x_i = z_i, \qquad 1 \leqslant i \leqslant m,$$

$$x_j = z_j - \varphi_j(\psi_1(z_1, ..., z_m), ..., \psi_m(z_1, ..., z_m)), \qquad m + 1 \leqslant j \leqslant n,$$

it is clear that

$$\det \left(\frac{\partial x_i}{\partial y_l}\right)_{1 \leqslant i,\ l \leqslant m} \neq 0, \qquad \det \left(\frac{\partial x_j}{\partial z_k}\right)_{1 \leqslant j,\ k \leqslant n} \neq 0.$$

Therefore $\{x_1, ..., x_n\}$ gives the desired coordinate system.

A generalization is given by Theorem 15.5.

2. Transformation of Vector Fields

Let M and N be C^∞ manifolds and Φ a differentiable mapping of M into N. Let X and Y be vector fields on M and N, respectively; X and Y are called *Φ-related* if

$$d\Phi_p(X_p) = Y_{\Phi(p)} \qquad \text{for all } p \in M. \tag{3}$$

It is easy to see that (3) is equivalent to

$$(Yf) \circ \Phi = X(f \circ \Phi) \qquad \text{for all } f \in C^\infty(N). \tag{4}$$

It is convenient to write $d\Phi \cdot X = Y$ or $X^\Phi = Y$ instead of (3).

Proposition 3.3.

(i) *Suppose* $d\Phi \cdot X_i = Y_i$ $(i = 1, 2)$. *Then*

$$d\Phi \cdot [X_1, X_2] = [Y_1, Y_2].$$

(ii) *Suppose Φ is a diffeomorphism of M onto itself and put* $f^\Phi = f \circ \Phi^{-1}$ *for* $f \in C^\infty(M)$. *Then if* $X \in \mathfrak{D}^1(M)$,

$$(fX)^\Phi = f^\Phi X^\Phi, \qquad (Xf)^\Phi = X^\Phi f^\Phi.$$

Proof. From (4) we have $(Y_1(Y_2 f)) \circ \Phi = X_1(Y_2 f \circ \Phi) = X_1(X_2(f \circ \Phi))$, so (i) follows. The last relation in (ii) is also an immediate consequence of (4). As to the first one, we have for $g \in C^\infty(M)$

$$((fX)^\Phi g) \circ \Phi = (fX)(g \circ \Phi) = f((X^\Phi g) \circ \Phi),$$

so

$$(fX)^\Phi g = f^\Phi (X^\Phi g).$$

Remark. Since $X^{\Phi}f = (Xf^{\Phi^{-1}})^{\Phi}$ it is natural to make the following definition. Let Φ be a diffeomorphism of M onto M and A a mapping of $C^{\infty}(M)$ into itself. The mapping A^{Φ} is defined by $A^{\Phi}f = (Af^{\Phi^{-1}})^{\Phi}$ for $f \in C^{\infty}(M)$. We also write $[Af](p)$ for the value of the function Af at $p \in M$. If Φ and Ψ are two diffeomorphisms of M, then $f^{\Phi\Psi} = (f^{\Psi})^{\Phi}$ and $A^{\Phi\Psi} = (A^{\Psi})^{\Phi}$.

3. Effect on Differential Forms

Let M and N be C^{∞} manifolds and $\Phi : M \to N$ a differentiable mapping. Let ω be an r-form on N. Then we can define an r-form $\Phi^*\omega$ on M which satisfies

$$\Phi^*\omega(X_1, ..., X_r) = \omega(Y_1, ..., Y_r) \circ \Phi$$

whenever the vector fields X_i and Y_i $(1 \leqslant i \leqslant r)$ are Φ-related. It suffices to put

$$(\Phi^*\omega)_p(A_1, ..., A_r) = \omega_{\Phi(p)}(d\Phi_p(A_1), ..., d\Phi_p(A_r))$$

for each $p \in M$, and $A_i \in M_p$. If $f \in C^{\infty}(N)$, we put $\Phi^*f = f \circ \Phi$ and by linearity $\Phi^*\theta$ is defined for each $\theta \in \mathfrak{A}(M)$. Then the following formulas hold:

$$\Phi^*(\omega_1 \wedge \omega_2) = \Phi^*(\omega_1) \wedge \Phi^*(\omega_2), \qquad \omega_1, \omega_2 \in \mathfrak{A}(M); \tag{5}$$

$$d(\Phi^*\omega) = \Phi^*(d\omega). \tag{6}$$

In fact, (5) follows from (4), § 2, and (6) is proved below. In the same way we can define Φ^*T for an arbitrary covariant tensor field $T \in \mathfrak{D}_*(M)$. If $M = N$ and Φ is a diffeomorphism of M onto itself such that $\Phi^*T = T$, we say that T is *invariant* under Φ.

The computation of $\Phi^*\omega$ in coordinates is very simple. Suppose U and V are open sets in M and N, respectively, where the coordinate systems

$$\xi: q \to (x_1(q), ..., x_m(q)), \qquad \eta: r \to (y_1(r), ..., y_n(r))$$

are valid. Assume $\Phi(U) \subset V$. On U, Φ has a coordinate expression

$$y_j = \varphi_j(x_1, ..., x_m) \qquad (1 \leqslant j \leqslant n).$$

If $\omega \in \mathfrak{A}(N)$, the form ω_V has an expression

$$\omega_V = \sum g_{j_1 ... j_s}\, dy_{j_1} \wedge ... \wedge dy_{j_s} \tag{7}$$

where $g_{j_1 ... j_s} \in C^{\infty}(V)$. The form $\Phi^*\omega$ induces the form $(\Phi^*\omega)_U$ on U, which has an expression

$$(\Phi^*\omega)_U = \sum f_{i_1 ... i_r}\, dx_{i_1} \wedge ... \wedge dx_{i_r}.$$

This expression is obtained just by substituting

$$y_j = \varphi_j(x_1, ..., x_m), \qquad dy_j = \sum_{i=1}^{m} \frac{\partial \varphi_j}{\partial x_i} dx_i \qquad (1 \leqslant j \leqslant n)$$

into (7). This follows from (5) if we observe that (2) implies

$$\Phi^*(dy_j) = \sum_{i=1}^{m} \left(\frac{\partial \varphi_j}{\partial x_i} \circ \xi \right) dx_i.$$

This proves (6) if ω is a function, hence, by (7), in general.

§ 4. Affine Connections

Definition. An *affine connection* on a manifold M is a rule ∇ which assigns to each $X \in \mathfrak{D}^1(M)$ a linear mapping ∇_X of the vector space $\mathfrak{D}^1(M)$ into itself satisfying the following two conditions:

(∇_1) $$\nabla_{fX+gY} = f\nabla_X + g\nabla_Y;$$

(∇_2) $$\nabla_X(fY) = f\nabla_X(Y) + (Xf)\, Y$$

for $f, g \in C^\infty(M)$, $X, Y \in \mathfrak{D}^1(M)$. The operator ∇_X is called *covariant differentiation* with respect to X. For a motivation see Exercises F.1-F.3.

Lemma 4.1. *Suppose M has the affine connection $X \to \nabla_X$ and let U be an open submanifold of M. Let $X, Y \in \mathfrak{D}^1(M)$. If X or Y vanishes identically on U, then so does $\nabla_X(Y)$.*

Proof. Suppose Y vanishes on U. Let $p \in U$ and $g \in C^\infty(M)$. To prove that $(\nabla_X(Y)g)(p) = 0$, we select $f \in C^\infty(M)$ such that $f(p) = 0$ and $f = 1$ outside U (Lemma 1.2). Then $fY = Y$ and

$$\nabla_X(Y)g = \nabla_X(fY)g = (Xf)(Yg) + f(\nabla_X(Y)g)$$

which vanishes at p. The statement about X follows similarly.

An affine connection ∇ on M induces an affine connection ∇_U on an arbitrary open submanifold U of M. In fact, let X, Y be two vector fields on U. For each $p \in U$ there exist vector fields X', Y' on M which agree with X and Y in an open neighborhood V of p. We then put $((\nabla_U)_X(Y))_q = (\nabla_{X'}(Y'))_q$ for $q \in V$. By Lemma 4.1, the right-hand side of this equation is independent of the choice of X', Y'. It follows immediately that the rule ∇_U: $X \to (\nabla_U)_X$ $(X \in \mathfrak{D}^1(U))$ is an affine connection on U.

In particular, suppose U is a coordinate neighborhood where a coordinate system $\varphi : q \to (x_1(q), \ldots, x_m(q))$ is valid. For simplicity, we write ∇_i instead of $(\nabla_U)_{\partial/\partial x_i}$. We define the functions $\Gamma_{ij}{}^k$ on U by

$$\nabla_i\left(\frac{\partial}{\partial x_j}\right) = \sum_k \Gamma_{ij}{}^k \frac{\partial}{\partial x_k}. \tag{1}$$

For simplicity of notation we write also $\Gamma_{ij}{}^k$ for the function $\Gamma_{ij}{}^k \circ \varphi^{-1}$. If $\{y_1, \ldots, y_m\}$ is another coordinate system valid on U, we get another set of functions $\Gamma'_{\alpha\beta}{}^\gamma$ by

$$\nabla_\alpha\left(\frac{\partial}{\partial y_\beta}\right) = \sum_\gamma \Gamma'_{\alpha\beta}{}^\gamma \frac{\partial}{\partial y_\gamma}.$$

Using the axioms ∇_1 and ∇_2 we find easily

$$\Gamma'_{\alpha\beta}{}^\gamma = \sum_{i,j,k} \frac{\partial x_i}{\partial y_\alpha} \frac{\partial x_j}{\partial y_\beta} \frac{\partial y_\gamma}{\partial x_k} \Gamma_{ij}{}^k + \sum_j \frac{\partial^2 x_j}{\partial y_\alpha \partial y_\beta} \frac{\partial y_\gamma}{\partial x_j}. \tag{2}$$

On the other hand, suppose there is given a covering of a manifold M by open coordinate neighborhoods U and in each neighborhood a system of functions $\Gamma_{ij}{}^k$ such that (2) holds whenever two of these neighborhoods overlap. Then we can define ∇_i by (1) and thus we get an affine connection ∇_U in each coordinate neighborhood U. We finally define an affine connection $\tilde{\nabla}$ on M as follows: Let $X,\ Y \in \mathfrak{D}^1(M)$ and $p \in M$. If U is a coordinate neighborhood containing p, let

$$(\tilde{\nabla}_X(Y))_p = ((\nabla_U)_{X_1}(Y_1))_p$$

if X_1 and Y_1 are the vector fields on U induced by X and Y, respectively. Then $\tilde{\nabla}$ is an affine connection on M which on each coordinate neighborhood U induces the connection ∇_U.

Lemma 4.2. *Let $X,\ Y \in \mathfrak{D}^1(M)$. If X vanishes at a point p in M, then so does $\nabla_X(Y)$.*

Let $\{x_1, \ldots, x_m\}$ be a coordinate system valid on an open neighborhood U of p. On the set U we have $X = \sum_i f_i(\partial/\partial x_i)$ where $f_i \in C^\infty(U)$ and $f_i(p) = 0$, $(1 \leqslant i \leqslant m)$. Using Lemma 4.1 we find $(\nabla_X(Y))_p = \sum_i f_i(p)\,(\nabla_i(Y))_p = 0$.

Remark. Thus if $v \in M_p$, $\nabla_v(Y)$ is a well-defined vector in M_p.

Definition. Suppose ∇ is an affine connection on M and that Φ is a diffeomorphism of M. A new affine connection ∇' can be defined on M by

$$\nabla'_X(Y) = (\nabla_{X^\Phi}(Y^\Phi))^{\Phi^{-1}}, \qquad X,\ Y \in \mathfrak{D}^1(M).$$

That ∇' is indeed an affine connection on M is best seen from Prop. 3.3.

The affine connection ∇ is called *invariant* under Φ if $\nabla' = \nabla$. In this case Φ is called an *affine transformation* of M. Similarly one can define an affine transformation of one manifold onto another.

§ 5. Parallelism

Let M be a C^∞ manifold. A *curve* in M is a regular mapping of an open interval $I \subset \boldsymbol{R}$ into M. The restriction of a curve to a closed subinterval is called a *curve segment*. The curve segment is called finite if the interval is finite.

Let $\gamma : t \rightarrow \gamma(t)$ $(t \in I)$ be a curve in M. Differentiation with respect to the parameter will often be denoted by a dot $(\cdot)$. In particular, $\dot{\gamma}(t)$ stands for the tangent vector $d\gamma(d/dt)_t$. Suppose now that to each $t \in I$ is associated a vector $Y(t) \in M_{\gamma(t)}$. Assuming $Y(t)$ to vary differentiably with t, we shall now define what it means for the family $Y(t)$ to be parallel with respect to γ. Let J be a compact subinterval of I such that the finite curve segment $\gamma_J : t \rightarrow \gamma(t)$ $(t \in J)$ has no double points and such that $\gamma(J)$ is contained in a coordinate neighborhood U. Owing to the regularity of γ each point of I is contained in such an interval J with nonempty interior. Let $\{x_1, ..., x_m\}$ be a coordinate system on U.

Lemma 5.1. *Let $g(t)$ be a differentiable function on an open interval containing J. Then there exists a function $G \in C^\infty(M)$ such that*

$$G(\gamma(t)) = g(t) \qquad (t \in J).$$

Proof. Fix $t_0 \in J$. There exists an index i such that the mapping $t \rightarrow x_i(\gamma(t))$ has nonzero derivative when $t = t_0$. Thus there exists a function η_i of one variable, differentiable in a neighborhood of $x_i(\gamma(t_0))$, such that $t = \eta_i(x_i(\gamma(t)))$ for all t in an interval around t_0. The function $q \rightarrow g(\eta_i(x_i(q)))$ is defined and differentiable for all $q \in U$ sufficiently near $\gamma(t_0)$. Select $G^* \in C^\infty(U)$ such that $G^*(q) = g(\eta_i(x_i(q)))$ for all q in some neighborhood of $\gamma(t_0)$. Then

$$G^*(\gamma(t)) = g(t)$$

for all t in some interval around t_0. Owing to the compactness of J there exist finitely many relatively compact open subsets $U_1, ..., U_n$ of U covering $\gamma(J)$ and functions $G_i \in C^\infty(U)$ such that $G_i(\gamma(t)) = g(t)$ if $\gamma(t) \in U_i$ $(1 \leqslant i \leqslant n)$. Since U has a countable base it is paracompact and the sequence $U_1, ..., U_n$ can be completed to a locally finite covering $\{U_\alpha\}_{\alpha \in A}$ of U. We may assume that each U_α is relatively compact and that $U_\alpha \cap \gamma(J) = \emptyset$ if α is none of the numbers $1, ..., n$. Let $\{\varphi_\alpha\}_{\alpha \in A}$

be a partition of unity subordinate to this covering. Then the function $G' = \Sigma_{i=1}^{n} G_i \varphi_i$ belongs to $C^\infty(U)$ and $G'(\gamma(t)) = g(t)$ for each $t \in J$. Finally, let ψ be a function in $C^\infty(M)$ of compact support contained in U such that $\psi = 1$ on $\gamma(J)$. The function G given by $G(q) = \psi(q)\, G'(q)$ if $q \in U$ and $G(q) = 0$ if $q \notin U$ then has the required properties.

We put $X(t) = \dot{\gamma}(t)$ $(t \in I)$. Using Lemma 5.1 it is easy to see that there exist vector fields $X, Y \in \mathfrak{D}^1(M)$ such that ($Y(t)$ being as before)

$$X_{\gamma(t)} = X(t), \qquad Y_{\gamma(t)} = Y(t) \qquad (t \in J).$$

Given an affine connection ∇ on M, the family $Y(t)$ $(t \in J)$ is said to be *parallel* with respect to γ_J (or parallel along γ_J) if

$$\nabla_X(Y)_{\gamma(t)} = 0 \qquad \text{for all } t \in J. \tag{1}$$

To show that this definition is independent of the choice of X and Y, we express (1) in the coordinates $\{x_1, ..., x_m\}$. There exist functions X^i, Y^j $(1 \leqslant i, j \leqslant m)$ on U such that

$$X = \sum_i X^i \frac{\partial}{\partial x_i}, \qquad Y = \sum_j Y^j \frac{\partial}{\partial x_j} \qquad \text{on } U.$$

For simplicity we put $x_i(t) = x_i(\gamma(t))$, $X^i(t) = X^i(\gamma(t))$, and $Y^i(t) = Y^i(\gamma(t))$ $(t \in J)$ $(1 \leqslant i \leqslant m)$. Then $X^i(t) = \dot{x}_i(t)$ and since

$$\nabla_X(Y) = \sum_k \Big(\sum_i X^i \frac{\partial Y^k}{\partial x_i} + \sum_{i,j} X^i Y^j \Gamma_{ij}{}^k \Big) \frac{\partial}{\partial x_k} \qquad \text{on } U$$

we obtain

$$\frac{dY^k}{dt} + \sum_{i,j} \Gamma_{ij}{}^k \frac{dx_i}{dt} Y^j = 0 \qquad (t \in J). \tag{2}$$

This equation involves X and Y only through their values *on* the curve. Consequently, condition (1) for parallelism is independent of the choice of X and Y. It is now obvious how to define parallelism with respect to any finite curve segment γ_J and finally with respect to the entire curve γ.

Definition. Let $\gamma : t \to \gamma(t)$ $(t \in I)$ be a curve in M. The curve γ is called a *geodesic* if the family of tangent vectors $\dot{\gamma}(t)$ is parallel with respect to γ. A geodesic γ is called maximal if it is not a proper restriction of any geodesic.

Suppose γ_J is a finite geodesic segment without double points contained in a coordinate neighborhood U where the coordinates $\{x_1, \ldots, x_m\}$ are valid. Then (2) implies

$$\frac{d^2x_k}{dt^2} + \sum_{i,j} \Gamma_{ij}{}^k \frac{dx_i}{dt} \frac{dx_j}{dt} = 0 \qquad (t \in J). \tag{3}$$

If we change the parameter on the geodesic and put $t = f(s)$, $(f'(s) \neq 0)$, then we get a new curve $s \to \gamma_J(f(s))$. This curve is a geodesic if and only if f is a linear function, as (3) shows.

Proposition 5.2. *Let p, q be two points in M and γ a curve segment from p to q. The parallelism τ with respect to γ induces an isomorphism of M_p onto M_q.*

Proof. Without loss of generality we may assume that γ has no double points and lies in a coordinate neighborhood U. Let $\{x_1, \ldots, x_m\}$ be a system of coordinates on U. Suppose the curve segment γ is given by the mapping $t \to \gamma(t)$ $(a \leqslant t \leqslant b)$ such that $\gamma(a) = p$, $\gamma(b) = q$. As before we put $x_i(t) = x_i(\gamma(t))$ $(a \leqslant t \leqslant b)$ $(1 \leqslant i \leqslant m)$.

Consider the system (2). From the theory of systems of ordinary, linear differential equations of first order we can conclude:

There exist m functions $\varphi_i(t, y_1, \ldots, y_m)$ $(1 \leqslant i \leqslant m)$ defined and differentiable† for $a \leqslant t \leqslant b$, $-\infty < y_i < \infty$ such that

(i) For each m-tuple $(y_1, \ldots, y_m)$, the functions $Y^i(t) = \varphi_i(t, y_1, \ldots, y_m)$ satisfy the system (2).

(ii) $\varphi_i(a, y_1, \ldots, y_m) = y_i \qquad (1 \leqslant i \leqslant m)$.

The functions φ_i are uniquely determined by these properties.

The properties (i) and (ii) show that the family of vectors $Y(t) = \Sigma_i\, Y^i(t)\, (\partial/\partial x_i)$ $(a \leqslant t \leqslant b)$ is parallel with respect to γ and that $Y(a) = \Sigma_i\, y_i(\partial/\partial x_i)_p$. The mapping $Y(a) \to Y(b)$ is a linear mapping of M_p into M_q since the functions φ_i are linear in the variables $y_1, \ldots, y_m$. This mapping is one-to-one owing to the uniqueness of the functions φ_i. Consequently, it is an isomorphism.

Proposition 5.3. *Let M be a differentiable manifold with an affine connection. Let p be any point in M and let $X \neq 0$ in M_p. Then there exists a unique maximal geodesic $t \to \gamma(t)$ in M such that*

$$\gamma(0) = p, \qquad \dot{\gamma}(0) = X. \tag{4}$$

† A function on a closed interval I is called differentiable on I if it is extendable to a differentiable function on some open interval containing I.

Proof. Let $\varphi : q \to (x_1(q), ..., x_m(q))$ be a system of coordinates on a neighborhood U of p such that $\varphi(U)$ is a cube $\{(x_1, ..., x_m) : |x_i| < c\}$ and $\varphi(p) = 0$. Then X can be written $X = \sum_i \alpha_i(\partial/\partial x_i)_p$ where $\alpha_i \in \boldsymbol{R}$. We consider the system of differential equations

$$\frac{dx_i}{dt} = z_i \qquad (1 \leqslant i \leqslant m), \tag{5}$$

$$\frac{dz_k}{dt} = -\sum_{i,j=1}^{m} \Gamma_{ij}{}^k (x_1, ..., x_m)\, z_i z_j \qquad (1 \leqslant k \leqslant m), \tag{5'}$$

with the initial conditions

$$(x_1, ..., x_m, z_1, ..., z_m)_{t=0} = (0, ..., 0, \alpha_1, ..., \alpha_m).$$

Let c_1, K satisfy $0 < c_1 < c$, $0 < K < \infty$. In the interval $|x_i| < c_1$, $|z_i| < K$ $(1 \leqslant i \leqslant m)$, the right-hand sides of the foregoing equations satisfy a Lipschitz condition.

From the existence and uniqueness theorem (see, e.g., Miller and Murray [1], p. 42) for a system of ordinary differential equations we conclude:

There exists a constant $b_1 > 0$ and differentiable functions $x_i(t)$, $z_i(t)$ $(1 \leqslant i \leqslant m)$ in the interval $|t| \leqslant b_1$ such that

(i) $\dfrac{dx_i(t)}{dt} = z_i(t) \qquad (1 \leqslant i \leqslant m), \qquad |t| < b_1,$

$$\frac{dz_k(t)}{dt} = -\sum_{i,j=1}^{m} \Gamma_{ij}{}^k(x_1(t), ..., x_m(t))\, z_i(t)\, z_j(t) \qquad (1 \leqslant k \leqslant m), \qquad |t| < b_1;$$

(ii) $(x_1(t), ..., x_m(t), z_1(t), ..., z_m(t))_{t=0} = (0, ..., 0, \alpha_1, ..., \alpha_m);$

(iii) $|x_i(t)| < c_1, |z_i(t)| < K \qquad (1 \leqslant i \leqslant m), \qquad |t| < b_1;$

(iv) $x_i(t), z_i(t) \quad (1 \leqslant i \leqslant m)$ is the only set of functions satisfying the conditions (i), (ii), and (iii).

This shows that there exists a geodesic $t \to \gamma(t)$ in M satisfying (4) and that two such geodesics coincide in some interval around $t = 0$. Moreover, we can conclude from (iv) that if two geodesics $t \to \gamma_1(t)$ $(t \in I_1)$, $t \to \gamma_2(t)$ $(t \in I_2)$ coincide in some open interval, then they coincide for all $t \in I_1 \cap I_2$. Proposition 5.3 now follows immediately.

Definition. The geodesic with the properties in Prop. 5.3 will be denoted γ_X. If $X = 0$, we put $\gamma_X(t) = p$ for all $t \in \boldsymbol{R}$.

§ 6. The Exponential Mapping

Suppose again M is a C^∞ manifold with an affine connection. Let $p \in M$. We use the notation from the proof of Prop. 5.3. We shall now study the solutions of (5) and (5′) and their dependence on the initial values. From the existence and uniqueness theorem (see, e.g., Miller and Murray [1], p. 64) for the system (5), (5′), we can conclude:

There exists a constant b $(0 < b < c)$ and differentiable functions $\varphi_i(t, \xi_1, ..., \xi_m, \zeta_1, ..., \zeta_m)$ for $|t| \leqslant 2b$, $|\xi_i| \leqslant b$, $|\zeta_j| \leqslant b\,(1 \leqslant i, j \leqslant m)$ such that:

(i) For each fixed set $(\xi_1, ..., \xi_m, \zeta_1, ..., \zeta_m)$ the functions

$$x_i(t) = \varphi_i(t, \xi_1, ..., \xi_m, \zeta_1, ..., \zeta_m)$$

$$z_i(t) = \left[\frac{\partial \varphi_i}{\partial t}\right](t, \xi_1, ..., \xi_m, \zeta_1, ..., \zeta_m), \qquad 1 \leqslant i \leqslant m, \qquad |t| \leqslant 2b,$$

satisfy (5) and (5′) and $|x_i(t)| < c_1$, $|z_i(t)| < K$.

(ii) $(x_1(t), ..., x_m(t), z_1(t), ..., z_m(t))_{t=0} = (\xi_1, ..., \xi_m, \zeta_1, ..., \zeta_m)$.

(iii) The functions φ_i are uniquely determined by the above properties.

Theorem 6.1. *Let M be a manifold with an affine connection. Let p be any point in M. Then there exists an open neighborhood N_0 of 0 in M_p and an open neighborhood N_p of p in M such that the mapping $X \to \gamma_X(1)$ is a diffeomorphism of N_0 onto N_p.*

Proof. Using the notation above, we put

$$\psi_i(t, \zeta_1, ..., \zeta_m) = \varphi_i(t, 0, ..., 0, \zeta_1, ..., \zeta_m)$$

for $1 \leqslant i \leqslant m$, $|t| \leqslant 2b$, $|\zeta_i| \leqslant b$. Then

$$\psi_i(0, \zeta_1, ..., \zeta_m) = 0,$$

$$\left[\frac{\partial \psi_i}{\partial t}\right](0, \zeta_1, ..., \zeta_m) = \zeta_i.$$

Since $\gamma_X(st) = \gamma_{sX}(t)$, the uniqueness (iii) implies

$$\psi_i(st, \zeta_1, ..., \zeta_m) = \psi_i(t, s\zeta_1, ..., s\zeta_m) \tag{1}$$

for $|s| \leqslant 1$, $|t| \leqslant 2b$, $|\zeta_i| \leqslant b$. Now let D_i denote partial derivative with respect to the ith argument; from (1) we get by differentiating with respect to s,

$$t[D_1\psi_i]\,(st, \zeta_1, ..., \zeta_m) = \sum_{k=1}^{m} \zeta_k[D_{k+1}\psi_i]\,(t, s\zeta_1, ..., s\zeta_m),$$

$$t^2[D_1^2\psi_i]\,(st, \zeta_1, ..., \zeta_m) = \sum_{j,k=1}^{m} \zeta_j\zeta_k[D_{j+1}D_{k+1}\psi_i]\,(t, s\zeta_1, ..., s\zeta_m).$$

From Taylor's formula we find

$$\psi_i(b, \zeta_1, ..., \zeta_m) = \zeta_i b + \tfrac{1}{2}[D_1^2\psi_i]\,(b^*, \zeta_1, ..., \zeta_m)\, b^2 \qquad (0 \leqslant b^* \leqslant b),$$

so

$$\psi_i(b, \zeta_1, ..., \zeta_m) = \zeta_i b + \tfrac{1}{2}\sum_{j,k=1}^{m} \zeta_j\zeta_k[D_{j+1}D_{k+1}\psi_i]\left(b, \frac{b^*}{b}\zeta_1, ..., \frac{b^*}{b}\zeta_m\right).$$

This shows that the mapping

$$\Psi : (\zeta_1, ..., \zeta_m) \rightarrow (\psi_1(b, \zeta_1, ..., \zeta_m), ..., \psi_m(b, \zeta_1, ..., \zeta_m))$$

has Jacobian at the origin equal to b^m. The mapping Ψ is just the mapping $X \rightarrow \gamma_X(b)$ expressed in coordinates (§ 3, No. 1). Since $\gamma_{bX}(1) = \gamma_X(b)$, the theorem follows.

Definition. The mapping $X \rightarrow \gamma_X(1)$ described in Theorem 6.1 is called the *Exponential mapping* at p and will be denoted by Exp (or Exp_p).

Definition. Let M be a manifold with an affine connection and p a point in M. An open neighborhood N_0 of the origin in M_p is said to be *normal* if: (1) the mapping Exp is a diffeomorphism of N_0 onto an open neighborhood N_p of p in M; (2) if $X \in N_0$, and $0 \leqslant t \leqslant 1$, then $tX \in N_0$.

The last condition means that N_0 is "*star-shaped.*" A neighborhood N_p of p in M is called a *normal neighborhood* of p if $N_p = \text{Exp}\, N_0$ where N_0 is a normal neighborhood of 0 in M_p. Assuming this to be the case, and letting $X_1, ..., X_m$ denote some basis of M_p, the inverse mapping†

$$\text{Exp}_p\,(a_1X_1 + ... + a_mX_m) \rightarrow (a_1, ..., a_m)$$

of N_p into $\boldsymbol{R}^m$ is called a system of *normal coordinates* at p.

We shall now prove a useful refinement of Theorem 6.1.

† Here and sometimes in the sequel we allow ourselves to denote the inverse of a one-to-one mapping $X \rightarrow \phi(X)$ by $\phi(X) \rightarrow X$.

Theorem 6.2. *Let M be a C^∞ manifold with an affine connection. Then each point $p \in M$ has a normal neighborhood N_p which is a normal neighborhood of each of its points. (In particular, two arbitrary points in N_p can be joined by exactly one*† *geodesic segment contained in N_p.)*

We shall use the notation from the proof of Prop. 5.3 and consider again the functions $\varphi_i(t, \xi_1, ..., \xi_m, \zeta_1, ..., \zeta_m)$ above. If $q \in U$ and $0 < \delta \leqslant c - \max_i | x_i(q) |$, then the subset of U given by

$$V_\delta(q) = \left\{ r \in U : \sum_{i=1}^{m} (x_i(r) - x_i(q))^2 < \delta^2 \right\}$$

will be called a *spherical neighborhood* of q with radius δ.

Now consider an m-tuple $(\xi_1, ..., \xi_m)$ where $| \xi_i | < b$, $1 \leqslant i \leqslant m$. Let $q \in U$ be determined by $x_i(q) = \xi_i$ $(1 \leqslant i \leqslant m)$. By the proof above, the mapping

$$\Phi : (\zeta_1, ..., \zeta_m) \to (\varphi_1(b, \xi_1, ..., \xi_m, \zeta_1, ..., \zeta_m), ..., \varphi_m(b, \xi_1, ..., \xi_m, \zeta_1, ..., \zeta_m))$$

has Jacobian b^m at the origin $(\zeta_1, ..., \zeta_m) = (0, ..., 0)$. Hence $\varphi^{-1} \circ \Phi$ is a diffeomorphism of a neighborhood $\zeta_1^2 + ... + \zeta_m^2 < r^2$ $(r \leqslant b)$ of the origin in $\boldsymbol{R}^m$ onto an open neighborhood N_q of q in M. We can suppose r taken as large as possible with this property. For reasons of continuity there exists a $\delta_0 > 0$ such that, if $\xi_1^2 + ... + \xi_m^2 < \delta_0^2$, then the corresponding N_q all have a spherical neighborhood of p in common. By taking δ_0 small enough we may assume that this spherical neighborhood is $V_{4\delta_0}(p)$. Since N_q is normal, this proves:

Lemma 6.3. *There exists a number $\delta_0 > 0$ such that for each $q \in V_{\delta_0}(p)$, the spherical neighborhood $V_{2\delta_0}(q)$ is contained in a normal neighborhood of q.*

The neighborhood $V_{\delta_0}(p)$ therefore has the following property: For each pair of points $q_1, q_2 \in V_{\delta_0}(p)$ there exists *at most* one geodesic segment contained in $V_{\delta_0}(p)$, joining q_1 and q_2. A neighborhood with this property will be called *simple*. It is obvious that if $0 < \delta \leqslant \delta_0$, then $V_\delta(p)$ is also simple. We shall now show that for all sufficiently small δ, $V_\delta(p)$ is also *convex*, that is, two arbitrary points in $V_\delta(p)$ can be joined by a geodesic segment contained in $V_\delta(p)$.

Let δ^* be a number satisfying the following two conditions:

(i) $0 < \delta^* < \delta_0$;

(ii) The matrix $(\delta_{ij} - \Sigma_k x_k \Gamma_{ij}{}^k)$ is strictly positive definite for $\Sigma_k x_k^2 \leqslant (\delta^*)^2$. (Here $\delta_{ij} = 1$ if $i = j$, 0 otherwise.)

† Except for a linear change of parameter on the geodesic.

It is obvious that such a number δ^* does exist. Theorem 6.2 is contained in the following lemma.

Lemma 6.4. *If* $0 < \delta \leqslant \delta^*$, *the neighborhood* $V_\delta(p)$ *is a normal neighborhood of each of its points. In particular,* $V_\delta(p)$ *is simple and convex.*

Proof. The boundary D of $V_\delta(p)$ is a submanifold of $V_{\delta_0}(p)$. We first prove that if a geodesic $\gamma : t \to \gamma(t)$ is tangent to D at a point $q_0 = \gamma(t_0)$, then for all $t \neq t_0$ sufficiently close to t_0, the point $\gamma(t)$ lies outside D. As before, we put $x_k(t) = x_k(\gamma(t))$ $(1 \leqslant k \leqslant m)$. Then the functions $x_k(t)$ satisfy (3), § 5, in a neighborhood of t_0. In Taylor's formula

$$F(t_0 + \Delta t) = F(t_0) + \Delta t \dot{F}(t_0) + \tfrac{1}{2}(\Delta t)^2 \ddot{F}(t_0) + 0(\Delta t)^3$$

for the function $F(t) = \sum_{k=1}^{m} (x_k(t))^2 - \delta^2$, we have

$$\dot{F}(t_0) = 2\sum_{k=1}^{m} x_k(t_0)\, \dot{x}_k(t_0) = 0, \qquad F(t_0) = 0,$$

$$\ddot{F}(t) = 2\sum_{k=1}^{m} (\dot{x}_k(t)\, \dot{x}_k(t) + x_k(t)\, \ddot{x}_k(t))$$

$$= 2\sum_{i,j=1}^{m} \Big(\delta_{ij} - \sum_k x_k(t)\, \Gamma_{ij}{}^k\Big)\, \dot{x}_i(t)\, \dot{x}_j(t).$$

Using (ii), it follows that $F(t_0 + \Delta t) > 0$ provided Δt is sufficiently small and $\neq 0$. This proves the statement concerning γ.

For a pair $P, Q \in V_\delta(p)$ we have therefore only two possibilities:

1) There is no geodesic segment inside $V_\delta(p)$ which joins P and Q. In this case, the unique geodesic segment inside N_P which joins P and Q will contain points outside the boundary D.

2) There exists a geodesic segment inside $V_\delta(p)$ which joins P and Q. In this case P and Q are said to be *mutually visible* inside $V_\delta(p)$.

Let S denote the subset of $V_\delta(p) \times V_\delta(p)$ consisting of all point-pairs which are mutually visible inside $V_\delta(p)$. The set S is nonempty and we shall now show that S is open and closed in the relative topology of $V_\delta(p) \times V_\delta(p)$. In view of the connectedness of $V_\delta(p)$, this will prove Lemma 6.4.

I. *S is closed.* Let (p_n, q_n) be a sequence in S which converges to $(p^*, q^*) \in V_\delta(p) \times V_\delta(p)$. We join p_n and q_n by a geodesic segment $t \to \gamma_n(t)$ in $V_\delta(p)$ such that $\gamma_n(0) = p_n$, $\gamma_n(b) = q_n$. Similarly, p^* and q^* are joined by a geodesic segment $\gamma^*(t)$ $(0 \leqslant t \leqslant b)$ inside N_{p^*}. Consider the mapping Φ above for $(\xi_1^{(n)}, \ldots, \xi_m^{(n)}) = (x_1(p_n), \ldots, x_m(p_n))$.

Under this mapping the point q_n corresponds to a certain m-tuple $(\zeta_1^{(n)}, ..., \zeta_m^{(n)})$. Since these m-tuples are bounded ($|\zeta_i| \leqslant b$), we can, passing to a subsequence if necessary, assume that $(\zeta_1^{(n)}, ..., \zeta_m^{(n)})$ converges to a limit $(\zeta_1^*, ..., \zeta_m^*)$ as $n \to \infty$. Then for $1 \leqslant i \leqslant m$ and $0 \leqslant t \leqslant b$ the sequence

$$\varphi_i(t, \xi_1^{(n)}, ..., \xi_m^{(n)}, \zeta_1^{(n)}, ..., \zeta_m^{(n)})$$

converges to

$$\varphi_i(t, x_1(p^*), ..., x_m(p^*), \zeta_1^*, ..., \zeta_m^*)$$

which represents a geodesic inside $V_{2\delta_0}(p^*)$ joining p^* to q^*. Owing to the uniqueness, it follows that

$$x_i(\gamma^*(t)) = \varphi_i(t, x_1(p^*), ..., x_m(p^*), \zeta_1^*, ..., \zeta_m^*)$$

for $0 \leqslant t \leqslant b$, $1 \leqslant i \leqslant m$; in other words $\gamma_n(t) \to \gamma^*(t)$ for $0 \leqslant t \leqslant b$. Since $\gamma_n(t) \in V_\delta(p)$ $(0 \leqslant t \leqslant b)$ it follows that γ^* contains no points outside the boundary D. Owing to 1) above, we have $(p^*, q^*) \in S$; hence S is closed.

II. *S is open.* In fact, the same argument as in I shows that the complement $(V_\delta(p) \times V_\delta(p)) - S$ is closed.

Definition. Let M be a manifold with an affine connection. Let p be a point in M and N_p a normal neighborhood of p. Let $X \in M_p$ and for each $q \in N_p$ put $(X^*)_q = \tau_{pq}X$ where τ_{pq} is the parallel translation along the unique geodesic segment in N_p which joins p and q. It is clear from (2), § 5, that $(X^*)_q$ depends differentiably on q. The vector field X^* on N_p, thus defined, is said to be *adapted* to the tangent vector X. As before, let $\theta(X^*)$ denote the Lie derivative with respect to X^*.

Definition. An affine connection ∇ on an analytic manifold M is called *analytic* if for each $p \in M$, $\nabla_X(Y)$ is analytic at p whenever the vector fields X and Y are analytic at p.

Theorem 6.5. *Let M be an analytic manifold with an analytic affine connection ∇. Let $p \in M$ and $X \neq 0$ in M_p. Then there exists an $\epsilon > 0$ such that the differential of Exp (= Exp_p) is given by*

$$(d\,\mathrm{Exp})_{tX}(Y) = \left\{\frac{1 - e^{\theta(-tX^*)}}{\theta(tX^*)}(Y^*)\right\}_{\mathrm{Exp}\,tX}, \qquad Y \in M_p,$$

for $|t| < \epsilon$.

Here $(1 - e^{-A})/A$ stands for $\sum_0^\infty (-A)^m/(m+1)!$ and as usual (Remark, § 2, No. 1) M_p is identified with its tangent space at each point.

Proof. The mapping Exp is analytic at the origin in M_p. Let f be an anlytic function at p. Then there exists a star-shaped neighborhood U_0 of 0 in M_p such that

$$f(\operatorname{Exp} Z) = P(z_1, ..., z_m), \qquad Z \in U_0,$$

where P is an absolutely convergent power series and $z_1, ..., z_m$ are the coordinates of Z with respect to some basis of M_p. It follows that for fixed $Z \in U_0$

$$f(\operatorname{Exp} tZ) = P(tz_1, ..., tz_m) = \sum_0^\infty \frac{1}{n!} a_n t^n \qquad (a_n \in \boldsymbol{R})$$

for $0 \leqslant t \leqslant 1$. If t is sufficiently small

$$[Z^*f](\operatorname{Exp} tZ) = \left\{\frac{d}{du} f(\operatorname{Exp}(t+u) Z)\right\}_{u=0} = \frac{d}{dt} f(\operatorname{Exp} tZ)$$

and by induction

$$[(Z^*)^n f](\operatorname{Exp} tZ) = \frac{d^n}{dt^n} f(\operatorname{Exp} tZ).$$

On putting $t = 0$ we find that $[(Z^*)^n f](p) = a_n$; hence

$$f(\operatorname{Exp} Z) = \sum_0^\infty \frac{1}{n!} [(Z^*)^n f](p) \qquad (Z \in U_0). \tag{2}$$

Now suppose $Y \in M_p$. Then

$$d \operatorname{Exp}_{tX}(Y) f = Y_{tX}(f \circ \operatorname{Exp}) = \left\{\frac{d}{du} f(\operatorname{Exp}(tX + uY))\right\}_{u=0}.$$

If t and u are sufficiently small we get from (2)

$$f(\operatorname{Exp}(tX + uY)) = \sum_0^\infty \frac{1}{r!} [(tX^* + uY^*)^r f](p) = \sum_{m,n \geq 0} \frac{t^n u^m}{(n+m)!} [S_{n,m} f](p) \tag{3}$$

where $S_{n,m}$ is the coefficient to $t^n u^m$ in $(tX^* + uY^*)^{n+m}$. In particular,

$$S_{n,1} = (X^*)^n Y^* + (X^*)^{n-1} Y^* X^* + ... + Y^*(X^*)^n.$$

We differentiate the expansion (3) with respect to u and put $u = 0$. We obtain

$$d \operatorname{Exp}_{tX}(Y) f = \sum_0^\infty \frac{t^n}{(n+1)!} [((X^*)^n Y^* + ... + Y^*(X^*)^n) f](p).$$

Let N_p be a normal neighborhood of p. Let $\boldsymbol{D}(N_p)$ denote the algebra of operators on $C^\infty(N_p)$ generated by the vector fields Z^*, as Z varies through M_p. Let L_{X^*} and R_{X^*} denote the linear transformations of $\boldsymbol{D}(N_p)$ given by $L_{X^*}: A \to X^*A$ and $R_{X^*}: A \to AX^*$. Since $\theta(X^*)\,Y^* = X^*Y^* - Y^*X^*$, we put $\theta(X^*)\,A = X^*A - AX^*$ for each $A \in \boldsymbol{D}(N_p)$. Then $\theta(X^*) = L_{X^*} - R_{X^*}$ so $\theta(X^*)$ and L_{X^*} commute; hence we have

$$(R_{X^*})^m = (L_{X^*} - \theta(X^*))^m = \sum_{p=0}^{m} (-1)^p \binom{m}{p} (L_{X^*})^{m-p}(\theta(X^*))^p.$$

On using the relation

$$\sum_{p=0}^{n-k} \binom{n-p}{k} = \binom{n+1}{k+1}$$

we find

$$S_{n,1} = \sum_{p=0}^{n} (X^*)^p \sum_{k=0}^{n-p} (-1)^k \binom{n-p}{k} (X^*)^{n-p-k}\, \theta(X^*)^k\, (Y^*)$$

$$= \sum_{k=0}^{n} \binom{n+1}{k+1} (X^*)^{n-k}\, \theta(-X^*)^k\, (Y^*)$$

so

$$d\,\mathrm{Exp}_{tX}(Y)f = \sum_{n=0}^{\infty} \Big[\sum_{k=0}^{n} \Big\{ \frac{(tX^*)^{n-k}}{(n-k)!} \frac{\theta(-tX^*)^k}{(k+1)!} (Y^*) \Big\} f \Big] (p).$$

For sufficiently small t, the right-hand side can be rewritten by the formula

$$\sum_{n=0}^{\infty} \Big[\sum_{k=0}^{n} \Big\{ \frac{(tX^*)^{n-k}}{(n-k)!} \frac{(\theta(-tX^*))^k}{(k+1)!} (Y^*) \Big\} f \Big] (p) \tag{4}$$

$$= \sum_{r=0}^{\infty} \Big[\frac{(tX^*)^r}{r!} \Big(\sum_{m=0}^{\infty} \Big\{ \frac{(\theta(-tX^*))^m}{(m+1)!} (Y^*) \Big\} f \Big) \Big] (p).$$

In order to justify (4) we first prove the statements (i) and (ii) below.

(i) There exists an interval $I_\delta : -\delta < t < \delta$ and an open neighborhood U of p such that the series

$$G(t, q) = \sum_{m=0}^{\infty} \Big\{ \frac{\theta(-tX)^m}{(m+1)!} (Y^*) \Big\}_q f \tag{5}$$

converges absolutely and represents an analytic function for $(t, q) \in I_\delta \times U$, and such that the operator X^* can be applied to the series (5) term by term.

(ii) There exists a subinterval $I_{\delta_1}: -\delta_1 < t < \delta_1$ of I_δ such that the series

$$\sum_{r=0}^{\infty} \frac{t^r}{r!} [(X^*)^r G](t, p)$$

converges uniformly for $t \in I_{\delta_1}$.

In order to prove (i) and (ii) we can assume that in a suitable coordinate system $\{x_1, ..., x_m\}$ valid near p we have $x_1(p) = ... = x_m(p) = 0$ and $X^* = \partial/\partial x_1$ (see Exercise A.7 in this chapter). We may also assume that $Y^* = g\partial/\partial x_2$ where g is analytic. Then

$$\theta(X^*)^m(Y^*) = \frac{\partial^m g}{\partial x_1^m} \frac{\partial}{\partial x_2}$$

and the series (5) reduces to

$$\sum_{s=0}^{\infty} \frac{(-t)^s}{(s+1)!} \frac{\partial^s g}{\partial x_1^s} \frac{\partial f}{\partial x_2}.$$

Since g is analytic its derivatives of sth order are bounded by $Cs!$ (C = constant) uniformly in a neighborhood of the origin. Hence (i) follows. For (ii) we expand $G(t, q)$ in a power series

$$G(t, q) = \sum a_{i_0 i_1 \ldots i_m} t^{i_0} x_1^{i_1} \ldots x_m^{i_m}.$$

Then

$$[(X^*)^r G](t, p) = \sum_{i=0}^{\infty} (r!\, a_{i,r,0\ldots 0})\, t^i.$$

Let $\rho > 0$ be a number such that the series for $G(t, q)$ converges for $(t, x_1, ..., x_m) = (\rho, \rho, ..., \rho)$. Then there exists a constant K such that

$$| a_{i_0 i_1 \ldots i_m} | \leqslant K\rho^{-(i_0 + \ldots + i_m)}.$$

It follows that

$$| [(X^*)^r G](t, p) | \leqslant K\rho^{-r} r! \sum_{i=0}^{\infty} (t/\rho)^i \leqslant 2K\rho^{-r} r!$$

provided $| t | \leqslant \frac{1}{2}\rho$. Statement (ii) is now obvious.

We put now

$$a_{m+r,r} = \left[\left\{\frac{(X^*)^r}{r!}\frac{\theta(-X^*)^m}{(m+1)!}(Y^*)\right\}f\right](p).$$

Then from (i) and (ii) we know that there exists a disk D around 0 in the complex plane such that the series

$$\sum_{n=r}^{\infty} a_{n,r}t^n \quad \text{and} \quad \sum_{r=0}^{\infty}\left(\sum_{n=r}^{\infty} a_{n,r}t^n\right)$$

converge absolutely and uniformly for $t \in D$. By Weierstrass' theorem on double series we can interchange the summation so that we have

$$\sum_{r=0}^{\infty}\sum_{n=r}^{\infty} a_{n,r}t^n = \sum_{n=0}^{\infty}\sum_{r=0}^{n} a_{n,r}t^n.$$

This, however, is precisely the relation (4). In view of (2), we have proved: There exists a number $\epsilon_f > 0$ such that

$$d\,\mathrm{Exp}_{tX}(Y)f = \left\{\frac{1-e^{-\theta(tX^*)}}{\theta(tX^*)}(Y^*)\right\}_{\mathrm{Exp}\,tX} f$$

for $|t| < \epsilon_f$. Using this relation on the coordinate functions $f = x_1, \dots, f = x_m$, the theorem follows with $\epsilon = \min(\epsilon_{x_1}, \dots, \epsilon_{x_m})$.

§ 7. Covariant Differentiation

In § 5, parallelism was defined by means of the covariant differentiation ∇_X. Theorem 7.1 below shows that it is also possible to go the other way and describe the covariant derivative by means of parallel translation. This makes it possible to define the covariant derivative of other objects.

Definition. Let X be a vector field on a manifold M. A curve $s \to \varphi(s)$ $(s \in I)$ is called an *integral curve* of X if

$$\dot{\varphi}(s) = X_{\varphi(s)}, \qquad s \in I. \tag{1}$$

Assuming $0 \in I$, let $p = \varphi(0)$ and let $\{x_1, \dots, x_m\}$ be a system of coordinates valid in a neighborhood U of p. There exist functions

$X^i \in C^\infty(U)$ such that $X = \sum_i X^i \, \partial/\partial x_i$ on U. For simplicity let $x_i(s) = x_i(\varphi(s))$ and write X^i instead of $(X^i)^*$ (§ 2, No. 1). Then (1) is equivalent to

$$\frac{dx_i(s)}{ds} = X^i(x_1(s), \ldots, x_m(s)) \qquad (1 \leqslant i \leqslant m). \tag{2}$$

Therefore if $X_p \neq 0$ there exists an integral curve of X through p.

Theorem 7.1. *Let M be a manifold with an affine connection. Let $p \in M$ and let X, Y be two vector fields on M. Assume $X_p \neq 0$. Let $s \to \varphi(s)$ be an integral curve of X through $p = \varphi(0)$ and τ_t the parallel translation from p to $\varphi(t)$ with respect to the curve φ. Then*

$$(\nabla_X(Y))_p = \lim_{s\to 0} \frac{1}{s} (\tau_s^{-1} Y_{\varphi(s)} - Y_p).$$

Proof. We shall use the notation introduced above. Consider a fixed $s > 0$ and the family $Z_{\varphi(t)}$ $(0 \leqslant t \leqslant s)$ which is parallel with respect to the curve φ such that $Z_{\varphi(0)} = \tau_s^{-1} Y_{\varphi(s)}$. We can write

$$Z_{\varphi(t)} = \sum_i Z^i(t) \left(\frac{\partial}{\partial x_i}\right)_{\varphi(t)}, \qquad Y_{\varphi(t)} = \sum_i Y^i(t) \left(\frac{\partial}{\partial x_i}\right)_{\varphi(t)},$$

and have the relations

$$\dot{Z}^k(t) + \sum_{i,j} \Gamma_{ij}{}^k \, \dot{x}_i(t) \, Z^j(t) = 0 \qquad (0 \leqslant t \leqslant s)$$

$$Z^k(s) = Y^k(s) \qquad (1 \leqslant k \leqslant m).$$

By the mean value theorem

$$Z^k(s) = Z^k(0) + s\dot{Z}^k(t^*)$$

for a suitable number t^* between 0 and s. Hence the kth component of $(1/s)\,(\tau_s^{-1} Y_{\varphi(s)} - Y_p)$ is

$$\begin{aligned} \frac{1}{s}(Z^k(0) - Y^k(0)) &= \frac{1}{s}\{Z^k(s) - s\dot{Z}^k(t^*) - Y^k(0)\} \\ &= \sum_{i,j} \Gamma_{ij}{}^k(\varphi(t^*)) \, \dot{x}_i(t^*) \, Z^j(t^*) + \frac{1}{s}(Y^k(s) - Y^k(0)). \end{aligned}$$

As $s \to 0$ this expression has the limit

$$\frac{dY^k}{ds} + \sum_{i,j} \Gamma_{ij}{}^k \frac{dx_i}{ds} Y^j.$$

Let this last expression be denoted by A_k. It was shown earlier that

$$\nabla_X(Y)_p = \sum_k A_k \left(\frac{\partial}{\partial x_k}\right)_p.$$

This proves the theorem.

By using Theorem 7.1 it is now possible to define covariant derivatives of arbitrary tensor fields. Let p and q be two points in M and γ a curve segment in M from p to q. Let τ denote the parallel translation along γ. If $F \in M_p^*$ we define $\tau \cdot F \in M_q^*$ by the formula $(\tau \cdot F)(A) = F(\tau^{-1} \cdot A)$ for each $A \in M_q$. If T is a tensor field on M of type (r, s) where $r + s > 0$, we define $\tau \cdot T_p \in \mathfrak{D}_s^r(q)$ by

$$(\tau \cdot T_p)(F_1, ..., F_r, A_1, ..., A_s) = T_p(\tau^{-1}F_1, ..., \tau^{-1}F_r, \tau^{-1}A_1, ..., \tau^{-1}A_s)$$

for $A_i \in M_q$, $F_j \in M_q^*$. Now, let $X \in \mathfrak{D}^1(M)$ and let p be any point in M where $X_p \neq 0$. With the notation of Theorem 7.1 we put

$$(\nabla_X T)_p = \lim_{s \to 0} \frac{1}{s} (\tau_s^{-1} T_{\varphi(s)} - T_p). \tag{3}$$

For each point $q \in M$ where $X_q = 0$ we put $(\nabla_X T)_q = 0$ in accordance with Lemma 4.2. For a function $f \in C^\infty(M)$ we put

$$(\nabla_X f)_p = \lim_{s \to 0} \frac{1}{s} (f(\varphi(s)) - f(p)),$$

if $X_p \neq 0$, otherwise we put $(\nabla_X f)_p = 0$. Then we have $\nabla_X f = Xf$. Finally ∇_X is extended to a linear mapping of $\mathfrak{D}$ into $\mathfrak{D}$.

Proposition 7.2. *The operator ∇_X has the following properties:*

(i) *∇_X is a derivation of the mixed tensor algebra $\mathfrak{D}(M)$ (considered as an algebra over **R**).*

(ii) *∇_X preserves type of tensors.*

(iii) *∇_X commutes with all contractions $C^i{}_j$.*

The verification of these properties is quite straightforward. For a simple application, let $X, Y \in \mathfrak{D}^1(M)$, $\omega \in \mathfrak{D}_1(M)$. Then by (i)

$$\nabla_X(Y \otimes \omega) = \nabla_X(Y) \otimes \omega + Y \otimes \nabla_X \omega, \tag{4}$$

so (iii) implies

$$(\nabla_X \omega)(Y) = X \cdot \omega(Y) - \omega(\nabla_X(Y)). \tag{5}$$

The tensor field $Y \otimes \omega$ can be regarded as an $\mathfrak{F}$-linear mapping of $\mathfrak{D}^1$ into itself given by

$$Y \otimes \omega : Z \to \omega(Z)\, Y \qquad (Z \in \mathfrak{D}^1).$$

If A, B are two mappings of $\mathfrak{D}^1$ into itself we put $[A, B] = AB - BA$. Then

$$[\nabla_X, Y \otimes \omega] = \nabla_X(Y \otimes \omega). \tag{6}$$

In fact,

$$[\nabla_X, Y \otimes \omega]\,(Z) = \nabla_X(\omega(Z)Y) - (Y \otimes \omega)\,(\nabla_X Z)$$
$$= \omega(Z)\,\nabla_X(Y) + (X \cdot \omega(Z))\,Y - \omega(\nabla_X Z)\,Y.$$

On the other hand, (4) and (5) imply

$$\nabla_X(Y \otimes \omega)\,(Z) = \omega(Z)\,\nabla_X(Y) + (\nabla_X\omega)\,(Z)\,Y$$
$$= \omega(Z)\,\nabla_X(Y) + (X \cdot \omega(Z))\,Y - \omega(\nabla_X(Z))\,Y,$$

proving (6). We have therefore

$$[\nabla_X, B] = \nabla_X \cdot B \tag{7}$$

if B is a tensor field of type (1,1). On the left-hand side, B is to be considered as the linear mapping of $\mathfrak{D}^1$ into $\mathfrak{D}^1$ given by

$$Z \to C^1{}_1(Z \otimes B),$$

where $C^1{}_1$ is the contraction of the first contravariant and first covariant index.

§ 8. The Structural Equations

Let M be a manifold with an affine connection ∇. We put

$$T(X, Y) = \nabla_X(Y) - \nabla_Y(X) - [X, Y],$$
$$R(X, Y) = \nabla_X\nabla_Y - \nabla_Y\nabla_X - \nabla_{[X,Y]}$$

for all $X, Y \in \mathfrak{D}^1$. Note that $T(X, Y) = -T(Y, X)$ and $R(X, Y) = -R(Y, X)$. It is easy to verify that $T(fX, gY) = fgT(X, Y)$ and $R(fX, gY) \cdot hZ = fghR(X, Y) \cdot Z$ for all $f, g, h \in C^\infty(M)$, $X, Y, Z \in \mathfrak{D}^1$. The mapping

$$(\omega, X, Y) \to \omega(T(X, Y))$$

is an $\mathfrak{F}$-multilinear mapping of $\mathfrak{D}_1 \times \mathfrak{D}^1 \times \mathfrak{D}^1$ into $\mathfrak{F}$ and therefore is an element of $\mathfrak{D}_2^1(M)$. This element is called the *torsion tensor field* and is also denoted by T. Similarly, the mapping

$$(\omega, Z, X, Y) \to \omega(R(X, Y) \cdot Z)$$

is an $\mathfrak{F}$-multilinear mapping of $\mathfrak{D}_1 \times \mathfrak{D}^1 \times \mathfrak{D}^1 \times \mathfrak{D}^1$ into $\mathfrak{F}$ and therefore is an element of $\mathfrak{D}_3^1(M)$. This element is called the *curvature tensor field* and is also denoted by R. The tensor fields T and R have type $(1, 2)$ and $(1, 3)$, respectively.

Let $p \in M$ and suppose $X_1, \ldots, X_m$ is a basis for the vector fields in some open neighborhood N_p of p, that is, each vector field X on N_p can be written $X = \sum_i f_i X_i$ where $f_i \in C^\infty(N_p)$. We define the functions $\Gamma_{ij}{}^k$, $T^k{}_{ij}$, $R^k{}_{lij}$ on N_p by the formulas

$$\nabla_{X_i}(X_j) = \sum_k \Gamma_{ij}{}^k X_k,$$

$$T(X_i, X_j) = \sum_k T^k{}_{ij} X_k,$$

$$R(X_i, X_j) \cdot X_l = \sum_k R^k{}_{lij} X_k.$$

Let ω^i, $\omega^i{}_j$ $(1 \leqslant i, j \leqslant m)$ be the 1-forms on N_p determined by

$$\omega^i(X_j) = \delta^i{}_j, \qquad \omega^i{}_j = \sum_k \Gamma_{kj}{}^i \omega^k.$$

It is clear that the forms $\omega^i{}_j$ determine the functions $\Gamma_{kj}{}^i$ on N_p and thereby the connection ∇. On the other hand, as the next theorem shows, the forms $\omega^i{}_j$ are described by the torsion and curvature tensor fields.

Theorem 8.1 (*the structural equations of Cartan*).

$$d\omega^i = -\sum_p \omega^i{}_p \wedge \omega^p + \tfrac{1}{2}\sum_{j,k} T^i{}_{jk}\omega^j \wedge \omega^k, \tag{1}$$

$$d\omega^i{}_l = -\sum_p \omega^i{}_p \wedge \omega^p{}_l + \tfrac{1}{2}\sum_{j,k} R^i{}_{ljk}\omega^j \wedge \omega^k. \tag{2}$$

Both sides of (1) represent a 2-form on N_p. We apply both sides of that equation to (X_j, X_k) and evaluate by means of the rules (4) and (9) in § 2. If we define the functions $c^i{}_{jk}$ by $[X_j, X_k] = \sum_i c^i{}_{jk} X_i$, the left-hand side of (1) is

$$d\omega^i(X_j, X_k) = \tfrac{1}{2}\{X_j \cdot \omega^i(X_k) - X_k \cdot \omega^i(X_j) - \omega^i([X_j, X_k])\}$$

$$= \tfrac{1}{2}\{0 - 0 - c^i{}_{jk}\}.$$

As for the right-hand side we have first

$$T(X_j, X_k) = \nabla_{X_j}(X_k) - \nabla_{X_k}(X_j) - [X_j, X_k]$$

$$= \sum_i (\Gamma_{jk}{}^i - \Gamma_{kj}{}^i - c^i{}_{jk}) X_i$$

so

$$T^i{}_{jk} = \Gamma_{jk}{}^i - \Gamma_{kj}{}^i - c^i{}_{jk}. \tag{3}$$

Similarly, we find

$$R^k{}_{lij} = \sum_p (\Gamma_{jl}{}^p \Gamma_{ip}{}^k - \Gamma_{il}{}^p \Gamma_{jp}{}^k) + X_i \cdot \Gamma_{jl}{}^k - X_j \cdot \Gamma_{il}{}^k - \sum_p c^p{}_{ij} \Gamma_{pl}{}^k; \tag{4}$$

furthermore,

$$\Big(-\sum_p \omega^i{}_p \wedge \omega^p\Big)(X_j, X_k) = -\sum_p \tfrac{1}{2}\{\omega^i{}_p(X_j)\, \omega^p(X_k) - \omega^p(X_j)\, \omega^i{}_p(X_k)\}$$

$$= \tfrac{1}{2}(\Gamma_{kj}{}^i - \Gamma_{jk}{}^i),$$

$$\Big(\tfrac{1}{2}\sum_{r,s} T^i{}_{rs}\omega^r \wedge \omega^s\Big)(X_j, X_k) = \tfrac{1}{4}\sum_{r,s} (\Gamma_{rs}{}^i - \Gamma_{sr}{}^i - c^i{}_{rs})(\delta_{jr}\delta_{ks} - \delta_{js}\delta_{kr})$$

$$= \tfrac{1}{2}(\Gamma_{jk}{}^i - \Gamma_{kj}{}^i - c^i{}_{jk}),$$

so (1) follows immediately; in the same way (2) follows if we use formula (4).

Suppose $Y_1, \ldots, Y_m$ is a basis for the tangent space M_p. Let N_0 be a normal neighborhood of the origin in M_p and let N_p denote the normal neighborhood Exp N_0 of p in M. Let $Y_1^*, \ldots, Y_m^*$ be the vector fields on N_p that are adapted to the tangent vectors $Y_1, \ldots, Y_m$. Then $Y_1^*, \ldots, Y_m^*$ is a basis for the vector fields on N_p, due to Prop. 5.2. Suppose ω^i, $\omega^i{}_l$, $\Gamma_{ij}{}^k$, $R^k{}_{lij}$, and $T^k{}_{ij}$ are defined by means of this basis. Let V be the set of points $(t, a_1, \ldots, a_m) \in \boldsymbol{R} \times \boldsymbol{R}^m$ for which $ta_1Y_1 + \ldots + ta_mY_m \in N_0$. We consider now the mapping $\Phi : V \to N_p$ given by

$$\Phi : (t, a_1, \ldots, a_m) \to \operatorname{Exp}(ta_1Y_1 + \ldots + ta_mY_m). \tag{5}$$

We shall then prove that the dual forms $\Phi^*\omega^i$ and $\Phi^*\omega^i{}_l$ are given by the formulas

$$\Phi^*\omega^i = a_i dt + \bar{\omega}^i, \qquad \Phi^*\omega^i{}_l = \bar{\omega}^i{}_l, \tag{5'}$$

where $\bar{\omega}^i$ and $\bar{\omega}^i{}_l$ are 1-forms in $da_1, ..., da_m$ (and do not contain dt). In fact, we can write

$$\Phi^*\omega^i = f_i(t, a_1, ..., a_m)\, dt + \bar{\omega}^i,$$

$$\Phi^*\omega^i{}_l = g_{il}(t, a_1, ..., a_m)\, dt + \bar{\omega}^i{}_l.$$

For a fixed point $(a_1, ..., a_m) \in \boldsymbol{R}^m$ we consider the mapping

$$\tau : t \to \operatorname{Exp}(ta_1 Y_1 + ... + ta_m Y_m),$$

which maps an open subset of $\boldsymbol{R}$ into M. It is easy to see that

$$\tau^*\omega^i = f_i(t, a_1, ..., a_m)\, dt, \qquad \tau^*\omega^i{}_l = g_{il}(t, a_1, ..., a_m)\, dt,$$

$$\dot{\tau}(t) = \Big(\sum_j a_j Y_j^*\Big)_{\tau(t)}.$$

By using the duality of τ^* and $d\tau$ we obtain

$$f_i(t, a_1, ..., a_m) = a_i,$$

$$g_{il}(t, a_1, ..., a_m) = \sum_q \Gamma_{ql}{}^i\, a_q.$$

This last sum, however, vanishes, because $Y^* = \Sigma_j\, a_j Y_j^*$ is the tangent vector field to the geodesic $t \to \operatorname{Exp}(ta_1 Y_1 + ... + ta_m Y_m)$ and consequently the expression

$$\nabla_{Y^*}(Y_l^*) = \sum_k \Big(\sum_q a_q\, \Gamma_{ql}{}^k\Big)\, Y_k^*$$

vanishes along that geodesic. This proves (5′).

The forms $\bar{\omega}^i$ vanish for $t = 0$. In fact, let A be the point $(0, a_1, ..., a_m)$ in V. Then

$$\bar{\omega}^i{}_A\Big(\frac{\partial}{\partial a_j}\Big) = \omega^i\Big(d\Phi_A\Big(\frac{\partial}{\partial a_j}\Big)\Big),$$

and if f is differentiable in a neighborhood of p

$$d\Phi_A\Big(\frac{\partial}{\partial a_j}\Big) f = \Big\{\frac{\partial}{\partial a_j}(f \circ \Phi)\Big\}_{t=0} = \frac{\partial}{\partial a_j}\Big((f \circ \Phi)_{t=0}\Big) = 0.$$

Similarly, the forms $\bar{\omega}^i{}_l$ vanish for $t = 0$.

For the exterior derivatives of the forms (5′) we have

$$\Phi^*(d\omega^i) = d(\Phi^*\omega^i) = da_i \wedge dt + dt \wedge \frac{\partial \bar{\omega}^i}{\partial t} + \dots$$

$$\Phi^*(d\omega^i{}_l) = d(\Phi^*\omega^i{}_l) = dt \wedge \frac{\partial \bar{\omega}^i{}_l}{\partial t} + \dots,$$

where the terms which are not written do not contain dt. On the other hand, since Φ^* is a homomorphism with respect to exterior products ((5), § 3), we can evaluate $\Phi^*(d\omega^i)$ and $\Phi^*(d\omega^i{}_l)$ by means of the structural equations. Equating the coefficients to dt, we obtain the system of differential equations on V:

$$\frac{\partial \bar{\omega}^i}{\partial t} = da_i + \sum_k a_k \bar{\omega}^i{}_k + \sum_{j,k} T^i{}_{jk} a_j \bar{\omega}^k, \qquad \bar{\omega}^i(t; a_j; da_k)_{t=0} = 0; \tag{6}$$

$$\frac{\partial \bar{\omega}^i{}_l}{\partial t} = \sum_{j,k} R^i{}_{ljk} a_j \bar{\omega}^k, \qquad \bar{\omega}^i{}_l(t; a_j; da_k)_{t=0} = 0, \tag{7}$$

which will be useful later. In the derivation of (6) and (7) the antisymmetry of R and T in the two last indices was used. Note that in (6) and (7) we have written for simplicity $T^i{}_{jk}$ and $R^i{}_{ljk}$ in place of $(T^i{}_{jk} \circ \Phi)$ and $(R^i{}_{ljk} \circ \Phi)$. These equations, which represent the structural equations in "polar coordinates," are particularly important in É. Cartan's treatment of Riemannian geometry (É. Cartan [22]). A simple application is given in Exercise C.3.

§ 9. The Riemannian Connection

Definition. Let M be a C^∞-manifold. A *pseudo-Riemannian structure* on M is a tensor field g of type (0, 2) which satisfies

(a) $g(X, Y) = g(Y, X)$ for all $X, Y \in \mathfrak{D}^1(M)$.

(b) For each $p \in M$, g_p is a nondegenerate bilinear form on $M_p \times M_p$.

A *pseudo-Riemannian manifold* is a *connected* C^∞-manifold with a pseudo-Riemannian structure. If (and only if) g_p is positive definite for each $p \in M$, we drop the prefix "pseudo" and speak of a Riemannian structure and Riemannian manifold. A Riemannian structure on a manifold induces in an obvious manner a Riemannian structure on any submanifold. The analogous statement does not hold for a general pseudo-Riemannian structure.

Theorem 9.1. *On a pseudo-Riemannian manifold there exists one and only one affine connection satisfying the following two conditions*:

(i) *The torsion tensor T is* 0.

(ii) *The parallel displacement preserves the inner product on the tangent spaces.*

Proof. Conditions (i) and (ii) can be written:

$$\text{(i}'\text{)} \quad \nabla_X Y - \nabla_Y X = [X, Y], \qquad X, Y \in \mathfrak{D}^1;$$
$$\text{(ii}'\text{)} \quad \nabla_Z g = 0, \qquad Z \in \mathfrak{D}^1.$$

We apply the derivation ∇_Z to the tensor field $X \otimes Y \otimes g$ and use the fact that ∇_Z commutes with contractions. In view of (ii′) we obtain

$$Z\,g(X, Y) = g(\nabla_Z X, Y) + g(X, \nabla_Z Y),$$
$$Z\,g(X, Y) = g(\nabla_X Z, Y) + g(X, \nabla_Z Y) + g([Z, X], Y). \tag{1}$$

In (1) we permute the letters cyclically and eliminate ∇_X and ∇_Y from the obtained relations. This gives

$$\begin{aligned} 2g(X, \nabla_Z Y) = Z\,g(X, Y) + g(Z, [X, Y]) + Y\,g(X, Z) + g(Y, [X, Z]) \\ - X\,g(Y, Z) - g(X, [Y, Z]), \end{aligned} \tag{2}$$

and this relation shows (g being nondegenerate) that there can be at most one affine connection satisfying (i) and (ii). On the other hand, we can define $\nabla_Z Y$ by (2) and a routine computation shows that the axioms ∇_1 and ∇_2 for an affine connection are satisfied. Moreover, carrying out the computations above in reverse order, one verifies (i′) and (ii′) on the basis of (2).

The connection ∇ given by (2) is called the *pseudo-Riemannian* (*or Riemannian*) connection. If M is analytic and the tensor field g is analytic, M is called an *analytic pseudo-Riemannian manifold.* In this case, the pseudo-Riemannian connection is analytic.

Suppose now M is a Riemannian manifold. Then g_p is positive definite for each $p \in M$. If $X \in M_p$, we sometimes write $\| X \|$ instead of $g_p(X, X)^{1/2}$. There exists a basis $Y_1, \ldots, Y_m$ of M_p such that $g_p(Y_i, Y_j) = \delta_{ij}$ $(1 \leqslant i, j \leqslant m)$. Let N_0 denote a normal neighborhood of 0 in M_p and let $N_p = \operatorname{Exp} N_0$. We shall now apply (6) and (7) in § 8 to the adapted basis $Y_1^*, \ldots, Y_m^*$ of vector fields on the normal neighborhood N_p. Since g is invariant under parallelism, we have

$$g(Y_i^*, Y_j^*) = g(Y_i, Y_j) = \delta_{ij}$$

on N_p. Since $\omega^i(Y_j^*) = \delta^i{}_j$, we have

$$g = \sum_i (\omega^i)^2 \qquad \text{on } N_p,$$

where the symmetric product $\alpha\beta$ of two 1-forms is given by $\alpha\beta = \frac{1}{2}(\alpha \otimes \beta + \beta \otimes \alpha)$.

Let S denote the unit sphere $\| X \| = 1$ in M_p, and let U denote the set of all pairs $(t, X) \in \boldsymbol{R} \times S$ for which $tX \in N_0$. Then U is open in $\boldsymbol{R} \times S$. Let Ψ denote the mapping $\Psi = \Phi \circ I$ where Φ is the mapping (5), § 8, and I is the identity mapping of U into $\boldsymbol{R} \times M_p$, M_p being identified with $\boldsymbol{R}^m$ by means of the basis $Y_1, \ldots, Y_m$. Thus $\Psi(t, X) = \operatorname{Exp} tX$ if $(t, X) \in U$. Then Ψ^*g (§ 3, No. 3) is an element of $\mathfrak{D}_2^0(U)$ and we have

$$\Psi^*g = \sum_i (\Psi^*\omega^i)^2.$$

Lemma 9.2. *The tensor field Ψ^*g on U is given by*

$$\Psi^*g = (dt)^2 + \sum_{i=1}^{m} (\bar{\omega}^i)^2.$$

Proof. For the Riemannian connection we have $\omega^l{}_i = -\omega^i{}_l$ so (by (5′), § 8) $\bar{\omega}^i{}_l = -\bar{\omega}^l{}_i$. In fact, this is an immediate consequence of the relation

$$g(\nabla_{Y_i^*}(Y_j^*), Y_k^*) + g(Y_j^*, \nabla_{Y_i^*}(Y_k^*)) = Y_i^* g(Y_j^*, Y_k^*) = 0.$$

Using (5′), § 8, we have

$$\Phi^*g = \sum_i (\Phi^*\omega^i)^2 = \sum_i a_i^2 (dt)^2 + \sum_i (\bar{\omega}^i)^2 + 2\sum_i a_i\bar{\omega}^i dt.$$

From (6), § 8, we obtain

$$\frac{\partial}{\partial t}\left(\sum_i a_i\bar{\omega}^i\right) = \sum_i a_i da_i + \sum_{i,k} a_i a_k \bar{\omega}^i{}_k = \sum_i a_i da_i$$

due to the skew symmetry of $\bar{\omega}^i{}_k$. Moreover, $I^*(\sum_i a_i da_i) = 0$. Now, N_0 is star-shaped. Therefore, if $X \in S$, the set of $t \in \boldsymbol{R}$ such that $(t, X) \in U$ is an interval containing $t = 0$. Inasmuch as $\bar{\omega}^i = 0$ for $t = 0$, we obtain $I^*(\sum_i a_i\bar{\omega}^i) = 0$. Moreover

$$I^*\left(\sum_i a_i^2\right) = 1,$$

so, using $\Psi^* = I^* \circ \Phi^*$ the lemma follows.

We shall now introduce the Riemannian metric on the Riemannian manifold M. Let $t \to \gamma(t)$ $(\alpha \leqslant t \leqslant \beta)$ be a curve segment in M. The arc length of γ is defined by

$$L(\gamma) = \int_\alpha^\beta \{g_{\gamma(t)}(\dot{\gamma}(t), \dot{\gamma}(t))\}^{1/2}\, dt. \tag{3}$$

It is clear from (3) that two curve segments which are the same except for a change of parameter have the same arc length.

For a Riemannian manifold, we write for simplicity "geodesic" instead of "geodesic segment." It will also be convenient not always to distinguish between two curves which coincide after a change of parameter.

Lemma 9.3. *Let M be a Riemannian manifold and p any point in M. Let N_0 be any normal neighborhood of 0 in M_p and put $N_p = Exp\, N_0$. For each $q \in N_p$, let γ_{pq} denote the unique geodesic in N_p joining p to q. Then*

$$L(\gamma_{pq}) < L(\gamma)$$

for each curve segment $\gamma \neq \gamma_{pq}$ in N_p which joins p to q. If, in particular, the normal neighborhood N_0 is an open ball $0 \leqslant \| X \| < \delta$ in M_p, the inequality $L(\gamma_{pq}) < L(\gamma)$ holds for each curve segment $\gamma \neq \gamma_{pq}$ in M which joins p to q.

Proof. Let $s \to \gamma(s)$ $(0 \leqslant s \leqslant 1)$ be any curve segment in N_p joining p to q. For the purpose of proving the inequality above, we can assume that $\gamma(s) \neq p$ for $s \neq 0$. We can then write $\gamma = \Psi \circ \gamma_0$ where γ_0 is a curve segment in $\boldsymbol{R} \times S$,

$$\gamma_0 : s \to (t(s), X(s)) \qquad (0 \leqslant s \leqslant 1),$$

such that $t(s) > 0$ for $0 < s \leqslant 1$. The curve segment γ_0 is contained in the set U of Lemma 9.2. We have

$$\dot{\gamma}_0(s) = d\gamma_0\left(\frac{d}{ds}\right) = \left(\dot{t}(s)\frac{d}{dt}, \dot{X}(s)\right),$$

t denoting the coordinate on $\boldsymbol{R}$. Consequently,

$$g(\dot{\gamma}(s), \dot{\gamma}(s)) = (\Psi^*g)\,(\dot{\gamma}_0(s), \dot{\gamma}_0(s)) = (\dot{t}(s))^2 + \sum_{i=1}^m (\bar{\omega}^i(\dot{X}(s)))^2. \tag{4}$$

Then

$$L(\gamma) = \int_0^1 \Big\{ \dot{t}(s))^2 + \sum_i (\bar{\omega}^i(\dot{X}(s)))^2 \Big\}^{1/2} ds \geqslant \int_0^1 |\, \dot{t}(s)\, |\, ds \geqslant L(\gamma_{pq}). \tag{5}$$

If the equality signs hold we have $\bar{\omega}^i(\dot{X}(s)) = 0$ for all i and all s. In view of (5′), § 8, this is equivalent to $\dot{X}(s) = 0$ or $X(s) =$ constant, which means that $\gamma = \gamma_{pq}$ (up to change of parameter).

Finally, let us consider the case when N_0 is an open ball $0 \leqslant \| X \| < \delta$ in M_p. Let $s \to \gamma(s)$ be a curve segment in M joining p to q such that γ does not lie in N_p. Let X_1 be the element in N_0 such that Exp $X_1 = q$ and suppose δ^* satisfies the inequalities $\| X_1 \| < \delta^* < \delta$. Put

$$N^* = \{\text{Exp}\, X : X \in M_p, \| X \| < \delta^*\}.$$

Let s_0 be the infimum of the set of parameter values s for which $\gamma(s) \notin N^*$. Then the point $q_0 = \gamma(s_0)$ lies on the boundary of N^* and, by the first part of the proof, the length of γ from p to q_0 is $\geqslant \delta^*$. Since $L(\gamma_{pq}) = \| X_1 \|$, it follows that

$$L(\gamma) > L(\gamma_{pq}).$$

This proves the lemma.

The Riemannian manifold M can now be turned into a metric space. Since M is connected, each pair of points $p, q \in M$ can be joined by a curve segment. The *distance* of p and q is defined by

$$d(p, q) = \inf_{\gamma} L(\gamma), \tag{6}$$

where γ runs over all curve segments joining p and q. Then we have

(a) $d(p, q) = d(q, p)$,
(b) $d(p, q) \leqslant d(p, r) + d(r, q)$,
(c) $d(p, q) = 0$ if and only if $p = q$.

The two first are obvious and the last one is a direct consequence of Lemma 9.3. Thus d is a metric on the set M. For $p \in M$ we put

$$B_r(p) = \{q \in M : d(p, q) < r\}, \qquad 0 \leqslant r \leqslant \infty,$$

$$S_r(p) = \{q \in M : d(p, q) = r\}, \qquad 0 \leqslant r < \infty;$$

$B_r(p)$ is called the *open ball* around p with radius r and $S_r(p)$ is called the *sphere* around p with radius r.

Proposition 9.4. *Suppose that the open ball*

$$V_r(0) = \{X \in M_p : 0 \leqslant \| X \| < r\}$$

is a normal neighborhood of 0 *in* M_p. *Then*

$$B_r(p) = \text{Exp}\, V_r(0).$$

Proof. It is obvious that Exp $V_r(0) \subset B_r(p)$. On the other hand, if q were a point in $B_r(p)$, not belonging to Exp $V_r(0)$, then each curve segment joining p to q must intersect the boundary of each set Exp $V_\rho(p)$ $(\rho < r)$. Lemma 9.3 then implies that $d(p, q) \geqslant \rho$ for each $\rho < r$. Hence $d(p, q) \geqslant r$. This contradiction shows that $B_r(p) =$ Exp $V_r(0)$.

Corollary 9.5. *The topology of M given by the metric d coincides with the original topology of M.*

In fact, the sets $B_r(p)$ $(r > 0)$, form a fundamental system of neighborhoods of p in the metric topology of M. On the other hand, Theorem 6.1 shows that in the original topology of M, the sets Exp $V_r(0)$ $(r > 0)$ form a fundamental system of neighborhoods of the point p.

Proposition 9.6. *Every Riemannian manifold is separable.*

This proposition is just a special case of Theorem 15.4 in the Appendix: *A connected, locally compact metric space is separable.*

Definition. Under the assumption of Prop. 9.4, the neighborhoods $B_r(p)$ and $V_r(0)$ are called *spherical normal neighborhoods* of p in M and of 0 in M_p, respectively.

Suppose V is a connected submanifold of a Riemannian manifold M. The Riemannian structure on M induces a Riemannian structure on V. Let d_M and d_V be the distances in M and V given by the Riemannian structures. Then $d_V(p, q) \geqslant d_M(p, q)$ for each pair $p, q \in V$. Examples show that in general we do not have equality sign; however, in the case when V is an open submanifold of M and $p \in V$, the equality sign holds, owing to Prop. 9.4, if q is sufficiently close to p.

Suppose $B_\delta(p)$ is a spherical normal neighborhood of p; if $r < \delta$, then $S_r(p)$ is the image of the sphere $\| X \| = r$ in M_p under the diffeomorphism Exp_p. Thus $S_r(p)$ is a submanifold of M.

Lemma 9.7. *Let $r < \delta$. Then each geodesic γ emanating from p is perpendicular to $S_r(p)$ at the first point of intersection.*

Proof. Assuming the geodesic γ parametrized by its arc length measured from p, let X denote its tangent vector at p. Then $\| X \| = 1$ and the segment of γ from p to the first point of intersection with $S_r(p)$ is $\gamma(s) =$ Exp sX $(0 \leqslant s \leqslant r)$. Let Y be any tangent vector to $S_r(p)$ at the point $\gamma(r)$. Then there exists a unique tangent vector Y_0 to S at X such that $(d\Psi)_{(r,X)}(0, Y_0) = Y$. Then

$$g\left(d\gamma_r\left(\frac{d}{ds}\right), Y\right) = (\Psi^* g)\left(\left(\frac{d}{ds}, 0\right), (0, Y_0)\right) = 0,$$

owing to Lemma 9.2.

Remark. In general the geodesic γ will intersect $S_r(p)$ $(r < \delta)$ more than once. The intersection does not always take place at a right angle. An example is provided by an everywhere dense geodesic on a flat two-dimensional torus. (For the definition of "flat" see Chapter V, § 6.)

Lemma 9.8. *Let M be a Riemannian manifold with metric d given by (6). Let p and q be two points in M and γ_{pq} a curve segment joining p and q. If $L(\gamma_{pq}) = d(p, q)$, then γ_{pq} is a geodesic.*

Proof. There exists a finite sequence of points $r_0, r_1, ..., r_n$ (where $r_0 = p$, $r_n = q$) on γ_{pq} such that each segment $\gamma_{r_i r_{i+1}}$ lies in a spherical normal neighborhood $B_{\delta_i}(r_i)$ (see Theorem 6.2). Then

$$\sum_{i=0}^{n-1} L(\gamma_{r_i r_{i+1}}) = L(\gamma_{pq}), \qquad d(p, q) \leqslant \sum_{i=0}^{n-1} d(r_i, r_{i+1})$$

and $d(r_i, r_{i+1}) \leqslant L(\gamma_{r_i r_{i+1}})$. Assuming now $L(\gamma_{pq}) = d(p, q)$ it follows that

$$d(r_i, r_{i+1}) = L(\gamma_{r_i r_{i+1}}).$$

By Lemma 9.3, $\gamma_{r_i r_{i+1}}$ is a geodesic and the lemma follows.

Remark. The conclusion of the lemma holds even if γ_{pq} is only assumed piecewise differentiable.

The following result which combines Lemmas 6.4 and 9.3 will often be useful.

Theorem 9.9. *Let M be a Riemannian manifold with metric d. To each $p \in M$ corresponds a number $r(p) > 0$ such that if $0 < \rho \leqslant r(p)$, then $B_\rho(p)$ has the properties:*

(A) *$B_\rho(p)$ is a normal neighborhood of each of its points.*

(B) *Let $a, b \in B_\rho(p)$ and let γ_{ab} be the unique geodesic in $B_\rho(p)$ joining a and b. Then γ_{ab} is the only curve segment in M of length $d(a, b)$ which joins a and b.*

Proof. Let $X_1, ..., X_m$ be an orthonormal basis of M_p and let $x_1, ..., x_m$ be normal coordinates at p with respect to this basis, valid on a neighborhood U of p. If $\delta > 0$ is sufficiently small we have

$$B_\delta(p) = \left\{ q \in U : \sum_{i=1}^{m} x_i(q)^2 < \delta^2 \right\}.$$

Using Lemma 6.4 we conclude that there exists a number $\delta^* > 0$ such that for $0 < \delta \leqslant \delta^*$, $B_\delta(p)$ is a normal neighborhood of each of its points. We put $r(p) = \frac{1}{4}\delta^*$. If $0 < \rho \leqslant r(p)$, then the neighborhood $B_\rho(p)$ clearly has property (A). It has also property (B). In fact, since $B_{\delta^*}(p)$ is a normal neighborhood of a, γ_{ab} is the only shortest curve segment in $B_{\delta^*}(p)$ which joins a and b. On the other hand, a curve segment which joins a and b but does not lie entirely in $B_{\delta^*}(p)$ has obviously length $> 3\rho$. Since $L(\gamma_{ab}) \leqslant d(a, p) + d(p, b) \leqslant 2\rho$, property (B) is also verified.

Definition. A ball $B_\rho(p)$ which is a normal neighborhood of each of its points will be called a *convex normal ball*. It will be called *minimizing* if it also has the property (B) above.

Remark. It is easy to show by examples that a convex normal ball is not necessarily minimizing.

Proposition 9.10. *In the notation of Theorem 9.9 let A and B be the unique points in M_p satisfying the relations*

$$\mathrm{Exp}_p\, A = a, \qquad \mathrm{Exp}_p\, B = b, \qquad \| A \| < r(p), \qquad \| B \| < r(p).$$

Then

$$\frac{\| A - B \|}{d(a, b)} \to 1$$

as $(a, b) \to (p, p)$.

Proof. We may assume that the straight line segment joining A and B does not pass through the origin. Consider now Eqs. (6) and (7) in § 8. The forms $\bar{\omega}^i - t\,da_i$ and $\bar{\omega}^i{}_l$ and their first derivatives with respect to t all vanish for $t = 0$. Using Taylor's formula with remainder we conclude that

$$\bar{\omega}^i = t\,da_i + t^2\theta^i, \qquad \bar{\omega}^i{}_l = t^2\theta^i{}_l,$$

where θ^i and $\theta^i{}_l$ are 1-forms. Now let $\Gamma : s \to \Gamma(s)$ $(0 \leqslant s \leqslant 1)$ be any curve segment in $B_{r(p)}(p)$ joining a and b and not passing through p. Let Γ_0 be the curve segment in the ball $\| X \| < r(p)$ in M_p joining A and B such that $\Gamma = \mathrm{Exp} \circ \Gamma_0$. Then $\Gamma_0(s)$ can be written

$$\Gamma_0(s) = t(s)\, X(s) \qquad (0 \leqslant s \leqslant 1),$$

where $t(s) > 0$ for all s and $s \to X(s)$ is a curve segment on S. Then we have as before

$$g(\dot{\Gamma}(s), \dot{\Gamma}(s)) = \dot{t}(s)^2 + \sum_{i=1}^{m} (\bar{\omega}^i(\dot{X}(s)))^2$$

and

$$L(\Gamma) = \int_0^1 \Big\{ \dot{t}(s)^2 + t(s)^2 \sum_i [da_i(\dot{X}(s)) + t(s)\, \theta^i(\dot{X}(s))]^2 \Big\}^{1/2} ds. \tag{7}$$

For the Riemannian manifold M_p we have $R = 0$ and $\bar{\omega}^i = t da_i$. Hence

$$L(\Gamma_0) = \int_0^1 \Big\{ \dot{t}(s)^2 + (t(s))^2 \sum_i (da_i(X(s)))^2 \Big\}^{1/2} ds. \tag{8}$$

If Γ_0 is the straight line joining A and B, then $t(s) \to 0$ uniformly in s as $(a, b) \to (p, p)$. It follows from (7) and (8) that

$$\lim_{(a,b)\to(p,p)} \frac{L(\Gamma_0)}{L(\Gamma)} = 1. \tag{9}$$

This relation holds for the same reason for γ_{ab}, the unique geodesic in $B_{r(p)}(p)$ joining a and b, and γ_{AB}, the corresponding curve segment in the ball $\| X \| < r(p)$. In other words,

$$\lim_{(a,b)\to(p,p)} \frac{L(\gamma_{AB})}{L(\gamma_{ab})} = 1. \tag{10}$$

Now $L(\gamma_{AB}) \geqslant L(\Gamma_0)$ and $L(\Gamma) \geqslant L(\gamma_{ab})$ so

$$\frac{L(\gamma_{AB})}{L(\gamma_{ab})} \geqslant \frac{L(\Gamma_0)}{L(\gamma_{ab})} \geqslant \frac{L(\Gamma_0)}{L(\Gamma)}\,.$$

Since $L(\Gamma_0) = \| A - B \|$ and $L(\gamma_{ab}) = d(a, b)$, the proposition follows from (9) and (10).

Remark. The hyperbolic plane (Exercise G) is an illuminating example of all the notions we shall develop for Riemannian manifolds.

§ 10. Complete Riemannian Manifolds

Definition. A Riemannian manifold M is said to be *complete* if every Cauchy sequence in M is convergent.

Lemma 10.1. *For each point p_0 in a Riemannian manifold M there exists a convex normal ball $B_\rho(p_0)$ around p_0 with the following property: Let p and q be two points in $B_\rho(p_0)$, γ the unique geodesic in $B_\rho(p_0)$ joining p and q. Let $L(\gamma)$ denote the length of γ. Let Y and Z denote the unit tangent vectors to γ at p and q, respectively.*

Then:

(i) *If* (p', Y') *is sufficiently close to* (p, Y) *and* Y' *is a unit vector in* $M_{p'}$, *there exists a geodesic in* $B_\rho(p_0)$ *of length* $L(\gamma)$, *starting at* p' *with tangent vector* Y'.

(ii) *The pair* (q, Z) *depends differentiably on* p, Y *and* $L(\gamma)$.

This lemma follows directly from the existence and uniqueness theorem stated in the beginning of § 6.

Lemma 10.2. *Let* p *be a point in a Riemannian manifold* M *and* $\gamma_n(t)$ $(t \in I_n)$ *a sequence of geodesics emanating from* p, t *being the arc length measured from* p. *Let* X_n *be a tangent vector to* γ_n *at* p *and suppose the sequence* (X_n) *converges to* $X \in M_p$. *Let* $\gamma_X(t)$ $(t \in I)$ *be the maximal geodesic tangent to* X *such that* $\gamma_X(0) = p$. *Assume* $t^* \in I$ *is a limit* $t^* = \lim t_n$ *where* $t_n \in I_n$. *Then* $\gamma_X(t^*) = \lim_n \gamma_n(t_n)$.

In fact, the segment $\gamma_X(t)$, $0 \leqslant t \leqslant t^*$, can be broken into finitely many segments each of which lies in a convex normal ball B_i with the property of Lemma 10.1. The first part of Lemma 10.1 implies that for sufficiently large n, all $\gamma_n(t)$, $0 \leqslant t \leqslant t_n$, lie in the union of the balls B_i. Now Lemma 10.2 follows by repeated application of the second part of Lemma 10.1.

The following two theorems show clearly the importance of the completeness condition.

Theorem 10.3. *Let* M *be a Riemannian manifold. The following conditions are equivalent.*

(i) M *is complete.*

(ii) *Each bounded closed subset of* M *is compact.*

(iii) *Each maximal geodesic in* M *has the form* $\gamma_X(t)$, $-\infty < t < \infty$ (*"has infinite length"*).

Theorem 10.4. *In a complete Riemannian manifold* M *with metric* d *each pair* $p, q \in M$ *can be joined by a geodesic of length* $d(p, q)$.

These two theorems will be proved simultaneously.

(i) $\Rightarrow$ (iii). Let $\gamma_X(t)$ $(t \in I)$ be a maximal geodesic in M, $|t|$ being the arc length measured from the point of origin of X. If t_0 were a boundary point of the (open) interval I, say on the right, select a sequence $(t_n) \subset I$ converging to t_0. Then $(\gamma_X(t_n))$ is a Cauchy sequence in M, hence converges to a limit $p \in M$, which is clearly independent of the choice of (t_n). Let $B_\rho(p)$ be a convex normal ball around p and let $\{x_1, ..., x_m\}$ be a system of normal coordinates at p valid on $B_\rho(p)$. Let $J = \{t \in I : \gamma_X(t) \in B_\rho(p)\}$, put $x_i(t) = x_i(\gamma_X(t))$ $(t \in J)$ and $x_i(t_0) = \lim_{t \to t_0} x_i(t)$, knowing that this limit exists.

Now we have

$$\ddot{x}_i(t) + \sum_{j,k} \Gamma_{jk}{}^i\, \dot{x}_j(t)\, \dot{x}_k(t) = 0 \tag{1}$$

for $t \in J$. The functions $\dot{x}_i(t)$ are bounded ($\Sigma_i\, \dot{x}_i(t)^2 = 1$) and the differential equation shows that each $\ddot{x}_i(t)$ is bounded. In particular, $\dot{x}_i(t)$ is uniformly continuous near t_0 and thus has a limit as $t \to t_0$. From the mean value theorem we have

$$(x_i(t) - x_i(t_0))/(t - t_0) = \dot{x}_i(t_0 + \theta(t - t_0)) \qquad (0 \leqslant \theta \leqslant 1)$$

which implies for the left derivative $\dot{x}_i(t_0)$,

$$\dot{x}_i(t_0) = \lim_{t \to t_0} \dot{x}_i(t) \qquad (1 \leqslant i \leqslant m). \tag{2}$$

From the differential equation (1) follows the existence of the limit $\lim_{t \to t_0} \ddot{x}_i(t)$; the mean value theorem again implies for the left derivative

$$\ddot{x}_i(t_0) = \lim_{t \to t_0} \ddot{x}_i(t). \tag{3}$$

The vector $Z = (\dot{x}_1(t_0), ..., \dot{x}_m(t_0))$ in the tangent space M_p has length 1 and we can form the geodesic $\gamma_Z(t)$ for $t_0 \leqslant t < t_0 + \rho$. The mapping $t \to \Gamma(t)$ where

$$\Gamma(t) = \begin{cases} \gamma_X(t), & t \in I, \\ \gamma_Z(t), & t_0 \leqslant t < t_0 + \rho, \end{cases}$$

satisfies (1) for $t \in J$ and for $t_0 < t < t_0 + \rho$. Moreover, (1) is satisfied for the *right derivatives* at $t = t_0$. Equations (2) and (3) show that (1) is also satisfied for the left derivatives at $t = t_0$. Thus, $\Gamma(t)$ is a geodesic, contradicting the maximality of $\gamma_X(t)$, $t \in I$.

(ii) ⇒ (i). Let (x_n) be a Cauchy sequence in M. The closure of the set (x_n) is bounded, hence compact by (ii). Thus a subsequence of (x_n) is convergent; being a Cauchy sequence, the sequence (x_n) itself is convergent.

Using a procedure of de Rham [1] we next prove that if the condition (iii) is satisfied, then each pair p, $q \in M$ can be joined by a geodesic of length $d(p, q)$. For each $r \geqslant 0$ let $\bar{B}_r$ denote the closure of the open ball $B_r(p)$ and let E_r denote the set of points x in $\bar{B}_r$ which can be joined to p by a geodesic of length $d(p, x)$. It suffices to prove

$$E_r = \bar{B}_r \tag{4}$$

for each $r \geqslant 0$. In view of Lemma 10.2, (iii) implies separability of M and compactness of E_r. The relation (4) is valid for $r = 0$, and if it is valid for $r = r_0 > 0$, it is obviously valid for $r < r_0$. On the other hand, if (4) holds for $r < r_0$, it is also valid for $r = r_0$; in fact, each $x \in \bar{B}_{r_0}$ is a limit of a sequence of points x_n each of which has distance from p less than r_0. Thus, $x_n \in E_{r_0}$ and $x \in E_{r_0}$, the set E_{r_0} being closed. Thus, it suffices to prove that if (4) holds for $r = R$, it holds also for some larger value $r = R + \rho$. For this we may assume $B_R \neq M$.

By compactness of $E_R = \bar{B}_R$ there exist finitely many points $x_1, ..., x_N \in E_R$ and positive numbers $\rho_1, ..., \rho_N$ such that the balls $B_{\rho_i}(x_i)$ $(1 \leqslant i \leqslant N)$ cover E_R and such that each $B_{2\rho_i}(x_i)$ is a relatively compact, minimizing convex normal ball. Since the set $\bigcup_{i=1}^N B_{\rho_i}(x_i)$ is relatively compact there exists a point in its complement at shortest distance from p. Since this distance is $> R$, there exists a number ρ such that $0 < \rho < \min(\rho_1, ..., \rho_N)$ and such that $\bar{B}_{R+\rho} \subset \bigcup_{i=1}^N B_{\rho_i}(x_i)$.

Suppose now y is a point in M such that $R < d(p, y) \leqslant R + \rho$. Let x be a point on the (compact) sphere $S_R(p)$ at smallest distance from y. Since every curve segment joining p and y must intersect $S_R(p)$, it follows that

$$d(p, x) + d(x, y) = d(p, y).$$

Consequently, $d(x, y) = d(p, y) - d(p, x) \leqslant R + \rho - R = \rho$ so y and x lie in the same ball $B_{2\rho_i}(x_i)$. Combining the shortest curve joining x and y with a curve of shortest length joining p and x, we obtain a curve of length $d(p, y)$ joining p and y. By Lemma 9.8 this curve is a geodesic so (4) is proved for all $r \geqslant 0$.

(iii) $\Rightarrow$ (ii). Let S be a bounded closed subset of M. Let (q_n) be a sequence of points in S, and p any fixed point in M. We know now that M is separable and that there exists a geodesic γ_n of length $d(p, q_n)$ joining p and q_n. Passing to a subsequence if necessary, we can assume that the unit tangent vectors to γ_n at p form a convergent sequence and that the sequence $(d(p, q_n))$ converges. Lemma 10.2 now shows that (iii) $\Rightarrow$ (ii). This concludes the proof of Theorems 10.3 and 10.4.

Remark. Call a Riemannian manifold M *complete at a point* $p \in M$ if Exp_p is defined on the entire M_p. Then the proof above shows that completeness at a single point $p \in M$ implies completeness of M.

Proposition 10.5. *For each point p in a complete Riemannian manifold M the Exponential mapping* Exp_p *is a differentiable mapping of* M_p *onto M. If M is analytic, then* Exp_p *is analytic.*

Proof. Lemma 10.2 expresses the continuity of Exp_p. The differentiability is proved in the same way (applying Lemma 10.1) since we now know that Exp_p is defined on the entire M_p. If M is analytic we first observe that the existence and uniqueness theorem in § 6 also holds for the analytic case giving analytic solutions. Thus we can replace "differentiably" in Lemma 10.1 by "analytically" and proceed as before.

Definition. Let M be a complete Riemannian manifold, p a point in M and Exp the Exponential mapping at p. Let $C(p)$ denote the set of vectors $X \in M_p$ for which the linear mapping $d\,\mathrm{Exp}_X$ is singular. A point in M (or in M_p) is said to be *conjugate* to p if it lies in Exp $C(p)$ (or $C(p)$).

In view of Prop. 10.5 the set $C(p)$ is a closed subset of M_p. It plays an important role in global differential geometry. In general, Exp $C(p)$ is not a submanifold of M. For the sphere S^2, the set Exp $C(p)$ consists of two antipodal points. In Chapter VII we shall prove the inequality

$$\dim(\mathrm{Exp}\, C(p)) \leqslant \dim M - 2$$

for a symmetric space of the compact type, dim denoting topological dimension. However, this inequality fails to hold for general Riemannian manifolds.

Let M and N be connected and locally connected spaces and $\pi:M \to N$ a continuous mapping. The pair (M, π) is called a *covering space* of N if each point $n \in N$ has an open neighborhood U such that each component of $\pi^{-1}(U)$ is homeomorphic to U under π.

Suppose N is a differentiable manifold and that (M, π) is a covering space of N. Then there is a unique differentiable structure on M such that the mapping π is regular. If M is given this differentiable structure, we say that (M, π) is a *covering manifold* of N.

We shall require the following standard theorem from the theory of covering spaces. We state it only for manifolds although it holds under suitable local connectedness hypotheses.

Let (M, π) be a covering manifold of N and let $\Gamma : [a, b] \to N$ be a path in N. If m is any point in M such that $\pi(m) = \Gamma(a)$, there exists a unique path $\Gamma^* : [a, b] \to M$ such that $\Gamma^*(a) = m$ and $\pi \circ \Gamma^* = \Gamma$.

The path Γ^* is called the *lift* of Γ through m.

Proposition 10.6. *Let N be a Riemannian manifold with a Riemannian structure g. Let (M, π) be a covering manifold of N. Then π^*g is a Riemannian structure on M. Moreover, M is complete if and only if N is complete.*

Proof. The mapping π is regular, so obviously π^*g is a Riemannian structure on M. If γ is a curve segment in M, then $\pi \circ \gamma$ is a curve segment in N. Using the characterization of geodesics by means of differential equations (3), (§ 5), it is clear that γ is a geodesic if and only if $\pi \circ \gamma$ is a geodesic. But completeness is equivalent to the infiniteness of each maximal geodesic. The proposition follows immediately.

Remark. Let N be a Riemannian manifold, (M, π) a covering manifold of N, M taken with Riemannian structure induced by π. Given a point $a \in N$, there exists a convex, minimizing normal ball $B_r(a)$ such that each component B of $\pi^{-1}(B_r(a))$ is diffeomorphic to $B_r(a)$ under π. Since the geodesics correspond under π, it follows that π is a distance-preserving mapping of B onto $B_r(a)$.

Definition. A geodesic $\gamma(t)$, $0 \leqslant t < \infty$, in a Riemannian manifold is called a *ray* if it realizes the shortest distance between any two of its points. The point $\gamma(0)$ is called the *initial point* of the ray.

Proposition 10.7. *Let o be a point in a complete, noncompact Riemannian manifold M. Then M contains a ray with initial point o.*

It follows from Theorem 10.3 that M is not bounded. Let (p_n) be a sequence in M such that $d(o, p_n) \to \infty$. Let γ_n be a geodesic of length $d(o, p_n)$ joining o and p_n. We parametrize γ_n by arc length measured from o. Let X_n be the unit tangent vector to γ_n at o. We can assume that the sequence (X_n) converges to a limit $X \in M_o$. Then $t \to \gamma_X(t)$ $(0 \leqslant t < \infty)$ is a ray. In fact, let $t_0 \geqslant 0$. There exists an integer N such that $d(o, p_n) > t_0$ for $n \geqslant N$. We have from Lemma 10.2

$$\lim_{n > N, n \to \infty} \gamma_n(t_0) = \gamma_X(t_0),$$

and consequently

$$\lim_{n > N, n \to \infty} d(o, \gamma_n(t_0)) = d(o, \gamma_X(t_0)).$$

On the other hand, $t_0 = d(o, \gamma_n(t_0))$ for $n \geqslant N$. The last relation therefore shows that $t_0 = d(o, \gamma_X(t_0))$. This implies that $d(\gamma_X(t_1), \gamma_X(t_0)) = t_0 - t_1$ for $0 \leqslant t_1 \leqslant t_0$, proving the proposition.

§ 11. Isometries

Definition. Let M and N be two C^∞ manifolds with pseudo-Riemannian structures g and h, respectively. Let φ be a mapping of M into N.

(i) φ is called an *isometry* if φ is a diffeomorphism of M onto N and $\varphi^*h = g$.

(ii) φ is called a *local isometry* if for each $p \in M$ there exist open neighborhoods U of p and V of $\varphi(p)$ such that φ is an isometry of U onto V.

It is obvious that if φ is an isometry of a Riemannian manifold M onto itself, then φ preserves distances, i. e., $d(\varphi(p), \varphi(q)) = d(p, q)$ for $p, q \in M$. On the other hand, we have:

Theorem 11.1. *Let M be a Riemannian manifold and φ a distance-preserving mapping of M onto itself. Then φ is an isometry.*

Proof. Let p be an arbitrary point in M and put $q = \varphi(p)$. Let $B_r(p)$ and $B_r(q)$ be spherical normal neighborhoods of $p \in M$ and $q \in M$, respectively. Then φ gives a one-to-one mapping of $B_r(p)$ onto $B_r(q)$. For each $X \in M_p$ we consider the geodesic Exp tX $(-r/\| X \| < t < r/\| X \|)$. The image $\Gamma(t) = \varphi\,(\text{Exp}\, tX)$ lies in $B_r(q)$ and has the property that $d(\Gamma(t), \Gamma(t')) = | t - t' |\, \| X \|$ for all t, t' in the interval of definition. To see that Γ is a geodesic we consider the point $q = \Gamma(0)$ and an arbitrary point Q on Γ. They can be joined by a unique geodesic γ of length $d(q, Q)$. Let $B_R(Q)$ be a spherical normal neighborhood of Q and let m be any point of Γ between q and Q such that $m \in B_R(Q)$. Then $d(q, m) + d(m, Q) = d(q, Q)$. If we join q and m by the shortest geodesic, and then join m and Q by the shortest geodesic, we get a broken curve of length $d(q, Q)$ joining q and Q. This curve must coincide with γ due to Lemma 9.8 (and the subsequent remark). Since Q was arbitrary on Γ, this proves that Γ is a geodesic; in particular, Γ is differentiable. Let X' denote the tangent vector to Γ at the point q. We have obtained a mapping $X \to X'$ of M_p into M_q. Denoting this mapping by φ' we have $\| X \| = \| \varphi'(X) \|$ and $\varphi'(\alpha X) = \alpha\varphi'(X)$ for $\alpha \in \boldsymbol{R}$, $X \in M_p$. Let $A, B \in M_p$ and select ρ such that $\| \rho A \|$ and $\| \rho B \|$ are both less than r. Let $a_t = \text{Exp}\, tA$, $b_t = \text{Exp}\, tB$ for $0 \leqslant t \leqslant \rho$. Then Prop. 9.10 shows that

$$\frac{2g_p(A, B)}{\| A \|\, \| B \|} = \frac{\| A \|^2 + \| B \|^2}{\| A \|\, \| B \|} - \frac{\| tA - tB \|^2}{\| tA \|\, \| tB \|}$$

$$= \frac{\| A \|^2 + \| B \|^2}{\| A \|\, \| B \|} - \lim_{t \to 0} \frac{d(a_t, b_t)^2}{\| tA \|\, \| tB \|}.$$

Since the right-hand side is preserved by the mapping φ, it follows that

$$g_p(A, B) = g_q(\varphi' A, \varphi' B).$$

But $A + B$ is determined by the quantities $\| A \|$, $\| B \|$, and $g_p(A, B)$, all of which are preserved by φ'. It follows that $\varphi'(A + B) = \varphi' A + \varphi' B$ which together with the previous properties of φ' shows that it is a diffeomorphism of M_p onto M_q. On $B_r(p)$ we have

$$\varphi = \text{Exp}_q \circ \varphi' \circ \text{Exp}_p^{-1} \tag{1}$$

and the theorem follows.

Lemma 11.2. *Let M be a Riemannian manifold, φ and ψ two isometries of M onto itself. Suppose there exists a point $p \in M$ for which $\varphi(p) = \psi(p)$ and $d\varphi_p = d\psi_p$. Then $\varphi = \psi$.*

Proof. Considering $\varphi \circ \psi^{-1}$ it is clear that we may assume that $\varphi(p) = p$ and that $d\varphi_p$ is the identity mapping. It is then obvious that all points in an arbitrary normal neighborhood of p are left fixed by φ. Since M is connected, each point $q \in M$ can be connected to p by a chain of overlapping normal neighborhoods. It follows that $\varphi(q) = q$.

Let M and N be Riemannian manifolds, U_i a domain in M ($i = 1, 2$), and φ_i an isometry of U_i onto a domain in N ($i = 1, 2$). It follows from Lemma 11.2 that if $(d\varphi_1)_p = (d\varphi_2)_p$ and $\varphi_1(p) = \varphi_2(p)$ for some point $p \in U_1 \cap U_2$, then φ_1 and φ_2 coincide on the component of p in $U_1 \cap U_2$. The isometries φ_1 and φ_2 are called *immediate continuations* if $U_1 \cap U_2 \neq \emptyset$ and $\varphi_1 = \varphi_2$ on $U_1 \cap U_2$.

Let φ be an isometry of a domain $U \subset M$ onto a domain in N. Let $\gamma(t)$, $0 \leqslant t \leqslant 1$, be a continuous curve in M such that $\gamma(0) \in U$. The isometry φ is said to be *extendable along* γ if for each t ($0 \leqslant t \leqslant 1$) there exists an isometry φ_t of a domain U_t containing $\gamma(t)$ onto an open subset of N such that $\varphi_0 = \varphi$ and $U_0 = U$ and such that φ_t and $\varphi_{t'}$ are immediate continuations whenever $|t - t'|$ is sufficiently small. The family φ_t, $0 \leqslant t \leqslant 1$, is called a *continuation of φ along γ*. It follows that the differential $(d\varphi_t)_{\gamma(t)}$ as well as $\varphi_t(\gamma(t))$ depends continuously on t. Suppose now ψ_t, $0 \leqslant t \leqslant 1$, is another continuation of φ along γ and V_t the domain of definition of ψ_t. From the foregoing remarks it follows that the set of t for which $\varphi_t(\gamma(t)) = \psi_t(\gamma(t))$ and $(d\varphi_t)_{\gamma(t)} = (d\psi_t)_{\gamma(t)}$ is an open and closed subset of the unit interval and contains $t = 0$. Thus, for each t, φ_t and ψ_t coincide in the component of $\gamma(t)$ in $U_t \cap V_t$. Roughly speaking, we can therefore say in analogy with analytic continuation of holomorphic functions: the continuation of an isometry along a curve is unique whenever possible.

Proposition 11.3. *Let M and N be analytic and complete Riemannian manifolds, φ an isometry of a domain $U \subset M$ onto a domain in N. Let $\gamma(t)$, $0 \leqslant t \leqslant 1$, be a continuous curve in M such that $\gamma(0) \in U$. Then φ is extendable along γ.*

Proof. Let $p \in M$, $q \in N$, and suppose $B_\rho(p)$ and $B_\rho(q)$ are spherical normal neighborhoods of p and q, respectively. Suppose $r < \rho$ and suppose ψ is an isometry of $B_r(p)$ into N such that $\psi(p) = q$. Then ψ can be extended uniquely to an isometry ψ' of $B_\rho(p)$ onto $B_\rho(q)$. To see this, we note first that the expression of ψ in normal coordinates at p and $\psi(p)$ is a linear mapping which we can use to define ψ'. If g and h

denote the metric tensors on M and N, respectively, and if X and Y are analytic vector fields on $B_\rho(p)$, then $h(X^{\psi'}, Y^{\psi'}) \circ \psi' = g(X, Y)$ on $B_r(p)$ by the assumption; this relation also holds on $B_\rho(p)$ due to the analyticity of ψ' (see Lemma 4.3, Chapter VI). Hence ψ' is an isometry.

Without restriction of generality we can assume $\gamma(t)$ differentiable. Let s^* be the supremum of the parameter values s such that a continuation φ_t of φ exists along the curve $\gamma(t)$, $0 \leqslant t \leqslant s$. We put $p_t = \gamma(t)$ for $0 \leqslant t \leqslant 1$, and $q_t = \varphi_t(p_t)$ for $0 \leqslant t < s^*$. It is clear from the completeness assumption that the limit $q^* = \lim_{t \to s^*} \varphi_t(p_t)$ exists. We select $\rho > 0$ such that $B_{3\rho}(p_{s^*})$ and $B_{3\rho}(q^*)$ are convex normal balls. Select $s' < s^*$ such that

$$p_t \in B_\rho(p_{s^*}), \qquad q_t \in B_\rho(q^*) \qquad \text{for } s' \leqslant t < s^*.$$

Then $B_{2\rho}(p_{s'})$ and $B_{2\rho}(q_{s'})$ are normal neighborhoods of $p_{s'}$ and $q_{s'}$, respectively. As shown above, $\varphi_{s'}$ can be extended to an isometry of $B_{2\rho}(p_{s'})$ onto $B_{2\rho}(q_{s'})$. Since $p^* \in B_{2\rho}(p_{s'})$, this shows that $s^* = 1$ and that φ has a continuation along γ.

Proposition 11.4. *Let the assumptions be as in Prop. 11.3 and suppose $\delta(t)$, $0 \leqslant t \leqslant 1$, is a continuous curve in M, homotopic to γ. Let φ_t and ψ_t $(0 \leqslant t \leqslant 1)$ be continuations of φ along γ and δ, respectively. Then φ_1 and ψ_1 coincide in a neighborhood of $\gamma(1) = \delta(1)$.*

Proof. Since γ and δ are homotopic, there exists a continuous mapping α of the closed unit square into M such that

$$\alpha(0, t) = \gamma(t), \qquad 0 \leqslant t \leqslant 1,$$
$$\alpha(1, t) = \delta(t), \qquad 0 \leqslant t \leqslant 1;$$
$$\alpha(s, 0) = \gamma(0), \qquad \alpha(s, 1) = \gamma(1) \qquad \text{for all } 0 \leqslant s \leqslant 1.$$

For a fixed s, let α^s denote the continuous curve $t \to \alpha(s, t)$ $(0 \leqslant t \leqslant 1)$, and let φ_t^s $(0 \leqslant t \leqslant 1)$ denote the continuation of φ along α^s. Let σ denote the supremum of the values s^* such that for each s satisfying $0 \leqslant s \leqslant s^*$, φ_1^s coincides with φ_1 in a neighborhood of $\gamma(1)$. Consider now the continuous curve α^σ. The mapping $t \to \varphi_t^\sigma(\alpha^\sigma(t))$ is also a continuous curve. Hence there exists a number $r > 0$ such that for each t, $0 \leqslant t \leqslant 1$, the balls $B_{2r}(\alpha^\sigma(t))$ and $B_{2r}(\varphi_t^\sigma(\alpha^\sigma(t)))$ are normal neighborhoods of their centers. Now, there exists an $\epsilon > 0$ such that $d(\alpha^\sigma(t), \alpha^s(t)) < r$ for $0 \leqslant t \leqslant 1$, and $| \sigma - s | < \epsilon$. For such s, the family φ_t^σ gives a continuation of φ along α^s. Now as remarked before, the continuation of an isometry along a given curve is unique. It follows that if $| s - \sigma | < \epsilon$, the isometries φ_1^σ and φ_1^s coincide in a neighborhood of $\gamma(1)$. This shows firstly that $\sigma = 1$ and secondly that if $0 \leqslant s \leqslant 1$, then φ_1^1 and φ_1^s coincide in a neighborhood of $\gamma(1)$. This proves the proposition.

§ 12. Sectional Curvature

In this section we shall exhibit the classical geometric significance of the curvature tensor for a Riemannian manifold.

Let F be a Riemannian manifold of dimension 2 and let p be a point in F. Let $V_r(0)$ denote the open ball in the tangent plane F_p with center 0 and radius r. Suppose r is so small that Exp_p is a diffeomorphism of $V_r(0)$ onto the open ball $B_r(p)$. Let $A_0(r)$ and $A(r)$ denote the (two-dimensional) areas of $V_r(0)$ and $B_r(p)$, respectively.

Definition. The *curvature*† of F at p is defined as the limit

$$K = \lim_{r \to 0} 12 \frac{A_0(r) - A(r)}{r^2 A_0(r)}.$$

The existence of this limit is contained in the following lemma, which at the same time facilitates the computation of K.

Lemma 12.1. *Let f denote the "Radon-Nikodym derivative" of Exp_p on $V_r(0)$. Then*

$$K = -\frac{3}{2}[\Delta f](0)$$

where Δ is the Laplacian on the metric space F_p, that is, $\Delta = \partial^2/\partial x_1^2 + \partial^2/\partial x_2^2$, x_1 and x_2 being coordinates with respect to some orthonormal basis.

Proof. The definition of $f(X)$ is expressed in the formula $\text{Exp}_p^* (dq) = f\, dX$ if dq and dX denote the surface elements (Chapter VIII, §2) of $B_r(p)$ and $V_r(0)$, respectively. Hence

$$A(r) = \int_{V_r(0)} f(X)\, dX.$$

The differentiable function $f(X)$ can be expanded in a finite Taylor series around the point $X = 0$. If this series is integrated over the disk $V_r(0)$ one finds

$$A(r) = A_0(r) \{f(0) + \tfrac{1}{8} r^2 [\Delta f](0) + O(r^3)\}.$$

The lemma now follows immediately, $f(0)$ being equal to 1.

† It is known from the differential geometry of surfaces that if F is a surface, then K is equal to the Gaussian curvature of F at p, but we shall not need this fact.

Let M be a Riemannian manifold and p a point in M. Let N_0 be a normal neighborhood of 0 in M_p and let $N_p = \text{Exp } N_0$. Let S be a two-dimensional vector subspace of M_p. Then $\text{Exp}(N_0 \cap S)$ is a connected submanifold of M of dimension 2 and has a Riemannian structure induced by that of M. The curvature of $\text{Exp}(N_0 \cap S)$ at p is called the *sectional curvature* of M at p along the *plane section* S.

Theorem 12.2. *Let M be a Riemannian manifold with curvature tensor field R and Riemannian structure g. Let p be a point in M and S a two-dimensional vector subspace of the tangent space M_p. The sectional curvature of M at p along the section S is then*

$$K(S) = -\frac{g_p(R_p(Y, Z)\, Y, Z)}{|\, Y \vee Z\,|^2}.$$

Here Y and Z are any linearly independent vectors in S; $Y \vee Z$ denotes the parallelogram spanned by these vectors and $|\, Y \vee Z\,|$ the area.

Proof. We shall first assume that M and g are analytic in order to apply Theorem 6.5. We also assume temporarily that Y and Z are orthonormal vectors in S. Let $X_1, \ldots, X_m$ be an orthonormal basis of M_p such that $X_1 = Y$ and $X_2 = Z$. Then each $X \in S$ can be written $X = x_1X_1 + x_2X_2$, $x_1, x_2 \in \boldsymbol{R}$. The Laplacian Δ on S is

$$\Delta = \frac{\partial^2}{\partial x_1^2} + \frac{\partial^2}{\partial x_2^2}.$$

Let N_0 be a normal neighborhood of 0 in M_p and put $N_p = \text{Exp } N_0$. A curve in the manifold $M_S = \text{Exp}(S \cap N_0)$ has the same length regardless whether it is measured by means of the Riemannian structure on M or by means of the induced structure on M_S. If $q \in M_S$, the geodesic in N_p from p to q is the shortest curve in M_S joining p and q. It follows that the Exponential mappings at p for M and M_S, respectively, coincide on $S \cap N_0$. Let $X_1^*, \ldots, X_m^*, X^*$ denote the vector fields on N_p adapted to the tangent vectors $X_1, \ldots, X_m, X$. If $X \in S \cap N_0$, we put

$$v_1 = d\,\text{Exp}_X(X_1), \qquad v_2 = d\,\text{Exp}_X(X_2),$$

and define the functions $c^k{}_{ij}$ by

$$[X_i^*, X_j^*] = \sum_k c^k{}_{ij} X_k^*. \tag{1}$$

The mapping $\text{Exp}(x_1X_1 + x_2X_2) \rightarrow (x_1, x_2)$ is a system of coordinates on the manifold $M_S = \text{Exp}(S \cap N_0)$ and v_1 and v_2 are tangent vectors to M_S. If a and b are two vectors in a metric vector space, we denote by $a \vee b$ the parallelogram spanned by a and b and by $| a \vee b |$ the area. Let f denote the ratio of the surface elements in M_S and S (at the points Exp X and X). In other words,

$$f(X) = \frac{| v_1 \vee v_2 |}{| X_1 \vee X_2 |} = | v_1 \vee v_2 |.$$

The vectors v_1 and v_2 can be expressed

$$v_1 = \sum_{i=1}^{m} f_i X_i^*, \qquad v_2 = \sum_{j=1}^{m} g_j X_j^*,$$

where f_i and g_j are analytic functions of (x_1, x_2). These functions are determined by Theorem 6.5 for small (x_1, x_2). In fact, we have

$$v_1 = X_1^* - \tfrac{1}{2}[X^*, X_1^*] + \tfrac{1}{6}[X^*, [X^*, X_1^*]] - \ldots \tag{2}$$

$$v_2 = X_2^* - \tfrac{1}{2}[X^*, X_2^*] + \tfrac{1}{6}[X^*, [X^*, X_2^*]] - \ldots. \tag{2'}$$

The vectors $(X_i^*)_{\text{Exp } X}$ $(1 \leqslant i \leqslant m)$ form an orthonormal basis of $M_{\text{Exp } X}$ (Theorem 9.1 (ii)). The projection of $v_1 \vee v_2$ into the 2-plane spanned by the vectors $(X_i^*)_{\text{Exp } X}$ and $(X_j^*)_{\text{Exp } X}$ has area $| f_i g_j - g_i f_j |$. It follows that

$$| v_1 \vee v_2 |^2 = \sum_{i<j} (f_i g_j - f_j g_i)^2. \tag{3}$$

We denote this quantity by F. The relation $f = F^{1/2}$ implies

$$2f\Delta f = \Delta F - \frac{1}{2f^2}\left\{\left(\frac{\partial F}{\partial x_1}\right)^2 + \left(\frac{\partial F}{\partial x_2}\right)^2\right\}.$$

We have to evaluate this expression for $(x_1, x_2) = (0, 0)$. Since the torsion T vanishes, we have

$$[X_i^*, X_j^*] = \nabla_{X_i^*}(X_j^*) - \nabla_{X_j^*}(X_i^*)$$

and consequently the functions $c^k{}_{ij}$ vanish at p; in other words, the restrictions of $c^k{}_{ij}$ to M_S vanish for $(x_1, x_2) = (0, 0)$. From (2) and (2′) we obtain expansions for f_i, g_j:

$$f_i = \delta_{1i} - \frac{x_2}{2} c^i{}_{21} + \frac{x_1 x_2}{6} (X_1^* \cdot c^i{}_{21}) + \frac{x_2^2}{6} (X_2^* \cdot c^i{}_{21}) + \ldots,$$

$$g_j = \delta_{2j} - \frac{x_1}{2} c^j{}_{12} + \frac{x_1 x_2}{6} (X_2^* \cdot c^j{}_{12}) + \frac{x_1^2}{6} (X_1^* \cdot c^j{}_{12}) + \ldots,$$

where the terms which are not written vanish for $(x_1, x_2) = (0, 0)$ of higher than second order. It follows easily that $\partial F/\partial x_1$ and $\partial F/\partial x_2$ vanish for $(x_1, x_2) = (0, 0)$ and

$$2[\Delta f](0) = [\Delta F](0) = [\Delta (f_1 g_2)^2](0).$$

Omitting again terms of higher than 2nd order we have

$$(f_1 g_2)^2$$
$$= 1 - x_1 c^2{}_{12} - x_2 c^1{}_{21} + \frac{x_1^2}{3}(X_1^* c^2{}_{12}) + \frac{x_2^2}{3}(X_2^* c^1{}_{21}) + \frac{x_1 x_2}{3}(X_1^* c^1{}_{21} + X_2^* c^2{}_{12}).$$

Since

$$X_1 c^2{}_{12} = [X_1^* c^2{}_{12}](0) = \left[\frac{\partial}{\partial x_1} c^2{}_{12}\right](0), \quad \text{etc.,}$$

we obtain

$$2[\Delta f](0) = -\frac{4}{3}(X_1 c^2{}_{12} + X_2 c^1{}_{21}),$$

and by Lemma 12.1

$$K(S) = X_1 \cdot g([X_1^*, X_2^*], X_2^*) + X_2 \cdot g([X_2^*, X_1^*], X_1^*). \tag{4}$$

On the other hand, using $[X_i^*, X_j^*]_p = 0$ and the formulas from Theorem 9.1, we get

$$-g_p(R_p(Y, Z)\, Y, Z) = g_p(\nabla_{X_2^*} \nabla_{X_1^*} \cdot X_1^*, X_2^*) - g_p(\nabla_{X_1^*} \nabla_{X_2^*} \cdot X_1^*, X_2^*)$$
$$= X_2 \cdot g(\nabla_{X_1^*} \cdot X_1^*, X_2^*) - X_1 \cdot g(\nabla_{X_2^*} \cdot X_1^*, X_2^*)$$
$$- g_p(\nabla_{X_1^*} \cdot X_1^*, \nabla_{X_2^*} \cdot X_2^*) + g_p(\nabla_{X_2^*} \cdot X_1^*, \nabla_{X_1^*} \cdot X_2^*).$$

The two last terms vanish since in general $(\nabla_{X_i^*} \cdot X_j^*)_p = 0$. For the two other terms we use (2), § 9. Since $g(X_j^*, X_k^*)$ is constant we obtain

$$-g_p(R_p(Y, Z)\, Y, Z) = X_2 \cdot g([X_2^*, X_1^*], X_1^*) + X_1 \cdot g([X_1^*, X_2^*], X_2^*).$$

In view of (4) this proves Theorem 12.2 in the analytic case for Y, Z orthonormal. If Y, Z are linearly independent but not necessarily orthonormal, we can write $A = y_1 Y + z_1 Z$, $B = y_2 Y + z_2 Z$ where A, B are orthonormal vectors in S. Anticipating the first and third relations of Lemma 12.5 we find

$$K(S) = -g_p(R_p(A, B)\, A, B)$$
$$= -g_p(R_p(y_1 Y + z_1 Z, y_2 Y + z_2 Z) \cdot (y_1 Y + z_1 Z), y_2 Y + z_2 Z)$$
$$= -(y_1 z_2 - y_2 z_1)^2 g_p(R_p(Y, Z)\, Y, Z)$$
$$= -\frac{g_p(R_p(Y, Z)\, Y, Z)}{|Y \vee Z|^2}.$$

Finally we drop the analyticity assumption. Let $\{x_1, ..., x_m\}$ be a system of coordinates valid on an open neighborhood U of p, such that $(\partial/\partial x_1)_p = Y$, $(\partial/\partial x_2)_p = Z$. Consider the function g_{ij} defined by

$$g_{ij} = g\left(\frac{\partial}{\partial x_i}, \frac{\partial}{\partial x_j}\right).$$

Then for each $q \in U$, the matrix $(g_{ij})_q$ is symmetric and strictly positive definite. There exist analytic functions $\gamma_{ij} = \gamma_{ji}$ on an open set $V(p \in V \subset U)$ whose derivatives of order $0 \leqslant k \leqslant 3$ approximate those of g_{ij} as well as we please. For sufficiently good approximation the matrix $(\gamma_{ij})_q$ will be symmetric and strictly positive definite for each $q \in V$; we get an analytic Riemannian structure γ on V by requiring

$$\gamma\left(\frac{\partial}{\partial x_i}, \frac{\partial}{\partial x_j}\right) = \gamma_{ij} \qquad (1 \leqslant i, j \leqslant m).$$

The sectional curvature and the curvature tensor field for γ approximate the corresponding sectional curvature and the curvature tensor field for g. Since Theorem 12.2 holds for the Riemannian manifold induced by the Riemannian structure γ on V, the theorem holds also for g.

The next proposition shows, that in a certain sense, the sectional curvature determines the curvature tensor.

Proposition 12.3. *Let M be a Riemannian manifold, p a point in M. Let g and g' be two Riemannian structures on M, R and R' the corresponding curvature tensors, and $K(S)$ and $K'(S)$ the corresponding sectional curvatures at p along a plane section $S \subset M_p$.*

Suppose that $g_p = g'_p$. If $K(S) = K'(S)$ for all plane sections $S \subset M_p$, then $R_p = R'_p$.

We first prove two simple lemmas.

Lemma 12.4. *Let A be a ring with identity element e such that $6a \neq 0$ for $a \neq 0$ in A. Let E be a module over A. Suppose a mapping $B: E \times E \times E \times E \rightarrow A$ is quadrilinear and satisfies the identities*

(a) $B(X, Y, Z, T) = -B(Y, X, Z, T)$.

(b) $B(X, Y, Z, T) = -B(X, Y, T, Z)$.

(c) $B(X, Y, Z, T) + B(Y, Z, X, T) + B(Z, X, Y, T) = 0$.

Then

(d) $B(X, Y, Z, T) = B(Z, T, X, Y)$.

If, in addition to (*a*), (*b*), and (*c*), *B satisfies*

(e) $B(X, Y, X, Y) = 0$ *for all* $X, Y \in E$,

then $B = 0$.

Proof. First we interchange T in (c) with X, Y, Z, respectively, and add the four obtained relations. Using (a) and (b) one obtains

$$(f) \quad B(T, X, Y, Z) + B(T, Y, Z, X) + B(T, Z, X, Y) = 0.$$

From (c) and (a) it follows that

$$B(Z, X, T, Y) = + B(T, X, Z, Y) - B(T, Z, X, Y).$$

On substituting in (f), relation (d) follows. In particular, $B(X, Y, X, T)$ is symmetric in Y and T. Thus (e) implies $B(X, Y, X, T) \equiv 0$. In view of (a) and (b) this implies that B is alternate; then (c) shows at once that $B \equiv 0$.

Lemma 12.5. *Let M be a manifold with an affine connection ∇. Let R and T denote the curvature tensor and torsion tensor, respectively. Then R satisfies the following identities*

$$R(X, Y) = - R(Y, X).$$

If $T = 0$, then

$$R(X, Y) \cdot Z + R(Y, Z) \cdot X + R(Z, X) \cdot Y = 0 \quad \textit{(Bianchi's identity)}.$$

If g is a pseudo-Riemannian structure on M and ∇ is the corresponding pseudo-Riemannian connection, then

$$g(R(X, Y)Z, V) = - g(R(X, Y)V, Z),$$
$$g(R(X, Y)Z, V) = g(R(Z, V)X, Y).$$

Proof. The first identity $R(X, Y) = - R(Y, X)$ is obvious. For the second, we use $T = 0$ and obtain

$$(\nabla_X\nabla_Y - \nabla_Y\nabla_X)Z + (\nabla_Y\nabla_Z - \nabla_Z\nabla_Y)X + (\nabla_Z\nabla_X - \nabla_X\nabla_Z)Y$$
$$- \nabla_{[X,Y]}Z - \nabla_{[Y,Z]}X - \nabla_{[Z,X]}Y$$
$$= \nabla_X[Y, Z] - \nabla_{[Y,Z]}X + \nabla_Y[Z, X] - \nabla_{[Z,X]}Y + \nabla_Z[X, Y] - \nabla_{[X,Y]}Z$$
$$= [X, [Y, Z]] + [Y, [Z, X]] + [Z, [X, Y]] = 0$$

by the Jacobi identity for vector fields (§2, No. 1). For the third identity we can assume the vector fields X, Y, Z, V are adapted to their values at some point p. From (2), §9, we find in this case

$$g(\nabla_X Z, Z) = 0.$$

For an arbitrary vector field W we have

$$Wg(X, Y) = g(\nabla_W X, Y) + g(X, \nabla_W Y)$$

from §9; also $(\nabla_W(Z))_p = 0$ since Z is adapted to Z_p. Hence

$$\begin{aligned} g_p(R_p(X, Y) Z, Z) &= g_p(\nabla_X \nabla_Y Z, Z) - g_p(\nabla_Y \nabla_X Z, Z) \\ &= X_p g(\nabla_Y Z, Z) - Y_p g(\nabla_X Z, Z) = 0. \end{aligned}$$

This proves the third identity above. The last follows from Lemma 12.4.

Returning now to Prop. 12.3, we use Lemma 12.4 on the quadrilinear function on $M_p \times M_p \times M_p \times M_p$ given by

$$B(X, Y, Z, T) = g_p(R_p(X, Y) Z, T) - g'_p(R'_p(X, Y) Z, T).$$

Since $g_p = g'_p$ the parallelogram $X \vee Y$ has the same area whether measured by means of g or g'. Now $K(S) = K'(S)$ implies $B(X, Y, X, Y) = 0$ for all $X, Y \in M_p$ so by Lemma 12.4, $B \equiv 0$. But g_p is nondegenerate so $R_p = R'_p$.

§ 13. Riemannian Manifolds of Negative Curvature

The local Riemannian geometry developed in §9 was mostly based on properties of the forms $(\mathrm{Exp})^* \omega^i$, which are 1-forms in a normal neighborhood in the tangent space to the manifold at some fixed point. The forms ω^i are only defined locally and the same is therefore the case with the forms $(\mathrm{Exp})^* \omega^i$. However, we shall now see that these last forms (in contrast to the ω^i) can be extended to the entire tangent space or at any rate to the part of the tangent space where Exp is defined and regular. No assumption will be made about the curvature for the time being.

Let M be a Riemannian manifold, p a point in M, M_p the tangent space at p. Let Exp stand for the mapping Exp_p. Let N_0 be any open subset of M_p star-shaped with respect to 0 such that Exp is a regular mapping of N_0 into M. Note that N_0 is not assumed to be a normal neighborhood of 0. Let $Y_1, ..., Y_m$ be any orthonormal basis of M_p. If $X \in N_0$, there exists an open neighborhood N_X of X in N_0 which Exp maps diffeomorphically onto an open neighborhood B of Exp X in M. For each $Y \in N_X$ let $(Y_i^*)_{\mathrm{Exp}\ Y}$ denote the tangent vector at Exp Y which is obtained by parallel translating Y_i along the geodesic Exp tY $(0 \leqslant t \leqslant 1)$. Then $Y_1^*, ..., Y_m^*$ is a basis for the vector fields on B. Let $\Gamma_{ij}{}^k$, $R^k{}_{lij}$ be the corresponding functions on B as defined in §8 and let the 1-forms ω^i, $\omega^i{}_l$ $(1 \leqslant i, l \leqslant m)$ on B be determined by $\omega^i(Y_j^*) = \delta^i{}_j$, $\omega^i{}_l = \Sigma_k \Gamma_{kl}{}^i \omega^k$. Let V_X be the set of points $(t, a_1, ..., a_m)$ $\in \boldsymbol{R} \times \boldsymbol{R}^m$ for which $ta_1Y_1 + ... + ta_mY_m \in N_X$. Consider the mapping $\Phi : V_X \to B$ given by $\Phi : (t, a_1, ..., a_m) \to \operatorname{Exp}(ta_1Y_1 + ... + ta_mY_m)$. Just as in §8 one proves that the 1-forms $\Phi^*(\omega^i)$ and $\Phi^*(\omega^i{}_l)$ are given by

$$\Phi^*(\omega^i) = a_i dt + \bar{\omega}^i, \qquad \Phi^*(\omega^i{}_l) = \bar{\omega}^i{}_l,$$

where $\bar{\omega}^i$ and $\bar{\omega}^i{}_l$ are 1-forms on V_X not containing dt. If X' is another point in N_0 we can as above construct 1-forms $\bar{\omega}^i$ and $\bar{\omega}^i{}_l$ on $V_{X'}$. It is clear from the definition that these forms agree on $V_X \cap V_{X'}$ with the ones previously constructed. Thus, if V denotes the set of points $(t, a_1, ..., a_m)$ in $\boldsymbol{R} \times \boldsymbol{R}^m$ such that $ta_1Y_1 + ... + ta_mY_m \in N_0$, it follows that the forms $\bar{\omega}^i$ and $\bar{\omega}^i{}_l$ can be defined on the entire set V. They satisfy the differential equations

$$\frac{\partial \bar{\omega}^i}{\partial t} = da_i + \sum_k a_k \bar{\omega}^i{}_k, \qquad \bar{\omega}^i(t, a_j; da_k)_{t=0} = 0, \tag{1}$$

$$\frac{\partial \bar{\omega}^i{}_l}{\partial t} = \sum_{j,k} R^i{}_{ljk} a_j \bar{\omega}^k, \qquad \bar{\omega}^i{}_l(t, a_j; da_k)_{t=0} = 0, \tag{2}$$

in each V_X where $R^i{}_{ljk}$ stands for $R^i{}_{ljk} \circ \Phi$; hence (1) and (2) hold in the entire set V. Combining (1) and (2) we obtain the system

$$\frac{\partial^2 \bar{\omega}^i}{\partial t^2} = \sum_{j,k,l} R^i{}_{ljk} a_l a_j \bar{\omega}^k, \qquad \bar{\omega}^i(t, a_j; da_k)_{t=0} = 0, \tag{3}$$

on V. This system is a generalization of the so-called Jacobi equation for surfaces.

Let S denote the unit sphere $\| X \| = 1$ in M_p and let U denote the set of all pairs $(t, X) \in \boldsymbol{R} \times S$ such that $tX \in N_0$. Consider the mapping $\Psi : (t, X) \to \operatorname{Exp} tX$ of U into M. If we identify M_p and $\boldsymbol{R}^m$ by means of the basis $Y_1, ..., Y_m$, the sphere S is identified with the submanifold $a_1^2 + ... + a_m^2 = 1$ of $\boldsymbol{R}^m$. Thus the forms $\bar{\omega}^i$ and $\bar{\omega}^i{}_l$ induce 1-forms on U which we denote by the same symbol. Using the fact that N_0 is star-shaped, one finds that Lemma 9.2, stating that

$$\Psi^* g = (dt)^2 + \sum_{i=1}^m (\bar{\omega}^i)^2, \qquad \text{on } U, \tag{4}$$

is still valid. If ξ denotes the mapping $(t, X) \to tX$ of U into N_0, then $\Psi = \operatorname{Exp} \circ \xi$. We can use (4) in the special case when $M = M_p$. Here $R = 0$, $\bar{\omega}^i = t da_i$ and $\Psi = \xi$.

Hence

$$\xi^*(g_p) = (dt)^2 + t^2 \sum_{i=1}^{m} (da_i)^2. \tag{5}$$

Let A be any fixed vector field on the sphere $a_1^2 + ... + a_m^2 = 1$ in $\boldsymbol{R}^m$ and put

$$\alpha_i = \bar{\omega}^i(A), \qquad \gamma_i = da_i(A) \qquad (1 \leqslant i \leqslant m).$$

The functions $\alpha_i = \alpha_i(t, a_1, ..., a_m)$ are defined on the set U and satisfy the equation

$$\sum_{i=1}^{m} \alpha_i \frac{\partial^2 \alpha_i}{\partial t^2} = \sum_{i,j,k,l} R^i{}_{ljk} a_l a_j \alpha_i \alpha_k, \qquad \alpha_i(0, a_1, ..., a_m) = 0. \tag{6}$$

We assume now that the sectional curvature of M is $\leqslant 0$ along each plane section at each point of M. For simplicity we express this condition by saying that *M has negative curvature.*

Consider now a *fixed* point $(t, a_1, ..., a_m) \in U$ and the corresponding numbers $\alpha_1, ..., \alpha_m$. Now the point $X = t(a_1 Y_1 + ... + a_m Y_m)$ lies in N_0 and the neighborhood N_X above is diffeomorphic to B. Let $q = \operatorname{Exp} X$, $Y = \sum_{i=1}^m a_i Y_i^*$, $Z = \sum_{i=1}^m \alpha_i Y_i^*$. Then $q \in B$ and Y and Z are vector fields on B. It is clear that

$$g_q(R_q(Y_q, Z_q)\, Y_q, Z_q) = \sum_{i,j,k,l} R^i{}_{ljk} a_l a_j \alpha_i \alpha_k$$

so (6) implies

$$\sum_i \alpha_i \frac{\partial^2 \alpha_i}{\partial t^2} \geqslant 0 \qquad \text{on } U, \tag{7}$$

due to the curvature assumption.

Again fix a point $(a_1, ..., a_m) \in S$. We put $h(t) = (\sum_i \alpha_i^2)^{1/2}$ for all $t \geqslant 0$ for which $(t, a_1, ..., a_m) \in U$. We assume temporarily that A does not vanish at the point $(a_1, ..., a_m)$. Then $h(0) = 0$ and $h(t) > 0$ for $t > 0$. Since

$$\left(\frac{\partial \bar{\omega}^i}{\partial t}\right)_{t=0} = da_i, \qquad \left(\frac{\partial \alpha_i}{\partial t}\right)_{t=0} = \gamma_i,$$

it follows that $h'(0) = (\sum_i \gamma_i^2)^{1/2}$. From (7) and the identity

$$h(t)^3\, h''(t) = h(t)^2 \sum_i \alpha_i \frac{\partial^2 \alpha_i}{\partial t^2} + \left(\sum_i \alpha_i^2 \sum_i \left(\frac{\partial \alpha_i}{\partial t}\right)^2 - \left(\sum_i \alpha_i \frac{\partial \alpha_i}{\partial t}\right)^2\right)$$

it follows that

$$h''(t) \geqslant 0, \qquad h'(t) \geqslant h'(0), \qquad \text{for } t > 0.$$

Consequently, $h(t) \geqslant th'(0)$, so

$$\sum_{i=1}^{m} (\bar{\omega}^i(A))^2 \geqslant t^2 \sum_{i=1}^{m} (da_i(A))^2 \tag{8}$$

at the point $(a_1, ..., a_m)$. If A vanishes at the point $(a_1, ..., a_m) \in S$, (8) holds trivially. Hence (8) holds for an arbitrary vector field A on S and for all points in U for which $t \geqslant 0$. Using now (4), (5), and the fact that $\Psi = \text{Exp} \circ \xi$, $\Psi^* = \xi^* \circ \text{Exp}^*$, we obtain

$$\| d \, \text{Exp}_X (Y) \| \geqslant \| Y \|,$$

if X is any point in N_0 and Y is any tangent vector to N_0 at X. This proves the following theorem.

Theorem 13.1. *Let M be a Riemannian manifold of negative curvature and p any point in M. Let N_0 be any open subset of M_p star-shaped with respect to 0 such that Exp (the Exponential mapping at p) is a regular mapping of N_0 into M. Then*

$$\| d \, \text{Exp}_X (Y) \| \geqslant \| Y \|,$$

if X is any point in N_0 and Y is any tangent vector to N_0 at X. In particular,

$$L(\text{Exp} \circ \Gamma) \geqslant L(\Gamma)$$

for any curve segment Γ in N_0, L denoting arc length.

Corollary 13.2. *Suppose M is a Riemannian manifold of negative curvature and V a minimizing convex normal ball in M. Let ABC be a triangle inside V whose angles are A, B, C and whose sides are geodesics of lengths a, b, and c. Then*

(i) $a^2 + b^2 - 2ab \cos C \leqslant c^2$;
(ii) $A + B + C \leqslant \pi$.

In fact, let us use Theorem 13.1 on Exp_C. Let Γ_a, Γ_b, and Γ_c denote the geodesics forming the sides of the triangle and let γ_a, γ_b, and γ_c denote the corresponding curve segments in the tangent space M_C ($\text{Exp}_C(\gamma_a) = \Gamma_a$, etc.). Let γ_0 denote the straight line in M_C joining the end points of γ_c, and put $\Gamma_0 = \text{Exp}_C(\gamma_0)$.

Then

$$a = L(\gamma_a) = L(\Gamma_a),$$

$$b = L(\gamma_b) = L(\Gamma_b),$$

$$L(\gamma_0) \leqslant L(\gamma_c), \qquad L(\gamma_0)^2 = a^2 + b^2 - 2ab \cos C,$$

since the angle between γ_a and γ_b is C.

Suppose now that the sectional curvature is everywhere $\leqslant 0$. Then $L(\gamma_c) \leqslant L(\Gamma_c)$ and (i) follows. For (ii) we first observe that $c = d(A, B)$, etc., and consequently each length a, b, or c is majorized by the sum of the two others. We can therefore find an ordinary plane triangle with sides a, b, c. Denoting its angles by A', B', C' we have by (i): $A \leqslant A'$, $B \leqslant B'$, $C \leqslant C'$. Since $A' + B' + C' = \pi$, relation (ii) follows.

Theorem 13.3.

(i) *Let M be a complete Riemannian manifold of negative curvature and p any point in M. Then M contains no points conjugate to p.*

(ii) *Let M be a complete Riemannian manifold and suppose there exists a point $p \in M$ such that M contains no point conjugate to p. Then the pair (M_p, Exp_p) is a covering manifold of M. In particular, if M is simply connected, Exp_p is a diffeomorphism of M_p onto M.*

Proof. (i) As before, we write Exp instead of Exp_p and denote by $C(p)$ the (closed) set of points in M_p which are conjugate to p. If $C(p)$ were not empty, let X be a point in $C(p)$ at minimum distance from the origin. Then there exists a vector $Y \neq 0$ in M_p such that $d\,\text{Exp}_X(Y) = 0$. On the other hand, Theorem 13.1 implies that $\| d\,\text{Exp}_{tX}(Y) \| \geqslant \| Y \|$ for $0 \leqslant t < 1$. By continuity, $\| d\,\text{Exp}_X(Y) \| \geqslant \| Y \|$, which is a contradiction. Thus $C(p) = \emptyset$.

In order to prove the latter statement of the theorem we follow a suggestion of I. Singer and consider the tensor field $g^* = \text{Exp}^* g$ on M_p, g denoting the Riemannian structure on M. Owing to the regularity of Exp, g^* is a Riemannian structure on M_p. The space M_p with the Riemannian structure g^* is complete; in fact, the geodesics through the origin in M_p are straight lines. Thus M_p is complete at p (in the sense of the remark following Theorem 10.4), hence complete. Theorem 13.3 now follows from the next lemma.

Lemma 13.4.† *Let V and W be Riemannian manifolds, V complete, and φ a differentiable mapping of V onto W. Assume that $d\varphi_v$ is an isometry for each $v \in V$. Then (V, φ) is a covering space of W.*

† Ambrose [1], p. 360.

Proof (Due to Palais, from Hicks [1]). Let $w \in W$, let N_0 be a normal neighborhood of 0 in W_w of the form $\| X \| < r$, and put $N_w = \operatorname{Exp}_w N_0$. Let v be any point in $\varphi^{-1}(w)$. Let ψ denote the inverse of the mapping $\operatorname{Exp}_w : N_0 \to N_w$. Since V is complete, the mapping $f = \operatorname{Exp}_v \circ (d\varphi_v)^{-1} \circ \psi$ is a well-defined mapping of N_w onto a subset N_v of V and it is obvious that $\varphi \circ f$ is the identity mapping of N_w onto itself. Similarly, $f \circ \varphi$ is the identity mapping of N_v onto itself. Since $(d\varphi_v)^{-1}(N_0)$ is the ball $\| X \| < r$ in V_v, it is clear that N_v is contained in the open ball $B_r(v)$. On the other hand, $B_r(v) \subset N_v$ due to Theorem 10.4. Thus $N_v = B_r(v)$ and φ is a diffeomorphism of N_v onto N_w. Now suppose $v_1, v_2 \in \varphi^{-1}(w)$, $v_1 \neq v_2$. Then the balls $B_r(v_1)$ and $B_r(v_2)$ are disjoint because otherwise there would be a point $v^* \in B_r(v_2) \cap B_r(v_1)$ such that w and $\varphi(v^*)$ are joined by geodesics of different length lying inside N_w. Moreover,

$$\bigcup_{v \in \varphi^{-1}(w)} B_r(v) = \varphi^{-1}(N_w),$$

because each point in $\varphi^{-1}(N_w)$ can be joined to a point in $\varphi^{-1}(w)$ by means of a geodesic of length $< r$. This proves that (V, φ) is a covering manifold of W.

The next theorem, due to É. Cartan, has an important application to Lie groups (see Chapter VI); in fact, it leads to the only known proof of the conjugacy of maximal compact subgroups of a semisimple Lie group.

Theorem 13.5. *Let M be a complete simply connected Riemannian manifold of negative curvature. Let K be a compact Lie transformation group*† *of M whose elements are isometries of M. Then the members of K have a common fixed point.*

Proof. Let d denote the distance function on M and let dk denote the Haar measure on K, normalized by $\int_K dk = 1$. Select a point $p \in M$ and consider the real function J on M given by

$$J(q) = \int_K d^2(q, k \cdot p) \, dk.$$

Then J is a nonnegative continuous function on M. Since the orbit of p is compact, there exists a ball $B_r(p)$ such that $J(q) > J(p)$ for $q \notin B_r(p)$. The closure of $B_r(p)$ is compact, and contains therefore a minimum point q_0 for J. Then q_0 is also a minimum for J on M. It is

† See the definition in Chapter II, § 3.

clear that $J(k \cdot q_0) = J(q_0)$ for $k \in K$, so in order to prove the existence of the fixed point, it suffices to prove that

$$J(q) > J(q_0) \qquad \text{if } q \neq q_0. \tag{9}$$

Now, due to Theorem 13.3, any two points in M can be joined by a unique geodesic and its length is the distance between the points. Thus, due to Cor. 13.2, the cosine inequality

$$a^2 + b^2 - 2ab \cos C \leqslant c^2$$

is valid for an arbitrary geodesic triangle in M. Suppose now $q \neq q_0$ and let $t \to q_t$ $(0 \leqslant t \leqslant d(q_0, q))$ denote the geodesic joining q_0 to q. If $k \cdot p \neq q_t$, let $\alpha_t(k)$ denote the angle between the geodesics (q_t, q) and $(k \cdot p, q_t)$. In view of Lemma 13.6 we have

$$\frac{d}{dt} d^2(q_t, k \cdot p) = \begin{cases} 2d(q_t, k \cdot p) \cos \alpha_t(k), & \text{if } k \cdot p \neq q_t, \\ 0, & \text{if } k \cdot p = q_t. \end{cases} \tag{10}$$

We shall now prove that the function

$$F(t, k) = \frac{d}{dt} d^2(q_t, k \cdot p) \qquad (0 \leqslant t \leqslant d(q_0, q), k \in K),$$

is continuous at each point $(0, k)$, $k \in K$. Then F is clearly continuous everywhere. Let K_1 denote the (closed) set of elements k in K such that $k \cdot p = q_0$; let K_2 denote the complement $K - K_1$. Now, the mapping $k \to k \cdot p$ of K into M is differentiable; using Lemma 10.1 successively, it follows quickly that the function $(t, k) \to \cos \alpha_t(k)$ is continuous at $(0, k_0)$ if $k_0 \in K_2$. Next, let $k_0 \in K_1$ and suppose the sequence (t_n, k_n) converges to $(0, k_0)$. By (10) we have

$$| F(t, k) | \leqslant 2d(q_t, k \cdot p),$$

and since $d(q_{t_n}, k_n \cdot p) \to d(q_0, k_0 \cdot p) = 0$, it follows that $F(t_n, k_n) \to F(0, k_0)$. This proves the continuity of F. Thus the function $t \to J(q_t)$ is differentiable and its derivative can be obtained by differentiating under the integral sign. Since the minimum occurs for $t = 0$, we obtain from (10)

$$\int_{K_2} d(q_0, k \cdot p) \cos \alpha_0(k) \, dk = 0.$$

The cosine inequality above shows that

$$d^2(q, k \cdot p) \geqslant d^2(q_0, k \cdot p) + d^2(q_0, q) - 2d(q_0, q)\, d(q_0, k \cdot p) \cos (\pi - \alpha_0(k))$$

for $k \in K_2$. By integration we obtain

$$\int_{K_2} d^2(q, k \cdot p)\, dk \geqslant \int_{K_2} d^2(q_0, k \cdot p)\, dk + d^2(q_0, q) \int_{K_2} dk.$$

The similar inequality for K_1 is trivial; adding these inequalities we get

$$J(q) \geqslant J(q_0) + d^2(q_0, q),$$

which proves the theorem.

Lemma 13.6. *Let M be as in Theorem 13.5 and let $p \in M$. Let $\gamma : t \to q_t$ $(0 \leqslant t \leqslant L)$ be a curve segment not containing p, t being the arc parameter. Then*

$$\left[\frac{d}{dt}\, d(q_t, p)\right]_{t=0} = \cos \alpha,$$

where α denotes the angle between the segment γ and the geodesic (pq_0) at q_0 (see Fig. 1).

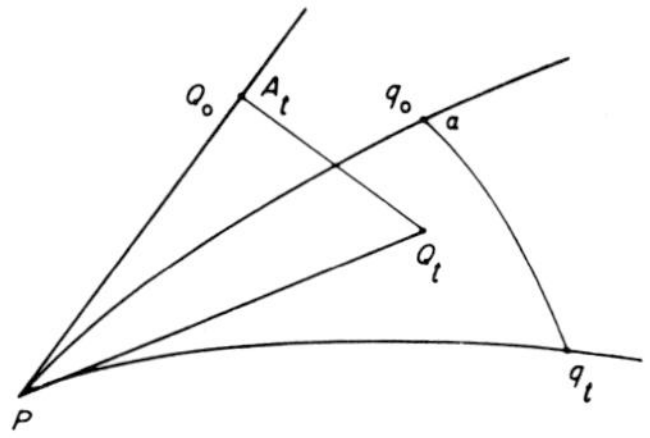

FIG. 1

Proof. Let Exp stand for Exp_p and determine $Q_t \in M_p$ such that $\text{Exp}\, Q_t = q_t$. Let the distance in M_p also be denoted by d. Then

$$\lim_{t \to 0} \frac{1}{t} (d(q_t, p) - d(q_0, p)) = \lim_{t \to 0} \frac{1}{2d(q_0, p)\, t} (d^2(q_t, p) - d^2(q_0, p))$$

$$= \lim_{t \to 0} \frac{1}{2d(Q_0, p)\, t} (d^2(Q_t, p) - d^2(Q_0, p)).$$

Now

$$d^2(Q_t, p) - d^2(Q_0, p) = d^2(Q_0, Q_t) + 2d(Q_0, p)\, d(Q_0, Q_t) \cos A_t,$$

where A_t is the angle between the straight lines (pQ_0) and (Q_0Q_t). Since Exp is a diffeomorphism, the mapping $t \to Q_t$ $(0 \leqslant t \leqslant L)$ is a curve segment. Let Y denote its tangent vector at Q_0, L_t its arc length measured from Q_0 to Q_t, and A the angle between Y and (pQ_0). Then

$$\lim_{t\to 0} \frac{d(Q_0, Q_t)}{L_t} = 1, \qquad \lim_{t\to 0} \frac{L_t}{t \,||\, Y \,||} = 1. \tag{11}$$

It follows at once that $(1/t)\,(d^2(Q_0, Q_t)) \to 0$ as $t \to 0$. Consequently,

$$\left[\frac{d}{dt}\, d(q_t, p)\right]_{t=0} = \lim_{t\to 0} \frac{1}{t}\, d(Q_0, Q_t) \cos A_t = ||\, Y \,|| \cos A.$$

Now $Y = Y_0 + Y_1$ where Y_0 has the direction of (pQ_0) and Y_1 is perpendicular to that direction. In view of Lemma 9.7, $d\,\mathrm{Exp}_{Q_0}(Y_1)$ is perpendicular to the geodesic (pq_0). Since $||\, d\,\mathrm{Exp}_{Q_0} Y_0 \,|| = ||\, Y_0 \,||$ we have

$$||\, Y \,|| \cos A = ||\, d\,\mathrm{Exp}_{Q_0}(Y) \,|| \cos \alpha = \cos \alpha,$$

and the lemma is proved.

§ 14. Totally Geodesic Submanifolds

Let M be a differentiable manifold, S a submanifold. Let $m = \dim M$, $s = \dim S$. A curve in S is of course a curve in M, but a curve in M contained in S is not necessarily a curve in S, because it may not even be continuous. However, we have:

Lemma 14.1. *Let φ be a differentiable mapping of a manifold V into the manifold M such that $\varphi(V)$ is contained in the submanifold S. If the mapping $\varphi : V \to S$ is continuous it is also differentiable.*

Let $p \in V$. In view of Prop. 3.2, there exists a coordinate system $\{x_1, ..., x_m\}$ valid on an open neighborhood N of $\varphi(p)$ in M such that the set

$$N_S = \{r \in N : x_j(r) = 0 \text{ for } s < j \leqslant m\}$$

together with the restrictions of $(x_1, ..., x_s)$ to N_S form a local chart on S containing $\varphi(p)$. By the continuity of φ there exists a local chart (W, ψ) around p such that $\varphi(W) \subset N_S$. The coordinates $x_j(\varphi(q))$ $(1 \leqslant j \leqslant m)$ depend differentiably on the coordinates of $q \in W$. In particular, this holds for the coordinates $x_j(\varphi(q))$ $(1 \leqslant j \leqslant s)$ so the mapping $\varphi : V \to S$ is differentiable.

As an immediate consequence of this lemma we have the following statement: Suppose that V and S are submanifolds of M and $V \subset S$. If S has the relative topology of M, then V is a submanifold of S.

In the remainder of this section we shall assume that M is a Riemannian manifold and S a connected submanifold. The Riemannian structure on M induces a Riemannian structure on S. Let d_M and d_S denote the distance functions in M and S, respectively. It is obvious that

$$d_M(p, q) \leqslant d_S(p, q)$$

for $p, q \in S$. In order to distinguish between geodesics in M and S we shall call them M-geodesics and S-geodesics, respectively.

Lemma 14.2. *Let γ be a curve in S, and suppose γ is an M-geodesic. Then γ is an S-geodesic.*

Proof. Let o and p be any points on γ, say $o = \gamma(r_0)$ and $p = \gamma(r)$; let N_o be a spherical normal neighborhood of o in M. If r is sufficiently close to r_0 the geodesic segment

$$\gamma_{op} : t \to \gamma(t), \qquad | t - r_0 | \leqslant | r - r_0 |,$$

is contained in N_o. In view of Lemma 9.3, the length of γ_{op} satisfies

$$L(\gamma_{op}) = d_M(o, p) \leqslant d_S(o, p) \leqslant L(\gamma_{op}).$$

Consequently $L(\gamma_{op}) = d_S(o, p)$; thus γ_{op} is a curve of shortest length in S joining o and p, hence an S-geodesic.

Definition. Let M be a Riemannian manifold and S a connected submanifold of M. Let $p \in S$. The submanifold S is said to be *geodesic at p* if each M-geodesic which is tangent to S at p is a curve in S. The submanifold S is called *totally geodesic* if it is geodesic at each of its points.

Lemma 14.3. *Suppose S is a submanifold of M, geodesic at a point $p \in S$. If γ is an S-geodesic through p, then γ is also an M-geodesic. If M is complete, then S is complete.*

Proof. Let Γ be the maximal M-geodesic tangent to γ at p. Then $\Gamma \subset S$ so by Lemma 14.2, Γ is an S-geodesic. Hence $\gamma \subset \Gamma$. Now suppose M is complete and let Exp_M and Exp_S denote the Exponential mapping at p for M and S, respectively. By assumption Exp_M is defined on the entire M_p. Since S is geodesic at p, Exp_S is the restriction of Exp_M to S_p, in particular, S is complete at p in the sense of the remark following Theorem 10.4. By that remark, S is complete.

Proposition 14.4. *Suppose S is a totally geodesic submanifold of M, and let I denote the identity mapping of S into M. For each $p \in S$ there exists an open neighborhood U_p of p in S on which I is distance preserving, that is,*

$$d_S(q_1, q_2) = d_M(q_1, q_2) \qquad \textit{for } q_1, q_2 \in U_p. \tag{1}$$

Proof. Let $B_\rho(p)$ be a minimizing convex normal ball around p in M. Since I is continuous, the intersection $B_\rho(p) \cap S$ is an open subset of S. Let U_p be a minimizing convex normal ball around p in S such that $U_p \subset B_\rho(p)$. Let q, r be arbitrary points in U_p and γ_{qr} the S-geodesic inside U_p joining q and r. Then $d_S(q, r) = L(\gamma_{qr})$. Consider the maximal M-geodesic Γ such that Γ and γ_{qr} have the same tangent vector at q. Since S is totally geodesic, $\Gamma \subset S$; from Lemma 14.2 follows that $\gamma_{qr} \subset \Gamma$, so γ_{qr} is an M-geodesic. Since $\gamma_{qr} \subset B_\rho(p)$ it follows that $d_M(q, r) = L(\gamma_{qr})$. This proves the proposition.

Remark. Examples are easily constructed (e.g., geodesics on a cone) which show that relation (1) does not in general hold for all $q_1, q_2 \in S$.

Theorem 14.5. *Let M be a Riemannian manifold and S a connected, complete submanifold of M. Then S is totally geodesic if and only if M-parallel translation along curves in S always transports tangents to S into tangents to S.*

Proof. Let $s = \dim S$, $m = \dim M$ and let o be an arbitrary point in S. In view of Prop. 3.2 there exists an open neighborhood N of o in M on which a coordinate system $\{x_1, ..., x_m\}$ is valid such that the set

$$U = \{q \in N : x_j(q) = 0 \quad \text{for} \quad s + 1 \leqslant j \leqslant m\}$$

is a normal neighborhood of o in S and such that the restrictions of $(x_1, ..., x_s)$ to U form a coordinate system on U.

Let $p \in U$ and let $\gamma : t \to \gamma(t)$ be a curve in U such that $p = \gamma(0)$. Let $Y(t)$ be a family of tangent vectors to M which is M-parallel along the curve γ and such that $Y(0) \in S_p$. Writing $Y(t) = \sum_{a=1}^m Y^a(t)\, \partial/\partial x_a$ the coefficients $Y^a(t)$ satisfy the equations

$$\dot{Y}^a(t) + \sum_{b,c=1}^{m} \Gamma_{bc}{}^a\, \dot{x}_b(t)\, Y^c(t) = 0, \qquad 1 \leqslant a, b, c \leqslant m, \tag{2}$$

$$Y^a(0) = 0, \qquad x_a(t) \equiv 0, \qquad s < a \leqslant m.$$

Let $\pi : t \to \pi(t)$ be an M-geodesic tangent to S at p, t being the arc parameter measured from p. If t is sufficiently small and we write $x_a(t)$ for $x_a(\pi(t))$,

$$\ddot{x}_a(t) + \sum_{b,c=1}^{m} \Gamma_{bc}{}^a\, \dot{x}_b(t)\, \dot{x}_c(t) = 0, \qquad 1 \leqslant a, b, c, \leqslant m, \tag{3}$$

$$\dot{x}_a(0) = 0, \qquad s < a \leqslant m.$$

In the computations below we adopt the following range of indices:

$$1 \leqslant i, j, k \leqslant s,$$

$$s + 1 \leqslant \alpha, \beta, \gamma \leqslant m.$$

Suppose now S is totally geodesic. Then the M-geodesic π above is a curve in S, hence an S-geodesic. For small t, $\pi(t)$ lies in U so $x_\alpha(t) \equiv 0$. Since every S-geodesic is now an M-geodesic, (3) implies

$$\Gamma_{jk}^{\alpha}(p) = 0, \qquad p \in U. \tag{4}$$

For the curve γ above we have $\dot{x}_\alpha(t) \equiv 0$. In view of (4) we obtain

$$\dot{Y}^{\alpha}(t) + \sum_{j,\beta} \Gamma_{j\beta}^{\alpha}\, \dot{x}_j(t)\, Y^{\beta}(t) = 0 \qquad \text{on } \gamma. \tag{5}$$

Now, $Y(0) \in S_p$ so $Y^{\beta}(0) = 0$. Owing to the uniqueness theorem for the system (5) of linear differential equations we have $Y^{\beta}(t) \equiv 0$. Consequently, the family $Y(t)$ is tangent to S. Finally, let $\beta : t \to \beta(t)$ $(t \in J)$ be an arbitrary curve in S and let $Z(t)$ be an M-parallel family along β such that $Z(t_0) \in S_{\beta(t_0)}$ for some $t_0 \in J$. The set of $t \in J$ such that $Z(t) \in S_{\beta(t)}$ is clearly closed in J. The argument above shows that this set is open in J. Thus $Z(t) \in S_{\beta(t)}$ for all $t \in J$ and the first half of the theorem is proved.

To prove the converse, suppose that for each curve as above, the relation $Y(0) \in S_p$ implies $Y(t) \in S_{\gamma(t)}$ for each t. In (2) we have therefore

$$Y^{\alpha}(t) \equiv 0 \qquad \dot{x}_\alpha(t) \equiv 0$$

and (4) follows. Now substitute (4) into (3). Since $\Gamma_{bc}^{a} = \Gamma_{cb}^{a}$ (torsion is 0), we obtain

$$\ddot{x}_\alpha(t) + 2 \sum_{j,\beta} \Gamma_{j\beta}^{\alpha}\, \dot{x}_j(t)\, \dot{x}_\beta(t) + \sum_{\beta,\gamma} \Gamma_{\beta\gamma}^{\alpha}\, \dot{x}_\beta(t)\, \dot{x}_\gamma(t) = 0. \tag{6}$$

Since $\dot{x}_\alpha(0) = 0$ we conclude from the uniqueness theorem for the nonlinear system (6) that $x_\alpha(t)$ is constant, that is, $x_\alpha(t) = 0$ for all t in a certain interval around 0. The functions $x_j(t)$ are differentiable; consequently, a piece π' of π containing p is a curve in U, hence a curve in S. Let π^* be the maximal S-geodesic tangent to π at p, parametrized by the arc length t^* measured from p. Since S is complete, t^* runs from $-\infty$ to ∞. Now $\pi^*(t) = \pi(t)$ if t is sufficiently small; moreover, the set of t-values for which $\pi(t) = \pi^*(t)$ is open and closed. Thus $\pi \subset \pi^*$, π is a curve in S and the theorem is proved.

Theorem 14.6. *Let M be a simply connected, complete Riemannian manifold of negative curvature. Let S be a closed totally geodesic submanifold of M. For each $p \in S$, the geodesics in M which are perpendicular to S at p make up a submanifold $S^{\perp}(p)$ of M and M is the disjoint union:*

$$M = \bigcup_{p \in S} S^{\perp}(p).$$

Proof. Let Exp_p denote the Exponential mapping of M_p into M and let T_p denote the orthogonal complement of the tangent space S_p in M_p. Since S is complete and since S-geodesics are M-geodesics, it follows that

$$S = \text{Exp}_p\,(S_p).$$

Moreover, we have by definition

$$S^{\perp}(p) = \text{Exp}_p\,(T_p).$$

Since Exp_p is a diffeomorphism (Theorem 13.3), $S^{\perp}(p)$ is a submanifold of M.

Now, let q be an arbitrary point in M lying outside S; S being closed there exists a point $p_0 \in S$ at shortest distance from q. The unique geodesic Γ from q to p_0 is perpendicular to S. In fact, if $\sigma : t \to s_t$ is a curve in S, the derivative $(d/dt)\, d(s_t, q)$ equals $\cos \alpha_t$ where α_t is the angle between σ and the geodesic from q to s_t (Lemma 13.6). On the other hand, if a geodesic connecting q to S is perpendicular to S, then this geodesic must coincide with Γ since the sum of the angles in a geodesic triangle in M is $\leqslant \pi$. This shows that each point $q \in M$ lies in exactly one of the manifolds $S^{\perp}(p)$ and the theorem is proved.

§ 15. Appendix

In this section we collect some tools which have been used in this chapter and prove some supplemental results mentioned in the text.

1. Topology

Proposition 15.1. *Let M be a locally compact Hausdorff space which has a countable basis for the open sets. Then M is paracompact.*

Proof. Let $U_1, U_2, \ldots$ be a countable base for the open subsets of M. Since M is locally compact, we may assume that the U_i are relatively compact. Now define inductively

$$V_1 = U_1, \qquad V_2 = U_1 \cup \ldots \cup U_{i_1}, \qquad \text{and} \qquad V_{k+1} = U_1 \cup \ldots \cup U_{i_k}$$

where $i_0 = 1$, and i_k the smallest integer $> i_{k-1}$ such that

$$\bar{V}_k \subset \bigcup_{i=1}^{i_k} U_i.$$

Then

$$M = \bigcup_1^\infty V_k, \qquad \bar{V}_k \subset V_{k+1}.$$

Now suppose $\{U_\alpha\}_{\alpha \in A}$ is an arbitrary covering of M. By the compactness of each $\bar{V}_k$ we can choose finite subcoverings of the coverings

$$\{U_\alpha \cap V_3\}_{\alpha \in A} \qquad \text{of} \quad \bar{V}_2;$$

$$\{U_\alpha \cap (V_{k+1} - \bar{V}_{k-2})\}_{\alpha \in A} \qquad \text{of} \quad \bar{V}_k - V_{k-1} \quad (k \geqslant 3).$$

The members of these subcoverings constitute a locally finite covering of M which clearly is a refinement of $\{U_\alpha\}_{\alpha \in A}$.

Proposition 15.2. *Every paracompact space M is normal.*

Proof. Let A and B be two closed disjoint subsets of M. Fix $p \in A$. We first show that p and B can be separated by disjoint open sets, $U(p)$ and $V(p)$ with $p \in U(p)$, $B \subset V(p)$. For each $q \in B$ there exist, by the Hausdorff axiom, two disjoint open sets U_q and V_q with $p \in U_q$, $q \in V_q$. The sets $M - B$ and V_q $(q \in B)$ constitute a covering of M which, by paracompactness, has a locally finite refinement $\{W_\alpha\}$. Then the set

$$V(p) = \bigcup_{W_\alpha \cap B \neq \emptyset} W_\alpha \tag{1}$$

is an open set containing B. By the local finiteness, p has an open neighborhood N which intersects only finitely many W_α in (1), say $W_1, \ldots, W_n$. For each of these W_i choose q_i such that $W_i \subset V_{q_i}$; then the set

$$U(p) = \Big(\bigcap_i U_{q_i}\Big) \cap N$$

and $V(p)$ above have the desired property.

The covering $M - A$, $U(p)$ $(p \in A)$ of M has a locally finite refinement $\{N_\beta\}$. The set

$$U = \bigcup_{N_\beta \cap A \neq \emptyset} N_\beta \tag{2}$$

is an open set containing A. Each $q \in B$ is contained in an open neighborhood B_q which intersects only finitely many N_β in (2), say $N_1^q, ..., N_b^q$. Each of these N_j^q is contained in some $U(p)$, say $U(p_j)$. The intersection

$$\Big(\bigcap_j V(p_j)\Big) \cap B_q$$

then is an open neighborhood of q disjoint from U. The union V of these neighborhoods as q runs through B has the desired property.

Proposition 15.3. *Let M be a normal topological space and $\{U_\alpha\}_{\alpha \in A}$ a locally finite covering of M. Then the sets U_α can be shrunk to sets V_α such that $\bar{V}_\alpha \subset U_\alpha$ for each $\alpha \in A$ and such that $\{V_\alpha\}_{\alpha\in A}$ is still a covering of M.*

Proof. Let Φ denote the set of all functions φ on the index set A satisfying

(i) $\varphi(\alpha) = U_\alpha$ or

$$\varphi(\alpha) = \text{open set } V_\alpha \text{ satisfying } \bar{V}_\alpha \subset U_\alpha;$$

(ii) $\{\varphi(\alpha)\}_{\alpha\in A}$ is a covering of M.

We give Φ a partial ordering: $\varphi < \varphi'$ means that $\varphi(\alpha) = \varphi'(\alpha)$ whenever $\varphi(\alpha) = V_\alpha$. Let $\Psi \subset \Phi$ be a totally ordered subset and put

$$\psi^*(\alpha) = \bigcap_{\psi\in\Psi} \psi(\alpha).$$

Then we claim $\psi^* \in \Phi$. In fact, since Ψ is totally ordered, the family $\{\psi(\alpha) : \psi \in \Psi\}$ consists of at most two members, so (i) is obvious; for (ii) let $p \in M$ and $U_{\alpha_1}, ..., U_{\alpha_n}$ the *finitely* many members of the covering $\{U_\alpha\}_{\alpha\in A}$ which contain p. Define $\psi_p(\alpha) = \psi^*(\alpha)$ for $\alpha = \alpha_1, ..., \alpha_n$; otherwise $\psi_p(\alpha) = U_\alpha$. Then $\{\psi_p(\alpha)\}_{\alpha\in A}$ is obtained from $\{U_\alpha\}_{\alpha\in A}$ by shrinking at most finitely many U_α, each shrinking leading to a new covering. Thus $\psi_p \in \Phi$, so in particular, (ii) holds for ψ^*, that is, $\psi^* \in \Phi$. The definition of ψ^* shows $\psi^* = \sup \Psi$. Thus we can apply Zorn's lemma and conclude that Φ has a maximal element φ^*.

It remains to prove that $\varphi^*(\alpha) = V_\alpha$ for every α. But if $\varphi^*(\beta) = U_\beta (\neq \bar{U}_\beta)$ for some $\beta \in A$, we consider the subset

$$M_\beta = M - \bigcup_{\alpha\neq\beta} \varphi^*(\alpha).$$

Then M_β is a closed subset of U_β (since φ^* satisfies (ii)), so by normality there exists an open set V_β such that $M_\beta \subset V_\beta \subset \bar{V}_\beta \subset U_\beta$. But then φ_0 defined by

$$\varphi_0(\alpha) = \varphi^*(\alpha), \qquad \varphi_0(\beta) = V_\beta$$

would contradict the maximality of φ^*.

Theorem 15.4. *A connected, locally compact metric space M is separable.*

Proof. Consider the open balls $B_r(p)$ in M. For each $p \in M$, $B_r(p)$ is relatively compact if r is sufficiently small; let $r(p)$ be the supremum of $r \in \boldsymbol{R}$ for which $B_r(p)$ is relatively compact. If $r(p) = \infty$ for some p, there is nothing to prove since a compact metric space is clearly separable.

Suppose therefore $r(p_0) < \infty$ for all $p_0 \in M$. Then $r(p)$ depends continuously on p; in fact

$$r(q) \leqslant r(p) + d(p, q) \qquad (p, q \in M). \tag{3}$$

To see this, suppose to the contrary that $r(q) = r(p) + d(p, q) + 2\epsilon$ for some $p, q \in M$ and $\epsilon > 0$. Then by the triangle inequality,

$$B_{r(p)+\epsilon}(p) \subset B_{r(q)-\epsilon}(q),$$

contradicting the maximality of $r(p)$. Consider now the closed balls

$$V(p) = \{q \in M : d(p, q) \leqslant \tfrac{1}{2}r(p)\}$$

and put

$$C_0 = V(p_0), \qquad C_{n+1} = \bigcup_{p \in C_n} V(p), \qquad n = 0, 1, \ldots.$$

Then the union $C = \bigcup_{k=0}^{\infty} C_k$ is obviously open in M; it is also closed as a consequence of (3); in fact if (p_k) is a sequence in C converging to $q \in M$, we have $d(p_k, q) \leqslant \frac{1}{3}r(q)$ for k sufficiently large, whence by (3), $q \in V(p_k)$. Thus $q \in C_{n_k+1}$ if $p_k \in C_{n_k}$. By the connectedness of M, the family $\{C_n\}_{n \in \boldsymbol{Z}}$ therefore has union M.

To conclude the proof of the theorem it suffices to prove that each C_k is compact. This is so for $k = 0$ and assuming it true for $k \leqslant n$ we shall prove C_{n+1} compact. So let (p_i) be a sequence in C_{n+1}. Then we choose $q_i \in C_n$ such that $p_i \in V(q_i)$. Passing to a subsequence if necessary we may, by the compactness of C_n, assume (q_i) converges to a limit

$q \in C_n$. Then $r(q_i)$ converges to $r(q)$ and since $p_i \in V(q_i)$ we conclude that if $\frac{1}{2} < \mu < 1$, then p_i belongs to the closed ball

$$\{p \in M : d(q, p) \leqslant \mu r(q)\}$$

for i sufficiently large. This last ball being compact, (p_i) has a convergent subsequence, so the theorem is proved.

2. Mappings of Constant Rank

Theorem 15.5. *Let M and N be C^∞ manifolds and $\Phi : M \to N$ a differentiable mapping. Let $p \in M$ and suppose the linear transformation $d\Phi_q : M_q \to N_{\Phi(q)}$ has constant rank k for all q in a neighborhood of p. Then there exists a coordinate system $\xi = \{x_1, ..., x_m\}$ near $p \in M$ and a coordinate system $\eta = \{y_1, ..., y_n\}$ near $\Phi(p) \in N$ such that $\xi(p) = 0$, $\eta(\Phi(p)) = 0$, and the expression of Φ, that is, the map $\eta \circ \Phi \circ \xi^{-1}$, is given by*

$$\eta \circ \Phi \circ \xi^{-1} : (x_1, ..., x_m) \to (x_1, ..., x_k, 0, ..., 0).$$

Proof. Let $\mu = \{u_1, ..., u_m\}$ be coordinates near p and $\nu = \{v_1, ..., v_n\}$ coordinates near $\Phi(p)$ such that $\mu(p) = 0$, $\nu(\Phi(p)) = 0$. If $\nu \circ \Phi \circ \mu^{-1}$ is given by

$$v_i = \varphi_i(u_1, ..., u_m) \qquad (1 \leqslant i \leqslant n),$$

we can assume the indexing done such that

$$\left(\det \left(\frac{\partial \varphi_\alpha}{\partial u_\beta}\right)_{1 \leqslant \alpha, \beta \leqslant k}\right)(\mu(p)) \neq 0.$$

Thus, introducing the functions

$$x_\alpha = v_\alpha \circ \Phi, \qquad 1 \leqslant \alpha \leqslant k, \tag{4}$$

$$x_j = u_j, \qquad k < j \leqslant m, \tag{5}$$

we have $x_\alpha(q) = \varphi_\alpha(u_1(q), ..., u_m(q))$ $(1 \leqslant \alpha \leqslant k)$ so the mapping $(u_1, ..., u_m) \to (x_1, ..., x_m)$ has a nonsingular Jacobian at the origin. Thus the mapping $\xi : q \to (x_1(q), ..., x_m(q))$ is a local coordinate system near $p \in M$. Writing now

$$\nu \circ \Phi \circ \xi^{-1}(x_1, ..., x_m) = (\psi_1, ..., \psi_n),$$

each ψ_i is a C^∞ function of $(x_1, ..., x_m)$ and because of (4), $\psi_\alpha = x_\alpha$ $(1 \leqslant \alpha \leqslant k)$, so the Jacobian has the form

$$\left(\frac{\partial \psi_i}{\partial x_j}\right) = \left(\begin{array}{c|ccc} I_k & & 0 & \\ \hline & D_{k+1}\psi_{k+1} & \cdots & D_m\psi_{k+1} \\ * & \vdots & & \\ & D_{k+1}\psi_n & & D_m\psi_n \end{array}\right), \tag{6}$$

where I_k is the unit matrix of order k and $D_j = \partial/\partial x_j$ $(1 \leqslant j \leqslant m)$. Since the matrix (6) has rank k, the lower right-hand block must vanish, that is, each ψ_i $(k < i \leqslant n)$ is given by a function

$$\psi_i = \psi_i(x_1, ..., x_k) \qquad (k < i \leqslant n),$$

which is independent of x_j $(j > k)$. Finally, we introduce the functions

$$y_\alpha = v_\alpha, \qquad 1 \leqslant \alpha \leqslant k,$$

$$y_i = v_i - \psi_i(v_1, ..., v_k), \qquad k < i \leqslant n,$$

and let η denote the mapping $r \to (y_1(r), ..., y_n(r))$. Then

$$\eta \circ \nu^{-1}(v_1, ..., v_n) = (v_1, ..., v_k, v_{k+1} - \psi_{k+1}(v_1, ..., v_k), ..., v_n - \psi_n(v_1, ..., v_k)),$$

whence

$$\begin{aligned} \eta \circ \Phi \circ \xi^{-1}(x_1, ..., x_m) &= \eta \circ \nu^{-1} \circ (\nu \circ \Phi \circ \xi^{-1})(x_1, ..., x_m) \\ &= \eta \circ \nu^{-1}(x_1, ..., x_k, \psi_{k+1}(x_1, ..., x_k), ..., \psi_n(x_1, ..., x_k)) \\ &= (x_1, ..., x_k, 0, ..., 0), \end{aligned}$$

and the proof is finished.

If $d\Phi_q(M_q) = N_{\Phi(q)}$, Φ is called a *submersion at* q; it is called a submersion if it is a submersion at each point.

Corollary 15.6. *With the notation of Theorem 15.5 suppose $d\Phi_q$ has rank k for all $q \in M$. Let $r \in \Phi(M)$. Then the closed subset $\Phi^{-1}(r) \subset M$ with the topology induced by that of M has a unique differentiable structure with which it is a submanifold of M of dimension $m - k$.*

For this let $S = \Phi^{-1}(r)$, $p \in S$, and consider the coordinate systems ξ and η above. Then the mapping $\sigma : q \to (x_{k+1}(q), ..., x_m(q))$ is an open chart on a neighborhood of p in S. If p' is another point in S and ξ', σ' the associated charts, then $\sigma' \circ \sigma^{-1}$ is differentiable, being a restriction of $\xi' \circ \xi^{-1}$ to an open set in $\boldsymbol{R}^{m-k}$. Hence S is a manifold; and since the identity map $I : S \to M$ has the expression

$$\xi \circ I \circ \sigma^{-1} : (x_{k+1}, ..., x_m) \to (0, ..., 0, x_{k+1}, ..., x_m),$$

S is a submanifold of M. The uniqueness is immediate from Lemma 14.1.

EXERCISES AND FURTHER RESULTS

A. Manifolds

1. Let M be a paracompact manifold, A and B disjoint closed subsets of M. Then there exists a function $f \in C^\infty(M)$ such that $f \equiv 1$ on A, $f \equiv 0$ on B.

2. Let M be a connected manifold and p, q two points in M. Then there exists a diffeomorphism Φ of M onto itself such that $\Phi(p) = q$.

3. Let M be a Hausdorff space and let δ and δ' be two differentiable structures on M. Let $\mathfrak{F}$ and $\mathfrak{F}'$ denote the corresponding sets of C^∞ functions. Then $\delta = \delta'$ if and only if $\mathfrak{F} = \mathfrak{F}'$.

Deduce that the real line $\boldsymbol{R}$ with its ordinary topology has infinitely many different differentiable structures.

4. Let Φ be a differentiable mapping of a manifold M onto a manifold N. A vector field X on M is called *projectable* (Koszul [1]) if there exists a vector field Y on N such that $d\Phi \cdot X = Y$.

(i) Show that X is projectable if and only if $X\mathfrak{F}_0 \subset \mathfrak{F}_0$ where $\mathfrak{F}_0 = \{f \circ \Phi : f \in C^\infty(N)\}$.

(ii) A necessary condition for X to be projectable is that

$$d\Phi_p(X_p) = d\Phi_q(X_q) \tag{1}$$

whenever $\Phi(p) = \Phi(q)$. If, in addition, $d\Phi_p(M_p) = N_{\Phi(p)}$ for each $p \in M$, this condition is also sufficient.

(iii) Let $M = \boldsymbol{R}$ with the usual differentiable structure and let N be the topological space $\boldsymbol{R}$ with the differentiable structure obtained by requiring the homeomorphism $\psi : x \to x^{1/3}$ of M onto N to be a diffeomorphism. In this case the identity mapping $\Phi : x \to x$ is a differentiable mapping of M onto N. The vector field $X = \partial/\partial x$ on M is not projectable although (1) is satisfied.

5. Deduce from §3.1 that diffeomorphic manifolds have the same dimension.

6. Using Exercise A.1 prove the following variation of Theorem 1.3. Let M be a paracompact manifold and $\{U_\alpha\}_{\alpha\in A}$ a locally finite covering of M. Then there exists a system $\{\varphi_\alpha\}_{\alpha\in A}$ of differentiable functions on M such that: (i) Each φ_α has support contained in U_α; (ii) $\varphi_\alpha \geqslant 0$, $\sum_{\alpha\in A} \varphi_\alpha = 1$.

7. Let M be a manifold, $p \in M$, and X a vector field on M such that $X_p \neq 0$. Then there exists a local chart $\{x_1, ..., x_m\}$ on a neighborhood U of p such that $X = \partial/\partial x_1$ on U. Deduce that the differential equation $Xu = f$ $(f \in C^\infty(M))$ has a solution u in a neighborhood of p.

8. Let M be a manifold and X, Y two vector fields both $\neq 0$ at a point $o \in M$. For p close to o and s, $t \in \boldsymbol{R}$ sufficiently small let $\varphi_s(p)$ and $\psi_t(p)$ denote the integral curves through p of X and Y, respectively. Let

$$\gamma(t) = \psi_{-\sqrt{t}}(\varphi_{-\sqrt{t}}(\psi_{\sqrt{t}}(\varphi_{\sqrt{t}}(o)))).$$

Prove that

$$[X, Y]_o = \lim_{t\to 0} \dot{\gamma}(t)$$

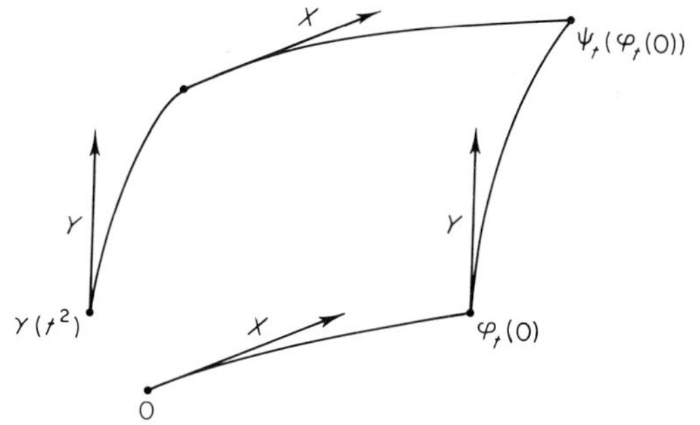

(Hint: The curves $t \to \varphi_t(\varphi_s(p))$ and $t \to \varphi_{t+s}(p)$ must coincide; deduce $(X^n f)(p) = [d^n/dt^n f(\varphi_t \cdot p)]_{t=0}$).

B. The Lie Derivative and the Interior Product

1. Let M be a manifold, X a vector field on M. The Lie derivative $\theta(X) : Y \to [X, Y]$ which maps $\mathfrak{D}^1(M)$ into itself can be extended uniquely to a mapping of $\mathfrak{D}(M)$ into itself such that:

(i) $\theta(X)f = Xf$ for $f \in C^\infty(M)$.

(ii) $\theta(X)$ is a derivation of $\mathfrak{D}(M)$ preserving type of tensors.

(iii) $\theta(X)$ commutes with contractions.

2. Let Φ be a diffeomorphism of a manifold M onto itself. Then Φ induces a unique type-preserving automorphism $T \to \Phi \cdot T$ of the tensor algebra $\mathfrak{D}(M)$ such that:

(i) The automorphism commutes with contractions.
(ii) $\Phi \cdot X = X^{\Phi}$, $(X \in \mathfrak{D}^1(M))$, $\Phi \cdot f = f^{\Phi}$, $(f \in C^\infty(M))$.

Prove that $\Phi \cdot \omega = (\Phi^{-1})^* \, \omega$ for $\omega \in \mathfrak{D}_*(M)$.

3. Let g_t be a one-parameter Lie transformation group of M and denote by X the vector field on M induced by g_t (Chapter II, §3). Then

$$\theta(X)T = \lim_{t\to 0} \frac{1}{t}(T - g_t \cdot T)$$

for each tensor field T on M ($g_t \cdot T$ is defined in Exercise 2).

4. The Lie derivative $\theta(X)$ on a manifold M has the following properties:

(i) $\theta([X, Y]) = \theta(X)\,\theta(Y) - \theta(Y)\,\theta(X)$, $\quad X, Y \in \mathfrak{D}^1(M)$.

(ii) $\theta(X)$ commutes with the alternation $A : \mathfrak{D}_*(M) \to \mathfrak{A}(M)$ and therefore induces a derivation of the Grassmann algebra of M.

(iii) $\theta(X)\, d = d\theta(X)$, that is, $\theta(X)$ commutes with exterior differentiation.

5. For $X \in \mathfrak{D}^1(M)$ there is a unique linear mapping $i(X) : \mathfrak{A}(M) \to \mathfrak{A}(M)$, the *interior product*, satisfying:

(i) $i(X)f = 0$ for $f \in C^\infty(M)$.
(ii) $i(X)\omega = \omega(X)$ for $\omega \in \mathfrak{A}_1(M)$.
(iii) $i(X) : \mathfrak{A}_r(M) \to \mathfrak{A}_{r-1}(M)$ and

$$i(X)(\omega_1 \wedge \omega_2) = i(X)(\omega_1) \wedge \omega_2 + (-1)^r \omega_1 \wedge i(X)(\omega_2)$$

if $\omega_1 \in \mathfrak{A}_r(M)$, $\omega_2 \in \mathfrak{A}(M)$.

6. (cf. H. Cartan [1]). Prove that if $X, Y \in \mathfrak{D}^1(M)$, $\omega_1, \ldots, \omega_r \in \mathfrak{A}_1(M)$,

(i) $i(X)^2 = 0$.

(ii) $$i(X)(\omega_1 \wedge \ldots \wedge \omega_r) = \sum_{1\leqslant k\leqslant r} (-1)^{k+1}\, \omega_k(X)\, \omega_1 \wedge \ldots \wedge \hat{\omega}_k \wedge \ldots \wedge \omega_r; \qquad \omega_i \in \mathfrak{A}_1(M).$$

(iii) $i([X, Y]) = \theta(X)\, i(Y) - i(Y)\, \theta(X)$.
(iv) $\theta(X) = i(X)\, d + d\, i(X)$.

C. Affine Connections

1. Let M be a connected manifold with a countable basis. Using partition of unity show that M has a Riemannian structure. On the other hand, a Riemannian manifold has a countable basis (Prop. 9.6).

2. Let ∇ be the affine connection on $\boldsymbol{R}^n$ determined by $\nabla_X(Y) = 0$ for $X = \partial/\partial x_i$, $Y = \partial/\partial x_j$, $1 \leqslant i, j \leqslant n$. Find the corresponding affine transformations.

3. Let M be a manifold with an affine connection ∇ satisfying $R = 0$, $T = 0$. Deduce from §8 that for each $p \in M$, Exp_p induces an affine transformation of a normal neighborhood of 0 in M_p onto a normal neighborhood of p in M.

4. Let M be a manifold with a torsion-free affine connection ∇. Suppose $X_1, \ldots, X_m$ is a basis for the vector fields on an open subset U of M. Let the forms $\omega^1, \ldots, \omega^m$ on U be determined by $\omega^i(X_j) = \delta^i{}_j$. Prove the formula

$$d\theta = \sum_{i=1}^{m} \omega^i \wedge \nabla_{X_i}(\theta)$$

for each differential form θ on U.

5. In Prop. 10.7 it was proved that a complete noncompact Riemannian manifold M always contains a ray. Does M always contain a straight line, that is, a geodesic $\gamma(t)$ $(-\infty < t < \infty)$ which realizes the shortest distance between any two of its points?

6. Let M and N be analytic, complete, simply connected Riemannian manifolds. Suppose that an open subset of M is isometric to an open subset of N. Using results from §11 show that M and N are isometric (Myers-Rinow).

D. Submanifolds

1. Let M and N be differentiable manifolds and Φ a differentiable mapping of M into N. Consider the mapping $\varphi : m \to (m, \Phi(m))$ $(m \in M)$ and the graph

$$G_\Phi = \{(m, \Phi(m)) : m \in M\}$$

of Φ with the topology induced by the product space $M \times N$. Then φ is a homeomorphism of M onto G_Φ and if the differentiable structure of M is transferred to G_Φ by φ, the graph G_Φ becomes a closed submanifold of $M \times N$.

2. Let N be a manifold and M a topological space, $M \subset N$ (as sets).

Show that there exists at most one differentiable structure on the topological space M such that M is a submanifold of N.

3. Using the figure 8 as a subset of $\boldsymbol{R}^2$ show that

(i) A closed connected submanifold of a connected manifold does not necessarily carry the relative topology.

(ii) A subset M of a connected manifold N may have two different topologies and differentiable structures such that in both cases M is a submanifold of N.

4. Let M be a submanifold of a manifold N and suppose $M = N$ (as sets). Assuming M to have a countable basis for the open sets, prove that $M = N$ (as manifolds). (Use Prop. 3.2 and Lemma 3.1, Chapter II.)

5. Let N be a manifold with a countable basis and M a closed submanifold of N. Then each $g \in C^\infty(M)$ can be extended to a $G \in C^\infty(N)$. (Proceed as in the proof of Lemma 5.1.)

6. Let M be a Riemannian manifold and S a connected, complete submanifold of M. Show that S is totally geodesic if and only if M-parallel translation of tangent vectors to S along curves in S always coincides with the S-parallel translation (see (2), Chapter V, §6).

E. Curvature

1. Let M be a Riemannian manifold of dimension 2, p a point in M, $r(q)$ the distance $d(p, q)$. Show that the curvature K of M at p satisfies

$$K = -3 \lim_{r\to 0} \Delta (\log r)$$

where Δ is the Laplace-Beltrami operator on M

$$\Delta : f \to \frac{1}{\sqrt{\bar{g}}} \sum_k \partial_k \Big(\sum_i g^{ik} \sqrt{\bar{g}} \partial_i f \Big), \qquad f \in C^\infty(M),$$

where

$$\partial_k = \frac{\partial}{\partial x_k}, \qquad g_{ij} = g(\partial_i, \partial_j), \qquad \sum_j g_{ij} g^{jk} = \delta_{ik}, \qquad \bar{g} = |\det(g_{ij})|.$$

2. Let M be a manifold with an affine connection ∇. Let $p \in M$ and $Z \in M_p$. Fix linearly independent vectors X_p, $Y_p \in M_p$. Let $\{x_1, ..., x_n\}$ be a coordinate system near p such that

$$x_i(p) = 0, \quad 1 \leqslant i \leqslant n, \qquad \Big(\frac{\partial}{\partial x_1}\Big)_p = X_p, \quad \Big(\frac{\partial}{\partial x_2}\Big)_p = Y_p.$$

Let γ_ϵ denote the "counter clockwise" boundary of the "square" $0 \leqslant x_1 \leqslant \epsilon$, $0 \leqslant x_2 \leqslant \epsilon$, $x_i = 0$ $(i > 2)$ in M and let $Z_\epsilon \in M_p$ denote the vector obtained by parallel translating Z around γ_ϵ. Prove that the curvature tensor R satisfies

$$R_p(X_p, Y_p)Z = \lim_{\epsilon \to 0} \frac{1}{\epsilon^2}(Z - Z_\epsilon).$$

Thus R measures the dependence of parallelism on the path and is closely related to the holonomy group described in the introduction to Chapter IV.

F. Surfaces

1. Let S be a surface in $\boldsymbol{R}^3$, X and Y two vector fields on S. Let $s \in S$, $X_s \neq 0$ and $t \to \gamma(t)$ a curve on S through s such that $\dot{\gamma}(t) = X_{\gamma(t)}$, $\gamma(0) = s$. Viewing $Y_{\gamma(t)}$ as a vector in $\boldsymbol{R}^3$ and letting $\pi_s : \boldsymbol{R}^3 \to S_s$ denote the orthogonal projection put

$$\nabla'_X(Y)_s = \pi_s\left(\lim_{t \to 0} \frac{1}{t}(Y_{\gamma(t)} - Y_s)\right).$$

Prove that this defines an affine connection on S.

2. (Minding) Let S be an orientable surface in $\boldsymbol{R}^3$, oriented by means of a continuous family of unit normal vectors $\xi_s (s \in S)$. Let $t \to \gamma_S(t)$ be a curve in S, t being the arc-parameter. The triple vector product $(\xi \times \dot{\gamma}_S \cdot \ddot{\gamma}_S)(s)$ is called the *geodesic curvature* of γ_S at s. Show that the geodesic curvature can be expressed in terms of $\dot{\gamma}_S$, $\ddot{\gamma}_S$, the Riemannian structure of S, and its derivatives with respect to local coordinates. Deduce that the geodesic curvature is invariant under orientation-preserving isometries.

3*. (Levi-Civita) Suppose a surface S in $\boldsymbol{R}^3$ rolls without slipping on a plane π. Let the point of contact run through a curve γ_S on S and a curve γ_π on π. Using the result of Exercise F.2 show that the Euclidean parallelism along γ_π corresponds to the parallelism along γ_S in the sense of the Riemannian connection on S.

G. The Hyperbolic Plane

1. Let D be the open disk $|z| < 1$ in $\boldsymbol{R}^2$ with the usual differentiable structure but given the Riemannian structure

$$g(u, v) = \frac{(u, v)}{(1 - |z|^2)^2} \qquad (u, v \in D_z)$$

$(\ ,\ )$ denoting the usual inner product on $\boldsymbol{R}^2$.

(i) Show that the angle between u and v in the Riemannian structure g coincides with the Euclidean angle.

(ii) Show that the Riemannian structure can be written

$$g = \frac{dx^2 + dy^2}{(1 - x^2 - y^2)^2} \qquad (z = x + iy).$$

(iii) Show that the arc length L satisfies

$$L(\gamma_0) \leqslant L(\gamma)$$

if γ is any curve joining the origin 0 and x $(0 < x < 1)$ and $\gamma_0(t) = tx$ $(0 \leqslant t \leqslant 1)$.

(iv) Show that the transformation

$$\varphi : z \to \frac{az + b}{\bar{b}z + \bar{a}} \qquad (|\,a\,|^2 - |\,b\,|^2 = 1)$$

is an isometry of D.

(v) Deduce from (iii) and (iv) that the geodesics in D are the circular arcs perpendicular to the boundary $|\,z\,| = 1$.

(vi) Prove from (iii) that

$$d(0, z) = \frac{1}{2} \log \frac{1 + |\,z\,|}{1 - |\,z\,|} \qquad (z \in D)$$

and using (iv) that

$$d(z_1, z_2) = \frac{1}{2} \log \left(\frac{z_1 - b_2}{z_1 - b_1} : \frac{z_2 - b_2}{z_2 - b_1} \right) \qquad (z_1, z_2 \in D)$$

with b_1 and b_2 as in the figure.

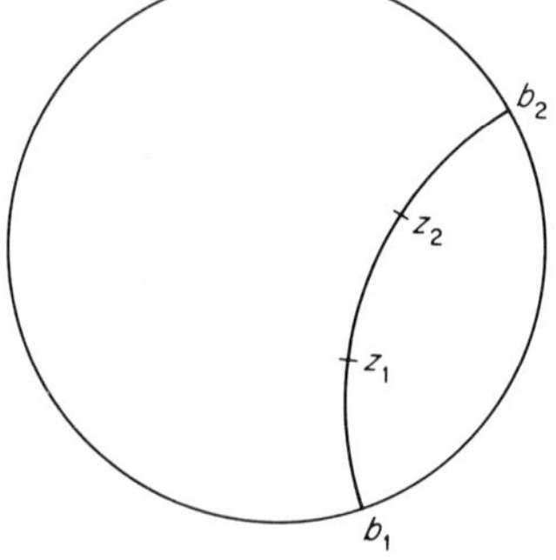

(vii) Show that the maps φ in (iv) together with the complex conjugation $z \to \bar{z}$ generate the group of all isometries of D.

(viii) Show that in geodesic polar coordinates (cf. Exercise E.1)

$$g = dr^2 + \tfrac{1}{4}\sinh^2(2r)\,d\theta^2.$$

Deduce from this and Lemma 12.1 that D has constant curvature $K = -4$.

(ix) The Cayley transform

$$c : z \to w = -i\,\frac{z+i}{z-i}$$

is an isometry of D onto the upper half plane $v > 0$ with the metric

$$h = \frac{du^2 + dv^2}{4v^2}, \qquad w = u + iv.$$

NOTES

§1-§3. The treatment of differentiable manifolds M given here is similar to Chevalley's development of analytic manifolds in [2]. In particular, a tangent vector (and hence a vector field and forms) is defined in terms of the function algebra $C^\infty(M)$ rather than as an equivalence class of curves. Since we have not given any motivation of Cartan's exterior derivative d, it is of interest to recall Palais' characterization [4] of d as the only linear operator (up to a constant factor) from p-forms to $(p+1)$-forms which commutes with mappings in the sense of (6), §3. Partition of unity (Theorem 1.3) which is a standard tool in reducing global problems to local ones, seems to be an outgrowth of work of S. Bochner, J. Dieudonné and W. Hurewicz.

§4-§5. Levi-Civita's concept of parallelism with respect to a curve on a surface is explained in Exercises F.1-3; this concept lies behind the definition of an affine connection. There are several ways of formulating this definition. Classically, an affine connection is defined in terms of the Christoffel symbols Γ^k_{ij} as indicated in §4. Cartan's "method of moving frames" uses the "connection forms" ω^i_j and the structural equations (1), (2), §8 instead. Koszul replaced the conditions for Γ^k_{ij} by axioms ∇_1 and ∇_2 in §4 (cf. Nomizu [2]) and Ehresmann [2] defined a connection on M in terms of the frame bundle over M. For the equivalence of these two last definitions see Nomizu [4], Chapter III, §4. Useful as the frame bundle definition is in global differential geometry, we have nevertheless preferred the vector field definition because the spaces with which we are mainly concerned are coset spaces G/K and for these the natural bundle to consider is the group G itself; the frame bundle would only be extra baggage.

§6. Theorem 6.2 is due to Whitehead [2]. The proof in the text is a slight simplification of Whitehead's proof using stronger differentiability hypotheses. Theorem 6.5 was proved by the author [4].

§9-§10. The treatment of local Riemannian geometry given here is partly based on É. Cartan [22], Chapter X. The equivalence of the completeness conditions (Theorem 10.3) for a two-dimensional Riemannian manifold is due to Hopf-Rinow [1]. A simplified proof was given by de Rham [1]. See also Myers [1] and Whitehead [3].

§11. Theorem 11.1 is due to Myers-Steenrod [1]. Their proof was simplified by Palais [1]. The remaining results in §11 which deal with continuations of local isometries and are useful in Chapter IV are based on de Rham's paper [1]. See also Rinow [1].

§12. Going back to Riemann's original lecture [1], the formula of Theorem 12.2 is the customary definition of sectional curvature. We have used instead an intrinsic definition in terms of areas. Lemma 12.1 and the ensuing proof of Theorem 12.2 are from Helgason [4].

§13-§14. The treatment of Riemannian manifolds of negative curvature is based on É. Cartan's book [22], Note III. In the two-dimensional case, Theorem 13.3 is due to Hadamard [1]. The concept of a totally geodesic submanifold is due to Hadamard [2]. Theorem 14.5 is proved in Cartan [22], p. 115. Cartan also proves, [22], p. 232, that if every submanifold of dimension $s \geqslant 2$ which is geodesic at a point is also totally geodesic then the manifold has constant sectional curvature. Theorem 14.6 can be regarded as a generalization of a decomposition theorem due to Mostow for a semisimple Lie group (Theorem 1.4, Chapter VI).

§15. The notion of paracompactness and Propositions 15.1-15.3 are from Dieudonné [1]. Theorem 15.4 is from Alexandroff [1]; cf. also Pfluger [1]. In the proof of the rank theorem (Theorem 15.5) we have used Spivak [1].

CHAPTER II

LIE GROUPS AND LIE ALGEBRAS

A Lie group is, roughly speaking, an analytic manifold with a group structure such that the group operations are analytic. Lie groups arise in a natural way as transformation groups of geometric objects. For example, the group of all affine transformations of a connected manifold with an affine connection and the group of all isometries of a pseudo-Riemannian manifold are known to be Lie groups in the compact open topology. However, the group of all diffeomorphisms of a manifold is too big to form a Lie group in any reasonable topology.

The tangent space $\mathfrak{g}$ at the identity element of a Lie group G has a rule of composition $(X, Y) \rightarrow [X, Y]$ derived from the bracket operation on the left invariant vector fields on G. The vector space $\mathfrak{g}$ with this rule of composition is called the Lie algebra of G. The structures of $\mathfrak{g}$ and G are related by the exponential mapping exp: $\mathfrak{g} \rightarrow$ G which sends straight lines through the origin in $\mathfrak{g}$ onto one-paramater subgroups of G. Several properties of this mapping are developed already in §1 because they can be derived as special cases of properties of the Exponential mapping for a suitable affine connection on G. Although the structure of $\mathfrak{g}$ is determined by an arbitrary neighborhood of the identity element of G, the exponential mapping sets up a far-reaching relationship between $\mathfrak{g}$ and the group G in the large. We shall for example see in Chapter VII that the center of a compact simply connected Lie group G is explicitly determined by the Lie algebra $\mathfrak{g}$. In §2 the correspondence (induced by exp) between subalgebras and subgroups is developed. This correspondence is of basic importance in the theory in spite of its weakness that the subalgebra does not in general decide whether the corresponding subgroup will be closed or not, an important distinction when coset spaces are considered.

In §4 we investigate the relationship between homogeneous spaces and coset spaces. It is shown that if a manifold M has a separable transitive Lie transformation group G acting on it, then M can be identified with a coset space G/H (H closed) and therefore falls inside the realm of Lie group theory. Thus, one can, for example, conclude that if H is compact, then M has a G-invariant Riemannian structure.

Let G be a connected Lie group with Lie algebra $\mathfrak{g}$. If $\sigma \in G$, the inner automorphism $g \rightarrow \sigma g \sigma^{-1}$ induces an automorphism Ad (σ) of $\mathfrak{g}$ and the mapping $\sigma \rightarrow$ Ad (σ) is an analytic homomorphism of G onto an analytic subgroup Ad (G) of $\boldsymbol{GL}(\mathfrak{g})$, the adjoint group. The group Ad (G) can be defined by $\mathfrak{g}$ alone and since its Lie algebra is isomorphic to $\mathfrak{g}/\mathfrak{z}$ ($\mathfrak{z}$ = center of $\mathfrak{g}$), one can, for example, conclude that a semisimple Lie algebra over $\boldsymbol{R}$ is isomorphic to the Lie algebra of a Lie group. This fact holds for arbitrary Lie algebras over $\boldsymbol{R}$ but will not be needed in this book in that generality.

Section 6 deals with some preliminary results about semisimple Lie groups. The main result is Weyl's theorem stating that the universal covering group of a compact semisimple Lie group is compact. In §7 we discuss invariant forms on G and their determination from the structure of $\mathfrak{g}$.

§1. The Exponential Mapping

1. The Lie Algebra of a Lie Group

Definition. A *Lie group* is a group G which is also an analytic manifold such that the mapping $(\sigma, \tau) \rightarrow \sigma\tau^{-1}$ of the product manifold $G \times G$ into G is analytic.

Examples. 1. Let G be the group of all isometries of the Euclidean plane $\boldsymbol{R}^2$ which preserve the orientation. If $\sigma \in G$, let $(x(\sigma), y(\sigma))$ denote the coordinates of the point $\sigma \cdot 0$ ($0 =$ origin of $\boldsymbol{R}^2$) and let $\theta(\sigma)$ denote the angle between the x-axis l and the image of l under σ. Then the mapping $\varphi : \sigma \rightarrow (x(\sigma), y(\sigma), \theta(\sigma))$ maps G in a one-to-one fashion onto the product manifold $\boldsymbol{R}^2 \times \boldsymbol{S}^1$ ($\boldsymbol{S}^1 = \boldsymbol{R} \bmod 2\pi$). We can turn G into an analytic manifold by requiring φ to be an analytic diffeomorphism. An elementary computation shows that for $\sigma, \tau \in G$

$$x(\sigma\tau^{-1}) = x(\sigma) - x(\tau) \cos(\theta(\sigma) - \theta(\tau)) + y(\tau) \sin(\theta(\sigma) - \theta(\tau));$$
$$y(\sigma\tau^{-1}) = y(\sigma) - x(\tau) \sin(\theta(\sigma) - \theta(\tau)) - y(\tau) \cos(\theta(\sigma) - \theta(\tau));$$
$$\theta(\sigma\tau^{-1}) = \theta(\sigma) - \theta(\tau) \pmod{2\pi}.$$

Since the functions sin and cos are analytic, it follows that G is a Lie group.

2. Let $\tilde{G}$ be the group of all isometries of $\boldsymbol{R}^2$. If s is the symmetry of $\boldsymbol{R}^2$ with respect to a line, then $\tilde{G} = G \cup sG$ (disjoint union). We can turn sG into an analytic manifold by requiring the mapping $\sigma \rightarrow s\sigma$ ($\sigma \in G$) to be an analytic diffeomorphism of G onto sG. This makes $\tilde{G}$ a Lie group.

On the other hand, if G_1 and G_2 are two components of a Lie group G and $x_1 \in G_1$, $x_2 \in G_2$, then the mapping $g \rightarrow x_2x_1^{-1}g$ is an analytic diffeomorphism of G_1 onto G_2.

Remark. A Lie group is always paracompact. In fact, let G be a topological group which is *locally Euclidean*, that is, has a neighborhood of the identity e, homeomorphic to a Euclidean space. Let G_0 denote the *identity component* of G (that is, the component of G containing e). Then G_0 is a connected topological group and as such it is generated by any neighborhood of the identity. It follows that G_0 has a countable base, in particular G_0 is paracompact. The same statement follows for G by the definition of paracompactness.

Let G be a connected topological group. A *covering group* of G is a pair $(\tilde{G}, \pi)$ where $\tilde{G}$ is a topological group and π is a homomorphism of $\tilde{G}$ into G such that $(\tilde{G}, \pi)$ is a covering space of G. In the case when G is a Lie group, then $\tilde{G}$ has clearly an analytic structure such that $\tilde{G}$ is a Lie group, π analytic and $(\tilde{G}, \pi)$ a covering manifold of G.

Definition. A homomorphism of a Lie group into another which is also an analytic mapping is called an *analytic homomorphism*. An isomorphism of one Lie group onto another which is also an analytic diffeomorphism is called an *analytic isomorphism*.

Let G be a Lie group. If $\rho \in G$, the left translation $L_\rho : g \rightarrow \rho g$ of G onto itself is an analytic diffeomorphism. A vector field Z on G is called left invariant if $dL_\rho Z = Z$ for all $\rho \in G$. Given a tangent vector $X \in G_e$ there exists exactly one left invariant vector field $\tilde{X}$ on G such that $\tilde{X}_e = X$ and this $\tilde{X}$ is analytic. In fact, $\tilde{X}$ can be defined by

$$[\tilde{X}f](\rho) = X f^{L_{\rho^{-1}}} = \left\{\frac{d}{dt} f(\rho\gamma(t))\right\}_{t=0}$$

if $f \in C^\infty(G)$, $\rho \in G$, and $\gamma(t)$ is any curve in G with tangent vector X for $t = 0$. If $X, Y \in G_e$, then the vector field $[\tilde{X}, \tilde{Y}]$ is left invariant due to Prop. 3.3, Chapter I. The tangent vector $[\tilde{X}, \tilde{Y}]_e$ is denoted by $[X, Y]$. The vector space G_e, with the rule of composition $(X, Y) \rightarrow [X, Y]$ we denote by $\mathfrak{g}$ (or $\mathfrak{L}(G)$) and call the *Lie algebra of G*.

More generally, let $\mathfrak{a}$ be a vector space over a field K (of characteristic 0). The set $\mathfrak{a}$ is called a *Lie algebra over K* if there is given a rule of composition $(X, Y) \rightarrow [X, Y]$ in $\mathfrak{a}$ which is bilinear and satisfies (a) $[X, X] = 0$ for all $X \in \mathfrak{a}$; (b) $[X, [Y, Z]] + [Y, [Z, X]] + [Z, [X, Y]] = 0$ for all $X, Y, Z \in \mathfrak{a}$. The identity (b) is called the *Jacobi identity*.

The Lie algebra of G above is clearly a Lie algebra over $\boldsymbol{R}$.

If $\mathfrak{a}$ is a Lie algebra over K and $X \in \mathfrak{a}$, the linear transformation $Y \rightarrow [X, Y]$ of $\mathfrak{a}$ is denoted by $\operatorname{ad} X$ (or $\operatorname{ad}_{\mathfrak{a}} X$ when a confusion would otherwise be possible). Let $\mathfrak{b}$ and $\mathfrak{c}$ be two vector subspaces of $\mathfrak{a}$. Then $[\mathfrak{b}, \mathfrak{c}]$ denotes the vector subspace of $\mathfrak{a}$ generated by the set of elements $[X, Y]$ where $X \in \mathfrak{b}$, $Y \in \mathfrak{c}$. A vector subspace $\mathfrak{b}$ of $\mathfrak{a}$ is called a *subalgebra* of $\mathfrak{a}$ if $[\mathfrak{b}, \mathfrak{b}] \subset \mathfrak{b}$ and an *ideal* in $\mathfrak{a}$ if $[\mathfrak{b}, \mathfrak{a}] \subset \mathfrak{b}$. If $\mathfrak{b}$ is an ideal in $\mathfrak{a}$ then the factor space $\mathfrak{a}/\mathfrak{b}$ is a Lie algebra with the bracket operation inherited from $\mathfrak{a}$. Let $\mathfrak{a}$ and $\mathfrak{b}$ be two Lie algebras over the same field K and σ a linear mapping of $\mathfrak{a}$ into $\mathfrak{b}$. The mapping σ is called a *homomorphism* if $\sigma([X, Y]) = [\sigma X, \sigma Y]$ for all $X, Y \in \mathfrak{a}$. If σ is a homomorphism then $\sigma(\mathfrak{a})$ is a subalgebra of $\mathfrak{b}$ and the kernel $\sigma^{-1}\{0\}$ is an ideal in $\mathfrak{a}$. If $\sigma^{-1}\{0\} = \{0\}$, then σ is called an *isomorphism* of $\mathfrak{a}$ into $\mathfrak{b}$. An isomorphism of a Lie algebra onto itself is called an *automorphism*. If $\mathfrak{a}$ is a Lie algebra and $\mathfrak{b}$, $\mathfrak{c}$ subsets of $\mathfrak{a}$, the *centralizer* of $\mathfrak{b}$ in $\mathfrak{c}$ is $\{X \in \mathfrak{c} : [X, \mathfrak{b}] = 0\}$. If $\mathfrak{b} \subset \mathfrak{a}$ is a subalgebra, its *normalizer* in $\mathfrak{a}$ is $\mathfrak{n} = \{X \in \mathfrak{a} : [X, \mathfrak{b}] \subset \mathfrak{b}\}$; $\mathfrak{b}$ is an ideal in $\mathfrak{n}$.

Let V be a vector space over a field K and let $\mathfrak{gl}(V)$ denote the vector space of all endomorphisms of V with the bracket operation $[A, B] = AB - BA$. Then $\mathfrak{gl}(V)$ is a Lie algebra over K. Let $\mathfrak{a}$ be a Lie algebra over K. A homomorphism of $\mathfrak{a}$ into $\mathfrak{gl}(V)$ is called a *representation* of $\mathfrak{a}$ on V. In particular, since $\text{ad}([X, Y]) = \text{ad}\, X\, \text{ad}\, Y - \text{ad}\, Y\, \text{ad}\, X$, the linear mapping $X \to \text{ad}\, X$ $(X \in \mathfrak{a})$ is a representation of $\mathfrak{a}$ on $\mathfrak{a}$. It is called the *adjoint representation* of $\mathfrak{a}$ and is denoted ad (or $\text{ad}_{\mathfrak{a}}$ when a confusion would otherwise be possible). The kernel of $\text{ad}_{\mathfrak{a}}$ is called the *center* of $\mathfrak{a}$. If the center of $\mathfrak{a}$ equals $\mathfrak{a}$, $\mathfrak{a}$ is said to be *abelian*. Thus $\mathfrak{a}$ is abelian if and only if $[\mathfrak{a}, \mathfrak{a}] = \{0\}$.

Let $\mathfrak{a}$ and $\mathfrak{b}$ be two Lie algebras over the same field K. The vector space $\mathfrak{a} \times \mathfrak{b}$ becomes a Lie algebra over K if we define

$$[(X, Y), (X', Y')] = ([X, X'], [Y, Y']).$$

This Lie algebra is called the *Lie algebra product* of $\mathfrak{a}$ and $\mathfrak{b}$. The sets $\{(X, 0) : X \in \mathfrak{a}\}$, $\{(0, Y) : Y \in \mathfrak{b}\}$ are ideals in $\mathfrak{a} \times \mathfrak{b}$ and $\mathfrak{a} \times \mathfrak{b}$ is the direct sum of these ideals.

In the following a Lie algebra shall always mean a finite-dimensional Lie algebra unless the contrary is stated.

2. The Universal Enveloping Algebra

Let $\mathfrak{a}$ be a Lie algebra over a field K. The rule of composition $(X, Y) \to [X, Y]$ is rarely associative; we shall now assign to $\mathfrak{a}$ an associative algebra with unit, the *universal enveloping algebra* of $\mathfrak{a}$, denoted $U(\mathfrak{a})$. This algebra is defined as the factor algebra $T(\mathfrak{a})/J$ where $T(\mathfrak{a})$ is the tensor algebra over $\mathfrak{a}$ (considered as a vector space) and J is the two-sided ideal in $T(\mathfrak{a})$ generated by the set of all elements of the form $X \otimes Y - Y \otimes X - [X, Y]$ where $X, Y \in \mathfrak{a}$. If $X \in \mathfrak{a}$, let X^* denote the image of X under the canonical mapping π of $T(\mathfrak{a})$ onto $U(\mathfrak{a})$. The identity element in $U(\mathfrak{a})$ will be denoted by 1. Then $1 \neq 0$ if $\mathfrak{a} \neq \{0\}$. Proposition 1.9 (b) motivates the consideration of $U(\mathfrak{a})$.

Proposition 1.1. *Let V be a vector space over K. There is a natural one-to-one correspondence between the set of all representations of $\mathfrak{a}$ on V and the set of all representations of $U(\mathfrak{a})$ on V. If ρ is a representation of $\mathfrak{a}$ on V and ρ^* is the corresponding representation of $U(\mathfrak{a})$ on V, then*

$$\rho(X) = \rho^*(X^*) \qquad \textit{for } X \in \mathfrak{a}. \tag{1}$$

Proof. Let ρ be a representation of $\mathfrak{a}$ on V. Then there exists a unique representation $\tilde{\rho}$ of $T(\mathfrak{a})$ on V satisfying $\tilde{\rho}(X) = \rho(X)$ for all $X \in \mathfrak{a}$. The mapping $\tilde{\rho}$ vanishes on the ideal J because

$$\tilde{\rho}(X \otimes Y - Y \otimes X - [X, Y]) = \rho(X)\,\rho(Y) - \rho(Y)\,\rho(X) - \rho([X, Y]) = 0.$$

Thus we can define a representation ρ^* of $U(\mathfrak{a})$ on V by the condition $\rho^* \circ \pi = \tilde{\rho}$. Then (1) is satisfied and determines ρ^* uniquely. On the other hand, suppose σ is a representation of $U(\mathfrak{a})$ on V. If $X \in \mathfrak{a}$ we put $\rho(X) = \sigma(X^*)$. Then the mapping $X \to \rho(X)$ is linear and in fact a representation of $\mathfrak{a}$ on V, because

$$\begin{aligned}\rho([X, Y]) = \sigma([X, Y]^*) &= \sigma(\pi(X \otimes Y - Y \otimes X)) \\ &= \sigma(X^*Y^* - Y^*X^*) = \rho(X)\rho(Y) - \rho(Y)\rho(X)\end{aligned}$$

for $X, Y \in \mathfrak{a}$. This proves the proposition.

Let $X_1, \ldots, X_n$ be a basis of $\mathfrak{a}$ and put $X^*(t) = \sum_{i=1}^n t_i X_i^*$ $(t_i \in K)$. Let $M = (m_1, \ldots, m_n)$ be an ordered set of integers $m_i \geqslant 0$. We shall call M a *positive integral n-tuple*. We put $|M| = m_1 + \ldots + m_n$, $t^M = t_1^{m_1} \ldots t_n^{m_n}$. Considering $t_1, \ldots, t_n$ as indeterminates the various t^M are linearly independent over K and for $|M| > 0$ we can define $X^*(M) \in U(\mathfrak{a})$ as the coefficient of t^M in the expansion of $(|M|!)^{-1}(X^*(t))^{|M|}$. Put $X^*(M) = 1$ if $|M| = 0$.

Proposition 1.2. *The smallest vector subspace of $U(\mathfrak{a})$ containing all the elements $X^*(M)$ (where M is a positive integral n-tuple) is $U(\mathfrak{a})$ itself.*

Proof. It suffices to prove that each element $X_{i_1}^* X_{i_2}^* \ldots X_{i_p}^*$ $(1 \leqslant i_1, \ldots, i_p \leqslant n)$ can be expressed as a finite sum $\sum_{|M| \leqslant p} a_M X^*(M)$ where $a_M \in K$. Consider the element

$$u_p = \frac{1}{p!} \sum_{\sigma} X_{i_{\sigma(1)}}^* \ldots X_{i_{\sigma(p)}}^*,$$

where σ runs over all permutations of the set $\{1, 2, \ldots, p\}$. It is clear that $u_p = cX^*(M)$, where $c \in K$ and M is a suitable positive integral n-tuple. Using the relation $X_j^* X_k^* - X_k^* X_j^* = [X_j, X_k]^*$ we see that

$$X_{i_1}^* \ldots X_{i_p}^* - X_{i_{\sigma(1)}}^* \ldots X_{i_{\sigma(p)}}^*$$

is a linear combination (with coefficients in K) of elements of the form $X_{j_1}^* \ldots X_{j_{p-1}}^*$ $(1 \leqslant j_1 \ldots j_{p-1} \leqslant n)$ where each X_{j_q} $(1 \leqslant q \leqslant p-1)$ belongs to the subalgebra of $\mathfrak{a}$ generated by $X_{i_1}, \ldots, X_{i_p}$. The formula

$$X_{i_1}^* X_{i_2}^* \ldots X_{i_p}^* = \sum_{|M| \leqslant p} a_M X^*(M)$$

now follows by induction on p.

Corollary 1.3. *Let $\mathfrak{b}$ be a subalgebra of $\mathfrak{a}$. Suppose $\mathfrak{b}$ has dimension $n - r$ and let the basis $X_1, \ldots, X_n$ of $\mathfrak{a}$ be chosen in such a way that the*

$n - r$ last elements lie in $\mathfrak{b}$. Let $\mathfrak{B}$ denote the vector subspace of $U(\mathfrak{a})$ spanned by all elements $X^*(M)$ where M varies over all positive integral n-tuples of the form $(0, ..., 0, m_{r+1}, ..., m_n)$. Then $\mathfrak{B}$ is a subalgebra of $U(\mathfrak{a})$.

In fact, the proof above shows that the product $X^*_{i_1} ... X^*_{i_p}$ $(r < i_1, ..., i_p \leqslant n)$ can be written as a linear combination of elements $X^*(M)$ for which $m_1 = ... = m_r = 0$.

3. Left Invariant Affine Connections

Let G be a Lie group, and ∇ an affine connection on G; ∇ is said to be *left invariant* if each L_σ $(\sigma \in G)$ is an affine transformation of G. Let $X_1, ..., X_n$ be a basis of the Lie algebra $\mathfrak{g}$ of G and let $\tilde{X}_1, ..., \tilde{X}_n$ denote the corresponding left invariant vector fields on G. Then if ∇ is left invariant, the vector fields $\nabla_{\tilde{X}_i}(\tilde{X}_j)$ $(1 \leqslant i, j \leqslant n)$ are obviously left invariant. On the other hand we can define an affine connection ∇ on G by requiring the $\nabla_{\tilde{X}_i}(\tilde{X}_j)$ to be any left invariant vector fields. Let Z, Z' be arbitrary vector fields in $\mathfrak{D}^1$. Then $Z = \Sigma_i f_i \tilde{X}_i$, $Z' = \Sigma_j g_j \tilde{X}_j$ where $f_i, g_j \in C^\infty(G)$. Using the axioms ∇_1 and ∇_2 and Prop. 3.3 in Chapter I we find easily that $\nabla_{dL_\sigma Z}(dL_\sigma Z') = dL_\sigma \nabla_Z(Z')$ for each $\sigma \in G$ so ∇ is left invariant.

Proposition 1.4. *There is a one-to-one correspondence between the set of left invariant affine connections ∇ on G and the set of bilinear functions α on $\mathfrak{g} \times \mathfrak{g}$ with values in $\mathfrak{g}$ given by*

$$\alpha(X, Y) = (\nabla_{\tilde{X}}(\tilde{Y}))_e.$$

Let $X \in \mathfrak{g}$. The following statements are then equivalent:

(i) $\alpha(X, X) = 0$;

(ii) *The geodesic $t \to \gamma_X(t)$ is an analytic homomorphism of **R** into G.*

Proof. Given a bilinear mapping $\alpha : \mathfrak{g} \times \mathfrak{g} \to \mathfrak{g}$, we define the affine connection ∇ by the requirement

$$\nabla_{\tilde{X}_i}(\tilde{X}_j) = \alpha(X_i, X_j)^\sim \qquad (1 \leqslant i, j \leqslant n).$$

By the remark above, ∇ is left invariant, and the correspondence follows. Also, ∇ is analytic.

Next let $X \in \mathfrak{g}$ and let $\tilde{X}$ be the corresponding left invariant vector field on G. Locally there exist integral curves to the vector field $\tilde{X}$ (Chapter I, §7). In other words, there exists a number $\epsilon > 0$ and a curve segment $\Gamma : t \to \Gamma(t)$ $(0 \leqslant t \leqslant \epsilon)$ in G such that

$$\Gamma(0) = e, \qquad \dot{\Gamma}(s) = \tilde{X}_{\Gamma(s)} \tag{2}$$

for $0 \leqslant s \leqslant \epsilon$. Using induction we define $\Gamma(t)$ for all $t \geqslant 0$ by the requirement

$$\Gamma(t) = \Gamma(n\epsilon)\, \Gamma(t - n\epsilon), \qquad \text{if} \qquad n\epsilon \leqslant t \leqslant (n+1)\,\epsilon,$$

n being a nonnegative integer. On the interval $n\epsilon \leqslant t \leqslant (n+1)\,\epsilon$ we have $\Gamma \circ L_{-n\varepsilon} = L_{\Gamma(n\varepsilon)^{-1}} \circ \Gamma$. We use both sides of this equation on the tangent vector $(d/dt)_t$ $(n\epsilon \leqslant t \leqslant (n+1)\,\epsilon)$. From (2) we obtain

$$\begin{aligned} \dot{\Gamma}(t) = d\Gamma\left(\frac{d}{dt}\right)_t &= dL_{\Gamma(n\varepsilon)} \circ d\Gamma \circ dL_{-n\varepsilon}\left(\frac{d}{dt}\right)_t \\ &= dL_{\Gamma(n\varepsilon)} \cdot \tilde{X}_{\Gamma(t-n\varepsilon)} \\ &= \tilde{X}_{\Gamma(t)}. \end{aligned}$$

Thus (2) holds for all $s \geqslant 0$ (including the points $n\epsilon$).

Assume now $\alpha(X, X) = 0$. Then, due to the left invariance of the corresponding affine connection ∇, we have $\nabla_{\tilde{X}}(\tilde{X}) = 0$. Hence the curve segment $\Gamma(t)$ $(t \geqslant 0)$ is a geodesic segment, and by the uniqueness of such, we have $\Gamma(t) = \gamma_X(t)$ for all $t \geqslant 0$. For any affine connection, $\gamma_{-X}(t) = \gamma_X(-t)$. Since $\alpha(-X, -X) = 0$, it follows that $\gamma_X(t)$ is defined for all $t \in \boldsymbol{R}$. Now let $s \geqslant 0$. Then the curves $t \to \gamma_X(s+t)$ and $t \to \gamma_X(s)\,\gamma_X(t)$ are both geodesics in G (since ∇ is left invariant) passing through $\gamma_X(s)$. These geodesics have tangent vectors $\dot{\gamma}_X(s)$ and $dL_{\gamma_X(s)} \cdot X$, respectively, at the point $\gamma_X(s)$. These are equal since (2) holds for all $s \geqslant 0$. We conclude that

$$\gamma_X(s+t) = \gamma_X(s)\,\gamma_X(t) \tag{3}$$

for $s \geqslant 0$ and all t. Using again $\gamma_{-X}(t) = \gamma_X(-t)$, we see that (3) holds for all s and t. This proves that (i) $\Rightarrow$ (ii).

Suppose now θ is any analytic homomorphism of $\boldsymbol{R}$ into G such that $\dot{\theta}(0) = X$. Then from $\theta(s+t) = \theta(s)\,\theta(t)$, $(t, s \in \boldsymbol{R})$, follows that

$$\theta(0) = e, \qquad \dot{\theta}(s) = \tilde{X}_{\theta(s)} \qquad \text{for all } s \in \boldsymbol{R}. \tag{4}$$

In particular, if γ_X is an analytic homomorphism, we have $\nabla_{\tilde{X}}(\tilde{X}) = 0$ on the curve γ_X; hence $\alpha(X, X) = (\nabla_{\tilde{X}}(\tilde{X}))_e = 0$.

Corollary 1.5. *Let $X \in \mathfrak{g}$. There exists a unique analytic homomorphism θ of $\boldsymbol{R}$ into G such that $\dot{\theta}(0) = X$.*

Proof. Let ∇ be any affine connection on G for which $\alpha(X, X) = 0$. Then $\theta = \gamma_X$ is a homomorphism with the required properties. For the uniqueness we observe that (4), in connection with $\alpha(X, X) = 0$, shows, that any homomorphism θ with the required properties must be a geodesic; by the uniqueness of geodesics (Prop. 5.3, Chapter I), $\theta = \gamma_X$.

Definition. For each $X \in \mathfrak{g}$, we put $\exp X = \theta(1)$ if θ is the homomorphism of Cor. 1.5. The mapping $X \rightarrow \exp X$ of $\mathfrak{g}$ into G is called the *exponential mapping*.

We have the formula

$$\exp (t + s) X = \exp tX \exp sX$$

for all $s, t \in \boldsymbol{R}$ and all $X \in \mathfrak{g}$. This follows immediately from the fact that if $\alpha(X, X) = 0$, then $\theta(t) = \gamma_X(t) = \gamma_{tX}(1) = \exp tX$.

Definition. A *one-parameter subgroup* of a Lie group G is an analytic homomorphism of $\boldsymbol{R}$ into G.

We have seen above that the one-parameter subgroups are the mappings $t \rightarrow \exp tX$ where X is an element of the Lie algebra.

We see from Prop. 1.4 and the corollary that the exponential mapping agrees with the mapping Exp_e (from Chapter I) for all left invariant affine connections on G satisfying $\alpha(X, X) = 0$ for all $X \in \mathfrak{g}$. The classical examples (Cartan and Schouten [1]) are $\alpha \equiv 0$ (the $(-)$-connection), $\alpha(X, Y) = \frac{1}{2}[X, Y]$ (the (0)-connection) and $\alpha(X, Y) = [X, Y]$ (the $(+)$-connection).

From Theorem 6.1, Chapter I, we deduce the following statement.

Proposition 1.6. *There exists an open neighborhood N_0 of 0 in $\mathfrak{g}$ and an open neighborhood N_e of e in G such that exp is an analytic diffeomorphism of N_0 onto N_e.*

Let $X_1, \ldots, X_n$ be a basis of $\mathfrak{g}$. The mapping

$$\exp (x_1X_1 + \ldots + x_nX_n) \rightarrow (x_1, \ldots, x_n)$$

of N_e onto N_0 is a coordinate system on N_e, called a system of *canonical coordinates* with respect to $X_1, \ldots, X_n$. The set N_e is called a *canonical coordinate neighborhood.* Note that N_0 is not required to be star-shaped.

4. Taylor's Formula and the Differential of the Exponential Mapping

Let G be a Lie group with Lie algebra $\mathfrak{g}$. Let $X \in \mathfrak{g}$, $g \in G$, and $f \in C^\infty(G)$. Since the homomorphism $\theta(t) = \exp tX$ satisfies $\dot{\theta}(0) = X$ we obtain

$$\tilde{X}_g f = X(f \circ L_g) = \left\{\frac{d}{dt} f(g \exp tX)\right\}_{t=0}, \tag{5}$$

It follows that the value of $\tilde{X}f$ at $g \exp uX$ is

$$[\tilde{X}f](g \exp uX) = \left\{\frac{d}{dt} f(g \exp uX \exp tX)\right\}_{t=0} = \frac{d}{du} f(g \exp uX)$$

and by induction

$$[\tilde{X}^n f]\,(g \exp uX) = \frac{d^n}{du^n} f(g \exp uX).$$

Suppose now that f is analytic at g. Then there exists a star-shaped neighborhood N_0 of 0 in $\mathfrak{g}$ such that

$$f(g \exp X) = P(x_1, ..., x_n) \qquad (X \in N_0),$$

where P denotes an absolutely convergent power series and $(x_1, ..., x_n)$ are the coordinates of X with respect to a fixed basis of $\mathfrak{g}$. Then we have for a fixed $X \in N_0$

$$f(g \exp tX) = P(tx_1, ..., tx_n) = \sum_0^\infty \frac{1}{m!} a_m t^m \qquad (a_m \in \boldsymbol{R}),$$

for $0 \leqslant t \leqslant 1$. It follows that each coefficient a_m equals the mth derivative of $f(g \exp tX)$ for $t = 0$; consequently

$$a_m = [\tilde{X}^m f]\,(g).$$

This proves the "Taylor formula";

$$f(g \exp X) = \sum_0^\infty \frac{1}{n!} [\tilde{X}^n f]\,(g) \tag{6}$$

for $X \in N_0$.

Theorem 1.7. *Let G be a Lie group with Lie algebra $\mathfrak{g}$. The exponential mapping of the manifold $\mathfrak{g}$ into G has the differential*

$$d \exp_X = d(L_{\exp X})_e \circ \frac{1 - e^{-\operatorname{ad} X}}{\operatorname{ad} X} \qquad (X \in \mathfrak{g}).$$

As usual, $\mathfrak{g}$ is here identified with the tangent space $\mathfrak{g}_X$.

Proof. We consider the left invariant affine connection on G given by $\alpha(X, Y) = 0$ for all $X, Y \in \mathfrak{g}$ (Prop. 1.4). Then the left invariant vector field $\nabla_{\tilde{X}}(\tilde{Y})$ vanishes identically on G. It follows that the vector field X^* adapted to X (Chapter I, §6) coincides with $\tilde{X}$ in a neighborhood of e in G. From Theorem 6.5, Chapter I we conclude that the equation

$$d(L_{\exp - tX}) \circ d \exp_{tX}(Y) = \frac{1 - e^{-\operatorname{ad} tX}}{\operatorname{ad} tX}(Y)$$

holds for all t in some interval $|t| < \delta$. Both sides of this equation are analytic functions on $\boldsymbol{R}$ with values in $\mathfrak{g}$. Since they agree for $|t| < \delta$, they must agree for all $t \in \boldsymbol{R}$; this proves the theorem.

The following result will be needed later.

Lemma 1.8. *Let G be a Lie group with Lie algebra $\mathfrak{g}$, and let exp be the exponential mapping of $\mathfrak{g}$ into G. Then, if $X, Y \in \mathfrak{g}$,*

(i) $\exp tX \exp tY = \exp \{t(X + Y) + \frac{t^2}{2}[X, Y] + O(t^3)\}$,

(ii) $\exp(-tX) \exp(-tY) \exp tX \exp tY = \exp \{t^2[X, Y] + O(t^3)\}$,

(iii) $\exp tX \exp tY \exp(-tX) = \exp \{tY + t^2[X, Y] + O(t^3)\}$.

In each case $O(t^3)$ denotes a vector in $\mathfrak{g}$ with the property: there exists an $\epsilon > 0$ such that $(1/t^3)\, O(t^3)$ is bounded and analytic for $|t| < \epsilon$.

We first prove (i). Let f be analytic at e. Then using the formula

$$[\tilde{X}^n f](g \exp tX) = \frac{d^n}{dt^n} f(g \exp tX)$$

twice we obtain

$$[\tilde{X}^n \tilde{Y}^m f](e) = \left[\frac{d^n}{dt^n} \frac{d^m}{ds^m} f(\exp tX \exp sY)\right]_{s=0,\, t=0}.$$

Therefore, the Taylor series for $f(\exp tX \exp sY)$ is

$$f(\exp tX \exp sY) = \sum_{m,n \geqslant 0} \frac{t^n}{n!} \frac{s^m}{m!} [\tilde{X}^n \tilde{Y}^m f](e) \tag{7}$$

for sufficiently small t and s. On the other hand,

$$\exp tX \exp tY = \exp Z(t)$$

for sufficiently small t where $Z(t)$ is a function with values in $\mathfrak{g}$, analytic at $t = 0$. We have $Z(t) = tZ_1 + t^2 Z_2 + O(t^3)$ where Z_1 and Z_2 are fixed vectors in $\mathfrak{g}$. Then if f is any of the canonical coordinate functions $\exp(x_1X_1 + \ldots + x_nX_n) \to x_i$ we have

$$f(\exp Z(t)) = f(\exp(tZ_1 + t^2Z_2)) + O(t^3)$$

$$= \sum_0^\infty \frac{1}{n!} [(t\tilde{Z}_1 + t^2\tilde{Z}_2)^n f](e) + O(t^3). \tag{8}$$

If we compare (7) for $t = s$ and (8) we find $Z_1 = X + Y$, $\frac{1}{2}\tilde{Z}_1^2 + \tilde{Z}_2 = \frac{1}{2}\tilde{X}^2 + \tilde{X}\tilde{Y} + \frac{1}{2}\tilde{Y}^2$. Consequently

$$Z_1 = X + Y, \qquad Z_2 = \tfrac{1}{2}[X, Y],$$

which proves (i). The relation (ii) is obtained by applying (i) twice. To prove (iii), let again f be analytic at e; then for small t

$$f(\exp tX \exp tY \exp(-tX)) = \sum_{m,n,p\geq 0} \frac{t^m}{m!}\frac{t^n}{n!}\frac{t^p}{p!}[\tilde{X}^m\tilde{Y}^n(-\tilde{X})^p f](e) \tag{9}$$

and

$$\exp tX \exp tY \exp(-tX) = \exp S(t)$$

where $S(t) = tS_1 + t^2S_2 + O(t^3)$ and $S_1, S_2 \in \mathfrak{g}$. If f is any canonical coordinate function, then

$$f(\exp S(t)) = f(\exp(tS_1 + t^2S_2)) + O(t^3)$$

$$= \sum_0^\infty \frac{1}{n!}[(t\tilde{S}_1 + t^2\tilde{S}_2)^n f](e) + O(t^3), \tag{10}$$

and we find by comparing coefficients in (9) and (10), $S_1 = Y$, $S_2 = [X, Y]$, which proves (iii).

Remark. The relation (ii) gives a geometric interpretation of the bracket $[X, Y]$; in fact, it shows that $[X, Y]$ is the tangent vector at e to the C^1 curve segment

$$s \to \exp(-\sqrt{s}\,X)\exp(-\sqrt{s}\,Y)\exp\sqrt{s}\,X\exp\sqrt{s}\,Y \qquad (s \geqslant 0).$$

Note also that this is a special case of Exercise A.8, Chapter I.

Let $\boldsymbol{D}(G)$ denote the algebra of operators on $C^\infty(G)$ generated by all the left invariant vector fields on G and I (the identity operator on

$C^\infty(G)$). If $X \in \mathfrak{g}$ we shall also denote the corresponding left invariant vector field on G by X. Similarly the operator $\tilde{X}_1 \cdot \tilde{X}_2 \dots \tilde{X}_k$ ($X_i \in \mathfrak{g}$) will be denoted by $X_1 \cdot X_2 \dots X_k$ for simplicity. Let $X_1, \dots, X_n$ be any basis of $\mathfrak{g}$ and put $X(t) = \Sigma_{i=1}^n t_i X_i$. Let $M = (m_1, \dots, m_n)$ be a positive integral n-tuple, let $t^M = t_1^{m_1} \cdots t_n^{m_n}$ and let $X(M)$ denote the coefficient of t^M in the expansion of $(|M|!)^{-1}(X(t))^{|M|}$. If $|M| = 0$ put $X(M) = I$. It is clear that $X(M) \in \boldsymbol{D}(G)$.

Proposition 1.9.

(a) *As M varies through all positive integral n-tuples the elements $X(M)$ form a basis of $\boldsymbol{D}(G)$ (considered as a vector space over $\boldsymbol{R}$).*

(b) *The universal enveloping algebra $U(\mathfrak{g})$ is isomorphic to $\boldsymbol{D}(G)$.*

Proof. Let f be an analytic function at $g \in G$; we have by (6)

$$f(g \exp X(t)) = \sum_M t^M [X(M)f](g), \tag{11}$$

if the t_i are sufficiently small. If we compare this formula with the ordinary Taylor formula for the function F defined by $F(t_1, \dots, t_n) = f(g \exp X(t))$, we obtain

$$[X(M)f](g) = \frac{1}{m_1! \dots m_n!} \left\{ \frac{\partial^{|M|}}{\partial t_1^{m_1} \dots \partial t_n^{m_n}} f(g \exp X(t)) \right\}_{t_1 = \dots = t_n = 0}. \tag{12}$$

It follows immediately that the various $X(M)$ are linearly independent. The Lie algebra $\mathfrak{g}$ has a representation ρ on $C^\infty(G)$ if we associate to each $X \in \mathfrak{g}$ the corresponding left invariant vector field. The representation ρ^* from Prop. 1.1 gives a homomorphism of $U(\mathfrak{g})$ into $\boldsymbol{D}(G)$ such that $\rho^*(X^*) = \rho(X)$ for $X \in \mathfrak{g}$. The mapping ρ^* sends the element $X_{i_1}^* \dots X_{i_p}^* \in U(\mathfrak{g})$ into $X_{i_1} \dots X_{i_p} \in \boldsymbol{D}(G)$; thus $\rho^*(U(\mathfrak{g})) = \boldsymbol{D}(G)$. Moreover, ρ^* sends the element $X^*(M) \in U(\mathfrak{g})$ into $X(M) \in \boldsymbol{D}(G)$. Since the elements $X(M)$ are linearly independent, the proposition follows from Prop. 1.2.

Corollary 1.10. *With the notation above, the elements $X_1^{e_1} \dots X_n^{e_n}$ ($e_i \geqslant 0$) form a basis of $\boldsymbol{D}(G)$.*

Since $X_i X_j - X_j X_i = [X_i, X_j]$ it is clear that each $X(M)$ can be written as a real linear combination of elements $X_1^{e_1} \dots X_n^{e_n}$ where $e_1 + \dots + e_n \leqslant |M|$. On the other hand, as noted in the proof of Prop. 1.2, each $X_1^{e_1} \dots X_n^{e_n}$ can be written as a real linear combination of elements $X(M)$ for which $|M| \leqslant e_1 + \dots + e_n$. Since the number of elements $X(M)$, $|M| \leqslant e_1 + \dots + e_n$ equals the number of elements

$X_1^{f_1} \dots X_n^{f_n}$ $(f_1 + \dots + f_n \leqslant e_1 + \dots + e_n)$, the corollary follows from Prop. 1.9.

This corollary shows quickly that $\boldsymbol{D}(G)$ has no divisors of 0.

Definition. Let G and G' be two Lie groups with identity elements e and e'. These groups are said to be *isomorphic* if there exists an analytic isomorphism of G onto G'. The groups G and G' are said to be *locally isomorphic* if there exist open neighborhoods U and U' of e and e', respectively, and an analytic diffeomorphism f of U onto U' satisfying:

(a) If $x, y, xy \in U$, then $f(xy) = f(x)f(y)$.

(b) If $x', y', x'y' \in U'$, then $f^{-1}(x'y') = f^{-1}(x')f^{-1}(y')$.

Theorem 1.11. *Two Lie groups are locally isomorphic if and only if their Lie algebras are isomorphic.*

Proof. Let G be a Lie group with Lie algebra $\mathfrak{g}$. Let $X_1, \dots, X_n$ be a basis of $\mathfrak{g}$. Owing to Prop. 1.9 we can legitimately write $X(M)$ instead of $X^*(M)$; there exist uniquely determined constants $C^P{}_{MN} \in \boldsymbol{R}$ such that

$$X(M)\, X(N) = \sum_P C^P{}_{MN}\, X(P),$$

M, N, and P denoting positive integral n-tuples. Owing to Prop. 1.9, the constants $C^P{}_{MN}$ depend only on the Lie algebra $\mathfrak{g}$. If N_e is a canonical coordinate neighborhood of $e \in G$ and $g \in N_e$ let $g_1, \dots, g_n$ denote the canonical coordinates of g. Then if $x, y, xy \in N_e$, we have

$$x = \exp(x_1X_1 + \dots + x_nX_n), \qquad y = \exp(y_1X_1 + \dots + y_nX_n),$$

$$xy = \exp((xy)_1X_1 + \dots + (xy)_nX_n).$$

We also put

$$x^M = x_1^{m_1} \dots x_n^{m_n}, \qquad y^M = y_1^{m_1} \dots y_n^{m_n}.$$

Using (7) on the function $f: x \to x_k$ we find for sufficiently small x_i, y_j

$$(xy)_k = \sum_{M,N} x^My^N[X(M)\, X(N)\, x_k]\,(e). \tag{13}$$

From (12) it follows that

$$[X(P)\, x_k]\,(e) = \begin{cases} 1 \text{ if } P = (\delta_{k1}, \delta_{k2}, \dots, \delta_{kn}), \\ 0 \text{ otherwise.} \end{cases}$$

Putting $[k] = (\delta_{1k}, \dots, \delta_{nk})$, relation (13) becomes

$$(xy)_k = \sum_{M,N} C^{[k]}{}_{MN}\, x^My^N, \tag{14}$$

if x_i, y_j $(1 \leqslant i, j \leqslant n)$ are sufficiently small. This last formula shows that the group law is determined in a neighborhood of e by the Lie algebra. In particular, Lie groups with isomorphic Lie algebras are locally isomorphic. Before proving the converse of Theorem 1.11 we prove a general lemma about homomorphisms.

Lemma 1.12. *Let H and K be Lie groups with Lie algebras $\mathfrak{h}$ and $\mathfrak{k}$, respectively. Let φ be an analytic homomorphism of H into K. Then $d\varphi_e$ is a homomorphism of $\mathfrak{h}$ into $\mathfrak{k}$ and*

$$\varphi(\exp X) = \exp d\varphi_e(X) \qquad (X \in \mathfrak{h}). \tag{15}$$

Proof. Let $X \in \mathfrak{h}$. The mapping $t \to \varphi(\exp tX)$ is an analytic homomorphism of $\boldsymbol{R}$ into K. If we put $X' = d\varphi_e(X)$, Cor. 1.5 implies that $\varphi(\exp tX) = \exp tX'$ for all $t \in \boldsymbol{R}$. Since φ is a homomorphism, we have $\varphi \circ L_\sigma = L_{\varphi(\sigma)} \circ \varphi$ for $\sigma \in H$. It follows that

$$(d\varphi_\sigma) \circ dL_\sigma \cdot X = dL_{\varphi(\sigma)} \cdot X'.$$

This means that the left invariant vector fields $\tilde{X}$ and $\tilde{X}'$ are φ-related. Hence, by Prop. 3.3, Chapter I, $d\varphi_e$ is a homomorphism and the lemma is proved.

To finish the proof of Theorem 1.11, we suppose now that the Lie groups G and G' are locally isomorphic. Let $\mathfrak{g}$ and $\mathfrak{g}'$ denote their respective Lie algebras. There is no restriction of generality in assuming G and G' connected. Let $(\tilde{G}, \pi)$ and $(\tilde{G}', \pi')$ be the universal covering groups of G and G', respectively. It follows from Lemma 1.12 that the mappings $d\pi_e$ and $d\pi'_e$ are Lie algebra isomorphisms. From the first part of the proof it follows that π and π' are local isomorphisms. The given local isomorphism between G and G' therefore induces a local isomorphism θ between $\tilde{G}$ and $\tilde{G}'$. Now $\tilde{G}$ is simply connected. Owing to a well-known theorem on topological groups, see e.g. Chevalley [2], p. 49, there exists a continuous homomorphism $\tilde{\theta}$ of $\tilde{G}$ into $\tilde{G}'$ which coincides with θ in a neighborhood of the identity; in particular, $\tilde{\theta}$ is analytic. On interchanging $\tilde{G}$ and $\tilde{G}'$ we see that $\tilde{\theta}$ is an isomorphism of $\tilde{G}$ onto $\tilde{G}'$. From Lemma 1.12 it follows that $d\tilde{\theta}_e$ is an isomorphism between the corresponding Lie algebras. This in turn gives the desired isomorphism of the Lie algebras $\mathfrak{g}$ and $\mathfrak{g}'$.

Example. Let $\boldsymbol{GL}(n, \boldsymbol{R})$ denote the group of all real nonsingular $n \times n$ matrices and let $\mathfrak{gl}(n, \boldsymbol{R})$ denote the Lie algebra of all real $n \times n$ matrices, the bracket being $[A, B] = AB - BA$, $A, B \in \mathfrak{gl}(n, \boldsymbol{R})$. If we consider the matrix $\sigma = (x_{ij}(\sigma)) \in \boldsymbol{GL}(n, \boldsymbol{R})$ as the set of coordinates of a point in $\boldsymbol{R}^{n^2}$ then $\boldsymbol{GL}(n, \boldsymbol{R})$ can be regarded as an open submanifold of $\boldsymbol{R}^{n^2}$. With this analytic structure $\boldsymbol{GL}(n, \boldsymbol{R})$ is a Lie group;

this is obvious by considering the expression of $x_{ij}(\sigma\tau^{-1})$ $(\tau, \sigma \in \boldsymbol{GL}(n, \boldsymbol{R}))$ in terms of $x_{kl}(\sigma)$, $x_{pq}(\tau)$, given by matrix multiplication.

Let X be an element of $\mathfrak{L}(\boldsymbol{GL}(n, \boldsymbol{R}))$ and let $\tilde{X}$ denote the left invariant vector field on $\boldsymbol{GL}(n, \boldsymbol{R})$ such that $\tilde{X}_e = X$. Let $(a_{ij}(X))$ denote the matrix $(\tilde{X}_e x_{ij})$. We shall prove that the mapping $\varphi : X \to (a_{ij}(X))$ is an isomorphism of $\mathfrak{L}(\boldsymbol{GL}(n, \boldsymbol{R}))$ onto $\mathfrak{gl}(n, \boldsymbol{R})$. The mapping φ is linear and one-to-one; in fact, the relation $(a_{ij}(X)) = 0$ implies $\tilde{X}_e f = 0$ for all differentiable functions f, hence $\tilde{X} = 0$. Considering the dimensions of the Lie algebras we see that the range of φ is $\mathfrak{gl}(n, \boldsymbol{R})$. Next we consider $[\tilde{X}x_{ij}]\,(\sigma) = (dL_\sigma X)\, x_{ij} = X(x_{ij} \circ L_\sigma)$. If $\tau \in \boldsymbol{GL}(n, \boldsymbol{R})$, then

$$(x_{ij} \circ L_\sigma)\,(\tau) = x_{ij}(\sigma\tau) = \sum_{k=1}^{n} x_{ik}(\sigma)\, x_{kj}(\tau). \tag{16}$$

Hence

$$[\tilde{X}x_{ij}]\,(\sigma) = \sum_{k=1}^{n} x_{ik}(\sigma)\, a_{kj}(X). \tag{17}$$

It follows that

$$[(\tilde{X}\tilde{Y} - \tilde{Y}\tilde{X})\, x_{ij}]\,(e) = \sum_{k=1}^{n} a_{ik}(X)\, a_{kj}(Y) - a_{ik}(Y)\, a_{kj}(X)$$

$$= [\varphi(X), \varphi(Y)]_{ij}.$$

Consequently, the Lie algebra of $\boldsymbol{GL}(n, \boldsymbol{R})$ can be identified with $\mathfrak{gl}(n, \boldsymbol{R})$ so we now write X_{ij} instead of $a_{ij}(X)$ above. Using the general formula

$$[\tilde{X}f]\,(\exp tX) = \frac{d}{dt} f(\exp tX)$$

for a differentiable function f, we obtain from (16) and (17)

$$\frac{d}{dt} x_{ij}(\exp tX) = \sum_{k=1}^{n} x_{ik}(\exp tX)\, X_{kj}.$$

Thus the matrix function $Y(t) = \exp tX$ satisfies the differential equation

$$\frac{dY(t)}{dt} = Y(t)\, X, \qquad Y(0) = I.$$

Since this equation is also satisfied by the matrix exponential function

$$Y(t) = e^{tX} = I + tX + \frac{t^2X^2}{2!} + \ldots,$$

we conclude that $\exp X = e^X$ for all $X \in \mathfrak{gl}(n, \boldsymbol{R})$.

Let V be an n-dimensional vector space over $\boldsymbol{R}$. Let $\mathfrak{gl}(V)$ be the Lie algebra of all endomorphisms of V and let $\boldsymbol{GL}(V)$ be the group of invertible endomorphisms of V. Fix a basis $e_1, \ldots, e_n$ of V. To each $\sigma \in \mathfrak{gl}(V)$ we associate the matrix $(x_{ij}(\sigma))$ given by

$$\sigma e_j = \sum_{i=1}^{n} x_{ij}(\sigma)\, e_i.$$

The mapping $J_e : \sigma \to (x_{ij}(\sigma))$ is an isomorphism of $\mathfrak{gl}(V)$ onto $\mathfrak{gl}(n, \boldsymbol{R})$ whose restriction to $\boldsymbol{GL}(V)$ is an isomorphism of $\boldsymbol{GL}(V)$ onto $\boldsymbol{GL}(n, \boldsymbol{R})$. This isomorphism turns $\boldsymbol{GL}(V)$ into a Lie group with Lie algebra isomorphic to $\mathfrak{gl}(V)$. If $f_1, \ldots, f_n$ is another basis of V, we get another isomorphism $J_f : \mathfrak{gl}(V) \to \mathfrak{gl}(n, \boldsymbol{R})$. If $A \in \boldsymbol{GL}(V)$ is determined by $Ae_i = f_i$ $(1 \leqslant i \leqslant n)$, then J_f and J_e are connected by the equation $J_e(\sigma) = J_f(A) J_f(\sigma) J_f(A^{-1})$. Since the mapping $g \to J_f(A) g J_f(A^{-1})$ is an analytic isomorphism of $\boldsymbol{GL}(n, \boldsymbol{R})$ onto itself, we conclude: (1) The analytic structure of $\boldsymbol{GL}(V)$ is independent of the choice of basis. (2) There is an isomorphism of $\mathfrak{L}(\boldsymbol{GL}(V))$ onto $\mathfrak{gl}(V)$ (namely, $J_e^{-1} \circ dJ_e$) which is independent of the choice of basis of V.

§ 2. Lie Subgroups and Subalgebras

Definition. Let G be a Lie group. A submanifold H of G is called a *Lie subgroup* if

(i) H is a subgroup of the (abstract) group G;

(ii) H is a topological group.

A Lie subgroup is itself a Lie group; in order to see this, consider the analytic mapping $\alpha : (x, y) \to xy^{-1}$ of $G \times G$ into G. Let α_H denote the restriction of α to $H \times H$. Then the mapping $\alpha_H : H \times H \to G$ is analytic, and by (ii) the mapping $\alpha_H : H \times H \to H$ is continuous. In view of Lemma 14.1, Chapter I, the mapping α_H is an analytic mapping of $H \times H$ into H so H is a Lie group.

A connected Lie subgroup is often called an *analytic subgroup.*

Theorem 2.1. *Let G be a Lie group. If H is a Lie subgroup of G, then the Lie algebra $\mathfrak{h}$ of H is a subalgebra of $\mathfrak{g}$, the Lie algebra of G. Each subalgebra of $\mathfrak{g}$ is the Lie algebra of exactly one connected Lie subgroup of G.*

Proof. If I denotes the identity mapping of H into G, then by Lemma 1.12 dI_e is a homomorphism of $\mathfrak{h}$ into $\mathfrak{g}$. Since H is a submanifold of G, dI_e is one-to-one. Thus $\mathfrak{h}$ can be regarded as a subalgebra of $\mathfrak{g}$.

Let $\exp_{\mathfrak{h}}$ and $\exp_{\mathfrak{g}}$, respectively, denote the exponential mappings of $\mathfrak{h}$ into H and of $\mathfrak{g}$ into G. From Cor. 1.5 we get immediately

$$\exp_{\mathfrak{h}}(X) = \exp_{\mathfrak{g}}(X), \qquad X \in \mathfrak{h}. \tag{1}$$

We can therefore drop the subscripts and write exp instead of $\exp_{\mathfrak{h}}$ and $\exp_{\mathfrak{g}}$. If $X \in \mathfrak{h}$, then the mapping $t \to \exp tX$ ($t \in \boldsymbol{R}$) is a curve in H. On the other hand, suppose $X \in \mathfrak{g}$ such that the mapping $t \to \exp tX$ is a path in H, that is, a continuous curve in H. By Lemma 14.1, Chapter I, the mapping $t \to \exp tX$ is an analytic mapping of $\boldsymbol{R}$ into H. Thus $X \in \mathfrak{h}$, so we have

$$\mathfrak{h} = \{X \in \mathfrak{g} : \text{the map } t \to \exp tX \text{ is a path in } H\}. \tag{2}$$

To prove the second statement of Theorem 2.1, suppose $\mathfrak{h}$ is any subalgebra of $\mathfrak{g}$. Let H be the smallest subgroup of G containing $\exp \mathfrak{h}$. Let $(X_1, ..., X_n)$ be a basis of $\mathfrak{g}$ such that (X_i) $(r < i \leqslant n)$ is a basis of $\mathfrak{h}$. Then we know from Cor. 1.3 (and Prop. 1.9) that all real linear combinations of elements $X(M)$, where the n-tuple M has the form $(0, ..., 0, m_{r+1}, ..., m_n)$, actually form a subalgebra of $U(\mathfrak{g})$. Let $|X| = (x_1^2 + ... + x_n^2)^{1/2}$ if $X = x_1X_1 + ... + x_nX_n$ $(x_i \in \boldsymbol{R})$. Choose $\delta > 0$ such that exp is a diffeomorphism of the open ball $B_\delta = \{X : |X| < \delta\}$ onto an open neighborhood N_e of e in G and such that (14), §1, holds for $x, y, xy \in N_e$. Denote the subset $\exp(\mathfrak{h} \cap B_\delta)$ of N_e by V. The mapping

$$\exp(x_{r+1}X_{r+1} + ... + x_nX_n) \to (x_{r+1}, ..., x_n)$$

is a coordinate system on V with respect to which V is a connected manifold. Since $\mathfrak{h} \cap B_\delta$ is a submanifold of B_δ, V is a submanifold of N_e; hence V is a submanifold of G. Now suppose $x, y \in V$, $xy \in N_e$, and consider the canonical coordinates of xy as given by (14), §1. Since $x_k = y_k = 0$, if $1 \leqslant k \leqslant r$, we find (using the remark above about $X(M)$) that $(xy)_k = 0$ if $1 \leqslant k \leqslant r$. Thus we have

$$VV \cap N_e \subset V. \tag{3}$$

Let $\mathscr{V}$ denote the family of subsets of H containing a neighborhood of e in V. Let us verify that $\mathscr{V}$ satisfies the following six axioms for a topological group (Chevalley [2], Chapter II, § II).

I. The intersection of any two sets of $\mathscr{V}$ lies in $\mathscr{V}$.

II. The intersection of all sets of $\mathscr{V}$ is $\{e\}$.

III. Any subset of H containing a set in $\mathscr{V}$ lies in $\mathscr{V}$.

IV. If $U \in \mathscr{V}$, there exists a set $U_1 \in \mathscr{V}$ such that $U_1U_1 \subset U$.

V. If $U \in \mathscr{V}$, then $U^{-1} \in \mathscr{V}$.

VI. If $U \in \mathscr{V}$ and $h \in H$, then $hUh^{-1} \in \mathscr{V}$.

Of these axioms, I, II, III, and V are obvious and IV follows from (3). For VI, let $U \in \mathscr{V}$ and $h \in H$. Let log denote the inverse of the mapping $\exp : B_\delta \to N_e$. Then log maps V onto $\mathfrak{h} \cap B_\delta$. If $X \in \mathfrak{g}$, there exists a unique vector $X' \in \mathfrak{g}$ such that $h \exp tX\, h^{-1} = \exp tX'$ for all $t \in \boldsymbol{R}$. The mapping $X \to X'$ is an automorphism of $\mathfrak{g}$ (Lemma 1.12); it maps $\mathfrak{h}$ into itself as is easily seen from (3) by using a decomposition $h = \exp Z_1 \ldots \exp Z_p$ where each Z_i belongs to $B_\delta \cap \mathfrak{h}$. Consequently, we can select δ_1 $(0 < \delta_1 < \delta)$ such that the open ball B_{δ_1} satisfies

$$h \exp (B_{\delta_1} \cap \mathfrak{h})\, h^{-1} \subset V,$$

$$h (\exp B_{\delta_1})\, h^{-1} \subset N_e.$$

The mapping $X \to \log (h \exp Xh^{-1})$ of $B_{\delta_1} \cap \mathfrak{h}$ into $B_\delta \cap \mathfrak{h}$ is regular so the image of $B_{\delta_1} \cap \mathfrak{h}$ is a neighborhood of 0 in $\mathfrak{h}$. Applying the mapping exp we see that $h \exp (B_{\delta_1} \cap \mathfrak{h})\, h^{-1}$ is a neighborhood of e in V. This shows that $hUh^{-1} \in \mathscr{V}$. Axioms I-VI are therefore satisfied. Hence there exists a topology on H such that H is a topological group and such that $\mathscr{V}$ is the family of neighborhoods of e in H. In particular V is a neighborhood of e in H.

For each $z \in G$, consider the mapping

$$\Phi_z : z \exp (x_1 X_1 + \ldots + x_n X_n) \to (x_1, \ldots, x_n),$$

which maps zN_e onto B_δ. Let φ_z denote the restriction of Φ_z to zV. If $z \in H$ then φ_z maps the neighborhood zV (of z in H) onto the open subset $B_\delta \cap \mathfrak{h}$ in Euclidean space $\boldsymbol{R}^{n-r}$. Moreover, if $z_1, z_2 \in H$ the mapping $\varphi_{z_1} \circ \varphi_{z_2}^{-1}$ is the restriction of $\Phi_{z_1} \circ \Phi_{z_2}^{-1}$ to an open subset of $\mathfrak{h}$, hence analytic. The space H with the collection of maps φ_z, $z \in H$, is therefore an analytic manifold.

Now V is a submanifold of G. Since left translations are diffeomorphisms of H it follows that H is a submanifold of G. Hence H is a Lie subgroup of G.

We know that $\dim H = \dim \mathfrak{h}$. For $i > r$ the mapping $t \to \exp tX_i$ is a curve in H. This in view of (2) proves that H has Lie algebra $\mathfrak{h}$. Moreover, H is connected since it is generated by $\exp \mathfrak{h}$ which is a connected neighborhood of e in H.

Finally, in order to prove uniqueness, suppose H_1 is any connected Lie subgroup of G with $(H_1)_e = \mathfrak{h}$. From (1) we see that $H = H_1$ (set theoretically). Since exp is an analytic diffeomorphism of a neighborhood of 0 in $\mathfrak{h}$ onto a neighborhood of e in H and H_1, it is clear that the Lie groups H and H_1 coincide.

Corollary 2.2. *Suppose H_1 and H_2 are two Lie subgroups of a Lie group G such that $H_1 = H_2$ (as topological groups). Then $H_1 = H_2$ (as Lie groups).*

Relation (2) shows, in fact, that H_1 and H_2 have the same Lie algebra. By Theorem 2.1, their identity components coincide as Lie groups. Since left translations on H_1 and H_2 are analytic, it follows at once that the Lie groups H_1 and H_2 coincide.

Theorem 2.3. *Let G be a Lie group with Lie algebra $\mathfrak{g}$ and H an (abstract) subgroup of G. Suppose H is a closed subset of G. Then there exists a unique analytic structure on H such that H is a topological Lie subgroup of G.*

We begin by proving a simple lemma.

Lemma 2.4. *Suppose $\mathfrak{g}$ is a direct sum $\mathfrak{g} = \mathfrak{m} + \mathfrak{n}$ where $\mathfrak{m}$ and $\mathfrak{n}$ are two vector subspaces of $\mathfrak{g}$. Then there exist bounded, open, connected neighborhoods $U_{\mathfrak{m}}$ and $U_{\mathfrak{n}}$ of 0 in $\mathfrak{m}$ and $\mathfrak{n}$, respectively, such that the mapping $\Phi : (A, B) \to \exp A \exp B$ is a diffeomorphism of $U_{\mathfrak{m}} \times U_{\mathfrak{n}}$ onto an open neighborhood of e in G.*

Proof. Let $X_1, ..., X_n$ be a basis of $\mathfrak{g}$ such that $X_i \in \mathfrak{m}$ for $1 \leqslant i \leqslant r$, $X_j \in \mathfrak{n}$ for $r < j \leqslant n$. Let $\{t_1, ..., t_n\}$ denote the canonical coordinates of the element $\exp(x_1X_1 + ... + x_rX_r) \exp(x_{r+1}X_{r+1} + ... + x_nX_n)$ with respect to this basis. Then $t_j = \varphi_j(x_1, ..., x_n)$, $1 \leqslant j \leqslant n$, where the functions φ_j are analytic at $(0, ..., 0)$. If $x_i = \delta_{ij}s$, then $t_i = \delta_{ij}s$ and the Jacobian determinant $\partial(\varphi_1, ..., \varphi_n)/\partial(x_1, ..., x_n)$ equals 1 for $x_1 = ... = x_n = 0$. This proves the lemma (Prop. 3.1, Chapter I).

Remark. The lemma generalizes immediately to an arbitrary direct decomposition $\mathfrak{g} = \mathfrak{m}_1 + ... + \mathfrak{m}_s$ of $\mathfrak{g}$ into subspaces.

Turning now to the proof of Theorem 2.3, let $\mathfrak{h}$ denote the subset of $\mathfrak{g}$ given by

$$\mathfrak{h} = \{X : \exp tX \in H \text{ for all } t \in \boldsymbol{R}\}.$$

We shall prove that $\mathfrak{h}$ is a subalgebra of $\mathfrak{g}$. First we note that $X \in \mathfrak{h}$, $s \in \boldsymbol{R}$ implies $sX \in \mathfrak{h}$. Next, suppose $X, Y \in \mathfrak{h}$. By Lemma 1.8 we have for a given $t \in \boldsymbol{R}$,

$$\left(\exp \frac{t}{n} X \exp \frac{t}{n} Y\right)^n = \exp \left\{t(X + Y) + \frac{t^2}{2n} [X, Y] + O\left(\frac{1}{n^2}\right)\right\},$$

$$\left(\exp \left(-\frac{t}{n} X\right) \exp \left(-\frac{t}{n} Y\right) \exp \frac{t}{n} X \exp \frac{t}{n} Y\right)^{n^2} = \exp \left\{t^2[X, Y] + O\left(\frac{1}{n}\right)\right\}.$$

The left-hand sides of these equations belong to H; since H is closed, the limit as $n \to \infty$ also belongs to H. Thus $t(X + Y) \in \mathfrak{h}$ and $t^2[X, Y] \in \mathfrak{h}$ as desired.

Let H^* denote the connected Lie subgroup of G with Lie algebra $\mathfrak{h}$. Then $H^* \subset H$. We shall now prove, that if H is given the relative topology of G, and H_0 is the identity component of H, then $H^* = H_0$ (as topological groups). For this we now prove that if N is a neighborhood of e in H^*, then N is a neighborhood of e in H. If N were not a neighborhood of e in H, there would exist a sequence $(c_k) \subset H - N$ such that $c_k \to e$ (in the topology of G). Using Lemma 2.4 for $\mathfrak{h} = \mathfrak{n}$ and $\mathfrak{m}$ any complementary subspace we can assume that $c_k = \exp A_k \exp B_k$ where $A_k \in U_{\mathfrak{m}}$, $B_k \in U_{\mathfrak{n}}$, and $\exp B_k \in N$. Then

$$A_k \neq 0, \qquad \lim A_k = 0.$$

Since $A_k \neq 0$, there exists an integer $r_k > 0$ such that

$$r_k A_k \in U_{\mathfrak{m}}, \qquad (r_k + 1) A_k \notin U_{\mathfrak{m}}.$$

Now, $U_{\mathfrak{m}}$ is bounded, so we can assume, passing to a subsequence if necessary, that the sequence $(r_k A_k)$ converges to a limit $A \in \mathfrak{m}$. Since $(r_k + 1) A_k \notin U_{\mathfrak{m}}$ and $A_k \to 0$, we see that A lies on the boundary of $U_{\mathfrak{m}}$; in particular $A \neq 0$.

Let p, q be any integers ($q > 0$). Then we can write $pr_k = qs_k + t_k$ where s_k, t_k are integers and $0 \leqslant t_k < q$. Then $\lim (t_k/q) A_k = 0$, so

$$\exp \frac{p}{q} A = \lim_k \exp \frac{pr_k}{q} A_k = \lim_k (\exp A_k)^{s_k},$$

which belongs to H. By continuity, $\exp tA \in H$ for each $t \in \boldsymbol{R}$, so $A \in \mathfrak{h}$. This contradicts the fact that $A \neq 0$ and $A \in \mathfrak{m}$.

We have therefore proved: (1) H_0 is open in H (taking $N = H^*$); (2) H_0 (and therefore H) has an analytic structure compatible with the relative topology of G in which it is a submanifold of G, hence a Lie subgroup of G. The uniqueness statement of Theorem 2.3 is immediate from Cor. 2.2.

Remark. The subgroup H above is discrete if and only if $\mathfrak{h} = 0$.

Lemma 2.5. *Let G be a Lie group and H a Lie subgroup. Let $\mathfrak{g}$ and $\mathfrak{h}$ denote the corresponding Lie algebras. Suppose H is a topological subgroup of G. Then there exists an open neighborhood V of 0 in $\mathfrak{g}$ such that:*

(i) *exp maps V diffeomorphically onto an open neighborhood of e in G.*

(ii) *$\exp (V \cap \mathfrak{h}) = (\exp V) \cap H$.*

Proof. First select a neighborhood W_0 of 0 in $\mathfrak{g}$ such that exp is one-to-one on W_0. Then select an open neighborhood N_0 of 0 in $\mathfrak{h}$ such that $N_0 \subset W_0$ and such that exp is a diffeomorphism of N_0 onto an open neighborhood N_e of e in H. Now, since H is a topological subspace of G there exists a neighborhood U_e of e in G such that $U_e \cap H = N_e$. Finally select an open neighborhood V of 0 in $\mathfrak{g}$ such that $V \subset W_0$, $V \cap \mathfrak{h} \subset N_0$ and such that exp is a diffeomorphism of V onto an open subset of G contained in U_e. Then V satisfies (i). Condition (ii) is also satisfied. In fact, let $X \in V$ such that $\exp X \in H$. Since $\exp X \in U_e \cap H = N_e$ there exists a vector $X_\mathfrak{h} \in N_0$ such that $\exp X_\mathfrak{h} = \exp X$. Since $X, X_\mathfrak{h} \in W_0$ we have $X = X_\mathfrak{h}$ so $\exp X \in \exp(V \cap \mathfrak{h})$. This proves $(\exp V) \cap H \subset \exp(V \cap \mathfrak{h})$. The converse inclusion being obvious the lemma is proved.

Theorem 2.6. *Let G and H be Lie groups and φ a continuous homomorphism of G into H. Then φ is analytic.*

Proof. Let the Lie algebras of G and H be denoted by $\mathfrak{g}$ and $\mathfrak{h}$, respectively. The product manifold $G \times H$ is a Lie group whose Lie algebra is the product $\mathfrak{g} \times \mathfrak{h}$ as defined in §1, No. 1. The graph of φ is the subset of $G \times H$ given by $K = \{(g, \varphi(g)) : g \in G\}$. It is obvious that K is a closed subgroup of $G \times H$. As a result of Theorem 2.3, K has a unique analytic structure under which it is a topological Lie subgroup of $G \times H$. Its Lie algebra is given by

$$\mathfrak{k} = \{(X, Y) \in \mathfrak{g} \times \mathfrak{h} : (\exp tX, \exp tY) \in K \text{ for } t \in \boldsymbol{R}\}. \tag{4}$$

Let N_0 be an open neighborhood of 0 in $\mathfrak{h}$ such that exp maps N_0 diffeomorphically onto an open neighborhood N_e of e in H. Let M_0 and M_e be chosen similarly for G. We may assume that $\varphi(M_e) \subset N_e$. In view of Lemma 2.5 we can also assume that exp is a diffeomorphism of $(M_0 \times N_0) \cap \mathfrak{k}$ onto $(M_e \times N_e) \cap K$. We shall now show that for a given $X \in \mathfrak{g}$ there exists a unique $Y \in \mathfrak{h}$ such that $(X, Y) \in \mathfrak{k}$. The uniqueness is obvious from (4); in fact, if (X, Y_1) and (X, Y_2) belong to $\mathfrak{k}$, then $(0, Y_1 - Y_2) \in \mathfrak{k}$ so by (4), $(e, \exp t(Y_1 - Y_2)) \in K$ for all $t \in \boldsymbol{R}$. By the definition of K, $\exp t(Y_1 - Y_2) = \varphi(e) = e$ for $t \in \boldsymbol{R}$ so $Y_1 - Y_2 = 0$. In order to prove the existence of Y, select an integer $r > 0$ such that the vector $X_r = (1/r)X$ lies in M_0. Since $\varphi(\exp X_r) \in N_e$, there exists a unique vector $Y_r \in N_0$ such that $\exp Y_r = \varphi(\exp X_r)$ and a unique $Z_r \in (M_0 \times N_0) \cap \mathfrak{k}$ such that

$$\exp Z_r = (\exp X_r, \exp Y_r).$$

Now exp is one-to-one on $M_0 \times N_0$ so this relation implies $Z_r = (X_r, Y_r)$

and we can put $Y = rY_r$. The mapping $\psi : X \to Y$ thus obtained is clearly a homomorphism of $\mathfrak{g}$ into $\mathfrak{h}$. Relation (4) shows that

$$\varphi(\exp tX) = \exp t\psi(X), \qquad X \in \mathfrak{g}. \tag{5}$$

Let $X_1, ..., X_n$ be a basis of $\mathfrak{g}$. Then by (5)

$$\varphi((\exp t_1X_1)(\exp t_2X_2) \ldots (\exp t_nX_n)) = (\exp t_1\psi(X_1))(\exp t_2\psi(X_2)) \ldots (\exp t_n\psi(X_n)). \tag{6}$$

The remark following Lemma 2.4 shows that the mapping $(\exp t_1X_1) \ldots (\exp t_nX_n) \to (t_1, ..., t_n)$ is a coordinate system on a neighborhood of e in G. But then by (6), φ is analytic at e, hence everywhere on G.

We shall now see that a simple countability assumption makes it possible to sharpen relation (2) and Cor. 2.2 substantially.

Proposition 2.7. *Let G be a Lie group and H a Lie subgroup. Let $\mathfrak{g}$ and $\mathfrak{h}$ denote the corresponding Lie algebras. Assume that the Lie group H has at most countably many components. Then*

$$\mathfrak{h} = \{X \in \mathfrak{g}: \exp tX \in H \textit{ for all } t \in \boldsymbol{R}\}.$$

Proof. We use Lemma 2.4 for $\mathfrak{n} = \mathfrak{h}$ and $\mathfrak{m}$ any complementary subspace to $\mathfrak{h}$ in $\mathfrak{g}$. Let V denote the set $\exp U_{\mathfrak{m}} \exp U_{\mathfrak{h}}$ (from Lemma 2.4) with the relative topology of G and put

$$\mathfrak{a} = \{A \in U_{\mathfrak{m}}: \exp A \in H\}.$$

Then

$$H \cap V = \bigcup_{A \in \mathfrak{a}} \exp A \exp U_{\mathfrak{h}}$$

and this is a disjoint union due to Lemma 2.4. Each member of this union is a neighborhood in H. Since H has a countable basis the set $\mathfrak{a}$ must be countable. Consider now the mapping π of V onto $U_{\mathfrak{m}}$ given by $\pi(\exp X \exp Y) = X$ $(X \in U_{\mathfrak{m}}, Y \in U_{\mathfrak{h}})$. This mapping is continuous and maps $H \cap V$ onto $\mathfrak{a}$. The component of e in $H \cap V$ (in the topology of V) is mapped by π onto a connected countable subset of $U_{\mathfrak{m}}$, hence the single point 0. Since $\pi^{-1}(0) = \exp U_{\mathfrak{h}}$ we conclude that $\exp U_{\mathfrak{h}}$ is the component of e in $H \cap V$ (in the topology of V).

Now let $X \in \mathfrak{g}$ such that $\exp tX \in H$ for all $t \in \boldsymbol{R}$. The mapping $\varphi : t \to \exp tX$ of $\boldsymbol{R}$ into G is continuous. Hence there exists a connected neighborhood U of 0 in $\boldsymbol{R}$ such that $\varphi(U) \subset V$. Then $\varphi(U) \subset H \cap V$ and since $\varphi(U)$ is connected, $\varphi(U) \subset \exp U_{\mathfrak{h}}$. But $\exp U_{\mathfrak{h}}$ is an arbitrarily small neighborhood of e in H so the mapping φ is a continuous mapping of $\boldsymbol{R}$ into H. By (2) we have $X \in \mathfrak{h}$ and the proposition is proved.

Remark. The countability assumption is essential in Prop. 2.7 as is easily seen by considering a Lie subgroup with the discrete topology.

Corollary 2.8. *Let G be a Lie group and let H_1 and H_2 be two Lie subgroups each having countably many components. Suppose that $H_1 = H_2$ (set theoretically). Then $H_1 = H_2$ (as Lie groups).*

In fact, Prop. 2.7 shows that H_1 and H_2 have the same Lie algebra.

Corollary 2.9. *Let G be a Lie group and let K and H be two analytic subgroups of G. Assume $K \subset H$. Then the Lie group K is an analytic subgroup of the Lie group H.*

Let $\mathfrak{k}$ and $\mathfrak{h}$ denote the Lie algebras of K and H. Then $\mathfrak{k} \subset \mathfrak{h}$ by Prop. 2.7. Let K^* denote the analytic subgroup of H with Lie algebra $\mathfrak{k}$. Then the analytic subgroups K and K^* of G have the same Lie algebra. By Theorem 2.1 the Lie groups K and K^* coincide.

Let $\boldsymbol{S}^1$ denote the unit circle and $\boldsymbol{T}$ the group $\boldsymbol{S}^1 \times \boldsymbol{S}^1$. Let $t \to \gamma(t)$ ($t \in \boldsymbol{R}$) be a continuous one-to-one homomorphism of $\boldsymbol{R}$ into $\boldsymbol{T}$. If we carry the analytic structure of $\boldsymbol{R}$ over by the homomorphism we obtain a Lie subgroup $\Gamma = \gamma(\boldsymbol{R})$ of $\boldsymbol{T}$. This Lie subgroup is neither closed in $\boldsymbol{T}$ nor a topological subgroup of $\boldsymbol{T}$. We shall now see that these anomalies go together.

Theorem 2.10. *Let G be a Lie group and H a Lie subgroup of G.*

(i) *If H is a topological subgroup of G then H is closed in G.*

(ii) *If H has at most countably many components and is closed in G then H is a topological subgroup of G.*

Part (i) is contained in a more general result.

Proposition 2.11. *Let G be a topological group and H a subgroup which in the relative topology is locally compact. Then H is closed in G. In particular, if H, in the relative topology, is discrete, then it is closed.*

Proof. We first construct, without using the group structure, an open set $V \subset G$ whose intersection with the closure $\bar{H}$ is H. Let $h \in H$, let U_h be a compact neighborhood of h in H, and V_h a neighborhood of h in G such that $V_h \cap H = U_h$. Let $\mathring{V}_h$ be the interior of V_h. If $g \in \mathring{V}_h \cap \bar{H}$ and N_g is any neighborhood of g in G, then N_g intersects the closed set $V_h \cap H$ $(N_g \cap (V_h \cap H) \supset (N_g \cap \mathring{V}_h) \cap H \neq \emptyset)$, whence $g \in V_h \cap H$. Thus $\mathring{V}_h \cap \bar{H} \subset V_h \cap H$ so $\mathring{V}_h \cap \bar{H} = \mathring{V}_h \cap H$. Taking $V = \bigcup_{h \in H} \mathring{V}_h$, we have $H = \bar{H} \cap V$.

Let $b \in H$ and W be a neighborhood of e in G such that $bW \subset V$. If $a \in \bar{H}$, $aW^{-1} \cap H$ contains an element c; hence $bc^{-1}a \subset bW \subset V$. Also $bc^{-1} \in H$, so $bc^{-1}a \in \bar{H}$. Thus $bc^{-1}a \in V \cap \bar{H} = H$, so $a \in H$. Q.E.D.

(ii) H being closed in G it has by Theorem 2.3 an analytic structure in which it is a topological Lie subgroup of G. Let H' denote this Lie subgroup. Then the identity mapping $I : H \to H'$ is continuous. Each component of H lies in a component of H'. Since H has countably many components the same holds for H'. Now (ii) follows from Cor. 2.8.

§ 3. Lie Transformation Groups

Let M be a Hausdorff space and G a topological group such that to each $g \in G$ is associated a homeomorphism $p \to g \cdot p$ of M onto itself such that

(1) $g_1 g_2 \cdot p = g_1 \cdot (g_2 \cdot p)$ for $p \in M$, $g_1, g_2 \in G$;

(2) the mapping $(g, p) \to g \cdot p$ is a continuous mapping of the product space $G \times M$ onto M.

The group G is then called a *topological transformation group* of M. From (1) follows that $e \cdot p = p$ for all $p \in M$. If e is the only element of G which leaves each $p \in M$ fixed then G is called *effective* and is said to act effectively on M.

Example. Suppose A is a topological group and F a closed subgroup of A. The system of left cosets aF, $a \in A$ is denoted A/F; let π denote the natural mapping of A onto A/F. The set A/F can be given a topology, the *natural topology*, which is uniquely determined by the condition that π is a continuous and open mapping. This makes A/F a Hausdorff space and it is not difficult to see that if to each $a \in A$ we assign the mapping $\tau(a) : bF \to abF$, then A is a topological transformation group of A/F. The group A is effective if and only if F contains no normal subgroup of A. The coset space A/F is a *homogeneous space*, that is, has a transitive group of homeomorphisms, namely $\tau(A)$. Theorem 3.2 below deals with the converse question, namely that of representing a homogeneous space by means of a coset space.

Lemma 3.1 (the category theorem). *If a locally compact space M is a countable union*

$$M = \bigcup_{n=1}^{\infty} M_n,$$

where each M_n is a closed subset, then at least one M_n contains an open subset of M.

Proof. Suppose no M_n contains an open subset of M. Let U_1 be an open subset of M whose closure $\bar{U}_1$ is compact. Select successively

$a_1 \in U_1 - M_1$ and a neighborhood U_2 of a_1 such that $\bar{U}_2 \subset U_1$ and $\bar{U}_2 \cap M_1 = \emptyset$;

$a_2 \in U_2 - M_2$ and a neighborhood U_3 of a_2 such that $\bar{U}_3 \subset U_2$ and $\bar{U}_3 \cap M_2 = \emptyset$, etc.

Then $\bar{U}_1, \bar{U}_2, \ldots$ is a decreasing sequence of compact sets $\neq \emptyset$. Thus there is a point $b \in M$ in common to all $\bar{U}_n$. But this implies $b \notin M_n$ for each n which is a contradiction.

Theorem 3.2. *Let G be a locally compact group with a countable base. Suppose G is a transitive topological transformation group of a locally compact Hausdorff space M. Let p be any point in M and H the subgroup of G which leaves p fixed. Then H is closed and the mapping*

$$gH \to g \cdot p$$

is a homeomorphism of G/H onto M.

Proof. Since the mapping $\varphi : g \to g \cdot p$ of G onto M is continuous, it follows that $H = \varphi^{-1}(p)$ is closed in G. The natural mapping $\pi : G \to G/H$ is open and continuous. Thus, in order to prove Theorem 3.2, it suffices to prove that φ is open. Let V be an open subset of G and g a point in V. Select a compact neighborhood U of e in G such that $U = U^{-1}$, $gU^2 \subset V$. There exists a sequence $(g_n) \subset G$ such that $G = \bigcup_n g_n U$. The group G being transitive, this implies $M = \bigcup_n g_n U \cdot p$. Each summand is compact, hence a closed subset of M. By the lemma above, some summand, and therefore $U \cdot p$, contains an inner point $u \cdot p$. Then p is an inner point of $u^{-1}U \cdot p \subset U^2 \cdot p$ and consequently $g \cdot p$ is an inner point of $V \cdot p$. This shows that the mapping φ is open.

Definition. The group H is called the *isotropy* group at p (or the isotropy subgroup of G at p).

Corollary 3.3. *Let G and X be two locally compact groups. Assume G has a countable base. Then every continuous homomorphism ψ of G onto X is open.*

In fact, if we associate to each $g \in G$ the homeomorphism $x \to \psi(g)\,x$ of X onto itself, then G becomes a transitive topological transformation group of X. If f denotes the identity element of X, the proof above shows that the mapping $g \to \psi(g)f$ of G onto X is open.

Let G be a Lie group and M a differentiable manifold. Suppose G is a topological transformation group of M; G is said to be a *Lie transformation group* of M if the mapping $(g, p) \to g \cdot p$ is a differentiable mapping of $G \times M$ onto M. It follows that for each $g \in G$ the mapping $p \to g \cdot p$ is a diffeomorphism of M onto itself.

Let G be a Lie transformation group of M. To each X in $\mathfrak{g}$, the Lie algebra of G, we can associate a vector field X^+ on M by the formula

$$[X^+f](p) = \lim_{t\to 0} \frac{f(\exp tX \cdot p) - f(p)}{t}$$

for $f \in C^\infty(M)$, $p \in M$. The existence of X^+ follows from the fact that the mapping $(g, p) \to g \cdot p$ is a differentiable mapping of $G \times M$ onto M. It is also easy to check that X^+ is a derivation of $C^\infty(M)$. It is called the vector field on M *induced* by the one-parameter subgroup $\exp tX$, $t \in \boldsymbol{R}$.

Theorem 3.4. *Let G be a Lie transformation group of M. Let X, Y be in $\mathfrak{g}$, the Lie algebra of G, and let X^+, Y^+ be the vector fields on M induced by $\exp tX$ and $\exp tY$ ($t \in \boldsymbol{R}$). Then*†

$$[X^+, Y^+] = -[X, Y]^+. \tag{1}$$

We first prove a lemma which also shows what would have happened had we used right translation $R_\rho : g \to g\rho$ instead of left translation in the definition of the Lie algebra.

Lemma 3.5. *Let $\tilde{X}$ and $\tilde{Y}$ denote the right invariant vector fields on G such that $\tilde{X}_e = X$, $\tilde{Y}_e = Y$. Then*

$$[\tilde{X}, \tilde{Y}] = -[X, Y]^\sim. \tag{2}$$

Proof. In analogy with (5), §1 we have

$$(\tilde{X}f)(g) = \left\{\frac{d}{dt} f(\exp(tX)g)\right\}_{t=0} \tag{3}$$

for $f \in C^\infty(G)$. Then if J denotes the diffeomorphism $g \to g^{-1}$ of G,

$$dJ_g(\tilde{X}_g)f = \left\{\frac{d}{dt}(f \circ J)(g \exp tX)\right\}_{t=0} = -X(f \circ R_{g^{-1}}) = -\tilde{X}_{g^{-1}}f.$$

Thus $dJ(\tilde{X}) = -\tilde{X}$, so (2) follows from Prop. 3.3, Chapter I.

For (1) fix $p \in M$ and the map $\Phi : g \in G \to g \cdot p \in M$. Then

$$d\Phi_g(\tilde{X}_g)f = \tilde{X}_g(f \circ \Phi) = \left\{\frac{d}{dt} f(\exp(tX)g \cdot p)\right\}_{t=0}.$$

Thus $d\Phi(\tilde{X}) = X^+$, so $d\Phi([\tilde{X}, \tilde{Y}]) = [X^+, Y^+]$ and (2) implies (1).

† See §8 for an application.

§ 4. Coset Spaces and Homogeneous Spaces

Let G be a Lie group and H a closed subgroup. The group H will always be given the analytic structure from Theorem 2.3. Let $\mathfrak{g}$ and $\mathfrak{h}$ denote the Lie algebras of G and H, respectively, and let $\mathfrak{m}$ denote some vector subspace of $\mathfrak{g}$ such that $\mathfrak{g} = \mathfrak{m} + \mathfrak{h}$ (direct sum). Let π be the natural mapping of G onto the space G/H of left cosets gH, $g \in G$. As usual we give G/H the natural topology determined by the requirement that π should be continuous and open. We put $p_0 = \pi(e)$ and let ψ denote the restriction of exp to $\mathfrak{m}$.

Lemma 4.1. *There exists a neighborhood U of 0 in $\mathfrak{m}$ which is mapped homeomorphically under ψ and such that π maps $\psi(U)$ homeomorphically onto a neighborhood of p_0 in G/H.*

Proof. Let $U_{\mathfrak{m}}$, $U_{\mathfrak{h}}$ have the property described in Lemma 2.4 for $\mathfrak{h} = \mathfrak{n}$. Then since H has the relative topology of G, we can select a neighborhood V of e in G such that $V \cap H = \exp U_{\mathfrak{h}}$. Let U be a compact neighborhood of 0 in $U_{\mathfrak{m}}$ such that $\exp(-U)\exp U \subset V$. Then ψ is a homeomorphism of U onto $\psi(U)$. Moreover, π is one-to-one on $\psi(U)$ because if $X', X'' \in U$ satisfy $\pi(\exp X') = \pi(\exp X'')$, then $\exp(-X'')\exp X' \subset V \cap H$ so $\exp X' = \exp X'' \exp Z$ where $Z \in U_{\mathfrak{h}}$. From Lemma 2.4 we can conclude that $X' = X''$, $Z = 0$; consequently, π is one-to-one on $\psi(U)$, hence a homeomorphism. Finally, $U \times U_{\mathfrak{h}}$ is a neighborhood of $(0, 0)$ in $U_{\mathfrak{m}} \times U_{\mathfrak{h}}$; hence $\exp U \exp U_{\mathfrak{h}}$ is a neighborhood of e in G and since π is an open mapping, the set $\pi(\exp U \exp U_{\mathfrak{h}}) = \pi(\psi(U))$ is a neighborhood of p_0 in G/H. This proves the lemma. The set $\psi(U)$ will be referred to as a *local cross section.*

Let N_0 denote the interior of the set $\pi(\psi(U))$ and let $X_1, \ldots, X_r$ be a basis of $\mathfrak{m}$. If $g \in G$, then the mapping

$$\pi(g \exp(x_1 X_1 + \ldots + x_r X_r)) \rightarrow (x_1, \ldots, x_r)$$

is a homeomorphism of the open set $g \cdot N_0$ onto an open subset of $\boldsymbol{R}^r$. It is easy to see† that with these charts, G/H is an analytic manifold. Moreover, if $x \in G$, the mapping $\tau(x) : yH \rightarrow xyH$ is an analytic diffeomorphism of G/H.

Theorem 4.2. *Let G be a Lie group, H a closed subgroup of G, G/H the space of left cosets gH with the natural topology. Then G/H has a unique analytic structure with the property that G is a Lie transformation group of G/H.*

We use the notation above and let $B = \psi(\mathring{U})$ where $\mathring{U}$ is the interior of U. Remembering that the mapping Φ in Lemma 2.4 is a diffeo-

† See Exercise C.4.

morphism, the set B is a submanifold of G. The mappings in the diagram

$$\begin{array}{ccc} G \times B & \xrightarrow{\Phi} & G \\ {\scriptstyle I \times \pi} \downarrow & & \downarrow {\scriptstyle \pi} \\ G \times N_0 & & G/H \end{array} \quad \text{are:} \quad \begin{array}{ll} I \times \pi : (g, x) \to (g, xH), & g \in G, x \in B; \\ \Phi : (g, x) \to gx, & g \in G, x \in B. \end{array}$$

Then the mapping $(g, xH) \to gxH$ of $G \times N_0$ onto G/H can be written $\pi \circ \Phi \circ (I \times \pi)^{-1}$ which is analytic. Thus G is a Lie transformation group of G/H. The uniqueness results from the following proposition which should be compared with Theorem 3.2.

Proposition 4.3. *Let G be a transitive Lie transformation group of a C^∞ manifold M. Let p_0 be a point in M and let G_{p_0} denote the subgroup of G that leaves p_0 fixed. Then G_{p_0} is closed. Let α denote the mapping $gG_{p_0} \to g \cdot p_0$ of G/G_{p_0} onto M.*

(a) *If α is a homeomorphism, then it is a diffeomorphism (G/G_{p_0} having the analytic structure defined above).*

(b) *Suppose α is a homeomorphism and that M is connected. Then G_0, the identity component of G, is transitive on M.*

Proof. (a) We put $H = G_{p_0}$ and use Lemma 4.1. Let B and N_0 have the same meaning as above. Then B is a submanifold of G, diffeomorphic to N_0 under π. Let i denote the identity mapping of B into G and let β denote the mapping $g \to g \cdot p_0$ of G onto M. By assumption, α_{N_0}, the restriction of α to N_0, is a homeomorphism of N_0 onto an open subset of M. The mapping α is differentiable since $\alpha_{N_0} = \beta \circ i \circ \pi^{-1}$. To show that α^{-1} is differentiable, we begin by showing that the Jacobian of β at $g = e$ has rank r_β equal to dim M.

$$\begin{array}{ccc} B & \xrightarrow{i} & G \\ {\scriptstyle \pi} \downarrow & & \downarrow {\scriptstyle \beta} \\ N_0 & \xrightarrow{\alpha} & M \end{array}$$

The mapping $d\beta_e$ is a linear mapping of $\mathfrak{g}$ into M_{p_0}. Suppose X is in the kernel of $d\beta_e$. Then if $f \in C^\infty(M)$, we have

$$0 = (d\beta_e X) f = X(f \circ \beta) = \left\{ \frac{d}{dt} f(\exp tX \cdot p_0) \right\}_{t=0}. \tag{1}$$

Let $s \in \mathbf{R}$; we use (1) on the function $f^*(q) = f(\exp sX \cdot q)$, $q \in M$. Then

$$0 = \left\{ \frac{d}{dt} f^*(\exp tX \cdot p_0) \right\}_{t=0} = \left\{ \frac{d}{dt} f(\exp tX \cdot p_0) \right\}_{t=s},$$

which shows that $f(\exp sX \cdot p_0)$ is constant in s. Since f is arbitrary, we have $\exp sX \cdot p_0 = p_0$ for all s so $X \in \mathfrak{h}$. On the other hand, it is obvious that $d\beta_e$ vanishes on $\mathfrak{h}$ so $\mathfrak{h} = \text{kernel}\,(d\beta_e)$. Hence $r_\beta = \dim \mathfrak{g} - \dim \mathfrak{h}$. But α is a homeomorphism, so the topological invariance of dimension implies $\dim G/H = \dim M$. Thus $r_\beta = \dim M$, α^{-1} is differentiable at p and, by translation, on M. This proves (a).

(b) If α is a homeomorphism, β above is an open mapping. There exists a subset $\{x_\gamma : \gamma \in I\}$ of G such that $G = \bigcup_{\gamma \in I} G_0 x_\gamma$. Each orbit $G_0 x_\gamma \cdot p_0$ is an open subset of M; two orbits $G_0 x_\gamma \cdot p_0$ and $G_0 x_{\gamma'} \cdot p_0$ are either disjoint or equal. Therefore, since M is connected, all orbits must coincide and (b) follows.

Definition. In the sequel the coset space G/H (G a Lie group, H a closed subgroup) will always be taken with the analytic structure described in Theorem 4.2. If $x \in G$, the diffeomorphism $yH \to xyH$ of G/H onto itself will be denoted by $\tau(x)$. The group H is called the *isotropy group*. The group H^* of linear transformations $(d\tau(h))_{\pi(e)}$, $(h \in H)$, is called the *linear isotropy group*.

Let N be a Lie subgroup of G. Then the subset $N \cap H$ is closed in N and the coset space $N/N \cap H$ is in one-to-one correspondence with the orbit of $\pi(e)$ in G/H under N. If $\mathfrak{n}$ and $\mathfrak{h}_1$ denote the Lie algebras of N and $N \cap H$, respectively, then $\mathfrak{h}_1 = \mathfrak{h} \cap \mathfrak{n}$ by (2), §2, and Theorem 2.3.

Proposition 4.4.

(a) *The orbit $N/N \cap H$ is a submanifold of G/H.*

(b) *If N is a topological subgroup of G, and if H is compact, then the submanifold $N/N \cap H$ is a closed topological subspace of G/H.*

Proof. (a) We consider the commutative diagram

$$\begin{array}{ccc} N & \xrightarrow{\;\;i\;\;} & G \\ {\scriptstyle \pi_1}\big\downarrow & & \big\downarrow{\scriptstyle \pi} \\ N/N \cap H & \xrightarrow{\;\;I\;\;} & G/H \end{array}$$

where π_1 and π are the natural mappings of N onto $N/N \cap H$ and of G onto G/H, respectively. The identity mapping of N into G is denoted by i and I denotes the mapping $n(N \cap H) \to i(n)H$ of $N/N \cap H$ into G/H. Let $\mathfrak{n}_1$ be a complementary subspace of $\mathfrak{h}_1$ in $\mathfrak{n}$ and $\mathfrak{g}_1$ a complementary subspace of $\mathfrak{h} + \mathfrak{n}_1$ in $\mathfrak{g}$. We use Lemma 4.1 on the decompositions $\mathfrak{n} = \mathfrak{h}_1 + \mathfrak{n}_1$ and $\mathfrak{g} = \mathfrak{h} + (\mathfrak{n}_1 + \mathfrak{g}_1)$. We can then get submanifolds $B_N \subset N$, $B_G \subset G$ through e which π_1 and π map diffeomorphically onto open neighborhoods of $\pi_1(e)$ in $N/N \cap H$ and of $\pi(e)$ in G/H, respectively. Here we can take B_N as a submanifold of B_G. Put $V_1 = \pi_1(B_N)$, $V = \pi(B_G)$. The restriction of I to V_1, say I_{V_1}, can be written $I_{V_1} = \pi \circ i \circ \pi_1^{-1}$. The Jacobian of I_{V_1} at $\pi_1(e)$ therefore has rank equal to dim $(N/N \cap H)$. Hence I is regular, proving (a).

(b) Since N is now a topological subgroup of G, the diagram above shows that I is a homeomorphism of $N/N \cap H$ into G/H. In order to show that $N/N \cap H$ is closed, let (p_k) be a sequence in $N/N \cap H$ converging to a point $q \in G/H$. Select $g \in G$ such that $\pi(g) = q$. We may assume that all p_k belong to the neighborhood $g \cdot V$ of q. Hence there is a unique element $g_k \in gB_G$ such that $\pi(g_k) = p_k$. Since π is a homeomorphism of gB_G onto $g \cdot V$, we have $\lim g_k = g$.

On the other hand, for each index k there exists an element $n_k \in N$ such that $\pi_1(n_k) = p_k$. Thus there exists an element $h_k \in H$ such that $g_k = n_k h_k$. Since H is compact we may assume that (h_k) is a convergent sequence. It follows that the sequence (n_k) is convergent; the limit n^* lies in N since N is closed in G. Consequently, $\pi_1(n^*) = q$ so the orbit $N/N \cap H$ is closed.

Remark. Prop. 4.4 holds in greater generality; see Exercise C.5 following this chapter and Exercise A.5, Chapter IV.

§ 5. The Adjoint Group

Let $\mathfrak{a}$ be a Lie algebra over $\boldsymbol{R}$. Let $\boldsymbol{GL}(\mathfrak{a})$ as usual denote the group of all nonsingular endomorphisms of $\mathfrak{a}$. We recall that an endomorphism of a vector space V (in particular of a Lie algebra) simply means a linear mapping of V into itself. The Lie algebra $\mathfrak{gl}(\mathfrak{a})$ of $\boldsymbol{GL}(\mathfrak{a})$ consists of the vector space of all endomorphisms of $\mathfrak{a}$ with the bracket operation $[A, B] = AB - BA$. The mapping $X \to \operatorname{ad} X$, $X \in \mathfrak{a}$ is a homomorphism of $\mathfrak{a}$ onto a subalgebra ad $(\mathfrak{a})$ of $\mathfrak{gl}(\mathfrak{a})$. Let Int $(\mathfrak{a})$ denote the analytic subgroup of $\boldsymbol{GL}(\mathfrak{a})$ whose Lie algebra is ad $(\mathfrak{a})$; Int $(\mathfrak{a})$ is called the *adjoint group* of $\mathfrak{a}$.

The group Aut $(\mathfrak{a})$ of all automorphisms of $\mathfrak{a}$ is a closed subgroup of $\boldsymbol{GL}(\mathfrak{a})$. Thus Aut $(\mathfrak{a})$ has a unique analytic structure in which it is a topological Lie subgroup of $\boldsymbol{GL}(\mathfrak{a})$. Let $\partial(\mathfrak{a})$ denote the Lie algebra of Aut $(\mathfrak{a})$. From §2 we know that $\partial(\mathfrak{a})$ consists of all endomorphisms D of $\mathfrak{a}$ such that $e^{tD} \in$ Aut $(\mathfrak{a})$ for each $t \in \boldsymbol{R}$. Let $X, Y \in \mathfrak{a}$. The relation $e^{tD}[X, Y] = [e^{tD}X, e^{tD}Y]$ for all $t \in \boldsymbol{R}$ implies

$$D[X, Y] = [DX, Y] + [X, DY]. \tag{1}$$

An endomorphism D of $\mathfrak{a}$ satisfying (1) for all $X, Y \in \mathfrak{a}$ is called a *derivation* of $\mathfrak{a}$. By induction we get from (1)

$$D^k[X, Y] = \sum_{i+j=k} \frac{k!}{i!j!} [D^iX, D^jY], \qquad i \geqslant 0, j \geqslant 0, \tag{2}$$

where D^0 means the identity mapping of $\mathfrak{a}$. From (2) follows that $e^{tD}[X, Y] = [e^{tD}X, e^{tD}Y]$ and thus $\partial(\mathfrak{a})$ *consists of all derivations of* $\mathfrak{a}$. Using the Jacobi identity we see that $\mathrm{ad}(\mathfrak{a}) \subset \partial(\mathfrak{a})$ and therefore $\mathrm{Int}(\mathfrak{a}) \subset \mathrm{Aut}(\mathfrak{a})$. The elements of $\mathrm{ad}(\mathfrak{a})$ and $\mathrm{Int}(\mathfrak{a})$, respectively, are called the *inner derivations* and *inner automorphisms* of $\mathfrak{a}$. Since $\mathrm{Aut}(\mathfrak{a})$ is a topological subgroup of $\boldsymbol{GL}(\mathfrak{a})$ the identity mapping of $\mathrm{Int}(\mathfrak{a})$ into $\mathrm{Aut}(\mathfrak{a})$ is continuous. In view of Lemma 14.1, Chapter I, $\mathrm{Int}(\mathfrak{a})$ is a Lie subgroup of $\mathrm{Aut}(\mathfrak{a})$. We shall now prove that $\mathrm{Int}(\mathfrak{a})$ is a normal subgroup of $\mathrm{Aut}(\mathfrak{a})$. Let $s \in \mathrm{Aut}(\mathfrak{a})$. Then the mapping $\sigma : g \to sgs^{-1}$ is an automorphism of $\mathrm{Aut}(\mathfrak{a})$, and $(d\sigma)_e$ is an automorphism of $\partial(\mathfrak{a})$. If A, B are endomorphisms of a vector space and A^{-1} exists, then $Ae^BA^{-1} = e^{ABA^{-1}}$. Considering Lemma 1.12 we have

$$(d\sigma)_e D = sDs^{-1} \qquad \text{for } D \in \partial(\mathfrak{a}).$$

If $X \in \mathfrak{a}$, we have $s\,\mathrm{ad}\,X\,s^{-1} = \mathrm{ad}(s \cdot X)$, so

$$(d\sigma)_e\,\mathrm{ad}\,X = \mathrm{ad}(s \cdot X),$$

and consequently

$$\sigma \cdot e^{\mathrm{ad}\,X} = e^{\mathrm{ad}(s \cdot X)} \qquad (X \in \mathfrak{a}).$$

Now, the group $\mathrm{Int}(\mathfrak{a})$ is connected, so it is generated by the elements $e^{\mathrm{ad}\,X}$, $X \in \mathfrak{a}$. It follows that $\mathrm{Int}(\mathfrak{a})$ is a normal subgroup of $\mathrm{Aut}(\mathfrak{a})$ and the automorphism s of $\mathfrak{a}$ induces the analytic isomorphism $g \to sgs^{-1}$ of $\mathrm{Int}(\mathfrak{a})$ onto itself.

More generally, if s is an isomorphism of a Lie algebra $\mathfrak{a}$ onto a Lie algebra $\mathfrak{b}$ (both Lie algebras over $\boldsymbol{R}$) then the mapping $g \to sgs^{-1}$ is an isomorphism of $\mathrm{Aut}(\mathfrak{a})$ onto $\mathrm{Aut}(\mathfrak{b})$ which maps $\mathrm{Int}(\mathfrak{a})$ onto $\mathrm{Int}(\mathfrak{b})$.

Let G be a Lie group. If $\sigma \in G$, the mapping $I(\sigma) : g \to \sigma g \sigma^{-1}$ is an analytic isomorphism of G onto itself. We put $\mathrm{Ad}(\sigma) = dI(\sigma)_e$. Sometimes we write $\mathrm{Ad}_G(\sigma)$ instead of $\mathrm{Ad}(\sigma)$ when a misunderstanding might otherwise arise. The mapping $\mathrm{Ad}(\sigma)$ is an automorphism of $\mathfrak{g}$, the Lie algebra of G. We have by Lemma 1.12

$$\exp \mathrm{Ad}(\sigma) X = \sigma \exp X \sigma^{-1} \qquad \text{for } \sigma \in G,\ X \in \mathfrak{g}. \tag{3}$$

The mapping $\sigma \to \mathrm{Ad}(\sigma)$ is a homomorphism of G into $\boldsymbol{GL}(\mathfrak{g})$. This homomorphism is called the *adjoint representation* of G. Let us prove that this homomorphism is analytic. For this it suffices to prove that for each $X \in \mathfrak{g}$ and each linear function ω on $\mathfrak{g}$ the function $\sigma \to \omega(\mathrm{Ad}(\sigma) X)$ $(\sigma \in G)$, is analytic at $\sigma = e$. Select $f \in C^\infty(G)$ such that f is analytic at $\sigma = e$ and such that $Yf = \omega(Y)$ for all $Y \in \mathfrak{g}$.

Then, using (3), we obtain

$$\omega(\mathrm{Ad}\,(\sigma)\,X) = (\mathrm{Ad}\,(\sigma)\,X)f = \left\{\frac{d}{dt}f(\sigma \exp tX\sigma^{-1})\right\}_{t=0},$$

which is clearly analytic at $\sigma = e$.

Next, let X and Y be arbitrary vectors in $\mathfrak{g}$. From Lemma 1.8 (iii) we have

$$\exp\,(\mathrm{Ad}\,(\exp tX)\,tY) = \exp\,(tY + t^2[X, Y] + O(t^3)).$$

It follows that

$$\mathrm{Ad}\,(\exp tX)\,Y = Y + t[X, Y] + O(t^2). \tag{4}$$

The differential $d\,\mathrm{Ad}_e$ is a homomorphism of $\mathfrak{g}$ into $\mathfrak{gl}(\mathfrak{g})$ and due to (4) we have

$$d\,\mathrm{Ad}_e(X) = \mathrm{ad}\,X, \qquad X \in \mathfrak{g}.$$

Applying the exponential mapping on both sides we obtain (Lemma 1.12)

$$\mathrm{Ad}\,(\exp X) = e^{\mathrm{ad}\,X}, \qquad X \in \mathfrak{g}. \tag{5}$$

Let G be a connected Lie group and H an analytic subgroup. Let $\mathfrak{g}$ and $\mathfrak{h}$ denote the corresponding Lie algebras. Relations (3) and (5) show that H is a normal subgroup of G if and only if $\mathfrak{h}$ is an ideal in $\mathfrak{g}$.

Lemma 5.1. *Let G be a connected Lie group with Lie algebra $\mathfrak{g}$ and let φ be an analytic homomorphism of G into a Lie group X with Lie algebra $\mathfrak{x}$. Then:*

(i) *The kernel $\varphi^{-1}(e)$ is a topological Lie subgroup of G. Its Lie algebra is the kernel of $d\varphi$ $(= d\varphi_e)$.*

(ii) *The image $\varphi(G)$ is a Lie subgroup of X with Lie algebra $d\varphi(\mathfrak{g}) \subset \mathfrak{x}$.*

(iii) *The factor group $G/\varphi^{-1}(e)$ with its natural analytic structure is a Lie group and the mapping $g\varphi^{-1}(e) \to \varphi(g)$ is an analytic isomorphism of $G/\varphi^{-1}(e)$ onto $\varphi(G)$. In particular, the mapping $\varphi : G \to \varphi(G)$ is analytic.*

Proof. (i) According to Theorem 2.3, $\varphi^{-1}(e)$ has a unique analytic structure with which it is a topological Lie subgroup of G. Moreover, its Lie algebra contains a vector $Z \in \mathfrak{g}$ if and only if $\varphi(\exp tZ) = e$ for all $t \in \boldsymbol{R}$. Since $\varphi(\exp tZ) = \exp td\varphi(Z)$, the condition is equivalent to $d\varphi(Z) = 0$.

(ii) Let X_1 denote the analytic subgroup of X with Lie algebra $d\varphi(\mathfrak{g})$. The group $\varphi(G)$ is generated by the elements $\varphi(\exp Z)$, $Z \in \mathfrak{g}$. The group X_1 is generated by the elements $\exp\,(d\varphi(Z))$, $Z \in \mathfrak{g}$. Since $\varphi(\exp Z) = \exp d\varphi(Z)$ it follows that $\varphi(G) = X_1$.

(iii) Let H be any closed normal subgroup of G. Then H is a topological Lie subgroup and the factor group G/H has a unique analytic structure such that the mapping $(g, xH) \to g\,xH$ is an analytic mapping of $G \times G/H$ onto G/H. In order to see that G/H is a Lie group in this analytic structure we use the local cross section $\psi(U)$ from Lemma 4.1. Let $B = \psi(\mathring{U})$ where $\mathring{U}$ is the interior of U. In the commutative diagram

$$\begin{array}{ccc} G \times G/H & \xrightarrow{\ \Phi\ } & G/H \\ & {\scriptstyle \pi \times I} \searrow \quad \nearrow {\scriptstyle \alpha} & \\ & G/H \times G/H & \end{array}$$

the symbols Φ, $\pi \times I$, and α denote the mappings:

$$\begin{aligned} \Phi &: (g, xH) \to g^{-1}xH, && x, g \in G; \\ \pi \times I &: (g, xH) \to (gH, xH), && x, g \in G; \\ \alpha &: (gH, xH) \to g^{-1}xH, && x, g \in G. \end{aligned}$$

The mapping α is well defined since H is a normal subgroup of G. Let g_0, x_0 be arbitrary two points in G. The restriction of $\pi \times I$ to $(g_0B) \times (G/H)$ is an analytic diffeomorphism of $(g_0B) \times (G/H)$ onto a neighborhood N of (g_0H, x_0H) in $G/H \times G/H$. On N we have $\alpha = \Phi \circ (\pi \times I)^{-1}$ which shows that α is analytic. Hence G/H is a Lie group.

Now choose for H the group $\varphi^{-1}(e)$ and let $\mathfrak{h}$ denote the Lie algebra of H. Then $\mathfrak{h} = d\varphi^{-1}(0)$ so $\mathfrak{h}$ is an ideal in $\mathfrak{g}$. By (ii) the Lie algebra of G/H is $d\pi(\mathfrak{g})$ which is isomorphic to the algebra $\mathfrak{g}/\mathfrak{h}$. On the other hand, the mapping $Z + \mathfrak{h} \to d\varphi(Z)$ is an isomorphism of $\mathfrak{g}/\mathfrak{h}$ onto $d\varphi(\mathfrak{g})$. The corresponding local isomorphism between G/H and $\varphi(G)$ coincides with the (algebraic) isomorphism $gH \to \varphi(g)$ on some neighborhood of the identity. It follows that this last isomorphism is analytic at e, hence everywhere.

Corollary 5.2. *Let G be a connected Lie group with Lie algebra $\mathfrak{g}$. Let Z denote the center of G. Then:*

(i) *Ad_G is an analytic homomorphism of G onto Int $(\mathfrak{g})$ with kernel Z.*

(ii) *The mapping $gZ \to Ad_G(g)$ is an analytic isomorphism of G/Z onto Int $(\mathfrak{g})$.*

In fact $\mathrm{Ad}_G(G) = \mathrm{Int}(\mathfrak{g})$ due to (5) and $\mathrm{Ad}_G^{-1}(e) = Z$ due to (3). The remaining statements are contained in Lemma 5.1.

Corollary 5.3. *Let $\mathfrak{g}$ be a Lie algebra over $\boldsymbol{R}$ with center $\{0\}$. Then the center of Int $(\mathfrak{g})$ consists of the identity element alone.*

In fact, let $G' = \text{Int}(\mathfrak{g})$ and let Z denote the center of G'. Let ad denote the adjoint representation of $\mathfrak{g}$ and let Ad′ and ad′ denote the adjoint representation of G' and ad ($\mathfrak{g}$), respectively. The mapping

$$\theta : gZ \to \text{Ad}'(g), \qquad g \in G',$$

is an isomorphism of G'/Z onto Int (ad ($\mathfrak{g}$)). On the other hand, the mapping $s : X \to \text{ad}\, X$ ($X \in \mathfrak{g}$) is an isomorphism of $\mathfrak{g}$ onto ad ($\mathfrak{g}$) and consequently the mapping $S : g \to s \circ g \circ s^{-1}$ ($g \in G'$) is an isomorphism of G' onto Int (ad ($\mathfrak{g}$)). Moreover, if $X \in \mathfrak{g}$, we obtain from (5)

$$S(e^{\text{ad}\, X}) = s \circ e^{\text{ad}\, X} \circ s^{-1} = e^{(\text{ad}'(\text{ad}\, X))} = \text{Ad}'(e^{\text{ad}\, X}),$$

ad ($\mathfrak{g}$) being the Lie algebra of G'. It follows that $S^{-1} \circ \theta$ is an isomorphism of G'/Z onto G', mapping gZ onto g ($g \in G'$). Obviously Z must consist of the identity element alone.

Remark. The conclusion of Cor. 5.3 does not hold in general, if $\mathfrak{g}$ has nontrivial center. Let, for example, $\mathfrak{g}$ be the three-dimensional Lie algebra $\mathfrak{g} = \boldsymbol{R}X_1 + \boldsymbol{R}X_2 + \boldsymbol{R}X_3$ with the bracket defined by: $[X_1, X_2] = X_3$, $[X_1, X_3] = [X_2, X_3] = 0$. Here $\mathfrak{g}$ is nonabelian, whereas Int ($\mathfrak{g}$) is abelian and has dimension 2.

Definition. Let $\mathfrak{g}$ be a Lie algebra over $\boldsymbol{R}$. Let $\mathfrak{k}$ be a subalgebra of $\mathfrak{g}$ and K^* the analytic subgroup of Int ($\mathfrak{g}$) which corresponds to the subalgebra $\text{ad}_{\mathfrak{g}}(\mathfrak{k})$ of $\text{ad}_{\mathfrak{g}}(\mathfrak{g})$. The subalgebra $\mathfrak{k}$ is called a *compactly imbedded subalgebra of* $\mathfrak{g}$ if K^* is compact. The Lie algebra $\mathfrak{g}$ is said to be *compact* if it is compactly imbedded in itself or equivalently if Int ($\mathfrak{g}$) is compact.

It should be observed that the topology of K^* might *a priori* differ from the relative topology of the group Int ($\mathfrak{g}$) which again might differ from the relative topology of $\boldsymbol{GL}(\mathfrak{g})$. The next proposition clarifies this situation.

Proposition 5.4. *Let $\tilde{K}$ denote the abstract group K^* with the relative topology of $\boldsymbol{GL}(\mathfrak{g})$. Then K^* is compact if and only if $\tilde{K}$ is compact.*

The identity mapping of K^* into $\boldsymbol{GL}(\mathfrak{g})$ is analytic, in particular, continuous. Thus $\tilde{K}$ is compact if K^* is compact. On the other hand, if $\tilde{K}$ is compact, then it is closed in $\boldsymbol{GL}(\mathfrak{g})$; by Theorem 2.10, K^* and $\tilde{K}$ are homeomorphic.

Remark. Suppose G is any connected Lie group with Lie algebra $\mathfrak{g}$. Let K be the analytic subgroup of G with Lie algebra $\mathfrak{k}$. Then the group K^* above coincides with $\text{Ad}_G(K)$; in fact, both groups are generated by $\text{Ad}_G(\exp X)$, $X \in \mathfrak{k}$.

§ 6. Semisimple Lie Groups

Let $\mathfrak{g}$ be a Lie algebra over a field of characteristic 0. Denoting by Tr the trace of a vector space endomorphism we consider the bilinear form $B(X, Y) = \text{Tr}(\text{ad } X \text{ ad } Y)$ on $\mathfrak{g} \times \mathfrak{g}$. The form B is called the *Killing form* of $\mathfrak{g}$. It is clearly symmetric.

For a subspace $\mathfrak{a} \subset \mathfrak{g}$ let $\mathfrak{a}^\perp = \{X \in \mathfrak{g} : B(X, \mathfrak{a}) = 0\}$. The map $X \to X^*$ of $\mathfrak{g} \to \mathfrak{g}^\wedge$ (dual of $\mathfrak{g}$) given by $X^*(Y) = B(X, Y)$ has kernel $\mathfrak{g}^\perp$, so by $\dim \mathfrak{a}^\perp = \dim \mathfrak{g} - \dim \mathfrak{a}^*$,

$$\dim \mathfrak{a} + \dim \mathfrak{a}^\perp = \dim \mathfrak{g} + \dim(\mathfrak{a} \cap \mathfrak{g}^\perp). \tag{1}$$

If σ is an automorphism of $\mathfrak{g}$, then $\text{ad}(\sigma X) = \sigma \circ \text{ad } X \circ \sigma^{-1}$ so by $\text{Tr}(AB) = \text{Tr}(BA)$, we have

$$B(\sigma X, \sigma Y) = B(X, Y), \qquad \sigma \in \text{Aut}(\mathfrak{g}),$$
$$B(X, [Y, Z]) = B(Y, [Z, X]) = B(Z, [X, Y]), \qquad X, Y, Z \in \mathfrak{g}. \tag{2}$$

Suppose $\mathfrak{a}$ is an ideal in $\mathfrak{g}$. Then it is easily verified that the Killing form of $\mathfrak{a}$ coincides with the restriction of B to $\mathfrak{a} \times \mathfrak{a}$.

Definition. A Lie algebra $\mathfrak{g}$ over a field of characteristic 0 is called *semisimple* if the Killing B of $\mathfrak{g}$ is nondegenerate. We shall call a Lie algebra $\mathfrak{g} \neq \{0\}$ *simple*† if it is semisimple and has no ideals except $\{0\}$ and $\mathfrak{g}$. A Lie group is called semisimple (simple) if its Lie algebra is semisimple (simple).

Proposition 6.1. *Let $\mathfrak{g}$ be a semisimple Lie algebra, $\mathfrak{a}$ an ideal in $\mathfrak{g}$. Let $\mathfrak{a}^\perp$ denote the set of elements $X \in \mathfrak{g}$ which are orthogonal to $\mathfrak{a}$ with respect to B. Then $\mathfrak{a}$ is semisimple, $\mathfrak{a}^\perp$ is an ideal and*

$$\mathfrak{g} = \mathfrak{a} + \mathfrak{a}^\perp \quad (\textit{direct sum}).$$

Proof. The fact that $\mathfrak{a}^\perp$ is an ideal is obvious from (2). Since B is nondegenerate, (1) implies $\dim \mathfrak{a} + \dim \mathfrak{a}^\perp = \dim \mathfrak{g}$. If $Z \in \mathfrak{g}$ and $X, Y \in \mathfrak{a} \cap \mathfrak{a}^\perp$, we have $B(Z, [X, Y]) = B([Z, X], Y) = 0$ so $[X, Y] = 0$. Hence $\mathfrak{a} \cap \mathfrak{a}^\perp$ is an abelian ideal in $\mathfrak{g}$. Let $\mathfrak{b}$ be any subspace of $\mathfrak{g}$ complementary to $\mathfrak{a} \cap \mathfrak{a}^\perp$. If $Z \in \mathfrak{g}$ and $T \in \mathfrak{a} \cap \mathfrak{a}^\perp$, then the endomorphism $\text{ad } T \text{ ad } Z$ maps $\mathfrak{a} \cap \mathfrak{a}^\perp$ into $\{0\}$, and $\mathfrak{b}$ into $\mathfrak{a} \cap \mathfrak{a}^\perp$. In particular, $\text{Tr}(\text{ad } T \text{ ad } Z) = 0$. It follows that $\mathfrak{a} \cap \mathfrak{a}^\perp = \{0\}$ and we get the direct decomposition $\mathfrak{g} = \mathfrak{a} + \mathfrak{a}^\perp$. Since the Killing form of $\mathfrak{a}$ is the restriction of B to $\mathfrak{a} \times \mathfrak{a}$, the semisimplicity of $\mathfrak{a}$ is obvious.

† This definition of a simple Lie algebra is convenient for our purposes but is formally different from the usual one: A Lie algebra $\mathfrak{g}$ is simple if it is nonabelian and has no ideals except $\{0\}$ and $\mathfrak{g}$. However, the two definitions are equivalent, cf. Exercise B.8, Chapter III.

Corollary 6.2. *A semisimple Lie algebra has center* $\{0\}$.

Corollary 6.3. *A semisimple Lie algebra* $\mathfrak{g}$ *is the direct sum*

$$\mathfrak{g} = \mathfrak{g}_1 + \dots + \mathfrak{g}_r,$$

where $\mathfrak{g}_i$ $(1 \leqslant i \leqslant r)$ *are all the simple ideals in* $\mathfrak{g}$. *Each ideal* $\mathfrak{a}$ *of* $\mathfrak{g}$ *is the direct sum of certain* $\mathfrak{g}_i$.

In fact, Prop. 6.1 implies that $\mathfrak{g}$ can be written as a direct sum of simple ideals $\mathfrak{g}_i$ $(1 \leqslant i \leqslant s)$ such that $\mathfrak{a}$ is the direct sum of certain of these $\mathfrak{g}_i$. If $\mathfrak{b}$ were a simple ideal which does not occur among the ideals $\mathfrak{g}_i$ $(1 \leqslant i \leqslant s)$, then $[\mathfrak{g}_i, \mathfrak{b}] \subset \mathfrak{g}_i \cap \mathfrak{b} = \{0\}$ for $1 \leqslant i \leqslant s$. This contradicts Cor. 6.2.

Proposition 6.4. *If* $\mathfrak{g}$ *is semisimple, then* $ad\,(\mathfrak{g}) = \partial(\mathfrak{g})$, *that is, every derivation is an inner derivation.*

Proof. The algebra $\text{ad}\,(\mathfrak{g})$ is isomorphic to $\mathfrak{g}$, hence semisimple. If D is a derivation of $\mathfrak{g}$ then $\text{ad}\,(DX) = [D, \text{ad}\, X]$ for $X \in \mathfrak{g}$, hence $\text{ad}\,(\mathfrak{g})$ is an ideal in $\partial(\mathfrak{g})$. Its orthogonal complement, say $\mathfrak{a}$, is also an ideal in $\partial(\mathfrak{g})$. Then $\mathfrak{a} \cap \text{ad}\,(\mathfrak{g})$ is orthogonal to $\text{ad}\,(\mathfrak{g})$ also with respect to the Killing form of $\text{ad}\,(\mathfrak{g})$, hence $\mathfrak{a} \cap \text{ad}\,(\mathfrak{g}) = \{0\}$. Consequently $D \in \mathfrak{a}$ implies $[D, \text{ad}\, X] \in \mathfrak{a} \cap \text{ad}\,(\mathfrak{g}) = \{0\}$. Thus $\text{ad}\,(DX) = 0$ for each $X \in \mathfrak{g}$, hence $D = 0$. Thus, $\mathfrak{a} = \{0\}$ so, by (1), $\text{ad}\,(\mathfrak{g}) = \partial(\mathfrak{g})$.

Corollary 6.5. *For a semisimple Lie algebra* $\mathfrak{g}$ *over* $\boldsymbol{R}$, *the adjoint group* $Int\,(\mathfrak{g})$ *is the identity component of* $Aut\,(\mathfrak{g})$. *In particular,* $Int\,(\mathfrak{g})$ *is a closed topological subgroup of* $Aut\,(\mathfrak{g})$.

Remark. If $\mathfrak{g}$ is not semisimple, the group $\text{Int}\,(\mathfrak{g})$ is not necessarily closed in $\text{Aut}\,(\mathfrak{g})$ (see Exercise D.3 for this chapter).

Proposition 6.6.

(i) *Let* $\mathfrak{g}$ *be a semisimple Lie algebra over* $\boldsymbol{R}$. *Then* $\mathfrak{g}$ *is compact if and only if the Killing form of* $\mathfrak{g}$ *is strictly negative definite.*

(ii) *Every compact Lie algebra* $\mathfrak{g}$ *is the direct sum* $\mathfrak{g} = \mathfrak{z} + [\mathfrak{g}, \mathfrak{g}]$ *where* $\mathfrak{z}$ *is the center of* $\mathfrak{g}$ *and the ideal* $[\mathfrak{g}, \mathfrak{g}]$ *is semisimple and compact.*

Proof. Suppose $\mathfrak{g}$ is a Lie algebra over $\boldsymbol{R}$ whose Killing form is strictly negative definite. Let $\boldsymbol{O}(B)$ denote the group of all linear transformations of $\mathfrak{g}$ which leave B invariant. Then $\boldsymbol{O}(B)$ is compact in the relative topology of $\boldsymbol{GL}(\mathfrak{g})$. We have $\text{Aut}\,(\mathfrak{g}) \subset \boldsymbol{O}(B)$, so by Cor. 6.5, $\text{Int}\,(\mathfrak{g})$ is compact.

Suppose now $\mathfrak{g}$ is an arbitrary compact Lie algebra. The Lie subgroup $\text{Int}\,(\mathfrak{g})$ of $\boldsymbol{GL}(\mathfrak{g})$ is compact; hence it carries the relative topology of

$GL(\mathfrak{g})$. There exists a strictly positive definite quadratic form Q on $\mathfrak{g}$ invariant under the action of the compact linear group Int $(\mathfrak{g})$. There exists a basis $X_1, ..., X_n$ of $\mathfrak{g}$ such that $Q(X) = \Sigma_{i=1}^n x_i^2$ if $X = \Sigma_{i=1}^n x_i X_i$. By means of this basis, each $\sigma \in$ Int $(\mathfrak{g})$ is represented by an orthogonal matrix and each ad X, $(X \in \mathfrak{g})$, by a skew symmetric matrix, say $(a_{ij}(X))$. Now the center $\mathfrak{z}$ of $\mathfrak{g}$ is invariant under Int $(\mathfrak{g})$, that is $\sigma \cdot \mathfrak{z} \subset \mathfrak{z}$ for each $\sigma \in$ Int $(\mathfrak{g})$. The orthogonal complement $\mathfrak{g}'$ of $\mathfrak{z}$ in $\mathfrak{g}$ with respect to Q is also invariant under Int $(\mathfrak{g})$ and under ad $(\mathfrak{g})$. Hence $\mathfrak{g}'$ is an ideal in $\mathfrak{g}$. This being so, the Killing form B' of $\mathfrak{g}'$ is the restriction to $\mathfrak{g}' \times \mathfrak{g}'$ of the Killing form B of $\mathfrak{g}$. Now, if $X \in \mathfrak{g}$

$$B(X, X) = \text{Tr}\,(\text{ad}\, X\, \text{ad}\, X) = \sum_{i,j} a_{ij}(X)\, a_{ji}(X) = -\sum_{i,j} a_{ij}(X)^2 \leqslant 0.$$

The equality sign holds if and only if ad $X = 0$, that is, if and only if $X \in \mathfrak{z}$. This proves that $\mathfrak{g}'$ is semisimple and compact. The decomposition in Cor. 6.3 shows that $[\mathfrak{g}', \mathfrak{g}'] = \mathfrak{g}'$. Hence $\mathfrak{g}' = [\mathfrak{g}, \mathfrak{g}]$ and the proposition is proved.

Corollary 6.7. *A Lie algebra $\mathfrak{g}$ over $\boldsymbol{R}$ is compact if and only if there exists a compact Lie group G with Lie algebra isomorphic to $\mathfrak{g}$.*

For this corollary one just has to remark that every abelian Lie algebra is isomorphic to the Lie algebra of a torus $\boldsymbol{S}^1 \times ... \times \boldsymbol{S}^1$.

Proposition 6.8. *Let $\mathfrak{g}$ be a Lie algebra over $\boldsymbol{R}$ and let $\mathfrak{z}$ denote the center of $\mathfrak{g}$. Suppose $\mathfrak{k}$ is a compactly imbedded subalgebra of $\mathfrak{g}$. If $\mathfrak{k} \cap \mathfrak{z} = \{0\}$ then the Killing form of $\mathfrak{g}$ is strictly negative definite on $\mathfrak{k}$.*

Proof. Let B denote the Killing form of $\mathfrak{g}$, and let K denote the analytic subgroup of the adjoint group Int $(\mathfrak{g})$ with Lie algebra $\text{ad}_{\mathfrak{g}}\,(\mathfrak{k})$. Owing to our assumptions, K is a compact Lie subgroup of $GL(\mathfrak{g})$; hence it carries the relative topology of $GL(\mathfrak{g})$. There exists a strictly positive definite quadratic form Q on $\mathfrak{g}$ invariant under K. There exists a basis of $\mathfrak{g}$ such that each endomorphism $\text{ad}_{\mathfrak{g}}\,(T)$ $(T \in \mathfrak{k})$ is expressed by means of a skew symmetric matrix, say $(a_{ij}(T))$. Then

$$B(T, T) = \sum_{i,j} a_{ij}(T)\, a_{ji}(T) = -\sum_{i,j} a_{ij}(T)^2 \leqslant 0,$$

and equality sign holds only if $T \in \mathfrak{z} \cap \mathfrak{k} = \{0\}$.

Theorem 6.9. *Let G be a compact, connected semisimple Lie group. Then the universal covering group G^* of G is compact.*

Proof. Let $\mathfrak{g}$ denote the Lie algebra of G (and G^*), and let B be the

Killing form of $\mathfrak{g}$. There exists unique left invariant Riemannian structures Q and Q^* on G and G^*, respectively, such that $Q_e = Q^*_{e*} = -B$. Here e and e^* denote the identity elements in G and G^*, respectively. Since

$$B(\mathrm{Ad}\,(g)\,X, \mathrm{Ad}(g)\,Y) = B(X, Y), \qquad X, Y \in \mathfrak{g}, g \in G,$$

it follows that Q and Q^* are also invariant under right translations on G and G^*. Let π denote the covering mapping of G^* onto G. Then $Q^* = \pi^*Q$ and since G is complete, the covering manifold G^* is also complete (Prop. 10.6, Chapter I). From (2) we have

$$Q^*([\tilde{Z}, \tilde{X}], \tilde{Y}) + Q^*(\tilde{X}, [\tilde{Z}, \tilde{Y}]) = 0,$$

where $\tilde{X}$, $\tilde{Y}$, and $\tilde{Z}$ are the left invariant vector fields on G^* which have values X, Y, Z at e. Let ∇ denote the Riemannian connection on G^* induced by Q^* (Theorem 9.1, Chapter I). Then we see from (2), §9, Chapter I, that $\nabla_{\tilde{X}}(\tilde{X}) = 0$ for all $X \in \mathfrak{g}$. From Prop. 1.4 we deduce that the geodesics in G^* through e^* are the one-parameter subgroups. This implies again that G^* is complete.

Suppose now the theorem were false for G. Then, due to Prop. 10.7, Chapter I, G^* contains a ray emanating from e^*. Let γ be the one-parameter subgroup containing this ray. Then γ is a "straight line" in G^*, that is, it realizes the shortest distance in G^* between any two of its points. In fact, any pair of points on γ can be moved by a left translation on a pair of points on the ray. We parametrize γ by arc length t measured from the point $e^* = \gamma(0)$. The set $\pi(\gamma)$ is a one-parameter subgroup of G; its closure in G is a compact, abelian, connected subgroup, hence a torus. By the classical theorem of Kronecker, there exists a sequence $(t_n) \subset \boldsymbol{R}$ such that $t_n \to \infty$ and $\pi(\gamma(t_n)) \to e$. We can assume that all $\pi(\gamma(t_n))$ lie in a minimizing convex normal ball $B_r(e)$ and that each component C of $\pi^{-1}(B_r(e))$ is diffeomorphic to $B_r(e)$ under π. Then the mapping $\pi : C \to B_r(e)$ is distance-preserving; hence there exists an element $z_n \in G^*$ such that

$$\pi(z_n) = e, \tag{3}$$
$$d(z_n, \gamma(t_n)) = d(e, \pi(\gamma(t_n))).$$

Here d denotes the distance in G as well as in G^*. Since (G^*, π) is a covering group of G, the kernel of π is contained in the center Z of G^*. Hence by (3), we have $z_n \in Z$. We intend to show $\gamma \subset Z$.

Now for a given element $a \in G^*$, consider the one-parameter subgroup $\delta : t \to a\gamma(t)a^{-1}$ $(t \in \boldsymbol{R})$. Since left and right translations on G^* are isometries, δ is a "straight line" and $|t|$ is the arc parameter measured from e^*. Since $z_n \in Z$ we have $d(\delta(t_n), z_n) = d(\gamma(t_n), z_n)$ and this shows that

$$\lim_{n\to\infty} d(\gamma(t_n), \delta(t_n)) = 0. \tag{4}$$

Suppose now $\delta(t) \neq \gamma(t)$ for some $t \neq 0$. Then the angle between the vectors $\dot\gamma(0)$ and $\dot\delta(0)$ is different from 0 (possibly 180°). In any case, we have from Lemma 9.8, Chapter I and subsequent remark

$$d(\gamma(-1), \delta(+1)) < d(e^*, \gamma(-1)) + d(e^*, \delta(1)) = 2.$$

From (4) we can determine an integer N such that

$$t_N > 1, \quad d(\gamma(t_N), \delta(t_N)) < 2 - d(\gamma(-1), \delta(+1)).$$

We consider now the following broken geodesic ζ : from $\gamma(-1)$ to $\delta(+1)$ along a shortest geodesic, from $\delta(+1)$ to $\delta(t_N)$ on δ, from $\delta(t_N)$ to $\gamma(t_N)$ along a shortest geodesic. The curve ζ joins $\gamma(-1)$ to $\gamma(t_N)$ and has length

$$d(\gamma(-1), \delta(+1)) + (t_N - 1) + d(\delta(t_N), \gamma(t_N)),$$

which is strictly smaller than $t_N + 1 = d(\gamma(-1), \gamma(t_N))$. This contradicts the property of γ being a straight line.

It follows that $\delta(t) = \gamma(t)$ for all $t \in \boldsymbol{R}$. Since $a \in G^*$ was arbitrary it follows that $\gamma \subset Z$. But then $\mathfrak{z}$, the Lie algebra of Z, is $\neq \{0\}$, and this contradicts the semisimplicity of $\mathfrak{g}$.

Proposition 6.10. *Let G be a connected Lie group with compact Lie algebra $\mathfrak{g}$. Then the mapping* $\exp : \mathfrak{g} \to G$ *is surjective.*

The proof is contained in the first part of the proof of Theorem 6.9. In fact, Ad(G) being compact, there exists a strictly positive definite quadratic form on $\mathfrak{g}$ invariant under Ad(G). In the corresponding left and right invariant Riemannian metric on G the geodesics through e are the one-parameter subgroups. Thus G is complete and the result follows from Theorem 10.4, Chapter I.

§ 7. Invariant Differential Forms

Let G be a Lie group with Lie algebra $\mathfrak{g}$. A differential form ω on G is called *left invariant* if $L(x)^* \omega = \omega$ for all $x \in G$, $L(x)$ denoting the left translation $g \to xg$ on G. Similarly we define *right invariant* differential forms on G. A form is called *bi-invariant* if it is both left and right invariant.

Let ω be a left invariant p-form on G. Then if $X_1, \ldots, X_{p+1} \in \mathfrak{g}$ are arbitrary, $\tilde{X}_i$ the corresponding left invariant vector fields on G, we have by (9), Chapter I, §2,

$$(p+1)\, d\omega(\tilde{X}_1, \ldots, \tilde{X}_{p+1})$$

$$= \sum_{i<j} (-1)^{i+j}\, \omega([\tilde{X}_i, \tilde{X}_j], \tilde{X}_1, \ldots, \hat{X}_i, \ldots, \hat{X}_j, \ldots, \tilde{X}_{p+1}). \tag{1}$$

Lemma 7.1. *Let ω be a left invariant form on G. If ω is right invariant then ω is closed, that is, $d\omega = 0$.*

Proof. Let ω be a left invariant p-form and let $X \in \mathfrak{g}$. We have

$$(\tilde{X}_i^{R(\exp tX)})_e = \mathrm{Ad}(\exp(-tX))(X_i),$$

so by the formula ((4), §5)

$$\left\{\frac{d\,\mathrm{Ad}(\exp tX)}{dt}\right\}_{t=0} = \mathrm{ad}\, X$$

the right invariance of ω implies

$$\sum_1^p \omega(\tilde{X}_1, \ldots, [\tilde{X}, \tilde{X}_i], \tilde{X}_{i+1}, \ldots, \tilde{X}_p) = 0, \tag{2}$$

so by (1), and some manipulation, or by Exercise E.1, $d\omega = 0$.

Remark. Even for G connected, a closed left invariant form is not necessarily right invariant. The group G of the mappings $T_{a,b} : x \to ax + b$ $(x \in \boldsymbol{R})$ where a and b are real numbers, $a > 0$, has a Lie algebra $\mathfrak{g} = \boldsymbol{R}e_1 + \boldsymbol{R}e_2$ where $[e_1, e_2] = e_2$. Let $\omega \neq 0$ be an arbitrary left invariant 2-form on G. Then $d\omega = 0$, whereas if ω were right invariant, (2) would imply $\omega(e_1, e_2) = 0$ contradicting $\omega \neq 0$.

Let V be a finite-dimensional vector space and $Z_1, \ldots, Z_n$ a basis of V. In order that a bilinear map $(X, Y) \to [X, Y]$ of $V \times V$ into V turn V into a Lie algebra it is necessary and sufficient that the *structural constants* $\gamma^i{}_{jk}$ given by

$$[Z_j, Z_k] = \sum_1^n \gamma^i{}_{jk} Z_i$$

satisfy the conditions

$$\gamma^i{}_{jk} + \gamma^i{}_{kj} = 0$$

$$\sum_{j=1}^n \gamma^i{}_{jl}\gamma^j{}_{km} + \gamma^i{}_{jm}\gamma^j{}_{lk} + \gamma^i{}_{jk}\gamma^j{}_{ml} = 0.$$

Proposition 7.2. *Let $X_1, \ldots, X_n$ be a basis of $\mathfrak{g}$ and $\omega_1, \ldots, \omega_n$ the 1-forms on G determined by $\omega_i(\tilde{X}_j) = \delta_{ij}$. Then*

$$d\omega_i = -\tfrac{1}{2} \sum_{j,k=1}^{n} c^i{}_{jk}\omega_j \wedge \omega_k \tag{3}$$

if $c^i{}_{jk}$ are the structural constants given by

$$[X_j, X_k] = \sum_{i=1}^{n} c^i{}_{jk} X_i.$$

Equations (3) are known as the Maurer-Cartan equations. They follow immediately from (1). They also follow from Theorem 8.1, Chapter I if we give G the left invariant affine connection for which α in Prop. 1.4 is identically 0. Note that the Jacobi identity for $\mathfrak{g}$ is reflected in the relation $d^2 = 0$.

Example. Consider as in §1 the general linear group $\boldsymbol{GL}(n, \boldsymbol{R})$ with the usual coordinates $\sigma \to (x_{ij}(\sigma))$. Writing $X = (x_{ij})$, $dX = (dx_{ij})$, the matrix

$$\Omega = X^{-1}\,dX,$$

whose entries are 1-forms on G, is invariant under left translations $X \to \sigma X$ on G. Writing

$$dX = X\Omega,$$

we can derive

$$0 = (dX) \wedge \Omega + X \wedge d\Omega,$$

where $\wedge$ denote the obvious wedge product of matrices. Multiplying by X^{-1}, we obtain

$$d\Omega + \Omega \wedge \Omega = 0, \tag{4}$$

which is an equivalent form of (3).

More generally, consider for each x in the Lie group G the mapping

$$dL(x^{-1})_x : G_x \to \mathfrak{g}$$

and let Ω denote the family of these maps. In other words,

$$\Omega_x(v) = dL(x^{-1})(v) \qquad \text{if} \quad v \in G_x.$$

Then Ω is a 1-form on G with values in $\mathfrak{g}$. Moreover, if $x, y \in G$, then

$$\Omega_{xy} \circ dL(x)_y = \Omega_y,$$

so Ω is left invariant. Thus $\Omega_x = \Sigma_{i=1}^n (\theta_i)_x X_i$ in terms of the basis $X_1, ..., X_n$ in Prop. 7.2, $\theta_1, ..., \theta_n$ being left invariant 1-forms on G. But applying Ω_x to the vectors $(\tilde{X}_j)_x$ it is clear that $\theta_j = \omega_j$ $(1 \leqslant j \leqslant n)$. Hence we write

$$\Omega = \sum_{i=1}^{n} \omega_i X_i, \qquad d\Omega = \sum_{i=1}^{n} d\omega_i X_i.$$

If θ is any $\mathfrak{g}$-valued 1-form on a manifold X, we can define $[\theta, \theta]$ as the 2-form with values in $\mathfrak{g}$ given by

$$[\theta, \theta]_x (v_1, v_2) = [\theta_x(v_1), \theta_x(v_2)], \qquad x \in X, \quad v_1, v_2 \in X_x.$$

Then Prop. 7.2 can be reformulated as follows.

Proposition 7.3. *Let Ω denote the unique left invariant $\mathfrak{g}$-valued 1-form on G such that Ω_e is the identity mapping of G_e into $\mathfrak{g}$. Then*

$$d\Omega + \tfrac{1}{2}[\Omega, \Omega] = 0.$$

In fact, since $c^i{}_{jk}$ is skew in (j, k)

$$[\Omega, \Omega]_x (v_1, v_2) = \Big[\sum_j \omega_j(v_1) X_j, \sum_k \omega_k(v_2) X_k\Big]$$

$$= \sum_{i,j,k} \omega_j(v_1) \omega_k(v_2) c^i{}_{jk} X_i = \sum_{i,j,k} c^i{}_{jk}(\omega_j \wedge \omega_k)(v_1, v_2) X_i$$

$$= -2(d\Omega)_x (v_1, v_2).$$

We shall now determine the Maurer-Cartan forms ω_i explicitly in terms of the structural constants $c^l{}_{jk}$. Since exp is a C^∞ map from $\mathfrak{g}$ into G, the forms $\exp^* \omega_i$ can be expressed in terms of the Cartesian coordinates $(x_1, ..., x_n)$ of $\mathfrak{g}$ with respect to the basis $X_1, ..., X_n$,

$$(\exp^*(\omega_i))_X (X_j) = A_{ij}(x_1, ..., x_n), \tag{5}$$

where $X = \Sigma_i x_i X_i$ and $A_{ij} \in C^\infty(\boldsymbol{R}^n)$. Now let N_0 be an open star-shaped neighborhood of 0 in $\mathfrak{g}$ which exp maps diffeomorphically onto an open neighborhood N_e of e in G. Then $(x_1, ..., x_n)$ are canonical coordinates of $x = \exp X$ $(X \in N_0)$ with respect to the basis $X_1, ..., X_n$. Then, if $f \in C^\infty(G)$,

$$d \exp_X(X_j) f = (X_j)_X (f \circ \exp) = \Big(\frac{d}{dt} f(\exp(X + tX_j))\Big)_{t=0},$$

whence

$$d \exp_X(X_j) = \frac{\partial}{\partial x_j}.$$

Consequently,

$$(\omega_i)_x \left(\frac{\partial}{\partial x_j}\right) = \omega_i(d \exp_X(X_j)) = \exp^*(\omega_i)_X (X_j),$$

so

$$(\omega_i)_x = \sum_{j=1}^{n} A_{ij}(x_1, ..., x_n)\, dx_j. \tag{6}$$

Thus by Theorem 1.7 and the left invariance of ω_i,

$$A_{ij}(x_1, ..., x_n) = (\omega_i)_x (d \exp_X(X_j)) = (\omega_i)_e \left(\frac{1 - e^{-\text{ad}\, X}}{\text{ad}\, X} (X_j)\right).$$

Summarizing, we have proved the following result.

Theorem 7.4. *Let $X_1, ..., X_n$ be a basis of $\mathfrak{g}$ and the left-invariant 1-forms ω_i determined by $\omega_i(\tilde{X}_j) = \delta_{ij}$. Then the functions A_{ij} in* (5) *and* (6) *are given by the structural constants as follows. For $X = \sum_i x_i X_i$ in $\mathfrak{g}$ let $A(X)$ be defined by*

$$A(X)(X_j) = \sum_i A_{ij}(x_1, ..., x_n)\, X_i \qquad (1 \leqslant j \leqslant n).$$

Then

$$A(X) = \frac{1 - e^{-\text{ad}\, X}}{\text{ad}\, X} \tag{7}$$

and

$$\text{ad}\, X(X_j) = \sum_k \left(\sum_i x_i c^k{}_{ij}\right) X_k.$$

Theorem 7.5. (the third theorem of Lie) *Let $c^i{}_{jk} \in \boldsymbol{R}$ be constants $(1 \leqslant i, j, k \leqslant n)$ satisfying the relations*

$$c^i{}_{jk} + c^i{}_{kj} = 0 \tag{8}$$

$$\sum_{j=1}^{n} (c^i{}_{jl} c^j{}_{km} + c^i{}_{jm} c^j{}_{lk} + c^i{}_{jk} c^j{}_{ml}) = 0. \tag{9}$$

Then there exist an open neighborhood N of 0 in $\boldsymbol{R}^n$ and a basis $Y_1, ..., Y_n$ of $\mathfrak{D}^1(N)$ over $C^\infty(N)$ satisfying the relations

$$[Y_j, Y_k] = \sum_{i=1}^{n} c^i{}_{jk} Y_i. \tag{10}$$

Proof. We shall find a basis $\omega_1, ..., \omega_n$ of $\mathfrak{D}_1(N)$ over $C^\infty(N)$ satisfying the relations

$$d\omega_i = -\tfrac{1}{2} \sum_{j,k=1}^{n} c^i{}_{jk} \omega_j \wedge \omega_k. \tag{11}$$

Then the $Y_1, ..., Y_n$ can be chosen as the basis dual to $\omega_1, ..., \omega_n$ (Lemma 2.3, Chapter I), and (10) follows from (11).

A natural method would be to define A_{ij} by (7) and to define ω_i by (6). Then (11) amounts to the following "integrability condition":

$$2\left(\frac{\partial A_{ip}}{\partial x_q} - \frac{\partial A_{iq}}{\partial x_p}\right) = \sum_{j,k} c^i{}_{jk}(A_{jp}A_{kq} - A_{jq}A_{kp}). \tag{12}$$

Since (9) would have to enter into its verification, we adopt a simpler method, motivated by §8, Chapter I. Since the structural equation (1) there becomes formula (11) above for a special left invariant affine connection on a Lie group ($\alpha = 0$ in Prop. 1.4) and since (1) $\Rightarrow$ (6) in §8, Chapter I, we start by *defining* 1-forms

$$\theta_i = \theta_i(t, a_1, ..., a_n) = \sum_{j=1}^{n} f_{ij}(t, a_1, ..., a_n)\, da_j$$

as solutions to the differential equations

$$\frac{\partial \theta_i}{\partial t} = da_i - \sum_{j,k} c^i{}_{jk} a_j \theta_k, \qquad \theta_i(0, a_1, ..., a_n) = 0. \tag{13}$$

This amounts to a linear inhomogeneous constant coefficient system of differential equations for the functions f_{ij}, so these functions are uniquely determined for $(t, a_1, ..., a_n) \in \boldsymbol{R}^{n+1}$. Using (13) we get

$$\begin{aligned} d\theta_i &= \sum_j \frac{\partial f_{ij}}{\partial t}\, dt \wedge da_j + \sum_{j,k} \frac{\partial f_{ij}}{\partial a_k}\, da_k \wedge da_j \\ &= \Big(-da_i + \sum_{j,k} c^i{}_{jk} a_j \theta_k\Big) \wedge dt + \sum_{j,k} \frac{\partial f_{ij}}{\partial a_k}\, da_k \wedge da_j. \end{aligned}$$

We write this formula

$$d\theta_i = \alpha_i \wedge dt + \beta_i, \tag{14}$$

where the α_i and β_i are 1-forms and 2-forms, respectively, which do not contain dt. Next we put

$$\sigma_i = \beta_i + \tfrac{1}{2} \sum_{j,k} c^i{}_{jk}\theta_j \wedge \theta_k$$

and since we would by §8, Chapter I expect the forms $\theta_i(t, a_1, ..., a_n)_{t=1}$ to satisfy (11) we now try to prove that $\sigma_i = 0$. Using (8), and writing $\cdots$ for terms which do not contain dt, we have

$$\begin{aligned} d\sigma_i &= d\beta_i + \sum_{j,k} c^i{}_{jk} d\theta_j \wedge \theta_k \\ &= -dt \wedge d\alpha_i - dt \wedge \sum_{j,k} c^i{}_{jk}\alpha_j \wedge \theta_k + ... \end{aligned}$$

Using the expression for α_i and (14), this becomes

$$\begin{aligned} &-dt \wedge \sum_{j,k} c^i{}_{jk}(da_j \wedge \theta_k + a_j\beta_k + \alpha_j \wedge \theta_k) + ... \\ &= -dt \wedge \sum_{j,k} c^i{}_{jk} \Big(\sum_{pq} c^j{}_{pq} a_p \theta_q \wedge \theta_k + a_j\beta_k\Big) + ... \end{aligned}$$

But since $\theta_q \wedge \theta_k = -\theta_k \wedge \theta_q$, we have

$$\sum_{j,k,q} c^i{}_{jk} c^j{}_{pq} \theta_q \wedge \theta_k = \tfrac{1}{2} \sum_{j,k,q} (c^i{}_{jk} c^j{}_{pq} - c^i{}_{jq} c^j{}_{pk}) \theta_q \wedge \theta_k,$$

which by (8) and (9) equals

$$-\tfrac{1}{2} \sum_{j,k,q} c^i{}_{jp} c^j{}_{qk} \theta_q \wedge \theta_k.$$

This proves

$$\begin{aligned} d\sigma_i &= -dt \wedge \Big(\sum_{j,k} c^i{}_{jk} a_j \beta_k - \tfrac{1}{2} \sum_{j,k,p,q} a_p c^i{}_{jp} c^j{}_{qk} \theta_q \wedge \theta_k\Big) + ... \\ &= -dt \wedge \sum_{j,k} c^i{}_{jk} \Big(-a_k\beta_j - \tfrac{1}{2} \sum_{qr} a_k c^j{}_{qr} \theta_q \wedge \theta_r\Big) + ... \end{aligned}$$

so

$$d\sigma_i = dt \wedge \sum_{jk} c^i{}_{jk} a_k \sigma_j + ... \tag{15}$$

This amounts to

$$\frac{\partial \sigma_i}{\partial t} = \sum_{jk} c^i{}_{jk} a_k \sigma_j$$

which, since the σ_j all vanish for $t = 0$, implies that each σ_i vanishes identically. Thus we see from (14) that the forms $\omega_i = \theta_i\,(1, a_1, \ldots, a_n)$ will satisfy (11). Finally, (13) implies that

$$\theta_i(t, 0, \ldots, 0) = t\,da_i,$$

so the forms ω_i are linearly independent at $(a_1, \ldots, a_n) = (0, \ldots, 0)$, hence also in a suitable neighborhood of the origin. This concludes the proof.

Now let G be a compact connected Lie group. Let dg denote the Haar measure on G normalized by $\int_G dg = 1$, let Q be a fixed positive definite quadratic form on $\mathfrak{g}$ invariant under $\mathrm{Ad}(G)$, and fix a basis $X_1, \ldots, X_n$ of $\mathfrak{g}$ orthonormal with respect to Q. Let $\omega_1, \ldots, \omega_n$ be the left invariant 1-forms on G given by $\omega_i(\tilde{X}_j) = \delta_{ij}$ and put $\theta = \omega_1 \wedge \cdots \wedge \omega_n$. Then θ is left invariant and also right invariant because $\det \mathrm{Ad}(g) \equiv 1$ by the compactness of G. Also each n-form ω on G can be written $\omega = f\theta$ where $f \in C^\infty(G)$ is unique, so we can define

$$\int_G \omega = \int_G f(g)\,dg. \tag{16}$$

Lemma 7.6. *Let ω be an $(n-1)$-form on G. Then*

$$\int_G d\omega = 0.$$

This is a special case of Stokes's theorem and can be proved quickly as follows. We have $d\omega = h\theta$ where $h \in C^\infty(G)$. By (16) and the bi-invariance of dg and θ, we have, since d commutes with mappings and integration with respect to another variable,

$$\int_G d\omega = \int_G \int_G \int_G R(x)^*\,L(y)^*\,(d\omega)\,dx\,dy$$

$$= \int_G d\left(\int_{G\times G} R(x)^*\,L(y)^*\omega\,dx\,dy\right) = 0,$$

the last equality following from Lemma 7.1.

Next we recall the $*$ operator which maps $\mathfrak{A}(G)$ onto itself, $\mathfrak{A}_p(G)$ onto $\mathfrak{A}_{n-p}(G)$ $(0 \leqslant p \leqslant n)$. Let $\sigma_1, \ldots, \sigma_n$ be the basis of the dual space $\mathfrak{g}^*$, dual to (X_i), $\mathfrak{A}(e)$ the Grassmann algebra of $\mathfrak{g} = G_e$, and $* : \mathfrak{A}(e) \to \mathfrak{A}(e)$ the mapping determined by linearity and the condition

$$*(\sigma_{i_1} \wedge \ldots \wedge \sigma_{i_p}) = \pm\sigma_{j_1} \wedge \ldots \wedge \sigma_{j_{n-p}}, \tag{17}$$

where $\{i_1, \ldots, i_p, j_1, \ldots, j_{n-p}\}$ is a permutation of $\{1, \ldots, n\}$, the sign being

$+$ or $-$ depending on whether the permutation is even or odd. We shall use the following simple fact from linear algebra (for proofs, see e.g. Flanders [1], Chapter 2):

(i) If (X_i) is replaced by another orthonormal basis (X'_j) where $X'_j = \sum_{i=1}^n g_{ij} X_i$ with $\det(g_{ij}) = 1$, then the definition of $*$ does not change.

(ii) If $i_1, < \cdots < i_p$, then

$$\sigma_{i_1} \wedge \ldots \wedge \sigma_{i_p} \wedge *(\sigma_{i_1} \wedge \ldots \wedge \sigma_{i_p}) = \sigma_1 \wedge \ldots \wedge \sigma_n.$$

(iii) $**\sigma = (-1)^{p(n-p)}\sigma$ if $\sigma \in \mathfrak{A}_p(e)$.

From (i) we have since $\det \operatorname{Ad}(g) = 1$, $\operatorname{Ad}(g)* = *\operatorname{Ad}(g)$ $(g \in G)$. Thus we can define $* : \mathfrak{A}(g) \to \mathfrak{A}(g)$ as the map $L(g^{-1})^* * L(g)^*$ or as the map $R(g^{-1})^* * R(g)^*$. Finally, the mapping $* : \mathfrak{A}(G) \to \mathfrak{A}(G)$ is defined by the condition

$$(*\omega)_g = *(\omega_g), \qquad \omega \in \mathfrak{A}(G), \quad g \in G.$$

Then $*$ commutes with $L(x)^*$ and $R(y)^*$ for all $x, y \in G$.

Next we define the linear operator $\delta : \mathfrak{A}(G) \to \mathfrak{A}(G)$ which maps p-forms into $(p-1)$-forms according to the formula

$$\delta\omega = (-1)^{np+n+1} * d * \omega, \qquad \omega \in \mathfrak{A}_p(G).$$

We then introduce an inner product $\langle\ ,\ \rangle$ on $\mathfrak{A}(G)$ by

$$\langle \omega, \eta \rangle = 0 \qquad \text{if} \quad \omega \in \mathfrak{A}_p(G), \quad \eta \in \mathfrak{A}_q(G) \quad (p \neq q),$$

$$\langle \omega, \eta \rangle = \int_G \omega \wedge *\eta \qquad \text{if} \quad \omega, \eta \in \mathfrak{A}_p(G)$$

and the requirement of bilinearity. This inner product is strictly positive definite; in fact we can write

$$\omega = \sum_{i_1 < \ldots < i_p} a_{i_1 \ldots i_p} \omega_{i_1} \wedge \ldots \wedge \omega_{i_p}$$

and then

$$\omega \wedge *\omega = \Big(\sum_{i_1 < \ldots < i_p} a_{i_1 \ldots i_p}^2 \Big) \theta$$

so the statement follows. Moreover d and δ are adjoint operators, that is,

$$\langle d\omega, \eta \rangle = \langle \omega, \delta\eta \rangle, \qquad \omega, \eta \in \mathfrak{A}(G). \tag{18}$$

It suffices to verify this when $\omega \in \mathfrak{A}_{p-1}(G)$, $\eta \in \mathfrak{A}_p(G)$. But then

$$d(\omega \wedge *\eta) = d\omega \wedge *\eta + (-1)^{p-1}\omega \wedge d * \eta = d\omega \wedge *\eta - \omega \wedge *\delta\eta,$$

since $** = (-1)^{p(n-p)}$ on $\mathfrak{A}_p(G)$. Integrating this over G and using Lemma 7.6, we derive (18). We consider now the operator $\Delta = -d\delta - \delta d$ on $\mathfrak{A}(G)$ which maps each $\mathfrak{A}_p(G)$ into itself. A form ω satisfying $\Delta\omega = 0$ is called a *harmonic* form.

Lemma 7.7. *A form ω on G is harmonic if and only if $d\omega = 0$ and $\delta\omega = 0$.*

In fact,

$$-\langle \Delta\omega, \omega\rangle = \langle \delta\omega, \delta\omega\rangle + \langle d\omega, d\omega\rangle$$

so the result follows.

Theorem 7.8. (Hodge) *The harmonic forms on a compact connected Lie group G are precisely the bi-invariant forms.*

A bi-invariant form ω satisfies $d\omega = 0$ (Lemma 7.1); and since $*$ commutes with left and right translations, $\delta\omega = 0$. Conversely, suppose $\Delta\omega = 0$, so by Lemma 7.7, $d\omega = \delta\omega = 0$. Let $X \in \mathfrak{g}$ and let $\tilde{X}$ denote the left invariant vector field on G such that $\tilde{X}_e = X$. By Exercise B.6, Chapter I we have $\theta(\tilde{X})\omega = i(\tilde{X})\, d\omega + di(\tilde{X})\omega = di(\tilde{X})\omega$. Then

$$\langle \theta(\tilde{X})\omega, \theta(\tilde{X})\omega\rangle = \langle \delta\theta(\tilde{X})\omega, i(\tilde{X})\omega\rangle = 0,$$

since $\theta(\tilde{X})\omega$ is harmonic. Hence $\theta(\tilde{X})\omega = 0$, so ω is right invariant (Exercise B.3, Chapter I). Left invariance follows in the same way.

Q.E.D.

§ 8. Perspectives

This section contains some informal comments whose purpose it is to put some of the topics of this chapter in better perspective.

First we explain how Theorem 3.4 is connected with the original foundation of Lie group theory. Inspired by Galois' theory for algebraic equations, Lie raised the following question in his paper [2]: *How can the knowledge of a stability group for a differential equation be utilized toward its integration* ? (A point transformation is said to leave a differential equation *stable* if it permutes the solutions.) Lie proved in [2] that a one-parameter transformation group φ_t of $\boldsymbol{R}^2$ with induced vector field

$$\Phi_p = \left(\frac{d(\varphi_t \cdot p)}{dt}\right)_{t=0} = \xi(p)\frac{\partial}{\partial x} + \eta(p)\frac{\partial}{\partial y}$$

leaves a differential equation $dy/dx = Y(x, y)/X(x, y)$ stable if and only if the vector field $Z = X\,\partial/\partial x + Y\,\partial/\partial y$ satisfies $[\Phi, Z] = \lambda Z$ where λ is a function; in this case $(X\eta - Y\xi)^{-1}$ is an integrating factor for the equation $X\,dy - Y\,dx = 0$.

Example.

$$\frac{dy}{dx} = \frac{y + x(x^2 + y^2)}{x - y(x^2 + y^2)}\,.$$

This equation can be written

$$\left(\frac{dy}{dx} - \frac{y}{x}\right)\Big/\left(1 + \frac{y}{x}\frac{dy}{dx}\right) = x^2 + y^2;$$

and since the left-hand side is $\tan\alpha$ where α is the angle between the integral curve through (x, y) and the radius vector, it is clear that the integral curves intersect each circle around $(0, 0)$ under a fixed angle. Thus the rotation group

$$\varphi_t : (x, y) \to (x \cos t - y \sin t,\ x \sin t + y \cos t),$$

for which $\Phi = -y\,\partial/\partial x + x\,\partial/\partial y$, leaves the equation stable and Lie's theorem gives the solution $y = x\tan(\frac{1}{2}(x^2 + y^2) + C)$, C a constant.

Generalizing (φ_t) above, Lie considered transformations $(x_1, \ldots, x_n) \to (x_1', \ldots, x_n')$ given by

$$T : x_i' = f_i(x_1, \ldots, x_n; t_1, \ldots, t_r) \tag{1}$$

depending effectively on r parameters t_k, i.e., the f_i are C^∞ functions and the matrix $(\partial f_i/\partial t_k)$ has rank r. We assume that the identity transformation is given by $t_1 = \cdots = t_r = 0$ and that if a transformation S corresponds to the parameters $(s_1, \ldots, s_r)$ then TS^{-1} is for sufficiently small t_i, s_j given by

$$TS^{-1} : x_i' = f_i(x_1, \ldots, x_n; u_1, \ldots, u_r) \tag{2}$$

where the u_k are analytic functions of the t_i and s_j. Generalizing Φ above, Lie introduced the vector fields

$$T_k = \sum_{i=1}^{n} \left(\frac{\partial f_i}{\partial t_k}\right)_{t=0} \frac{\partial}{\partial x_i} \qquad (1 \leqslant k \leqslant r)$$

and, as a result of the group property (2), proved the fundamental formula

$$[T_k, T_l] = \sum_{p=1}^{r} c^p{}_{kl} T_p, \tag{3}$$

where the $c^p{}_{kl}$ are *constants* satisfying

$$c^p{}_{kl} = -c^p{}_{lk}, \qquad \sum_{q=1}^{r} (c^l{}_{kq} c^q{}_{lm} + c^p{}_{mq} c^q{}_{kl} + c^p{}_{lq} c^q{}_{mk}) = 0. \tag{4}$$

Independently of Lie, Killing had through geometric investigations been led to concepts close to T_k and relations (3) and he attacked the algebraic problem of classifying all solutions to (4). See notes to Chapter X.

We shall now show how (3) follows from Theorem 3.4. So the (x_i) are coordinates on M and the (t_k) coordinates near e in G. We first assume that the (t_k) are "canonical coordinates of the second kind," i.e., for a suitable basis $X_1, \ldots, X_r$ of $\mathfrak{g}$,

$$\begin{aligned}\exp(t_1X_1) \ldots \exp(t_rX_r) \cdot (x_1, \ldots, x_n)\\ = f_1(x_1, \ldots, x_n; t_1, \ldots, t_r), \ldots, f_n(x_1, \ldots, x_n; t_1, \ldots, t_r).\end{aligned}$$

Such coordinates exist by the remark following Lemma 2.4. Then $X_k = (\partial/\partial t_k)_e$ $(1 \leqslant k \leqslant r)$ and

$$(X_k^+ f)(p) = \sum_{i=1}^{n} (X_k^+ x_i)(p) \left(\frac{\partial}{\partial x_i}\right)_p (f),$$

$$\begin{aligned}(X_k^+ x_i)(p) &= \left\{\frac{d}{ds} x_i(\exp(sX_k) \cdot (x_1, \ldots, x_n))\right\}_{s=0}\\ &= \left\{\frac{\partial}{\partial s} f_i(x_1, \ldots, x_n; 0, \ldots, s, \ldots, 0)\right\}_{s=0} = \left(\frac{\partial f_i}{\partial t_k}\right)_{t=0}.\end{aligned}$$

It follows that $X_+^k = T_k$, so Theorem 3.4 implies (3) for this coordinate system $\{t_1, \ldots, t_r\}$. But if $\{s_1, \ldots, s_r\}$ is another coordinate system on a neighborhood of e in G, with $s_1(e) = \cdots = s_r(e) = 0$, then

$$f_i(x_1, \ldots, x_n; s_1, \ldots, s_r) = f'_i(x_1, \ldots, x_n; t_1, \ldots, t_r)$$

where the f'_i are obtained by changing to the canonical coordinates $(t_1, \ldots, t_k)$ of the second kind. Then

$$\frac{\partial f_i}{\partial s_k} = \sum_{l=1}^{r} \frac{\partial f'_i}{\partial t_l} \frac{\partial t_l}{\partial s_k}$$

so the $(\partial f_i/\partial s_k)_{s=0}$ are certain constant linear combinations of the T_l; thus (3) holds in general.

A major problem in the early theory of Lie groups was to establish the existence of a "local Lie group" with a given Lie algebra. This was solved by "Lie's three theorems" which were proved by means of differential equations. The third of these is Theorem 7.5. The second theorem is equivalent to the integrability conditions (12), §7. The first theorem then amounts to, that when the vector fields $Y_1, ..., Y_n$ of Theorem 7.5 are integrated to give local one-parameter transformation groups, these generate a local group for which the Y_i are left invariant vector fields.

Nowadays the existence of a (global) Lie group with a given Lie algebra can be proved without recourse to differential equations, as we shall indicate below. Besides semisimple Lie groups whose theory is intertwined with the theory of symmetric spaces, solvable Lie groups (cf. Chapter III, §2) are the other fundamental class of Lie groups. They include the abelian Lie groups and the nilpotent Lie groups. Each Lie algebra $\mathfrak{g}$ has a "Levi decomposition" $\mathfrak{g} = \mathfrak{r} + \mathfrak{s}$ where $\mathfrak{r}$ is a solvable ideal, $\mathfrak{s}$ a semisimple subalgebra and $\mathfrak{r} \cap \mathfrak{s} = (0)$. These properties determine $\mathfrak{r}$ uniquely and determine $\mathfrak{s}$ up to conjugacy. From this decomposition it is easy to prove the existence of a Lie group with Lie algebra $\mathfrak{g}$. (In fact, the existence of a Lie group R with Lie algebra $\mathfrak{r}$ follows by induction on the dimension, the group $S = \text{Int}(\mathfrak{s})$ has Lie algebra $\mathfrak{s}$ and the desired group G can be taken as a semidirect product R^*S^*, $*$ denoting the universal covering group.)

The analog of Prop. 6.10 fails to hold for semisimple Lie groups in general (cf. Exercise B.1) and fails to hold for solvable Lie groups in general (cf. Exercise B.4). However it does hold for nilpotent Lie groups as will be proved in Chapter VI.

EXERCISES AND FURTHER RESULTS

A. On the Geometry of Lie Groups

1. Let G be a Lie group, $L(x)$ and $R(x)$, respectively the left translation $g \to xg$, and the right translation $g \to gx$. Prove:

(i) $\text{Ad}(x) = dR(x^{-1})_x \circ dL(x)_e = dL(x)_{x^{-1}} \circ dR(x^{-1})_e$.

(ii) If J is the map $g \to g^{-1}$ then

$$dJ_x = -dL(x^{-1})_e \circ dR(x^{-1})_x = -dR(x^{-1})_e \circ dL(x^{-1})_x.$$

(iii) If Φ is the mapping $(g, h) \to gh$ of $G \times G$ into G, then if $X \in G_g$, $Y \in G_h$,

$$d\Phi_{(g,h)}(X, Y) = dL(g)_h(Y) + dR(h)_g(X).$$

2. Let $\gamma(t)$ $(t \in \boldsymbol{R})$ be a one-parameter subgroup of a Lie group. Assume that γ intersects itself. Then γ is a "closed" one-parameter subgroup, that is, there exists a number $L > 0$ such that $\gamma(t + L) = \gamma(t)$ for all $t \in \boldsymbol{R}$.

3. Let $\gamma(t)$, $\delta(t)$ $(t \in \boldsymbol{R})$ be two one-parameter subgroups of a Lie group. If $\gamma(L) = \delta(L)$ for some $L > 0$, then the curve $\sigma(t) = \gamma(t)\,\delta(-t)$ $(0 \leqslant t \leqslant L)$ is smooth at e, that is, $\dot{\sigma}(e) = \dot{\sigma}(L)$ (Goto and Jakobsen [1]).

4. Let G be a locally compact group, H a closed subgroup. Prove that the space G/H is complete in any G-invariant metric.

5. Let G be a connected Lie group with Lie algebra $\mathfrak{g}$. Let B be a nondegenerate symmetric bilinear form on $\mathfrak{g} \times \mathfrak{g}$. Then there exists a unique left invariant pseudo-Riemannian structure Q on G such that $Q_e = B$. Show, using Prop. 1.4 and (2), §9, Chapter I, that the following conditions are equivalent:

(i) The geodesics through e are the one-parameter subgroups.

(ii) $B(X, [X, Y]) = 0$, for all $X, Y \in \mathfrak{g}$.

(iii) $B(X, [Y, Z]) = B([X, Y], Z)$ for all $X, Y, Z \in \mathfrak{g}$.

(iv) Q is invariant under all right translations on G.

(v) Q is invariant under the mapping $g \to g^{-1}$ of G onto itself.

6. Let G be a connected Lie group with Lie algebra $\mathfrak{g}$. Then there exists a unique affine connection ∇ on G invariant under all left and right translations and under the map $J : g \to g^{-1}$. Let $X, Y \in \mathfrak{g}$. Prove that:

(i) The parallel translate of X along the curve $\gamma(t) = \exp tY$ $(0 \leqslant t \leqslant 1)$ is given by

$$dL(\exp \tfrac{1}{2}Y)\, dR(\exp \tfrac{1}{2}Y)X.$$

(ii) $\nabla_{\tilde{X}}(\tilde{Y}) = \frac{1}{2}[\tilde{X}, \tilde{Y}]$ where $\tilde{X}$ and $\tilde{Y}$ are the left invariant vector fields with $\tilde{X}_e = X$, $\tilde{Y}_e = Y$.

(iii) The geodesics are the translates of one-parameter subgroups.

(iv) The torsion T and curvature R of ∇ are given by

$$T = 0, \qquad R(X, Y) = -\tfrac{1}{4}\,\mathrm{ad}([X, Y]).$$

B. The Exponential Mapping

1. Let $SL(2, \boldsymbol{R})$ denote the group of all real 2×2 matrices with determinant 1. Its Lie algebra $\mathfrak{sl}(2, \boldsymbol{R})$ consists of all real 2×2 matrices of trace 0.

(i) Let $X \in \mathfrak{sl}(2, \boldsymbol{R})$, $I =$ unit matrix. Show that

$$e^X = \cosh(-\det X)^{1/2}\, I + \frac{\sinh(-\det X)^{1/2}}{(-\det X)^{1/2}}\, X \qquad \text{if} \quad \det X < 0$$

$$e^X = \cos(\det X)^{1/2}\, I + \frac{\sin(\det X)^{1/2}}{(\det X)^{1/2}}\, X \qquad \text{if} \quad \det X > 0$$

$$e^X = I + X \qquad \text{if} \quad \det X = 0.$$

(ii) Let us consider one-parameter subgroups the same if they have proportional tangent vectors at e. Then the matrix

$$\begin{pmatrix} \lambda & 0 \\ 0 & \lambda^{-1} \end{pmatrix} \in SL(2, \boldsymbol{R}) \qquad (\lambda \neq 1)$$

lies on exactly one one-parameter subgroup if $\lambda > 0$, on infinitely many one-parameter subgroups if $\lambda = -1$ and one no one-parameter subgroup if $\lambda < 0$, $\lambda \neq -1$.

2. Show that the group $SL(2, \boldsymbol{R})$ admits a bi-invariant pseudo-Riemannian structure. This pseudo-Riemannian manifold is *complete* in the sense that the geodesics are indefinitely extendable (Exercise A.5). Show that:

(i) Two points in $SL(2, \boldsymbol{R})$ can not in general be joined by a geodesic.

(ii) Two points in $SL(2, \boldsymbol{R})$ can always be joined by a singly broken geodesic.

(iii) On any connected manifold M with an affine connection an arbitrary pair of points can always be joined by a finitely broken geodesic. The number of breaks necessary may be unbounded even if M is complete (Hicks [2]).

3. The Lie group $GL(n, \boldsymbol{C})$ has Lie algebra $\mathfrak{gl}(n, \boldsymbol{C})$ and the mapping

$$\exp : \mathfrak{gl}(n, \boldsymbol{C}) \to GL(n, \boldsymbol{C})$$

is surjective. (Use the Jordan canonical form (Prop. 1.1, Chapter III and Lemma 4.5, Chapter VI.)

4. Let G denote the subgroup of $\boldsymbol{GL}(n, \boldsymbol{R})$ given by

$$\begin{pmatrix} \cos\gamma & \sin\gamma & 0 & \alpha \\ -\sin\gamma & \cos\gamma & 0 & \beta \\ 0 & 0 & 1 & \gamma \\ 0 & 0 & 0 & 1 \end{pmatrix} \qquad (\alpha, \beta, \gamma \in \boldsymbol{R})$$

Describe its Lie algebra $\mathfrak{g} \subset \mathfrak{gl}(n, \boldsymbol{R})$, show that $\mathfrak{g}$ is solvable, but that the mapping $\exp : \mathfrak{g} \to G$ is neither injective nor surjective.

5. Using the exponential mapping show that each Lie group G contains a neighborhood of e containing no subgroup $\neq \{e\}$. ("A Lie group has no small subgroups.")

C. Subgroups and Transformation Groups

1. Verify the description of the Lie algebras of the various subgroups of $\boldsymbol{GL}(n, \boldsymbol{C})$ listed in Chapter X, §2.

2. Show that a commutative connected Lie group is isomorphic to a product group of the form $\boldsymbol{R}^n \times \boldsymbol{T}^m$ where $\boldsymbol{T}^m$ is an m-dimensional torus. Deduce that a one-parameter subgroup γ of a Lie group H is either closed or has compact closure.

3. Let $H \subset G$ be connected Lie groups. Suppose the identity mapping $I : H \to G$ is continuous. Then H is a Lie subgroup of G.

4. (The analytic structure of G/H) With the notation prior to Theorem 4.2 let $g, g' \in G$ and consider the two homeomorphisms

$$\psi_g : g \exp(x_1 X_1 + \ldots + x_r X_r) \cdot p_0 \to (x_1, \ldots, x_r) \qquad \text{of } g \cdot N_0 \text{ into } \boldsymbol{R}^r;$$

$$\psi_{g'} : g' \exp(y_1 X_1 + \ldots + y_r X_r) \cdot p_0 \to (y_1, \ldots, y_r) \qquad \text{of } g' \cdot N_0 \text{ into } \boldsymbol{R}^r.$$

Prove that the mapping $\psi_{g'} \circ \psi_g^{-1}$ is an analytic mapping of

$$\psi_g(g \cdot N_0 \cap g' \cdot N_0)$$

onto

$$\psi_{g'}(g \cdot N_0 \cap g' \cdot N_0).$$

5. Let G be a Lie transformation group of a manifold M. Then each orbit $G \cdot p$ is a submanifold of M, diffeomorphic to G/G_p. (Proceed as in the proof of Prop. 4.3.)

6. Let G be a locally connected topological group. Suppose the identity component G_0 has an analytic structure compatible with the topology in which it is a Lie group. Show that G has the same property. (Hint: Use Theorem 2.6.)

This shows that the definition of a Lie group adopted here is equivalent to that of Chevalley [2].

7*. Suppose an abstract subgroup H of a connected Lie group G has a manifold structure in which it is a submanifold of G with at most countably many components. Then H is a Lie subgroup of G. (cf. Freudenthal [4]; see also Kobayashi and Nomizu [1], I, p. 275 or F. Warner [1], p. 95, and Chevalley [2], p. 96).

8. Let G be a connected Lie group, $H \subset G$ a closed subgroup. The action of G on the manifold $M = G/H$ is called *imprimitive* if there exists a connected submanifold N of M $(0 < \dim N < \dim M)$ such that for each $g \in G$ either $g \cdot N = N$ or $g \cdot N \cap N = \emptyset$. Show that this is equivalent to the existence of a Lie subgroup L, $H \subset L \subset G$, such that $\dim H < \dim L < \dim G$.

9*. Let G be a Lie transformation group of a manifold M, M/G the orbit space topologized by the finest topology for which the natural mapping $\pi : M \to M/G$ is continuous. Let

$$D = \{(p, q) \in M \times M : p = g \cdot q \text{ for some } g \in G\}.$$

Prove that:

(i) M/G is a Hausdorff space if and only if the subset $D \subset M \times M$ is closed.

(ii) There exists a differentiable structure on the topological space M/G such that $\pi : M \to M/G$ is a submersion if and only if the topological subspace $D \subset M \times M$ is a closed submanifold.

In this case the differentiable structure is unique and all the G-orbits in M have the same dimension (see, e.g., Dieudonné [2], Chapitre XVI).

D. Closed Subgroups

1. Let Γ be a discrete subgroup of $\boldsymbol{R}^2$ such that $\boldsymbol{R}^2/\Gamma$ is compact. Show that an analytic subgroup of $\boldsymbol{R}^2$ is always closed but that its image in $\boldsymbol{R}^2/\Gamma$ under the natural mapping is not necessarily closed.

2. Let $\mathfrak{g}$ be a Lie algebra such that $\mathrm{Int}(\mathfrak{g})$ has compact closure in $\boldsymbol{GL}(\mathfrak{g})$. Then $\mathrm{Int}(\mathfrak{g})$ is compact. (Hint: Repeat the proof of Prop. 6.6 and use Theorem 6.9.)

3. Let G denote the five-dimensional manifold $\boldsymbol{C} \times \boldsymbol{C} \times \boldsymbol{R}$ with multiplication defined as follows (van Est [1], Hochschild [1]):

$$(c_1, c_2, r)(c_1', c_2', r') = (c_1 + e^{2\pi i r}c_1', c_2 + e^{2\pi i h r}c_2', r + r'),$$

where h is a fixed irrational number and $c_1, c_2, c_1', c_2' \in \boldsymbol{C}$, $r, r' \in \boldsymbol{R}$. Then G is a Lie group.

(i) Let $s, t \in \boldsymbol{R}$ and define the mapping $\alpha_{s,t} : G \to G$ by $\alpha_{s,t}(c_1, c_2, r) = (e^{2\pi i s} c_1, e^{2\pi i t} c_2, r)$. Show that $\alpha_{s,t}$ is an analytic isomorphism.

(ii) If $t = hs + hn$ where n is an integer, then $\alpha_{s,t}$ coincides with the inner automorphism

$$(c_1, c_2, r) \to (0, 0, s+n)(c_1, c_2, r)(0, 0, s+n)^{-1}.$$

(iii) Let $\mathfrak{g}$ denote the Lie algebra of G and let $A_{s,t}$ denote the automorphism $d\alpha_{s,t}$ of $\mathfrak{g}$. If $s_n \to s_0$, $t_n \to t_0$ then $A_{s_n,t_n} \to A_{s_0,t_0}$ in Aut $(\mathfrak{g})$.

(iv) Show that $A_{0,1/3} \notin \text{Int}(\mathfrak{g})$. Deduce from (iii) that Int $(\mathfrak{g})$ is not closed in Aut $(\mathfrak{g})$.

4*. Let G be a connected Lie group and H an analytic subgroup. Let $\mathfrak{g}$ and $\mathfrak{h}$ denote the corresponding Lie algebras.

(i) Assume G simply connected. If $\mathfrak{h}$ is an ideal in $\mathfrak{g}$ then H is closed in G (Chevalley [2], p. 127).

(ii) Assume G simply connected. If $\mathfrak{h}$ is semisimple then H is closed in G (Mostow [2], p. 615).

(iii) Assume G compact. If $\mathfrak{h}$ is semisimple then H is closed in G (Mostow [2], p. 615).

(iv) Assume $G = \boldsymbol{GL}(n, \boldsymbol{C})$. If $\mathfrak{h}$ is semisimple then H is closed in G (Goto [1], Yosida [1]).

(v) Suppose H is not closed in G. Then there exists a one-parameter subgroup γ of H whose closure (in G) is not contained in H (Goto [1]).

(vi) H is closed if exp $\mathfrak{h}$ is closed. This follows from (v).

(vii) Assume G solvable and simply connected. Then H is closed and simply connected (Chevalley [8]).

(viii) Suppose $G = \boldsymbol{SO}(n)$ and that H acts irreducibly on $\boldsymbol{R}^n$. Then H is closed in G (Borel and Lichnèrowicz [2], Kobayashi and Nomizu [1], I, p. 277).

E. Invariant Differential Forms

1. Let G be a connected Lie group with Lie algebra $\mathfrak{g}$. Let $(X_i)_{1 \leqslant i \leqslant n}$ be a basis of $\mathfrak{g}$, $\tilde{X}_i$ $(1 \leqslant i \leqslant n)$ the corresponding left invariant vector fields, and ω_j $(1 \leqslant j \leqslant n)$ the dual forms given by $\omega_j(\tilde{X}_i) = \delta_{ij}$. From (1), §7 or Exercise C4, Chapter I deduce the formula (cf. Koszul [4])

$$2\, d\omega = \sum_{k=1}^{n} \omega_k \wedge \theta(\tilde{X}_k)\omega \qquad (\omega \text{ left invariant})$$

where $\theta(\tilde{X}_k)$ is the Lie derivative (Exercise B.1, Chapter I). Show that if $\omega = \omega_i$, this formula reduces to the Maurer-Cartan equations (3), §7.

2. Prove that for the orthogonal group $\boldsymbol{O}(n)$ the matrix of 1-forms $\Omega = g^{-1}\,dg$ $(g \in \boldsymbol{O}(n))$ satisfies

$$d\Omega + \Omega \wedge \Omega = 0, \qquad \Omega + {}^t\Omega = 0,$$

${}^tA$ denoting the transpose of a matrix A. Generalize these relations to $\boldsymbol{U}(n)$ and $\boldsymbol{Sp}(n)$.

3. Using the method of Exercise E2 show that the group of matrices

$$g = \begin{pmatrix} 1 & x & z \\ 0 & 1 & y \\ 0 & 0 & 1 \end{pmatrix} \qquad (x, y, z \in \boldsymbol{R})$$

has a basis of left invariant 1-forms given by

$$\omega_1 = dx, \qquad \omega_2 = dy, \qquad \omega_3 = dz - x\,dy$$

and that the Maurer-Cartan equations are

$$d\omega_1 = 0, \qquad d\omega_2 = 0, \qquad d\omega_3 = -\omega_1 \wedge \omega_2.$$

NOTES

In the early days of Lie group theory, the late nineteenth century, the notion of a Lie group had, in the hands of S. Lie, W. Killing, and É. Cartan, a primarily local character. For information about this early period see for example Lie [1], Vol. III, Mostow [7], Bourbaki [2], Chapters II-III and Helgason [10]. Global Lie groups were not emphasized until during the 1920's through the work of H. Weyl, É. Cartan, and O. Schreier. These two viewpoints, the infinitesimal method and the integral method, were not completely coordinated until É. Cartan proved in 1930 ([16], [20]) that every Lie algebra over $\boldsymbol{R}$ is the Lie algebra of a Lie group. The book of Chevalley [2] gives a systematic exposition of Lie group theory from the global point of view.

§1-§3. No generality is gained by replacing the analyticity requirement in the definition of a Lie group by differentiability. This was stated without proof in Lie [3]; it was proved by F. Schur [1-3] along with Theorem 7.4 where the functions A_{ij} automatically turn out to be analytic functions (cf. Pontrjagin [1], §56, Satz 88). Even the differentiability assumption is not essential; a locally Euclidean topological group has an analytic structure (actually unique) in which the group operations are analytic. This theorem, which after many partial steps was proved by A. Gleason, D. Montgomery, and L. Zippin in 1952, constitutes an affirmative solution to a problem posed by Hilbert in 1900 (see Montgomery and Zippin [1]). In addition it was proved by Gleason [1] for finite-dimensional groups and by Yamabe [2] in general that a locally compact group without small subgroups is a Lie group.

The universal enveloping algebra was defined by Poincaré [1], Birkhoff [1], and Witt [1]; starting with Harish-Chandra's work [3] it plays an important role in infinite-dimensional representation theory of Lie groups. See Dixmier [1] for a wide-ranging exposition. The isomorphism $U(\mathfrak{g}) \approx \boldsymbol{D}(G)$ in Prop. 1.9 is proved in Harish-Chandra [7] and is attributed to L. Schwartz in Godement [1]. The injectivity of the mapping $X \to X^*$ of $\mathfrak{g}$ into $U(\mathfrak{g})$ (the Poincaré-Birkhoff-Witt theorem) is an obvious consequence.

Various invariant affine connections on a Lie group were introduced by Cartan and Schouten [1]; Prop. 1.4 is proved by Nomizu [2]. The formula in Theorem 1.7 for the differential of the exponential mapping was proved in Helgason [2]; it also gives a proof of Theorem 7.4, originally proved by F. Schur [1-3].

The treatment of Lie subgroups and subalgebras in §2 is primarily based on Chevalley [2]. Several simplifications have been possible since the exponential mapping is available. In particular, the proofs of the basic Theorems 1.11, 2.1 and 2.3 are from Bruhat [2]. The proof of Theorem 2.6 also occurs there (although oversimplified) and in Freudenthal-deVries [1], §11. The use of the graph of a homomorphism also occurs in another context in Chevalley [2], p. 112. Theorem 2.3 on closed subgroups was originally proved by von Neumann [1] for matrix groups and generalized by É. Cartan [16] to arbitrary Lie groups. Theorem 3.2 was proved by Arens [1], but Cor. 3.3 is older (see Pontrjagin [1]).

§5-§6. The adjoint group goes back to Lie. The existence of a positive definite quadratic form invariant under a given compact linear group is one of H. Weyl's important applications of the invariant measure. The fundamental Theorem 6.9 is also due to Weyl [1], Kap IV, Satz 2. For a fuller exposition of Weyl's proof see Pontrjagin [1], §64. The proof given in the text is due to Samelson [1]. Cartan's proof is contained in Theorem 6.1, Chapter VII. For other proofs see Harish–Chandra [2], Chevalley–Eilenberg [1], Séminaire Sophus Lie [1] (also in Serre [2], Varadarajan [1]), and Seifert [1].

§7-§8. In [1] Maurer proves integrability conditions (12), §7 which are equivalent to the Maurer-Cartan equations (3). Theorem 7.5 is proved in Lie [1], Vol. 2, Chapter XVII (cf. Pontrjagin [1], §56). The proof in the text is Cartan's [25, 26]; we have followed the exposition in Flanders [1], Chapter VII. For a more explicit construction of the local group from the forms ω_i see Chevalley [2], Chapter V. Theorem 7.8 is proved in Hodge [1], §56.2, by essentially the same method.

CHAPTER III

STRUCTURE OF SEMISIMPLE LIE ALGEBRAS

As will become clear in Chapter V, the study of symmetric spaces leads quickly and rather surprisingly to semisimple Lie algebras. This chapter is devoted to a preliminary study of these Lie algebras. The central result is Theorem 6.3, asserting that any complex semisimple Lie algebra has a compact real form. The proof is based on the root space decomposition of a complex semisimple Lie algebra with respect to a Cartan subalgebra. However, the existence of a Cartan subalgebra is not a trivial matter. It is based on Lie's theorem on solvable Lie algebras, proved in §2. In the last section we determine the Cartan subalgebras, and the root pattern for the complex classical Lie algebras $\mathfrak{a}_n$, $\mathfrak{b}_n$, $\mathfrak{c}_n$, and $\mathfrak{d}_n$.

§ 1. Preliminaries

Let K be a field and V a finite-dimensional vector space over K. We shall recall some facts concerning endomorphisms of V. Let Hom (V, V) denote the ring of all endomorphisms of V. Let $e_1, ..., e_n$ be a basis of V. To each $A \in$ Hom (V, V) we associate the matrix

$$(\alpha_{ij}) = \begin{pmatrix} \alpha_{11} & \alpha_{12} & \cdots & \alpha_{1n} \\ \alpha_{21} & \alpha_{22} & \cdots & \alpha_{2n} \\ \cdots & & \cdots & \\ \alpha_{n1} & \alpha_{n2} & \cdots & \alpha_{nn} \end{pmatrix}$$

which is determined by the condition $Ae_j = \sum_{i=1}^{n} \alpha_{ij}e_i$ $(1 \leqslant j \leqslant n)$. We shall call the matrix (α_{ij}) the *expression* of A in terms of the basis $e_1, ..., e_n$. The mapping $A \to (\alpha_{ij})$ is an isomorphism of Hom (V, V) onto the ring $M_n(K)$ of all $n \times n$ matrices with entries in K. If $f_1, ..., f_n$ is a basis dual to $e_1, ..., e_n$, then the endomorphism ${}^tA : V^* \to V^*$ has matrix expression (α_{ji}), the transpose of (α_{ij}). A matrix (α_{ij}) for which $\alpha_{ij} = 0$ if $i > j$ is called an *upper triangular* matrix, a matrix (β_{ij}) for which $\beta_{ij} = 0$ if $i < j$ is called a *lower triangular* matrix. A matrix which is both upper and lower triangular is called a *diagonal matrix*.

If $\lambda \in K$, let V_λ denote the set of elements $e \in V$ such that $Ae = \lambda e$. If $V_\lambda \neq \{0\}$, then λ is called an *eigenvalue* of A and V_λ is called the *eigenspace* of A for the eigenvalue λ. A vector $v \neq 0$ in V which belongs to some eigenspace of A is called an *eigenvector* of A. The equation in λ,

$$\det(\lambda I - A) = 0 \qquad (I = \text{identity endomorphism})$$

is called the *characteristic equation* of A. Those solutions of this equation which lie in K coincide with the eigenvalues of A. The left-hand side of the equation is called the *characteristic polynomial* of A.

An endomorphism $N \in \text{Hom}(V, V)$ is called *nilpotent* if $N^k = 0$ for some integer $k > 0$. If $V \neq \{0\}$ and if $N \in \text{Hom}(V, V)$ is nilpotent, then N has exactly one eigenvalue, namely 0. Let $e_1 \neq 0$ be a vector in V such that $Ne_1 = 0$. If E_1 denotes the one-dimensional subspace of V spanned by e_1, N induces an endomorphism N_1 of the factor space V/E_1. This endomorphism N_1 is again nilpotent and if $\dim V/E_1 \neq 0$, we can select $e_2 \neq 0$ in V such that the vector $(e_2 + E_1) \in V/E_1$ is an eigenvector of N_1. By a continuation of this process we obtain a basis $e_1, \ldots, e_n$ of V such that

$$Ne_1 = 0, \; Ne_p = 0 \bmod (e_1, \ldots, e_{p-1}), \; 2 \leqslant p \leqslant n.$$

Here $(e_1, \ldots, e_{p-1})$ denotes the subspace of V spanned by the vectors $e_1, \ldots, e_{p-1}$. The matrix (n_{ij}) expressing N in terms of the basis $e_1, \ldots, e_n$ has 0 on and below the diagonal. On the other hand, let (n_{ij}) be an $n \times n$ matrix with 0 on and below the diagonal. If $f_1, \ldots, f_n$ is any basis of V, the endomorphism N given by $Nf_j = \sum_{i=1}^n n_{ij} f_i$ is nilpotent. *Thus N is nilpotent if and only if it has a matrix expression with 0 on and below the diagonal.*

Consider now a subset $\mathfrak{S} \subset \text{Hom}(V, V)$. A subspace W of V is called *invariant* (under $\mathfrak{S}$) if $SW \subset W$ for each $S \in \mathfrak{S}$. The space V is called *irreducible* if its only invariant subspaces are $\{0\}$ and V. The set $\mathfrak{S}$ is called *semisimple* if each invariant subspace (under $\mathfrak{S}$) has a complementary invariant subspace. In this case V can be written as a direct sum $V = \sum_i V_i$ where the spaces V_i are invariant and irreducible (under $\mathfrak{S}$). If $\mathfrak{S}$ is a commutative family of endomorphisms, then $\mathfrak{S}$ is semisimple if and only if each $S \in \mathfrak{S}$ is semisimple.

Each $A \in \text{Hom}(V, V)$ can be uniquely decomposed:

$$A = S + N, \; S \text{ semisimple}, \; N \text{ nilpotent}, \; SN = NS. \tag{1}$$

In this decomposition, S and N are polynomials in A, and are called the *semisimple part* and *nilpotent part* of A respectively. Suppose λ is an eigenvalue of A. Select a vector $e \neq 0$ such that $Ae = \lambda e$. Since N is a polynomial in A, but has 0 as the only eigenvalue, it follows that $Ne = 0$ and $Se = \lambda e$. On the other hand, let λ be an eigenvalue of S

and let V_λ be the eigenspace of S for the eigenvalue λ. Since $AS = SA$, we have $AV_\lambda \subset V_\lambda$ and the restriction of $A - \lambda I$ to V_λ is nilpotent. In particular, λ is an eigenvalue of A. Thus, A and S have the same eigenvalues, say $\lambda_1, \cdots, \lambda_r$.

Suppose now that K is algebraically closed. Let $\mathfrak{S}$ be a semisimple commutative family of endomorphisms of V. Then each nonzero invariant irreducible subspace (under $\mathfrak{S}$) is one-dimensional, so V has a basis in terms of which all $S \in \mathfrak{S}$ are expressed by diagonal matrices. In particular, this is the case when $\mathfrak{S}$ consists of the single endomorphism S above. Thus V can be written as a direct sum

$$V = \sum_{i=1}^{r} V_i,$$

where V_i is the eigenspace of S for the eigenvalue λ_i. Let $v \neq 0$ be a vector in V and $\lambda \in K$ such that

$$(A - \lambda I)^k v = 0$$

for some integer k. Taking k as small as possible, we see that λ is an eigenvalue of A, let us say $\lambda = \lambda_1$. We can write $v = \sum_i v_i$ where $v_i \in V_i$, and since V_i is invariant under A, we obtain

$$(A - \lambda_1 I)^k v_i = 0 \qquad \text{for } 1 \leqslant i \leqslant r.$$

Now, the endomorphism $N_i = A - \lambda_i I$ is nilpotent on V_i, whereas the equation

$$(N_i + (\lambda_i - \lambda_1) I)^k v_i = 0$$

shows that if $v_i \neq 0$, N_i has the eigenvalue $\lambda_1 - \lambda_i$ on V_i. It follows that $v_i = 0$ if $i > 1$; hence $v \in V_1$. Summarizing, we get:

Proposition 1.1. *Let V be a finite-dimensional vector space over an algebraically closed field K. Let $A \in \mathrm{Hom}(V, V)$, and let $\lambda_1, \ldots, \lambda_r \in K$ be the different eigenvalues of A. Put*

$$V_i = \{v \in V : (A - \lambda_i I)^k v = 0 \text{ for } k \text{ sufficiently large}\}.$$

Then

(i) $$V = \sum_{i=1}^{r} V_i \qquad (\textit{direct sum}).$$

(ii) *Each V_i is invariant under A.*

(iii) *The semisimple part of* A *is given by*

$$S\left(\sum_{i=1}^{r} v_i\right) = \sum_{i=1}^{r} \lambda_i v_i \qquad (v_i \in V_i).$$

(iv) *The characteristic polynomial of* A *is*

$$\det(\lambda I - A) = (\lambda - \lambda_1)^{d_1} \ldots (\lambda - \lambda_r)^{d_r},$$

where $d_i = \dim V_i$ $(1 \leqslant i \leqslant r)$.

§ 2. Theorems of Lie and Engel

Throughout this section, K denotes a field of characteristic 0 and $\tilde{K}$ its algebraic closure. Let $\mathfrak{g}$ be a Lie algebra over K. The vector space spanned by all elements $[X, Y]$, $X, Y \in \mathfrak{g}$, is an ideal in $\mathfrak{g}$, called the *derived* algebra of $\mathfrak{g}$. The derived algebra will be denoted $\mathfrak{D}\mathfrak{g}$ and the nth derived algebra $\mathfrak{D}^n\mathfrak{g}$ of $\mathfrak{g}$ is defined inductively by $\mathfrak{D}^0\mathfrak{g} = \mathfrak{g}$ and $\mathfrak{D}^n\mathfrak{g} = \mathfrak{D}(\mathfrak{D}^{n-1}\mathfrak{g})$. Each $\mathfrak{D}^n\mathfrak{g}$ is an ideal in $\mathfrak{g}$.

Definition. The Lie algebra $\mathfrak{g}$ is called *solvable* if there exists an integer $n \geqslant 0$ such that $\mathfrak{D}^n\mathfrak{g} = \{0\}$. A Lie group is called solvable if its Lie algebra is solvable.

Let $\mathfrak{g}$ be a solvable Lie algebra $\neq \{0\}$ and let n be the smallest integer for which $\mathfrak{D}^n\mathfrak{g} = \{0\}$. Then $\mathfrak{D}^{n-1}\mathfrak{g}$ is a nonzero abelian ideal in $\mathfrak{g}$. Hence $\mathfrak{g}$ is not semisimple (Prop. 6.1, Chapter II).

Definition. A Lie algebra $\mathfrak{g}$ is said to satisfy the *chain condition* if for each ideal $\mathfrak{h} \neq \{0\}$ in $\mathfrak{g}$ there exists an ideal $\mathfrak{h}_1$ of $\mathfrak{h}$ of codimension 1.

Lemma 2.1. *A Lie algebra* $\mathfrak{g}$ *is solvable if and only if it satisfies the chain condition.*

Proof. If $\mathfrak{g}$ is solvable and $\neq \{0\}$, then $\mathfrak{D}\mathfrak{g} \neq \mathfrak{g}$. Hence there exists a subspace $\mathfrak{h}$ of $\mathfrak{g}$ of codimension 1, containing $\mathfrak{D}\mathfrak{g}$. Then $\mathfrak{h}$ is an ideal in $\mathfrak{g}$. Each ideal (even each subalgebra) of a solvable Lie algebra is solvable; consequently, $\mathfrak{g}$ satisfies the chain condition. On the other hand, suppose $\mathfrak{g}$ is a Lie algebra satisfying the chain condition. Then there exists a sequence

$$\mathfrak{g} = \mathfrak{g}_0 \supset \mathfrak{g}_1 \supset \cdots \supset \mathfrak{g}_{n-1} \supset \mathfrak{g}_n = \{0\},$$

where $\mathfrak{g}_r$ is an ideal in $\mathfrak{g}_{r-1}$ of codimension 1 $(1 \leqslant r \leqslant n)$, and thus $\mathfrak{D}(\mathfrak{g}_{r-1}) \subset \mathfrak{g}_r$. By induction, $\mathfrak{g}$ is solvable.

Theorem 2.2. (Lie) *Let $\mathfrak{g}$ be a solvable Lie algebra over K. Let $V \neq \{0\}$ be a finite-dimensional vector space over $\tilde{K}$, the algebraic closure of K. Let π be a homomorphism of $\mathfrak{g}$ into $\mathfrak{gl}(V)$. Then there exists a vector $v \neq 0$ in V which is an eigenvector of all the members of $\pi(\mathfrak{g})$.*

Proof. We shall prove the theorem by induction on $\dim \mathfrak{g}$. If $\dim \mathfrak{g} = 1$, the theorem is a consequence of Prop. 1.1; we assume now that the theorem holds for all solvable Lie algebras over K of dimension $< \dim \mathfrak{g}$. Let $\mathfrak{h}$ be an ideal in $\mathfrak{g}$ of codimension 1. Then $\mathfrak{h}$ is solvable, so by assumption there exists a vector $e_0 \neq 0$ in V and a linear function $\lambda : \mathfrak{h} \to \tilde{K}$ such that

$$\pi(H)\, e_0 = \lambda(H)\, e_0 \qquad \text{for all } H \in \mathfrak{h}.$$

Select $X \in \mathfrak{g}$ such that $X \notin \mathfrak{h}$ and put

$$e_{-1} = 0, \qquad e_p = \pi(X)^p\, e_0, \qquad p = 1, 2, \ldots .$$

The subspace W of V spanned by all e_p $(p \geqslant 0)$ is invariant under $\pi(X)$. We shall now prove by induction that

$$\pi(H)\, e_p \equiv \lambda(H)\, e_p \quad \mod (e_0, \ldots, e_{p-1}) \quad \text{for all } H \in \mathfrak{h},\ p \geqslant 0. \tag{1}$$

In fact, (1) holds for $p = 0$, and, assuming it for p, we have

$$\pi(H)\, e_{p+1} = \pi(H)\, \pi(X)\, e_p = \pi([H, X])\, e_p + \pi(X)\, \pi(H)\, e_p$$

$$\equiv \lambda([H, X])\, e_p + \pi(X)\, \lambda(H)\, e_p \qquad \mod (e_0, \ldots, e_{p-1}, \pi(X)\, e_0, \ldots, \pi(X)\, e_{p-1})$$

so

$$\pi(H)\, e_{p+1} \equiv \lambda(H)\, e_{p+1} \qquad \mod (e_0, e_1 \ldots e_p) \qquad \text{for } H \in \mathfrak{h}.$$

It follows that W is invariant under $\pi(\mathfrak{g})$ and $\mathrm{Tr}_W\, \pi(H) = \lambda(H) \dim W$. Now $\pi([H, X]) = \pi(H)\, \pi(X) - \pi(X)\, \pi(H)$ so $\mathrm{Tr}_W\, \pi([H, X]) = 0$. Since $\dim W > 0$ we obtain $\lambda([H, X]) = 0$. Now, we have

$$\pi(H)\, e_{p+1} = \pi([H, X])\, e_p + \pi(X)\, \pi(H)\, e_p$$

and the relation

$$\pi(H)\, e_p = \lambda(H)\, e_p \qquad (H \in \mathfrak{h}, \text{ all } p \geqslant 0)$$

follows by induction on p. This shows that for $H \in \mathfrak{h}$, $\pi(H) = \lambda(H)\, I$ on W. Since $\pi(X)$ leaves W invariant it has an eigenvector $v \neq 0$ in W. This vector has the properties stated in the theorem.

Corollary 2.3. *Let $\mathfrak{g}$ be a solvable Lie algebra over a field K and π a representation of $\mathfrak{g}$ on a finite-dimensional vector space $V \neq \{0\}$ over $\tilde{K}$,* *the algebraic closure of K. Then there exists a basis $e_1, \ldots, e_n$ of V, in terms of which all the endomorphisms $\pi(X)$, $X \in \mathfrak{g}$, are expressed by upper triangular matrices.*

In fact, we can apply Theorem 2.2. Let $e_1 \neq 0$ be a common eigenvector of all $\pi(X)$, $X \in \mathfrak{g}$, and consider the subspace E_1 of V spanned by e_1. The representation π induces a representation π_1 of $\mathfrak{g}$ on the factor space V/E_1, so if $\dim V/E_1 \neq 0$ we can select $e_2 \in V$ such that the vector $(e_2 + E_1) \in V/E_1$ is an eigenvector for all $\pi_1(X)$. Continuing in this manner we find a basis $e_1, \ldots, e_n$ of V such that for each $X \in \mathfrak{g}$

$$\pi(X)\, e_i \equiv 0 \qquad \mod (e_1, e_2 \ldots e_i).$$

This means that the matrix representing $\pi(X)$ has zeros below the main diagonal.

Definition. A Lie algebra $\mathfrak{g}$ over K is said to be *nilpotent* if for each $Z \in \mathfrak{g}$, $\mathrm{ad}_{\mathfrak{g}}\, Z$ is a nilpotent endomorphism of $\mathfrak{g}$. A Lie group is called nilpotent if its Lie algebra is nilpotent.

Theorem 2.4. (Engel) *Let V be a nonzero finite-dimensional vector space over K, and let $\mathfrak{g}$ be a subalgebra of $\mathfrak{gl}(V)$ consisting of nilpotent elements. Then*

(i) *$\mathfrak{g}$ is nilpotent.*

(ii) *There exists a vector $v \neq 0$ in V such that $Zv = 0$ for all $Z \in \mathfrak{g}$.*

(iii) *There exists a basis $e_1, \ldots, e_n$ of V in terms of which all the endomorphisms $X \in \mathfrak{g}$ are expressed by matrices with zeros on and below the diagonal.*

Proof. (i) For $Z \in \mathfrak{gl}(V)$ consider the endomorphisms L_Z and R_Z on $\mathfrak{gl}(V)$ given by $L_Z X = ZX$, $R_Z X = XZ$ $(X \in \mathfrak{gl}(V))$. Then L_Z and R_Z commute and if ad denotes the adjoint representation of $\mathfrak{gl}(V)$, we have $\mathrm{ad}\, Z = L_Z - R_Z$. It follows that for $X \in \mathfrak{g}$ and any integer $p \geqslant 0$

$$(\mathrm{ad}\, Z)^p\,(X) = \sum_{i=0}^{p} (-1)^i \binom{p}{i} Z^{p-i} X Z^i. \tag{2}$$

Suppose $Z \in \mathfrak{g}$. Then Z is nilpotent and by relation (2) ad Z is nilpotent on $\mathfrak{gl}(V)$. Since $\mathrm{ad}_{\mathfrak{g}}\, Z$ is the restriction of ad Z to $\mathfrak{g}$, it follows that $\mathrm{ad}_{\mathfrak{g}}\, Z$ is nilpotent.

For the second part of the theorem let $r = \dim \mathfrak{g}$. We shall use induction on r. If $r = 1$, (ii) is trivial. Assume now that (ii) holds for algebras of dimension $< r$. Let $\mathfrak{h}$ be a proper subalgebra of $\mathfrak{g}$ of maximum dimension. If $H \in \mathfrak{h}$, then by (i), $\mathrm{ad}_{\mathfrak{g}}\, H$ is a nilpotent endomorphism of $\mathfrak{g}$

and maps $\mathfrak{h}$ into itself, hence $\text{ad}_{\mathfrak{g}}\, H$ induces a nilpotent endomorphism H^* on the vector space $\mathfrak{g}/\mathfrak{h}$. The set $\{H^* : H \in \mathfrak{h}\}$ is a subalgebra of $\mathfrak{gl}(\mathfrak{g}/\mathfrak{h})$ having dimension $< r$ and consisting of nilpotent elements. Using the induction hypothesis we conclude that there exists an element $X \in \mathfrak{g}$, $X \notin \mathfrak{h}$, such that $\text{ad}_{\mathfrak{g}}\, H(X) \in \mathfrak{h}$ for all $H \in \mathfrak{h}$. The subspace $\mathfrak{h} + KX$ of $\mathfrak{g}$ is therefore a subalgebra of $\mathfrak{g}$ which, due to the maximality of $\mathfrak{h}$, must coincide with $\mathfrak{g}$. Thus $\mathfrak{h}$ is an ideal in $\mathfrak{g}$.

Now let W denote the subspace of V given by

$$W = \{e \in V : He = 0 \text{ for all } H \in \mathfrak{h}\}.$$

Owing to the induction hypothesis, $W \neq \{0\}$. Moreover, if $e \in W$ we have

$$HXe = [H, X]\, e + XHe = 0$$

so $X \cdot W \subset W$. The restriction of X to W is nilpotent and there exists a vector $v \neq 0$ in W such that $Xv = 0$. This vector v has the property required in (ii).

To prove (iii) let e_1 be any vector in V such that $e_1 \neq 0$ and $Ze_1 = 0$ for all $Z \in \mathfrak{g}$. Let E_1 be the subspace of V spanned by e_1. Then each $Z \in \mathfrak{g}$ induces a nilpotent endomorphism Z^* of the vector space V/E_1. If $V/E_1 \neq \{0\}$ we can select $e_2 \in V$, $e_2 \notin E_1$ such that $e_2 + E_1 \in V/E_1$ is annihilated by all Z^* $(Z \in \mathfrak{g})$. Continuing in this manner we find a basis $e_1, ..., e_n$ of V such that for each $Z \in \mathfrak{g}$

$$Ze_1 = 0, \qquad Ze_i \equiv 0 \quad \text{mod}\,(e_1, ..., e_{i-1}), \qquad 2 \leqslant i \leqslant n. \tag{3}$$

The matrix expressing Z in terms of the basis $e_1, ..., e_n$ has zeros on and below the diagonal.

Corollary 2.5. *In the notation of Theorem* 2.4 *we have*

$$X_1 X_2 ... X_s = 0$$

if $s \geqslant \dim V$ *and* $X_i \in \mathfrak{g}$ $(1 \leqslant i \leqslant s)$.

In fact, this is an immediate consequence of (3).

Corollary 2.6. *A nilpotent Lie algebra* $\mathfrak{g}$ *is solvable.*

In fact, the algebra $\text{ad}_{\mathfrak{g}}(\mathfrak{g})$ is a subalgebra of $\mathfrak{gl}(\mathfrak{g})$ and consists of nilpotent endomorphisms of $\mathfrak{g}$. The product of s such endomorphisms is 0 if $s \geqslant \dim \mathfrak{g}$ (Cor. 2.5). In particular, $\mathfrak{g}$ is solvable.

Definition. For a Lie algebra $\mathfrak{l}$, we define

$$\mathscr{C}^0\mathfrak{l} = \mathfrak{l}, \qquad \mathscr{C}^{p+1}\mathfrak{l} = [\mathfrak{l}, \mathscr{C}^p\mathfrak{l}], \qquad p = 0, 1, ...$$

The series

$$\mathscr{C}^0\mathfrak{l} \supset \mathscr{C}^1\mathfrak{l} \supset \mathscr{C}^2\mathfrak{l} \supset \ldots$$

is called the *central decending series* of $\mathfrak{l}$.

Corollary 2.7. *A Lie algebra* $\mathfrak{l}$ *over* K *is nilpotent if and only if* $\mathscr{C}^m\mathfrak{l} = \{0\}$ *for* $m \geqslant$ *dim* $\mathfrak{l}$.

In fact, if $\mathfrak{l}$ is nilpotent we can use Cor. 2.5 on the Lie algebra $\mathfrak{g} = \text{ad}(\mathfrak{l})$ and deduce $\mathscr{C}^m\mathfrak{l} = \{0\}$ if $m \geqslant \dim \mathfrak{l}$. The converse is trivial.

Corollary 2.8. *A nilpotent Lie algebra* $\mathfrak{l} \neq \{0\}$ *has nonzero center.*

In fact, if $\mathscr{C}^m\mathfrak{l} = \{0\}$ then $\mathscr{C}^{m-1}\mathfrak{l}$ lies in the center of $\mathfrak{l}$.

§ 3. Cartan Subalgebras

In this section $\mathfrak{g}$ denotes an arbitrary fixed semisimple Lie algebra over the complex numbers C. The adjoint representation of $\mathfrak{g}$ will be denoted ad.

Definition. A *Cartan subalgebra* of $\mathfrak{g}$ is a subalgebra $\mathfrak{h}$ of $\mathfrak{g}$ satisfying the following conditions:

(i) $\mathfrak{h}$ is a maximal abelian subalgebra of $\mathfrak{g}$.

(ii) For each $H \in \mathfrak{h}$, the endomorphism ad H of $\mathfrak{g}$ is semisimple.

In this section we shall prove that every semisimple Lie algebra $\mathfrak{g}$ over C has a Cartan subalgebra. Later on we shall see that for any two Cartan subalgebras $\mathfrak{h}_1$ and $\mathfrak{h}_2$ of $\mathfrak{g}$, there exists an automorphism σ of $\mathfrak{g}$ such that $\sigma \cdot \mathfrak{h}_1 = \mathfrak{h}_2$.

Let H be any element in $\mathfrak{g}$ and let $0 = \lambda_0, \lambda_1, \ldots, \lambda_r$ be the different eigenvalues of ad H. For each $\lambda \in C$ we consider the subspace

$$\mathfrak{g}(H, \lambda) = \{X \in \mathfrak{g} : (\text{ad } H - \lambda I)^k X = 0 \text{ for some } k\}.$$

Then, according to Prop. 1.1 we have $\mathfrak{g}(H, \lambda) = 0$ unless $\lambda = \lambda_i$ for some i; also

$$\mathfrak{g} = \sum_{i=0}^{r} \mathfrak{g}(H, \lambda_i) \qquad \text{(direct sum)}.$$

Definition. The element $H \in \mathfrak{g}$ is called *regular* if

$$\dim \mathfrak{g}(H, 0) = \min_{X \in \mathfrak{g}} (\dim \mathfrak{g}(X, 0)).$$

We shall now prove the following theorem, which ensures the existence of Cartan subalgebras.

Theorem 3.1. *Let H_0 be a regular element in $\mathfrak{g}$. Then $\mathfrak{g}(H_0, 0)$ is a Cartan subalgebra of $\mathfrak{g}$.*

This will be proved by means of the theorems of Lie and Engel. We put $\mathfrak{h} = \mathfrak{g}(H_0, 0)$.

Lemma 3.2. *If $Z \in \mathfrak{g}$, then $[\mathfrak{g}(Z, \lambda), \mathfrak{g}(Z, \mu)] \subset \mathfrak{g}(Z, \lambda + \mu)$. In particular, $\mathfrak{h}$ is a subalgebra of $\mathfrak{g}$.*

Let $X_\lambda \in \mathfrak{g}(Z, \lambda)$, $X_\mu \in \mathfrak{g}(Z, \mu)$. Then

$$(\text{ad } Z - (\lambda + \mu) I) [X_\lambda, X_\mu] = [(\text{ad } Z - \lambda I) X_\lambda, X_\mu] + [X_\lambda, (\text{ad } Z - \mu I) X_\mu],$$

and by induction

$$(\text{ad } Z - (\lambda + \mu) I)^n [X_\lambda, X_\mu] = \sum_{i=0}^{n} \binom{n}{i} [(\text{ad } Z - \lambda I)^i X_\lambda, (\text{ad } Z - \mu I)^{n-i} X_\mu].$$

The lemma follows immediately.

Lemma 3.3. *The algebra $\mathfrak{h}$ is nilpotent.*

Let $0 = \lambda_0, \lambda_1, ..., \lambda_r$ be the different eigenvalues of ad H_0 and let $\mathfrak{g}'$ denote the subspace $\sum_{i=1}^{r} \mathfrak{g}(H_0, \lambda_i)$ of $\mathfrak{g}$. Then $[\mathfrak{h}, \mathfrak{g}'] \subset \mathfrak{g}'$ due to Lemma 3.2. For each $H \in \mathfrak{h}$ let H' denote the restriction of ad H to $\mathfrak{g}'$. Let $d(H) = \det H'$. Then the function $H \to d(H)$ is a polynomial function on $\mathfrak{h}$ and since H_0' has only nonzero eigenvalues we have $d(H_0) \neq 0$. Now if a polynomial function vanishes on an open set it must vanish identically. We conclude that the subset S of $\mathfrak{h}$ consisting of all points $H \in \mathfrak{h}$ for which $d(H) \neq 0$ is a dense subset of $\mathfrak{h}$. If H is any element in S, the endomorphism H' of $\mathfrak{g}'$ has all its eigenvalues $\neq 0$; it follows that $\mathfrak{g}(H, 0) \subset \mathfrak{h}$. Since H_0 is regular we conclude that $\mathfrak{g}(H, 0) = \mathfrak{h}$. This means that the restriction of ad H to $\mathfrak{h}$ is nilpotent. This restriction is $\text{ad}_\mathfrak{h} H$, so if $l = \dim \mathfrak{h}$ we have

$$(\text{ad}_\mathfrak{h} H)^l = 0 \qquad \text{for each } H \in S. \tag{1}$$

Since S is dense in $\mathfrak{h}$, relation (1) follows by continuity for all $H \in \mathfrak{h}$; thus $\mathfrak{h}$ is nilpotent.

The definition of a regular element and the proof of Lemma 3.3 holds for any complex Lie algebra $\mathfrak{g}$. In the next lemma, however, we make use of the semisimplicity of $\mathfrak{g}$.

Lemma 3.4. *The algebra $\mathfrak{h}$ is abelian and in fact a maximal abelian subalgebra of $\mathfrak{g}$.*

As usual, let B denote the Killing form of $\mathfrak{g}$. If $X \in \mathfrak{g}(H_0, \lambda)$ and

$H \in \mathfrak{h}$, then $\operatorname{ad} X \operatorname{ad} H$ maps the subspace $\mathfrak{g}(H_0, \mu)$ into $\mathfrak{g}(H_0, \lambda + \mu)$; choosing a basis of $\mathfrak{g}$ composed of bases of the spaces $\mathfrak{g}(H_0, \lambda_i)$ we see that

$$\operatorname{Tr}(\operatorname{ad} X \operatorname{ad} H) = B(X, H) = 0 \qquad \text{if } H \in \mathfrak{h},\ X \in \mathfrak{g}'. \tag{2}$$

Now $\mathfrak{h}$, being nilpotent, is solvable (Cor. 2.6) so, by Cor. 2.3, there exists a basis of $\mathfrak{g}$ with respect to which all the endomorphisms $\operatorname{ad} H$ $(H \in \mathfrak{h})$ are expressed by upper triangular matrices. If A, B, C are upper triangular matrices, then ABC and BAC have the same diagonal elements; hence $\operatorname{Tr}(ABC) = \operatorname{Tr}(BAC)$. In particular we have

$$\operatorname{Tr}(\operatorname{ad}[H_1, H_2] \operatorname{ad} H) = 0 \qquad \text{if } H_1, H_2, H \in \mathfrak{h}.$$

Combining this with (2) we see that $[H_1, H_2]$ is orthogonal to $\mathfrak{g}$ (with respect to B). Making now use of the semisimplicity of $\mathfrak{g}$ we conclude that $\mathfrak{h}$ is abelian. The maximality is immediate from the definition of $\mathfrak{h}$.

Passing now to the proof of Theorem 3.1, let λ be one of the nonzero eigenvalues $\lambda_1, \ldots, \lambda_r$ of $\operatorname{ad} H_0$. Each endomorphism $\operatorname{ad} H$ $(H \in \mathfrak{h})$ leaves $\mathfrak{g}(H_0, \lambda)$ invariant, so if $\operatorname{ad}_\lambda H$ denotes the restriction of $\operatorname{ad} H$ to $\mathfrak{g}(H_0, \lambda)$, the mapping $\operatorname{ad}_\lambda : H \to \operatorname{ad}_\lambda H$ is a representation of $\mathfrak{h}$ on $\mathfrak{g}(H_0, \lambda)$. Since the family $\operatorname{ad}_\lambda(\mathfrak{h})$ is commutative there exists a basis $e_1, \ldots, e_s$ of $\mathfrak{g}(H_0, \lambda)$ such that each $\operatorname{ad}_\lambda(H)$ is upper triangular and has the diagonal for its semisimple part. The diagonal elements $\alpha_1(H), \ldots, \alpha_s(H)$ are linear functions on $\mathfrak{h}$ and $\alpha_1(H_0) = \ldots = \alpha_s(H_0) = \lambda$. Let β be any linear function on $\mathfrak{h}$ such that $\beta(H_0) = \lambda$. Let V_β be the subspace of $\mathfrak{g}(H_0, \lambda)$ spanned by the basis vectors e_j for which $\alpha_j(H) \equiv \beta(H)$ for all $H \in \mathfrak{h}$. Then V_β is the set of vectors $X \in \mathfrak{g}(H_0, \lambda)$ such that

$$(\operatorname{ad} H - \beta(H) I)^k X = 0 \tag{3}$$

for all $H \in \mathfrak{h}$ and some fixed k (independent of H). On the other hand, since $\beta(H_0) = \lambda$, no vector $X \in \mathfrak{g}$ which does not lie in $\mathfrak{g}(H_0, \lambda)$ could satisfy (3). It follows that

$$V_\beta = \{X \in \mathfrak{g} : (\operatorname{ad} H - \beta(H) I)^k X = 0 \text{ for all } H \in \mathfrak{h} \text{ and some fixed } k\}. \tag{4}$$

Given any linear function β on $\mathfrak{h}$ we can define a subset V_β of $\mathfrak{g}$ by (4). Then V_β is a subspace of $\mathfrak{g}$ and $V_0 = \mathfrak{h}$. Moreover, the relation

$$[V_\alpha, V_\beta] \subset V_{\alpha+\beta} \tag{5}$$

is proved just as Lemma 3.2.

It has been shown above that $\mathfrak{g}$ is a direct sum of certain of the spaces V_β, say

$$\mathfrak{g} = \sum_i V_{\beta_i}.$$

Then, if $H, H' \in \mathfrak{h}$ we have

$$B(H, H') = \operatorname{Tr}(\operatorname{ad} H \operatorname{ad} H') = \sum_i \beta_i(H)\,\beta_i(H') \dim V_{\beta_i}. \tag{6}$$

The endomorphism ad H can be decomposed

$$\operatorname{ad} H = S + N,$$

where S is semisimple, N is nilpotent, and $SN = NS$. Since S is a polynomial in ad H, S leaves each V_β invariant and (4) shows that $SX = \beta(H)\,X$ for all $X \in V_\beta$. From (5) it now follows quickly that S is a derivation of $\mathfrak{g}$. From Prop. 6.4, Chapter II we know that every derivation of $\mathfrak{g}$ is inner; in other words, there exists an element $Z \in \mathfrak{g}$ such that $S = \operatorname{ad} Z$. Now S, being a polynomial in ad H, commutes with all elements of ad $\mathfrak{h}$. Since $\mathfrak{h}$ is maximal abelian in $\mathfrak{g}$, this implies $Z \in \mathfrak{h}$. Now, ad $Z(X) = \beta(H)\,X$ for $X \in V_\beta$, so by the definition of V_β we conclude that $\beta(H) = \beta(Z)$ for every linear function β on $\mathfrak{h}$ for which $V_\beta \neq \{0\}$. But then (6) and (2) show that $Z - H$ is orthogonal to $\mathfrak{g}$ (with respect to B). Consequently, $Z = H$ so ad H is semisimple. This concludes the proof of Theorem 3.1.

§ 4. Root Space Decomposition

The structure theory of semisimple Lie algebras is based on the following theorem of which a proof was given in §3.

Theorem 4.1. *Every semisimple Lie algebra over* **C** *contains a Cartan subalgebra.*

In §4 *and* §5 *let* $\mathfrak{g}$ *be an arbitrary semisimple Lie algebra over* **C** *and let* $\mathfrak{h}$ *denote an arbitrary fixed Cartan subalgebra.*

Let α be a linear function on the complex vector space $\mathfrak{h}$. Let $\mathfrak{g}^\alpha$ denote the linear subspace of $\mathfrak{g}$ given by

$$\mathfrak{g}^\alpha = \{X \in \mathfrak{g} : [H, X] = \alpha(H)\,X \text{ for all } H \in \mathfrak{h}\}.$$

The linear function α is called a *root* (of $\mathfrak{g}$ with respect to $\mathfrak{h}$) if $\mathfrak{g}^\alpha \neq \{0\}$. In that case, $\mathfrak{g}^\alpha$ is called a *root subspace*. Since $\mathfrak{h}$ is a maximal abelian subalgebra of $\mathfrak{g}$ we have $\mathfrak{g}^0 = \mathfrak{h}$. From the Jacobi identity we obtain easily the relation

$$[\mathfrak{g}^\alpha, \mathfrak{g}^\beta] \subset \mathfrak{g}^{\alpha+\beta} \tag{1}$$

for any pair α, β of C-linear functions on $\mathfrak{h}$.

Let Δ denote the set of all nonzero roots and as before let B denote the Killing form on $\mathfrak{g} \times \mathfrak{g}$.

Theorem 4.2.

(i) $\mathfrak{g} = \mathfrak{h} + \sum_{\alpha \in \Delta} \mathfrak{g}^\alpha$ *(direct sum).*

(ii) *dim* $\mathfrak{g}^\alpha = 1$ *for each* $\alpha \in \Delta$.

(iii) *Let* α, β *be two roots such that* $\alpha + \beta \neq 0$. *Then* $\mathfrak{g}^\alpha$ *and* $\mathfrak{g}^\beta$ *are orthogonal under* B.

(iv) *The restriction of* B *to* $\mathfrak{h} \times \mathfrak{h}$ *is nondegenerate. For each linear form* α *on* $\mathfrak{h}$ *there exists a unique element* $H_\alpha \in \mathfrak{h}$ *such that*

$$B(H, H_\alpha) = \alpha(H) \qquad \text{for all } H \in \mathfrak{h}.$$

We put $\langle \lambda, \mu \rangle = B(H_\lambda, H_\mu)$.

(v) *If* $\alpha \in \Delta$, *then* $-\alpha \in \Delta$ *and*

$$[\mathfrak{g}^\alpha, \mathfrak{g}^{-\alpha}] = CH_\alpha, \qquad \alpha(H_\alpha) \neq 0.$$

Proof. (i) We prove first that the sum is direct. If this were not so we would have a relation

$$H^* + \sum_i X_{\alpha_i} = 0,$$

where $H^* \in \mathfrak{h}$, $X_{\alpha_i} \neq 0$ in $\mathfrak{g}^{\alpha_i}$ where the roots α_i are different and not 0. Then we can select $H \in \mathfrak{h}$ such that the numbers $\alpha_i(H)$ are all different and not zero. In fact, the subset N of $\mathfrak{h}$ where all α_i are different and nonzero is the complement of the union of finitely many hyperplanes. In particular N is not empty. Then H^* and the vectors X_{α_i} lie in different eigenspaces of ad H and are therefore linearly independent. On the other hand, since the set $\mathrm{ad}_{\mathfrak{g}}(\mathfrak{h})$ is semisimple, we can write $\mathfrak{g} = \sum_i \mathfrak{g}_i$ (direct sum), where each $\mathfrak{g}_i$ is a one-dimensional subspace of $\mathfrak{g}$, invariant under $\mathrm{ad}_{\mathfrak{g}}(\mathfrak{h})$. This means that $\mathfrak{g}_i \subset \mathfrak{g}^\alpha$ for a suitable root α, and (i) follows. As a consequence of (i) we have

$$\text{If} \qquad \alpha(H_0) = 0 \qquad \text{for each } \alpha \in \Delta, \text{ then } H_0 = 0. \tag{2}$$

In fact, (i) shows that $[H_0, X] = 0$ for all $X \in \mathfrak{g}$; hence $H_0 = 0$, the center of $\mathfrak{g}$ being $\{0\}$. In order to prove (iii), select any $X \in \mathfrak{g}^\alpha$, $Y \in \mathfrak{g}^\beta$. Then ad X ad Y maps $\mathfrak{g}^\gamma$ into $\mathfrak{g}^{\gamma+\alpha+\beta}$; since $\alpha + \beta \neq 0$, it is clear that $\mathfrak{g}^\gamma \cap \mathfrak{g}^{\gamma+\alpha+\beta} = \{0\}$. Therefore, if the endomorphism ad X ad Y is expressed by means of a basis, each of whose elements lies in a root subspace $\mathfrak{g}^\gamma$, it is obvious that Tr (ad X ad Y) $= 0$. We next prove (iv). If $H_0 \in \mathfrak{h}$ satisfies $B(H_0, H) = 0$ for all $H \in \mathfrak{h}$, then by (iii), $B(H_0, X) = 0$ for all $X \in \mathfrak{g}$. Thus $H_0 = 0$. The latter part of (iv) is a consequence of the first. To prove (v) let $\alpha \in \Delta$. If $\mathfrak{g}^{-\alpha}$ were $\{0\}$, then by (iii) each $X_\alpha \in \mathfrak{g}^\alpha$

would satisfy $B(X_\alpha, X) = 0$ for all $X \in \mathfrak{g}$ which is impossible. Let $H, X_\alpha, X_{-\alpha}$ be arbitrary elements in $\mathfrak{h}, \mathfrak{g}^\alpha, \mathfrak{g}^{-\alpha}$, respectively. Then

$$B([X_\alpha, X_{-\alpha}], H) = B(X_\alpha, [X_{-\alpha}, H]) = B(X_\alpha, X_{-\alpha})\, B(H_\alpha, H).$$

Using (iv) we obtain therefore

$$[X_\alpha, X_{-\alpha}] = B(X_\alpha, X_{-\alpha})\, H_\alpha. \tag{3}$$

Next consider any element $E_{-\alpha} \neq 0$ in $\mathfrak{g}^{-\alpha}$. Since B induces a non-degenerate bilinear form on $\mathfrak{g}^\alpha \times \mathfrak{g}^{-\alpha}$, there exists a vector $E_\alpha \in \mathfrak{g}^\alpha$ such that

$$B(E_\alpha, E_{-\alpha}) = +1.$$

Let β be any root and put $\mathfrak{g}^* = \Sigma_{n\in N}\, \mathfrak{g}^{\beta+n\alpha}$ where N is the set of all integers n for which $\beta + n\alpha$ is a root. Owing to (1), the subspace $\mathfrak{g}^*$ is invariant under ad $E_{-\alpha}$, ad E_α, ad H_α. We compute the trace of ad H_α in two ways. Since $[E_\alpha, E_{-\alpha}] = H_\alpha$ it follows that

$$\mathrm{Tr}_{\mathfrak{g}^*}\, \mathrm{ad}\, H_\alpha = -\, \mathrm{Tr}_{\mathfrak{g}^*}\, \mathrm{ad}\, E_{-\alpha}\, \mathrm{ad}\, E_\alpha + \mathrm{Tr}_{\mathfrak{g}^*}\, \mathrm{ad}\, E_\alpha\, \mathrm{ad}\, E_{-\alpha} = 0.$$

On the other hand, ad H_α leaves each space $\mathfrak{g}^{\beta+n\alpha}$ invariant, so

$$\mathrm{Tr}_{\mathfrak{g}^*}\, \mathrm{ad}\, H_\alpha = \sum_{n\in N} (\beta + n\alpha)\, (H_\alpha)\, \dim \mathfrak{g}^{\beta+n\alpha}.$$

Thus we have the relation

$$\beta(H_\alpha) \sum_{n\in N} \dim \mathfrak{g}^{\beta+n\alpha} = -\, \alpha(H_\alpha) \sum_{n\in N} n \dim \mathfrak{g}^{\beta+n\alpha} \tag{4}$$

for each $\alpha \in \Delta$ and each root β.

Since $\dim \mathfrak{g}^\beta > 0$ we deduce from (2) and (4) that $\alpha(H_\alpha) \neq 0$. This proves (v). To prove (ii), suppose $\dim \mathfrak{g}^\alpha$ were > 1. Then (E_α and $E_{-\alpha}$ being as above), there exists a vector $D_\alpha \neq 0$ in $\mathfrak{g}^\alpha$ such that

$$B(D_\alpha, E_{-\alpha}) = 0.$$

We put $D_{-1} = 0$, $D_n = (\mathrm{ad}\, E_\alpha)^n\, D_\alpha$, $n = 0, 1, 2, \ldots$. Then

$$[E_{-\alpha}, D_n] = -\, \frac{n(n+1)}{2}\, \alpha(H_\alpha)\, D_{n-1}, \qquad n = 0, 1, 2 \ldots. \tag{5}$$

For $n = 0$ this is clear from (3). Assuming (5) true for n, we have

$$[E_{-\alpha}, D_{n+1}] = [E_{-\alpha}, [E_\alpha, D_n]] = -[E_\alpha, [D_n, E_{-\alpha}]] - [D_n, [E_{-\alpha}, E_\alpha]]$$

$$= -\frac{n(n+1)}{2}\alpha(H_\alpha)[E_\alpha, D_{n-1}] + [D_n, H_\alpha]$$

$$= -\left(\frac{n(n+1)}{2} + (n+1)\right)\alpha(H_\alpha) D_n = -\frac{(n+1)(n+2)}{2}\alpha(H_\alpha) D_n,$$

where we have used the fact that $D_n \in \mathfrak{g}^{(n+1)\alpha}$. Since $D_0 \neq 0$, (5) shows that all $D_n \neq 0$, $n = 0, 1, 2 \ldots$ which is impossible. This proves (ii) so Theorem 4.2 is proved.

Let $\alpha \in \Delta$ and let β be any root. The *α-series* containing β is by definition the set of all roots of the form $\beta + n\alpha$ where n is an integer.

Theorem 4.3. *Let β be a root, and $\alpha \in \Delta$.*

(i) *The α-series containing β has the form $\beta + n\alpha$ ($p \leqslant n \leqslant q$) (the α-series is an uninterrupted string). Also*

$$-2\frac{\beta(H_\alpha)}{\alpha(H_\alpha)} = p + q.$$

(ii) *Let $X_\alpha \in \mathfrak{g}^\alpha$, $X_{-\alpha} \in \mathfrak{g}^{-\alpha}$, $X_\beta \in \mathfrak{g}^\beta$ where $\beta \neq 0$. Then*

$$[X_{-\alpha}, [X_\alpha, X_\beta]] = \frac{q(1-p)}{2}\alpha(H_\alpha) B(X_\alpha, X_{-\alpha}) X_\beta.$$

(iii) *The only roots proportional to α are* $-\alpha, 0, \alpha$.

(iv) *Suppose $\alpha + \beta \neq 0$. Then $[\mathfrak{g}^\alpha, \mathfrak{g}^\beta] = \mathfrak{g}^{\alpha+\beta}$.*

Proof. Let $E_{-\alpha}$, E_α be any elements in $\mathfrak{g}^{-\alpha}$ and $\mathfrak{g}^\alpha$ respectively satisfying

$$B(E_\alpha, E_{-\alpha}) = 1.$$

To prove (i), let r, s be two integers such that $\beta + n\alpha$ is a root for $r \leqslant n \leqslant s$ but neither for $n = r - 1$ nor $n = s + 1$. Such a set $\beta + n\alpha$, ($r \leqslant n \leqslant s$), we shall call a *maximal string*. The subspace

$$\mathfrak{g}^* = \sum_{n=r}^{s} \mathfrak{g}^{\beta+n\alpha}$$

is invariant under ad $E_{-\alpha}$, ad E_α, ad H_α. Since $H_\alpha = [E_\alpha, E_{-\alpha}]$ we have

$$\mathrm{Tr}_{\mathfrak{g}^*}(\mathrm{ad}\, H_\alpha) = 0.$$

On the other hand,

$$\mathrm{Tr}_{\mathfrak{g}^*}(\mathrm{ad}\, H_\alpha) = \sum_{n=r}^{s} (\beta + n\alpha)(H_\alpha)$$

and since $\alpha(H_\alpha) \neq 0$ we obtain

$$-2\frac{\beta(H_\alpha)}{\alpha(H_\alpha)} = r + s. \tag{6}$$

The α-series containing β is of course a union of maximal strings $\beta + n\alpha$, $r_i \leqslant n \leqslant s_i$. Since (6) holds for each maximal string, the α-series can only consist of one such string. This proves (i). For (iii) let $n\alpha$ ($p \leqslant n \leqslant q$) be the α-series containing 0. The subspace

$$\mathfrak{s} = \sum_{n=-1}^{q} \mathfrak{g}^{n\alpha}$$

is then invariant under ad $E_{-\alpha}$, ad E_α, and ad H_α. We find as before

$$0 = \mathrm{Tr}_{\mathfrak{s}}\, \mathrm{ad}\, H_\alpha = \sum_{n=-1}^{q} n\alpha(H_\alpha),$$

which implies $q = 1$. Using Theorem 4.2 (v), we see that $p = -1$. Now suppose there were a complex number c which is not an integer such that $c\alpha$ is a root. Using (i) on the root $\beta = c\alpha$ we see that $c = n + \frac{1}{2}$ where n is an integer. The α-series containing β will also contain $-(n + \frac{1}{2})\alpha$, and since this series consists of just one string it must contain $-\frac{1}{2}\alpha$ and $\frac{1}{2}\alpha$. But since $\alpha = 2(\frac{1}{2}\alpha)$, this contradicts the first part of the proof. We next prove (iv). We have $[\mathfrak{g}^\alpha, \mathfrak{g}^\beta] \subset \mathfrak{g}^{\alpha+\beta}$ so (iv) is obvious if $\alpha + \beta$ is not a root. Suppose $\alpha + \beta$ is a root and $\beta + n\alpha$, $p \leqslant n \leqslant q$, is the α-series containing β. Then $q \geqslant 1$. If $[\mathfrak{g}^\alpha, \mathfrak{g}^\beta]$ were $\{0\}$, then the subspace

$$\mathfrak{t} = \sum_{n=p}^{0} \mathfrak{g}^{\beta+n\alpha}$$

would be invariant under ad E_α, ad $E_{-\alpha}$, ad H_α. We find as before

$$0 = \mathrm{Tr}_{\mathfrak{t}}\, \mathrm{ad}\, H_\alpha = \sum_{n=p}^{0} (\beta + n\alpha)(H_\alpha)$$

or $-2\beta(H_\alpha) = \alpha(H_\alpha)\, p$. Then (i) implies that $q = 0$ which is a contradiction. Finally we prove (ii). First observe that $\beta + p\alpha \neq 0$. Select a vector $E_p \neq 0$ in $\mathfrak{g}^{\beta+p\alpha}$, and put $E_n = (\text{ad } E_\alpha)^{n-p} E_p$ for $n \geqslant p$. Then $E_n = 0$ for $n > q$; if $p \leqslant n \leqslant q$, then $E_n \neq 0$ as a consequence of (iv) and Theorem 4.2 (v). We shall now prove

$$[E_{-\alpha}, [E_\alpha, E_n]] = \frac{(q-n)(1-p+n)}{2} \alpha(H_\alpha) E_n \qquad (n \geqslant p). \tag{7}$$

Since X_β is a scalar multiple of E_0, (7) would imply (ii). We prove (7) by induction and consider first the case $n = p$. By the Jacobi identity

$$\begin{aligned}[E_{-\alpha}, [E_\alpha, E_p]] &= -[E_\alpha, [E_p, E_{-\alpha}]] - [E_p, [E_{-\alpha}, E_\alpha]] \\ &= 0 + [E_p, H_\alpha] = -(\beta + p\alpha)(H_\alpha) E_p,\end{aligned}$$

which by (i) equals $\frac{1}{2}(q-p)\alpha(H_\alpha) E_p$. Now assume (7); we have

$$[E_{-\alpha}, [E_\alpha, E_{n+1}]] = -[E_\alpha, [E_{n+1}, E_{-\alpha}]] - [E_{n+1}, [E_{-\alpha}, E_\alpha]]. \tag{8}$$

The first term on the right is $[E_\alpha, [E_{-\alpha}, [E_\alpha, E_n]]$ which by induction hypothesis equals

$$\frac{(q-n)(1-p+n)}{2} \alpha(H_\alpha) E_{n+1}.$$

The last term on the right-hand side of (8) is $-(\beta + (n+1)\alpha)(H_\alpha) E_{n+1}$ since $E_{n+1} \in \mathfrak{g}^{\beta+(n+1)\alpha}$. Using (i) we find that these two terms add up to

$$\begin{aligned}&\tfrac{1}{2}\alpha(H_\alpha) E_{n+1}\{(q-n)(1-p+n) + p + q - 2n - 2\} \\ &\qquad = \frac{(q-n-1)(n+2-p)}{2} \alpha(H_\alpha) E_{n+1}.\end{aligned}$$

This proves (7), and Theorem 4.3 is proved.

Theorem 4.4. *Let* $\mathfrak{h}_R = \sum_{\alpha \in \Delta} R H_\alpha$. *Then*

(i) *B is real and strictly positive definite on $\mathfrak{h}_R \times \mathfrak{h}_R$.*

(ii) *$\mathfrak{h} = \mathfrak{h}_R + i\mathfrak{h}_R$ (direct sum).*

Proof. We have for $H, H' \in \mathfrak{h}$

$$B(H, H') = \text{Tr}(\text{ad } H \text{ ad } H') = \sum_{\beta \in \Delta} \beta(H)\beta(H'). \tag{9}$$

From Theorem 4.3 (i) we know that

$$2\beta(H_\alpha) = -\alpha(H_\alpha)(p_{\beta,\alpha} + q_{\beta,\alpha}), \qquad p_{\beta,\alpha}, q_{\beta,\alpha} \text{ integers},$$

so

$$\alpha(H_\alpha) = B(H_\alpha, H_\alpha) = \tfrac{1}{4}\alpha(H_\alpha)^2 \sum_{\beta \in \Delta} (p_{\beta,\alpha} + q_{\beta,\alpha})^2.$$

Since $\alpha(H_\alpha) \neq 0$ this shows that $\alpha(H_\alpha)$ is real (and positive) and $\beta(H)$ is real for each $H \in \mathfrak{h}_R$. Using (2) and (9), part (i) follows. Moreover, (i) shows that $\mathfrak{h}_R \cap i\mathfrak{h}_R = \{0\}$. Finally, the spaces $\mathfrak{h}$ and $\Sigma_{\alpha \in \Delta} CH_\alpha$ must coincide; in fact, suppose the contrary were the case. Then there exists a linear function λ on $\mathfrak{h}$ which is not identically 0 but vanishes on the subspace $\Sigma_{\alpha \in \Delta} CH_\alpha$. There exists a unique element $H_\lambda \in \mathfrak{h}$ such that $B(H, H_\lambda) = \lambda(H)$ for all $H \in \mathfrak{h}$. In particular, $\alpha(H_\lambda) = 0$ for all $\alpha \in \Delta$, so by (2), $H_\lambda = 0$ and $\lambda \equiv 0$. This contradiction proves (ii).

§ 5. Significance of the Root Pattern

To a semisimple Lie algebra $\mathfrak{g}$ and a Cartan subalgebra $\mathfrak{h}$ we have associated a system of vectors H_α, $\alpha \in \Delta$. We shall now see that this system determines $\mathfrak{g}$ up to isomorphism.

Let S be any subset of Δ. The *hull* of S, denoted $\bar{S}$, is by definition the set of all roots in Δ of the form $\pm \alpha$, $\pm (\alpha + \beta)$ where α, β run through S. For each pair α, $-\alpha \in \bar{S}$ we select vectors $E_\alpha \in \mathfrak{g}^\alpha$, $E_{-\alpha} \in \mathfrak{g}^{-\alpha}$ such that

$$B(E_\alpha, E_{-\alpha}) = 1 \qquad \text{for } \alpha, -\alpha \in \bar{S}. \tag{1}$$

Let α, β be any elements in $\bar{S}$ such that $\alpha + \beta \neq 0$ and such that either $\alpha + \beta \in \bar{S}$ or $\alpha + \beta \notin \Delta$. We define the number $N_{\alpha,\beta}$ by

$$[E_\alpha, E_\beta] = N_{\alpha,\beta} E_{\alpha+\beta} \qquad \text{if } \alpha + \beta \in \bar{S},$$
$$N_{\alpha,\beta} = 0 \qquad \text{if } \alpha + \beta \notin \Delta.$$

Thus $N_{\alpha,\beta}$ is defined under the conditions:

(a) $\alpha, \beta \in \bar{S}$; (b) $\alpha + \beta \neq 0$; (c) $\alpha + \beta \in \bar{S}$, or $\alpha + \beta \notin \Delta$.

We have obviously $N_{\alpha,\beta} = - N_{\beta,\alpha}$.

Lemma 5.1. *Suppose $\alpha, \beta, \gamma \in \bar{S}$ and $\alpha + \beta + \gamma = 0$. Then*

$$N_{\alpha,\beta} = N_{\beta,\gamma} = N_{\gamma,\alpha} \qquad (\textit{all } N \textit{ are defined}).$$

Proof. We use the Jacobi identity on the vectors E_α, E_β, E_γ. In view of (3), §4, we obtain

$$N_{\beta,\gamma} H_\alpha + N_{\gamma,\alpha} H_\beta + N_{\alpha,\beta} H_\gamma = 0.$$

On the other hand, $-H_\gamma = H_\alpha + H_\beta$ and β is not proportional to α. The lemma follows.

Lemma 5.2. *Suppose* $\alpha, \beta \in S$, $\alpha + \beta \in \Delta$. *Let* $\beta + n\alpha$ $(p \leqslant n \leqslant q)$ *denote the* α*-series containing* β. *Then*

$$N_{\alpha,\beta}N_{-\alpha,-\beta} = -\frac{q(1-p)}{2}\alpha(H_\alpha).$$

Proof. Since $-\alpha + (\alpha + \beta) = \beta$, $N_{-\alpha,\alpha+\beta}$ is defined; using Theorem 4.3 (ii), we have

$$N_{\alpha,\beta}N_{-\alpha;\alpha+\beta} = \frac{q(1-p)}{2}\alpha(H_\alpha).$$

From Lemma 5.1 we have

$$N_{-\alpha,\alpha+\beta} = N_{\alpha+\beta,-\beta} = N_{-\beta,-\alpha}$$

and Lemma 5.2 follows.

Lemma 5.3. *Suppose* $\alpha, \beta, \gamma, \delta$ *are four roots in* S (*not necessarily distinct*), *no two of which have sum* 0. *If*

$$\alpha + \beta + \gamma + \delta = 0,$$

then

$$N_{\alpha,\beta}N_{\gamma,\delta} + N_{\beta,\gamma}N_{\alpha,\delta} + N_{\gamma,\alpha}N_{\beta,\delta} = 0 \qquad (\textit{all } N \textit{ are defined}).$$

Proof. Suppose first that $\beta + \gamma$ is a root. Then $\beta + \gamma \in \bar{S}$ and since $\alpha + (\beta + \gamma) = -\delta$, $N_{\alpha,\beta+\gamma}$ is defined. In this case, we have

$$[E_\alpha, [E_\beta, E_\gamma]] = N_{\beta,\gamma}N_{\alpha,\beta+\gamma}E_{-\delta}.$$

Applying Lemma 5.1 to α, $\beta + \gamma$, δ, we have $N_{\alpha,\beta+\gamma} = N_{\delta,\alpha}$ and therefore

$$[E_\alpha, [E_\beta, E_\gamma]] = -N_{\beta,\gamma}N_{\alpha,\delta}E_{-\delta}. \tag{2}$$

This relation holds also if $\beta + \gamma$ is not a root because then both sides are 0. In (2) we permute the letters α, β, γ cyclically and use Jacobi's identity. Since $E_{-\delta} \neq 0$, Lemma 5.3 follows.

A set M is said to be *ordered* (or totally ordered) by means of a relation $<$ if to any two elements a, b in the set exactly one of the following relations holds: $a < b$, $b < a$ or $a = b$. Moreover, it is assumed that whenever $a < b$ and $b < c$, then $a < c$. The relation $a > b$ is to mean the same as $b < a$. If the relation $a < b$ (or $a > b$) is defined only for certain pairs in M, the set M is called *partially ordered.*

Let V be a finite-dimensional vector space over $\boldsymbol{R}$; V is said to be an *ordered vector space* if it is an ordered set and the ordering relation $<$ satisfies the conditions: (1) $X > 0$, $Y > 0$ implies $X + Y > 0$.

(2) If $X > 0$ and a is a positive real number, then $aX > 0$. Note that (1) implies: $X > 0$ if and only if $-X < 0$. If $X_1, ..., X_n$ is a basis of V then V can be turned into an ordered vector space as follows: Let $X, Y \in V$. We say $X > Y$ if $X - Y = \sum_{i=1}^{n} a_i X_i$ and the first nonzero number in the sequence $a_1, a_2, ..., a_n$ is > 0. It is readily verified that V is an ordered vector space with this ordering, which is called the *lexicographic ordering* of V with respect to the basis $X_1, ..., X_n$. Let $V^\wedge$ denote the dual space of V. Let $\lambda, \mu \in V^\wedge$. We say $\lambda > \mu$ if the first nonzero number in the sequence $\lambda(X_1) - \mu(X_1), ..., \lambda(X_n) - \mu(X_n)$ is positive. This ordering of $V^\wedge$ is called the lexicographic ordering with respect to the basis $X_1, ..., X_n$ of $V(= (V^\wedge)^\wedge)$. The element $\lambda \in V^\wedge$ is called positive if $\lambda > 0$.

Let V and W be two vector spaces over $\mathbf{R}$ and $V^\wedge$ and $W^\wedge$ their duals. If φ is a linear mapping of V into W then ${}^t\varphi$ shall denote the dual mapping of $W^\wedge$ into $V^\wedge$ which is determined by

$$({}^t\varphi)(F)(v) = F(\varphi(v)) \qquad \text{if } v \in V, F \in W^\wedge.$$

The following important theorem shows that a semisimple Lie algebra over $\mathbf{C}$ is determined (up to isomorphism) by means of a Cartan subalgebra and the corresponding pattern of roots.

Theorem 5.4. *Let $\mathfrak{g}$ and $\mathfrak{g}'$ be two semisimple Lie algebras, $\mathfrak{h}$ and $\mathfrak{h}'$ Cartan subalgebras of $\mathfrak{g}$ and $\mathfrak{g}'$, respectively. Let Δ and Δ' denote the corresponding sets of nonzero roots and as usual let*

$$\mathfrak{h}_{\mathbf{R}} = \sum_{\alpha \in \Delta} \mathbf{R} H_\alpha, \qquad \mathfrak{h}'_{\mathbf{R}} = \sum_{\alpha' \in \Delta'} \mathbf{R} H_{\alpha'}.$$

Then Δ and Δ' can be considered as subsets of the dual space of $\mathfrak{h}_{\mathbf{R}}$ and $\mathfrak{h}'_{\mathbf{R}}$, respectively, since each $\beta \in \Delta$ ($\beta' \in \Delta'$) is real on $\mathfrak{h}_{\mathbf{R}}$ ($\mathfrak{h}'_{\mathbf{R}}$).

Suppose φ is a one-to-one $\mathbf{R}$-linear mapping of $\mathfrak{h}_{\mathbf{R}}$ onto $\mathfrak{h}'_{\mathbf{R}}$ such that ${}^t\varphi$ maps Δ' onto Δ. Then φ can be extended to an isomorphism $\tilde{\varphi}$ of $\mathfrak{g}$ onto $\mathfrak{g}'$.

Proof. The Killing forms B and B' induce positive definite metrics on $\mathfrak{h}_{\mathbf{R}}$ and $\mathfrak{h}'_{\mathbf{R}}$. We shall first prove that φ is an isometry, that is,

$$B(H_1, H_2) = B'(\varphi H_1, \varphi H_2) \qquad \text{for } H_1, H_2 \in \mathfrak{h}_{\mathbf{R}}.$$

It suffices to prove that

$$B(H_\alpha, H_\beta) = B'(H_{\alpha'}, H_{\beta'})$$

for all $\alpha, \beta \in \Delta$, where $\alpha = {}^t\varphi \cdot \alpha'$, $\beta = {}^t\varphi \cdot \beta'$. Using Theorem 4.3 (i), we obtain

$$\frac{\beta(H_\alpha)}{\alpha(H_\alpha)} = \frac{\beta'(H_{\alpha'})}{\alpha'(H_{\alpha'})}$$

for all $\alpha, \beta \in \Delta$. We write this relation as

$$\beta(H_\alpha) = c_\alpha \beta'(H_{\alpha'}).$$

Interchanging α and β, we find $c_\alpha = c$, independent of α. Since

$$B(H_\alpha, H_\beta) = \sum_{\gamma \in \Delta} \gamma(H_\alpha)\,\gamma(H_\beta) = c^2 \sum_{\gamma' \in \Delta'} \gamma'(H_{\alpha'})\,\gamma'(H_{\beta'}) = c^2 B(H_{\alpha'}, H_{\beta'})$$

we see that $c^2 = c$ so $c = 1$ as stated.

Now given any basis in $\mathfrak{h}_R$ we can introduce a lexicographic ordering in the dual space of $\mathfrak{h}_R$. Thus Δ becomes an ordered set and we shall prove Theorem 5.4 by induction with respect to this ordering. We select for each $\alpha \in \Delta$ an element $E_\alpha \in \mathfrak{g}^\alpha$ such that (1) holds for all $\alpha \in \Delta$. The numbers $N_{\alpha,\beta}$ are then defined for all $\alpha, \beta \in \Delta$ for which $\alpha + \beta \neq 0$.

We shall show that to each $\alpha' \in \Delta'$ there exists an element $E_{\alpha'} \in \mathfrak{g}'^{(\alpha')}$ such that

$$B'(E_{\alpha'}, E_{-\alpha'}) = 1, \qquad \alpha \in \Delta, \tag{3}$$

and such that

$$[E_{\alpha'}, E_{\beta'}] = N_{\alpha,\beta} E_{\alpha'+\beta'} \qquad \text{if } \alpha, \beta, \alpha+\beta \in \Delta. \tag{4}$$

Let $\rho \in \Delta$ be a positive root and let Δ_ρ denote the set of roots $\alpha \in \Delta$ satisfying $-\rho < \alpha < \rho$. If there exists a root greater than ρ, then ρ^* will denote the smallest such root.

The induction hypothesis is that for each $\alpha \in \Delta_\rho$, the vector $E_{\alpha'}$ can be chosen in $\mathfrak{g}'^{(\alpha')}$ in such a way that (3) is satisfied for all $\alpha \in \Delta_\rho$ and (4) is satisfied if $\alpha, \beta, \alpha + \beta \in \Delta_\rho$. We shall then define $E_{\rho'}$ and $E_{-\rho'}$ in such a way that (3) is satisfied for all $\alpha \in \Delta_{\rho^*}$ and (4) is satisfied if $\alpha, \beta, \alpha + \beta \in \Delta_{\rho^*}$. If ρ^* does not exist, Δ_{ρ^*} is to mean Δ itself.

If ρ has no decomposition $\rho = \alpha + \beta$ with $\alpha, \beta \in \Delta_\rho$ then we just have to take for $E_{\rho'}$ an arbitrary nonzero element in $\mathfrak{g}'^{(\rho')}$ and then fix $E_{-\rho'}$ by the relation

$$B'(E_{\rho'}, E_{-\rho'}) = 1. \tag{5}$$

If ρ has a decomposition $\rho = \alpha + \beta$, $\alpha, \beta \in \Delta_\rho$ we select the particular decomposition for which α is as small as possible and define $E_{\rho'}$ by means of the equation

$$[E_{\alpha'}, E_{\beta'}] = N_{\alpha,\beta} E_{\rho'}. \tag{6}$$

Then $E_{\rho'} \neq 0$ and we can again define $E_{-\rho'}$ by (5). In order to prove that (4) holds for all $\alpha, \beta, \alpha + \beta \in \Delta_{\rho^*}$, we define the numbers $M_{\gamma,\delta}$ by means of the relation

$$[E_{\gamma'}, E_{\delta'}] = M_{\gamma,\delta} E_{\gamma'+\delta'} \qquad \text{if } \gamma, \delta, \gamma + \delta \in \Delta_{\rho^*}. \tag{7}$$

We also put $M_{\gamma,\delta} = 0$ if γ, δ satisfy the conditions $\gamma, \delta \in \Delta_{\rho^*}$, $\gamma + \delta \neq 0$, $\gamma + \delta \notin \Delta$. We shall prove that $N_{\gamma,\delta} = M_{\gamma,\delta}$ whenever $\gamma, \delta, \gamma + \delta \in \Delta_{\rho^*}$. We have to consider various possibilities:

1. $\gamma, \delta, \gamma + \delta \in \Delta_\rho$. Then $N_{\gamma,\delta} = M_{\gamma,\delta}$ by the induction hypothesis.

2. $\gamma + \delta = \rho$. Then $\gamma, \delta \in \Delta_\rho$. We can assume that the decomposition $\rho = \gamma + \delta$ differs from the decomposition $\rho = \alpha + \beta$. Then the roots $\alpha, \beta, -\gamma, -\delta$ satisfy the relation $\alpha + \beta + (-\gamma) + (-\delta) = 0$ and no two of these roots have sum 0. We can apply Lemmas 5.2 and 5.3 to $\mathfrak{g}$ for $S = \Delta$. We obtain

$$N_{\alpha,\beta} N_{-\gamma,-\delta} = -N_{\beta,-\gamma} N_{\alpha,-\delta} - N_{-\gamma,\alpha} N_{\beta,-\delta}, \tag{8}$$

$$N_{\gamma,\delta} N_{-\gamma,-\delta} = -\frac{l(1-k)}{2} \gamma(H_\gamma), \tag{8$'$}$$

if $\delta + n\gamma$ $(k \leqslant n \leqslant l)$ is the γ-series containing δ. We can also apply Lemmas 5.2 and 5.3 to $\mathfrak{g}'$ by taking for S the set of roots $\alpha', \beta', -\gamma', -\delta'$. To see that the lemmas can be applied, we note that ${}^t\varphi(\bar{S}) \subset \Delta_{\rho^*}$ so $E_{\mu'}$ is defined for each $\mu' \in \bar{S}$; also $M_{\mu,\nu}$ is defined under the required conditions (a) $\mu', \nu' \in \bar{S}$, (*b*) $\mu' + \nu' \neq 0$, (*c*) $\mu' + \nu' \in \bar{S}$ or $\mu' + \nu' \notin \Delta'$. We obtain

$$M_{\alpha,\beta} M_{-\gamma,-\delta} = -M_{\beta,-\gamma} M_{\alpha,-\delta} - M_{-\gamma,\alpha} M_{\beta,-\delta}, \tag{9}$$

$$M_{-\gamma,-\delta} M_{\gamma,\delta} = -\frac{l(1-k)}{2} (-\gamma'(H_{-\gamma'})), \tag{9$'$}$$

because $-\delta + n(-\gamma)$, $(k \leqslant n \leqslant l)$, is the $(-\gamma)$-series containing $-\delta$. From **1** we know that the right-hand sides of (9) and (8) are the same. From (6) and (7) we have $M_{\alpha,\beta} = N_{\alpha,\beta} \neq 0$. It follows that $N_{-\gamma,-\delta} = M_{-\gamma,-\delta}$. By the first part of the proof, $\gamma(H_\gamma) = \gamma'(H_{\gamma'})$ so by (8′) and (9′), $N_{\gamma,\delta} = M_{\gamma,\delta}$ as we wished to prove.

3. $\gamma + \delta = -\rho$. Then $(-\gamma) + (-\delta) = \rho$, $(-\gamma), (-\delta) \in \Delta_\rho$. By **2**, we have $N_{\gamma,\delta} = M_{\gamma,\delta}$.

4. One of the roots γ, δ equals $\pm \rho$. Suppose, for example, $\gamma = -\rho$. Then $\delta \neq \pm \rho$ and $\rho = \delta + (-\gamma - \delta)$ where $\delta, -\gamma - \delta \in \Delta_\rho$. Using **2** we have $N_{\delta,-\gamma-\delta} = M_{\delta,-\gamma-\delta}$. We apply Lemma 5.1 to $\mathfrak{g}$ for $S = \Delta$, and get

$$N_{\delta,-\gamma-\delta} = N_{-\gamma-\delta,\gamma} = N_{\gamma,\delta}.$$

Applying Lemma 5.1 to $\mathfrak{g}'$ for $S = \{\delta', -\gamma' - \delta', \gamma'\}$ we obtain (since ${}^t\varphi(\bar{S}) \subset \Delta_{\rho^*}$)

$$M_{\delta,-\gamma-\delta} = M_{-\gamma-\delta,\gamma} = M_{\gamma,\delta}$$

and $N_{\gamma,\delta} = M_{\gamma,\delta}$ follows. This proves relations (3) and (4).

Consider now the linear mapping $\tilde{\varphi} : \mathfrak{g} \to \mathfrak{g}'$ determined by

$$\tilde{\varphi}(H_\alpha) = \varphi(H_\alpha) = H_{\alpha'},$$

$$\tilde{\varphi}(E_\alpha) = E_{\alpha'}, \qquad \alpha \in \Delta.$$

Relations (3) and (4) then show that $\tilde{\varphi}$ is an isomorphism of $\mathfrak{g}$ onto $\mathfrak{g}'$.

Remark 1. The extension $\tilde{\varphi}$ is not in general unique; in fact, if $H \in \mathfrak{h}$, then $e^{\operatorname{ad} H}$ is an automorphism of $\mathfrak{g}$ whose restriction to $\mathfrak{h}$ is the identity mapping.

Remark 2. In the proof above the ordering of Δ was introduced in order to carry out an induction process. However, the ordering of Δ plays an important role in other contexts which will come up later. Therefore, it should be borne in mind that "ordering of Δ" shall always mean the ordering of the set Δ induced by some vector space ordering of the dual space of $\mathfrak{h}_{\boldsymbol{R}}$.

Theorem 5.5. *For each $\alpha \in \Delta$ a vector $X_\alpha \in \mathfrak{g}^\alpha$ can be chosen such that for all $\alpha, \beta \in \Delta$*

$$[X_\alpha, X_{-\alpha}] = H_\alpha, \quad [H, X_\alpha] = \alpha(H)\, X_\alpha \quad \text{for } H \in \mathfrak{h};$$

$$[X_\alpha, X_\beta] = 0 \qquad \text{if } \alpha + \beta \neq 0 \text{ and } \alpha + \beta \notin \Delta;$$

$$[X_\alpha, X_\beta] = N_{\alpha,\beta} X_{\alpha+\beta} \qquad \text{if } \alpha + \beta \in \Delta,$$

where the constants $N_{\alpha,\beta}$ satisfy

$$N_{\alpha,\beta} = - N_{-\alpha,-\beta}.$$

For any such choice

$$N_{\alpha,\beta}^2 = \frac{q(1-p)}{2}\, \alpha(H_\alpha),$$

where $\beta + n\alpha$ ($p \leqslant n \leqslant q$) is the α-series containing β.

Proof. Let φ denote the mapping $H \to -H$ of $\mathfrak{h}_{\boldsymbol{R}}$ onto itself. Then

$({}^t\varphi)(\lambda) = -\lambda$ for each real linear form λ on $\mathfrak{h}_R$. In particular, ${}^t\varphi$ maps the set Δ onto itself so from Theorem 5.4 we know that φ can be extended to an automorphism $\tilde{\varphi}$ of $\mathfrak{g}$. For each $\alpha \in \Delta$ we select $E_\alpha \in \mathfrak{g}^\alpha$ such that

$$B(E_\alpha, E_{-\alpha}) = 1 \qquad \text{for all } \alpha \in \Delta.$$

Since $\tilde{\varphi}(E_\alpha) \in \mathfrak{g}^{-\alpha}$ and $\dim \mathfrak{g}^{-\alpha} = 1$ there exists a complex number $c_{-\alpha}$ such that $\tilde{\varphi}(E_\alpha) = c_{-\alpha}E_{-\alpha}$. Since B is invariant under $\tilde{\varphi}$ we have $c_\alpha c_{-\alpha} = 1$. For each $\alpha \in \Delta$ one can select a number a_α such that

$$\left.\begin{aligned} a_\alpha^2 &= -c_\alpha, \\ a_\alpha a_{-\alpha} &= +1 \end{aligned}\right. \qquad \text{for } \alpha \in \Delta.$$

We put now

$$X_\alpha = a_\alpha E_\alpha, \qquad \alpha \in \Delta.$$

By (3), §4,

$$[X_\alpha, X_{-\alpha}] = B(X_\alpha, X_{-\alpha})\, H_\alpha = a_\alpha a_{-\alpha} B(E_\alpha, E_{-\alpha})\, H_\alpha = H_\alpha.$$

Also,

$$\tilde{\varphi}(X_\alpha) = a_\alpha \tilde{\varphi}(E_\alpha) = a_\alpha c_{-\alpha} E_{-\alpha} = -a_{-\alpha}E_{-\alpha} = -X_{-\alpha}.$$

If $\alpha, \beta, \alpha + \beta \in \Delta$ we define $N_{\alpha,\beta}$ by $[X_\alpha, X_\beta] = N_{\alpha,\beta}X_{\alpha+\beta}$. Then

$$-N_{\alpha,\beta}X_{-\alpha-\beta} = \tilde{\varphi}(N_{\alpha,\beta}X_{\alpha+\beta}) = \tilde{\varphi}[X_\alpha, X_\beta] = [-X_{-\alpha}, -X_{-\beta}] = N_{-\alpha,-\beta}X_{-\alpha-\beta},$$

so $N_{\alpha,\beta} = -N_{-\alpha,-\beta}$. The last relation of the theorem now follows from Lemma 5.2.

Definition. A root $\alpha > 0$ is called *simple* if it cannot be written as a sum $\alpha = \beta + \gamma$ where β and γ are positive roots.

Lemma 5.6. *Let α, β be simple roots, $\alpha \neq \beta$. Then $\beta - \alpha$ is not a root and $B(H_\alpha, H_\beta) \leqslant 0$.*

Proof. If $\beta - \alpha$ were a root, say γ, then $\gamma \in \Delta$. Writing $\beta = \alpha + \gamma$ if $\gamma > 0$ and $\alpha = \beta - \gamma$ if $\gamma < 0$ we get a contradiction to the simplicity of α and β. In the notation of Theorem 4.3, we have $p = 0$, $q \geqslant 0$. Since

$$-2B(H_\beta, H_\alpha) = B(H_\alpha, H_\alpha)\,(p + q)$$

the lemma follows.

Theorem 5.7. *Let* $\alpha_1, \ldots, \alpha_r$ *be the set of all simple roots. Then* $r = \dim \mathfrak{h}_R$ *and each* $\beta \in \Delta$ *has the form* $\beta = \sum_{i=1}^{r} n_i\alpha_i$ *where the* n_i *are integers which are either all positive or all negative.*

Proof. The simple roots are linearly independent. Otherwise there would be a relation (each $\alpha_i \neq$ each α_j)

$$\sum_i a_i\alpha_i = \sum_j b_j\alpha_j$$

with nonnegative real numbers a_i, b_j, not all zero. If $\gamma = \sum a_i\alpha_i$ and $H_\gamma \in \mathfrak{h}_R$ is determined by $B(H, H_\gamma) = \gamma(H)$ $(H \in \mathfrak{h})$, then

$$B(H_\gamma, H_\gamma) = \sum_{i,j} a_i b_j B(H_{\alpha_i}, H_{\alpha_j}).$$

The left-hand side is $\geqslant 0$ but the right-hand side is $\leqslant 0$ due to the lemma. Hence $\gamma = 0$ which is a contradiction.

Now let β be a root > 0. If β is not simple it can be written $\beta = \gamma + \delta$ where γ, δ are roots > 0. It follows by induction that $\beta = \sum_{i=1}^{r} n_i\alpha_i$ where each n_i is an integer $\geqslant 0$. It is now obvious that $r = \dim \mathfrak{h}_R$ and the theorem is proved.

The simple roots and their analogs for graded Lie algebras will be very useful in Chapter X.

§ 6. Real Forms

Let V be a vector space over $\boldsymbol{R}$ of finite dimension. A *complex structure* on V is an $\boldsymbol{R}$-linear endomorphism J of V such that $J^2 = -I$, where I is the identity mapping of V. A vector space V over $\boldsymbol{R}$ with a complex structure J can be turned into a vector space $\tilde{V}$ over $\boldsymbol{C}$ by putting

$$(a + ib)\, X = aX + bJX,$$

$$X \in V, \quad a, b \in \boldsymbol{R}.$$

In fact, $J^2 = -I$ implies $\alpha(\beta X) = (\alpha\beta)\, X$ for α, $\beta \in \boldsymbol{C}$ and $X \in V$. We have clearly $\dim_{\boldsymbol{C}} \tilde{V} = \frac{1}{2} \dim_{\boldsymbol{R}} V$ and consequently V must be even-dimensional. We call $\tilde{V}$ the *complex vector space associated to* V. Note that V and $\tilde{V}$ agree set theoretically.

On the other hand, if E is a vector space over $\boldsymbol{C}$ we can consider E as a vector space E^R over $\boldsymbol{R}$. The multiplication by i on E then becomes a complex structure J on E^R and it is clear that $E = (E^R)^{\sim}$.

A Lie algebra $\mathfrak{v}$ over $\boldsymbol{R}$ is said to have a complex structure J if J is a complex structure on the vector space $\mathfrak{v}$ and in addition

$$[X, JY] = J[X, Y], \qquad \text{for } X, Y \in \mathfrak{v}. \tag{1}$$

Condition (1) means $(\operatorname{ad} X) \circ J = J \circ \operatorname{ad} X$ for all $X \in \mathfrak{v}$, or equivalently, $\operatorname{ad}(JX) = J \circ \operatorname{ad} X$ for all $X \in \mathfrak{v}$. It follows from (1) that

$$[JX, JY] = -[X, Y].$$

The complex vector space $\tilde{\mathfrak{v}}$ becomes a Lie algebra over $\boldsymbol{C}$ with the bracket operation inherited from $\mathfrak{v}$. In fact

$$[(a + ib) X, (c + id) Y] = [aX + bJX, cY + dJY]$$
$$= ac[X, Y] + bcJ[X, Y] + adJ[X, Y] - bd[X, Y]$$

so

$$[(a + ib) X, (c + id) Y] = (a + ib)(c + id)[X, Y].$$

On the other hand, suppose $\mathfrak{e}$ is a Lie algebra over $\boldsymbol{C}$. The vector space $\mathfrak{e}^R$ has a complex structure J given by multiplication by i on $\mathfrak{e}$. With the bracket operation inherited from $\mathfrak{e}$, $\mathfrak{e}^R$ becomes a Lie algebra over $\boldsymbol{R}$ with the complex structure J.

Now suppose W is an arbitrary finite-dimensional vector space over $\boldsymbol{R}$. The product $W \times W$ is again a vector space over $\boldsymbol{R}$ and the endomorphism $J : (X, Y) \to (-Y, X)$ is a complex structure on $W \times W$. The complex vector space $(W \times W)^{\sim}$ is called the *complexification* of W and will be denoted W^C. We have of course $\dim_C W^C = \dim_R W$. The elements of W^C are the pairs (X, Y) where $X, Y \in W$ and since $(X, Y) = (X, 0) + i(Y, 0)$ we write $X + iY$ instead of (X, Y). Then since

$$(a + bJ)(X, Y) = a(X, Y) + b(-Y, X) = (aX - bY, aY + bX)$$

we have

$$(a + ib)(X + iY) = aX - bY + i(aY + bX).$$

On the other hand, each finite-dimensional vector space E over $\boldsymbol{C}$ is isomorphic to W^C for a suitable vector space W over $\boldsymbol{R}$; in fact, if (Z_i) is any basis of E, one can take W as the set of all vectors of the form $\sum_i a_i Z_i$, $a_i \in \boldsymbol{R}$.

Let $\mathfrak{l}_0$ be a Lie algebra over $\boldsymbol{R}$; owing to the conventions above, the complex vector space $\mathfrak{l} = (\mathfrak{l}_0)^C$ consists of all symbols $X + iY$, where $X, Y \in \mathfrak{l}_0$. We define the bracket operation in $\mathfrak{l}$ by

$$[X + iY, Z + iT] = [X, Z] - [Y, T] + i([Y, Z] + [X, T]),$$

and this bracket operation is bilinear over $\boldsymbol{C}$. It is clear that $\mathfrak{l} = (\mathfrak{l}_0)^C$ is a Lie algebra over $\boldsymbol{C}$; it is called the *complexification of the Lie algebra* $\mathfrak{l}_0$. The Lie algebra $\mathfrak{l}^R$ is a Lie algebra over $\boldsymbol{R}$ with a complex structure J derived from multiplication by i on $\mathfrak{l}$.

Lemma 6.1. *Let K_0, K, and K^R denote the Killing forms of the Lie algebras $\mathfrak{l}_0$, $\mathfrak{l}$, and $\mathfrak{l}^R$. Then*

$$K_0(X, Y) = K(X, Y) \qquad \textit{for } X, Y \in \mathfrak{l}_0,$$

$$K^R(X, Y) = 2 \operatorname{Re} (K(X, Y)) \qquad \textit{for } X, Y \in \mathfrak{l}^R \qquad (\operatorname{Re} = \textit{real part}).$$

The first relation is obvious. For the second, suppose X_i $(1 \leqslant i \leqslant n)$ is any basis of $\mathfrak{l}$; let $B + iC$ denote the matrix of ad X ad Y with respect to this basis, B and C being real. Then $X_1, \ldots, X_n, JX_1, \ldots, JX_n$ is a basis of $\mathfrak{l}^R$ and since the linear transformation ad X ad Y of $\mathfrak{l}^R$ commutes with J, it has the matrix expression

$$\begin{pmatrix} B & -C \\ C & B \end{pmatrix}$$

and the second relation above follows.

As a consequence of Lemma 6.1 we note that the algebras $\mathfrak{l}_0$, $\mathfrak{l}$, and $\mathfrak{l}^R$ are all semisimple if and only if one of them is.

Definition. Let $\mathfrak{g}$ be a Lie algebra over $\boldsymbol{C}$. A *real form* of $\mathfrak{g}$ is a subalgebra $\mathfrak{g}_0$ of the real Lie algebra $\mathfrak{g}^R$ such that

$$\mathfrak{g}^R = \mathfrak{g}_0 + J\mathfrak{g}_0 \qquad \text{(direct sum of vector spaces).}$$

In this case, each $Z \in \mathfrak{g}$ can be uniquely written

$$Z = X + iY, \qquad X, Y \in \mathfrak{g}_0.$$

Thus $\mathfrak{g}$ is isomorphic to the complexification of $\mathfrak{g}_0$. The mapping σ of $\mathfrak{g}$ onto itself given by $\sigma : X + iY \to X - iY$ $(X, Y \in \mathfrak{g}_0)$ is called the *conjugation* of $\mathfrak{g}$ with respect to $\mathfrak{g}_0$. The mapping σ has the properties

$$\sigma(\sigma(X)) = X, \qquad \sigma(X + Y) = \sigma(X) + \sigma(Y),$$

$$\sigma(\alpha X) = \bar{\alpha}\sigma(X), \qquad \sigma[X, Y] = [\sigma X, \sigma Y],$$

for $X, Y \in \mathfrak{g}$, $\alpha \in \boldsymbol{C}$. Thus σ is not an automorphism of $\mathfrak{g}$, but it is an automorphism of the real algebra $\mathfrak{g}^R$. On the other hand, let σ be a mapping of $\mathfrak{g}$ onto itself with the properties above. Then the set $\mathfrak{g}_0$ of fixed points of σ is a real form of $\mathfrak{g}$ and σ is the conjugation of $\mathfrak{g}$ with respect to $\mathfrak{g}_0$. In fact, $J\mathfrak{g}_0$ is the eigenspace of σ for the eigenvalue -1 and consequently $\mathfrak{g}^R = \mathfrak{g}_0 + J\mathfrak{g}_0$. If B is the Killing form on $\mathfrak{g} \times \mathfrak{g}$, it is easy to see from Lemma 6.1 that $B(\sigma X, \sigma Y)$ is the complex conjugate of $B(X, Y)$. Another useful remark in this connection is the following: Let $\mathfrak{g}_1$ and $\mathfrak{g}_2$ be two real forms of $\mathfrak{g}$ and σ_1 and σ_2 the corresponding conjugations. Then σ_1 leaves $\mathfrak{g}_2$ invariant if and only if σ_1 and σ_2 commute; in this case we have the direct decompositions

$$\mathfrak{g}_1 = \mathfrak{g}_1 \cap \mathfrak{g}_2 + \mathfrak{g}_1 \cap (i\mathfrak{g}_2),$$
$$\mathfrak{g}_2 = \mathfrak{g}_1 \cap \mathfrak{g}_2 + \mathfrak{g}_2 \cap (i\mathfrak{g}_1).$$

Lemma 6.2. *Suppose* $\mathfrak{g}$ *is a semisimple Lie algebra over* $\boldsymbol{C}$, $\mathfrak{g}_0$ *a real form of* $\mathfrak{g}$, *and* σ *the conjugation of* $\mathfrak{g}$ *with respect to* $\mathfrak{g}_0$. *Let ad denote the adjoint representation of* $\mathfrak{g}^R$ *and Int* $(\mathfrak{g}^R)$ *the adjoint group of* $\mathfrak{g}^R$. *If* G_0 *denotes the analytic subgroup of Int* $(\mathfrak{g}^R)$ *whose Lie algebra is ad* $(\mathfrak{g}_0)$, *then* G_0 *is a closed subgroup of Int* $(\mathfrak{g}^R)$ *and analytically isomorphic to Int* $(\mathfrak{g}_0)$.

Proof. Every automorphism s of $\mathfrak{g}^R$ gives rise to an automorphism $\tilde{s}$ of Int $(\mathfrak{g}^R)$ satisfying $\tilde{s}(e^{\text{ad}X}) = e^{\text{ad}(s \cdot X)}$ $(X \in \mathfrak{g}^R)$. In particular there exists an automorphism $\tilde{\sigma}$ of Int $(\mathfrak{g}^R)$ such that $(d\tilde{\sigma})_e$ (ad X) = ad $(\sigma \cdot X)$ for $X \in \mathfrak{g}^R$. Since ad is an isomorphism, this proves that ad $(\mathfrak{g}_0)$ is the set of fixed points of $(d\tilde{\sigma})_e$; thus G_0 is the identity component of the set of fixed points of $\tilde{\sigma}$. Now, let ad_0 denote the adjoint representation of $\mathfrak{g}_0$ and for each endomorphism A of $\mathfrak{g}^R$ leaving $\mathfrak{g}_0$ invariant, let A_0 denote its restriction to $\mathfrak{g}_0$. Then if $X \in \mathfrak{g}_0$, we have $(\text{ad } X)_0 = \text{ad}_0 X$ and the mapping $A \to A_0$ maps G_0 onto Int $(\mathfrak{g}_0)$. This mapping is an isomorphism of G_0 onto Int $(\mathfrak{g}_0)$. In fact, suppose $A \in G_0$ such that A_0 is the identity. Since A commutes with the complex structure J, it follows that A is the identity. Finally since the isomorphism is regular at the identity it is an analytic isomorphism.

The following theorem is of fundamental importance in the theory of semisimple Lie algebras and symmetric spaces.

Theorem 6.3. *Every semisimple Lie algebra* $\mathfrak{g}$ *over* $\boldsymbol{C}$ *has a real form which is compact.*

Proof. As always, let B denote the Killing form on $\mathfrak{g} \times \mathfrak{g}$. Let $\mathfrak{h}$ be a Cartan subalgebra of $\mathfrak{g}$, and Δ the corresponding set of nonzero roots. For each $\alpha \in \Delta$ we select $X_\alpha \in \mathfrak{g}^\alpha$ with the properties of Theorem 5.5. The first relation $[X_\alpha, X_{-\alpha}] = H_\alpha$ implies $B(X_\alpha, X_{-\alpha}) = 1$ by (3), § 4, and consequently

$$B(X_\alpha - X_{-\alpha}, X_\alpha - X_{-\alpha}) = -2,$$
$$B(i(X_\alpha + X_{-\alpha}), i(X_\alpha + X_{-\alpha})) = -2,$$
$$B(X_\alpha - X_{-\alpha}, i(X_\alpha + X_{-\alpha})) = 0,$$
$$B(iH_\alpha, iH_\alpha) = -\alpha(H_\alpha) < 0,$$

the last relation following from Theorem 4.4. Since $B(X_\alpha, X_\beta) = 0$ if $\alpha + \beta \neq 0$, it follows that B is strictly negative definite on the $\boldsymbol{R}$-linear subspace

$$\mathfrak{g}_k = \sum_{\alpha \in \Delta} \boldsymbol{R}(iH_\alpha) + \sum_{\alpha \in \Delta} \boldsymbol{R}(X_\alpha - X_{-\alpha}) + \sum_{\alpha \in \Delta} \boldsymbol{R}(i(X_\alpha + X_{-\alpha})). \tag{2}$$

Moreover, $\mathfrak{g} = \mathfrak{g}_k + i\mathfrak{g}_k$ (direct vector space sum). Using now (for the first time) the relation $N_{\alpha,\beta} = -N_{-\alpha,-\beta}$ (which in view of Lemma 5.2 implies that each $N_{\alpha,\beta}$ is real), we see that $X, Y \in \mathfrak{g}_k$ implies $[X, Y] \in \mathfrak{g}_k$. Thus $\mathfrak{g}_k$ is a real form of $\mathfrak{g}$. The Killing form of $\mathfrak{g}_k$ is strictly negative definite, being the restriction of B to $\mathfrak{g}_k \times \mathfrak{g}_k$. Thus $\mathfrak{g}_k$ is compact and the theorem is proved.

§ 7. Cartan Decompositions

Theorem 7.1. *Let $\mathfrak{g}_0$ be a semisimple Lie algebra over $\boldsymbol{R}$, $\mathfrak{g}$ its complexification, and $\mathfrak{u}$ any compact real form of $\mathfrak{g}$. Let σ and τ denote the conjugations of $\mathfrak{g}$ with respect to $\mathfrak{g}_0$ and $\mathfrak{u}$, respectively. Then there exists an automorphism φ of $\mathfrak{g}$ such that the compact real form $\varphi \cdot \mathfrak{u}$ is invariant under σ.*

Proof. The Hermitian form B_τ on $\mathfrak{g} \times \mathfrak{g}$ given by

$$B_\tau(X, Y) = - B(X, \tau Y), \qquad X, Y \in \mathfrak{g},$$

is strictly positive definite since $\mathfrak{u}$ is compact. The linear transformation $N = \sigma\tau$ is an automorphism of the complex algebra $\mathfrak{g}$ and hence leaves the Killing form invariant. Using $\sigma^2 = \tau^2 = I$, we obtain

$$B(NX, \tau Y) = B(X, N^{-1}\tau Y) = B(X, \tau NY)$$

or

$$B_\tau(NX, Y) = B_\tau(X, NY).$$

This shows that N is self-adjoint with respect to B_τ. Let $X_1, ..., X_n$ be a basis of $\mathfrak{g}$ with respect to which N is represented by a diagonal matrix. Then the endomorphism $P = N^2$ is represented by a diagonal matrix with positive diagonal elements $\lambda_1, ..., \lambda_n$. For each $t \in \boldsymbol{R}$, let P^t denote the linear transformation of $\mathfrak{g}$ represented by the diagonal matrix with diagonal elements $(\lambda_i)^t > 0$. Then each P^t commutes with N. Let ${c^k}_{ij}$ denote the constants determined by

$$[X_i, X_j] = \sum_{k=1}^{n} {c^k}_{ij} X_k$$

for $1 \leqslant i, j \leqslant n$. Since P is an automorphism, we have

$$\lambda_i \lambda_j c^k{}_{ij} = \lambda_k c^k{}_{ij} \qquad (1 \leqslant i, j, k \leqslant n).$$

This equation implies

$$(\lambda_i)^t (\lambda_j)^t c^k{}_{ij} = (\lambda_k)^t c^k{}_{ij} \qquad (t \in \boldsymbol{R}),$$

which shows that each P^t is an automorphism of $\mathfrak{g}$.

Consider now the mapping $\tau_1 = P^t \tau P^{-t}$ of $\mathfrak{g}$ into itself. The subspace $P^t \mathfrak{u}$ is a compact real form of $\mathfrak{g}$ and τ_1 is the conjugation of $\mathfrak{g}$ with respect to this form. Moreover we have $\tau N \tau^{-1} = N^{-1}$ so $\tau P \tau^{-1} = P^{-1}$. By a simple matrix computation the relation $\tau P = P^{-1}\tau$ implies $\tau P^t = P^{-t}\tau$ for all $t \in \boldsymbol{R}$. Consequently,

$$\sigma\tau_1 = \sigma P^t \tau P^{-t} = \sigma\tau P^{-2t} = NP^{-2t},$$

$$\tau_1 \sigma = (\sigma\tau_1)^{-1} = P^{2t}N^{-1} = N^{-1}P^{2t}.$$

If $t = \frac{1}{4}$, then $\sigma\tau_1 = \tau_1\sigma$. Thus the automorphism $\varphi = P^{1/4}$ has the desired properties.

Remark. The proof has shown that the automorphism φ can be chosen as $P^{1/4}$ where P^t ($t \in \boldsymbol{R}$) is a one-parameter group of semisimple, positive definite (for B_τ) automorphisms of $\mathfrak{g}$ satisfying $P^1 = (\sigma\tau)^2$.

Definition. Let $\mathfrak{g}_0$ be a semisimple Lie algebra over $\boldsymbol{R}$, $\mathfrak{g}$ its complexification, σ the conjugation of $\mathfrak{g}$ with respect to $\mathfrak{g}_0$. A direct decomposition $\mathfrak{g}_0 = \mathfrak{k}_0 + \mathfrak{p}_0$ of $\mathfrak{g}_0$ into a subalgebra $\mathfrak{k}_0$ and a vector subspace $\mathfrak{p}_0$ is called a *Cartan decomposition* if there exists a compact real form $\mathfrak{g}_k$ of $\mathfrak{g}$ such that

$$\sigma \cdot \mathfrak{g}_k \subset \mathfrak{g}_k, \qquad \mathfrak{k}_0 = \mathfrak{g}_0 \cap \mathfrak{g}_k, \qquad \mathfrak{p}_0 = \mathfrak{g}_0 \cap (i\mathfrak{g}_k). \tag{1}$$

It is an immediate consequence of Theorems 6.3 and 7.1 that each semisimple Lie algebra $\mathfrak{g}_0$ over $\boldsymbol{R}$ has a Cartan decomposition. The following theorem shows that any two Cartan decompositions are conjugate under an inner automorphism.

Theorem 7.2. *Let $\mathfrak{g}_0$ be a semisimple Lie algebra over $\boldsymbol{R}$. Suppose*

$$\mathfrak{g}_0 = \mathfrak{k}_1 + \mathfrak{p}_1, \qquad \mathfrak{g}_0 = \mathfrak{k}_2 + \mathfrak{p}_2$$

are two Cartan decompositions of $\mathfrak{g}_0$. Then there exists an element $\psi \in Int(\mathfrak{g}_0)$ such that

$$\psi \cdot \mathfrak{k}_1 = \mathfrak{k}_2, \qquad \psi \cdot \mathfrak{p}_1 = \mathfrak{p}_2.$$

Proof. Let $\mathfrak{g}$ be the complexification of $\mathfrak{g}_0$, σ the conjugation of $\mathfrak{g}$ with respect to $\mathfrak{g}_0$. Then there exist compact real forms $\mathfrak{u}_1$ and $\mathfrak{u}_2$ of $\mathfrak{g}$ such that

$$\sigma \cdot \mathfrak{u}_j \subset \mathfrak{u}_j, \qquad \mathfrak{k}_j = \mathfrak{g}_0 \cap \mathfrak{u}_j, \qquad \mathfrak{p}_j = \mathfrak{g}_0 \cap (i\mathfrak{u}_j) \qquad (j = 1, 2).$$

Let τ_1 and τ_2 denote the conjugations of $\mathfrak{g}$ with respect to $\mathfrak{u}_1$ and $\mathfrak{u}_2$, respectively. Then from Theorem 7.1 and the subsequent remark there exists a one-parameter group P^t of positive definite semisimple automorphisms of $\mathfrak{g}$ such that $P^1 = (\tau_1\tau_2)^2$ and $P^{1/4} \cdot \mathfrak{u}_2$ is a compact real form of $\mathfrak{g}$ invariant under τ_1. It follows that $P^{1/4} \cdot \mathfrak{u}_2$ is the direct sum of its intersections with $\mathfrak{u}_1$ and $i\mathfrak{u}_1$. Now B, the Killing form of $\mathfrak{g}$, is strictly positive definite on the subspace $i\mathfrak{u}_1$, and strictly negative definite on the compact form $P^{1/4}\mathfrak{u}_2$. Consequently, the intersection $P^{1/4}\mathfrak{u}_2 \cap i\mathfrak{u}_1$ reduces to $\{0\}$, so

$$\mathfrak{u}_1 = P^{1/4} \cdot \mathfrak{u}_2.$$

Since $\sigma\mathfrak{u}_j \subset \mathfrak{u}_j$, we have $\sigma\tau_j = \tau_j\sigma$ $(j = 1, 2)$ and consequently σ commutes with P^1. Now P^t is the unique positive definite tth power of P^1; hence σ commutes with each P^t so P^t leaves $\mathfrak{g}_0$ invariant. The restriction of the linear transformation P^t to $\mathfrak{g}_0$ gives rise to a one-parameter subgroup $\{\exp tX\}$ of Aut $(\mathfrak{g}_0)$. As a result of Cor. 6.5, Chapter II, we have $\{\exp tX\} \subset$ Int $(\mathfrak{g}_0)$. The theorem follows if we take $\psi = \exp \frac{1}{4} X$.

The proof has the following

Corollary 7.3. *If $\mathfrak{u}_1$ and $\mathfrak{u}_2$ are any compact real forms of a semisimple Lie algebra $\mathfrak{g}$ over $\boldsymbol{C}$ then there exists a one-parameter subgroup ψ^t $(t \in \boldsymbol{R})$ of automorphisms of $\mathfrak{g}$ such that $\psi^1\mathfrak{u}_1 = \mathfrak{u}_2$.*

Proposition 7.4. *Let $\mathfrak{g}_0$ be a semisimple Lie algebra over $\boldsymbol{R}$ which is the direct sum $\mathfrak{g}_0 = \mathfrak{k}_0 + \mathfrak{p}_0$ where $\mathfrak{k}_0$ is a subalgebra and $\mathfrak{p}_0$ a vector subspace. The following conditions are equivalent.*

(i) *$\mathfrak{g}_0 = \mathfrak{k}_0 + \mathfrak{p}_0$ is a Cartan decomposition of $\mathfrak{g}_0$.*

(ii) *The mapping $s_0 : T + X \to T - X$ $(T \in \mathfrak{k}_0,\ X \in \mathfrak{p}_0)$ is an automorphism of $\mathfrak{g}_0$ and the symmetric bilinear form*

$$B_{s_0}(X, Y) = -B(X, s_0Y)$$

is strictly positive definite (that is, $B < 0$ on $\mathfrak{k}_0$, $B > 0$ on $\mathfrak{p}_0$).

If these conditions are satisfied, $\mathfrak{k}_0$ is a maximal compactly imbedded subalgebra of $\mathfrak{g}_0$.

Definition. An involutive automorphism θ of a semisimple Lie algebra $\mathfrak{g}_0$ is called a *Cartan involution* if the bilinear form $B_\theta(X, Y) = -B(X, \theta Y)$ is strictly positive definite.

Proof. (ii) $\Rightarrow$ (i). Let $\mathfrak{g}$ denote the complexification of $\mathfrak{g}_0$, and let $\mathfrak{g}^R$ denote the Lie algebra $\mathfrak{g}$ when considered as a Lie algebra over $\boldsymbol{R}$. Since s_0 is an automorphism, we have $B(\mathfrak{k}_0, \mathfrak{p}_0) = 0$, $[\mathfrak{k}_0, \mathfrak{p}_0] \subset \mathfrak{p}_0$, and $[\mathfrak{p}_0, \mathfrak{p}_0] \subset \mathfrak{k}_0$. It follows that the subspace $\mathfrak{g}_k = \mathfrak{k}_0 + i\mathfrak{p}_0$ of $\mathfrak{g}^R$ is a subalgebra and in fact a compact real form of $\mathfrak{g}$, satisfying the relations (1). On the other hand, relations (1) show that (i) $\Rightarrow$ (ii). We know from Lemma 6.2 that the groups Int $(\mathfrak{g}_0)$ and Int $(\mathfrak{g}_k)$ can be regarded as closed subgroups of Int $(\mathfrak{g}^R)$. Now Int $(\mathfrak{g}_k)$ is compact and the same is true of Int $(\mathfrak{g}_0) \cap$ Int $(\mathfrak{g}_k)$. This last group is a Lie subgroup of Int $(\mathfrak{g}_0)$ and has Lie algebra $\mathfrak{g}_0 \cap \mathfrak{g}_k = \mathfrak{k}_0$. Thus $\mathfrak{k}_0$ is compactly imbedded in $\mathfrak{g}_0$. If $\mathfrak{k}_0$ were not maximal let $\mathfrak{k}_1$ be a compactly imbedded subalgebra of $\mathfrak{g}_0$, properly containing $\mathfrak{k}_0$. Then there exists an element $X \neq 0$ in $\mathfrak{k}_1 \cap \mathfrak{p}_0$. Let η denote the conjugation of $\mathfrak{g}$ with respect to $\mathfrak{g}_k$. Then $\eta\mathfrak{g}_0 \subset \mathfrak{g}_0$ and the bilinear form B_η on $\mathfrak{g}_0 \times \mathfrak{g}_0$ defined by

$$B_\eta(Y, Z) = -B(Y, \eta Z), \qquad Y, Z \in \mathfrak{g}_0,$$

is symmetric and strictly positive definite. Since

$$B([X, Y], \eta Z) = -B(Y, [X, \eta Z]) = B(Y, [\eta X, \eta Z])$$

we have

$$B_\eta(\text{ad } X(Y), Z = B_\eta(Y, \text{ad } X(Z)).$$

Thus ad X has all its eigenvalues real, and not all zero. But then the powers $e^{n \text{ ad } X}$ can not lie in a compact matrix group. This contradicts the fact that $\mathfrak{k}_1$ is a compactly imbedded subalgebra of $\mathfrak{g}_0$.

Corollary 7.5. *Let $\mathfrak{g}$ be a semisimple Lie algebra over $\boldsymbol{C}$ and let $\mathfrak{u}$ be any compact real form of $\mathfrak{g}$. Let $\mathfrak{g}^R$ denote the Lie algebra $\mathfrak{g}$ considered as a real Lie algebra and let J denote the complex structure of $\mathfrak{g}^R$ which corresponds to the multiplication by i on $\mathfrak{g}$. Then*

$$\mathfrak{g}^R = \mathfrak{u} + J\mathfrak{u}$$

is a Cartan decomposition of $\mathfrak{g}^R$.

In fact, let B^R, B, and B^C denote the Killing forms on $\mathfrak{g}^R$, $\mathfrak{g}$, and $(\mathfrak{g}^R)^C$. Then by Lemma 3.1 $B^C = B^R = 2 \text{ Re } B$ on $\mathfrak{g}^R \times \mathfrak{g}^R$. Since B

is strictly negative definite on $\mathfrak{u} \times \mathfrak{u}$ and strictly positive definite on $J\mathfrak{u} \times J\mathfrak{u}$, the same holds for B^C. Using Prop. 7.4 the corollary follows.

§ 8. Examples. The Complex Classical Lie Algebras

The complex classical Lie algebras are:

$\mathfrak{a}_n = \mathfrak{sl}(n+1, C)$ (the complex $(n+1) \times (n+1)$ matrices of trace 0).

$\mathfrak{b}_n = \mathfrak{so}(2n+1, C)$ (the complex skew symmetric matrices of order $2n+1$).

$\mathfrak{c}_n = \mathfrak{sp}(n, C)$ (the matrices

$$\begin{pmatrix} Z_1 & Z_2 \\ Z_3 & -{}^tZ_1 \end{pmatrix} \qquad \begin{matrix} Z_i \text{ complex } n \times n \text{ matrices} \\ Z_2 \text{ and } Z_3 \text{ symmetric}). \end{matrix}$$

$\mathfrak{d}_n = \mathfrak{so}(2n, C)$ (the complex skew symmetric matrices of order $2n$).

These are Lie algebras of certain matrix groups described in Chapter X. The groups corresponding to $\mathfrak{b}_n$ and $\mathfrak{d}_n$ are counterparts to the groups corresponding to $\mathfrak{c}_n$; while the former are invariance groups for a *symmetric* nondegenerate bilinear form, the latter is the invariance group for a *skew* bilinear form (which can only be nondegenerate in even dimensions).

Here we shall determine the Killing forms of these algebras, verify their semisimplicity, and determine their root pattern. Let E_{ij} denote a square matrix with entry 1 where the ith row and the jth column meet, all other entries being 0. Then $E_{ij}E_{kl} = \delta_{jk}E_{il}$ so

$$[E_{ij}, E_{kl}] = \delta_{jk}E_{il} - \delta_{li}E_{kj}. \tag{1}$$

The algebra $\mathfrak{a}_n = \mathfrak{sl}(n+1, C)$. Putting

$$H_i = E_{ii} - E_{i+1\,i+1} \quad (1 \leqslant i \leqslant n), \qquad \mathfrak{h} = \sum_i CH_i \tag{2}$$

we have the direct decomposition

$$\mathfrak{a}_n = \mathfrak{h} + \sum_{i \neq j} CE_{ij} \tag{3}$$

and $\mathfrak{h}$ consists of the diagonal matrices of trace 0. If $H \in \mathfrak{h}$ and $e_i(H)$

($1 \leqslant i \leqslant (n+1)$) are the diagonal elements, we have

$$[H, E_{ij}] = (e_i(H) - e_j(H))\, E_{ij}. \tag{4}$$

Hence the Killing form B satisfies

$$B(H, H) = \operatorname{Tr}((\operatorname{ad} H)^2) = \sum_{i,j} (e_i(H) - e_j(H))^2$$
$$= 2(n+1)\operatorname{Tr}(H^2) - 2\operatorname{Tr}(H)^2 = 2(n+1)\operatorname{Tr}(H^2).$$

Each $X \in \mathfrak{sl}(n+1, \boldsymbol{C})$ with all eigenvalues different can be diagonalized, so there exists a nonsingular matrix g such that $gXg^{-1} \in \mathfrak{h}$. By the invariance of B and Tr, $B(X, X) = 2(n+1)\operatorname{Tr}(X^2)$ and by continuity this holds for all $X \in \mathfrak{sl}(n+1, \boldsymbol{C})$. By polarization

$$B(X, Y) = 2(n+1)\operatorname{Tr}(XY), \qquad X, Y \in \mathfrak{a}_n. \tag{5}$$

This implies that $\mathfrak{sl}(n+1, \boldsymbol{C})$ is semisimple; then (3) and (4) show that $\mathfrak{h}$ is a Cartan subalgebra and that the roots are

$$e_i - e_j \qquad (1 \leqslant i, j \leqslant n+1). \tag{6}$$

The algebra $\mathfrak{d}_n = \mathfrak{so}(2n, \boldsymbol{C})$. Here we put

$$H_i = E_{2i-1,2i} - E_{2i,2i-1} \quad (1 \leqslant i \leqslant n), \qquad \mathfrak{h} = \sum_i \boldsymbol{C} H_i. \tag{7}$$

From (1) we deduce

$$[H_i, E_{2i-1,2j-1} \pm E_{2i2j}] = -E_{2i,2j-1} \pm E_{2i-1,2j} \qquad (i \neq j)$$
$$[H_i, E_{2i,2j-1} \mp E_{2i-1,2j}] = E_{2i-1,2j-1} \pm E_{2i,2j} \qquad (i \neq j)$$

so that, putting $F_{ij} = E_{ij} - E_{ji}$, the vectors

$$G_{jk}^{+} = F_{2j-1,2k-1} + F_{2j2k} + i(F_{2j-1,2k} - F_{2j,2k-1}) \qquad (1 \leqslant j \neq k \leqslant n)$$
$$G_{jk}^{-} = F_{2j-1,2k-1} - F_{2j2k} + i(F_{2j-1,2k} + F_{2j,2k-1}) \qquad (1 \leqslant j \neq k \leqslant n)$$

satisfy

$$[H, G_{jk}^{+}] = (e_j(H) - e_k(H))\, G_{jk}^{+} \qquad (1 \leqslant j \neq k \leqslant n) \tag{8}$$

$$[H, G_{jk}^{-}] = -(e_j(H) + e_k(H))\, G_{jk}^{-} \qquad (1 \leqslant j < k \leqslant n) \tag{8'}$$

$$[H, G_{jk}^{-}] = (e_j(H) + e_k(H))\, G_{jk}^{-} \qquad (1 \leqslant k < j \leqslant n) \tag{8''}$$

if e_j is the linear form on $\mathfrak{h}$ given by

$$e_j(H_k) = -i\,\delta_{jk}. \tag{9}$$

Furthermore we have the direct decomposition

$$\mathfrak{d}_n = \mathfrak{h} + \sum_{j\neq k} CG_{jk}^{+} + \sum_{j\neq k} CG_{jk}^{-}. \tag{10}$$

For $n = 2$ these matrices are

$$H_1 = \begin{pmatrix} 0 & 1 & 0 & 0 \\ -1 & 0 & 0 & 0 \\ 0 & 0 & 0 & 0 \\ 0 & 0 & 0 & 0 \end{pmatrix}, \qquad G_{12}^{+} = \begin{pmatrix} 0 & 0 & 1 & i \\ 0 & 0 & -i & 1 \\ -1 & i & 0 & 0 \\ -i & -1 & 0 & 0 \end{pmatrix},$$

$$H_2 = \begin{pmatrix} 0 & 0 & 0 & 0 \\ 0 & 0 & 0 & 0 \\ 0 & 0 & 0 & 1 \\ 0 & 0 & -1 & 0 \end{pmatrix}, \qquad G_{12}^{-} = \begin{pmatrix} 0 & 0 & 1 & i \\ 0 & 0 & i & -1 \\ -1 & -i & 0 & 0 \\ -i & 1 & 0 & 0 \end{pmatrix},$$

$$G_{21}^{+} = \begin{pmatrix} 0 & 0 & -1 & i \\ 0 & 0 & -i & -1 \\ 1 & i & 0 & 0 \\ -i & 1 & 0 & 0 \end{pmatrix}, \qquad G_{21}^{-} = \begin{pmatrix} 0 & 0 & -1 & i \\ 0 & 0 & i & 1 \\ 1 & -i & 0 & 0 \\ -i & -1 & 0 & 0 \end{pmatrix}.$$

The Killing form B satisfies

$$\begin{aligned} B(H, H) &= \mathrm{Tr}((\mathrm{ad}\, H)^2) \\ &= \sum_{j\neq k} (e_j(H) - e_k(H))^2 + (e_j(H) + e_k(H))^2 \\ &= (4n-4) \sum_{i=1}^{n} e_i(H)^2 = (2n-2)\,\mathrm{Tr}(H^2). \end{aligned}$$

Each skew symmetric matrix is conjugate to some element in $\mathfrak{h}$, so we deduce again $B(X, X) = (2n-2)\,\mathrm{Tr}(X^2)$ $(X \in \mathfrak{d}_n)$ and by polarization

$$B(X, Y) = (2n-2)\,\mathrm{Tr}(XY) \qquad (X, Y \in \mathfrak{d}_n). \tag{11}$$

Again this implies that $\mathfrak{d}_n$ is semisimple (for $n > 1$), that $\mathfrak{h}$ is a Cartan subalgebra, and that the roots are given by

$$e_i - e_j \quad (1 \leqslant i, j \leqslant n), \qquad \pm(e_j + e_k) \quad (1 \leqslant j < k \leqslant n). \tag{12}$$

The algebra $\mathfrak{b}_n = \mathfrak{so}(2n+1, C)$. Again let $\mathfrak{h}$ be defined by (7) and e_j by (9). Then the vectors

$$D_j^{\pm} = F_{2j-1,2n+1} \pm iF_{2j,2n+1} \qquad (1 \leqslant j \leqslant n)$$

satisfy

$$[H, D_j^{\pm}] = \mp e_j(H)\, D_j^{\pm} \tag{13}$$

and we have the direct decomposition

$$\mathfrak{b}_n = \mathfrak{h} + \sum_{j \neq k} CG_{jk}^{+} + \sum_{j \neq k} CG_{jk}^{-} + \sum_{j=1}^{n} CD_j^{+} + \sum_{j=1}^{n} CD_j^{-}. \tag{14}$$

Using the computation for $\mathfrak{d}_n$, the Killing form B now satisfies

$$B(H, H) = (4n-4) \sum_{i=1}^{n} e_i(H)^2 + 2 \sum_{i=1}^{n} e_i(H)^2 = (2n-1)\,\mathrm{Tr}(H^2),$$

so as before

$$B(X, Y) = (2n-1)\,\mathrm{Tr}(XY) \qquad (X, Y \in \mathfrak{b}_n). \tag{15}$$

Thus (11) and (15) are unified in the formula

$$B(X, Y) = (n-2)\,\mathrm{Tr}(XY), \qquad X, Y \in \mathfrak{so}(n, C). \tag{16}$$

Again (15) implies that $\mathfrak{b}_n$ is semisimple (for $n \geqslant 1$); then (8), (13), and (14) imply that $\mathfrak{h}$ is a Cartan subalgebra and that the roots are

$$\begin{gathered} \pm e_i \quad (1 \leqslant i \leqslant n), \qquad e_i - e_j \quad (1 \leqslant i, j \leqslant n), \\ \pm(e_i + e_j) \quad (1 \leqslant i < j \leqslant n). \end{gathered} \tag{17}$$

The algebra $\mathfrak{c}_n = \mathfrak{sp}(n, C)$. Here we let

$$H_i = E_{ii} - E_{n+i,n+i} \quad (1 \leqslant i \leqslant n), \qquad \mathfrak{h} = \sum_{i=1}^{n} CH_i, \tag{18}$$

and let e_j be the linear form on $\mathfrak{h}$ given by

$$e_j(H_i) = \delta_{ij}. \tag{19}$$

Then we have the direct decomposition

$$\mathfrak{c}_n = \mathfrak{h} + \sum_{i \leqslant j} C(E_{n+i,j} + E_{n+j,i}) + \sum_{i \leqslant j} C(E_{i,n+j} + E_{j,n+i})$$

$$+ \sum_{i \neq j} C(E_{ij} - E_{n+j,n+i})$$

and

$$[H, E_{n+i,j} + E_{n+j,i}] = -(e_i(H) + e_j(H))(E_{n+i,j} + E_{n+j,i}) \qquad (i \leqslant j)_i$$
$$[H, E_{i,n+j} + E_{j,n+i}] = (e_i(H) + e_j(H))(E_{i,n+j} + E_{j,n+i}) \qquad (i \leqslant j)$$
$$[H, E_{ij} - E_{n+j,n+i}] = (e_i(H) - e_j(H))(E_{ij} - E_{n+j,n+i}) \qquad (i \neq j).$$

These relations show that $\mathfrak{c}_n$ has center (0). Consider now the Lie algebra $\mathfrak{sp}(n)$ of matrices

$$\begin{pmatrix} U & V \\ -\bar{V} & \bar{U} \end{pmatrix} \qquad \begin{matrix} U \text{ skew Hermitian } n \times n \text{ matrix}, \\ V \text{ symmetric } n \times n \text{ matrix}. \end{matrix}$$

While $\mathfrak{sp}(n, C)$ is the Lie algebra of the group of matrices leaving invariant the exterior form

$$z_1 \wedge z_{n+1} + z_2 \wedge z_{n+2} + \dots + z_n \wedge z_{2n}, \tag{20}$$

$\mathfrak{sp}(n)$ is the Lie algebra of the group of unitary matrices leaving (20) invariant (cf. Chapter X). Moreover $\mathfrak{sp}(n)$ is a real form of $\mathfrak{sp}(n, C)$. Thus Prop. 6.6 (ii), Chapter II, implies that $\mathfrak{sp}(n, C)$ is semisimple. Then the relations above show that $\mathfrak{h}$ is a Cartan subalgebra and that the nonzero roots are ($\pm$ signs read independently)

$$\pm 2e_i \quad (1 \leqslant i \leqslant n), \qquad \pm e_i \pm e_j \quad (1 \leqslant i < j \leqslant n). \tag{21}$$

The Killing form B satisfies

$$B(H, H) = \text{Tr}(\text{ad } H \text{ ad } H) = 2 \sum_{i \leqslant j} (e_i(H) + e_j(H))^2 + \sum_{i,j} (e_i(H) - e_j(H))^2$$
$$= \sum_{i,j} (e_i(H) + e_j(H))^2 + \sum_i (2e_i(H))^2 + \sum_{i,j} (e_i(H) - e_j(H))^2$$
$$= (4n + 4) \sum_i e_i(H)^2 = (2n + 2) \text{ Tr}(HH).$$

Anticipating Cor. 6.6 in Chapter V, we have $B(X,X) = (2n + 2) \text{ Tr}(XX)$ for $X \in \mathfrak{c}_n$ regular, hence by continuity for all $X \in \mathfrak{c}_n$, whence by polarization

$$B(X, Y) = (2n + 2) \text{ Tr}(XY) \qquad (X, Y \in \mathfrak{c}_n). \tag{22}$$

Because of the complexity of the root space decompositions (10) and (14) for $\mathfrak{d}_n$ and $\mathfrak{b}_n$, one frequently uses a different description of these algebras (cf. Exercises B.5 and B.6) which at the same time brings out the analogy between $\mathfrak{c}_n$ and $\mathfrak{d}_n$.

Finally we refer to Chapter X for many examples of Cartan decompositions and the associated Cartan involutions.

EXERCISES AND FURTHER RESULTS

A. Solvable and Nilpotent Lie Algebras

1. A solvable Lie algebra has no semisimple subalgebra $\neq \{0\}$.

2. Let $\mathfrak{t}(n)$ denote the subalgebra of $\mathfrak{gl}(n, \boldsymbol{R})$ formed by all upper triangular $n \times n$ matrices and let $\mathfrak{n}(n)$ denote the subalgebra of matrices in $\mathfrak{t}(n)$ having all diagonal elements 0. Prove that:

(i) $\mathfrak{t}(n)$ is solvable, $\mathfrak{n}(n)$ is nilpotent and coincides with the derived algebra of $\mathfrak{t}(n)$.

(ii) The Lie algebras $\mathfrak{t}(n)$ and $\mathfrak{n}(n)$ both have centers of dimension 1.

(iii) Let B denote the Killing form of $\mathfrak{t}(n)$. Then $B(\mathfrak{t}(n), \mathfrak{n}(n)) = 0$.

3*. Let $\mathfrak{g}$ be a Lie algebra. Then $\mathfrak{g}$ is solvable $\Leftrightarrow$ $[\mathfrak{g}, \mathfrak{g}]$ is nilpotent $\Leftrightarrow$ $B(\mathfrak{g}, [\mathfrak{g}, \mathfrak{g}]) = 0$ (É. Cartan [1], p. 47; except for the second $\Leftarrow$ this is a simple consequence of the text).

4. Let G denote the group of all mappings $T_{a,b} : x \to ax + b$ $(x \in \boldsymbol{R})$, where a and b are real numbers, $a > 0$. Let G have the analytic structure determined by the condition that the mapping $(a, b) \to T_{a,b}$ is an analytic diffeomorphism. Show that G is a solvable Lie group and that its Lie algebra is the only noncommutative Lie algebra over $\boldsymbol{R}$ of dimension 2 (up to isomorphism).

5. Let $\mathfrak{g}$ be a solvable Lie algebra over $\boldsymbol{C}$ and let $\mathfrak{a}$ be a minimal proper ideal in $\mathfrak{g}$. Then $\mathfrak{a}$ has dimension 1. Formulate and prove an analogous result for solvable Lie algebras over $\boldsymbol{R}$.

B. Semisimple Lie Algebras

1. (i) Let $\mathfrak{g}$ be a semisimple Lie algebra over $\boldsymbol{C}$, $\mathfrak{h}$ a Cartan subalgebra. Let Δ denote the set of nonzero roots of $\mathfrak{g}$ with respect to $\mathfrak{h}$. Let Γ be a subset of Δ satisfying the conditions: If $\gamma \in \Gamma$, then $-\gamma \in \Gamma$; if $\gamma, \delta \in \Gamma$ and $\gamma + \delta \in \Delta$, then $\gamma + \delta \in \Gamma$. Let $\mathfrak{g}_\Gamma$ be the smallest subalgebra of $\mathfrak{g}$ containing all the root subspaces $\mathfrak{g}^\gamma$, $\gamma \in \Gamma$. Then $\mathfrak{g}_\Gamma$ is semisimple and $\mathfrak{h} \cap \mathfrak{g}_\Gamma$ is a Cartan subalgebra of $\mathfrak{g}_\Gamma$.

(ii) Deduce from (i) that if $H \in \mathfrak{h}$, its centralizer $\mathfrak{z}_H$ in $\mathfrak{g}$ is the direct sum $\mathfrak{z}_H = \mathfrak{c} + \mathfrak{g}_0$ where the ideals $\mathfrak{c}$ and $\mathfrak{g}_0$ are abelian and semisimple, respectively.

2. Let $\mathfrak{g}$ be a Lie algebra of dimension n. For $X \in \mathfrak{g}$ we write the characteristic polynomial of ad X, in the form

$$\det(\lambda I - \text{ad } X) = \lambda^n + a_{n-1}(X)\,\lambda^{n-1} + a_{n-2}(X)\,\lambda^{n-2} + \dots + a_0(X).$$

Prove that:

(i) $a_{n-1}(X) = -\text{Tr}(\text{ad } X)$.

(ii) $a_{n-2}(X) = \frac{1}{2}[(\text{Tr}(\text{ad } X))^2 - \text{Tr}((\text{ad } X)^2)]$.

(iii) $a_0(X) \equiv 0$.

(iv) If $\mathfrak{g}$ is semisimple, then $a_{n-1} \equiv 0$ so $a_{n-2}(X) = -\frac{1}{2} B(X, X)$.

(v) Let $\mathfrak{g}$ be semisimple over $\boldsymbol{C}$ and write

$$\det(\lambda I - \text{ad } X) = \lambda^n + \dots + a_l(X)\,\lambda^l \qquad (a_l \not\equiv 0).$$

Then X is regular if and only if $a_l(X) \neq 0$. Also if $\mathfrak{h} \subset \mathfrak{g}$ is any Cartan subalgebra,

$$a_l(H) = \prod_{\alpha \in \Delta} \alpha(H) \qquad (H \in \mathfrak{h})$$

(vi) Let $\mathfrak{g} = \mathfrak{sl}(2, \boldsymbol{C})$. Then $X \in \mathfrak{g}$ is regular if and only if $\det X \neq 0$.

(vii) Let $\mathfrak{g} = \mathfrak{sl}(n, \boldsymbol{C})$. Then $X \in \mathfrak{g}$ is regular if and only if the eigenvalues of X are all different.

3. A representation ρ of a Lie algebra $\mathfrak{g}$ (resp. a group G) on a finite-dimensional vector space V is called *semisimple* (or *completely reducible*) if each subspace of V invariant under $\rho(\mathfrak{g})$ (resp. $\rho(G)$) has a complementary invariant subspace.

(i) Any finite-dimensional representation of a compact topological group on a real or complex vector space V is semisimple.

(ii) (Weyl's unitary trick) Using a compact real form prove that any finite-dimensional representation π of a real semisimple Lie algebra $\mathfrak{g}$ on a real or complex vector space V is semisimple.

4. Let θ be a Cartan involution of a semisimple Lie algebra $\mathfrak{g}$ over $\boldsymbol{R}$ and σ an arbitrary involutive automorphism of $\mathfrak{g}$. Then there exists an automorphism φ of $\mathfrak{g}$ such that the Cartan involution $\varphi\theta\varphi^{-1}$ commutes with σ. (Hint: Repeat the proof of Theorem 7.1.)

5. Let $\mathfrak{d}_n$ denote the Lie algebra of the subgroup of $\boldsymbol{GL}(2n, \boldsymbol{C})$ leaving invariant the quadratic form

$$z_1 z_{n+1} + z_2 z_{n+2} + \dots + z_n z_{2n}.$$

Then $\mathfrak{d}_n$ is isomorphic to $\mathfrak{so}(2n, C)$. Let E_{ij}, F_{ij} be as in §8. Prove that:

(i) $\mathfrak{d}_n$ consists of the matrices

$$\begin{pmatrix} Z_1 & Z_2 \\ Z_3 & -{}^tZ_1 \end{pmatrix} \qquad \begin{array}{l} Z_i \text{ complex } n \times n \text{ matrices,} \\ Z_2, Z_3 \text{ skew symmetric.} \end{array}$$

(ii) The diagonal matrices in $\mathfrak{d}_n$ form a Cartan subalgebra $\mathfrak{a}$, and if e_i is the linear form on $\mathfrak{a}$ given by

$$e_i \begin{pmatrix} E_{jj} & 0 \\ 0 & -E_{jj} \end{pmatrix} = \delta_{ij},$$

then for $H \in \mathfrak{a}$

$$\left[H, \begin{pmatrix} E_{ij} & 0 \\ 0 & -E_{ji} \end{pmatrix}\right] = (e_i(H) - e_j(H)) \begin{pmatrix} E_{ij} & 0 \\ 0 & -E_{ji} \end{pmatrix} \qquad (1 \leqslant i, j \leqslant n)$$

$$\left[H, \begin{pmatrix} 0 & F_{ij} \\ 0 & 0 \end{pmatrix}\right] = (e_i(H) + e_j(H)) \begin{pmatrix} 0 & F_{ij} \\ 0 & 0 \end{pmatrix} \qquad (1 \leqslant i < j \leqslant n)$$

$$\left[H, \begin{pmatrix} 0 & 0 \\ F_{ij} & 0 \end{pmatrix}\right] = -(e_i(H) + e_j(H)) \begin{pmatrix} 0 & 0 \\ F_{ij} & 0 \end{pmatrix} \qquad (1 \leqslant i < j \leqslant n),$$

so the roots are given by

$$e_i - e_j \quad (1 \leqslant i, j \leqslant n), \qquad \pm(e_i + e_j) \quad (1 \leqslant i < j \leqslant n).$$

6. Let $\mathfrak{b}_n$ denote the Lie algebra of the subgroup of $\boldsymbol{GL}(2n + 1, \boldsymbol{C})$ which leaves invariant the quadratic form

$$z_0^2 + 2(z_1 z_{n+1} + z_2 z_{n+2} + \dots + z_n z_{2n}).$$

Then $\mathfrak{b}_n$ is isomorphic to $\mathfrak{so}(2n + 1, \boldsymbol{C})$. Let E_{ij}, F_{ij} be as in §8. Prove that:

(i) $\mathfrak{b}_n$ consists of the matrices

$$\begin{pmatrix} 0 & u & v \\ -{}^tv & Z_1 & Z_2 \\ -{}^tu & Z_3 & -{}^tZ_1 \end{pmatrix} \qquad \begin{array}{l} u, v \;\; 1 \times n \text{ complex matrices,} \\ Z_i \text{ complex } n \times n \text{ matrices,} \\ Z_2, Z_3 \text{ skew symmetric.} \end{array}$$

(ii) The diagonal matrices in (i) form a Cartan subalgebra $\mathfrak{a}$, the bracket relations in Exercise B.5 (ii) extend to $\mathfrak{b}_n$ and in addition we have the relations

$$[H, E_{1,i+1} - E_{n+1+i,1}] = -e_i(H)(E_{1,i+1} - E_{n+1+i,1}) \qquad (1 \leqslant i \leqslant n)$$
$$[H, E_{1,n+1+i} - E_{1+i,1}] = e_i(H)(E_{1,n+1+i} - E_{1+i,1}) \qquad (1 \leqslant i \leqslant n),$$

and the roots are given by

$$\pm e_i \quad (1 \leqslant i \leqslant n), \qquad (e_i - e_j) \quad (1 \leqslant i, j \leqslant n),$$
$$\pm(e_i + e_j) \quad (1 \leqslant i < j \leqslant n).$$

7. Let V be a finite-dimensional vector space over $\boldsymbol{C}$ and G a semisimple analytic subgroup of $\boldsymbol{GL}(V)$. Then:

(i) G has finite center (Cartan [12], p. 12).

(ii) G is closed in $\boldsymbol{GL}(V)$ (Yosida [1]).

(iii) Every continuous homomorphism ρ of G onto itself is an analytic isomorphism (Harish–Chandra [2]).

(iv) The semisimplicity assumption for G can be dropped neither in (i), (ii) nor in (iii).

8. Using the criterion that a derivation D of a finite-dimensional algebra A is nilpotent if $\mathrm{Tr}(DE) = 0$ for every derivation E of A prove that a Lie algebra $\mathfrak{g}$ over $\boldsymbol{R}$ of dimension > 1, with no ideals except 0 and $\mathfrak{g}$, is semisimple (cf. §6, Chapter II). The criterion is a special case of Theorem 17 in Chevalley [6], Vol. II, p. 182; cf. Varadarajan [1], p. 163.

C. Geometric Properties of the Root Pattern

1. Let $\mathfrak{g}$ be a semisimple Lie algebra and $\mathfrak{h}$ a Cartan subalgebra. Let Δ be the system of nonzero roots and put $\mathfrak{h}_{\boldsymbol{R}} = \sum_{\alpha\in\Delta} \boldsymbol{R}H_\alpha$. The system $\{H_\beta : \beta \in \Delta\}$ is invariant under the reflections

$$s_\alpha : H \to H - 2\frac{\alpha(H)}{\alpha(H_\alpha)} H_\alpha \qquad (H \in \mathfrak{h}_{\boldsymbol{R}})$$

in the hyperplanes $\alpha(H) = 0$ (Theorem 4.3 (i)).

2. Let α and β be nonproportional roots. Then:

(i) $\langle\alpha, \beta\rangle > 0 \Rightarrow \alpha - \beta$ is a root.

(ii) $\langle\alpha, \beta\rangle < 0 \Rightarrow \alpha + \beta$ is a root.

3. Let $\alpha \in \Delta$ and let β be any root. The α-series containing β contains at most four elements. Moreover the integer $a_{\beta,\alpha} = 2\langle\beta, \alpha\rangle/\langle\alpha, \alpha\rangle$

satisfies

$$|a_{\beta,\alpha}| \leqslant 3.$$

4. Putting $\langle\alpha, \beta\rangle = B(H_\alpha, H_\beta)$, formula (9) in §4 implies

$$\langle\alpha, \beta\rangle = \sum_{\gamma\in\Delta} \langle\alpha, \gamma\rangle\langle\gamma, \beta\rangle.$$

The matrix M with entries $\langle\alpha, \beta\rangle$ $(\alpha, \beta \in \Delta)$ therefore satisfies $M^2 = M$ and is symmetric. Deduce the formula

$$\sum_{\alpha\in\Delta} \langle\alpha, \alpha\rangle = \dim \mathfrak{h}$$

(cf. G. Brown [1]).

5. Let α and β be roots which are neither proportional nor orthogonal. Let θ $(0 < \theta < \pi)$ be the angle between H_α and H_β. Then

$$\cos\theta = \pm\tfrac{1}{2}\sqrt{m} \qquad \text{where} \quad m = 1, 2, 3$$

and if $\langle\beta, \beta\rangle \geqslant \langle\alpha, \alpha\rangle$,

$$\langle\beta, \beta\rangle = m\langle\alpha, \alpha\rangle.$$

6. Let $\alpha, \beta \in \Delta$ and $\beta + n\alpha$ $(p \leqslant n \leqslant q)$ be the α-series containg β. Assume $\beta + \alpha \in \Delta$. Then

$$\frac{\langle\beta + \alpha, \beta + \alpha\rangle}{\langle\beta, \beta\rangle} = \frac{-p + 1}{q}.$$

(Hint: Treat separately the cases $\langle\alpha, \alpha\rangle < \langle\beta, \beta\rangle$, $\langle\alpha, \alpha\rangle \geqslant \langle\beta, \beta\rangle$ and use Exercises C.2, C.3, and C.5.)

7. Let H_α, X_α $(\alpha \in \Delta)$ be as in Theorem 5.5, and put

$$H'_\alpha = \frac{2}{\langle\alpha, \alpha\rangle} H_\alpha, \qquad X'_\alpha = \left(\frac{2}{\langle\alpha, \alpha\rangle}\right)^{1/2} X_\alpha.$$

Using Exercise C.6 prove that the coefficients $N'_{\alpha,\beta}$ determined by

$$[X'_\alpha, X'_\beta] = N'_{\alpha,\beta}X'_{\alpha,\beta} \qquad (\alpha, \beta, \alpha + \beta \in \Delta)$$

satisfy

$$|N'_{\alpha,\beta}| = 1 - p$$

if $\beta + n\alpha$ $(p \leqslant n \leqslant q)$ is the α-series containing β. Thus $\mathfrak{g}$ has a basis for which the structural constants are integers (Chevalley [5], Théorème 1).

8. Let $\alpha, \beta \in \Delta$ and let $\beta + p\alpha, \ldots, \beta + q\alpha$ be the α-series containing β, $\alpha + p'\beta, \ldots, \alpha + q'\beta$ the β-series containing α. Prove that

$$\frac{\langle \alpha, \alpha \rangle}{\langle \beta, \beta \rangle} = \frac{p' + q'}{p + q} = \frac{q'(1 - p')}{q(1 - p)}.$$

NOTES

§2. Lie's theorem is proved in Lie [1], III, p. 678. The name "solvable" ("integrable" in Lie's terminology) is derived from the analogous concept for finite groups. In the theory of Picard-Vessiot extensions of differential fields solvable Lie groups play a role analogous to the role of solvable finite groups in Galois theory (Kolchin [1]). Concerning Engel's theorem, see É. Cartan [1], p. 46.

§3-§6. The root space decomposition is basic in the Killing-Cartan structure theory of simple Lie algebras over $\boldsymbol{C}$ (Killing [1], Cartan [1]). The roots are there introduced as zeros of the characteristic polynomial $\det(\lambda I - \operatorname{ad} X)$. Killing discovered some remarkable properties which the roots have under addition, for example Theorems 4.3 and 5.7. In Weyl [1, 2] the Cartan subalgebra is brought more to the foreground, the roots become linear functions, and many methods and simplifications are introduced. Among these is the ordering of the roots which is basic in Weyl's proof of Theorem 6.3 (existence of compact real forms ([1], III, Satz 6)). Exercise B.3 gives a significant application. Theorem 6.3 was known (but its full significance not realized) from Cartan's classification [2] of all real forms of simple Lie algebras over $\boldsymbol{C}$. The method of proof of Theorem 6.3 gives the general Theorem 5.4 (van der Waerden [1], Weyl [2], Chapter 2) which shows how $\mathfrak{g}$ is determined by the roots. This had been taken for granted by Killing and proved case-by-case by Cartan.

Later (in [12], p. 23) Cartan proposed the following method for proving Theorem 6.3: Let $\mathfrak{F}$ be the set of all bases $(e_1, \ldots, e_n)$ of $\mathfrak{g}$ such that $B(Z, Z) = -\Sigma_1^n z_i^2$ if $Z = \Sigma_1^n z_i e_i$ and let c_{ij}^k be the corresponding structure constants. Let f denote the function on $\mathfrak{F}$ defined by $f(e_1, \ldots, e_n) = \Sigma_{i,j,k} \mid c_{ij}^k \mid^2$. A simple argument (cf. Helgason [12], p. 28) shows that $\mathfrak{u} = \Sigma_1^n \boldsymbol{R} e_i$ is a compact real form of $\mathfrak{g}$ if and only if f has a minimum which is reached for c_{ij}^k real. A proof of the existence of $\mathfrak{u}$ along these lines was accomplished by Richardson [1]. As a corollary one has the existence of a Cartan subalgebra $\mathfrak{h}$ (without the theorems of Lie and Engel). In fact one can take $\mathfrak{h} = \mathfrak{t} + i\mathfrak{t}$ where $\mathfrak{t} \subset \mathfrak{u}$ is any maximal abelian subalgebra.

§7. Theorem 7.1 is proved by É. Cartan [12] and simplified by Mostow [1]. The proof in the text is modeled after Samelson (cf. [4]). Theorem 7.2 is also proved in Cartan [12]; the proof in the text is simpler, but possibly less instructive.

CHAPTER IV

SYMMETRIC SPACES

In this chapter we return to Riemannian geometry and begin a study of the Riemannian locally symmetric spaces. These are defined as Riemannian manifolds for which the curvature tensor is invariant under all parallel translations. É. Cartan set himself the problem of giving a complete classification of these spaces. In an ingeneous manner he gave the problem two different group-theoretic formulations [6]. One of these is particularly effective and strikingly enough reduces the problem to the classification of simple Lie algebras over $\boldsymbol{R}$, a problem which Cartan himself had solved already in 1914.

Cartan's first method was based on the so-called *holonomy group*. If o is a point in a Riemannian manifold M, then the holonomy group of M is the group of all linear transformations of the tangent space M_o obtained by parallel translation along closed curves starting at o. It is readily seen that different points of M give isomorphic holonomy groups. Of course each element of the holonomy group leaves the Riemannian structure g_o invariant; if M is locally symmetric the curvature tensor R_o is also left invariant. Hence it follows from the structural equations (6) and (7), §8, Chapter I, that each element of the holonomy group induces an isometry of a neighborhood of o in M onto itself leaving o fixed. This leads to algebraic relations between the Lie algebra $\mathfrak{k}$ of the identity component of the holonomy group and the tensors g_o and R_o, namely,

$$g_o(AX,Y) + g_o(X,AY) = 0, \qquad A \in \mathfrak{k},\ X,Y \in M_o;$$

$$[A,R_o(X,Y)] = R_o(AX,Y) + R_o(X,AY), \qquad A \in \mathfrak{k},\ X,Y \in M_o;$$

$$R_o(X,Y) \in \mathfrak{k}, \qquad X,Y \in M_o,$$

proved here in §5. Cartan now showed ([6], p. 225) that if for a given $\mathfrak{k}$ a tensor R_o of type (1,3) satisfies these formulas and in addition fulfills the general symmetry conditions for a curvature tensor (Lemma 12.5, Chapter I), then there exists a locally symmetric space for which it is the curvature tensor at o. In our treatment this proof is used in §5 to construct a globally symmetric space from a locally symmetric one. Since a symmetric space is determined locally from R_o and g_o (Lemma 1.2), the problem is reduced to two others: 1. Classify all possible holonomy groups of symmetric spaces. 2. Determine (up to a constant factor) the curvature tensor of a symmetric space by means of the holonomy group. For the second problem see Theorem 5.11, Chapter IX, and Cartan [6], pp. 221-224. For the first problem Cartan used his earlier work on finite-dimensional representations of Lie algebras, but the necessarily extensive calculations were not carried out in all details in [6] since a simpler method became available.

This second method of Cartan [7] (and the one which we shall follow) brings the problem of classifying locally symmetric spaces more directly into the realm of group theory. It is based on the fact that the invariance of the curvature tensor under parallelism is equivalent to the condition that the geodesic symmetry with respect to each point be a local isometry. (This explains the term "locally symmetric.") The consequences of this fact are more conveniently expressed for Riemannian globally symmetric spaces for which the geodesic symmetry by definition always extends to a global isometry. Such spaces have a transitive group of isometries and can be represented as coset spaces G/K where G is a connected Lie group with an involutive automorphism σ whose fixed point set is (essentially) K. The group G becomes semisimple after dividing out a direct factor which is a motion group of a Euclidean space. In this way the problem is reduced to the study of certain involutive automorphisms of semisimple Lie algebras.

In §1 the two definitions of locally symmetric spaces are considered and shown equivalent. The group of isometries of a Riemannian manifold is studied in §2; in §3 the results are applied to Riemannian globally symmetric spaces which then are represented as coset spaces of a special kind. The curvature tensor of a Riemannian globally symmetric space is computed in §4. The result is used in §5 for constructing a Riemannian globally symmetric space, a piece of which is isometric to a piece of a given locally symmetric space. In the last section it is shown how totally geodesic subspaces of a symmetric space are connected with Lie triple systems of the Lie algebra of the group of isometries.

§ 1. Affine Locally Symmetric Spaces

Let M be a C^∞ manifold with an affine connection ∇. Let p be a point in M and let N_0 be a normal neighborhood of the origin 0 in M_p, symmetric with respect to 0. As usual, put $N_p = \text{Exp}_p N_0$. For each $q \in N_p$, consider the geodesic $t \to \gamma(t)$ within N_p passing through p and q such that $\gamma(0) = p$, $\gamma(1) = q$. We put $q' = \gamma(-1)$. The mapping $q \to q'$ of N_p onto itself is called the *geodesic symmetry* with respect to p and will be denoted by s_p. In normal coordinates $\{x_1, ..., x_m\}$ at p, s_p has the expression $(x_1, ..., x_m) \to (-x_1, ..., -x_m)$. In particular, s_p is a diffeomorphism of N_p onto itself and $(ds_p)_p = -I$ where I denotes the identity mapping.

Definition. Let M be a manifold with an affine connection ∇ which has torsion tensor T and curvature tensor R; M is called *affine locally symmetric* if each point $m \in M$ has an open neighborhood N_m on which the geodesic symmetry s_m is an affine transformation.

Theorem 1.1. *A manifold M is affine locally symmetric if and only if $T = 0$ and $\nabla_Z R = 0$ for all $Z \in \mathfrak{D}^1(M)$.*

Proof. To begin with let M be any manifold with an affine connection ∇ and let φ be a diffeomorphism of M onto itself. We define the connection ∇' on M by

$$\nabla'_X(Y) = (\nabla_{X^\varphi}(Y^\varphi))^{\varphi^{-1}} \qquad \text{for } X, Y \in \mathfrak{D}^1(M). \qquad (1)$$

If we denote the mapping $X \to X^{\varphi}$ ($X \in \mathfrak{D}^1$) by φ, we have $\nabla'_X = \varphi^{-1} \circ \nabla_{\varphi X} \circ \varphi$ on $\mathfrak{D}^1$. Let T' and R' denote the corresponding torsion and curvature tensors. It is trivial to verify

$$T'(X, Y) = \varphi^{-1}(T(\varphi X, \varphi Y)), \tag{2}$$

$$R'(X, Y) = \varphi^{-1} \circ R(\varphi X, \varphi Y) \circ \varphi \tag{3}$$

for $X, Y \in \mathfrak{D}^1$. We shall now prove the relation

$$(\nabla'_Z R')(X, Y) = \varphi^{-1} \circ ((\nabla_{\varphi Z} R)(\varphi X, \varphi Y)) \circ \varphi \tag{4}$$

for $X, Y, Z \in \mathfrak{D}^1$. From relation (7) in Chapter I, §7, we have

$$[\nabla_Z, R(X, Y)] = \nabla_Z(R(X, Y)). \tag{5}$$

We now apply ∇_Z to the tensor field $X \otimes Y \otimes R$. Using (5) and the fact that ∇_Z commutes with contractions we obtain

$$(\nabla_Z R)(X, Y) = [\nabla_Z, R(X, Y)] - R(\nabla_Z X, Y) - R(X, \nabla_Z Y).$$

If we combine this with the similar formula for $(\nabla'_Z R')(X, Y)$, relation (4) follows easily.

Let m be an arbitrary point in M and let N_m be a normal neighborhood of m invariant under s_m. Suppose first that M is affine locally symmetric. Let $Z \in \mathfrak{D}^1(M)$ and γ a geodesic in N_m, passing through m with tangent vector Z_m at m. Let p and q be two points on γ, symmetric with respect to m. Let τ and τ_m denote the parallel translation (along γ) from p to q and m, respectively. Consider a vector $L \in M_p$. The vectors L and $\tau_m L$ are parallel with respect to the geodesic (mp). Since s_m is an affine transformation, the vectors $ds_m L$ and $ds_m \tau_m L$ are parallel with respect to the geodesic (mq). Since $ds_m \tau_m L = -\tau_m L$ it follows that

$$ds_m(L) = -\tau L. \tag{6}$$

Using (2) and (3) for $s_m = \varphi$, $R' = R$, $T' = T$, we obtain (§7, Chapter I)

$$\tau R_p = R_q, \qquad \tau T_p = -T_q. \tag{7}$$

Using the definition of $(\nabla_Z R)_m$ (§7, Chapter I) we deduce from (7) that $(\nabla_Z R)_m = 0$. Moreover, putting $p = q = m$, it follows from (7) that $T_m = 0$.

Next we prove the converse. The diffeomorphism s_m of N_m defines a new connection on N_m by (1). Since T and $\nabla_Z R$ vanish we find from (2) and (4) that $T' = 0$ and $\nabla'_Z R' = 0$. Finally we have $R'_m = R_m$

as a consequence of $(ds_m)_m = -I$. The theorem will thus follow from:

Lemma 1.2. *Let M and M' be two manifolds with affine connections ∇ and ∇'. Assume that*

$$\nabla_Z T = 0, \qquad \nabla_Z R = 0 \qquad \textit{for all } Z \in \mathfrak{D}^1(M),$$

$$\nabla'_{Z'} T' = 0, \qquad \nabla'_{Z'} R' = 0 \qquad \textit{for all } Z' \in \mathfrak{D}^1(M').$$

Let $p \in M$, $p' \in M'$ and let A be a linear one-to-one mapping of M_p onto $M'_{p'}$. Let $\tilde{A}$ denote the unique type-preserving isomorphism of the mixed tensor algebra $\mathfrak{D}(p)$ onto $\mathfrak{D}(p')$ extending A such that $\tilde{A}$ coincides with $({}^tA)^{-1}$ on the dual $(M_p)\hat{}$. Assume now that $\tilde{A} \cdot R_p = R'_{p'}$, $\tilde{A} \cdot T_p = T'_{p'}$. Then there exists an open neighborhood U_p of p in M and an affine transformation φ of U_p onto an open neighborhood $U_{p'}$ of p' in M' such that $\varphi(p) = p'$ and $d\varphi_p = A$.

Proof. Let N_0 and N'_0 be normal neighborhoods of the origin in M_p and $M'_{p'}$, respectively, and put $N_p = \text{Exp}\, N_0$, $N'_{p'} = \text{Exp}\, N'_0$. Let $Y_1, \ldots, Y_m$ be a basis of M_p and let $T^i{}_{jk}$ and $R^i{}_{ljk}$ be the coefficients of T and R in terms of the adapted vector field basis $Y_1^*, \ldots, Y_m^*$. These coefficients are then constants in N_p, due to the assumption. Now we express T' and R' in terms of the vector field basis on $N'_{p'}$, adapted to the basis $AY_1, \ldots, AY_m$ of $M'_{p'}$. Then the coefficients $T'^i{}_{jk}$ and $R'^i{}_{ljk}$ are *the same constants* as $T^i{}_{jk}$ and $R^i{}_{ljk}$.

Let V (respectively, V') denote the set of points $(t, a_1, \ldots, a_m) \in \boldsymbol{R} \times \boldsymbol{R}^m$ for which $ta_1Y_1 + \ldots + ta_mY_m \in N_0$, $(ta_1(AY_1) + \ldots + ta_m(AY_m) \in N'_0)$. The differential equations (6) and (7) in Chapter I, §8, are exactly the same for both connections ∇ and ∇'. For ∇ they hold on V and for ∇' they hold on V'. Since the equations are linear, the uniqueness of their solutions holds globally. Consequently, the solutions of the two sets of equations agree on $V \cap V'$. Let

$$U_p = \{\text{Exp}_p\,(t(a_1Y_1 + \ldots + a_mY_m)) : (t, a_1, \ldots, a_m) \in V \cap V'\};$$

$$U_{p'} = \{\text{Exp}_{p'}\,(t(a_1AY_1 + \ldots + a_mAY_m)) : (t, a_1, \ldots, a_m) \in V \cap V'\}.$$

Then U_p and $U_{p'}$ are normal neighborhoods of $p \in M$ and $p' \in M'$, respectively. If $q \in U_p$ and $q' \in U_{p'}$ have the same normal coordinates with respect to (Y_i) and (AY_i), respectively, the mapping $\varphi : q \rightarrow q'$ is an affine transformation of U_p onto $U_{p'}$ such that $\varphi(p) = p'$ and $d\varphi_p = A$.

Definition. A Riemannian manifold M is called a *Riemannian locally symmetric space* if for each $p \in M$ there exists a normal neighborhood of p on which the geodesic symmetry with respect to p is an isometry.

Theorem 1.3. *Let M be a Riemannian manifold. Then M is a Riemannian locally symmetric space if and only if the sectional curvature is invariant under all parallel translations.*

If M is locally symmetric, then the Riemannian structure g and the curvature tensor R are both invariant under all parallel translations. The invariance of the sectional curvature follows from Theorem 12.2, Chapter I. On the other hand, suppose the sectional curvatures invariant under all parallel translations. Let $p, q \in M$, γ a curve segment joining p to q, and τ the parallel translation from p to q along γ. Then if X, $Y \in M_p$, we have

$$g_p(R_p(X, Y)X, Y) = g_q(R_q(\tau X, \tau Y)\tau X, \tau Y),$$
$$g_p(R_p(X, Y)X, Y) = g_q(\tau(R_p(X, Y)X), \tau Y).$$

The quadrilinear form B given by

$$B(X, Y, Z, T) = g_q(R_q(\tau X, \tau Y)\tau Z, \tau T) - g_q(\tau(R_p(X, Y)Z), \tau T)$$

for $X, Y, Z, T \in M_p$, satisfies the conditions of Lemma 12.4, Chapter I. Consequently, $B \equiv 0$ so

$$\tau(R_p(X, Y)Z) = R_q(\tau X, \tau Y)\tau Z, \quad \text{that is,} \quad \tau \cdot R_p = R_q,$$

which shows that $\nabla_U R = 0$ for each $U \in \mathfrak{D}^1$. Since the geodesic symmetry s_m induces an isometry of M_m, the theorem now follows from Theorem 1.1 and:

Lemma 1.4. *Let φ be an affine transformation of a pseudo-Riemannian manifold M. Suppose that for some point $q \in M$, the mapping $d\varphi_q$: $M_q \to M_{\varphi(q)}$ is an isometry. Then φ is an isometry of M onto itself.*

Proof. Let $p \in M$ and $X, Y \in M_p$. We join p to q by a curve γ and let τ denote the parallel translation from p to q along γ. Then

$$g_p(X, Y) = g_q(\tau X, \tau Y) = g_{\varphi(q)}(d\varphi_q \tau X, d\varphi_q \tau Y).$$

This last quantity equals $g_{\varphi(p)}(d\varphi_p X, d\varphi_p Y)$ because φ, being an affine transformation, transforms vectors that are parallel along γ into vectors that are parallel along $\varphi \cdot \gamma$. This proves the lemma.

§ 2. Groups of Isometries

Let M be a Riemannian manifold and $I(M)$ the set of all isometries of M. Let $g_1, g_2 \in I(M)$. The composite mapping $g_1 \circ g_2$ is again an isometry. If we put $g_1 g_2 = g_1 \circ g_2$, $I(M)$ becomes a group. We shall always consider $I(M)$ with the *compact open topology*. This is defined as follows: Let C and U, respectively, be a compact and an open subset of M, and put

$$W(C, U) = \{g \in I(M) : g \cdot C \subset U\}.$$

The compact open topology on $I(M)$ is defined as the smallest topology on $I(M)$ for which all the sets $W(C, U)$ are open. It is obvious that $I(M)$ is a Hausdorff space. The identity component of $I(M)$ will be denoted $I_0(M)$.

Lemma 2.1. *The space $I(M)$ has a countable base.*

Proof. Since M is a separable metric space (Chapter I, §9) there exists a countable basis $O_1, \ldots, O_i, \ldots$ for the open subsets of M. Since M is locally compact, we can assume that the closure $\bar{O}_i$ is compact for each i. Let $C \subset M$ be compact, $U \subset M$ open, and f any element of $W(C, U)$. For each $p \in C$ there exists an index i and an index j such that $p \in O_i$, $f(O_i) \subset O_j \subset U$. We can find coverings $O_{i_1} \ldots O_{i_N}$ of C and $O_{j_1} \ldots O_{j_N}$ of $f(C)$ such that

$$f(\bar{O}_{i_k}) \subset O_{j_k} \subset U \qquad (1 \leqslant k \leqslant N).$$

It follows that

$$f \in \bigcap_{k=1}^{N} W(\bar{O}_{i_k}, O_{j_k}) \subset W(C, U).$$

This shows that the set Ω of all finite intersections of sets of the form $W(\bar{O}_i, O_j)$ forms a basis of the open sets of $I(M)$. Since Ω is countable, the lemma follows.

Theorem 2.2. *Let M be a Riemannian manifold and (f_n) a sequence in $I(M)$. Suppose there exists a point $o \in M$ such that the sequence $(f_n \cdot o)$ is convergent. Then there exists an element $f \in I(M)$ and a subsequence (f_{n_ν}) of (f_n) which converges to f in the compact open topology.*

We first prove a lemma.

Lemma 2.3. *Assume that a sequence (f_n) in $I(M)$ converges pointwise on a set $A \subset M$. Then (f_n) also converges pointwise on $\bar{A}$ (the closure of A).*

Proof. Let $p \in \bar{A}$ and choose $r > 0$ such that the open ball $B_r(p)$ has compact closure. Let ϵ be given, $0 < \epsilon < r$. We first select a point $p_1 \in A$ such that $d(p, p_1) < \epsilon/3$, then an integer N such that

$$d(f_n \cdot p_1, f_m \cdot p_1) < \epsilon/3 \qquad \text{for } n, m \geqslant N.$$

It follows that

$$d(f_n \cdot p, f_m \cdot p) \leqslant d(f_n \cdot p, f_n \cdot p_1) + d(f_n \cdot p_1, f_m \cdot p_1) + d(f_m \cdot p_1, f_m \cdot p) < \epsilon \tag{1}$$

if $m, n \geqslant N$. Therefore, all $f_n \cdot p$ $(n \geqslant N)$ lie inside the ball $B_\epsilon(f_N \cdot p)$ which has compact closure as well as $B_r(p)$. We can thus select a subsequence of $(f_n \cdot p)$ which converges to a limit, say p^*. From (1) we conclude that

$$d(f_n \cdot p, p^*) \leqslant \epsilon$$

for $n \geqslant N$ and the lemma is proved.

To prove Theorem 2.2, let S denote the set of points $q \in M$ for which the sequence $(f_n \cdot q)$ has compact closure. If (p_i) is a sequence in S and $(f_n{}^*)$ a subsequence of (f_n) we can, using a diagonal process, find a subsequence of $(f_n{}^*)$ which converges at each p_i. By Lemma 2.3 S is closed. We shall now prove that S is open. Since $o \in S$ and M is connected, this will prove that $S = M$.

Let $p^* \in S$ and choose $r > 0$ such that the ball $B_r(p^*)$ has compact closure. Let $p \in B_{r/4}(p^*)$ and let $(f_n{}^* \cdot p)$ be any subsequence of the sequence $(f_n \cdot p)$. There exists a subsequence (f_{n_μ}) of $(f_n{}^*)$ such that the sequence $(f_{n_\mu} \cdot p^*)$ converges to a limit q^* and such that the entire sequence $(f_{n_\mu} \cdot p^*)$ is contained in the ball $B_{r/4}(q^*)$. Then the sequence $(f_{n_\mu} \cdot p)$ is contained in the ball $B_{r/2}(q^*)$ which has compact closure since it is contained in each of the balls $f_{n_\mu} \cdot B_r(p^*)$. Hence $(f_n{}^* \cdot p)$ has a convergent subsequence. This proves that $S = M$.

Now M has a dense sequence of points, and using a diagonal process we can find a subsequence (φ_ν) of (f_n) which converges at all these points and by Lemma 2.3 on the entire M. The following lemma completes the proof of the theorem.

Lemma 2.4. *Let (φ_ν) be a sequence in $I(M)$ which converges pointwise on M to a mapping $f: M \rightarrow M$. Then $f \in I(M)$ and $\lim \varphi_\nu = f$ in the compact open topology.*

Proof. Let C be a compact subset of M and $\epsilon > 0$. We select points $p_1, ..., p_n$ such that each $p \in C$ has distance less than $\epsilon/3$ from some p_i. We can choose an integer N such that

$$d(\varphi_\nu \cdot p_i, f \cdot p_i) < \epsilon/3 \qquad \text{for } 1 \leqslant i \leqslant n, \qquad \nu > N.$$

If $p \in C$, we select p_j such that $d(p, p_j) < \epsilon/3$. Then, since f preserves distances, we have for $\nu > N$

$$d(\varphi_\nu \cdot p, f \cdot p) \leqslant d(\varphi_\nu \cdot p, \varphi_\nu \cdot p_j) + d(\varphi_\nu \cdot p_j, f \cdot p_j) + d(f \cdot p_j, f \cdot p) < \epsilon.$$

We finally prove that $f \cdot M = M$. Let $q \in f \cdot M$ and determine $q' \in M$ by $f(q') = q$. Then

$$0 = \lim d(q, \varphi_\nu \cdot q') = \lim d(\varphi_\nu^{-1} \cdot q, q').$$

Thus the sequence $\varphi_\nu^{-1} \cdot q$ converges to q'. We know from what is already proved, that there exists a subsequence (ψ_μ) of (φ_ν) such that $(\psi_\mu^{-1} \cdot p)$ converges for each $p \in M$. Let $p' = \lim (\psi_\mu^{-1} \cdot p)$. Then

$$\lim d(\psi_\mu \cdot p', p) = \lim d(p', \psi_\mu^{-1} \cdot p) = 0.$$

It follows that

$$p = \lim \psi_\mu p' = f \cdot p'.$$

Since $p \in M$ is arbitrary, we conclude that $fM = M$. From Theorem 11.1, Chapter I, we know that f is a diffeomorphism. Thus the lemma is proved.

Theorem 2.5.

(a) *Let M be a Riemannian manifold. The compact open topology of $I(M)$ turns $I(M)$ into a locally compact topological transformation group of M.*

(b) *Let $p \in M$ and let $\tilde{K}$ denote the subgroup of $I(M)$ which leaves p fixed. Then $\tilde{K}$ is compact.*

Proof. Let (f_n) be a sequence in $I(M)$ which converges to an element $f \in I(M)$. Then for each $p \in M$,

$$d(f_n^{-1} \cdot p, f^{-1} \cdot p) = d(p, f_n f^{-1} \cdot p) = d(f \cdot (f^{-1} \cdot p), f_n \cdot f^{-1} \cdot p),$$

which converges to 0 as $n \to \infty$. It follows from Lemma 2.4 that $\lim f_n^{-1} = f^{-1}$ in the compact open topology. Thus the inverse operation $f \to f^{-1}$ is continuous on $I(M)$. The continuity of the multiplication is proved similarly. Hence $I(M)$ is a topological group. Next, let $\lim f_n = f$ $(f_n \in I(M))$ and $\lim p_n = p$ $(p_n \in M)$. Let $\epsilon > 0$ be given and select an integer N such that $d(p_n, p) < \epsilon$ for $n \geqslant N$. The sequence $p_N, p_{N+1}, \ldots$ together with p form a compact set C. If $U = B_\epsilon(f \cdot p)$, then $f \in W(C, U)$. Let N_1 be an integer such that $f_n \in W(C, U)$ for $n \geqslant N_1$. Then $d(f_n \cdot p_n, f \cdot p) < \epsilon$ for $n \geqslant N_1 N$ so $I(M)$ is a topological transformation group of M. To show that $I(M)$ is locally compact and $\tilde{K}$ compact, let U be an open relatively compact neighborhood of p. Then $\tilde{K}$ is a closed subset of $W(\{p\}, U)$ and due to Theorem 2.2, $W(\{p\}, U)$ has compact closure. This finishes the proof.

§ 3. Riemannian Globally Symmetric Spaces

Let M be a Riemannian manifold with Riemannian structure Q; we recall that M is called an *analytic Riemannian manifold* if M and Q are both analytic. A mapping is called *involutive* if its square, but not the mapping itself, is the identity.

Definition. Let M be an analytic† Riemannian manifold; M is called *Riemannian globally symmetric* if each $p \in M$ is an isolated fixed point of an involutive isometry s_p of M.

Remark. It is obvious from the next lemma and Lemma 11.2, Chapter I, that there is only one such s_p.

Lemma 3.1. *Let M be Riemannian globally symmetric. For each $p \in M$ there exists a normal neighborhood N_p of $p \in M$ such that s_p is the geodesic symmetry on N_p.*

Let $A = (ds_p)_p$. Then $AM_p \subset M_p$ and $A^2 = I$. Writing

$$X = \tfrac{1}{2}(X - AX) + \tfrac{1}{2}(X + AX)$$

we see that $M_p = V^- + V^+$ (direct sum), where $V^{\pm} = \{X : AX = \pm X\}$. Suppose $X \neq 0$ belongs to V^+ and consider a geodesic γ tangent to X. Then s_p will leave γ pointwise fixed. This contradicts the assumption that p is an isolated fixed point. Thus $A = -I$ and the lemma follows.

Let M be a Riemannian globally symmetric space. Let γ be any geodesic in M. If $p \in \gamma$, $s_p \cdot \gamma$ gives an extension of γ, so that each maximal geodesic in M has infinite length. Thus M is complete and any two points $p, q \in M$ can be joined by a geodesic of length $d(p, q)$. If m is the midpoint of this geodesic then s_m interchanges p and q. In particular, the group $I(M)$ acts transitively on M. Owing to the theorems of §2, $I(M)$ has a countable base in the compact open topology and is a transitive, locally compact topological transformation group of M. Let $\tilde{K}$ be the subgroup of $I(M)$ which leaves some point p_0 of M fixed. Then $\tilde{K}$ is compact and due to Theorem 3.2 of Chapter II, $I(M)/\tilde{K}$ is homeomorphic to M under the mapping $g\tilde{K} \rightarrow g \cdot p_0$, $g \in I(M)$.

Lemma 3.2. *Let M be a Riemannian globally symmetric space. Then*

† This analyticity assumption is convenient; no loss of generality results from it as Prop. 5.5 shows.

$I(M)$ has an analytic structure compatible with the compact open topology in which it is a Lie transformation group of M.

Remark. The properties of $I(M)$ stated characterize its analytic structure uniquely. In fact, a topological group has at most one analytic structure compatible with its topology with respect to which it is a Lie group (Theorem 2.6, Chapter II).

Proof. Using the notation above, we consider the mapping $\sigma : k \rightarrow (dk)_{p_0}$ of $\tilde{K}$ into the orthogonal group $\boldsymbol{O}(M_{p_0})$. Let $X_1, ..., X_m$ be an orthonormal basis of M_{p_0}, $\{x_1, ..., x_m\}$ the normal coordinate system with respect to this basis, valid on a convex normal ball $B_r(p_0)$. The expression of the mapping k in coordinates $(x_1, ..., x_m)$ is the same as the expression of $(dk)_{p_0}$ in terms of the Cartesian coordinates on M_{p_0}. Thus σ is continuous. Owing to Lemma 11.2, Chapter I, σ is one-to-one, hence a homeomorphism. The linear isotropy group $K^* = \sigma(\tilde{K})$ is a compact subgroup of $\boldsymbol{O}(M_{p_0})$ and has therefore a unique differentiable structure compatible with the topology induced by $\boldsymbol{O}(M_{p_0})$ in which it is a Lie subgroup of $\boldsymbol{O}(M_{p_0})$. If we carry this differentiable structure over by σ^{-1}, $\tilde{K}$ becomes a compact Lie group.

Let π be the natural mapping $g \rightarrow g \cdot p_0$ of $I(M)$ onto M. We shall now construct a subset B of $I(M)$ (a certain local cross section), which π maps homeomorphically onto $B_r(p_0)$. Let $t \rightarrow p_t$ be a geodesic in $B_r(p_0)$ starting at p_0. For simplicity we put $s_{p_t} = s_t$. The mapping $T_t = s_{t/2}s_0$ is an isometry of M and sends p_0 into p_t. Owing to relation (6), §1, it is clear that $(dT_t)_{p_0}$ is the parallel translation from p_0 to p_t along the geodesic. Consider now the mapping $\psi : B_r(p_0) \rightarrow I(M)$ given by $\psi(p_t) = T_t$. The mapping ψ is of course one-to-one. In order to prove that ψ is continuous it suffices to prove that if a sequence $(q_n) \subset M$ converges to $q \in M$, then the corresponding symmetries s_{q_n} converge to s_q in $I(M)$. If p is sufficiently close to q then it is obvious that $(s_{q_n} \cdot p)$ converges to $s_q \cdot p$. Since an isometry is determined by its action on any open set it suffices, due to Lemma 2.4, to prove that the sequence $(s_{q_n} \cdot p)$ is convergent for each $p \in M$. Let S denote the set of points $p \in M$ for which $(s_{q_n} \cdot p)$ is convergent. The set S is open (Lemma 10.1, Chapter I) and not empty. It is also closed (Lemma 2.3), so $S = M$ and the continuity of ψ is established.

Let $B = \psi(B_r(p_0))$. The mapping π is one-to-one on B and $\pi \circ \psi = I$. Hence π is a homeomorphism of B onto $B_r(p_0)$. The set $B\tilde{K} = \{bk : b \in B, k \in \tilde{K}\}$ is the inverse image $\pi^{-1}(B_r(p_0))$, and is therefore an open subset of $I(M)$. Let $g \in B\tilde{K}$. Then $g = bk$ $(b \in B, k \in \tilde{K})$. It follows that $b = \psi(\pi(g))$ so the mapping $(b, k) \rightarrow bk$ is a homeomorphism of $B \times \tilde{K}$ onto $B\tilde{K}$. Hence if U is an open subset of $\tilde{K}$, the set BU

is open in $B\tilde{K}$, hence open in $I(M)$. In particular, let U be an open neighborhood of e in $\tilde{K}$ on which a system $\{y_1, ..., y_r\}$ of coordinates is valid. The mapping

$$\varphi_e : bu \to (x_1(\pi(b)), ..., x_m(\pi(b)), y_1(u), ..., y_r(u))$$

is a homeomorphism of BU onto an open subset of $\boldsymbol{R}^{m+r}$. For each $x \in I(M)$, the mapping $\varphi_x = \varphi_e \circ L_{x^{-1}}$ is a homeomorphism of xBU onto an open subset of $\boldsymbol{R}^{m+r}$. In order that this should give an analytic structure on $I(M)$, it suffices to verify that $\varphi_x \circ \varphi_e^{-1}$ is analytic on $\varphi_e(BU \cap xBU)$. This, and the fact that $I(M)$ is a Lie group, will follow if we can prove the following statement:

If $b_1, b_2 \in B$, $u_1, u_2 \in U$ such that

$$b_1u_1b_2u_2 = bu,$$

where $b \in B$, $u \in U$, then the coordinates $y_\alpha(u)$, $x_i(\pi(b))$ are analytic functions of the coordinates $x_j(\pi(b_1))$, $x_k(\pi(b_2))$, $y_\beta(u_1)$, $y_\gamma(u_2)$, $1 \leqslant i, j, k \leqslant m$, $1 \leqslant \alpha, \beta, \gamma \leqslant r$.

Let $k \in \tilde{K}$. Then the isometries ks_0k^{-1} and s_0 leave p_0 fixed and induce the same mapping of M_{p_0}. By Lemma 11.2, Chapter I, we have $ks_0k^{-1} = s_0$. In particular, if q is the midpoint of the geodesic from p_0 to $b_2 \cdot p_0$,

$$u_1b_2u_1^{-1} = u_1s_qs_0u_1^{-1} = u_1s_qu_1^{-1}s_0 = s_{q^*}s_0$$

if $q^* = u_1 \cdot q$. Hence $u_1b_2u_1^{-1}$ is an element b^* of B and the coordinates of b^* depend analytically on the coordinates of u_1 and b_2. We have now

$$b_1b^*u_1u_2 = bu$$

and

$$x_i(\pi(b_1b^*)) = x_i(\pi(b)), \qquad 1 \leqslant i \leqslant m.$$

Now, the point $b_1b^* \cdot p_0$ is determined from $b_1 \cdot p_0$ and $b^* \cdot p_0$ as follows: Let γ_1 and γ^* be the geodesics in $B_r(p_0)$ from p_0 to $b_1 \cdot p_0$ and from p_0 to $b^* \cdot p_0$. Let Y_1 and Y^* denote their unit tangent vectors at p_0. Let Y_3 be the parallel translate of Y^* along γ_1 and consider the geodesic γ_3 emanating from $b_1 \cdot p_0$ with tangent vector Y_3 and arc length $L(\gamma^*)$. The end point of γ_3 is $b_1b^* \cdot p_0$. This construction shows the analytic dependence of $x_i(\pi(b_1b^*))$ on $x_j(\pi(b_1))$ and $x_k(\pi(b^*))$, $1 \leqslant i, j, k \leqslant m$. Moreover, the coordinates of the tangent vector $d(u(u_1u_2)^{-1}) \cdot X_i = d(b^{-1}b_1b^*) \cdot X_i$ with respect to $X_1, ..., X_m$ depend analytically on the coordinates of b_1, b^*. Therefore, $\tilde{K}$ being a Lie group, it follows that the coordinates of u depend analytically on the coordinates of u_1, u_2, b_1, b^*. This shows that $I(M)$ is a Lie group. The argument above shows also that the mapping $(g, p) \to g \cdot p$ is an analytic mapping of $BU \times B_r(p_0)$ into M. Hence $I(M)$ is a Lie transformation group of M.

Theorem 3.3.

(i) *Let M be a Riemannian globally symmetric space and p_0 any point in M. If $G = I_0(M)$, and K is the subgroup of G which leaves p_0 fixed, then K is a compact subgroup of the connected group G and G/K is analytically diffeomorphic to M under the mapping $gK \to g \cdot p_0$, $g \in G$.*

(ii) *The mapping $\sigma : g \to s_{p_0} g s_{p_0}$ is an involutive automorphism of G such that K lies between the closed group K_σ of all fixed points of σ and the identity component of K_σ. The group K contains no normal subgroup of G other than $\{e\}$.*

(iii) *Let $\mathfrak{g}$ and $\mathfrak{k}$ denote the Lie algebras of G and K, respectively. Then $\mathfrak{k} = \{X \in \mathfrak{g} : (d\sigma)_e X = X\}$ and if $\mathfrak{p} = \{X \in \mathfrak{g} : (d\sigma)_e X = -X\}$ we have $\mathfrak{g} = \mathfrak{k} + \mathfrak{p}$ (direct sum). Let π denote the natural mapping $g \to g \cdot p_0$ of G onto M. Then $(d\pi)_e$ maps $\mathfrak{k}$ into $\{0\}$ and $\mathfrak{p}$ isomorphically onto M_{p_0}. If $X \in \mathfrak{p}$, then the geodesic emanating from p_0 with tangent vector $(d\pi)_e X$ is given by*

$$\gamma_{d\pi \cdot X}(t) = \exp tX \cdot p_0 \qquad (d\pi = (d\pi)_e).$$

Moreover, if $Y \in M_{p_0}$, then $(d \exp tX)_{p_0}(Y)$ is the parallel translate of Y along the geodesic.

Proof. The first part (i) follows from Prop. 4.3 in Chapter II. For (ii) and (iii) put $s_0 = s_{p_0}$. It is obvious that σ is an involutive automorphism of $I(M)$ and consequently maps the identity component G onto itself. If $k \in K$, the mappings k and $s_0 k s_0$ are isometries which induce the same mapping of M_{p_0}. From Lemma 11.2, Chapter I, we have $s_0 k s_0 = k$ for all $k \in K$. It follows that the automorphism $(d\sigma)_e$ of $\mathfrak{g}$ is identity on $\mathfrak{k}$. On the other hand, if $X \in \mathfrak{g}$ is left fixed by $(d\sigma)_e$, then $s_0 \exp tX\, s_0 = \exp tX$ for each $t \in \boldsymbol{R}$. This implies that $\exp tX \cdot p_0$ is a fixed point of s_0; since p_0 is an isolated fixed point of s_0 it follows that $\exp tX \cdot p_0 = p_0$ for all t so $X \in \mathfrak{k}$. Since $I(M)$ and G act effectively on G/K, K contains no normal subgroup $\neq \{e\}$ of G. This proves (ii).

The direct decomposition $\mathfrak{g} = \mathfrak{k} + \mathfrak{p}$ follows from the identity $X = \frac{1}{2}(X + d\sigma \cdot X) + \frac{1}{2}(X - d\sigma \cdot X)$. In the proof of Prop. 4.3, Chapter II, it is shown that $(d\pi)_e$ is a linear mapping of $\mathfrak{g}$ onto M_{p_0} with kernel $\mathfrak{k}$.

Finally, let $X \in \mathfrak{p}$; put $p_t = \gamma_{d\pi \cdot X}(t)$ and set as before $s_t = s_{p_t}$, $T_t = s_{t/2} s_0$. We have seen that $(dT_t)_{p_0}$ is the parallel translation along the geodesic from p_0 to p_t. Moreover, $s_\tau s_0 s_t = s_{\tau+t}$ because both sides of the equation are isometries which leave the point $p_{\tau+t}$ fixed and induce the same mapping in the tangent space $M_{p_{\tau+t}}$. It follows that $T_{2\tau+2t} = T_{2\tau}T_{2t}$ for all $\tau, t \in \boldsymbol{R}$. If t is sufficiently small, T_t lies in the local cross section B used in the proof of Lemma 3.2. Now π is an analytic diffeomorphism of B onto the normal neighborhood $B_r(p_0)$ and $\pi T_t = p_t$. Since the mapping $t \to p_t$ is analytic, it follows that $t \to T_t$ is a one-parameter subgroup. Hence $T_t = \exp tZ$ where $Z \in \mathfrak{g}$, $t \in \boldsymbol{R}$. Now $\sigma T_t = s_0 s_{t/2} = s_{-t/2} s_0 = T_{-t}$. Thus $d\sigma Z = -Z$ so $Z \in \mathfrak{p}$. But $\pi T_t = p_t$ so $d\pi Z = d\pi X$. Consequently, $X = Z$ and the theorem is proved

Remark. The isometries $\tau(\exp tX)$ are called *transvections.* They "slide" M along the geodesic $\gamma_{d\pi \cdot X}$ and realize the parallelism along this geodesic.

Definition. Let G be a connected Lie group and H a closed subgroup. The pair (G, H) is called a *symmetric pair* if there exists an involutive analytic automorphism σ of G such that $(H_\sigma)_0 \subset H \subset H_\sigma$, where H_σ is the set of fixed points of σ and $(H_\sigma)_0$ is the identity component of H_σ. If, in addition, the group† $\mathrm{Ad}_G(H)$ is compact, (G, H) is said to be a *Riemannian symmetric pair.*

As usual, we consider G as a Lie transformation group of the coset space G/H where each $g \in G$ gives rise to the diffeomorphism $\tau(g)$: $xH \to gxH$ of G/H onto itself.

Proposition 3.4. *Let (G, K) be a Riemannian symmetric pair. Let π denote the natural mapping of G onto G/K and put $o = \pi(e)$. Let σ be any analytic, involutive automorphism of G such that $(K_\sigma)_0 \subset K \subset K_\sigma$. In each G-invariant Riemannian structure Q on G/K (such Q exist) the manifold G/K is a Riemannian globally symmetric space. The geodesic symmetry s_o satisfies*

$$s_o \circ \pi = \pi \circ \sigma, \qquad \tau(\sigma(g)) = s_o \tau(g)\, s_o, \qquad g \in G;$$

in particular, s_o is independent of the choice of Q.

Remark 1. We shall see later, that the Riemannian connection on G/K is independent of the choice of Q.

Proof. Let σ be an arbitrary analytic involutive automorphism of G such that $(K_\sigma)_0 \subset K \subset K_\sigma$. For simplicity we shall write $d\sigma$ and $d\pi$ instead of $(d\sigma)_e$ and $(d\pi)_e$. Let $\mathfrak{g}$ and $\mathfrak{k}$ denote the Lie algebras of G and K, respectively, and put $\mathfrak{p} = \{X \in \mathfrak{g} : d\sigma X = -X\}$. Then $\mathfrak{g} = \mathfrak{k} + \mathfrak{p}$ (direct sum). If $X \in \mathfrak{p}$ and $k \in K$, then $\sigma(\exp \mathrm{Ad}(k)\, tX) = k \exp(-tX)k^{-1}$

† Here $\mathrm{Ad}_G(H)$ means the Lie subgroup of $\mathrm{Ad}_G(G)$ which is the image of H under Ad_G. As in Prop. 5.4, Chapter II, we see that if $\mathrm{Ad}_G(H)$ is compact, then it is compact in the ordinary matrix topology.

so $d\sigma \operatorname{Ad}(k) X = -\operatorname{Ad}(k) X$. Thus $\mathfrak{p}$ is invariant under $\operatorname{Ad}_G(K)$. The mapping $d\pi$ maps $\mathfrak{g}$ onto T_o, the tangent space to G/K at o, and the kernel of $d\pi$ is $\mathfrak{k}$. The resulting isomorphism of $\mathfrak{p}$ onto T_o commutes with the action of K, that is,

$$d\pi \cdot \operatorname{Ad}(k) X = d\tau(k) \cdot d\pi(X), \qquad k \in K,\ X \in \mathfrak{p}. \tag{1}$$

In fact, this formula is an immediate consequence of the relation

$$\pi(\exp \operatorname{Ad}(k)\, tX) = \pi(k \exp tX\, k^{-1}) = \tau(k)\, \pi(\exp tX).$$

Since $\operatorname{Ad}_G(K)$ is a compact group in the relative topology of $\boldsymbol{GL}(\mathfrak{g})$, there exists a strictly positive definite quadratic form B on $\mathfrak{p}$ invariant under $\operatorname{Ad}_G(K)$. Then the form $Q_o = B \circ (d\pi)^{-1}$ on T_o is invariant under all the mappings $d\tau(k)$, $k \in K$. Let the corresponding symmetric bilinear form on $T_o \times T_o$ also be denoted by Q_o. For each $p \in G/K$ we define the bilinear form Q_p on $(G/K)_p \times (G/K)_p$ by

$$Q_p(d\tau(g)\, X_0, d\tau(g)\, Y_0) = Q_o(X_0, Y_0), \qquad X_0, Y_0 \in T_o,$$

where $g \in G$ is chosen such that $g \cdot o = p$. The invariance of B under $\operatorname{Ad}_G(K)$ guarantees that Q_p is well defined. Since each $\tau(g)$, $g \in G$, is an analytic diffeomorphism of G/K it follows that $p \to Q_p$ is an analytic Riemannian structure on G/K, invariant under the action of G. On the other hand, each G-invariant Riemannian structure on G/K arises in this fashion from an invariant quadratic form on $\mathfrak{p}$.

We now define a mapping s_o of G/K onto itself by the condition $s_o \circ \pi = \pi \circ \sigma$. Then s_o is an involutive diffeomorphism of G/K onto itself and $(ds_o)_o = -I$. To see that s_o is an isometry, let $g \in G$, $p = \tau(g) \cdot o$, and $X, Y \in (G/K)_p$. Then the vectors $X_0 = d\tau(g^{-1})\, X$, $Y_0 = d\tau(g^{-1})\, Y$ belong to T_o. The formula $s_o \circ \pi = \pi \circ \sigma$ implies for each $x \in G$,

$$s_o \circ \tau(g)\, (xK) = \sigma(gx)\, K = \sigma(g)\, \sigma(x)\, K = (\tau(\sigma(g)) \circ s_o)\, (xK),$$

so $s_o \circ \tau(g) = \tau(\sigma(g)) \circ s_o$. Hence

$$\begin{aligned} Q(ds_o\, X, ds_o\, Y) &= Q(ds_o\, d\tau(g)\, X_0, ds_o d\tau(g)\, Y_0) \\ &= Q(ds_o\, X_0, ds_o\, Y_0) = Q(X_0, Y_0) = Q(X, Y). \end{aligned}$$

Thus s_o is an isometry and near o it must coincide with the geodesic symmetry. For an arbitrary point $p = \tau(g) \cdot o$ in G/K, the geodesic symmetry is given by

$$s_p = \tau(g) \circ s_o \circ \tau(g^{-1}).$$

This being an isometry, the space G/K is a Riemannian globally symmetric space. This finishes the proof.

The formula $s_o \circ \pi = \pi \circ \sigma$ shows that the geodesic symmetry on G/K is the same for all G-invariant metrics.

We shall now derive some further properties of the Riemannian symmetric pair (G, K). Let Z denote the center of G and let N denote the set of $n \in G$ for which $\tau(n)$ is the identity mapping of G/K. Then Z and N are closed normal subgroups of G and $N \subset K$. Due to Lemma 5.1, Chapter II, the group $K/K \cap Z$ and the linear group $\text{Ad}_G(K)$ are analytically isomorphic. Hence $K/K \cap Z$ is compact and since $K \cap Z \subset N$, K/N is compact. Let $I(G/K)$ denote the group of all isometries of G/K (with the Riemannian structure Q), and let $\tilde{K}$ denote the subgroup of $I(G/K)$ which leaves o fixed. Then, by Lemma 3.2 and Theorem 2.5, $I(G/K)$ and $\tilde{K}$ are Lie groups, $\tilde{K}$ compact.

Consider now the (algebraic) isomorphism $\beta : gN \rightarrow \tau(g)$ of G/N into $I(G/K)$. If a sequence (g_nN) converges to gN in G/N, then (g_nxN) converges to gxN for each $x \in G$ and therefore g_nxK converges to gxK in G/K. In view of Lemma 2.4 this proves the continuity of β. The restriction of β to K/N is a homeomorphism.

Remark 2. The group $\beta(G/N)$ is a closed subgroup of $I(G/K)$.

In fact, let $K_1 = \beta(K/N)$ and $G_1 = \beta(G/N)$. Then K_1 is a compact topological subgroup of $\tilde{K}$ and if the analytic structure of G/N is carried over on G_1 by β, G_1 is a Lie transformation group of G/K. Let (g_n) be a sequence in G_1 which converges in $I(G/K)$ to an element $g \in I(G/K)$. The sequence $(g_n \cdot o)$ converges to the point $p = g \cdot o$ in G/K. Select $g^* \in G_1$ such that $g^* \cdot o = p$. There exists a local cross section in G_1 through g^*, that is a submanifold B^* of G_1 containing g^* such that the natural mapping $x \rightarrow x \cdot o$ of G_1 onto G/K is a diffeomorphism of B^* onto an open neighborhood of p in G/K. If n is sufficiently large, there exists an element $k_n \in K_1$ such that $g_nk_n \in B^*$. It is clear that the sequence (g_nk_n) in B^* converges to g^*, and since K_1 is compact we may assume that the sequence (k_n) is convergent in K_1. The imbeddings $B^* \rightarrow G_1$ and $K_1 \rightarrow G_1$ being continuous the sequences (g_nk_n) and (k_n) converge in G_1. It follows that (g_n) converges in G_1. Finally, since the imbedding $G_1 \subset I(G/K)$ is continuous, $g \in G_1$, so G_1 is closed.

Now it follows from Theorem 2.3, Chapter II, that G_1 has a unique analytic structure in which it is a topological Lie subgroup G_2 of $I(G/K)$. The identity mapping $G_1 \rightarrow G_2$ is continuous, hence a homeomorphism (Cor. 3.3, Chapter II), hence an analytic isomorphism (Theorem 2.6, Chapter II). This means that β is an analytic isomorphism of G/N onto a closed, topological Lie subgroup of $I(G/K)$.

Let $\tilde{\mathfrak{g}}$ and $\tilde{\mathfrak{k}}$ denote the Lie algebras of $I(G/K)$ and $\tilde{K}$, respectively.

From Theorem 3.3 we obtain a subspace $\tilde{\mathfrak{p}} \subset \tilde{\mathfrak{g}}$ such that $\tilde{\mathfrak{g}} = \tilde{\mathfrak{k}} + \tilde{\mathfrak{p}}$ (direct sum) and

$$\tilde{\pi}(\exp X) = \operatorname{Exp} d\tilde{\pi}(X), \qquad X \in \tilde{\mathfrak{p}}, \tag{2}$$

$\tilde{\pi}$ denoting the natural mapping $g \to g \cdot o$ of $I(G/K)$ onto G/K. Let $\tilde{\sigma}$ denote the involutive automorphism $g \to s_o g s_o$ of $I(G/K)$ and let π_1 denote the natural mapping of G onto G/N. Then $\beta \circ \pi_1 = \tau$ and by Prop. 3.4

$$\beta(\pi_1(\sigma(g))) = \tau(\sigma(g)) = \tilde{\sigma}(\tau(g)) \qquad (g \in G),$$

so

$$d\beta \circ d\pi_1 \circ d\sigma = d\tilde{\sigma} \circ d\tau.$$

Consequently, the mapping $d\beta \circ d\pi_1$ maps $\mathfrak{p}$ onto $\tilde{\mathfrak{p}}$. Since $\tilde{\pi} \circ \beta \circ \pi_1 = \pi$ we get from (2) for $X \in \mathfrak{p}$

$$\begin{aligned}\pi(\exp X) &= \tilde{\pi}(\beta(\pi_1(\exp X)) = \tilde{\pi}(\exp d\beta(d\pi_1(X)) \\ &= \operatorname{Exp}(d\tilde{\pi}(d\beta(d\pi_1(X))\end{aligned}$$

so

$$\pi(\exp X) = \operatorname{Exp}(d\pi(X)), \qquad X \in \mathfrak{p}. \tag{3}$$

Thus the geodesics in G/K are still orbits of suitable one-parameter subgroups in G.

We shall now prove that under very general conditions, the automorphism σ is completely determined by its fixed points.

Proposition 3.5. *Let (G, K) be a Riemannian symmetric pair. Let $\mathfrak{k}$ denote the Lie algebra of K and let $\mathfrak{z}$ denote the Lie algebra of the center of G. Assume that $\mathfrak{k} \cap \mathfrak{z} = \{0\}$. Then there exists exactly one involutive, analytic automorphism σ of G such that $(K_\sigma)_0 \subset K \subset K_\sigma$.*

Proof. Let σ_1, σ_2 be two automorphisms with the described properties. Then the Lie algebra $\mathfrak{g}$ of G has direct decompositions $\mathfrak{g} = \mathfrak{k} + \mathfrak{p}_i$ where $\mathfrak{p}_i$ is the eigenspace for the eigenvalue -1 of the automorphism $d\sigma_i$ $(i = 1, 2)$. Since the Killing form B of $\mathfrak{g}$ is invariant under σ_i, it follows that $\mathfrak{k}$ is orthogonal to $\mathfrak{p}_i$ with respect to B $(i = 1, 2)$. Let $X_1 \in \mathfrak{p}_1$. Then there exists an element $X_2 \in \mathfrak{p}_2$ such that $X_1 = T + X_2$ where $T \in \mathfrak{k}$. It follows that T is orthogonal to $\mathfrak{k}$. Since $\mathfrak{k} \cap \mathfrak{z} = \{0\}$, B is strictly negative definite on $\mathfrak{k}$ (Prop. 6.8, Chapter II). Thus $T = 0$ and $\mathfrak{p}_1 = \mathfrak{p}_2$.

For further study of symmetric spaces it is important to express the symmetry conditions in terms of Lie algebras rather than in terms of the groups. In Theorem 3.3 we have seen that a Riemannian globally symmetric space gives rise to a pair $(\mathfrak{g}, s)$ where:

(i) $\mathfrak{g}$ is a Lie algebra over $\boldsymbol{R}$.

(ii) s is an involutive automorphism of $\mathfrak{g}$.

(iii) $\mathfrak{k}$, the set of fixed points of s, is a compactly imbedded subalgebra of $\mathfrak{g}$.

(iv) $\mathfrak{k} \cap \mathfrak{z} = \{0\}$ if $\mathfrak{z}$ denotes the center of $\mathfrak{g}$.

Definition. A pair $(\mathfrak{g}, s)$ with the properties (i), (ii), (iii) is called an *orthogonal symmetric Lie algebra*. It is said to be *effective* if, in addition, (iv) holds. A *pair* (G, K), where G is a connected Lie group with Lie algebra $\mathfrak{g}$, and K is a Lie subgroup of G with Lie algebra $\mathfrak{k}$, is said to be *associated* with the orthogonal symmetric Lie algebra $(\mathfrak{g}, s)$.

Remark 1. If $\mathfrak{k}$ is a compactly imbedded subalgebra of a Lie algebra $\mathfrak{g}$ with center $\mathfrak{z}$ and $\mathfrak{k} \cap \mathfrak{z} = \{0\}$, then by the proof of Prop. 3.5 there exists at most one involutive automorphism of $\mathfrak{g}$ whose fixed point set is $\mathfrak{k}$.

Remark 2. It is important to distinguish between a symmetric pair as defined earlier (before Prop. 3.4) and a pair associated with an orthogonal symmetric Lie algebra.

Proposition 3.6. *Let $(\mathfrak{g}, s)$ be an orthogonal symmetric Lie algebra, $\mathfrak{k}$ the set of fixed points of s. Let (G, K) and $(\tilde{G}, \tilde{K})$ be two pairs associated with $(\mathfrak{g}, s)$. Suppose K and $\tilde{K}$ are connected and $\tilde{G}$ simply connected. Then $\tilde{K}$ is closed and $(\tilde{G}, \tilde{K})$ is a Riemannian symmetric pair. If K is closed in G (this is the case if the center of $\mathfrak{g}$ is $\{0\}$), then G/K is Riemannian locally symmetric*† *for each G-invariant metric (such exist) and $\tilde{G}/\tilde{K}$ is the universal covering manifold of G/K.*

Proof. Since $\tilde{G}$ is simply connected, there exists an analytic homomorphism $\sigma : \tilde{G} \to \tilde{G}$ for which $(d\sigma)_e = s$. Since s is an involutive automorphism the same is true of σ. The group $\tilde{K}$ is the identity component of the group of fixed points of σ. In particular $\tilde{K}$ is closed in $\tilde{G}$. The space $\tilde{G}/\tilde{K}$ is simply connected. In fact, let $\gamma(t)$, $0 \leqslant t \leqslant 1$, be a continuous closed curve in $\tilde{G}/\tilde{K}$. Without loss of generality we can assume that $\gamma(0) = \gamma(1) = \tilde{\pi}(e)$ ($\tilde{\pi}$ being the natural mapping of $\tilde{G}$ onto $\tilde{G}/\tilde{K}$). Using local cross sections in $\tilde{G}$ (Lemma 4.1, Chapter II), it is easy to find a continuous curve $\tilde{\gamma}(t)$, $0 \leqslant t \leqslant 1$, in $\tilde{G}$ such that $\tilde{\pi}(\tilde{\gamma}(t)) = \gamma(t)$ for $0 \leqslant t \leqslant 1$. Then $\tilde{\gamma}(0)$ and $\tilde{\gamma}(1)$ belong to $\tilde{K}$ and can be joined by a continuous curve, $\tilde{K}$ being connected. The closed curve in $\tilde{G}$, so obtained, is homotopic to a point in $\tilde{G}$. It follows that the projection $\gamma(t)$ is also homotopic to a point in $\tilde{G}/\tilde{K}$.

† For an improvement, see Theorem 1.1, Chapter VI, and Exercise 10, Chapter VII.

The groups $\mathrm{Ad}_G(K)$ and $\mathrm{Ad}_{\tilde{G}}(\tilde{K})$ coincide because they are both analytic subgroups of Int $(\mathfrak{g})$ and have the same Lie algebra. They are compact (and thus carry the relative topology of $\boldsymbol{GL}(\mathfrak{g})$) since $\mathfrak{k}$ is compactly imbedded in $\mathfrak{g}$. The space $\mathfrak{p} = \{X \in \mathfrak{g} : s \cdot X = -X\}$ is invariant under $\mathrm{Ad}_G(K)$ and has a strictly positive definte quadratic form B invariant under $\mathrm{Ad}_G(K)$. As before, this form gives rise to a $\tilde{G}$-invariant Riemannian structure on $\tilde{G}/\tilde{K}$, and, if K is closed, to a G-invariant Riemannian structure on G/K. Let φ be the homomorphism of $\tilde{G}$ onto G such that $(d\varphi)_e$ is the identity mapping of $\mathfrak{g}$. Let $K^\natural$ denote the inverse image $\varphi^{-1}(K)$. Then $\tilde{K}$ is the identity component of $K^\natural$ and $\tilde{G}/\tilde{K}$ is a covering space (see Chevalley [2], p. 58) of $\tilde{G}/K^\natural$. If ψ denotes the mapping $g\tilde{K} \to \varphi(g)K$ of $\tilde{G}/\tilde{K}$ onto G/K, then the pair $(\tilde{G}/\tilde{K}, \psi)$ is the simply connected covering manifold of G/K (Lemma 13.4, Chapter I). Moreover, ψ is a local isometry. Since $\tilde{G}/\tilde{K}$ is globally symmetric, G/K is locally symmetric.

Let π denote the natural mapping of G onto G/K. Since one-parameter subgroups in $\tilde{G}$ and G correspond under φ and since geodesic in $\tilde{G}/\tilde{K}$ and G/K correspond under ψ, we obtain from (3),

$$\pi(\exp X) = \mathrm{Exp}\, d\pi X \qquad \text{for } X \in \mathfrak{p}, \tag{4}$$

exp and Exp referring to G and G/K, respectively. Relation (1) also holds here and allows us to identify $\mathfrak{p}$ and $(G/K)_{\pi(e)}$ whenever this is convenient.

Finally, let K^* denote the complete inverse image $\mathrm{Ad}_G^{-1}(\mathrm{Ad}_G(K))$. Since $\mathrm{Ad}_G(K)$ is closed in Int $(\mathfrak{g})$, K^* is closed in G. If $\mathfrak{k}^*$ denotes the Lie algebra of K^* we have $\mathrm{ad}_{\mathfrak{g}}(\mathfrak{k}^*) = \mathrm{ad}_{\mathfrak{g}}(\mathfrak{k})$. If $\mathfrak{g}$ has center $\{0\}$, it follows that $\mathfrak{k}^* = \mathfrak{k}$; consequently K is the identity component of K^*, hence closed.

§ 4. The Exponential Mapping and the Curvature

The notation in this section will be as follows: Let $(\mathfrak{g}, s)$ be an orthogonal symmetric Lie algebra, $\mathfrak{k}$ the set of fixed points of s, and $\mathfrak{p}$ the subspace $\{X \in \mathfrak{g} : sX = -X\}$. For $X \in \mathfrak{p}$, let T_X denote the restriction of $(\mathrm{ad}\, X)^2$ to $\mathfrak{p}$. Then $T_X\mathfrak{p} \subset \mathfrak{p}$. Suppose the pair (G, K) is associated with $(\mathfrak{g}, s)$ and suppose that K is connected and closed in G. Let π be the natural mapping of G onto G/K and put $o = \pi(e)$. For $g \in G$, let $\tau(g)$ denote the mapping $xK \to gxK$ of G/K onto itself. The subspace $\mathfrak{p}$ will be identified with the tangent space $(G/K)_o$ by means of the mapping $d\pi$. Let Q be any G-invariant Riemannian structure on G/K. Then G/K is complete and locally symmetric.

We shall now describe the geometric concepts Exponential mapping and curvature for G/K in group theoretic terms.

Theorem 4.1. *The Exponential mapping of* $\mathfrak{p}$ *into* G/K *is independent of the choice of* Q. *Its differential is given by*

$$d\,\mathrm{Exp}_X = d\tau(\exp X)_o \circ \sum_{n=0}^{\infty} \frac{(T_X)^n}{(2n+1)!}, \qquad X \in \mathfrak{p}.$$

Here $\mathfrak{p}$ *is considered as a manifold in the usual way and whose tangent space at each point is identified with* $\mathfrak{p}$ *itself.*

Proof. Let $X, Y \in \mathfrak{p}$. From (4), §3, we have $\pi(\exp X) = \mathrm{Exp}\, X$ so from Theorem 1.7, Chapter II, we obtain

$$d\,\mathrm{Exp}_X(Y) = d\pi \circ d\exp_X(Y) = d\pi \circ dL_{\exp X} \circ \frac{1 - e^{-\mathrm{ad}\, X}}{\mathrm{ad}\, X}(Y)$$

$$= d\tau\,(\exp X) \circ d\pi \circ \sum_{m=0}^{\infty} \frac{(-\,\mathrm{ad}\, X)^m}{(m+1)!}(Y).$$

From the relations $[\mathfrak{k}, \mathfrak{p}] \subset \mathfrak{p}$, $[\mathfrak{p}, \mathfrak{p}] \subset \mathfrak{k}$ it follows that $d\pi\,(\mathrm{ad}\, X)^m(Y)$ is equal to $T_X^n(Y)$ if $m = 2n$ and 0 if m is odd. This proves the theorem.

Theorem 4.2. *Let* R *denote the curvature tensor of the space* G/K *corresponding to the Riemannian structure* Q. *Then, at the point* $o \in G/K$,

$$R_o(X, Y)\,Z = -\,[[X, Y], Z] \qquad \text{for } X, Y, Z \in \mathfrak{p}.$$

Proof. First we evaluate the sectional curvature directly and then use results from §12, Chapter I, to find the curvature tensor.

Assuming of course that $\dim \mathfrak{p} > 1$, let S be a two-dimensional subspace of $\mathfrak{p}$ and let $X_1, X_2, \ldots, X_m$ be an orthonormal basis of $\mathfrak{p}$ such that X_1 and X_2 belong to S. Each $X \in S$ can be written $X = x_1 X_1 + x_2 X_2$ where $x_1, x_2 \in \boldsymbol{R}$ and the Laplacian Δ on S is given by

$$\frac{\partial^2}{\partial x_1^2} + \frac{\partial^2}{\partial x_2^2}.$$

We also put

$$A_X = \sum_{n=0}^{\infty} \frac{T_X^n}{(2n+1)!}, \qquad v_1 = A_X(X_1), \qquad v_2 = A_X(X_2).$$

If a and b are two vectors in a metric vector space, we denote by $a \vee b$ the parallelogram spanned by a and b and by $|\, a \vee b \,|$ the area.

Let N_0 be a normal neighborhood of 0 in the tangent space $\mathfrak{p}$. The submanifold $M_S = \operatorname{Exp}(S \cap N_0)$ of $M = G/K$ has a Riemannian structure induced by Q; a curve in M_S has the same arc length whether it is considered as a curve in M_S or M. If $p \in M_S$, the unique geodesic in $\operatorname{Exp} N_0$ from o to p is the shortest curve in M_S joining o and p. It follows that the Exponential mappings at o, for M_S and M, respectively, coincide on $S \cap N_0$. Thus, if $X \in S \cap N_0$, the vectors $d\tau(\exp X) \cdot v_1$ and $d\tau(\exp X) \cdot v_2$ are tangent vectors to M_S at $\operatorname{Exp} X$; the ratio of the surface elements in M_S and $S \cap N_0$ is therefore given by

$$f(X) = \frac{| d\tau(\exp X) v_1 \vee d\tau(\exp X) v_2 |}{| X_1 \vee X_2 |} = | v_1 \vee v_2 |,$$

since $\tau(\exp X)$ is an isometry of G/K. According to Lemma 12.1, Chapter I, the sectional curvature of M along the plane section S is given by

$$K(S) = -\frac{3}{2}\left[\left(\frac{\partial^2}{\partial x_1^2} + \frac{\partial^2}{\partial x_2^2}\right) f\right](0, 0).$$

Let (A_{ij}) be the matrix expressing A_X in terms of the basis $X_1, ..., X_m$ of $\mathfrak{p}$, $A_X X_j = \sum_i A_{ij} X_i$. Then

$$\begin{aligned} f(X) &= | (A_{11}X_1 + ... + A_{m1}X_m) \vee (A_{12}X_1 + ... + A_{m2}X_m) | \\ &= \left\{(A_{11}A_{22} - A_{12}A_{21})^2 + \sum_{1 < i < j \leq m} (A_{i1}A_{j2} - A_{j1}A_{i2})^2\right\}^{1/2}, \end{aligned}$$

since $| A_{i1}A_{j2} - A_{j1}A_{i2} |$ is the area of the projection of $v_1 \vee v_2$ on the (X_i, X_j)-plane. In computing $[\Delta f](0)$ from the expression for $f(X)$, we only have to consider terms of second order in x_1 or x_2. If $i \neq j$, A_{ij} only contains terms of second order and higher. Hence

$$[\Delta f](0) = [\Delta A_{11}A_{22}](0).$$

Now the matrix elements T_{ij} of $T_X = (\operatorname{ad}(x_1X_1 + x_2X_2))^2$ are of second order in x_1 and x_2. It follows that (writing Q for Q_o)

$$\begin{aligned} [\Delta f](0) &= \frac{1}{3!}[\Delta(T_{11} + T_{22})](0) = \frac{1}{3!}[\Delta\{Q(T_X X_1, X_1) + Q(T_X X_2, X_2)\}](0) \\ &= \frac{1}{3}\{Q(T_{X_1}X_1, X_1) + Q(T_{X_2}X_1, X_1) + Q(T_{X_1}X_2, X_2) + Q(T_{X_2}X_2, X_2)\}. \end{aligned}$$

In this expression the first and the last terms vanish. The two other terms are equal because of the invariance

$$Q_o(\operatorname{Ad}(k)X, \operatorname{Ad}(k)Y) = Q_o(X, Y), \qquad X, Y \in \mathfrak{p},$$

which implies

$$Q_o([Z, X], Y) + Q_o(X, [Z, Y]) = 0, \qquad X, Y \in \mathfrak{p},\ Z \in \mathfrak{k}. \tag{1}$$

We have thus proved

$$K(S) = -Q_o(T_{X_1}X_2, X_2) = Q_o(\text{ad}\,([X_1, X_2]) \cdot X_1, X_2).$$

To prove the formula for R we consider the quadrilinear form

$$B(X, Y, Z, T) = Q_o((R(X, Y) + \text{ad}\,[X, Y]) \cdot Z, T) \qquad (X, Y, Z, T \in \mathfrak{p}).$$

In view of Theorem 12.2, Chapter I, we know then that

$$B(X_1, X_2, X_1, X_2) = 0 \tag{2}$$

if X_1, X_2 are orthonormal vectors in $\mathfrak{p}$. We also have

$$B(X, Y, Z, T) = -B(Y, X, Z, T), \tag{3}$$

$$B(X, Y, Z, T) = -B(X, Y, T, Z), \tag{4}$$

$$B(X, Y, Z, T) + B(Y, Z, X, T) + B(Z, X, Y, T) = 0. \tag{5}$$

Here we have used the Jacobi identity and Lemma 12.5, Chapter I. Now if X, Y are arbitrary vectors in $\mathfrak{p}$, there exist orthonormal vectors X_1, X_2 in $\mathfrak{p}$ such that

$$X = x_1X_1 + x_2X_2, \qquad Y = y_1X_1 + y_2X_2, \qquad x_1, x_2, y_1, y_2 \in \boldsymbol{R}.$$

Then, using (3) and (4) we get

$$B(X, Y, X, Y) = (x_1y_2 - x_2y_1)^2\, B(X_1, X_2, X_1, X_2)$$

so

$$B(X, Y, X, Y) = 0 \qquad \text{for } X, Y \in \mathfrak{p}.$$

From Lemma 12.4, Chapter I we can conclude that $B \equiv 0$ and this proves Theorem 4.2.

The result shows that the curvature tensor is independent of the G-invariant Riemannian structure Q. In view of Lemma 1.2 we get the following:

Corollary 4.3. *The Riemannian connection on G/K is the same for all G-invariant Riemannian structures Q on G/K.*

§ 5. Locally and Globally Symmetric Spaces

Let M be a Riemannian manifold, p a point in M. In general it is impossible to find any neighborhood N of p which can be extended to a complete Riemannian manifold $\tilde{M}$. However, if M is locally symmetric then this turns out to be possible and $\tilde{M}$ can be taken globally symmetric. We shall also establish another relation between locally and globally symmetric spaces, namely that the universal covering manifold of a complete Riemannian locally symmetric space is globally symmetric.

Theorem 5.1. *Let M be a Riemannian locally symmetric space and p a point in M. There exists a Riemannian globally symmetric space $\tilde{M}$, an open neighborhood N_p of p in M and an isometry φ of N_p onto an open neighborhood of $\varphi(p)$ in $\tilde{M}$.*

The proof below is broken up into a few lemmas. Let the Riemannian structure of M be denoted by Q and let R denote the curvature tensor. Let $\mathfrak{p}$ denote the tangent space M_p. If A is an endomorphism of $\mathfrak{p}$, then A can be uniquely extended to the mixed tensor algebra $\mathfrak{D}(p)$ over $\mathfrak{p}$ as a derivation, preserving type of tensors and commuting with contractions. Denoting this extension again by A we have, if $X \in M_p$, $\omega \in M_p^*$,

$$A(\omega \otimes X) = A\omega \otimes X + \omega \otimes AX.$$

Applying contractions and noting that A annihilates scalars, we get

$$(A\omega)(X) = -\omega(AX).$$

It follows easily that

$$(A \cdot Q_p)(X, Y) = -Q_p(AX, Y) - Q_p(X, AY), \tag{1}$$

$$(A \cdot R_p)(X, Y) = [A, R_p(X, Y)] - R_p(AX, Y) - R_p(X, AY) \tag{2}$$

for $X, Y \in \mathfrak{p}$. Here the bracket $[E, F]$ of two endomorphisms denotes the endomorphism $EF - FE$.

Lemma 5.2. *Let $\mathfrak{k}$ denote the set of all endomorphisms of $\mathfrak{p}$, which, when extended to the mixed tensor algebra $\mathfrak{D}(p)$ as above, annihilate Q_p and R_p. Then $\mathfrak{k}$ is a Lie algebra with the bracket $[A, B] = AB - BA$; further, $R_p(X, Y) \in \mathfrak{k}$ for any $X, Y \in \mathfrak{p}$.*

Proof. By (1) and (2) above, $A \in \mathfrak{k}$ if and only if

$$Q_p(AX, Y) + Q_p(X, AY) = 0, \tag{3}$$

$$[A, R_p(X, Y)] = R_p(AX, Y) + R_p(X, AY) \tag{4}$$

for all $X, Y \in \mathfrak{p}$. We express (3) by saying that A is skew symmetric with respect to Q_p. Now, suppose $A, B \in \mathfrak{k}$; then

$$Q_p((AB - BA) X, Y) + Q_p(X, (AB - BA) Y)$$
$$= - Q_p(BX, AY) + Q_p(AX, BY) - Q_p(AX, BY) + Q_p(BX, AY) = 0,$$

so

$$Q_p([A, B] \cdot X, Y) + Q_p(X, [A, B] \cdot Y) = 0.$$

Similarly, by (4) and the Jacobi identity

$$\begin{aligned}[][A, B], R_p(X, Y)] &= - [[B, R_p(X, Y)], A] - [[R_p(X, Y), A], B] \\ &= [A, R_p(BX, Y) + R_p(X, BY)] - [B, R_p(AX, Y) + R_p(X, AY)] \\ &= R_p(ABX, Y) + R_p(BX, AY) + R_p(AX, BY) + R_p(X, ABY) \\ &\quad - R_p(BAX, Y) - R_p(AX, BY) - R_p(BX, AY) - R_p(X, BAY),\end{aligned}$$

so

$$[[A, B], R_p(X, Y)] = R_p([A, B] \cdot X, Y) + R_p(X, [A, B] \cdot Y).$$

Now, if $X, Y \in \mathfrak{p}$, the endomorphism $R_p(X, Y)$ of $\mathfrak{p}$ is skew symmetric with respect to Q_p (Lemma 12.5, Chapter I). Let X^* and Y^* be any vector fields on M such that $X_p^* = X$, $Y_p^* = Y$. In general, if D_1, D_2 are derivations of an algebra the same is true of $D_1D_2 - D_2D_1$. Thus the endomorphism

$$R(X^*, Y^*) = \nabla_{X^*}\nabla_{Y^*} - \nabla_{Y^*}\nabla_{X^*} - \nabla_{[X^*, Y^*]}$$

is a derivation of the mixed tensor algebra $\mathfrak{D}(M)$, preserving type of tensors and commuting with contractions. In addition, $R(fX^*, gY^*) \cdot hZ = fgh\, R(X^*, Y^*) \cdot Z$ for $f, g, h \in C^\infty(M)$ and $Z \in \mathfrak{D}^1(M)$. This implies that

$$(R(X^*, Y^*)\, T)_p$$

for a tensor field T only depends on the values $R_p(X, Y)$ and T_p. We can therefore put

$$R_p(X, Y) \cdot T_p = (R\,(X^*, Y^*)\, T)_p.$$

The mapping $T_p \to R_p(X, Y) \cdot T_p$ is the unique extension of the endomorphism $R_p(X, Y)$ of $\mathfrak{p}$ to a derivation of the mixed tensor algebra $\mathfrak{D}(p)$ commuting with contractions. Since M is locally symmetric, we have

$$R_p(X, Y) \cdot R_p = ((\nabla_{X^*}\nabla_{Y^*} - \nabla_{Y^*}\nabla_{X^*} - \nabla_{[X^*, Y^*]})\, R)_p = 0,$$

and this shows that $R_p(X, Y) \in \mathfrak{k}$, proving the lemma.

Consider now the direct sum $\mathfrak{g} = \mathfrak{k} + \mathfrak{p}$; we introduce a bracket operation in $\mathfrak{g}$ as follows:

For $X_1, X_2 \in \mathfrak{p}$, $\quad [X_1, X_2] = -R(X_1, X_2)$.

For $X \in \mathfrak{p}, T \in \mathfrak{k}$, $\quad [T, X] = -[X, T] = T \cdot X \quad$ (T operating on X).

For $T_1, T_2 \in \mathfrak{k}$, $\quad [T_1, T_2] = T_1T_2 - T_2T_1$.

The definition of the bracket $[X_1, X_2]$ is of course motivated by Theorem 4.2.

Lemma 5.3. *The bracket operation above turns* $\mathfrak{g}$ *into a Lie algebra.*

Proof. Since the bracket operation is skew symmetric, only the Jacobi identity

$$[Z_1, [Z_2, Z_3]] + [Z_2, [Z_3, Z_1]] + [Z_3, [Z_1, Z_2]] = 0 \qquad (5)$$

has to be verified. If all Z_i belong to $\mathfrak{k}$, (5) is just the Jacobi identity for $\mathfrak{k}$. If $Z_1, Z_2 \in \mathfrak{k}$, $Z_3 \in \mathfrak{p}$, (5) is immediate from the definition of the bracket. If $Z_1, Z_2 \in \mathfrak{p}$, $Z_3 \in \mathfrak{k}$, then (5) reduces to (4). Finally, if all Z_i belong to $\mathfrak{p}$, the Jacobi identity is the Bianchi identity (Lemma 12.5, Chapter I).

Now we have the relations

$$[\mathfrak{k}, \mathfrak{k}] \subset \mathfrak{k}, \qquad [\mathfrak{k}, \mathfrak{p}] \subset \mathfrak{p}, \qquad [\mathfrak{p}, \mathfrak{p}] \subset \mathfrak{k},$$

which show that the mapping $s : T + X \rightarrow T - X$ ($T \in \mathfrak{k}$, $X \in \mathfrak{p}$) is an involutive automorphism of $\mathfrak{g}$. The set of fixed points of s coincides with $\mathfrak{k}$.

Lemma 5.4. *The pair* $(\mathfrak{g}, s)$ *is an effective orthogonal symmetric Lie algebra.*

Proof. Suppose $\mathfrak{n}$ is an ideal of $\mathfrak{g}$ contained in $\mathfrak{k}$. Then if $T \in \mathfrak{n}$, $X \in \mathfrak{p}$ we have

$$[T, X] \in \mathfrak{p} \cap \mathfrak{n} = \{0\}$$

so $T \cdot X = 0$; hence $T = 0$ and $\mathfrak{n} = \{0\}$. In particular, if $\mathfrak{z}$ denotes the center of $\mathfrak{g}$, then $\mathfrak{k} \cap \mathfrak{z} = \{0\}$.

The adjoint group Int $(\mathfrak{g})$ has Lie algebra $\mathrm{ad}_{\mathfrak{g}}(\mathfrak{g})$. Let K denote the analytic subgroup of Int $(\mathfrak{g})$ whose Lie algebra is $\mathrm{ad}_{\mathfrak{g}}(\mathfrak{k})$. Each member of K leaves $\mathfrak{p}$ and $\mathfrak{k}$ invariant, so K is a Lie subgroup of the product group $\boldsymbol{GL}(\mathfrak{p}) \times \boldsymbol{GL}(\mathfrak{k})$. The mappings of $\boldsymbol{GL}(\mathfrak{p}) \times \boldsymbol{GL}(\mathfrak{k})$ onto $\boldsymbol{GL}(\mathfrak{p})$ and of $\boldsymbol{GL}(\mathfrak{p}) \times \boldsymbol{GL}(\mathfrak{k})$ onto $\boldsymbol{GL}(\mathfrak{k})$, obtained by restriction to $\mathfrak{p}$ and $\mathfrak{k}$, respectively, are analytic homomorphisms. The images $K_{\mathfrak{p}}$ and $K_{\mathfrak{k}}$

of K under these mappings are analytic subgroups of $\boldsymbol{GL}(\mathfrak{p})$ and $\boldsymbol{GL}(\mathfrak{k})$, respectively. Their Lie algebras are obtained by restricting the endomorphisms in $\text{ad}_{\mathfrak{g}}(\mathfrak{k})$ to $\mathfrak{p}$ and $\mathfrak{k}$, respectively. We see then that the Lie algebra of $K_{\mathfrak{p}}$ is exactly $\mathfrak{k}$ and the Lie algebra of $K_{\mathfrak{k}}$ is $\text{ad}_{\mathfrak{k}}(\mathfrak{k})$. Thus $K_{\mathfrak{k}} = \text{Int}(\mathfrak{k})$ (as Lie groups).

Now, each automorphism A of $\mathfrak{p}$ can be extended uniquely to a type preserving automorphism $\tilde{A}$ of the mixed tensor algebra $\mathfrak{D}(p)$ over $\mathfrak{p}$ such that $\tilde{A}$ coincides with $({}^tA)^{-1}$ on the dual space $\mathfrak{p}^\wedge$. Those automorphisms A of $\mathfrak{p}$ for which $\tilde{A}$ leaves invariant Q_p and R_p, form a compact Lie subgroup of $\boldsymbol{GL}(\mathfrak{p})$ with Lie algebra $\mathfrak{k}$. The identity component of this group must therefore coincide with $K_{\mathfrak{p}}$; thus the group $K_{\mathfrak{p}}$ is compact and so is its homomorphic image $K_{\mathfrak{k}}$.

Let (k_n) be a sequence in K. There exists a subsequence (k_ν) of (k_n) such that the corresponding sequences of restrictions to $\mathfrak{p}$ and $\mathfrak{k}$ are convergent; it follows that (k_ν) is convergent (in the relative topology of $\boldsymbol{GL}(\mathfrak{g})$) to an element $k \in K$. In particular, K is a closed subset of $\boldsymbol{GL}(\mathfrak{g})$; owing to Theorem 2.10, Chapter II, the Lie group K carries the relative topology of $\boldsymbol{GL}(\mathfrak{g})$. Thus K is a compact Lie group. This shows that $\mathfrak{k}$ is compactly imbedded in $\mathfrak{g}$ and the lemma is proved.

The center $\mathfrak{z}$ of $\mathfrak{g}$ is invariant under s; hence $\mathfrak{z} = (\mathfrak{k} \cap \mathfrak{z}) + (\mathfrak{p} \cap \mathfrak{z})$. Now, $\mathfrak{k} \cap \mathfrak{z} = \{0\}$ so $\mathfrak{z} \subset \mathfrak{p}$. Let $\mathfrak{p}'$ denote the orthogonal complement of $\mathfrak{z}$ in $\mathfrak{p}$ (with respect to Q_p). Then $[\mathfrak{k}, \mathfrak{p}'] \subset \mathfrak{p}'$ so $\mathfrak{k} + \mathfrak{p}'$ is an ideal of $\mathfrak{g}$, isomorphic to $\text{ad}_{\mathfrak{g}}(\mathfrak{g})$. Let $c = \dim \mathfrak{z}$. Then the product group

$$G = \text{Int}(\mathfrak{g}) \times \boldsymbol{R}^c$$

has Lie algebra $\text{ad}_{\mathfrak{g}}(\mathfrak{g}) \times \mathfrak{z}$; the Lie group K is a compact Lie subgroup of G with Lie algebra $\text{ad}_{\mathfrak{g}}(\mathfrak{k})$. The automorphism s of $\mathfrak{g}$ induces an automorphism σ of $\text{Int}(\mathfrak{g})$ such that

$$\sigma \cdot e^{\text{ad}\, X} = e^{\text{ad}(s.X)}, \qquad X \in \mathfrak{g}$$

(Chapter II, §5). We extend σ to an automorphism of G, also denoted σ, by putting $\sigma \cdot a = a^{-1}$ for $a \in \boldsymbol{R}^c$. Then K is the identity component of the set of fixed points of σ. Thus (G, K) is a Riemannian symmetric pair.

If $X \in \mathfrak{g}$, let $X_{\mathfrak{z}}$ denote the component of X in $\mathfrak{z}$ according to the direct decomposition $\mathfrak{g} = (\mathfrak{k} + \mathfrak{p}') + \mathfrak{z}$. Then the mapping $X \to (\text{ad}\, X, X_{\mathfrak{z}})$ is an isomorphism of $\mathfrak{g}$ onto $\text{ad}_{\mathfrak{g}}(\mathfrak{g}) \times \mathfrak{z}$, carrying $\mathfrak{k}$ onto $\text{ad}_{\mathfrak{g}}(\mathfrak{k})$ and such that the automorphisms s and $(d\sigma)_e$ correspond. Thus we can regard $\text{Ad}_G(K)$ as a group of automorphisms of $\mathfrak{g}$; this group leaves invariant $\mathfrak{p}$ and the quadratic form Q_p on $\mathfrak{p}$. Identifying $\mathfrak{p}$ with $(G/K)_p$ we see that there exists a unique G-invariant Riemannian structure $\tilde{Q}$ on G/K

such that $\tilde{Q}_p = Q_p$. With this Riemannian structure, G/K is globally symmetric (Prop. 3.4) and this is the desired space $\tilde{M}$. In fact, the curvature tensor $\tilde{R}$ of G/K is given by Theorem 4.2,

$$\tilde{R}_p(X, Y) \cdot Z = -[[X, Y], Z], \qquad X, Y, Z \in \mathfrak{p},$$

and by the definition of the bracket in $\mathfrak{g}$,

$$-[[X, Y], Z] = -[-R_p(X, Y), Z] = R_p(X, Y) \cdot Z.$$

Hence

$$\tilde{R}_p = R_p, \qquad \tilde{Q}_p = Q_p,$$

and the theorem now follows from Lemma 1.2 and Lemma 1.4.

Remark. In the proof above, the group G was constructed by means of the adjoint group Int ($\mathfrak{g}$). A minor shortcut could be taken by making use of the theorem, that for every Lie algebra $\mathfrak{a}$ over $\boldsymbol{R}$ there exists a Lie group A whose Lie algebra is isomorphic to $\mathfrak{a}$. However, this theorem is neither proved nor used in this book (cf. §8, Chapter II).

Proposition 5.5. *A Riemannian locally symmetric space M is an analytic Riemannian manifold.*

Proof. In view of Theorem 5.1, there exists a covering $\{B_\alpha\}_{\alpha \in A}$ of M with open balls $B_\alpha = B_{\rho(\alpha)}(p_\alpha)$ such that for each α, $B_{3\rho(\alpha)}(p_\alpha)$ is a normal neighborhood of p_α, isometric to an open set in a Riemannian globally symmetric space. We have to show that for any $\alpha, \beta \in A$, the normal coordinates at p_α and p_β are analytically related on $B_\alpha \cap B_\beta$. We may assume for the radii, that $\rho(\alpha) \geqslant \rho(\beta)$. Then if $B_\alpha \cap B_\beta \neq \emptyset$, the ball $B_{3\rho(\alpha)}(p_\alpha)$ contains B_β. Since the Riemannian structure on M is analytic on $B_{3\rho(\alpha)}(p_\alpha)$, the normal coordinates at p_α and p_β are analytically related on B_β, in particular on $B_\alpha \cap B_\beta$.

As a consequence we note that the analyticity assumption in the definition of a Riemannian globally symmetric space can be dropped.

We shall now establish another connection between locally and globally symmetric Riemannian spaces.

Theorem 5.6. *Let M be a complete, simply connected Riemannian locally symmetric space. Then M is Riemannian globally symmetric.*

Proof. Let $p \in M$ and let $B_\rho(p)$ be a spherical normal neighborhood such that the geodesic symmetry s_p is an isometry of $B_\rho(p)$ onto itself. We define a mapping Φ of M into M as follows. Let $q \in M$ and let $\gamma(t)$, $0 \leqslant t \leqslant 1$, be a continuous curve joining p and q. Let φ_t be a continuation of s_p along γ as defined in §11, Chapter I, and put $\Phi(q) =$

$\varphi_1(q)$. Since M is simply connected it follows from Prop. 11.4 that $\Phi(q)$ does not depend on the choice of γ. For the same reason Φ coincides with φ_1 in a neighborhood of q. Hence Φ is a differentiable mapping of M into M such that for each $q \in M$, $d\Phi_q$ is an isometry. Since Φ reverses the direction of each geodesic starting at p, it is clear that $\Phi(M) = M$ and $\Phi \circ \Phi$ is the identity mapping. Being involutive, Φ must be one-to-one and the theorem is proved.

Corollary 5.7. *Let M be a complete Riemannian locally symmetric space. Let g denote the Riemannian structure on M and let (M^*, π) be the universal covering manifold of M. Then M^*, with the Riemannian structure π^*g, is a Riemannian globally symmetric space.*

In fact, M^* satisfies the hypothesis of Theorem 5.6.

§ 6. Compact Lie Groups

A compact connected Lie group G can always be regarded as a Riemannian globally symmetric space. The mapping $\sigma : (g_1, g_2) \to (g_2, g_1)$ is an involutive automorphism of the product group $G \times G$. The fixed points of σ constitute the diagonal G^* of $G \times G$; the pair $(G \times G, G^*)$ is a Riemannian symmetric pair and the coset space $G \times G/G^*$ is diffeomorphic to the group G under the mapping

$$(g_1, g_2)\, G^* \to g_1 g_2^{-1}.$$

A Riemannian structure on $G \times G/G^*$ is $G \times G$-invariant if and only if the corresponding Riemannian structure on G is invariant under right and left translations. Thus by Prop. 3.4, G is a Riemannian globally symmetric space in each bi-invariant Riemannian structure. The natural mapping of $G \times G$ onto $G \times G/G^*$ now becomes the mapping $\pi : G \times G \to G$ given by $\pi(g_1, g_2) = g_1 g_2^{-1}$. Recalling that the geodesic symmetry s_e is given by $s_e \circ \pi = \pi \circ \sigma$, we see that $s_e(g) = g^{-1}$ for $g \in G$. More generally, $s_x(g) = xg^{-1}x$.

Let $\mathfrak{g}$ denote the Lie algebra of G. Then the product algebra $\mathfrak{g} \times \mathfrak{g}$ is the Lie algebra of $G \times G$ and the identity

$$(X, Y) = (\tfrac{1}{2}(X+Y), \tfrac{1}{2}(X+Y)) + (\tfrac{1}{2}(X-Y), -\tfrac{1}{2}(X-Y))$$

gives the decomposition of $\mathfrak{g} \times \mathfrak{g}$ into the two eigenspaces of $d\sigma$. Since $\pi(g_1, g_2) = g_1 g_2^{-1}$, it follows that

$$d\pi(X, Y) = X - Y, \qquad X, Y \in \mathfrak{g}.$$

We now denote,

exp* : the exponential mapping of $\mathfrak{g} \times \mathfrak{g}$ into $G \times G$;

exp : the exponential mapping of $\mathfrak{g}$ into G;

Exp : the Exponential mapping of $\mathfrak{g}$ into G (G being considered as a Riemannian globally symmetric space).

Formula (3) in §3 then implies

$$\pi(\exp^* (X, -X)) = \operatorname{Exp} (d\pi(X, -X)) \qquad (X \in \mathfrak{g}).$$

Hence $\exp X \cdot (\exp(-X))^{-1} = \operatorname{Exp} 2X$, so we have

$$\exp X = \operatorname{Exp} X, \qquad X \in \mathfrak{g}.$$

The geodesics in G through e are therefore just the one-parameter subgroups. This fact could also have been verified by using (2) §9, Chapter I, as was done in the special case considered in Theorem 6.9 in Chapter II.

The orthogonal symmetric Lie algebra associated with $(G \times G, G^*)$ is $(\mathfrak{g} \times \mathfrak{g}, \tau)$ where τ is the automorphism $(X, Y) \to (Y, X)$ of $\mathfrak{g} \times \mathfrak{g}$.

§ 7. Totally Geodesic Submanifolds. Lie Triple Systems

In contrast to general Riemannian manifolds, globally symmetric spaces contain plenty of totally geodesic submanifolds. We shall now describe these in Lie algebra terms.

Proposition 7.1. *Let M be a Riemannian manifold and S a totally geodesic submanifold of M. If M is locally symmetric, the same holds for S.*

The proof is straightforward because each geodesic symmetry on S is obtained from a geodesic symmetry of M by restriction.

Definition. Let $\mathfrak{g}$ be a Lie algebra over $\boldsymbol{R}$ and let $\mathfrak{m}$ be a subspace of $\mathfrak{g}$; $\mathfrak{m}$ is called a *Lie triple system* if $X, Y, Z \in \mathfrak{m}$ implies $[X, [Y, Z]] \in \mathfrak{m}$.

Theorem 7.2. *Let M be a Riemannian globally symmetric space and let the notation be as in Theorem 3.3. Identifying as usual the tangent space M_{p_0} with the subspace $\mathfrak{p}$ of the Lie algebra of $I(M)$, let $\mathfrak{s}$ be a Lie triple system contained in $\mathfrak{p}$. Put $S = \operatorname{Exp} \mathfrak{s}$. Then S has a natural differentiable structure in which it is a totally geodesic submanifold of M satisfying $S_{p_0} = \mathfrak{s}$.*

On the other hand, if S is a totally geodesic submanifold of M and $p_0 \in S$, then the subspace $\mathfrak{s} = S_{p_0}$ of $\mathfrak{p}$ is a Lie triple system.

Proof. Suppose first that S is a totally geodesic submanifold and let X, Y be two vectors in the tangent space $\mathfrak{s} = S_{p_0}$. For each $t \in \boldsymbol{R}$, the vector $A = d\,\mathrm{Exp}_{tY}(X)$ is a tangent vector to S at $\mathrm{Exp}\,(tY)$. As we have seen earlier, the vector $d\tau(\exp(-tY)) \cdot A$ is M-parallel to A along the curve $\exp tY$ $(t \in \boldsymbol{R})$. Using Theorem 14.5, Chapter I, we conclude that $d\tau(\exp(-tY)) \cdot A \in \mathfrak{s}$. In view of Theorem 4.1, this means that

$$\sum_0^\infty \frac{(T_{tY})^n}{(2n+1)!}(X) \in \mathfrak{s}$$

for all $t \in \boldsymbol{R}$. This implies that $T_Y(X) \in \mathfrak{s}$. Now,

$$T_{Y+Z} = T_Y + T_Z + \mathrm{ad}\,Y\,\mathrm{ad}\,Z + \mathrm{ad}\,Z\,\mathrm{ad}\,Y.$$

Combining this with the Jacobi identity, we obtain

$$2[Y, [Z, X]] + [X, [Y, Z]] \in \mathfrak{s} \qquad \text{for } X, Y, Z \in \mathfrak{s}. \tag{1}$$

Interchange of X, Y gives the equation

$$4[X, [Z, Y]] + 2[Y, [X, Z]] \in \mathfrak{s},$$

which, added to (1), shows that $[X, [Y, Z]] \in \mathfrak{s}$.

On the other hand, suppose $\mathfrak{s}$ is a Lie triple system. Still using the notation of Theorem 3.3, we have $[\mathfrak{s}, \mathfrak{s}] \subset [\mathfrak{p}, \mathfrak{p}] \subset \mathfrak{k}$. Moreover, the subspace $[\mathfrak{s}, \mathfrak{s}]$ is a subalgebra of $\mathfrak{k}$; this follows from the identity

$$[[X, Y], [U, V]] + [U, [V, [X, Y]]] + [V, [[X, Y], U]] = 0$$

combined with the fact that $\mathfrak{s}$ is a Lie triple system. It follows immediately that the subspace $\mathfrak{g}' = \mathfrak{s} + [\mathfrak{s}, \mathfrak{s}]$ is a subalgebra of $\mathfrak{g}$. Let G' denote the analytic subgroup of G with Lie algebra $\mathfrak{g}'$, let M' denote the orbit $G' \cdot p_0$ and let K' denote the subgroup of G' leaving the point p_0 fixed. Since the identity mapping of G' into G is continuous, K' is a closed subgroup of G'. Since M' is in one-to-one correspondence $g \cdot p_0 \rightarrow gK'$ with G'/K' we can carry the topology and differentiable structure of G'/K' over on M'; by Prop. 4.4, Chapter II, M' is then a submanifold of M. Furthermore, $(M')_{p_0} = \mathfrak{s}$. The M-geodesics through p_0 have the form $\exp tX \cdot p_0$ $(t \in \boldsymbol{R})$ where X is a general vector in $\mathfrak{p}$. This geodesic is tangent to M' at p_0 if and only if $X \in \mathfrak{s}$; it follows that the submanifold M' of M is geodesic at p_0. Since G' is a group of isometries of M and

M', and acts transitively on M', it follows that M' is geodesic at each of its points, hence totally geodesic. Obviously $M' = \text{Exp}\, \mathfrak{s}$, and the theorem is proved.

Remark. The automorphism $\sigma : g \to s_{p_0} g s_{p_0}$ has differential $d\sigma$ which leaves $\mathfrak{g}'$ invariant and has fixed point set $\mathfrak{k} \cap \mathfrak{g}'$ which is the Lie algebra of K'. Thus (G', K') is a symmetric pair and the manifold $M' = \text{Exp}\, \mathfrak{s}$ is a Riemannian globally symmetric space.

EXERCISES AND FURTHER RESULTS

A. Geometry of Homogeneous Spaces

1. Let (G, H) be a symmetric pair (§3) with respect to an involutive automorphism σ of G. Let s_o denote the diffeomorphism $gH \to \sigma(g)H$ of G/H onto itself. Then (cf. Nomizu [2]):

(i) G/H has a unique affine connection ∇ invariant under s_o and under the action of G.

(ii) Let $\mathfrak{g} = \mathfrak{h} + \mathfrak{m}$ be the decomposition of $\mathfrak{g}$ into the $(+1)$-eigenspace and (-1)-eigenspace of $d\sigma$. We identify $\mathfrak{m}$ and $(G/H)_o$ by means of the differential $d\pi$ of the natural map $\pi : G \to G/H$. The geodesics through the origin $o = \{H\}$ are the curves

$$\gamma_X : t \to \exp tX \cdot o \qquad (X \in \mathfrak{m}).$$

(iii) If $X, Y \in \mathfrak{m}$, the parallel translate of Y along γ_X is $(d\tau(\exp tX))_o(Y)$.

(iv) The torsion and curvature tensor of G/H are given by

$$T = 0, \qquad R_o(X, Y)(Z) = -[[X, Y], Z], \quad X, Y, Z \in \mathfrak{m}.$$

Moreover

$$\nabla_V(R) = 0 \qquad \text{for all} \quad V \in \mathfrak{D}^1(G/H).$$

2. Let G be a compact connected Lie group and $H \subset G$ a closed subgroup. Let $\mathfrak{h} \subset \mathfrak{g}$ denote the Lie algebras, let Q_o be an $\text{Ad}(G)$-invariant positive definite symmetric bilinear form on $\mathfrak{g}$ and let $\mathfrak{m}$ be the orthogonal complement of $\mathfrak{h}$ in $\mathfrak{g}$. Let Q be the corresponding G-invariant Riemannian structure on G/H. Then:

(i) The geodesics in G/H through $o = \{H\}$ are $\gamma_X(t) = \exp tX \cdot o$ $(X \in \mathfrak{m})$ (cf. Nomizu [2], Samelson [3]).

(ii) If X, $Y \in \mathfrak{m}$ are orthonormal, the corresponding sectional curvature of G/H is given by

$$K(\{X, Y\}) = \tfrac{1}{4} \|[X, Y]\|^2 + \tfrac{3}{4} \|[X, Y]_{\mathfrak{h}}\|^2,$$

where $\| Z \|^2 = Q_o(Z, Z)$ and $Z = Z_{\mathfrak{m}} + Z_{\mathfrak{h}}$ $(Z \in \mathfrak{g}, Z_{\mathfrak{m}} \in \mathfrak{m}, Z_{\mathfrak{h}} \in \mathfrak{h})$ (cf. Helgason [2]).

3. A compact semisimple Lie group G has a bi-invariant Riemannian structure Q such that Q_e is the negative of the Killing form of the Lie algebra $\mathfrak{g}$ of G. If G is considered as a symmetric space $(G \times G)/G^*$ as in §6, it acquires a bi-invariant Riemannian structure Q^* from the Killing form of $\mathfrak{g} \times \mathfrak{g}$. Show that $Q = 2Q^*$.

4. Show that any two complete simply connected Riemannian manifolds M_1, M_2 of the same dimension and of the same constant sectional curvature are isometric.

5. Let M be a connected locally compact metric space and $I(M)$ the group of distance-preserving mappings of M onto itself, topologized by the compact open topology. Let $H \subset I(M)$ be a closed subgroup. Then for each $p \in M$, the orbit $H \cdot p$ is closed.

6. Let M be a simply connected Riemannian globally symmetric space, $o \in M$ any point. Suppose $A : M_o \to M_o$ is a linear transformation leaving the metric tensor g_o and the curvature tensor R_o invariant. Then there exists an isometry φ of M onto M such that $\varphi(o) = o$ and $d\varphi_o = A$.

B. Cohomology of Symmetric Spaces

1. Let M be a Riemannian globally symmetric space, ω a differential form on M invariant under each member of $I_0(M)$. Then ω is closed, that is, $d\omega = 0$.

2. Let (G, K) be a symmetric pair, G and K compact and connected. Define the $*$ operator on $\mathfrak{A}(G/K)$ and harmonic form on G/K as in §7, Chapter II. Prove that the harmonic forms on G/K are precisely the G-invariant forms.

NOTES

§1. The material here is due to É. Cartan ([6] and [22], Chapter XI) for the Riemannian case. Concerning the affine case see Whitehead [1] for the local theory, Nomizu [2], Fedenko [1], Rozenfeld [2] and Berger [2] for the global theory. Riemannian manifolds for which the curvature is preserved by parallelism had independently been considered by H. Levy [1] without giving solutions beyond the spaces of constant curvature.

The results of §2 (which actually apply to all (separable) connected, locally compact metric spaces M) are due to van Dantzig and van der Waerden [1], see also Arens [1].

§3. Lemma 3.2 was proved by É. Cartan [6], p. 230, by the use of differential equations. It was extended by Myers and Steenrod [1] to all Riemannian manifolds. For a modern approach, using Palais [5], see Chu and Kobayashi [1]. In [16] É. Cartan points out the fact that the geodesics in M are orbits of one-parameter subgroups from $I(M)$. This property is further examined in Nomizu [2].

§4. The formula in Theorem 4.2 for the curvature tensor of a symmetric space is due to É. Cartan [6] and extended by Nomizu [2] to all reductive homogeneous spaces. The proof in the text is from Helgason [2] where the formula for the differential of Exp (Theorem 4.1) is also given.

§5. The relation between locally and globally symmetric spaces is not altogether clear from É. Cartan's work although his extensive paper [10] gives a global classification. Theorem 5.1 is a special case of a theorem of Nomizu [2] on reductive homogeneous spaces. The idea of the proof was already used by É. Cartan [6], p. 225, for the similar problem of constructing a locally symmetric space whose curvature tensor satisfies certain necessary conditions involving the holonomy group (see the introduction to this chapter). Theorem 5.6 is due to Borel and Lichnerowicz [1]; they outlined a proof based on results of Ehresmann [1] applied to the local group of local isometries of a Riemannian locally symmetric space. Ambrose has in [1] stated and proved an extension of Theorem 5.6 to arbitrary Riemannian manifolds.

§7. The connection between totally geodesic submanifolds and Lie triple systems is pointed out in É. Cartan [7], p. 133; see also Mostow [3].

CHAPTER V

DECOMPOSITION OF SYMMETRIC SPACES

In the previous chapter we have seen that a Riemannian globally symmetric space M gives rise to a pair $(\mathfrak{l},s)$ where $\mathfrak{l}$ is the Lie algebra of the group of isometries of $I(M)$ and s is an involutive automorphism of $\mathfrak{l}$ having a compactly imbedded subalgebra for the set of fixed points. This chapter is devoted to the study of such pairs. It is shown in §1 that they fall into three different categories, the compact type, the noncompact type, and the Euclidean type. An arbitrary pair $(\mathfrak{l},s)$ can be decomposed into three parts each of which is from the types above. Since the Euclidean type is uninteresting we are left with two types of pairs $(\mathfrak{l},s)$, the compact type and the noncompact type, both of which have $\mathfrak{l}$ semisimple. The symmetric spaces corresponding to these have positive sectional curvature and negative sectional curvature, respectively. There is a remarkable duality (§2) between the two types which for example provides two viewpoints of the classification problem and incidentally explains the formal analogy between spherical trigonometry and hyperbolic trigonometry.

The rank of a symmetric space M is an important invariant; it is defined as the maximum dimension of any flat totally geodesic subspace A of M. It is shown in §6 that each geodesic in M can be moved into A by an isometry of M. This means that for any two points in M one can speak of their complex distance; this is an l-tuple $(r_1, ..., r_l)$ of real numbers (l = rank of M) and has the property that two point-pairs in M are congruent under an isometry of M if and only if their complex distance is the same. Just as ordinary Euclidean distance is > 0, the l-tuple $(r_1, ..., r_l)$ is restricted to the fundamental domain of a certain discontinuous group. (For the noncompact type this is the Weyl group W (Chapter VII, §2); for the compact type the group is larger (Chapter VII, §8).)

§ 1. Orthogonal Symmetric Lie Algebras

We recall that an orthogonal symmetric Lie algebra is a pair $(\mathfrak{l}, s)$ where

(i) $\mathfrak{l}$ is a Lie algebra over $\boldsymbol{R}$.

(ii) s is an involutive† automorphism of $\mathfrak{l}$.

(iii) $\mathfrak{u}$, the set of fixed points of s, is a compactly imbedded subalgebra of $\mathfrak{l}$.

† That is, $s \neq I$ and $s^2 = I$.

If, in addition, $\mathfrak{u} \cap \mathfrak{z} = \{0\}$, where $\mathfrak{z}$ denotes the center of $\mathfrak{l}$, then $(\mathfrak{l}, s)$ is called *effective*. Two orthogonal symmetric Lie algebras $(\mathfrak{l}_1, s_1)$ and $(\mathfrak{l}_2, s_2)$ are called *isomorphic* if there exists an isomorphism φ of $\mathfrak{l}_1$ onto $\mathfrak{l}_2$ such that $\varphi \circ s_1 = s_2 \circ \varphi$.

Examples.

(a) Let $\mathfrak{l}$ be a compact semisimple Lie algebra and s any involutive automorphism of $\mathfrak{l}$. Then $(\mathfrak{l}, s)$ is an effective orthogonal symmetric Lie algebra.

(b) Let $\mathfrak{l}$ be a noncompact semisimple Lie algebra and let $\mathfrak{l} = \mathfrak{u} + \mathfrak{e}$ be any Cartan decomposition of $\mathfrak{l}$ (where $\mathfrak{u}$ is the subalgebra). Let s denote the automorphism of $\mathfrak{l}$ given by $s(T + X) = T - X$ $(T \in \mathfrak{u}, X \in \mathfrak{e})$. Then $(\mathfrak{l}, s)$ is an effective orthogonal symmetric Lie algebra (Prop. 7.4, Chapter III).

(c) Let $\mathfrak{e}$ be a finite-dimensional vector space over $\boldsymbol{R}$ and let $\mathfrak{u}$ be the Lie algebra of a compact Lie subgroup of $\boldsymbol{GL}(\mathfrak{e})$. Let $\mathfrak{l}$ denote the direct sum $\mathfrak{l} = \mathfrak{u} + \mathfrak{e}$; $\mathfrak{l}$ can be turned into a Lie algebra by defining

$$
\begin{array}{lll}
[X_1, X_2] = 0 & & \text{if } X_1, X_2 \in \mathfrak{e},\\
[T, X] = -[X, T] = T \cdot X & (T \text{ acting on } X) & \text{if } T \in \mathfrak{u}, X \in \mathfrak{e},\\
[T_1, T_2] = T_1T_2 - T_2T_1 & & \text{if } T_1, T_2 \in \mathfrak{u}.
\end{array}
$$

Then $\mathfrak{l}$ is a Lie algebra containing $\mathfrak{u}$ as a subalgebra. Assuming $\mathfrak{e} \neq \{0\}$, the mapping $s : T + X \to T - X$, $(T \in \mathfrak{u}, X \in \mathfrak{e})$, is an involutive automorphism of $\mathfrak{l}$. The pair $(\mathfrak{l}, s)$ is an effective orthogonal symmetric Lie algebra. The proof of this statement is the same as that of Lemma 5.4, Chapter IV.

Definition. Let $(\mathfrak{l}, s)$ be an effective orthogonal symmetric Lie algebra. Let $\mathfrak{l} = \mathfrak{u} + \mathfrak{e}$ be the decomposition of $\mathfrak{l}$ into the eigenspaces of s for the eigenvalue $+1$ and -1, respectively.

(a) If $\mathfrak{l}$ is compact and semisimple, $(\mathfrak{l}, s)$ is said to be of the *compact type*.

(b) If $\mathfrak{l}$ is noncompact and semisimple and $\mathfrak{l} = \mathfrak{u} + \mathfrak{e}$ is a Cartan decomposition of $\mathfrak{l}$, then $(\mathfrak{l}, s)$ is said to be of the *noncompact type*.

(c) If $\mathfrak{e}$ is an abelian ideal in $\mathfrak{l}$, then $(\mathfrak{l}, s)$ is said to be of the *Euclidean type*.

Definition. Let $(\mathfrak{l}, s)$ be an orthogonal symmetric Lie algebra and suppose the pair (L, U) is associated† with $(\mathfrak{l}, s)$. The pair (L, U) is

† See definition preceding Prop. 3.6, Chapter IV.

said to be of the compact type, noncompact type, or Euclidean type according to the type of $(\mathfrak{l}, s)$.

The next theorem shows that every effective orthogonal symmetric Lie algebra can be decomposed into three others, which are of the compact type, noncompact type, and Euclidean type, respectively.

Theorem 1.1. *Let $(\mathfrak{l}, s)$ be an effective, orthogonal symmetric Lie algebra. Then there exist ideals $\mathfrak{l}_0$, $\mathfrak{l}_-$, and $\mathfrak{l}_+$ in $\mathfrak{l}$ with the following properties:*

1. $\mathfrak{l} = \mathfrak{l}_0 + \mathfrak{l}_- + \mathfrak{l}_+$ *(direct sum).*

2. $\mathfrak{l}_0$, $\mathfrak{l}_-$ *and* $\mathfrak{l}_+$ *are invariant under* s *and orthogonal with respect to the Killing form of* $\mathfrak{l}$.

3. *Let s_0, s_-, and s_+ denote the restrictions of s to $\mathfrak{l}_0$, $\mathfrak{l}_-$, and $\mathfrak{l}_+$, respectively. The pairs $(\mathfrak{l}_0, s_0)$, $(\mathfrak{l}_-, s_-)$, and $(\mathfrak{l}_+, s_+)$, are effective orthogonal symmetric Lie algebras of the Euclidean type, compact type, and noncompact type, respectively.*

The proof of this theorem will be broken up into a sequence of lemmas. Let $\mathfrak{u}$ and $\mathfrak{e}$ denote the eigenspaces of s for the eigenvalues $+ 1$ and $- 1$, respectively. Then we have

$$\mathfrak{l} = \mathfrak{u} + \mathfrak{e} \text{ (direct sum)}, \quad [\mathfrak{u}, \mathfrak{u}] \subset \mathfrak{u}, \quad [\mathfrak{u}, \mathfrak{e}] \subset \mathfrak{e}, \quad [\mathfrak{e}, \mathfrak{e}] \subset \mathfrak{u}. \tag{1}$$

Let B denote the Killing form of $\mathfrak{l}$. Since B is invariant under each automorphism of $\mathfrak{l}$, in particular under s, it follows that the subspaces $\mathfrak{u}$ and $\mathfrak{e}$ are orthogonal with respect to B.

Lemma 1.2. *The Killing form B is strictly negative definite on* $\mathfrak{u}$.

This lemma is a special case of Prop. 6.8, Chapter II.

Now let U denote the analytic subgroup of the adjoint group $\mathrm{Int}(\mathfrak{l})$ with Lie algebra $\mathrm{ad}_{\mathfrak{l}}(\mathfrak{u})$. As a result of our assumptions, U is a compact Lie subgroup of $\boldsymbol{GL}(\mathfrak{l})$; thus it carries the relative topology of $\boldsymbol{GL}(\mathfrak{l})$. Since U is connected, relations (1) imply

$$u \cdot \mathfrak{u} \subset \mathfrak{u}, \qquad u \cdot \mathfrak{e} \subset \mathfrak{e} \qquad \text{for } u \in U.$$

Being a compact linear group, U leaves invariant a strictly positive definite, symmetric bilinear form Q on $\mathfrak{e} \times \mathfrak{e}$. There exists a basis $X_1, ..., X_n$ of $\mathfrak{e}$ and real numbers $\beta_1, ..., \beta_n$ such that

$$Q(X, X) = x_1^2 + ... + x_n^2,$$

$$B(X, X) = \beta_1 x_1^2 + ... + \beta_n x_n^2,$$

if $X = \sum_{i=1}^{n} x_i X_i$. Let

$$\mathfrak{e}_0 = \sum_{\beta_i = 0} \boldsymbol{R} X_i, \qquad \mathfrak{e}_- = \sum_{\beta_i < 0} \boldsymbol{R} X_i, \qquad \mathfrak{e}_+ = \sum_{\beta_i > 0} \boldsymbol{R} X_i.$$

Then $\mathfrak{e}$ is the direct sum of the subspaces $\mathfrak{e}_0$, $\mathfrak{e}_-$, $\mathfrak{e}_+$; moreover, these subspaces are orthogonal with respect to Q and B and each one is invariant under s. If we define the endomorphism b of $\mathfrak{e}$ by $bX_i = \beta_i X_i$ $(1 \leqslant i \leqslant n)$, then

$$Q(bX, Y) = B(X, Y)$$

for $X, Y \in \mathfrak{e}$. Since B and Q are invariant under U, the endomorphism b commutes with the restriction of each $u \in U$ to $\mathfrak{e}$. It follows that the spaces $\mathfrak{e}_0$, $\mathfrak{e}_-$, and $\mathfrak{e}_+$ are invariant under U and under $\mathrm{ad}_{\mathfrak{l}}(\mathfrak{u})$.

Lemma 1.3. *The subspaces $\mathfrak{e}_0$, $\mathfrak{e}_-$, and $\mathfrak{e}_+$ satisfy the following relations:*

(i) $\mathfrak{e}_0 = \{X \in \mathfrak{l} : B(X, Y) = 0$ *for all* $Y \in \mathfrak{l}\}$.

(ii) $[\mathfrak{e}_0, \mathfrak{e}] = \{0\}$ *and* $\mathfrak{e}_0$ *is an abelian ideal in* $\mathfrak{l}$.

(iii) $[\mathfrak{e}_-, \mathfrak{e}_+] = \{0\}$.

Proof. Let $\mathfrak{n}$ denote the set on the right-hand side in (i). Then $\mathfrak{n}$ is invariant under s, so $\mathfrak{n} = \mathfrak{n} \cap \mathfrak{u} + \mathfrak{n} \cap \mathfrak{e}$ (direct sum). Now $\mathfrak{n} \cap \mathfrak{u} = \{0\}$ due to Lemma 1.2 so $\mathfrak{n} \subset \mathfrak{e}$. But $\mathfrak{n} \cap \mathfrak{e}_- = \mathfrak{n} \cap \mathfrak{e}_+ = \{0\}$ so $\mathfrak{n} \subset \mathfrak{e}_0$. On the other hand, if $X \in \mathfrak{e}_0$, then $B(X, Y) = 0$ for all $Y \in \mathfrak{e}_0$, hence for all $Y \in \mathfrak{l}$. This proves (i). As a result of (i), $\mathfrak{e}_0$ is an ideal in $\mathfrak{l}$, but $[\mathfrak{e}_0, \mathfrak{e}] \subset \mathfrak{u}$ by (1). This proves (ii). In order to prove (iii) we observe that $[\mathfrak{e}_-, \mathfrak{e}_+] \subset \mathfrak{u}$, so, owing to Lemma 1.2, it suffices to prove

$$B(\mathfrak{u}, [\mathfrak{e}_-, \mathfrak{e}_+]) = 0.$$

But if $T \in \mathfrak{u}$, $X_\pm \in \mathfrak{e}_\pm$, then

$$B(T, [X_-, X_+]) = B([T, X_-], X_+) = 0$$

and the lemma is proved.

We define now $\mathfrak{u}_+ = [\mathfrak{e}_+, \mathfrak{e}_+]$, $\mathfrak{u}_- = [\mathfrak{e}_-, \mathfrak{e}_-]$ and let $\mathfrak{u}_0$ denote the orthogonal complement (with respect to B) of the subspace of $\mathfrak{u}$ spanned by $\mathfrak{u}_+$ and $\mathfrak{u}_-$.

Lemma 1.4. *The subspaces $\mathfrak{u}_0$, $\mathfrak{u}_+$, and $\mathfrak{u}_-$ are ideals in $\mathfrak{u}$, orthogonal with respect to B, and $\mathfrak{u} = \mathfrak{u}_0 + \mathfrak{u}_- + \mathfrak{u}_+$ (direct sum).*

Proof. Since $[\mathfrak{u}, \mathfrak{e}_+] \subset \mathfrak{e}_+$, we have by the Jacobi identity $[\mathfrak{u}_+, \mathfrak{u}] = [[\mathfrak{e}_+, \mathfrak{e}_+], \mathfrak{u}] \subset \mathfrak{u}_+$. Similarly $[\mathfrak{u}_-, \mathfrak{u}] \subset \mathfrak{u}_-$. Now, let $X_\pm \in \mathfrak{e}_\pm$, $Y_\pm \in \mathfrak{e}_\pm$. Then

$$B([X_+, Y_+], [X_-, Y_-]) = B(X_+, [Y_+, [X_-, Y_-]]) = 0$$

due to the Jacobi identity and Lemma 1.3 (iii). Hence $\mathfrak{u}_+$ and $\mathfrak{u}_-$ are orthogonal and the sum $\mathfrak{u}_- + \mathfrak{u}_+$ is an ideal in $\mathfrak{u}$. The orthogonal complement $\mathfrak{u}_0$ is also an ideal and the lemma now follows from Lemma 1.2.

Lemma 1.5. *The following commutation relations hold:*

(i) $[\mathfrak{u}_0, \mathfrak{e}_-] = [\mathfrak{u}_0, \mathfrak{e}_+] = \{0\}$.

(ii) $[\mathfrak{u}_-, \mathfrak{e}_0] = [\mathfrak{u}_-, \mathfrak{e}_+] = \{0\}$.

(iii) $[\mathfrak{u}_+, \mathfrak{e}_0] = [\mathfrak{u}_+, \mathfrak{e}_-] = \{0\}$.

Proof. (i) Let $T \in \mathfrak{u}_0$, $X, Y \in \mathfrak{e}_+$. Then

$$B([T, X], Y) = B(T, [X, Y]) = 0.$$

Thus $[\mathfrak{u}_0, \mathfrak{e}_+]$ is orthogonal to $\mathfrak{e}_+$. Since B is strictly positive definite on $\mathfrak{e}_+$ and since $[\mathfrak{u}_0, \mathfrak{e}_+] \subset \mathfrak{e}_+$, it follows that $[\mathfrak{u}_0, \mathfrak{e}_+] = \{0\}$. Similarly $[\mathfrak{u}_0, \mathfrak{e}_-] = \{0\}$. For (ii) we have $[\mathfrak{u}_-, \mathfrak{e}_0] = [[\mathfrak{e}_-, \mathfrak{e}_-], \mathfrak{e}_0] = \{0\}$ by Lemma 1.3 (ii). Moreover, $[\mathfrak{u}_-, \mathfrak{e}_+] = [[\mathfrak{e}_-, \mathfrak{e}_-], \mathfrak{e}_+] = \{0\}$ by Lemma 1.3 (iii). The last part (iii) follows in the same way.

In order to prove Theorem 1.1, we have to distinguish between two cases: $\mathfrak{e}_0 = \{0\}$ and $\mathfrak{e}_0 \neq \{0\}$. Suppose first $\mathfrak{e}_0 \neq \{0\}$. Then we put

$$\mathfrak{l}_0 = \mathfrak{u}_0 + \mathfrak{e}_0, \qquad \mathfrak{l}_- = \mathfrak{u}_- + \mathfrak{e}_-, \qquad \mathfrak{l}_+ = \mathfrak{u}_+ + \mathfrak{e}_+.$$

Then $\mathfrak{l}$ is the direct sum of the subspaces $\mathfrak{l}_0, \mathfrak{l}_-, \mathfrak{l}_+$; these subspaces are invariant under s and orthogonal with respect to B. Using the lemmas above, we have

$$\begin{aligned} [\mathfrak{l}_0, \mathfrak{l}] &= [\mathfrak{u}_0, \mathfrak{l}] + [\mathfrak{e}_0, \mathfrak{l}] \\ &= [\mathfrak{u}_0, \mathfrak{u}] + [\mathfrak{u}_0, \mathfrak{e}_0] + [\mathfrak{u}_0, \mathfrak{e}_+] + [\mathfrak{u}_0, \mathfrak{e}_-] + [\mathfrak{e}_0, \mathfrak{l}] \\ &\qquad \subset \mathfrak{u}_0 + \mathfrak{e}_0 + \{0\} + \{0\} + \mathfrak{e}_0 \end{aligned}$$

so $[\mathfrak{l}_0, \mathfrak{l}] \subset \mathfrak{l}_0$. Secondly

$$\begin{aligned} &[\mathfrak{l}_+, \mathfrak{l}] = [\mathfrak{u}_+, \mathfrak{l}] + [\mathfrak{e}_+, \mathfrak{l}] \\ &= [\mathfrak{u}_+, \mathfrak{u}] + [\mathfrak{u}_+, \mathfrak{e}_0] + [\mathfrak{u}_+, \mathfrak{e}_-] + [\mathfrak{u}_+, \mathfrak{e}_+] + [\mathfrak{e}_+, \mathfrak{u}] + [\mathfrak{e}_+, \mathfrak{e}_0] + [\mathfrak{e}_+, \mathfrak{e}_-] + [\mathfrak{e}_+, \mathfrak{e}_+] \\ &\qquad \subset \mathfrak{u}_+ + \{0\} + \{0\} + \mathfrak{e}_+ + \mathfrak{e}_+ + \{0\} + \{0\} + \mathfrak{u}_+ \end{aligned}$$

so $[\mathfrak{l}_+, \mathfrak{l}] \subset \mathfrak{l}_+$. Similarly $[\mathfrak{l}_-, \mathfrak{l}] \subset \mathfrak{l}_-$ so the subspaces $\mathfrak{l}_0, \mathfrak{l}_-, \mathfrak{l}_+$ are ideals in $\mathfrak{l}$. This being so, their Killing forms are obtained from B by restriction. Since B is strictly negative on $\mathfrak{l}_-$, it follows (Prop. 6.6, Chapter II) that $\mathfrak{l}_-$ is a semisimple compact Lie algebra. Since B is strictly negative definite on $\mathfrak{u}_+$, and strictly positive definite on $\mathfrak{e}_+$, $\mathfrak{l}_+$ is semisimple and it follows (Prop. 7.4, Chapter III) that the decomposition $\mathfrak{l}_+ = \mathfrak{u}_+ + \mathfrak{e}_+$ is a Cartan decomposition of $\mathfrak{l}_+$. Finally we consider $\mathfrak{l}_0$. Since the center $\mathfrak{z}_0$ of $\mathfrak{l}_0$ coincides with the center $\mathfrak{z}$ of $\mathfrak{l}$ we have $\mathfrak{u}_0 \cap \mathfrak{z}_0 \subset \mathfrak{u} \cap \mathfrak{z} = \{0\}$.

In order to show that $\mathfrak{u}_0$ is compactly imbedded in $\mathfrak{l}_0$, we make use of the following lemma which was communicated to the author by J. Hano.

Lemma 1.6. *Let G_0 be a Lie group and $\mathfrak{g}_0$ its Lie algebra. Suppose $\mathfrak{g}_0$ is the direct sum of two ideals $\mathfrak{g}_1$ and $\mathfrak{g}_2$. Let $\mathfrak{k}_i$ be a subalgebra of $\mathfrak{g}_i$ $(i = 1, 2)$, and put $\mathfrak{k}_0 = \mathfrak{k}_1 + \mathfrak{k}_2$. Then $\mathfrak{k}_0$ is a compactly imbedded subalgebra of $\mathfrak{g}_0$, if and only if $\mathfrak{k}_1$ and $\mathfrak{k}_2$ are compactly imbedded in $\mathfrak{g}_1$ and $\mathfrak{g}_2$, respectively.*

Proof. Without loss of generality, we can assume that G_0 is simply connected and is the direct product $G_0 = G_1 \times G_2$ where G_i is a simply connected Lie group with Lie algebra $\mathfrak{g}_i$ $(i = 1, 2)$. Let K_i denote the analytic subgroup of G_i with Lie algebra $\mathfrak{k}_i$ $(i = 0, 1, 2)$. Then $K_0 = K_1 \times K_2$. If Z_i denotes the center of G_i $(i = 0, 1, 2)$, then $Z_0 = Z_1 \times Z_2$. The mapping $(k_1(K_1 \cap Z_1), k_2(K_2 \cap Z_2)) \to k_1k_2(K_0 \cap Z_0)$ is a topological isomorphism of the product group

$$(K_1/K_1 \cap Z_1) \times (K_2/K_2 \cap Z_2) \quad \text{onto} \quad K_0/K_0 \cap Z_0.$$

In view of Lemma 5.1, Chapter II, the Lie group $K_i/K_i \cap Z_i$ is analytically isomorphic to the Lie subgroup $\mathrm{Ad}_{G_i}(K_i)$ of $\mathrm{Int}(\mathfrak{g}_i)$ $(i = 0, 1, 2)$. The lemma now follows immediately.

Returning now to Theorem 1.1, we first note that there exists a Lie group L whose Lie algebra is isomorphic to $\mathfrak{l}$. In fact, if $c = \dim \mathfrak{z}$, then the product group $L = \mathrm{Int}(\mathfrak{l}) \times \boldsymbol{R}^c$ has for Lie algebra the product $\mathrm{ad}_{\mathfrak{l}}(\mathfrak{l}) \times \mathfrak{z}$. To see that this Lie algebra is isomorphic to $\mathfrak{l}$, we observe that $\mathfrak{z} \subset \mathfrak{e}$ and denote by $\mathfrak{e}'$ the orthogonal complement (with respect to Q) of $\mathfrak{z}$ in $\mathfrak{e}$. Then $[\mathfrak{u}, \mathfrak{e}'] \subset \mathfrak{e}'$ and $[\mathfrak{e}', \mathfrak{e}'] \subset \mathfrak{u}$ so the subspace $\mathfrak{u} + \mathfrak{e}'$ is an ideal in $\mathfrak{l}$, isomorphic to $\mathrm{ad}_{\mathfrak{l}}(\mathfrak{l})$.

From Lemma 1.6 it now follows that $\mathfrak{u}_0$ is compactly imbedded in $\mathfrak{l}_0$. Moreover, $\mathfrak{e}_0$ is an abelian ideal in $\mathfrak{l}_0$ so $(\mathfrak{l}_0, s_0)$ is an orthogonal symmetric Lie algebra of the Euclidean type.

It remains to consider the case $\mathfrak{e}_0 = \{0\}$. In this case $\mathfrak{u}_0$ is an ideal in $\mathfrak{l}$. Hence its Killing form is strictly negative definite so $\mathfrak{u}_0$ is compact and semisimple. We put

$$\mathfrak{l}_0 = \{0\}, \qquad \mathfrak{l}_- = \mathfrak{u}_0 + \mathfrak{u}_- + \mathfrak{e}_-, \qquad \mathfrak{l}_+ = \mathfrak{u}_+ + \mathfrak{e}_+ \qquad \text{if } \mathfrak{e}_- \neq \{0\};$$
$$\mathfrak{l}_0 = \{0\}, \qquad \mathfrak{l}_- = \{0\}, \qquad \mathfrak{l}_+ = \mathfrak{u}_0 + \mathfrak{u}_+ + \mathfrak{e}_+ \qquad \text{if } \mathfrak{e}_- = \{0\}.$$

In each case, Theorem 1.1 follows easily.

Corollary 1.7. *Suppose $X \in \mathfrak{e}$ commutes elementwise with $\mathfrak{u}$. Then $X \in \mathfrak{e}_0$.*

In fact, we can write $X = X_0 + X_- + X_+$ where $X_0 \in \mathfrak{e}_0$, $X_- \in \mathfrak{e}_-$, and $X_+ \in \mathfrak{e}_+$. Then the hypothesis implies that $\mathrm{ad}_{\mathfrak{l}}(X_+)$ and $\mathrm{ad}_{\mathfrak{l}}(X_-)$

map $\mathfrak{u}$ into $\{0\}$. Thus $(\text{ad}_{\mathfrak{l}}(X_+))^2 = (\text{ad}_{\mathfrak{l}}(X_-))^2 = 0$ and therefore $B(X_+, X_+) = B(X_-, X_-) = 0$. Hence $X_+ = X_- = 0$ and $X \in \mathfrak{e}_0$.

A further decomposition of orthogonal symmetric Lie algebras into "irreducible" ones will be given later (cf. Chapter VIII, §5).

§ 2. The Duality

There is a remarkable and important duality between the compact type and the noncompact type. Let $(\mathfrak{l}, s)$ be an orthogonal symmetric Lie algebra and put $\mathfrak{l} = \mathfrak{u} + \mathfrak{e}$ as in (1), §1. Let $\mathfrak{l}^*$ denote the subset $\mathfrak{u} + i\mathfrak{e}$ of the complexification $\mathfrak{l}^C$ of $\mathfrak{l}$. With the bracket operation inherited from $\mathfrak{l}^C$, $\mathfrak{l}^*$ is a Lie algebra over $\boldsymbol{R}$. The mapping $s^* : T + iX \to T - iX$ ($T \in \mathfrak{u}$, $X \in \mathfrak{e}$) is an involutive automorphism of $\mathfrak{l}^*$. As will be verified presently, $(\mathfrak{l}^*, s^*)$ is an orthogonal symmetric Lie algebra, called the *dual* of $(\mathfrak{l}, s)$. Then $(\mathfrak{l}, s)$ is the dual of $(\mathfrak{l}^*, s^*)$.

Proposition 2.1. *Let* $(\mathfrak{l}, s)$ *be an orthogonal symmetric Lie algebra. Then:*

(i) *The pair* $(\mathfrak{l}^*, s^*)$ *is an orthogonal symmetric Lie algebra.*

(ii) *If* $(\mathfrak{l}, s)$ *is of the compact type, then* $(\mathfrak{l}^*, s^*)$ *is of the noncompact type and conversely.*

(iii) *If* $(\mathfrak{l}_1, s_1)$ *is isomorphic to* $(\mathfrak{l}_2, s_2)$, *then* $(\mathfrak{l}_1^*, s_{1^*})$ *is isomorphic to* $(\mathfrak{l}_2^*, s_{2^*})$.

Proof. (i) Let $(\mathfrak{l}^C)^R$ denote the Lie algebra $\mathfrak{l}^C$ when considered as a Lie algebra over $\boldsymbol{R}$. Then the Lie algebra $(\mathfrak{l}^C)^R$ has a complex structure J given by the multiplication by i on $\mathfrak{l}^C$. Each endomorphism A of $\mathfrak{l}$ or $\mathfrak{l}^*$ extends uniquely to a linear transformation of $(\mathfrak{l}^C)^R$ commuting with J. In this way the Lie groups $\boldsymbol{GL}(\mathfrak{l})$ and $\boldsymbol{GL}(\mathfrak{l}^*)$ become closed Lie subgroups of $\boldsymbol{GL}((\mathfrak{l}^C)^R)$. Consequently, the adjoint groups Int $(\mathfrak{l})$ and Int $(\mathfrak{l}^*)$ are Lie subgroups of $\boldsymbol{GL}((\mathfrak{l}^C)^R)$. Let U denote the analytic subgroup of Int $(\mathfrak{l})$ with Lie algebra $\text{ad}_{\mathfrak{l}}(\mathfrak{u})$. Now U is compact, so by Cor. 2.9, Chapter II, $\mathfrak{u}$ is compactly imbedded in $\mathfrak{l}^*$. For (ii) one just has to observe that $\mathfrak{l}$ and $\mathfrak{l}^*$ are real forms of $\mathfrak{l}^C$ so their Killing forms are obtained from the Killing form of $\mathfrak{l}^C$ by restriction. For (iii) suppose φ is an isomorphism of $\mathfrak{l}_1$ onto $\mathfrak{l}_2$ such that $\varphi \circ s_1 = s_2 \circ \varphi$. Then φ extends uniquely to an isomorphism $\tilde{\varphi}$ of $\mathfrak{l}_1^C$ onto $\mathfrak{l}_2^C$. The restriction of $\tilde{\varphi}$ to $\mathfrak{l}_1^*$ then sets up the required isomorphism between $(\mathfrak{l}_1^*, s_1^*)$ and $(\mathfrak{l}_2^*, s_2^*)$.

Let $\mathfrak{g}$ be a semisimple Lie algebra over $\boldsymbol{C}$. If $\mathfrak{l}$ runs through all compact real forms of $\mathfrak{g}$ and s runs through all involutive automorphisms of $\mathfrak{l}$, then $\mathfrak{l}^*$ runs through all noncompact real forms of $\mathfrak{g}$.

Proposition 2.2. *Let $\mathfrak{l}$ be a compact semisimple Lie algebra. Let s_1 and s_2 be two involutive automorphisms of $\mathfrak{l}$ and let $\mathfrak{l}_1^*$ and $\mathfrak{l}_2^*$ denote the corresponding real forms of $\mathfrak{l}^C$. Then s_1 and s_2 are conjugate within the group Aut $(\mathfrak{l})$ if and only if $\mathfrak{l}_1^*$ and $\mathfrak{l}_2^*$ are conjugate under an automorphism of $\mathfrak{l}^C$.*

Proof. Suppose first that there exists a $\sigma \in \text{Aut}(\mathfrak{l})$ such that $s_2 = \sigma s_1 \sigma^{-1}$. Let $\mathfrak{l} = \mathfrak{u}_1 + \mathfrak{e}_1$, $\mathfrak{l} = \mathfrak{u}_2 + \mathfrak{e}_2$ be the direct decompositions of $\mathfrak{l}$ into eigenspaces of s_1 and s_2, respectively. Then $\sigma \mathfrak{u}_1 = \mathfrak{u}_2$ and $\sigma \mathfrak{e}_1 = \mathfrak{e}_2$. Let Σ denote the unique extension of σ to a (complex) automorphism of $\mathfrak{l}^C$. Since $\mathfrak{l}_1^* = \mathfrak{u}_1 + i\mathfrak{e}_1$, $\mathfrak{l}_2^* = \mathfrak{u}_2 + i\mathfrak{e}_2$, it is obvious that $\Sigma \cdot \mathfrak{l}_1^* = \mathfrak{l}_2^*$.

For the converse (and nontrivial) part of Prop. 2.2 we shall use Theorem 7.2, Chapter III, stating that two Cartan decompositions of a semisimple Lie algebra are necessarily conjugate under an inner automorphism. Suppose then that there exists an automorphism Σ of $\mathfrak{l}^C$ such that $\Sigma \cdot \mathfrak{l}_1^* = \mathfrak{l}_2^*$. Then the two Cartan decompositions $\mathfrak{l}_2^* = \mathfrak{u}_2 + i\mathfrak{e}_2$ and $\mathfrak{l}_2^* = \Sigma \cdot \mathfrak{u}_1 + i\Sigma \cdot \mathfrak{e}_1$ are conjugate under an inner automorphism γ of $\mathfrak{l}_2^*$. Let Γ denote the unique extension of γ to an automorphism of $\mathfrak{l}^C$. Then the automorphism $\Gamma \circ \Sigma$ leaves $\mathfrak{l}$ invariant and its restriction to $\mathfrak{l}$ sets up the desired conjugacy of s_2 and s_1.

As shown in Chapter IV, any compact Lie group can be given the structure of a Riemannian globally symmetric space. We shall now see that the subclass of orthogonal symmetric Lie algebras of compact type, so obtained, corresponds, under the duality, to the class of orthogonal symmetric Lie algebras $(\mathfrak{l}, s)$ of noncompact type, where $\mathfrak{l}$ has complex structure and s is a conjugation.

Proposition 2.3. *Let $\mathfrak{l}_0$ be a compact semisimple Lie algebra and let s denote the automorphism $(X, Y) \to (Y, X)$ of the product algebra $\mathfrak{l} = \mathfrak{l}_0 \times \mathfrak{l}_0$. Then $(\mathfrak{l}, s)$ is an orthogonal symmetric Lie algebra of the compact type. If $(\mathfrak{l}^*, s^*)$ denotes the dual of $(\mathfrak{l}, s)$, then $\mathfrak{l}^*$ is isomorphic (as a real Lie algebra) to a complex subalgebra $\mathfrak{a}$ of $\mathfrak{l}^C$ in such a way that s^* corresponds to the conjugation of $\mathfrak{a}$ with respect to a compact real form of $\mathfrak{a}$.*

Proof. Let $\mathfrak{u} = \{(X, X) : X \in \mathfrak{l}_0\}$ and $\mathfrak{e} = \{(X, -X) : X \in \mathfrak{l}_0\}$. Then the direct decomposition $\mathfrak{l} = \mathfrak{u} + \mathfrak{e}$ is the usual decomposition of $\mathfrak{l}$ into eigenspaces of s. We have $\mathfrak{l}^* = \mathfrak{u} + i\mathfrak{e}$ and $\mathfrak{l}^C = \mathfrak{l} + i\mathfrak{l}$. The algebra $\mathfrak{a} = \mathfrak{u} + i\mathfrak{u}$ is a (complex) subalgebra of $\mathfrak{l}^C$ and the mapping

$$\varphi : (X, X) + \imath(Y, -Y) \to (X, X) + i(Y, Y), \qquad X, Y \in \mathfrak{l}_0,$$

is an isomorphism of $\mathfrak{l}^*$ onto $\mathfrak{a}$ (considered as real Lie algebras). Moreover, if μ denotes the conjugation of $\mathfrak{a}$ with respect to $\mathfrak{u}$, then $\varphi \circ s^* = \mu \circ \varphi$. Since $\mathfrak{u}$ is a compact real form of $\mathfrak{a}$, the proposition is proved.

Theorem 2.4. *Let* $(\mathfrak{l}, s)$ *be an orthogonal symmetric Lie algebra of the compact type and* $(\mathfrak{l}^*, s^*)$ *its dual. Then the Lie algebra* $\mathfrak{l}^*$ *has a complex structure if and only if* $\mathfrak{l}$ *can be written as a direct sum* $\mathfrak{l} = \mathfrak{l}_1 + \mathfrak{l}_2$, *where* $\mathfrak{l}_1$ *and* $\mathfrak{l}_2$ *are ideals in* $\mathfrak{l}$, *which are interchanged by* s.

Proof. Suppose first, that $\mathfrak{l}^*$ has a complex structure, which will be denoted by J in order to avoid confusion with the complex structure of $\mathfrak{l}^C$. Then J satisfies the relation $[X, JY] = J[X, Y]$ for $X, Y \in \mathfrak{l}^*$. Since $\mathfrak{l}^*$ can be considered as a semisimple Lie algebra over $\boldsymbol{C}$ (by means of J), it has a compact real form $\mathfrak{k}$. We have then the direct decomposition

$$\mathfrak{l}^* = \mathfrak{k} + J\mathfrak{k},$$

which is a Cartan decomposition of $\mathfrak{l}^*$. On the other hand, we have the decompositions

$$\mathfrak{l} = \mathfrak{u} + \mathfrak{e}, \qquad \mathfrak{l}^* = \mathfrak{u} + i\mathfrak{e}$$

into eigenspaces of s and s^*, respectively. Since all Cartan decompositions of $\mathfrak{l}^*$ are conjugate under an inner automorphism of $\mathfrak{l}^*$ there exists an element $\sigma \in \operatorname{Int}(\mathfrak{l}^*)$ such that $\sigma \cdot \mathfrak{k} = \mathfrak{u}$, $\sigma \cdot (J\mathfrak{k}) = i\mathfrak{e}$. Consider now the following mappings:

$$\mathfrak{e} \xrightarrow{i} i\mathfrak{e} \xrightarrow{\sigma^{-1}} J\mathfrak{k} \xrightarrow{-J} \mathfrak{k} \xrightarrow{\sigma} \mathfrak{u}$$

and put $\gamma(X) = -\sigma J \sigma^{-1} i X$ for $X \in \mathfrak{e}$. Then γ is a one-to-one linear mapping of $\mathfrak{e}$ onto $\mathfrak{u}$ and has the following properties:

$$\text{(a)} \quad [\gamma(X), \gamma(Y)] = [X, Y];$$

$$\text{(b)} \quad [X, \gamma(Y)] = [\gamma(X), Y];$$

$$\text{(c)} \quad \gamma([\gamma(X), Y]) = [X, Y]$$

for $X, Y \in \mathfrak{e}$. The last property is verified as follows:

$$\begin{aligned}\gamma[\gamma X, Y] &= \sigma J \sigma^{-1} i([\sigma J \sigma^{-1} i X, Y]) = \sigma J \sigma^{-1}([\sigma J \sigma^{-1} i X, iY]) \\ &= \sigma J[J\sigma^{-1} i X, \sigma^{-1} i Y] = -\sigma[\sigma^{-1} i X, \sigma^{-1} i Y] = [X, Y].\end{aligned}$$

Property (a) is proved in the same way and (b) follows from (*c*). We define now the subsets $\mathfrak{l}_1$ and $\mathfrak{l}_2$ of $\mathfrak{l}$ by

$$\mathfrak{l}_1 = \{X + \gamma(X) : X \in \mathfrak{e}\}, \qquad \mathfrak{l}_2 = \{X - \gamma(X) : X \in \mathfrak{e}\}.$$

Then the following statements hold:

(i) $\mathfrak{l}_1 \cap \mathfrak{l}_2 = \{0\}$.
(ii) s interchanges $\mathfrak{l}_1$ and $\mathfrak{l}_2$.
(iii) $\mathfrak{l}_1$ and $\mathfrak{l}_2$ are ideals in $\mathfrak{l}$.

The first statement is obvious because the relation $X + \gamma(X) = Y - \gamma(Y)$ implies $\gamma(X + Y) = Y - X \in \mathfrak{e} \cap \mathfrak{u} = \{0\}$. The second statement is obvious since $s(X + \gamma(X)) = - X + \gamma(X)$. Finally, properties (a), (b), and (c) above show that $\mathfrak{l}_1$ and $\mathfrak{l}_2$ are subalgebras of $\mathfrak{l}$ and $[\mathfrak{l}_1, \mathfrak{l}_2] = \{0\}$, proving (iii). This proves the first half of the theorem.

Remark. The mapping $X + \gamma(X) \rightarrow 2\gamma(X)$ is an isomorphism of $\mathfrak{l}_1$ onto $\mathfrak{u}$.

In order to prove the second half of the theorem, we consider a Lie algebra $\mathfrak{l}_0$ isomorphic to both $\mathfrak{l}_1$ and $\mathfrak{l}_2$. Let I_i denote an isomorphism of $\mathfrak{l}_i$ onto $\mathfrak{l}_0$ $(i = 1, 2)$. In the following let X denote an arbitrary element in $\mathfrak{l}_1$ and Y an arbitrary element in $\mathfrak{l}_2$. Let $\bar{\mathfrak{l}}$ denote the product algebra $\mathfrak{l}_0 \times \mathfrak{l}_0$ and let $\bar{s}$, s_0 denote the automorphisms of $\bar{\mathfrak{l}}$ given by

$$\bar{s} \cdot (I_1X, I_2Y) = (I_1sY, I_2sX),$$
$$s_0 \cdot (I_1X, I_2Y) = (I_2Y, I_1X).$$

Then we have the isomorphisms

$$(\mathfrak{l}, s) \xrightarrow{I_0} (\bar{\mathfrak{l}}, \bar{s}) \xrightarrow{\sigma} (\bar{\mathfrak{l}}, s_0),$$

where

$$I_0(X + Y) = (I_1X, I_2Y),$$
$$\sigma(I_1X, I_2Y) = (I_1X, I_1sY).$$

Let $(\bar{\mathfrak{l}}^*, \bar{s}^*)$ denote the dual of $(\bar{\mathfrak{l}}, \bar{s})$ and let $(\mathfrak{g}, s_0^*)$ denote the dual of $(\bar{\mathfrak{l}}, s_0)$. Then, as a result of Prop. 2.1, the orthogonal symmetric Lie algebras

$$(\mathfrak{l}^*, s^*), \qquad (\bar{\mathfrak{l}}^*, \bar{s}^*), \qquad (\mathfrak{g}, s_0^*)$$

are all isomorphic. But due to Prop. 2.3, $\mathfrak{g}$ has a complex structure. Using Prop. 2.2 it follows that $\bar{\mathfrak{l}}^*$, and therefore $\mathfrak{l}^*$, has a complex structure. This concludes the proof of the theorem.

Example I. *The coset spaces* $SO(p + q)/SO(p) \times SO(q)$ *and* $SO_0(p, q)/SO(p) \times SO(q)$.

Let $SO(p, q)$ denote the group of real quadratic matrices of determinant 1, leaving invariant the quadratic form

$$- x_1^2 - \ldots - x_p^2 + x_{p+1}^2 + \ldots + x_{p+q}^2 \qquad (p + q > 2).$$

Let I_n denote the unit matrix of order n and put

$$I_{p,q} = \begin{pmatrix} -I_p & 0 \\ 0 & I_q \end{pmatrix}.$$

Denoting by ${}^tA$ the transpose of a matrix A, we see that a matrix g of determinant 1 belongs to $\boldsymbol{SO}(p, q)$ if and only if ${}^tgI_{p,q}g = I_{p,q}$. Thus $\boldsymbol{SO}(p, q)$ is a closed subgroup of $\boldsymbol{GL}(p + q, \boldsymbol{R})$, hence a topological Lie subgroup. Its Lie algebra, denoted $\mathfrak{so}(p, q)$, is a subalgebra of $\mathfrak{gl}(p + q, \boldsymbol{R})$. According to formula (2) in Chapter II, §2 we have

$$X \in \mathfrak{so}(p,q) \text{ if and only if } e^{sX} \in \boldsymbol{SO}(p,q) \text{ for all } s \in \boldsymbol{R}.$$

Now $e^X \in \boldsymbol{SO}(p, q)$ if and only if ${}^t(e^X) = I_{p,q}\, e^{-X} I_{p,q}$; since ${}^t(e^X) = e^{{}^tX}$, it follows that

$$X \in \mathfrak{so}(p, q) \text{ if and only if } {}^tXI_{p,q} + I_{p,q}X = 0.$$

Thus $\mathfrak{so}(p, q)$ is the set of matrices

$$X = \begin{pmatrix} X_1 & X_2 \\ {}^tX_2 & X_3 \end{pmatrix}$$

where X_1 and X_3 are skew symmetric matrices of order p and q, respectively, and X_2 is an arbitrary $p \times q$ matrix. In particular, the Lie algebra $\mathfrak{so}(n)$ of the group $\boldsymbol{SO}(n)$ $(= \boldsymbol{SO}(n, 0))$ consists of all $n \times n$ skew symmetric matrices.

Now let $\mathfrak{l} = \mathfrak{so}(p + q)$ and let s denote the restriction to $\mathfrak{l}$ of the automorphism $\sigma_{p,q} : X \to I_{p,q}XI_{p,q}$ of $\mathfrak{gl}(p + q, \boldsymbol{C})$. Then $(\mathfrak{l}, s)$ is an orthogonal symmetric Lie algebra of the compact type. If $\mathfrak{l} = \mathfrak{u} + \mathfrak{e}$ is the usual decomposition into eigenspaces of s, it is easily seen that

$$\mathfrak{u} = \left\{ \begin{pmatrix} X_1 & 0 \\ 0 & X_3 \end{pmatrix} \middle| \begin{array}{l} X_1 : p \times p \text{ skew symmetric matrix} \\ X_3 : q \times q \text{ skew symmetric matrix} \end{array} \right\},$$

$$\mathfrak{e} = \left\{ \begin{pmatrix} 0 & X_2 \\ -{}^tX_2 & 0 \end{pmatrix} \middle| X_2 : p \times q \text{ arbitrary matrix} \right\}.$$

The pair $(\boldsymbol{SO}(p + q), \boldsymbol{SO}(p) \times \boldsymbol{SO}(q))$ is therefore associated with $(\mathfrak{l}, s)$. Let $(\mathfrak{l}^*, s^*)$ denote the dual of $(\mathfrak{l}, s)$. Then $\mathfrak{l}^*$ is the subalgebra of $\mathfrak{gl}(p+q,\boldsymbol{C})$ given by $\mathfrak{l}^* = \mathfrak{u} + i\mathfrak{e}$ and s^* is again the restriction of $\sigma_{p,q}$ to $\mathfrak{l}^*$. Now it is easy to verify that the mapping

$$\begin{pmatrix} X_1 & iX_2 \\ -i\,{}^tX_2 & X_3 \end{pmatrix} \to \begin{pmatrix} X_1 & X_2 \\ {}^tX_2 & X_3 \end{pmatrix} = \begin{pmatrix} -iI_p & 0 \\ 0 & I_q \end{pmatrix} \begin{pmatrix} X_1 & iX_2 \\ -i\,{}^tX_2 & X_3 \end{pmatrix} \begin{pmatrix} iI_p & 0 \\ 0 & I_q \end{pmatrix}$$

is an isomorphism of $\mathfrak{l}^*$ onto $\mathfrak{so}(p, q)$. Under this isomorphism, the automorphism s^* corresponds again to the automorphism $X \to I_{p,q}XI_{p,q}$ of $\mathfrak{so}(p, q)$. Let $\boldsymbol{SO}_0(p, q)$ denote the identity component of $\boldsymbol{SO}(p, q)$. Then the pair $(\boldsymbol{SO}_0(p, q), \boldsymbol{SO}(p) \times \boldsymbol{SO}(q))$ is associated with the orthogonal symmetric Lie algebra ($\mathfrak{so}(p, q)$, $\sigma_{p,q}$), which is isomorphic to the dual of ($\mathfrak{so}(p + q)$, s).

Example II. *The case* $p = 1$, $q = 3$.

Let Q denote the algebra of quaternions, Q_+ the subspace of pure quaternions, and G the multiplicative group of quaternions of norm 1. To each pair x, y from G we associate the endomorphism

$$T_{x,y} : u \to xuy^{-1}, \qquad u \in Q$$

of Q onto itself. Since $T_{x,y}$ is norm preserving, it belongs to the group of rotations of Q; since G is connected it follows that all $T_{x,y}$ belong to a connected part of the group of rotations. Hence $T_{x,y} \in \mathbf{SO}(4)$. Each endomorphism $T_{x,x}$ leaves the subspace Q_+ invariant; let τ_x denote the restriction of $T_{x,x}$ to Q_+. Then $\tau_x \in \mathbf{SO}(3)$. The following statements hold.

(a) *The mapping* $x \to \tau_x$ *is an analytic homomorphism of* G *onto* $\mathbf{SO}(3)$ *and the kernel consists of* e *and* $-e$, e *denoting the identity element in* Q.

(b) *The mapping* $(x, y) \to T_{x,y}$ *is an analytic homomorphism of the product* $G \times G$ *onto* $\mathbf{SO}(4)$ *and the kernel consists of* (e, e) *and* $(-e, -e)$.

The verification of (a) and (b) will be left to the reader. Passing to the Lie algebras we obtain an isomorphism φ of $\mathfrak{so}(3) \times \mathfrak{so}(3)$ onto $\mathfrak{so}(4)$. Let s_0 denote the automorphism $(X, Y) \to (Y, X)$ of $\mathfrak{so}(3) \times \mathfrak{so}(3)$. An elementary quaternion computation shows that $\sigma_{1,3}(T_{x,y}) = T_{y,x}(x, y \in G)$. Hence we obtain

(c) *The orthogonal symmetric Lie algebras*

$$(\mathfrak{so}(3) \times \mathfrak{so}(3), s_0) \qquad \textit{and} \qquad (\mathfrak{so}(4), \sigma_{1,3})$$

are isomorphic under φ.

Now by Prop. 2.1 (iii), the duals of these pairs are isomorphic. In view of Theorem 2.4 we can conclude that the Lie algebra $\mathfrak{so}(1, 3)$ has a complex structure. Since this Lie algebra has dimension 6, it is isomorphic to a semisimple, three dimensional complex Lie algebra (considered as a real Lie algebra). Such an algebra must have a Cartan subalgebra of dimension 1 and two nonzero roots α, $-\alpha$. In view of Theorem 5.4, Chapter III, there is therefore at most one three-dimensional complex semisimple Lie algebra. On the other hand, the Lie algebra $\mathfrak{sl}(2, \mathbf{C})$ of all complex 2×2 matrices of trace 0 is such an algebra. We can therefore conclude that $\mathfrak{so}(1, 3)$ is isomorphic to $\mathfrak{sl}(2, \mathbf{C})^R$.

§ 3. Sectional Curvature of Symmetric Spaces

The three classes of symmetric spaces can be distinguished by means of their curvature as shown in the following theorem.

Theorem 3.1. *Let* $(\mathfrak{l}, \mathfrak{s})$ *be an orthogonal symmetric Lie algebra and suppose that the pair* (L, U) *is associated with* $(\mathfrak{l}, s)$. *We assume that* U *is connected and closed.*† *Let* Q *be an arbitrary* L*-invariant Riemannian structure on* L/U *(such a* Q *exists).*

(i) *If* (L, U) *is of the compact type, then* L/U *has sectional curvature everywhere* $\geqslant 0$.

(ii) *If* (L, U) *is of the noncompact type, then* L/U *has sectional curvature everywhere* $\leqslant 0$.

(iii) *If* (L, U) *is of the Euclidean type, then* L/U *has sectional curvature everywhere* $= 0$.

Proof. The tangent space to L/U at the point $o = \{U\}$ can, as usual, be identified with $\mathfrak{e}$, the eigenspace of s for the eigenvalue -1. Let S be a two-dimensional subspace of $\mathfrak{e}$, and let X, Y be an orthonormal basis of S. Then, according to Theorem 4.2, Chapter IV, the curvature of L/U along the section S is given by

$$K(S) = -Q_o(R(X, Y)X, Y) = +Q_o([[X, Y], X], Y).$$

Part (iii) is now obvious, so we can assume that $\mathfrak{l}$ is semisimple. As in §1, let b denote the endomorphism of $\mathfrak{e}$ given by

$$Q_o(bX, Y) = B(X, Y), \qquad X, Y \in \mathfrak{e},$$

B denoting the Killing form of $\mathfrak{l}$. Since $Q_o(bX, Y) = Q_o(X, bY)$, the eigenvalues $\beta_1, \ldots, \beta_n$ of b are real. Let $\mathfrak{e}_1, \ldots, \mathfrak{e}_n$ be the corresponding eigenspaces of b. Then, if $i \neq j$, the spaces $\mathfrak{e}_i$ and $\mathfrak{e}_j$ are orthogonal with respect to B as well as Q_o. We shall prove that $[\mathfrak{e}_i, \mathfrak{e}_j] = \{0\}$. Let $\mathfrak{u}$ denote the Lie algebra of U. Then b commutes with each member of $\mathrm{ad}_{\mathfrak{l}}(\mathfrak{u})$. Hence $[\mathfrak{u}, \mathfrak{e}_i] \subset \mathfrak{e}_i$ for each i. Now let $X_i \in \mathfrak{e}_i$, $X_j \in \mathfrak{e}_j$, $T \in \mathfrak{u}$. Then $[X_i, X_j] \in \mathfrak{u}$ and

$$B(T, [X_i, X_j]) = B([T, X_i], X_j) = 0.$$

Owing to Lemma 1.2, B is strictly negative definite on $\mathfrak{u}$; consequently $[\mathfrak{e}_i, \mathfrak{e}_j] = \{0\}$.

† For the noncompact type this hypothesis is always satisfied (see Chapter VI). For the compact type, U is always closed but not necessarily connected (see Chapter VII).

Writing $X = \sum_{i=1}^n X_i$, $Y = \sum_{i=1}^n Y_i$ ($X_i, Y_i \in \mathfrak{e}_i$), we have

$$[X, Y] = \sum_{i=1}^{n} [X_i, Y_i], \qquad [[X_i, Y_i], X] = [[X_i, Y_i], X_i],$$

so

$$K(S) = \sum_{i=1}^{n} Q_o([[X_i, Y_i], X_i], Y_i) = \sum_{i=1}^{n} \frac{1}{\beta_i} B([[X_i, Y_i], X_i], Y_i)$$

and

$$K(S) = \sum_{i=1}^{n} \frac{1}{\beta_i} B([X_i, Y_i], [X_i, Y_i]). \tag{1}$$

Since $\beta_i < 0$ in case (i), $\beta_i > 0$ in case (ii), the theorem follows.

Remark 1. In case we take $Q = \epsilon B$, where $\epsilon = -1$ for the compact type, $\epsilon = +1$ for the noncompact type, we have by (1)

$$K(S) = \epsilon B([X, Y], [X, Y]). \tag{2}$$

Remark 2. Suppose $(\mathfrak{l}, s)$ is one of the types (i), (ii), or (iii). Then the curvature $K(S)$ along the section S is 0 if and only if S is an abelian subspace of $\mathfrak{e}$.

Example. As a special case of Example I, §2, we consider the spaces $\boldsymbol{SO}(p + 1)/\boldsymbol{SO}(p)$ and $\boldsymbol{SO}_0(p, 1)/\boldsymbol{SO}(p)$ which correspond to each other under the duality. Here the linear isotropy group at a point p acts transitively on the set of two-dimensional subspaces of the tangent space at p. Hence these spaces have constant sectional curvature. In particular, for $p = 2$, the spaces are the two-dimensional sphere and the two-dimensional non-Euclidean space of Lobatschevsky.

For the sphere $\boldsymbol{SO}(3)/\boldsymbol{SO}(2)$ the formulas

$$\frac{\sin a}{\sin A} = \frac{\sin b}{\sin B} = \frac{\sin c}{\sin C},$$

$$\cos a = \cos b \cos c + \sin b \sin c \cos A$$

hold for a geodesic triangle with angles A, B, C and sides of length a, b, c. For the two-dimensional Lobatschevsky space $\boldsymbol{SO}_0(2, 1)/\boldsymbol{SO}(2)$ the formulas are

$$\frac{\sinh a}{\sin A} = \frac{\sinh b}{\sin B} = \frac{\sinh c}{\sin C},$$

$$\cosh a = \cosh b \cosh c - \sinh b \sinh c \cos A.$$

Since $\sinh iz = i \sin z$ and $\cosh iz = \cos z$, the two sets of formulas correspond under the substitution $a \to ia$, $b \to ib$, $c \to ic$. The duality for symmetric spaces gives a general explanation of this formal analogy between spherical trigonometry and non-Euclidean trigonometry.

§ 4. Symmetric Spaces with Semisimple Groups of Isometries

Theorem 4.1. *Let (G, K) be a Riemannian symmetric pair. Suppose that G is semisimple and acts effectively on the coset space $M = G/K$. Let Q be any G-invariant Riemannian structure on M and let R denote the corresponding curvature tensor. Then:*

(i) $G = I_0(M)$ *(as Lie groups).*

(ii) *The linear isotropy subgroup of G at $o = \{K\}$ is a Lie subgroup K^* of $\mathbf{GL}(M_o)$, isomorphic to K. Its Lie algebra $\mathfrak{k}^*$ consists of all endomorphisms of M_o which, when extended to the mixed tensor algebra over M_o as derivations commuting with contractions, annihilate Q_o and R_o.*

(iii) *$\mathfrak{k}^*$ is spanned by the set $\{R_o(X, Y) : X, Y \in M_o\}$.*

Proof. According to Prop. 3.5, Chapter IV, there exists a unique analytic involutive automorphism σ of G such that $(K_\sigma)_0 \subset K \subset K_\sigma$. Here K_σ denotes the set of fixed points of σ and $(K_\sigma)_0$ is the identity component of K_σ. Let s_o denote the geodesic symmetry of G/K with respect to o. Then, as proved in Chapter IV (Prop. 3.4),

$$\sigma(g) = s_o g s_o, \qquad g \in G. \tag{1}$$

Let $\mathfrak{g} = \mathfrak{k} + \mathfrak{p}$ be the direct decomposition of the Lie algebra $\mathfrak{g}$ of G into the eigenspaces of $(d\sigma)_e$ for the eigenvalues $+1$ and -1, respectively. Let Z denote the center of G. According to Lemma 5.1, Chapter II, the group $\mathrm{Ad}_G(K)$ is analytically isomorphic to $K/K \cap Z$ which equals K, G being effective. Thus K is compact and isomorphic to the linear isotropy group K^*. Let $G' = I_0(M)$ and let K' denote the (compact) subgroup of G' leaving the point o fixed. Owing to Remark 2 following Prop. 3.4, Chapter IV, the group G is a closed subgroup of G'. Hence $\mathfrak{g}$ is a subalgebra of the Lie algebra $\mathfrak{g}'$ of G'. Let $\tilde{\sigma}$ denote the automorphism $g \to s_0 g s_0$ of G' and let

$$\mathfrak{g}' = \mathfrak{k}' + \mathfrak{p}'$$

be the decomposition of $\mathfrak{g}'$ into the eigenspaces of $(d\tilde{\sigma})_e$, $\mathfrak{k}'$ being the Lie algebra of K'. Here $\mathfrak{k} \subset \mathfrak{k}'$, $\mathfrak{p} = \mathfrak{p}'$. We now apply Theorem 1.1 and the terminology introduced there to the pair $(\mathfrak{g}', (d\tilde{\sigma})_e)$ which is an effective orthogonal symmetric Lie algebra. The subspace $(\mathfrak{p}')_0$ is an abelian ideal in $\mathfrak{g}'$ and $\mathfrak{g}$ (Lemma 1.3). Since $\mathfrak{g}$ is semisimple, $(\mathfrak{p}')_0 = \{0\}$. Hence $(\mathfrak{k}')_0$ is an ideal of $\mathfrak{g}'$ contained in $\mathfrak{k}'$; thus $(\mathfrak{k}')_0 = \{0\}$ and $[\mathfrak{p}', \mathfrak{p}'] = \mathfrak{k}'$. Since $\mathfrak{p} = \mathfrak{p}'$ and $[\mathfrak{p}, \mathfrak{p}] \subset \mathfrak{k}$, it follows that $\mathfrak{k}' = \mathfrak{k} = [\mathfrak{p}, \mathfrak{p}]$ and $\mathfrak{g} = \mathfrak{g}'$, proving (i). Moreover, the relation $[\mathfrak{p}, \mathfrak{p}] = \mathfrak{k}$ is equivalent to (iii), in view of the formula for the curvature tensor (Theorem 4.2, Chapter IV).

Finally, in order to prove (ii), let $\tilde{\mathfrak{k}}$ denote the Lie algebra of all endomorphisms of M_o which, when extended to the mixed tensor algebra over M_o as derivations commuting with contractions, annihilate Q_o and R_o. Then $\mathfrak{k} \subset \tilde{\mathfrak{k}}$ and the space $\tilde{g} = \tilde{\mathfrak{k}} + \mathfrak{p}$ is a Lie algebra if the bracket $[T, X]$ for $T \in \tilde{\mathfrak{k}}$, $X \in \mathfrak{p}$ is defined as $T \cdot X$ (T operating on X). As in the proof of $[\mathfrak{p}, \mathfrak{p}] = \mathfrak{k}'$ we see that $[\mathfrak{p}, \mathfrak{p}] = \tilde{\mathfrak{k}}$ so $\mathfrak{k} = \tilde{\mathfrak{k}}$ and (ii) follows.

Definition. Let M be a Riemannian globally symmetric space; M is said to be of the *compact type* or the *noncompact type* according to the type of the Riemannian symmetric pair $(I_0(M), K)$, K being the isotropy subgroup of $I_0(M)$ at some point in M. If $(\mathfrak{g}, \theta)$ is the corresponding orthogonal involutive Lie algebra, M is said to be *associated with* $(\mathfrak{g}, \theta)$.

Proposition 4.2. *Let M be a simply connected Riemannian globally symmetric space. Then M is a product*

$$M = M_0 \times M_- \times M_+,$$

where M_0 is a Euclidean space, M_- and M_+ are Riemannian globally symmetric of the compact and noncompact type, respectively.

Proof. Let $G = I_0(M)$ and let K denote the isotropy subgroup at some point o in M. Then $M = G/K$. Let $(\tilde{G}, \varphi)$ denote the universal covering group of G and let $\tilde{K}$ denote the identity component of $\varphi^{-1}(K)$. Then if ψ denotes the mapping $g\tilde{K} \to \varphi(g)K$ of $\tilde{G}/\tilde{K}$ onto G/K, the pair $(\tilde{G}/\tilde{K}, \psi)$ is a covering manifold of G/K. Since M is simply connected, $M = \tilde{G}/\tilde{K}$.

Let s denote the involutive automorphism of $\mathfrak{g}$, the Lie algebra of G (and $\tilde{G}$), which corresponds to the automorphism $g \to s_o g s_o$ of G. Then $(\mathfrak{g}, s)$ is an effective orthogonal symmetric Lie algebra. We decompose $\mathfrak{g}$ according to Theorem 1.1 and let $\tilde{G} = G_0 \times G_- \times G_+$ be the corresponding decomposition of $\tilde{G}$. If $\tilde{K} = K_0 \times K_- \times K_+$ is the decomposition induced on $\tilde{K}$, the spaces $M_0 = G_0/K_0$, $M_- = G_-/K_-$ and $M_+ = G_+/K_+$ have the required properties.

§ 5. Notational Conventions

In order to avoid repeated explanation of notation we shall now make some notational conventions which, with minor modifications, will be in force through Chapters VI, VII, and VIII.

The symbol $\mathfrak{g}_0$ shall denote an arbitrary semisimple Lie algebra over $\boldsymbol{R}$, and $\mathfrak{g}$ its complexification. Let $\mathfrak{g}_0 = \mathfrak{k}_0 + \mathfrak{p}_0$ be any Cartan decomposition of $\mathfrak{g}_0$ ($\mathfrak{k}_0$ the algebra), and let $\mathfrak{u}$ denote the compact real form $\mathfrak{k}_0 + i\mathfrak{p}_0$ of $\mathfrak{g}$. Let B denote the Killing form of $\mathfrak{g}$. Its restrictions to

$\mathfrak{g}_0 \times \mathfrak{g}_0$ and $\mathfrak{u} \times \mathfrak{u}$ are the Killing forms of $\mathfrak{g}_0$ and $\mathfrak{u}$, respectively. Let σ and τ denote the conjugations of $\mathfrak{g}$ with respect to $\mathfrak{g}_0$ and $\mathfrak{u}$, respectively, and put $\theta = \sigma\tau = \tau\sigma$. Then θ is an involutive automorphism of $\mathfrak{g}$. Subspaces of $\mathfrak{g}_0$ will usually be denoted by the subscript 0; the corresponding subspace of $\mathfrak{g}$ will then be denoted by the same letter but without the subscript. According to this convention, $\mathfrak{k}$ and $\mathfrak{p}$ denote the eigenspaces of the automorphism θ. If $\mathfrak{e}_0$ is a subspace of $\mathfrak{p}_0$, the subspace $i\mathfrak{e}_0$ of $i\mathfrak{p}_0$ will often be denoted by $\mathfrak{e}_*$.

The adjoint groups Int $(\mathfrak{g}_0)$, Int $(\mathfrak{u})$ are groups of endomorphisms of $\mathfrak{g}_0$ and $\mathfrak{u}$, which, however, can be extended to endomorphisms of $\mathfrak{g}$. The Lie algebras of Int $(\mathfrak{g}_0)$ and Int $(\mathfrak{u})$ will be identified with the subalgebras $\mathfrak{g}_0$ and $\mathfrak{u}$ of $\mathfrak{g}^R$. The analytic subgroup of Int $(\mathfrak{g}_0)$ whose Lie algebra is $\mathfrak{k}_0$ will be denoted by K^*. Then K^* is compact and a Lie subgroup of Int $(\mathfrak{u})$. We shall see in the following chapter that $K^* = \text{Int}(\mathfrak{g}_0) \cap \text{Int}(\mathfrak{u})$. All the groups K^*, Int $(\mathfrak{g}_0)$, Int $(\mathfrak{u})$ are closed, topological Lie subgroups of $GL(\mathfrak{g}^R)$.

Let (G, K_1) and (U, K_2) be Riemannian symmetric pairs associated with the orthogonal symmetric Lie algebras $(\mathfrak{g}_0, \theta)$ and $(\mathfrak{u}, \theta)$, respectively. In general, the space G/K_1, (and similarly U/K_2), will have many G-invariant Riemannian structures which are not proportional. However, the Riemannian connection is the same for all of these. Accordingly, we shall always (unless the contrary is specified) give G/K_1 the unique G-invariant Riemannian structure induced by the restriction of the Killing form B to $\mathfrak{p}_0 \times \mathfrak{p}_0$. Similarly, the space U/K_2 will be given the unique U-invariant Riemannian structure induced by the restriction of $- B$ to $i\mathfrak{p}_0 \times i\mathfrak{p}_0$.

§ 6. Rank of Symmetric Spaces

Definition. A Riemannian manifold is said to be *flat* if its curvature tensor vanishes identically.

Definition. Let M be a Riemannian globally symmetric space. The *rank* of M is the maximal dimension of a flat, totally geodesic submanifold of M.

Proposition 6.1. *Let M be a Riemannian globally symmetric space of the compact type or the noncompact type. Let o be any point in M and as usual identify the tangent space M_o with the subspace $\mathfrak{p}_0$ (or $i\mathfrak{p}_0$) of the Lie algebra of $I(M)$. Let $\mathfrak{s}_0$ be a Lie triple system contained in M_o. Then the totally geodesic submanifold $S = \text{Exp}\, \mathfrak{s}_0$ (with the differentiable structure from Theorem 7.2, Chapter IV) is flat if and only if $\mathfrak{s}_0$ is abelian.*

Proof. As we have seen in §7, Chapter IV, the manifold S is globally symmetric and can be written $S = G'/K'$ where G' is an analytic subgroup of $I(M)$, invariant under the automorphism $g \rightarrow s_o g s_o$ of $I(M)$. The proof of Theorem 7.2, Chapter IV, shows that the geodesics through o in S are $\exp tX \cdot o$ ($X \in \mathfrak{s}_0$), that is, (3), §3, Chapter IV, and therefore also Theorems 4.1 and 4.2, Chapter IV, are valid for G'/K' (although we neither claim K' connected nor its Lie algebra compactly imbedded in the Lie algebra of G'). The proposition now follows immediately from (1), §3.

Theorem 6.2. *Let M be a Riemannian globally symmetric space of the compact type or the noncompact type. Let l denote the rank of M and let A and A' denote two flat, totally geodesic submanifolds of M of dimension l.*

(i) *Let $q \in A$, $q' \in A'$. Then there exists an element $x \in I_0(M)$ such that $x \cdot A = A'$, $x \cdot q = q'$.*

(ii) *Let $X \in M_q$. Then there exists an element $k \in I_0(M)$ such that $k \cdot q = q$ and $dk(X) \in A_q$.*

(iii) *The manifolds A and A' are closed topological subspaces of M.*

In order to prove this theorem we begin with some general remarks about totally geodesic submanifolds. Let M be any manifold, S a connected submanifold. Let X and Y be vector fields on M such that $X_s, Y_s \in S_s$ for each $s \in S$. Then it follows easily from Prop. 3.2, Chapter I, that the families $s \rightarrow X_s$ ($s \in S$) and $s \rightarrow Y_s$ ($s \in S$) are vector fields on S. We denote these vector fields by $\bar{X}$ and $\bar{Y}$. It follows from Prop. 3.3, Chapter I, that $[X, Y]_s \subset S_s$ for each $s \in S$ and $[X, Y]^- = [\bar{X}, \bar{Y}]$.

Now suppose the manifold M is connected and has a Riemannian structure g. Let $\bar{g}$ denote the induced Riemannian structure on S. Let ∇ and $\bar{\nabla}$ denote the corresponding Riemannian connections. Let X, Y, Z be any vector fields on M for which $X_s, Y_s, Z_s \in S_s$ for each $s \in S$. Then we conclude from (2), §9, Chapter I, that

$$g(X, \nabla_Z(Y))(s) = \bar{g}(\bar{X}, \bar{\nabla}_{\bar{Z}}(\bar{Y}))(s) \qquad (s \in S). \tag{1}$$

Suppose now that S is totally geodesic. Then $\nabla_Z(Y)_s \in S_s$ ($s \in S$) by Theorem 14.5, Chapter I. Equation (1) therefore implies

$$\bar{\nabla}_{\bar{Z}}(\bar{Y}) = (\nabla_Z(Y))^- \tag{2}$$

Let R and $\bar{R}$ denote the curvature tensors of M and S, respectively. Then by (2)

$$\bar{R}(\bar{X}, \bar{Y}) \cdot \bar{Z} = (R(X, Y) \cdot Z)^- \tag{3}$$

and the sectional curvature along a two-dimensional subspace of S_s is the same for M and S.

After these preliminary remarks let us turn to the proof of Theorem 6.2. Let q be any point in the flat totally geodesic subspace A of M. As usual we identify the tangent space M_q with a subspace of the Lie algebra $\mathfrak{l}$ of $I(M)$. Let X and Y be two vectors in the tangent space A_q. By the preceding remarks the sectional curvature of M at q along the plane section spanned by X and Y is 0. Using (1), §3, it follows that $[X, Y] = 0$, the bracket being that of the Lie algebra $\mathfrak{l}$. Then by Prop. 6.1 A_q is a maximal abelian subspace of M_q. Let G' denote the analytic subgroup of $I(M)$ corresponding to the subalgebra A_q of $\mathfrak{l}$. Let K' denote the subgroup of G' leaving q fixed. The totally geodesic submanifold $\operatorname{Exp}_q(A_q)$ from Prop. 6.1 is the orbit $G' \cdot q$ with differentiable structure derived from G'/K'. Consider the automorphism $\sigma : g \to s_q g s_q$ of $I(M)$, s_q denoting the symmetry of M with respect to q. Then $M_q = \{X \in \mathfrak{l} : d\sigma(X) = -X\}$ so $\sigma(g) = g^{-1}$ for $g \in G'$. This relation also holds for the closure of G' which therefore has an abelian Lie algebra contained in M_q. On the other hand this Lie algebra contains A_q since G' is a Lie subgroup of its closure. By the maximality of A_q, G' is closed and thus carries the relative topology of $I(M)$. The group K' is therefore a closed subgroup of the isotropy subgroup of $I(M)$ at q, hence compact. Using now Prop. 4.4, Chapter II, we deduce that the submanifold $\operatorname{Exp}_q(A_q)$ is a closed topological subspace of M. The identity mapping of A into this submanifold is therefore continuous and, by Lemma 14.1, Chapter I, differentiable. We can therefore state that $\operatorname{Exp}_q(A_q)$ and A coincide as submanifolds of M. This proves (iii) and reduces (i) and (ii) to the following lemma (see §5 for notation).

Lemma 6.3. *Let $\mathfrak{a}_*$ and $\mathfrak{a}'_*$ be two maximal abelian subspaces of $\mathfrak{p}_*$. Then:*

(i) *There exists an element $H \in \mathfrak{a}_*$ whose centralizer in $\mathfrak{p}_*$ is $\mathfrak{a}_*$.*
(ii) *There exists an element $k \in K^*$ such that $k \cdot \mathfrak{a}_* = \mathfrak{a}'_*$.*
(iii) $\mathfrak{p}_* = \bigcup_{k \in K^*} k \cdot \mathfrak{a}_*$.

Proof. Let $P_* = \exp \mathfrak{p}_*$. Then $\operatorname{Int}(\mathfrak{u}) = P_* K^*$ since the geodesics $\operatorname{Exp} tX$ ($t \in \boldsymbol{R}$) cover the manifold $\operatorname{Int}(\mathfrak{u})/K^*$ as X varies through $\mathfrak{p}_*$. Let $\tilde{\theta}$ denote the involutive automorphism of $\operatorname{Int}(\mathfrak{u})$ which corresponds to the restriction of θ to $\mathfrak{u}$. Then

$$\tilde{\theta}((\exp X)\, k) = (\exp(-X))\, k$$

so

$$P_* = \{g\tilde{\theta}(g)^{-1} : g \in \operatorname{Int}(\mathfrak{u})\}.$$

Consequently, P_* is compact and closed in $\operatorname{Int}(\mathfrak{u})$.

Let A_* denote the closure of $\exp \mathfrak{a}_*$ in $\operatorname{Int}(\mathfrak{u})$. Then A_* is a torus, contained in P_*. Since $\tilde{\theta}(a) = a^{-1}$ for each $a \in A_*$, the Lie algebra of A_* is contained in $\mathfrak{p}_*$; being abelian this Lie algebra must coincide with $\mathfrak{a}_*$. Select $H \in \mathfrak{a}_*$ such that the one-parameter subgroup $\exp tH$ $(t \in \boldsymbol{R})$ is dense in A_*. Then the centralizer of H in $\mathfrak{p}_*$ is $\mathfrak{a}_*$.

In order to prove (iii) let X be an arbitrary element in $\mathfrak{p}_*$. The function $k \to B(H, k \cdot X)$ is a continuous function on the compact group K_*, and takes a minimum for $k = k_0$, say. If $T \in \mathfrak{k}_0$ we have therefore

$$\left\{\frac{d}{dt} B(H, (\exp tT)\, k_0 \cdot X)\right\}_{t=0} = 0.$$

This can be written

$$B(H, [T, k_0 \cdot X]) = 0.$$

Consequently, $B([k_0 \cdot X, H], T) = 0$ for all $T \in \mathfrak{k}_0$. Since $[k_0 \cdot X, H] \in \mathfrak{k}_0$ it follows that $[k_0 \cdot X, H] = 0$ and by (i), $k_0 \cdot X \in \mathfrak{a}_*$. This proves (iii).

Finally, using (iii) on $\mathfrak{a}_*'$, there exists a $k \in K^*$ such that $H \in k \cdot \mathfrak{a}_*'$. Each element in $k \cdot \mathfrak{a}_*'$ commutes with H; since $\mathfrak{a}_*$ is the centralizer of H in $\mathfrak{p}_*$ it follows that $k \cdot \mathfrak{a}_*' \subset \mathfrak{a}_*$. This finishes the proof of the lemma.

We shall now give two other applications of Lemma 6.3.

Theorem 6.4. *Let G be a connected, compact Lie group. Let $\mathfrak{g}$ denote the Lie algebra of G and let $\mathfrak{t}$ and $\mathfrak{t}'$ denote two maximal abelian subalgebras of $\mathfrak{g}$. Then*

(i) *There exists an element $H \in \mathfrak{t}$ whose centralizer in $\mathfrak{g}$ is $\mathfrak{t}$.*
(ii) *There exists an element $g \in G$ such that $Ad(g)\, \mathfrak{t} = \mathfrak{t}'$.*
(iii) $\mathfrak{g} = \bigcup_{g \in G} Ad(g)\, \mathfrak{t}$.

Proof. The group G can be written $G \times G/G^*$ where G^* is the diagonal in $G \times G$. If G is semisimple, the pair $(G \times G, G^*)$ is a Riemannian symmetric pair of the compact type with the involutive automorphism $\sigma : (x, y) \to (y, x)$ of $G \times G$. In this case Theorem 6.4 is a special instance of Lemma 6.3. If however, G is not semisimple a slight extension of Lemma 6.3 is necessary. In the proof of Lemma 6.3 the semisimplicity of $\mathfrak{k}_0 + \mathfrak{p}_*$ was never used and the form B could be replaced by any strictly negative definite bilinear form Q on $\mathfrak{u} \times \mathfrak{u}$ satisfying the invariance condition

$$Q(X, [Y, Z]) = Q([X, Y], Z)$$

for all $X, Y, Z \in \mathfrak{u}$. In the case of the Riemannian symmetric pair $(G \times G, G^*)$ such a bilinear form Q exists due to the compactness of G. Hence Theorem 6.4 holds for any connected compact Lie group G.

Theorem 6.5. *Let* $\mathfrak{g}$ *be a semisimple Lie algebra over* $\mathbf{C}$, $\mathfrak{h}_1$ *and* $\mathfrak{h}_2$ *two Cartan subalgebras of* $\mathfrak{g}$. *Then there exists an automorphism* σ *of* $\mathfrak{g}$ *such that* $\sigma\mathfrak{h}_1 = \mathfrak{h}_2$.

Proof. Each Cartan subalgebra $\mathfrak{h}$ of $\mathfrak{g}$ determines a compact real form $\mathfrak{u}$ of $\mathfrak{g}$ such that $i\mathfrak{h}_{\mathbf{R}} \subset \mathfrak{u}$ (Theorem 6.3, Chapter III). Let $\mathfrak{u}_1$ and $\mathfrak{u}_2$ denote two compact real forms of $\mathfrak{g}$ arising in this manner from $\mathfrak{h}_1$ and $\mathfrak{h}_2$, respectively. Then there exists an automorphism of $\mathfrak{g}$ carrying $\mathfrak{u}_1$ onto $\mathfrak{u}_2$. Hence we may assume $\mathfrak{u}_1 = \mathfrak{u}_2$ without loss of generality. The subspaces $i\mathfrak{h}_{1,\mathbf{R}}$ and $i\mathfrak{h}_{2,\mathbf{R}}$ are then maximal abelian subalgebras of $\mathfrak{u}_1$. By Theorem 6.4 these subalgebras are conjugate under an element $\sigma \in \mathrm{Int}(\mathfrak{u}_1)$. But σ extends uniquely to an automorphism of $\mathfrak{g}$ and $\mathfrak{h}_1$ and $\mathfrak{h}_2$ are conjugate under this automorphism.

Corollary 6.6. *Let* $\mathfrak{g}$ *be a semisimple Lie algebra over* $\mathbf{C}$, $\mathfrak{h} \subset \mathfrak{g}$ *a Cartan subalgebra, and* $X \in \mathfrak{g}$ *regular. Then* $\sigma X \in \mathfrak{h}$ *for some automorphism* σ *of* $\mathfrak{g}$.

In fact, X is by Theorem 3.1, Chapter III, contained in a Cartan subalgebra of $\mathfrak{g}$ which by Theorem 6.5 is conjugate to $\mathfrak{h}$.

We now write out the group-theoretic version of Theorem 6.2. Refinements of this result will be given later (Theorem 8.6, Chapter VII and Theorem 1.1, Chapter IX).

Theorem 6.7. *Let* $\mathfrak{g}_0$ *be a semisimple Lie algebra,* $\mathfrak{g}_0 = \mathfrak{k}_0 + \mathfrak{p}_0$ *a Cartan decomposition, and* $(\mathfrak{g}_0, \theta)$, $(\mathfrak{u}, \theta)$ *the corresponding dual orthogonal symmetric Lie algebras, where* $\mathfrak{u} = \mathfrak{k}_0 + i\mathfrak{p}_0$. *Let* (G, K_1) *and* (U, K_2) *be any Riemannian symmetric pairs associated with* $(\mathfrak{g}_0, \theta)$ *and* $(\mathfrak{u}, \theta)$, *respectively. Let* $\mathfrak{a}_0 \subset \mathfrak{p}_0$ *be any maximal abelian subspace and* $A_1 \subset G$, $A_2 \subset U$ *the analytic subgroups corresponding to* $\mathfrak{a}_0 \subset \mathfrak{g}_0$, $i\mathfrak{a}_0 \subset \mathfrak{u}$. *Then*

$$G = K_1 A_1 K_1 \qquad \textit{and} \qquad U = K_2 A_2 K_2.$$

Proof. If $(K_1)_0$ is the identity component of K_1 we have by the completeness, $G\,(K_1)_0 = \mathrm{Exp}(\mathfrak{p}_0)$. Using Lemma 6.3, we obtain for each $g \in G$ elements $H \in \mathfrak{p}_0$, $k \in K_1$ such that $g(K_1)_0 = \mathrm{Exp}\,\mathrm{Ad}(k)H$. Using now (4), §3, Chapter IV, this implies $g = k \exp Hk^{-1}k_1$ for some $k_1 \in K_1$, so the first relation above is proved. The same proof gives the second relation.

EXERCISES AND FURTHER RESULTS

1. If $X \in \mathfrak{so}(3)$ (that is, a 3×3 skew symmetric matrix) show that

$$e^X = I + \frac{\sin\rho}{\rho} X + \frac{1 - \cos\rho}{\rho^2} X^2,$$

where I is the identity matrix and $\rho^2 = -\frac{1}{2}\operatorname{Tr}(XX)$.

2. In §2 it was shown that $\mathfrak{so}(1, 3)$ is isomorphic to $\mathfrak{sl}(2, \boldsymbol{C})^{\boldsymbol{R}}$. Exhibit an explicit isomorphism.

3. Let $(\mathfrak{l}, s)$ be an orthogonal symmetric Lie algebra, $\mathfrak{l}$ semisimple. Then:

(i) $\mathfrak{u}$ equals its normalizer in $\mathfrak{l}$.

(ii) If $\mathfrak{u}$ contains no ideal in $\mathfrak{l}$, then $[\mathfrak{e}, \mathfrak{e}] = \mathfrak{u}$.

4. Let G be a connected Lie group which contains a compact subgroup of dimension $\geqslant 1$ but has center $\{e\}$. Show that G has a subgroup H such that (G, H) is a symmetric pair.

5. Let M be a Riemannian globally symmetric space, $I(M)$ its group of isometries. Prove that (i) M is of the compact type if and only if $I(M)$ is semisimple and compact; (ii) M is of the noncompact type if and only if the Lie algebra of $I(M)$ is semisimple and has no compact ideal $\neq \{0\}$.

6. Let M be a manifold with a pseudo-Riemannian structure g and curvature tensor R. Let $p \in M$, S a two-dimensional subspace in M_p such that g_p is nondegenerate on S. Let

$$K(S) = -\frac{g_p(R_p(X, Y)X, Y)}{g_p(X, X)\,g_p(Y, Y) - g_p(X, Y)^2},$$

X and Y being any linearly independent vectors in S.

Prove that the denominator is $\neq 0$, that $K(S)$ is independent of the choice of X and Y, and that it coincides with the sectional curvature in the case when g is positive definite.

7*. (The pseudo-Riemannian manifolds of constant sectional curvature) Consider the quadric Q_e in $\boldsymbol{R}^{p+q+1}$ given by

$$B_e(X) \equiv x_1^2 + \dots + x_p^2 - x_{p+1}^2 - x_{p+2}^2 - \dots + e x_{p+q+1}^2 = e \qquad (e = \pm 1).$$

with the pseudo-Riemannian structure g_e induced by B_e. Show that (cf. Helgason [3], [4]):

(i) $Q_{-1} \cong \boldsymbol{O}(p, q+1)/\boldsymbol{O}(p, q)$ (diffeomorphism), the pseudo-Riemannian structure g_{-1} has signature (p, q) and constant curvature -1.

(ii) $Q_{+1} \cong \boldsymbol{O}(p+1, q)/\boldsymbol{O}(p, q)$ (diffeomorphism), the pseudo-Riemannian structure g_+ has signature (p, q) and constant curvature $+1$.

(iii) Up to local isometry Q_{-1} and Q_{+1} exhaust the class of pseudo-Riemannian manifolds of constant curvature -1 and $+1$, respectively.

(iv) Q_{-1} and Q_{+1} are dual symmetric spaces (in the sense of a generalization of §2).

8. Interpret the relation $[\mathfrak{p}, \mathfrak{p}] = \mathfrak{k}$ (Theorem 4.1) for the sphere $\boldsymbol{SO}(n)/\boldsymbol{SO}(n-1) = \boldsymbol{S}^{n-1}$ in the following geometric fashion: Let $e_1, \ldots, e_n$ be the canonical basis of $\boldsymbol{R}^n$ and $\boldsymbol{SO}(n-1)$ the subgroup of $\boldsymbol{SO}(n)$ leaving e_n fixed. The rotations in the (e_i, e_n)-plane, that is, the group $\exp \boldsymbol{R}(E_{in} - E_{ni})$ induces the group of transvections along the geodesic $t \rightarrow \operatorname{Exp}_{e_n}(t(E_{in} - E_{ni}))$; we have

$$[E_{in} - E_{ni}, E_{jn} - E_{nj}] = E_{ji} - E_{ij} \qquad (1 \leqslant i < j < n)$$

and $\exp t(E_{ji} - E_{ij})$ is a rotation in the tangent space $(\boldsymbol{S}^{n-1})_{e_n}$.

NOTES

The results of §1-§4 are, for the most part, due to É. Cartan [16].

The conjugacy of maximal tori (or more precisely Theorem 6.4 (iii)) was first proved by Weyl [1], Kap. IV, Satz 1. The simple proof given here is due to Hunt [2] and this proof applies equally well to Lemma 6.3 first proved by É. Cartan in [10]. The conjugacy statement for Cartan subalgebras (Theorem 6.5) has been extended and sharpened by Chevalley [6], Chapter VI, §4, Théorème 4.

CHAPTER VI

SYMMETRIC SPACES OF THE NONCOMPACT TYPE

Having in the last chapter dealt with analogies and common properties of the two types of symmetric spaces we shall now study the two types separately and start with the noncompact type.

In §1 it is shown that for a given noncompact simple Lie algebra $\mathfrak{g}_0$ over $\boldsymbol{R}$ there exists a unique Riemannian globally symmetric space M of the noncompact type such that $I(M)$ has Lie algebra $\mathfrak{g}_0$. This M is diffeomorphic to a Euclidean space. Section 2 contains a proof of É. Cartan's conjugacy theorem for maximal compact subgroups. The relationship between $\mathfrak{g}_0$ and the geometry of M is further developed in the exercises. The subsequent sections deal with topics connected with the Iwasawa decomposition $G = KAN$ of a semisimple connected Lie group G into an (essentially) maximal compact subgroup K, an abelian group A, and a nilpotent group N.

§ 1. Decomposition of a Semisimple Lie Group

Let $\mathfrak{g}_0$ be a noncompact semisimple Lie algebra over $\boldsymbol{R}$ and let $\mathfrak{g}_0 = \mathfrak{k}_0 + \mathfrak{p}_0$ be a Cartan decomposition of $\mathfrak{g}_0$. The mapping $\theta : T + X \to T - X$ ($T \in \mathfrak{k}_0$, $X \in \mathfrak{p}_0$) is an involutive automorphism of $\mathfrak{g}_0$ and the pair $(\mathfrak{g}_0, \theta)$ is an orthogonal symmetric Lie algebra of the noncompact type. We recall that a pair (G, K) is said to be associated with $(\mathfrak{g}_0, \theta)$ if G is a connected Lie group with Lie algebra $\mathfrak{g}_0$ and K is a Lie subgroup of G with Lie algebra $\mathfrak{k}_0$. Such a pair is said to be of the noncompact type. This pair is said to be a Riemannian symmetric pair if K is closed, $\mathrm{Ad}_G(K)$ compact and there exists an analytic involutive automorphism $\tilde{\theta}$ of G such that $(K_{\tilde{\theta}})_0 \subset K \subset K_{\tilde{\theta}}$. Such a $\tilde{\theta}$ is necessarily unique and $d\tilde{\theta} = \theta$ (Prop. 3.5, Chapter IV). Finally, a Riemannian globally symmetric space M is said to be of the noncompact type if the pair $(I_0(M), H)$ is of the noncompact type, H being the isotropy subgroup of $I_0(M)$ at some point $o \in M$.

Theorem 1.1. *With the notation above, suppose* (G, K) *is any pair associated with* $(\mathfrak{g}_0, \theta)$. *Then:*

(i) *K is connected, closed, and contains the center Z of G. Moreover,*

K is compact if and only if Z is finite. In this case, K is a maximal compact subgroup of G.

(ii) *There exists an involutive, analytic automorphism θ of G whose fixed point set is K and whose differential at e is θ; the pair (G, K) is a Riemannian symmetric pair.*

(iii) *The mapping $\varphi : (X, k) \to (\exp X)k$ is a diffeomorphism of $\mathfrak{p}_0 \times K$ onto the group G and the mapping Exp is a diffeomorphism of $\mathfrak{p}_0$ onto the globally symmetric space G/K.*

Let Ad and ad denote the adjoint representations of G and $\mathfrak{g}_0$, respectively. Before starting on Theorem 1.1 we prove a simple lemma.

Lemma 1.2. *With respect to the positive definite symmetric bilinear form*

$$B_\theta(Y, Z) = -B(Y, \theta Z) \qquad (Y, Z \in \mathfrak{g}_0)$$

each ad X $(X \in \mathfrak{p}_0)$ *is symmetric and each* ad T $(T \in \mathfrak{k}_0)$ *is skew symmetric. Also* $\mathrm{Ad}(k)\theta = \theta\,\mathrm{Ad}(k)$ $(k \in K)$ *so* $\mathrm{Ad}(k)$ *is orthogonal.*

Proof. We have

$$\begin{aligned} B_\theta(\mathrm{ad}\, X(Y), Z) &= B([Y, X], \theta Z) \\ &= B(Y, [X, \theta Z]) \\ &= -B(Y, \theta[X, Z]) \\ &= B_\theta(Y, \mathrm{ad}\, X(Z)). \end{aligned}$$

The proof of the skew symmetry of ad T is similar.

Passing now to the proof of Theorem 1.1, let K_0 denote the identity component of K. Then owing to Prop. 3.6, Chapter IV, K_0 is closed in G, the coset space G/K_0 is Riemannian locally symmetric and $\pi(\exp X) = \mathrm{Exp}\, X$ for $X \in \mathfrak{p}_0$ if π denotes the natural mapping of G onto G/K_0. Thus G/K_0 is complete and consequently Exp maps $\mathfrak{p}_0$ onto G/K_0. It follows that φ maps $\mathfrak{p}_0 \times K_0$ onto G. To see that φ is one-to-one on $\mathfrak{p}_0 \times K$, suppose that $X_1, X_2 \in \mathfrak{p}_0$, $k_1, k_2 \in K$ such that

$$(\exp X_1)\, k_1 = (\exp X_2)\, k_2. \tag{1}$$

Applying Ad to this relation we obtain

$$e^{\mathrm{ad}\, X_1} \circ \mathrm{Ad}\,(k_1) = e^{\mathrm{ad}\, X_2} \circ \mathrm{Ad}\,(k_2). \tag{2}$$

The decomposition $m = po$ of a nonsingular matrix m, where p is symmetric and positive definite and o is orthogonal, is unique. Thus by (2)

$$e^{\operatorname{ad} X_1} = e^{\operatorname{ad} X_2},$$

$$\operatorname{Ad}(k_1) = \operatorname{Ad}(k_2).$$

The exponential mapping is one-to-one on the set of symmetric matrices; hence, ad $X_1 =$ ad X_2 and $X_1 = X_2$ since $\mathfrak{g}_0$ has center $\{0\}$. It follows from (1) that $k_1 = k_2$ so φ is one-to-one on $\mathfrak{p}_0 \times K$. Since we have proved that $\varphi(\mathfrak{p}_0 \times K_0) = \varphi(\mathfrak{p}_0 \times K)$, we conclude that $K_0 = K$. Now let as usual K^* denote the analytic subgroup of Int $(\mathfrak{g}_0)$ with Lie algebra $\mathfrak{k}_0$. Then the pair $(G, \operatorname{Ad}^{-1}(K^*))$ is associated with $(\mathfrak{g}_0, \theta)$; from what is already proved, the group $\operatorname{Ad}^{-1}(K^*)$ is connected. Having Lie algebra equal to $\mathfrak{k}_0$, it must coincide with K. Hence $Z \subset K$. Since $K^* = K/Z$ is compact, and Z is discrete (§2, Chapter II), it follows that K is compact if and only if Z is finite.

Let K_1 be a compact subgroup of G containing K. Then its Lie algebra is compactly imbedded in $\mathfrak{g}_0$ so by the maximality of $\mathfrak{k}_0$ (Prop. 7.4, Chapter III) K_1 and K have the same Lie algebra. By the proof above, $K = K_1$, so K is maximal.

The automorphism θ of $\mathfrak{g}_0$ induces an automorphism Θ of the universal covering group $\tilde{G}$ of G such that $d\Theta = \theta$. The fixed points of Θ form a subgroup $\tilde{K}$, which, by the above, must contain the center $\tilde{Z}$ of $\tilde{G}$. The kernel of the covering mapping of $\tilde{G}$ onto G is a discrete normal subgroup N of $\tilde{G}$ and must therefore belong to $\tilde{Z}$. The automorphism $\tilde{\theta}$ of $G = \tilde{G}/N$ induced by Θ then turns (G, K) into a Riemannian symmetric pair. Using (i) we see that the fixed point set of $\tilde{\theta}$ coincides with K.

Since the mapping φ is one-to-one, the mapping Exp is a one-to-one differentiable mapping of $\mathfrak{p}_0$ onto G/K. In order to prove that it is regular (and thus a diffeomorphism) it suffices, due to the formula for $d\,\mathrm{Exp}_X$ (Theorem 4.1, Chapter IV), to prove that

$$\det\left(\sum_0^\infty \frac{(T_X)^n}{(2n+1)!}\right) \neq 0 \qquad \text{for } X \in \mathfrak{p}_0. \tag{3}$$

The relations

$$B(T_X Y, Z) = B(Y, T_X Z),$$

$$B(T_X Y, Y) = -B([X, Y], [X, Y]) \geqslant 0,$$

valid for $X, Y, Z \in \mathfrak{p}_0$, show that T_X is symmetric and positive definite with respect to B. The validity of (3) is therefore obvious.

It remains only to prove that φ is everywhere regular. A general tangent vector to $\mathfrak{p}_0 \times K$ at the point (X, k) has the form $(Y, dL_k \cdot T)$ where $Y \in \mathfrak{p}_0$ and $T \in \mathfrak{k}_0$, L_x denoting left translation by the group element x.

Since

$$\varphi(X + tY, k) = \exp(X + tY)\,k = k \exp(\mathrm{Ad}(k^{-1})(X + tY))$$

and

$$\varphi(X, k \exp tT) = (\exp X)\,k(\exp tT)$$

it follows from Theorem 1.7, Chapter II, that

$$d\varphi_{(X,k)}(Y, dL_k \cdot T) = dL_{(\exp X)k} \cdot \Big(\frac{1 - e^{-\mathrm{ad}\,X'}}{\mathrm{ad}\,X'}(Y') + T\Big), \tag{4}$$

where $X' = \mathrm{Ad}(k^{-1})\,X$, $Y' = \mathrm{Ad}(k^{-1})\,Y$. The $\mathfrak{p}_0$-component of the vector $(1 - e^{-\mathrm{ad}\,X'})\,(\mathrm{ad}\,X')^{-1}\,(Y')$ is

$$\sum_0^\infty \frac{(T_{X'})^n}{(2n+1)!}(Y'),$$

which, due to (3), vanishes only if $Y' = 0$. Consequently the right-hand side of (4) is $\neq 0$ unless $T = Y = 0$. This shows that φ is regular and completes the proof of Theorem 1.1.

Corollary 1.3. *Let M and M' be two Riemannian globally symmetric spaces of the noncompact type such that the groups $I(M)$ and $I(M')$ have the same Lie algebra $\mathfrak{g}_0$. As usual,* (Chapter V, §5), *suppose the Riemannian structures on M and M' arise from the Killing form of $\mathfrak{g}_0$. Then M and M' are isometric.*

In fact, the spaces M and M' arise from two Cartan decompositions of $\mathfrak{g}_0$. If these Cartan decompositions coincide, the corollary follows from Theorem 1.1. Now, any two Cartan decompositions of $\mathfrak{g}_0$ are conjugate under an inner automorphism σ of $\mathfrak{g}_0$. It is easy to set up an isometry between M and M' by means of σ.

If the Lie algebra $\mathfrak{g}_0$ is simple then by the irreducibility (Chapter VIII, §5) the space $M = \mathrm{Int}(\mathfrak{g}_0)/K^*$ is, except for a multiplication of the metric by a constant factor, the unique Riemannian globally symmetric space for which $I(M)$ has Lie algebra $\mathfrak{g}_0$. For compact semisimple Lie algebras the situation is quite different as we shall see in the next chapter.

Let $\mathfrak{s}_0$ be a Lie triple system contained in $\mathfrak{p}_0$. Then, according to Theorem 7.2, Chapter IV, and Theorem 1.1, $\mathrm{Exp}\,\mathfrak{s}_0$ is a closed totally geodesic submanifold of G/K. We can therefore apply Theorem 14.6, Chapter I, and get the following extension of Theorem 1.1.

Theorem 1.4. *In the notation of Theorem* 1.1, *let* $\mathfrak{s}_0$ *be a Lie triple system contained in* $\mathfrak{p}_0$ *and let* $\mathfrak{t}_0$ *denote the orthogonal complement of* $\mathfrak{s}_0$ *in* $\mathfrak{p}_0$. *Let* $S_0 = \exp \mathfrak{s}_0$ *and* $T_0 = \exp \mathfrak{t}_0$ *be given the relative topology of* G. *Then* G *decomposes topologically*

$$G = S_0 \cdot T_0 \cdot K.$$

Proof. In view of Theorem 14.6, Chapter I, the mapping $(X, Y) \to \tau(\exp X) \cdot \operatorname{Exp} Y$, $(X \in \mathfrak{s}_0, Y \in \mathfrak{t}_0)$, is a continuous one-to-one mapping of $\mathfrak{s}_0 \times \mathfrak{t}_0$ onto G/K. Each bounded set in M is the image of a bounded set in $\mathfrak{s}_0 \times \mathfrak{t}_0$ (Cor. 13.2 (i), Chapter I). Hence the mapping is a homeomorphism. If we state this fact in terms of G and make use of Theorem 1.1, the present theorem follows.

§ 2. Maximal Compact Subgroups and Their Conjugacy

The symbols $\mathfrak{g}_0$, $\mathfrak{k}_0$, $\mathfrak{p}_0$, G, and K have the same meaning as in §1.

It was proved in Chapter III that all Cartan decompositions of a semisimple Lie algebra over $\boldsymbol{R}$ are conjugate under an inner automorphism. Using differential geometric results, we shall now prove a stronger theorem, namely, that all maximal compactly imbedded subalgebras of a semisimple Lie algebra over $\boldsymbol{R}$ are conjugate under an inner automorphism. Consequently, each maximal compactly imbedded subalgebra $\mathfrak{u}$ of a semisimple Lie algebra $\mathfrak{l}$ is a part of a Cartan decomposition and its orthogonal complement $\mathfrak{e}$ with respect to the Killing form of $\mathfrak{l}$ satisfies $[\mathfrak{e}, \mathfrak{e}] \subset \mathfrak{u}$.

Theorem 2.1. *Let* (G, K) *be a Riemannian symmetric pair of the noncompact type. Let* K_1 *be any compact subgroup of* G. *Then there exists an element* $x \in G$ *such that* $x^{-1} K_1 x \subset K$.

Proof. The relation $x^{-1}K_1x \subset K$ means that xK is a fixed point under the action of the group K_1 on the coset space G/K. Since the space G/K is a simply connected Riemannian manifold of negative curvature, the existence of the fixed point is assured by Theorem 13.5, Chapter I.

Theorem 2.2.

(i) *Let* (G, K) *be a Riemannian symmetric pair of the noncompact type. Then* K *has a unique maximal compact subgroup* K' *and this group is maximal compact in* G.

(ii) *All maximal compact subgroups of a connected semisimple Lie group* G *are connected and conjugate under an inner automorphism of* G.

(iii) *Let K' be any maximal compact subgroup of a connected semisimple Lie group G. Then there exists a submanifold E of G, diffeomorphic to a Euclidean space such that the mapping $(e, k) \to ek$ is a diffeomorphism of $E \times K'$ onto G.*

Proof. The group $\mathrm{Ad}_G(K)$ is compact and has Lie algebra $\mathfrak{k}_0$. Thus $\mathfrak{k}_0$ is a compact Lie algebra. According to Prop. 6.6, Chapter II, $\mathfrak{k}_0$ can be written as a direct sum $\mathfrak{k}_0 = \mathfrak{k}_s + \mathfrak{k}_a$ where the ideals $\mathfrak{k}_s$ and $\mathfrak{k}_a$ are semisimple and abelian, respectively. Let K_s and K_a denote the corresponding analytic subgroups of K. The group K_a is a direct product $K_a = T \times V$ of analytic subgroups T, V of G, where T is a torus and V is analytically isomorphic to Euclidean space. Now we put

$$K' = K_s T = \{kt : k \in K_s, t \in T\}.$$

As a result of Theorem 6.9, Chapter II, the group K_s, and therefore the group K', is compact. The groups K' and V commute elementwise; the group $K' \cap V$ is a compact subgroup of the Euclidean group V, hence $K' \cap V = \{e\}$. It follows that $K = K' \times V$ (direct product) and K' is the unique maximal compact subgroup of K. Combining this with Theorem 2.1, we see that each compact subgroup of G is conjugate to a subgroup of K'. This proves (i) and (ii).

Consider now the set

$$E = \{(\exp X)\, v : X \in \mathfrak{p}_0, v \in V\}.$$

Since $\mathfrak{p}_0 \times V$ is a submanifold of $\mathfrak{p}_0 \times K$, it follows from Theorem 1.1, that E is a submanifold of G, diffeomorphic to Euclidean space. Finally, the mappings

$$((\exp X)\, v, k) \to (\exp X, vk) \to \exp X\, kv,$$

$$E \times K' \to (\exp \mathfrak{p}_0) \times K \to G$$

yield the desired diffeomorphism of $E \times K'$ onto G.

§ 3. The Iwasawa Decomposition

We now combine the Cartan decomposition of a semisimple Lie algebra and the root space decomposition of its complexification. Among other things, this gives the so-called Iwasawa decomposition for which a more direct proof is given at the end of the section.

Lemma 3.1. *Let $\mathfrak{g}$ be a complex semisimple Lie algebra, $\mathfrak{g}_k$ any compact real form of $\mathfrak{g}$, and let η denote the conjugation of $\mathfrak{g}$ with respect to $\mathfrak{g}_k$. Let*

$\mathfrak{h}$ be a Cartan subalgebra of $\mathfrak{g}$ which is invariant under η. Let Δ denote the set of nonzero roots of $\mathfrak{g}$ with respect to $\mathfrak{h}$ and let $\mathfrak{h}_{\boldsymbol{R}} = \sum_{\alpha \in \Delta} \boldsymbol{R} H_\alpha$. Then:

(i) $\mathfrak{h}_{\boldsymbol{R}} \subset i\mathfrak{g}_k$.

(ii) *There exists a vector $E_\alpha \in \mathfrak{g}^\alpha$ such that for all $\alpha \in \Delta$*

$$(E_\alpha - E_{-\alpha}), \qquad i(E_\alpha + E_{-\alpha}) \in \mathfrak{g}_k,$$

$$[E_\alpha, E_{-\alpha}] = (2/\alpha(H_\alpha))\, H_\alpha.$$

Proof. The Killing form B of $\mathfrak{g}$ is strictly negative definite on $\mathfrak{g}_k \times \mathfrak{g}_k$, and moreover

$$B(X, \eta X) < 0 \qquad \text{for } X \neq 0 \text{ in } \mathfrak{g}.$$

For each $\alpha \in \Delta$ we can define the complex linear function α^η on $\mathfrak{h}$ by

$$\alpha^\eta(H) = \overline{(\alpha(\eta \cdot H))}, \qquad H \in \mathfrak{h},$$

the bar denoting complex conjugation. Then if $Y \in \mathfrak{g}^\alpha$, $H \in \mathfrak{h}$, we find

$$[\eta H, \eta Y] = \eta[H, Y] = \overline{\alpha(h)}\, \eta Y = \alpha^\eta(\eta H)\, \eta Y,$$

which shows that $\alpha^\eta \in \Delta$. Also, if $H \in \mathfrak{h}$ we have

$$B(\eta H_\alpha, H) = \overline{B(H_\alpha, \eta H)} = \overline{\alpha(\eta H)} = \alpha^\eta(H) = B(H_{\alpha^\eta}, H)$$

so

$$\eta H_\alpha - H_{\alpha^\eta}.$$

This shows that $\mathfrak{h}_{\boldsymbol{R}}$ is invariant under η. Since $H = \frac{1}{2}(H + \eta H) + \frac{1}{2}(H - \eta H)$ we have the direct decomposition $\mathfrak{h}_{\boldsymbol{R}} = \mathfrak{h}^+ + \mathfrak{h}^-$ where $\eta(H) = \pm H$ for $H \in \mathfrak{h}^\pm$. Here $\mathfrak{h}^+ = \{0\}$; in fact if $H \neq 0$ in $\mathfrak{h}^+$ then $B(H, H) = B(H, \eta H) < 0$ which contradicts the fact that B is strictly positive definite on $\mathfrak{h}_{\boldsymbol{R}} \times \mathfrak{h}_{\boldsymbol{R}}$ (Theorem 4.4, Chapter III). This proves (i) and $\alpha^\eta = -\alpha$.

To prove (ii) we turn the dual space of $\mathfrak{h}_{\boldsymbol{R}}$ into an ordered vector space. Since each $\alpha \in \Delta$ is real valued on $\mathfrak{h}_{\boldsymbol{R}}$, Δ has now become an ordered set. Let Δ^+ denote the set of positive roots. Since $\alpha(H_\alpha) > 0$ we can to each $\alpha \in \Delta^+$ select $E_\alpha \in \mathfrak{g}^\alpha$ such that

$$B(E_\alpha, \eta E_\alpha) = -\frac{2}{\alpha(H_\alpha)}.$$

Since $\alpha^\eta = -\alpha$ it is easy to see that $\eta E_\alpha \in \mathfrak{g}^{-\alpha}$ for $\alpha \in \Delta^+$. We put $E_{-\alpha} = -\eta E_\alpha$ for $\alpha \in \Delta^+$. Then $\eta E_\alpha = -E_{-\alpha}$ for all $\alpha \in \Delta$ and consequently the elements $(E_\alpha - E_{-\alpha})$ and $i(E_\alpha + E_{-\alpha})$ belong to $\mathfrak{g}_k$ for each $\alpha \in \Delta$. Finally

$$B(H, [E_\alpha, E_{-\alpha}]) = B([H, E_\alpha], E_{-\alpha}) = \frac{2\alpha(H)}{\alpha(H_\alpha)}$$

and since $[E_\alpha, E_{-\alpha}]$ is a scalar multiple of H_α (Theorem 4.2, Chapter III), it follows that $[E_\alpha, E_{-\alpha}] = (2/\alpha(H_\alpha))\, H_\alpha$.

Suppose now $\mathfrak{g}_0$ is a semisimple Lie algebra over $\boldsymbol{R}$ and that $\mathfrak{g}_0 = \mathfrak{k}_0 + \mathfrak{p}_0$ is a Cartan decomposition of $\mathfrak{g}_0$ as usual. Let $\mathfrak{g}$ be the complexification of $\mathfrak{g}_0$, put $\mathfrak{u} = \mathfrak{k}_0 + i\mathfrak{p}_0$ and let σ and τ denote the conjugations of $\mathfrak{g}$ with respect to $\mathfrak{g}_0$ and $\mathfrak{u}$, respectively. The automorphism $\sigma\tau$ of $\mathfrak{g}$ will be denoted by θ. Let ad denote the adjoint representation of $\mathfrak{g}$.

Let $\mathfrak{h}_{\mathfrak{p}_0}$ denote any maximal abelian subspace of $\mathfrak{p}_0$. Let $\mathfrak{h}_0$ be any maximal abelian subalgebra of $\mathfrak{g}_0$ containing $\mathfrak{h}_{\mathfrak{p}_0}$. The existence of $\mathfrak{h}_0$ is obvious from Zorn's lemma. If $X \in \mathfrak{h}_0$ and $Y \in \mathfrak{h}_{\mathfrak{p}_0}$ we have

$$[X - \theta X, Y] = [X, Y] - \theta[X, \theta Y] = [X, Y] + \theta[X, Y] = 0 + 0.$$

Since $X - \theta X \in \mathfrak{p}_0$ it follows, in view of the maximality of $\mathfrak{h}_{\mathfrak{p}_0}$, that $X - \theta X \in \mathfrak{h}_{\mathfrak{p}_0}$. Thus $\theta\mathfrak{h}_0 \subset \mathfrak{h}_0$ so we have the direct decomposition $\mathfrak{h}_0 = \mathfrak{h}_0 \cap \mathfrak{k}_0 + \mathfrak{h}_0 \cap \mathfrak{p}_0$. Obviously $\mathfrak{h}_{\mathfrak{p}_0} = \mathfrak{h}_0 \cap \mathfrak{p}_0$. We put $\mathfrak{h}_{\mathfrak{k}_0} = \mathfrak{h}_0 \cap \mathfrak{k}_0$. Let $\mathfrak{h}$, $\mathfrak{h}_\mathfrak{p}$, $\mathfrak{h}_\mathfrak{k}$, $\mathfrak{k}$, and $\mathfrak{p}$ denote the subspaces of $\mathfrak{g}$ generated by $\mathfrak{h}_0$, $\mathfrak{h}_{\mathfrak{p}_0}$, $\mathfrak{h}_{\mathfrak{k}_0}$, $\mathfrak{k}_0$ and $\mathfrak{p}_0$, respectively.

Lemma 3.2. *The algebra* $\mathfrak{h}$ *is a Cartan subalgebra of* $\mathfrak{g}$ *and* $\mathfrak{h}_{\boldsymbol{R}} = \mathfrak{h}_{\mathfrak{p}_0} + i\mathfrak{h}_{\mathfrak{k}_0}$.

It is obvious that $\mathfrak{h}$ is a maximal abelian subalgebra of $\mathfrak{g}$. Now, the Hermitian form $B_\tau(X, Y) = -B(X, \tau Y)$ on $\mathfrak{g} \times \mathfrak{g}$ is strictly positive definite and if $Z \in \mathfrak{u}$ we have

$$B_\tau([Z, X], Y) + B_\tau(X, [Z, Y]) = 0.$$

If ad Z leaves a subspace V of $\mathfrak{g}$ invariant then the orthogonal complement $V^\perp$ (with respect to B_τ) is also invariant and $\mathfrak{g} = V + V^\perp$ (direct sum). Hence ad Z is semisimple. Thus ad H is semisimple if $H \in \mathfrak{h}_\mathfrak{k} \cup \mathfrak{h}_\mathfrak{p}$. Since ad H_1 and ad H_2 commute if $H_1 \in \mathfrak{h}_\mathfrak{k}$, $H_2 \in \mathfrak{h}_\mathfrak{p}$, it follows that ad $(H_1 + H_2)$ is semisimple and $\mathfrak{h}$ is a Cartan subalgebra.

As a result of its definition, $\mathfrak{h}$ is invariant under σ and θ. Thus it is also invariant under τ. By Lemma 3.1 we have $\mathfrak{h}_{\boldsymbol{R}} \subset \mathfrak{h} \cap (i\mathfrak{u})$. But $\theta\mathfrak{h} \subset \mathfrak{h}$ implies $\mathfrak{h} \cap (i\mathfrak{u}) = \mathfrak{h} \cap i\mathfrak{k}_0 + \mathfrak{h} \cap \mathfrak{p}_0 = i\mathfrak{h}_{\mathfrak{k}_0} + \mathfrak{h}_{\mathfrak{p}_0}$. Since $\dim \mathfrak{h}_{\boldsymbol{R}} = \dim \mathfrak{h}_0$, the lemma follows.

Let V be a finite-dimensional vector space over $\boldsymbol{R}$, W a subspace of V. Let $V^\wedge$ and $W^\wedge$ denote their duals and suppose that $V^\wedge$ and $W^\wedge$

have been turned into ordered vector spaces. The orderings are said to be *compatible* (Harish-Chandra [8], p. 195) if $\lambda \in V^\wedge$ is positive whenever its restriction $\bar{\lambda}$ to W is positive. Compatible orderings can for example be constructed as follows: Let $X_1, \ldots, X_n$ be a basis of V such that $X_1, \ldots, X_m$ is a basis of W. Then the lexicographic orderings of $W^\wedge$ and $V^\wedge$ with respect to these bases are compatible.

Now select compatible orderings in the dual spaces of $\mathfrak{h}_{\mathfrak{p}_0}$ and $\mathfrak{h}_{\boldsymbol{R}}$, respectively. Since each root $\alpha \in \Delta$ is real valued on $\mathfrak{h}_{\boldsymbol{R}}$ we get in this way an ordering of Δ. Let Δ^+ denote the set of positive roots. Now for each $\alpha \in \Delta$ the linear functions α^τ, α^σ, and α^θ defined by

$$\alpha^\tau(H) = \overline{\alpha(\tau H)}, \quad \alpha^\sigma(H) = \overline{\alpha(\sigma H)}, \quad \alpha^\theta(H) = \alpha(\theta H) \qquad (H \in \mathfrak{h}),$$

are again members of Δ. The root α vanishes identically on $\mathfrak{h}_{\mathfrak{p}_0}$ if and only if $\alpha = \alpha^\theta$. Let $\Delta_{\mathfrak{p}}$ denote the set of roots which do not vanish identically on $\mathfrak{h}_{\mathfrak{p}_0}$. We divide the positive roots in two classes as follows:

$$P_+ = \{\alpha : \alpha \in \Delta^+, \alpha \neq \alpha^\theta\}, \qquad P_- = \{\alpha : \alpha \in \Delta^+, \alpha = \alpha^\theta\}.$$

Lemma 3.3.

(i) *If* $\alpha \in P_+$, *then* $-\alpha^\theta \in P_+$, $\alpha^\sigma \in P_+$, $\alpha^\tau = -\alpha$.

(ii) *If* $\beta \in P_-$, *then* $\beta^\theta = \beta$, $\beta^\sigma = -\beta$, $\beta^\tau = -\beta$, *and* $\mathfrak{g}^\beta + \mathfrak{g}^{-\beta} \in \mathfrak{k}$.

Proof. The restrictions to $\mathfrak{h}_{\mathfrak{p}_0}$ of α and α^θ have sum 0. By the compatibility of the orderings, we have $\alpha^\theta < 0$. Since α^σ and α agree on $\mathfrak{h}_{\mathfrak{p}_0}$ we have $\alpha^\sigma \in P_+$. The relations $\alpha^\tau = -\alpha$, $\beta^\tau = -\beta$ were established during the proof of Lemma 3.1. The relation $\beta^\theta = \beta$ implies $H_\beta \in \mathfrak{k} \cap \mathfrak{h}^* = i\mathfrak{h}_{\mathfrak{k}_0}$ so $\beta^\sigma = -\beta$. Since $\theta\mathfrak{g}^\beta = \mathfrak{g}^\beta$, $\theta^2 = 1$ and $\dim(\mathfrak{g}^\beta) = 1$, it is clear that $\theta Z = -Z$ or $\theta Z = Z$ for each $Z \in \mathfrak{g}^\beta$. If $\theta Z = -Z$, then $Z \in \mathfrak{p}$. For $H \in \mathfrak{h}_{\mathfrak{p}}$ we have $[H, Z] = \beta(H)Z = 0$ and since $\mathfrak{h}_{\mathfrak{p}}$ is a maximal abelian subspace of $\mathfrak{p}$ it follows that $Z = 0$ and $\mathfrak{g}^\beta \in \mathfrak{k}$. Similarly $\mathfrak{g}^{-\beta} \in \mathfrak{k}$.

Theorem 3.4. *Let* $\mathfrak{n} = \sum_{\alpha \in P_+} \mathfrak{g}^\alpha$, $\mathfrak{n}_0 = \mathfrak{g}_0 \cap \mathfrak{n}$, $\mathfrak{s}_0 = \mathfrak{h}_{\mathfrak{p}_0} + \mathfrak{n}_0$. *Then* $\mathfrak{n}$ *and* $\mathfrak{n}_0$ *are nilpotent Lie algebras,* $\mathfrak{s}_0$ *is a solvable Lie algebra, and*

$$\mathfrak{g}_0 = \mathfrak{k}_0 + \mathfrak{h}_{\mathfrak{p}_0} + \mathfrak{n}_0 \quad (\textit{direct vector space sum}).$$

Proof. Let $\alpha, \beta \in P_+$. If $\alpha + \beta \in \Delta$, then $\alpha + \beta \in P_+$ and $\mathfrak{n}$ is a subalgebra of $\mathfrak{g}$ which obviously is nilpotent. Hence $\mathfrak{n}_0$ is a nilpotent subalgebra of $\mathfrak{g}_0$. From the relation $[\mathfrak{n}_0 + \mathfrak{h}_{\mathfrak{p}_0}, \mathfrak{n}_0 + \mathfrak{h}_{\mathfrak{p}_0}] \subset \mathfrak{n}_0$ we see that $\mathfrak{h}_{\mathfrak{p}_0} + \mathfrak{n}_0$ is a solvable Lie algebra. To see that the sum $\mathfrak{k}_0 + \mathfrak{h}_{\mathfrak{p}_0} + \mathfrak{n}_0$ is direct, suppose $T \in \mathfrak{k}_0$, $H \in \mathfrak{h}_{\mathfrak{p}_0}$ and $X \in \mathfrak{n}_0$ such that $T + H + X = 0$.

Applying θ we find that $T - H + \theta X = 0$ so $2H + X - \theta X = 0$. Now by Lemma 3.3

$$\theta X \in \sum_{\alpha \in P_+} \mathfrak{g}^{-\alpha}$$

and since the sum $\mathfrak{h} + \sum_{\alpha \in \Delta} \mathfrak{g}^{\alpha} = \mathfrak{g}$ is direct we conclude that $H = 0$ and $X - \theta X = 0$. But $\mathfrak{k}_0 \cap \mathfrak{n}_0 = \{0\}$ so $X = T = 0$.

Now let $X \in \mathfrak{g}_0$. Since $X = \frac{1}{2}(X + \sigma X)$ it follows that X can be written

$$X = H + \sum_{\alpha \in \Delta} (X_\alpha + \sigma X_\alpha),$$

where $H \in \mathfrak{h}_0$, $X_\alpha \in \mathfrak{g}^\alpha$ for each $\alpha \in \Delta$. If α or $-\alpha$ belongs to P_-, then $X_\alpha + \sigma X_\alpha \in \mathfrak{k}_0$ due to Lemma 3.3. If $\alpha \in P_+$ then $X_\alpha + \sigma X_\alpha \in \mathfrak{n}_0$ by the same lemma. Finally, if $-\alpha \in P_+$ then $\tau(X_\alpha + \sigma X_\alpha) \in \mathfrak{g}^{-\alpha} + \mathfrak{g}^{\alpha^\theta} \subset \mathfrak{n}$ by Lemma 3.3. Consequently

$$X_\alpha + \sigma X_\alpha = \{(X_\alpha + \sigma X_\alpha) + \tau(X_\alpha + \sigma X_\alpha)\} - \tau(X_\alpha + \sigma X_\alpha) \in \mathfrak{u} \cap \mathfrak{g}_0 + \mathfrak{n} \cap \mathfrak{g}_0$$

so $X_\alpha + \sigma X_\alpha \in \mathfrak{k}_0 + \mathfrak{n}_0$. This proves the theorem.

Lemma 3.5. *There exists a basis* (X_i) *of* $\mathfrak{g}$ *such that the matrices representing ad* $(\mathfrak{g})$ *have the following properties*:

(i) *The matrices ad* $\mathfrak{u}$ *are skew Hermitian.*

(ii) The *matrices ad* $\mathfrak{n}$ *are lower triangular with zeros in the diagonal.*

(iii) *The matrices ad* $\mathfrak{h}_{\mathfrak{p}_0}$ *are diagonal matrices with a real diagonal.*

Proof. Let $\alpha_1 < \alpha_2 < \ldots$ be the roots in Δ^+ in increasing order. Let $H_1, \ldots, H_l$ be any basis of $\mathfrak{h}_R$, orthonormal with respect to B_τ. Select $E_{\alpha_i} \in \mathfrak{g}^{\alpha_i}$ such that $B_\tau(E_{\alpha_i}, E_{\alpha_i}) = 1$ $(i = 1, 2, \ldots)$. Since $\tau E_{\alpha_i} \in \mathfrak{g}^{-\alpha_i}$, the vectors $\ldots \tau E_{\alpha_2}, \tau E_{\alpha_1}, H_1, \ldots, H_l, E_{\alpha_1}, E_{\alpha_2}, \ldots$ form an orthonormal basis of $\mathfrak{g}$. This basis has the properties (i), (ii), and (iii).

Let $\mathfrak{m}_0$ denote the centralizer of $\mathfrak{h}_{\mathfrak{p}_0}$ in $\mathfrak{k}_0$. Let $\mathfrak{l}_0$ and $\mathfrak{q}_0$, respectively, denote the orthogonal complements of $\mathfrak{m}_0$ in $\mathfrak{k}_0$ and of $\mathfrak{h}_{\mathfrak{p}_0}$ in $\mathfrak{p}_0$. Here "orthogonal" refers to the positive definite form B_τ. Let $\mathfrak{m}$, $\mathfrak{l}$, and $\mathfrak{q}$ denote the subspaces of $\mathfrak{g}$ generated by $\mathfrak{m}_0$, $\mathfrak{l}_0$, and $\mathfrak{q}_0$, respectively.

Lemma 3.6. *The direct decompositions*

$$\mathfrak{g}_0 = \mathfrak{l}_0 + \mathfrak{m}_0 + \mathfrak{h}_{\mathfrak{p}_0} + \mathfrak{q}_0,$$

$$\mathfrak{g} = \mathfrak{l} + \mathfrak{m} + \mathfrak{h}_{\mathfrak{p}} + \mathfrak{q}$$

are orthogonal with respect to B_τ *and invariant under* θ. *Moreover, if* $X_\alpha \neq 0$ *is arbitrary in* $\mathfrak{g}^\alpha$ ($\alpha \in \Delta$),

$$\mathfrak{m} = \mathfrak{h}_{\mathfrak{k}} + \sum_{\alpha \in P_-} (\mathfrak{g}^\alpha + \mathfrak{g}^{-\alpha}),$$

$$\mathfrak{l} = \sum_{\alpha \in P_+} C(X_\alpha + \theta X_\alpha),$$

$$\mathfrak{q} = \sum_{\alpha \in P_+} C(X_\alpha - \theta X_\alpha).$$

Proof. Each $X \in \mathfrak{g}$ can be written

$$X = H^* + \sum_{\alpha \in \Delta} c_\alpha X_\alpha \qquad (H^* \in \mathfrak{h},\ c_\alpha \in \boldsymbol{C}).$$

Hence

$$[H, X] = \sum_{\alpha \in \Delta} c_\alpha \alpha(H)\, X_\alpha \tag{1}$$

for each $H \in \mathfrak{h}$. Thus X commutes with $H \in \mathfrak{h}$ if and only if $c_\alpha \alpha(H) = 0$ for all $\alpha \in \Delta$. It follows that $\mathfrak{h} + \sum_{\alpha \in P_-} (\mathfrak{g}^\alpha + \mathfrak{g}^{-\alpha})$ is the centralizer of $\mathfrak{h}_{\mathfrak{p}}$ in $\mathfrak{g}$. Since $\mathfrak{m}$ is the centralizer of $\mathfrak{h}_{\mathfrak{p}}$ in $\mathfrak{k}$, the expression for $\mathfrak{m}$ follows.

To prove the formula for $\mathfrak{l}$ let $H \in \mathfrak{h}_{\mathfrak{k}}$, $\alpha \in P_+$, and $\beta \in P_-$ or $-\beta \in P_-$. Then, using Theorem 4.2, Chapter III,

$$-B_\tau(X_\alpha + \theta X_\alpha, H) = B(X_\alpha, \tau H) + B(\theta X_\alpha, \tau H) = 0,$$

$$-B_\tau(X_\alpha + \theta X_\alpha, X_\beta) = B(X_\alpha, \tau X_\beta) + B(\theta X_\alpha, \tau X_\beta) = 0,$$

since $\tau X_\beta \in \mathfrak{g}^{-\beta}$, $\theta X_\alpha \in \mathfrak{g}^{\alpha^\theta}$, and $\alpha^\theta + \beta \neq 0$. This proves that

$$\sum_{\alpha \in P_+} C(X_\alpha + \theta X_\alpha) \subset \mathfrak{l}$$

and the inclusion

$$\sum_{\alpha \in P_+} C(X_\alpha - \theta X_\alpha) \subset \mathfrak{q}$$

is proved in exactly the same way. Now let α be a fixed element in P_+ and let $c \in \boldsymbol{C}$ be determined by $\theta X_{-\alpha} = c X_{-\alpha^\theta}$. Then

$$X_\alpha = \tfrac{1}{2}(X_\alpha + \theta X_\alpha) + \tfrac{1}{2}(X_\alpha - \theta X_\alpha) \in \mathfrak{l} + \mathfrak{q};$$

$$X_{-\alpha} = \tfrac{1}{2}(\theta X_{-\alpha} + \theta(\theta X_{-\alpha})) + \tfrac{1}{2}(\theta(\theta X_{-\alpha}) - \theta X_{-\alpha})$$

$$= \tfrac{1}{2} c(X_{-\alpha^\theta} + \theta X_{-\alpha^\theta}) - \tfrac{1}{2} c(X_{-\alpha^\theta} - \theta X_{-\alpha^\theta}) \in \mathfrak{l} + \mathfrak{q}.$$

Consequently the element X above belongs to $\mathfrak{m} + \mathfrak{h}_{\mathfrak{p}} + \sum_{\alpha \in P_+} C(X_\alpha + \theta X_\alpha) + \sum_{\alpha \in P_+} C(X_\alpha - \theta X_\alpha)$ and the lemma is proved.

The following corollary is an immediate consequence of (1).

Corollary 3.7. *If* $H \in \mathfrak{h}_{\mathfrak{p}}$ *and* $\alpha(H) \neq 0$ *for all* $\alpha \in P_+$, *then the centralizer of* H *in* $\mathfrak{g}$ *is* $\mathfrak{h}_{\mathfrak{p}} + \mathfrak{m}$.

We now give a more direct proof of Theorem 3.4 without invoking the complexifications. Let $\mathfrak{g}_0$ be a semisimple Lie algebra over $\boldsymbol{R}$, θ a Cartan involution, and $\mathfrak{g}_0 = \mathfrak{k}_0 + \mathfrak{p}_0$ the corresponding Cartan decomposition. Let $\mathfrak{h}_{\mathfrak{p}_0} \subset \mathfrak{p}_0$ be any maximal abelian subspace. Because of Lemma 1.2, each ad X ($X \in \mathfrak{p}_0$) can be diagonalized by means of a basis of $\mathfrak{g}_0$, so the commutative family ad H ($H \in \mathfrak{h}_{\mathfrak{p}_0}$) can be simultaneously diagonalized. For each real linear form λ on $\mathfrak{h}_{\mathfrak{p}_0}$ let

$$\mathfrak{g}_{0,\lambda} = \{X \in \mathfrak{g}_0 : [H, X] = \lambda(H)X \qquad \text{for all} \quad H \in \mathfrak{h}_{\mathfrak{p}_0}\}.$$

Then

$$\theta(\mathfrak{g}_{0,\lambda}) = \mathfrak{g}_{0,-\lambda}, \qquad [\mathfrak{g}_{0,\lambda}, \mathfrak{g}_{0,\mu}] \subset \mathfrak{g}_{0,\lambda+\mu}, \tag{2}$$

the latter relation following from the Jacobi identity. If $\lambda \neq 0$ and $\mathfrak{g}_{0,\lambda} \neq \{0\}$, then λ is called a *root* of $\mathfrak{g}_0$ with respect to $\mathfrak{h}_{\mathfrak{p}_0}$. If Σ denotes the set of all roots, the simultaneous diagonalization is expressed by

$$\mathfrak{g}_0 = \mathfrak{g}_{0,0} + \sum_{\lambda \in \Sigma} \mathfrak{g}_{0,\lambda} \qquad \text{(direct sum)}. \tag{3}$$

Now let the dual of $\mathfrak{h}_{\mathfrak{p}_0}$ be ordered in some way and let Σ^+ be the corresponding set of positive roots. Then by (2) the space $\mathfrak{n}_0' = \sum_{\lambda>0} \mathfrak{g}_{0,\lambda}$ is a subalgebra of $\mathfrak{g}_0$. Since $\mathfrak{h}_{\mathfrak{p}_0}$ is θ-invariant and maximal abelian in $\mathfrak{p}_0$ we have

$$\mathfrak{g}_{0,0} = (\mathfrak{g}_{0,0} \cap \mathfrak{k}_0) + \mathfrak{h}_{\mathfrak{p}_0} = \mathfrak{m}_0 + \mathfrak{h}_{\mathfrak{p}_0}. \tag{4}$$

Writing for $X \in \sum_{\lambda<0} \mathfrak{g}_{0,\lambda}$,

$$X = X + \theta X - \theta X,$$

we see that $X \in \mathfrak{k}_0 + \mathfrak{n}_0'$, so using (4) we obtain

$$\mathfrak{g}_0 = \mathfrak{k}_0 + \mathfrak{h}_{\mathfrak{p}_0} + \mathfrak{n}_0'. \tag{5}$$

Applying θ we conclude the directness of (5) from that of (3). Comparing this proof with that of Theorem 3.4 we conclude:

(i) Σ is the set of restrictions of $\Delta_{\mathfrak{p}}$ to $\mathfrak{h}_{\mathfrak{p}_0}$; and if this latter ordering of the dual of $\mathfrak{h}_{\mathfrak{p}_0}$ is compatible with the above ordering of the dual of $\mathfrak{h}_{\boldsymbol{R}}$, then Σ^+ is the set of restrictions of P_+ and $\mathfrak{n}_0' = \mathfrak{n}_0$.

(ii) If $\lambda \in \Sigma$, then, denoting restriction by a bar,

$$\dim \mathfrak{g}_{0,\lambda} = \text{number of } \alpha \in \Delta_{\mathfrak{p}} \text{ such that } \bar{\alpha} = \lambda. \tag{6}$$

Because of (i), the elements of Σ are also called *restricted roots*; and because of (ii), the number $m_\lambda = \dim \mathfrak{g}_{0,\lambda}$ is called the *multiplicity* of λ.

§ 4. Nilpotent Lie Groups

To begin with we establish certain facts concerning the exponential mapping of a nilpotent Lie group. We apply these to a more detailed study of the nilpotent Lie algebra which arises in the Iwasawa decomposition.

Let L be a Lie group with Lie algebra $\mathfrak{l}$. Let $T(\mathfrak{l})$ denote the tensor algebra over $\mathfrak{l}$ considered as a vector space. Let $X \to \bar{X}$ denote the identity mapping of $\mathfrak{l}$ into $T(\mathfrak{l})$ (this makes it unnecessary to denote the multiplication in $T(\mathfrak{l})$ by a separate symbol). Similarly, if $M = (m_1 ..., m_n)$ is a positive integral n-tuple we denote by $\bar{X}(M)$ the coefficient to $t_1^{m_1} ... t_n^{m_n}$ in the product $(|M|!)^{-1}(t_1\bar{X}_1 + ... + t_n\bar{X}_n)^{|M|}$, where $X_1, ..., X_n$ is a basis of $\mathfrak{l}$ and $|M| = m_1 + ... + m_n$. An element of the form $\sum a_{e_1...e_n} \bar{X}_1^{e_1} ... \bar{X}_n^{e_n}$ will be called a *canonical polynomial*. As before, let J denote the two-sided ideal in $T(\mathfrak{l})$ generated by the set of all elements of the form $\bar{X}\bar{Y} - \bar{Y}\bar{X} - ([X, Y])^-$, $X, Y \in \mathfrak{l}$. The factor algebra $T(\mathfrak{l})/J$ is the universal enveloping algebra $U(\mathfrak{l})$ of $\mathfrak{l}$. Let $X(M)$ be the image of $\bar{X}(M)$ under the canonical mapping of $T(\mathfrak{l})$ onto $U(\mathfrak{l})$. As proved earlier, the elements $X(M)$ form a basis of $U(\mathfrak{l})$. From Cor. 1.10, Chapter II, we obtain the following statement: To each $\bar{X}(M)$ corresponds a unique canonical polynomial $\bar{R}_M$ such that

$$\bar{X}(M) \equiv \bar{R}_M \qquad \text{mod } J.$$

Suppose now $\mathfrak{l}$ is nilpotent. In the central decending series $\mathscr{C}^0\mathfrak{l} \supset \mathscr{C}^1\mathfrak{l} \supset ...$ let $\mathscr{C}^{m-1}\mathfrak{l}$ denote the last nonzero term. The basis $X_1, ..., X_n$ of $\mathfrak{l}$ is said to be *compatible* with the central descending series if there exist integers $r_0 = 1 < r_1 < ... < r_m = n + 1$ such that $X_{r_i}, X_{r_i+1}, ..., X_{r_{i+1}-1}$ is a basis of a complementary subspace of $\mathscr{C}^{i+1}\mathfrak{l}$ in $\mathscr{C}^i\mathfrak{l}$ $(0 \leqslant i \leqslant m - 1)$. We put

$$w(\bar{X}_i) = p$$

if X_i lies in $\mathscr{C}^{p-1}\mathfrak{l}$ but not in $\mathscr{C}^p\mathfrak{l}$. In particular, $w(\bar{X}_i) = m$ if $X_i \in \mathscr{C}^{m-1}\mathfrak{l}$. We also put

$$w(c\bar{X}_{i_1} ... \bar{X}_{i_r}) = \sum_{k=1}^{r} w(\bar{X}_{i_k}), \qquad d(c\bar{X}_{i_1} ... \bar{X}_{i_r}) = r,$$

c being any real number $\neq 0$. We shall call w the *weight* and d the *degree*. The terms in $\bar{X}(M)$ all have the same weight, denoted $w(M)$.

Lemma 4.1. *Let L be a nilpotent Lie group with Lie algebra $\mathfrak{l}$. Let $X_1, ..., X_n$ be a basis of $\mathfrak{l}$ compatible with the central descending series*

$$\mathscr{C}^0\mathfrak{l} \supset \mathscr{C}^1\mathfrak{l} \supset ... \supset \mathscr{C}^m\mathfrak{l} = \{0\}, \qquad \mathscr{C}^{m-1}\mathfrak{l} \neq \{0\},$$

of $\mathfrak{l}$. Let the constants C^P_{MN} be determined by

$$X(M)\,X(N) = \sum_P C^P_{MN} X(P).$$

Let $[k]$ denote the n-tuple $(\delta_{k1}, ..., \delta_{kn})$. Then

$$C^{[k]}_{MN} = 0 \qquad \text{for each } k, \qquad 1 \leqslant k \leqslant n$$

provided $|M| + |N| > m$.

Proof. Consider the structural constants $c^k{}_{ij}$ defined by

$$[X_i, X_j] = \sum_{k=1}^{n} c^k{}_{ij} X_k, \qquad 1 \leqslant i, j \leqslant n.$$

Using $[\mathfrak{l}, \mathscr{C}^{q-1}\mathfrak{l}] = \mathscr{C}^q\mathfrak{l}$ and the Jacobi identity, we have

$$[\mathscr{C}^{p-1}\mathfrak{l}, \mathscr{C}^q\mathfrak{l}] \subset [\mathscr{C}^p\mathfrak{l}, \mathscr{C}^{q-1}] + [\mathfrak{l}, [\mathscr{C}^{p-1}\mathfrak{l}, \mathscr{C}^{q-1}\mathfrak{l}]].$$

The inclusion

$$[\mathscr{C}^{p-1}\mathfrak{l}, \mathscr{C}^{q-1}\mathfrak{l}] \subset \mathscr{C}^{p+q-1}\mathfrak{l}$$

thus follows by induction on q. Hence if $p = w(\bar{X}_i)$, $q = w(\bar{X}_j)$, we have by the choice of basis, $X_k \in \mathscr{C}^{p+q-1}\mathfrak{l}$ if $c^k{}_{ij} \neq 0$, whence

$$w(\bar{X}_k) \geqslant w(\bar{X}_i) + w(\bar{X}_j) \qquad \text{if} \quad c^k{}_{ij} \neq 0. \tag{1}$$

Now let M and N be any two positive integral n-tuples. Then, writing $\bar{X}^P = \bar{X}_1^{p_1} ... \bar{X}_n^{p_n}$, we have with some constants a_P,

$$\bar{X}(M)\,\bar{X}(N) \equiv \sum_P a_P \bar{X}^P \qquad (\text{mod } J) \tag{2}$$

$$\bar{X}(M)\,\bar{X}(N) \equiv \sum_Q C^Q_{MN} \bar{X}(Q) \qquad (\text{mod } J).$$

The expression on the right in (2) can be obtained by finitely many replacements

$$\bar{X}_i\bar{X}_j \to \bar{X}_j\bar{X}_i + \sum_{k=1}^{n} c^k{}_{ij}\bar{X}_k + u, \qquad u \in J,$$

which by (1) do not decrease the weight. Consequently,

$$w(\bar{X}^P) \geqslant w(M) + w(N) \qquad \text{if} \quad a_P \neq 0. \tag{3}$$

Next, we claim that

$$w(Q) \geqslant w(M) + w(N) \qquad \text{if} \quad C^Q_{MN} \neq 0. \tag{4}$$

In fact, let $k = \min\{w(Q) : C^Q_{MN} \neq 0\}$ and let us write $\equiv$ for congruence mod J. Then

$$\sum_{C^Q_{MN}\neq 0} C^Q_{MN}\bar{X}(Q) \equiv \sum_{w(Q')>k} C^{Q'}_{MN}\bar{X}(Q') + \sum_{w(Q_i)=k} C^{Q_i}_{MN}\bar{X}(Q_i).$$

For the first sum we have by the above

$$\sum_{w(Q')>k} C^{Q'}_{MN}\bar{X}(Q') \equiv \sum_{w(\bar{X}^P)>k} c_P\bar{X}^P.$$

For the second sum, let $r = \max_i |Q_i|$ and let this maximum be reached at exactly $Q_{i_1}, \ldots, Q_{i_m}$. Since $\bar{X}(Q_i) \equiv \bar{X}^{Q_i} + \sum_{|P|<r} a_{P,i}\bar{X}^P$, we see that

$$\sum_{w(Q_i)=k} C^{Q_i}_{MN}\bar{X}(Q_i) \equiv \sum_{s=1}^{m} C^{Q_{i_s}}_{MN}\bar{X}^{Q_{i_s}} + \sum_{w(\bar{X}^P)\geqslant k, |P|<r} d_P\bar{X}^P$$

Combining these congruences, we deduce

$$\bar{X}(M)\,\bar{X}(N) \equiv \sum_{w(\bar{X}^P)>k} c_P\bar{X}^P + \sum_{s=1}^{m} C^{Q_{i_s}}_{MN}\bar{X}^{Q_{i_s}} + \sum_{w(\bar{X}^P)\geqslant k, |P|<r} d_P\bar{X}^P$$

and since $w(\bar{X}^{Q_{i_s}}) = k$, $|Q_{i_s}| = r$, the middle sum must occur on the right-hand side in (2). Thus by (3), $k \geqslant w(M) + w(N)$, so (4) is proved.

Now suppose $|M| + |N| > m$. Then $w(M) + w(N) > m$. If $C^{[k]}_{MN}$ were $\neq 0$, then $\bar{X}_k$ would have weight $\geqslant w(M) + w(N) > m$. Since $\mathscr{C}^m\mathfrak{l} = \{0\}$ this is impossible and the lemma is proved.

Definition. Let V and W be two finite dimensional vector spaces over a field K. A *polynomial function* P on V is a function which can be put in the form $P = p(f_1, \ldots, f_n)$ where each f_i is a linear function on V with values in K and p is a polynomial (with coefficients in K). A mapping $\varphi : V \to W$ is said to be a *polynomial mapping* if $P \circ \varphi$ is a polynomial function on V whenever P is a polynomial function on W.

Suppose we have chosen bases for V and W. Then the mapping $\varphi : V \to W$ is a polynomial mapping if and only if the coordinates of $\varphi(X) \in W$ are polynomials p_i in the coordinates of $X \in V$. The highest degree of the polynomials p_i is a number, independent of the choice of bases. We shall call this number the *degree* of the polynomial mapping φ.

Theorem 4.2. *Let L be a Lie group with Lie algebra $\mathfrak{l}$. A necessary and sufficient condition for $\mathfrak{l}$ to be nilpotent is the existence of a polynomial mapping $P : \mathfrak{l} \times \mathfrak{l} \to \mathfrak{l}$ such that*

$$\exp X \exp Y = \exp P(X, Y) \qquad \textit{for } X, Y \in \mathfrak{l}. \tag{5}$$

In this case P has degree $\leqslant \dim \mathfrak{l}$.

Proof. Assume first that L is nilpotent and let $X_1, \ldots, X_n$ be a basis of $\mathfrak{l}$ compatible with the central descending series. Combining Lemma 4.1 with Cor. 2.7, Chapter III, we see that $C_{MN}^{[k]} = 0$ for all $k = 1, \ldots, n$ provided $|M| + |N| > \dim \mathfrak{l}$. Denoting canonical coordinates with subscripts we have relation (14) in Chapter II, §1:

$$(xy)_k = \sum_{M,N} C_{MN}^{[k]} x^M y^N,$$

where $x^M = x_1^{m_1} \ldots x_n^{m_n}$, etc. We can therefore state: There exist n polynomials $P_i(x_1, \ldots, x_n, y_1, \ldots, y_n)$, $1 \leqslant i \leqslant n$, of degree $\leqslant n$ and a number $a > 0$ such that

$$\exp(x_1X_1 + \ldots + x_nX_n) \exp(y_1X_1 + \ldots + y_nX_n) = \exp(P_1X_1 + \ldots + P_nX_n) \tag{6}$$

for $|x_i| < a$, $|y_i| < a$, $1 \leqslant i \leqslant n$. However, making use of the following lemma we conclude that (6) holds for all x_i and all y_i. Hence the condition of the theorem is necessary.

Lemma 4.3. *Let M and N be analytic manifolds, M connected. Let φ and ψ be two analytic mappings of M into N. Suppose $\varphi(p) = \psi(p)$ for all p in an open subset of M. Then $\varphi(p) = \psi(p)$ for all $p \in M$.*

Proof. Let $q \in M$. Let us say that φ and ψ have the same partial derivatives at q if (1) $\varphi(q) = \psi(q)$. (2) If $\{x_1, ..., x_m\}$ and $\{y_1, ..., y_n\}$ are coordinate systems valid near q and $\varphi(q)$, respectively, then the expressions of φ and ψ in these coordinates have the same partial derivatives (of all orders) at the point $(x_1(q), ..., x_m(q))$. Let M' be the subset of M consisting of all the points $q \in M$ such that φ and ψ have the same partial derivatives at q. Then M' is obviously closed in M. But M' is also open in M because the partial derivatives of an analytic function at a point determine the power series expansion of the function. Due to the connectedness of M we have $M' = M$ as desired.

To prove the second half of Theorem 4.2 suppose (5) holds for some polynomial mapping $P: \mathfrak{l} \times \mathfrak{l} \to \mathfrak{l}$. Let N_0 be a star-shaped open neighborhood of 0 in $\mathfrak{l}$ such that the mapping exp is a diffeomorphism of N_0 onto a neighborhood N_e of e in L. Let $\{x_1, ..., x_n\}$ be a system of canonical coordinates on N_e. Let $X, Y \in \mathfrak{l}$ and assume $X \in N_0$. Then

$$\exp X \exp tY = \exp (X + tZ_t) \qquad (t \in \boldsymbol{R}), \tag{7}$$

where $Z_t = Z_0 + tZ_1 + t^2Z_2 + ...$, each Z_i being of the form $Z_i = Q_i(X, Y)$ where Q_i is a polynomial mapping of $\mathfrak{l} \times \mathfrak{l}$ into $\mathfrak{l}$. Since

$$X + tZ_t - (X + tZ_0) = O(t^2)$$

it follows that

$$\left\{\frac{d}{dt} f(\exp (X + tZ_t))\right\}_{t=0} = \left\{\frac{d}{dt} f(\exp (X + tZ_0))\right\}_{t=0},$$

whenever f is one of the coordinate functions x_i. Using (7), we conclude

$$[dL_{\exp X} (Y)f] (\exp X) = \left\{\frac{d}{dt} f(\exp (X + tZ_0)\right\}_{t=0}.$$

On the other hand, we know from Theorem 1.7, Chapter II, that

$$\left\{\frac{d}{dt} f(\exp (X + tZ_0))\right\}_{t=0} = \left[\left\{dL_{\exp X} \circ \frac{1 - e^{-\operatorname{ad} X}}{\operatorname{ad} X} (Z_0)\right\} f\right] (\exp X).$$

Since $Z_0 = Q_0(X, Y)$ we get from these two equations

$$\frac{\operatorname{ad} X}{1 - e^{-\operatorname{ad} X}} (Y) = Q_0(X, Y)$$

if X is sufficiently small, Y arbitrary. Now if $| x | < 2\pi$ we have an absolutely convergent series $\Sigma_0^\infty a_n x^n$ satisfying

$$\left(\sum_0^\infty a_n x^n\right)\left(\sum_1^\infty \frac{(-1)^{n-1}}{n!} x^{n-1}\right) = 1 \tag{8}$$

and infinitely many a_n are $\neq 0$. Equation (8) simply amounts to an infinite set of equations for the coefficients a_n. If X is sufficiently small, the series $\Sigma_0^\infty a_n(\text{ad } X)^n$ and $\Sigma_1^\infty (-1)^{n-1}/n!\,(\text{ad } X)^{n-1}$ converge absolutely† and can be multiplied together, term by term. Thus relation (8) remains true if we replace x by ad X. It follows that

$$\sum_0^\infty a_n(\text{ad } X)^n (Y) = Q_0(X, Y).$$

Since infinitely many coefficients a_n are $\neq 0$, we conclude that ad X is nilpotent, hence $\mathfrak{l}$ is nilpotent.

Corollary 4.4. *Let L be a connected nilpotent Lie group with Lie algebra $\mathfrak{l}$. Then the exponential mapping is a regular mapping of $\mathfrak{l}$ onto L.*

If $X \in \mathfrak{l}$, then ad X is nilpotent so there exists a basis of $\mathfrak{l}$ such that the corresponding matrix expression for ad X has zeros on and below the diagonal. Consequently

$$\det\left(\frac{1 - e^{-\text{ad } X}}{\text{ad } X}\right) \neq 0.$$

In view of Theorem 1.7, Chapter II, this means that exp is regular everywhere on $\mathfrak{l}$. On the other hand, Theorem 4.2 shows that exp $(\mathfrak{l})$ is a subgroup of L, which, due to the regularity of exp, is an open subgroup. An open subgroup is always closed and due to the connectedness of L we find exp $(\mathfrak{l}) = L$.

Remark. This corollary can also be proved more directly and without using Theorem 4.2. If $\mathfrak{l} \neq \{0\}$ the center $\mathfrak{c}$ of $\mathfrak{l}$ is not zero and the factor algebra $\mathfrak{l}/\mathfrak{c}$ is again nilpotent and has dimension less than dim $\mathfrak{l}$. Corollary 4.4 can now be proved by induction. The details are left to the reader.

Let N be a nilpotent endomorphism of a finite-dimensional vector space V over $\boldsymbol{R}$. We put $\log(1 + N) = \Sigma_{n\geqslant 1}(-1)^{n-1} N^n/n$ (finite series). It is clear that $\log(1 + N)$ and $e^N - 1$ are also nilpotent.

† Let $|\ |$ be some norm on a finite-dimensional vector space V over $\boldsymbol{R}$. If A is an endomorphism of V, put $\|A\| = \sup(|Ax|/|x|)$. If A_n $(n = 0,1, \ldots)$ is an endomorphism of V, the series $\Sigma_0^\infty A_n$ is said to be absolutely convergent if $\Sigma_0^\infty \|A_n\|$ is convergent.

Lemma 4.5.

$$\log e^N = N, \qquad e^{\log(1+N)} = 1 + N.$$

Proof. Let x be a real number and let the coefficients $a_{m,n}$ be determined by

$$\Big(\sum_{r=1}^{\infty} \frac{1}{r!} x^r\Big)^n = \sum_{m=1}^{\infty} a_{m,n} x^m \qquad \text{for all } x. \tag{9}$$

If $|\exp x - 1| < 1$, then

$$x = \log((\exp x - 1) + 1) = \sum_{n=1}^{\infty} (-1)^{n-1}/n \Big(\sum_{r=1}^{\infty} \frac{1}{r!} x^r\Big)^n$$

$$= \sum_{n=1}^{\infty} (-1)^{n-1}/n \Big(\sum_{m=1}^{\infty} a_{m,n} x^m\Big).$$

Owing to Weierstrass' theorem on double series, the summations can be interchanged so

$$x = \sum_{m=1}^{\infty} \Big(\sum_{n=1}^{\infty} (-1)^{n-1}/n \, a_{m,n}\Big) x^m. \tag{10}$$

If A is any endomorphism of V, the series $\Sigma_1^{\infty} (1/r!) A^r$ is absolutely convergent. Relation (9) remains true if we replace x by A. Consequently,

$$\sum_{n=1}^{\infty} (-1)^{n-1}/n \Big(\sum_{r=1}^{\infty} \frac{1}{r!} N^r\Big)^n = \sum_{n=1}^{\infty} (-1)^{n-1}/n \sum_{m=1}^{\infty} a_{m,n} N^m.$$

Since N is nilpotent, the series are actually finite and the summations can of course be interchanged. Considering (10) it follows that

$$\log e^N = N.$$

The second relation can be proved in the same way.

§ 5. Global Decompositions

Theorem 5.1. *Let $\mathfrak{g}_0 = \mathfrak{k}_0 + \mathfrak{h}_{\mathfrak{p}_0} + \mathfrak{n}_0$ be an Iwasawa decomposition of a semisimple Lie algebra $\mathfrak{g}_0$ over* **R**. *Let G be any connected Lie group with Lie algebra $\mathfrak{g}_0$ and let K, $A_{\mathfrak{p}}$, N be the analytic subgroups of G with Lie algebras $\mathfrak{k}_0$, $\mathfrak{h}_{\mathfrak{p}_0}$, and $\mathfrak{n}_0$, respectively. Then the mapping*

$$\Phi : (k, a, n) \to kan \qquad (k \in K, a \in A_{\mathfrak{p}}, n \in N),$$

is an analytic diffeomorphism of the product manifold $K \times A_\mathfrak{p} \times N$ *onto* G. *The groups* $A_\mathfrak{p}$ *and* N *are simply connected.*

We begin by proving a general lemma (Harish-Chandra [4], p. 213), which will also be useful later.

Lemma 5.2. *Let* U *be a Lie group with Lie algebra* $\mathfrak{u}$. *Suppose* $\mathfrak{u}$ *is a direct sum* $\mathfrak{u} = \mathfrak{m} + \mathfrak{h}$ *where* $\mathfrak{m}$ *and* $\mathfrak{h}$ *are subalgebras of* $\mathfrak{u}$ *(not necessarily ideals). Let* M *and* H *be the analytic subgroups of* U *with Lie algebras* $\mathfrak{m}$ *and* $\mathfrak{h}$, *respectively. Then the mapping* $\alpha : (m, h) \to mh$ $(m \in M, h \in H)$ *of* $M \times H$ *into* U *is everywhere regular.*

Proof. As usual we denote by L_x the left translation by a group element x. The tangent vector to the curve $x \exp tX$ at x is $dL_x(X)$. We identify H and M, respectively, with the subgroups (e, H) and (M, e) of the product group $M \times H$. Also, the tangent space $(M \times H)_{(m,h)}$ is identified with the direct sum $M_m + H_h$ $(m \in M, h \in H)$.

Let $Y \in \mathfrak{m}$, $Z \in \mathfrak{h}$. Then

$$\alpha(m \exp tY, h) = mh \exp (t \operatorname{Ad}(h^{-1}) Y), \qquad t \in \boldsymbol{R},$$

$$\alpha(m, h \exp tZ) = mh \exp tZ.$$

It follows that

$$d\alpha_{(m,h)}(dL_m Y, dL_h Z) = dL_{mh}(\operatorname{Ad}(h^{-1}) Y + Z). \tag{1}$$

Now suppose $\operatorname{Ad}(h^{-1}) Y + Z = 0$; then $Y + \operatorname{Ad}(h) Z = 0$ and since $\operatorname{Ad}(h) Z \in \mathfrak{h}$ we have $Y = Z = 0$. This proves the lemma.

Let G_0 be the adjoint group of $\mathfrak{g}_0$. As usual we identify $\operatorname{ad}(\mathfrak{g}_0)$ and $\mathfrak{g}_0$. Let K_0, A_0, N_0, and S_0 denote the analytic subgroups of G_0 with Lie algebras $\mathfrak{k}_0$, $\mathfrak{h}_{\mathfrak{p}_0}$, $\mathfrak{n}_0$, and $\mathfrak{s}_0 = \mathfrak{h}_{\mathfrak{p}_0} + \mathfrak{n}_0$, respectively. We shall begin by proving Theorem 5.1 for the group G_0. The elements of G_0 are endomorphisms of $\mathfrak{g}_0$ which we extend to the complex algebra $\mathfrak{g}$. In terms of the basis (X_i) of $\mathfrak{g}$ from Lemma 3.5, the elements of K_0 are represented by unitary matrices, the group A_0 consists of positive diagonal matrices and the elements of N_0 are represented by lower triangular matrices with diagonal elements equal to 1. Now if a triangular matrix with positive diagonal elements is unitary, it must be the unit matrix. It follows that the mapping

$$(k, a, n) \to kan, \qquad k \in K_0,\ a \in A_0,\ n \in N_0,$$

of $K_0 \times A_0 \times N_0$ into G_0 is one-to-one. The group A_0 is obviously a simply connected closed subgroup of G_0. From Cor. 4.4 and Lemma 4.5 it follows immediately that the exponential mapping for matrices is an analytic diffeomorphism of the Lie algebra of all lower triangular matrices with zeros in the diagonal onto the Lie group of all lower triangular matrices with all diagonal elements equal to 1. This being so, it follows that N_0 is a closed, simply connected subgroup of G_0. The set $A_0N_0 = \{an : a \in A_0, n \in N_0\}$ is a subgroup of G_0. Since a represents the diagonal in the matrix an, and since A_0 is closed in G_0, it is obvious that the group A_0N_0 is a closed subgroup of G_0, hence an analytic subgroup of G_0. Now, by Lemma 5.2, the mapping $(a, n) \to an$ is a diffeomorphism of $A_0 \times N_0$ onto A_0N_0; hence $\dim A_0N_0 = \dim S_0$. But obviously $A_0N_0 \subset S_0$; hence $A_0N_0 = S_0$.

Consider now the Riemannian globally symmetric space $M = G_0/K_0$. Let R be the subgroup consisting of those elements in G_0 which leave every point of M fixed. Then we know (Remark 2, Prop. 3.4, Chapter IV) that the factor group G_0/R is a closed subgroup of the group $I(M)$ of all isometries of M. The natural mapping of G_0 onto G_0/R maps S_0 onto a subgroup S_* of G_0/R. Since S_0 is closed in G_0 and R compact, it follows that S_* is closed in G_0/R, hence a closed subgroup of $I(M)$. Let p denote the point in M given by the coset $\{K_0\}$. It is clear that $\dim S_* = \dim M$ so the orbit $S_* \cdot p$ is an open subset of M due to Lemma 4.1, Chapter II. But being the orbit of a closed subgroup of $I(M)$, $S_* \cdot p$ is a closed subset of M (Theorem 2.2, Chapter IV).† Hence $S_* \cdot p = M$. In terms of G_0, this result means that each $g \in G_0$ can be written $g = sk$, where $s \in S_0$, $k \in K_0$. Taking inverses it follows that each $g_1 \in G_0$ can be written $g_1 = k_1s_1$ $(k_1 \in K_0,\ s_1 \in S_0)$. The mapping $(k, a, n) \to kan$ of $K_0 \times A_0 \times N_0$ into G_0 is therefore one-to-one, onto and regular, hence a diffeomorphism.

To prove Theorem 5.1 in full generality, let π denote the natural mapping of G onto G_0. The kernel of π is the center Z of G. Since Z is discrete, (G, π) is a covering group of G_0. The identity component of the groups $\pi^{-1}(A_0)$ and $\pi^{-1}(N_0)$ coincides with $A_\mathfrak{p}$ and N, respectively. The groups A_0 and N_0 are simply connected and have covering groups $(A_\mathfrak{p}, \pi)$ and (N, π). Hence $A_\mathfrak{p} \cap Z = N \cap Z = \{e\}$ and $A_\mathfrak{p}$ and N are simply connected. If we put $\tilde{K} = \pi^{-1}(K_0)$ we have evidently $G = \tilde{K}A_\mathfrak{p}N$ and each g can be written uniquely $g = kan$ $(k \in \tilde{K},\ a \in A_\mathfrak{p},\ n \in N)$. Here a and n depend continuously on g, because π is a homeomorphism of $A_\mathfrak{p}$ onto A_0 and of N onto N_0. Hence k depends continuously on g; hence $A_\mathfrak{p}N$ and $KA_\mathfrak{p}N$ are closed in G. The regularity of Φ follows by applying Lemma 5.2 twice: first on the subgroups $A_\mathfrak{p}$ and N of $A_\mathfrak{p}N$, next on the subgroups K and $A_\mathfrak{p}N$ of G. Thus $KA_\mathfrak{p}N$ is open and closed in G so $G = KA_\mathfrak{p}N$; this finishes the proof of Theorem 5.1.

Proposition 5.3. *In the notation of Theorem* 5.1, *let* $S = A_\mathfrak{p}N$, $P = \exp \mathfrak{p}_0$. *Then* S *is a closed solvable subgroup of* G, P *is a closed*

† Or by Prop. 4.4, Chapter II.

submanifold of G. Let $\tilde{\theta}$ *denote the automorphism of G for which* $d\tilde{\theta} = \theta$. *Then the mapping*

$$\psi : s \to \tilde{\theta}(s)\, s^{-1}, \qquad s \in S,$$

is a diffeomorphism of S onto P.

Proof. Only the last statement has not been proved already. Each $g \in G$ can be written $g = pk$, $p \in P$, $k \in K$. Then $\tilde{\theta}(g) = p^{-1}k$ so $\tilde{\theta}(g)g^{-1} = p^{-2} \in P$. In particular, $\psi(S) \subset P$. The mapping ψ is one-to-one. In fact, if $\tilde{\theta}(s_1)\, s_1^{-1} = \tilde{\theta}(s_2)s_2^{-1}$, $(s_1, s_2 \in S)$, then $\tilde{\theta}(s_2^{-1}s_1) = s_2^{-1}s_1$; hence $s_2^{-1}s_1 \in K \cap S = \{e\}$ so $s_1 = s_2$. Furthermore, $\psi(S) = P$; in fact, given $p \in P$ there exists a unique $X \in \mathfrak{p}_0$ such that $p = \exp X$. By Theorem 5.1 there exist unique elements $k \in K$ and $s \in S$ such that $\exp \frac{1}{2} X = ks^{-1}$. Then $p = \tilde{\theta}(s)\, s^{-1}$ as desired. This mapping $\psi^{-1} : p \to s$ is differentiable because it is composed of the mappings

$$p \xrightarrow{\exp^{-1}} X \to \frac{X}{2} \to \exp \frac{X}{2} \to s.$$

§ 6. The Complex Case

It will be convenient later to have the Iwasawa decomposition (Theorems 3.4 and 5.1) restated for the case when the semisimple Lie algebra in question has a complex structure.

Suppose $\mathfrak{g}_0$ is a semisimple Lie algebra over $\boldsymbol{R}$ with complex structure J. This simply means (Chapter III, §6) that there exists an endomorphism J of $\mathfrak{g}_0$ such that

$$J^2 = -I,$$
$$(\mathrm{ad}_{\mathfrak{g}_0} X)\, J = J\, \mathrm{ad}_{\mathfrak{g}_0} X, \qquad X \in \mathfrak{g}_0.$$

As shown in §6, Chapter III, the set $\mathfrak{g}_0$ can be regarded as a Lie algebra $\tilde{\mathfrak{g}}_0$ over $\boldsymbol{C}$. The Lie algebra $\mathfrak{g}_0$ is obtained from $\tilde{\mathfrak{g}}_0$ by restricting the scalars to $\boldsymbol{R}$, in other words $(\tilde{\mathfrak{g}}_0)^R = \mathfrak{g}_0$. Let $\mathfrak{c}$ be any compact real form of the semisimple Lie algebra $\tilde{\mathfrak{g}}_0$. Then

$$\mathfrak{g}_0 = \mathfrak{c} + J\mathfrak{c}$$

is a Cartan decomposition of $\mathfrak{g}_0$ (Cor. 7.5, Chapter III). We can therefore carry through the construction in §3 for $\mathfrak{k}_0 = \mathfrak{c}$ and $\mathfrak{p}_0 = J\mathfrak{c}$. The maximal abelian subspace $\mathfrak{h}_{\mathfrak{p}_0}$ has the form $J\mathfrak{a}_0$ where $\mathfrak{a}_0$ is a maximal abelian subalgebra of $\mathfrak{c}$. Then $\mathfrak{h}_0 = \mathfrak{a}_0 + J\mathfrak{a}_0$ is a maximal abelian subalgebra of $\mathfrak{g}_0$. Since $J\mathfrak{h}_0 \subset \mathfrak{h}_0$, the Lie algebra $\mathfrak{h}_0$ has a complex structure inherited from $\mathfrak{g}_0$. The corresponding Lie algebra $\tilde{\mathfrak{h}}_0$ over C is a Cartan subalgebra of $\tilde{\mathfrak{g}}_0$. Let $\tilde{\Delta}$ denote the set of nonzero roots of $\tilde{\mathfrak{g}}_0$ with respect to $\tilde{\mathfrak{h}}_0$. In accordance with previous terminology we put

$$\tilde{\mathfrak{h}}_{0,R} = \sum_{\alpha \in \tilde{\Delta}} RH_\alpha.$$

Owing to Lemma 3.1 we have $(\tilde{\mathfrak{h}}_0)^* = J\mathfrak{a}_0$. Suppose now that the dual of the (real) vector space $J\mathfrak{a}_0$ has been turned into an ordered vector space. Let $(\tilde{\Delta})^+$ denote the set of positive roots in $\tilde{\Delta}$ according to this ordering. We put

$$\tilde{\mathfrak{n}}_+ = \sum_{\alpha \in (\tilde{\Delta})^+} \tilde{\mathfrak{g}}_0^\alpha \qquad \text{and} \qquad \mathfrak{n}_+ = (\tilde{\mathfrak{n}}_+)^R.$$

Let $\mathfrak{g} = \mathfrak{g}_0 + i\mathfrak{g}_0$ be the complexification of the real Lie algebra $\mathfrak{g}_0$ and extend J to a (C-linear) endomorphism of $\mathfrak{g}$. The algebra $\mathfrak{h} = \mathfrak{h}_0 + i\mathfrak{h}_0$ is a Cartan subalgebra of $\mathfrak{g}$ (Lemma 3.2). Let Δ denote the set of nonzero roots of $\mathfrak{g}$ with respect to $\mathfrak{h}$ and as before we put

$$\mathfrak{h}_R = \sum_{\alpha \in \Delta} RH_\alpha.$$

Then according to Lemma 3.2 we have $\mathfrak{h}_R = J\mathfrak{a}_0 + i\mathfrak{a}_0$. It is obviously possible to turn the dual space of $\mathfrak{h}_R$ into an ordered vector space such that the orderings in the duals of $\tilde{\mathfrak{h}}_{0,R}$ and $\mathfrak{h}_R$ are compatible. Let Δ^+ denote the set of positive roots in Δ with respect to this ordering. In §3 we have divided the set Δ^+ into two subsets P_- and P_+, P_- containing exactly those roots in Δ^+ which vanish identically on $\mathfrak{h}_{\mathfrak{p}_0}$.

Lemma 6.1. *In the present case, $\Delta^+ = P_+$ so P_- is empty.*

Proof. The equation $J[X, Y] = [JX, Y]$ holds for all $X, Y \in \mathfrak{g}_0$, hence for all $X, Y \in \mathfrak{g}$, J being C-linear. Now let α be a nonzero root of $\mathfrak{g}$ with respect to $\mathfrak{h}$ and select a nonzero vector $X_\alpha \in \mathfrak{g}^\alpha$. Let $H \in \mathfrak{h}$. The equation $[H, X_\alpha] = \alpha(H) X_\alpha$ implies $[H, JX_\alpha] = \alpha(H) JX_\alpha$. Hence $JX_\alpha = cX_\alpha$ where c is a complex number. On the other hand, $J\mathfrak{h} \subset \mathfrak{h}$ and $\alpha(JH) X_\alpha = [JH, X_\alpha] = [H, JX_\alpha] = c\alpha(H) X_\alpha$. Hence $\alpha(JH) = c\alpha(H)$ for all $H \in \mathfrak{h}$. From this equation it is obvious that α cannot vanish identically on the space $J\mathfrak{a}_0$ (which plays the role of $\mathfrak{h}_{\mathfrak{p}_0}$). This proves the lemma.

Lemma 6.2. *Let $\mathfrak{n} = \sum_{\alpha \in \Delta^+} \mathfrak{g}^\alpha$. Then $\mathfrak{n} \cap \mathfrak{g}_0 = \mathfrak{n}_+$.*

Proof. Let $\gamma \in (\tilde{\Delta})^+$ and $Z_\gamma \neq 0$ in $\tilde{\mathfrak{g}}_0^\gamma$. Then $[H, Z_\gamma] = \gamma(H) Z_\gamma$ for all $H \in \mathfrak{h}_0$. If we extend γ to a C-linear function γ^* on $\mathfrak{h}$ then $[H, Z_\gamma] =$

$\gamma^*(H) Z_\gamma$ for all $H \in \mathfrak{h}$. Thus $\gamma^* \in \Delta$ and due to the compatibility of the orderings we have $\gamma^* \in \Delta^+$. Hence $Z_\gamma \in \mathfrak{n} \cap \mathfrak{g}_0$ and $\mathfrak{n}_+ \subset \mathfrak{n} \cap \mathfrak{g}_0$. On the other hand, the number of elements in Δ^+ is twice the number of elements in $(\tilde{\Delta})^+$. It follows that

$$\dim_R \mathfrak{n}_+ = \dim_C \mathfrak{n} = \dim_R \mathfrak{n} \cap \mathfrak{g}_0.$$

This proves the lemma. Theorems 3.4 and 5.1 can now be restated (in simplified notation).

Theorem 6.3. *Let $\mathfrak{g}$ be a semisimple Lie algebra over $\boldsymbol{C}$, $\mathfrak{g}^R$ the Lie algebra $\mathfrak{g}$ considered as a Lie algebra over $\boldsymbol{R}$. Let J be the complex structure on $\mathfrak{g}^R$ which corresponds to multiplication by i on $\mathfrak{g}$. Let $\mathfrak{u}$ be any compact real form of $\mathfrak{g}$ and let $\mathfrak{a}$ be any maximal abelian subalgebra of $\mathfrak{u}$. Then the algebra $\mathfrak{h} = \mathfrak{a} + i\mathfrak{a}$ is a Cartan subalgebra of $\mathfrak{g}$. Let Δ be the set of roots of $\mathfrak{g}$ with respect to $\mathfrak{h}$ and let Δ^+ be the set of positive roots with respect to some ordering of Δ. If $\mathfrak{n}_+$ denotes the space $\sum_{\alpha\in\Delta^+} \mathfrak{g}^\alpha$ considered as a real subspace of $\mathfrak{g}^R$ the following direct decomposition is valid*

$$\mathfrak{g}^R = \mathfrak{u} + J\mathfrak{a} + \mathfrak{n}_+.$$

Let G_c be any connected Lie group with Lie algebra $\mathfrak{g}^R$ and let U, A^*, and N_+ denote the analytic subgroups of G_c with Lie algebras $\mathfrak{u}$, $J\mathfrak{a}$, and $\mathfrak{n}_+$, respectively. Then the mapping

$$(u, a, n) \to uan, \qquad u \in U,\ a \in A^*,\ n \in N_+,$$

is an analytic diffeomorphism of $U \times A^* \times N_+$ onto G_c. The groups A^* and N_+ are simply connected.

EXERCISES AND FURTHER RESULTS

A. Geometric Features of the Cartan Decomposition

1. Let G be connected semisimple Lie group whose Lie algebra has a complex structure. Show that G has finite center.

2. (See Chapter V, §2 for the notation.) The group $\boldsymbol{SO}(p) \times \boldsymbol{SO}(2)$ is a maximal compact subgroup of $\boldsymbol{SO}_0(p, 2)$. Deduce from Theorem 1.1 that the universal covering group of $\boldsymbol{SO}_0(p, 2)$ has infinite center ($p \geqslant 1$).

3. Let G, $\mathfrak{g}_0$, θ and K be as in Theorem 1.1.

(i) Prove that K equals its normalizer in G.

(ii) Prove that $\mathrm{Ad}(G)$ is invariant under the transpose $g \to {}^t g$ with respect to B_θ, and that ${}^t g = \theta g^{-1} \theta$.

(iii) Assuming G simple, prove that K is a maximal closed proper subgroup of G. (Use Prop. 5.1, Chapter VIII.)

(iv) Assuming G simple, prove that K is a maximal proper subgroup of G (cf. Brauer [1], Brun [1]).

4. Let A be a set of isometries of a complete simply connected Riemannian manifold M of negative curvature. Then

(i) If not empty, the set of fixed points under A forms a connected totally geodesic submanifold of M.

(ii) Suppose M is a Riemannian globally symmetric space of the noncompact type, $M = G/K$ ($G = I_0(M)$), and that A is a closed connected subgroup of K with Lie algebra $\mathfrak{a}$. Show that the fixed point set of A is Exp $\mathfrak{b}$ where

$$\mathfrak{b} = \{X \in \mathfrak{p}_0 : \mathrm{Ad}_G(\exp(-X))\,\mathfrak{a} \subset \mathfrak{k}_0\}.$$

(iii) Show that the set $\mathfrak{b}$ in (ii) can be written

$$\mathfrak{b} = \{X \in \mathfrak{p}_0 : [\boldsymbol{R}X, \mathfrak{a}] = \{0\}\}.$$

5. Let M be a Riemannian globally symmetric space of the noncompact type, o any point in M. Let σ denote the automorphism $g \to s_0 g s_0$ of $I_0(M)$. Let $I_0(M)$ be given the pseudo-Riemannian structure induced by the Killing form. Then the mapping

$$p \to g\sigma(g^{-1}) \qquad (g \cdot o = p),$$

is a diffeomorphism of M onto a closed totally geodesic submanifold S of $I_0(M)$. Under this mapping the action of an element $x \in I_0(M)$ on M corresponds to the diffeomorphism $s \to xs\sigma(x^{-1})$ of S.

6. Let $\mathfrak{g}_0$ be a semisimple Lie algebra, $K \subset \mathrm{Int}(\mathfrak{g}_0)$ a fixed maximal compact subgroup, and $M = \mathrm{Int}(\mathfrak{g}_0)/K$ the associated symmetric space. The mapping

$$p = gK \to gKg^{-1} \qquad (= \text{isotropy group } K_p \text{ at } p)$$

is a bijection of M onto the set of all maximal compact subgroups of $\mathrm{Int}(\mathfrak{g}_0)$. Using the fact that $K_p \cap K_q$ is the group of "rotations" of M around the geodesic γ_{pq}, deduce that K_p and K_q are conjugate under a $g \in \mathrm{Int}(\mathfrak{g}_0)$ commuting elementwise with $K_p \cap K_q$.

7. Let $\mathfrak{g}_0$ be a semisimple Lie algebra, σ any automorphism of $\mathfrak{g}_0$.

(i) Show that there is a unique automorphism Σ of $\mathrm{Int}(\mathfrak{g}_0)$ such that $d\Sigma = \sigma$. Prove that $\Sigma(g) = \sigma g \sigma^{-1}$ for $g \in \mathrm{Int}(\mathfrak{g}_0)$.

(ii) Σ permutes the maximal compact subgroups of $\mathrm{Int}(\mathfrak{g}_0)$. Show that the corresponding mapping I_σ of $M = \text{Int}(\mathfrak{g}_0)/K$ (cf. Exercise A.6) is an isometry.

(iii) Suppose $\mathfrak{g}_0$ has no compact ideals. Then the mapping $\sigma \to I_\sigma$ is an isomorphism of $\text{Aut}(\mathfrak{g}_0)$ onto $I(M)$.

8. (i) Let $\mathfrak{g}_0$ be a semisimple Lie algebra over $\boldsymbol{R}$ and M a compact group of automorphisms of $\mathfrak{g}_0$ leaving invariant a compactly imbedded subalgebra $\mathfrak{s}_0 \subset \mathfrak{g}_0$. Then M leaves invariant a Cartan decomposition $\mathfrak{g}_0 = \mathfrak{k}_0 + \mathfrak{p}_0$ such that $\mathfrak{s}_0 \subset \mathfrak{k}_0$. (Use the method of Theorem 2.1; cf. Borel [9].)

In particular, an automorphism of finite order always leaves some Cartan decomposition invariant (compare Exercise B.4, Chapter III).

(ii) If $\mathfrak{g}_0' \subset \mathfrak{g}_0$ is a semisimple subalgebra and $\mathfrak{g}_0' = \mathfrak{k}_0' + \mathfrak{p}_0'$ any Cartan decomposition, then there exists a Cartan decomposition $\mathfrak{g}_0 = \mathfrak{k}_0 + \mathfrak{p}_0$ such that $\mathfrak{k}_0' \subset \mathfrak{k}_0$, $\mathfrak{p}_0' \subset \mathfrak{p}_0$ (Mostow [3]; compare Lemma 2.2 in Chapter IX). (Hint: Use (i) on the conjugation σ of $\mathfrak{g}$ with respect to $\mathfrak{g}_0$.)

9. Let G be a connected semisimple Lie group, $\mathfrak{g}_0 = \mathfrak{k}_0 + \mathfrak{p}_0$ a Cartan decomposition of its Lie algebra, and θ the corresponding Cartan involution. Consider the left invariant Riemannian structure given by the positive definite bilinear form B_θ on $\mathfrak{g}_0 \times \mathfrak{g}_0$ and let ∇ denote the corresponding affine connection. Then ∇ and its geodesics have the following properties (H. C. Wang [2], and personal communication).

(i) If $Y, Z \in \mathfrak{g}_0$, then the left invariant vector fields $\tilde{Y}$, $\tilde{Z}$ satisfy

$$2\nabla_{\tilde{Y}}(\tilde{Z}) = [\tilde{Y}, \tilde{Z}] + [\theta\tilde{Y}, \tilde{Z}] + [\theta\tilde{Z}, \tilde{Y}].$$

(ii) Let $s \to \gamma(s)$ be a geodesic in G, $\gamma(0) = e$, and s the arc parameter. For each $s \in \boldsymbol{R}$ fix $T(s) \in \mathfrak{k}_0$, $X(s) \in \mathfrak{p}_0$ such that

$$\dot{\gamma}(s) = (T(s) + X(s))^{\sim}_{\gamma(s)}.$$

Then given $Y \in \mathfrak{g}_0$ the function

$$f_Y(s) = B_\theta(\dot{\gamma}(s), \tilde{Y}_{\gamma(s)})$$

satisfies

$$\frac{df_Y}{ds} = -B_\theta([T(s), X(s)], Y - \theta Y).$$

(iii) Deduce from (ii) that

$$\frac{dT(s)}{ds} = 0, \qquad \frac{dX(s)}{ds} = -2[T(s), X(s)]$$

so that, writing T_0 for $T(0)$, X_0 for $X(0)$,

$$\dot{\gamma}(s) = (T_0 + e^{-2\,\text{ad}(sT_0)}X_0)^{\sim}_{\gamma(s)}.$$

(iv) Deduce from (iii) that the geodesic γ is given by

$$\gamma(s) = \exp(s(X_0 - T_0)) \exp 2sT_0.$$

(v) Deduce from Exercise A.3, Chapter II that a geodesic in G which intersects itself is a closed geodesic.

10*. Let M be a Riemannian globally symmetric space of the noncompact type and L a (not necessarily connected) effective Lie transformation group of M. If $L \supset I_0(M)$, then $I(M) \supset L$ (cf. Ochiai [1]). For the compact M an entirely different situation prevails, cf. $\boldsymbol{SL}(2, \boldsymbol{C})$ acting on $\boldsymbol{S}^2$. For a general study, see Nagano [2].

B. The Iwasawa Decomposition

1. Using the Gram-Schmidt orthonormalization process, prove directly the Iwasawa decomposition for $\boldsymbol{SL}(n, \boldsymbol{R})$.

2. With the notation of Theorem 5.1 prove that (cf. Helgason [6], [7]):

(i) $K \cap (N(\theta N)) = \{e\}$.

(ii) $N(\theta N) \cap MA_{\mathfrak{p}} = \{e\}$.

(iii) The orbits $A_{\mathfrak{p}} \cdot o$ and $N \cdot o$ are perpendicular at o, the origin in G/K.

(iv) The distance function d on G/K satisfies

$$d(o, a \cdot o) < d(o, an \cdot o) \qquad \text{for} \quad a \in A, \quad n \neq e \text{ in } N.$$

(v) Let $H \in \mathfrak{h}_{\mathfrak{p}_0}$ be such that $\alpha(H) > 0$ for all $\alpha \in P_+$. Then

$$N = \{z \in G : \lim_{t \to +\infty} \exp(-tH)\, z \exp tH = e\}.$$

3. Give explicit Iwasawa decompositions for $\mathfrak{sl}(n, \boldsymbol{C})$, $\mathfrak{so}(n, \boldsymbol{C})$, and $\mathfrak{sp}(n, \boldsymbol{C})$.

4. Show that if $\mathfrak{g}$ is a semisimple Lie algebra over $\boldsymbol{C}$, then the real Lie algebra $\mathfrak{g}^{\boldsymbol{R}}$ has all its restricted roots of multiplicity 2.

5. With the notation of §3, assume $\mathfrak{k}$ contains no ideal $\neq 0$ in $\mathfrak{g}$. Then (cf. Helgason [9], p. 50)

$$\mathfrak{m} + \mathfrak{h}_{\mathfrak{p}} = \sum_{\alpha \in P_+} [\mathfrak{g}^{\alpha}, \mathfrak{g}^{-\alpha}].$$

C. The Displacement Function

Let M be a metric space with distance function d. If f is an isometry of M the *displacement function* δ_f is defined by

$$\delta_f(x) = d(x, f(x)) \qquad (x \in M).$$

The isometry f is called a *Clifford-Wolf isometry* if δ_f is constant on M.

1*. A Riemannian globally symmetric space M of the noncompact type has no isometry $f \neq I$ with δ_f bounded; in particular M has no Clifford-Wolf isometry (cf. Freudenthal [5], Tits [2]).

2*. Let M be a simply connected complete Riemannian manifold with sectional curvature $\kappa \leqslant 0$, and $\text{Crit}(f)$ the set of critical points of the function δ_f^2 (that is, where $d(\delta_f^2) = 0$). Then (cf. Ozols [1]):

(i) $\text{Crit}(f)$ is a totally geodesic submanifold.

(ii) $x \in \text{Crit}(f)$ if and only f preserves the unique geodesic joining x to $f(x)$ and induces a translation along this geodesic.

(iii) If $\kappa < 0$ and f is without fixed points, then either $\text{Crit}(f) = \emptyset$ or consists of a single geodesic.

3*. Let M be a Riemannian globally symmetric space of the noncompact type, and $g \in I(M)$. Then we have the following results (Mostow [6]):

(i) Let $S \subset M$ be a totally geodesic submanifold invariant under g. Then

$$\inf_{m \in M} \delta_g(m) = \inf_{s \in S} \delta_g(s).$$

(Let for $m \in M$, s be the foot of the perpendicular from m to S; use Cor. 13.2, Chapter I, on the triangles $(s, m, g \cdot m)$ and $(s, g \cdot m, g \cdot s)$ to deduce $\delta_f(s) \leqslant \delta_f(m)$.)

(ii) The $\inf_{m \in M} \delta_g(m)$ is reached if and only if g is semisimple. The points where the minimum is reached form a totally geodesic submanifold of M on which the centralizer of g in $I(M)$ acts transitively.

4*. Let M be a complete locally symmetric space and M^* its universal covering (cf. Theorem 5.6, Chapter IV), and $\Gamma \subset I(M^*)$ the corresponding group of covering transformations of M^* such that $M = M^*/\Gamma$. Then the following conditions are equivalent (cf. Wolf [2]):

(i) M is homogeneous;

(ii) Γ consists of Clifford-Wolf isometries;

(iii) $Z(\Gamma)$, the centralizer of Γ in $I(M^*)$, is transitive on M^*.

NOTES

§1-§2. Most of the results here are due to É. Cartan [10]; see also Mostow [1, 3]. Theorem 1.4 was first proved by Mostow in [3]; his proof differs somewhat from that of the text and as several results of the same paper it is based on the imbedding $\varphi: gK \to \mathrm{Ad}_G(g\sigma(g^{-1}))$ of G/K into the space P of positive definite symmetric matrices. This imbedding is also used in Mostow's profound study [6] of rigidity where he proves that a compact irreducible locally symmetric space of nonpositive curvature and dimension >2 is uniquely determined up to isometry (and a normalizing factor) by its fundamental group. (This fails for dimension 2: compact Riemann surfaces of the same genus need not be conformally equivalent.) Theorem 1.4 is applied to pseudo-Riemannian symmetric spaces by Flensted-Jensen [1, 2]. For further studies of these see, e.g., Koh [1], Loos [1], Shapiro [1], Rossmann [1], Oshima and Sekiguchi [1], in addition to the references to Chapter IV.

Other imbeddings of symmetric spaces and resulting compactifications have been studied by Satake [3], Furstenberg [1], Moore [1], Karpelevič [1], Oshima [1]. Cartan's conjugacy theorem (Theorem 2.1) has been extended by Iwasawa [1] to all connected locally compact groups.

§3-§5. Theorem 3.4 and its global analog Theorem 5.1 are from Iwasawa [1] (the results were extended from complex groups to real groups by Chevalley). In these theorems the Cartan subalgebra $\mathfrak{h}$ of $\mathfrak{g}$ is chosen such that $\mathfrak{h} \cap \mathfrak{p}$ has maximum dimension. The theorems have been extended by Harish-Chandra [8], p. 212 to other Cartan subalgebras; see also Satake [3], §3. Concerning Theorem 4.2 see Chen [1]. The proof of Lemma 4.1 incorporates Namioka's correction of the earlier version in [13].

CHAPTER VII

SYMMETRIC SPACES OF THE COMPACT TYPE

In contrast to the situation for the noncompact type there may be several Riemannian globally symmetric spaces U/K associated with the same orthogonal symmetric Lie algebra $(\mathfrak{u},\theta)$ of the compact type. These spaces are finite in number and have the same universal covering space. They can all be described by means of the center of the universal covering group of U (Theorem 9.1).

The entire chapter centers around the maximal abelian subspace $\mathfrak{h}_{\mathfrak{p}_*}$ together with the root system $\Delta_{\mathfrak{p}}$. These define the diagram $D(U,K)$ in which certain information about the space U/K is contained, for example, the location of its conjugate points and position of its closed geodesics. Elementary properties of the Weyl group $W(U, K)$ and the diagram are developed in §2 and §3, and some of these are extended to the affine Weyl group Γ_Σ in §8. For the isotropy action we prove the following identification of the orbit spaces: $K\backslash\mathfrak{p}_* = W(U, K)\backslash\mathfrak{h}_{\mathfrak{p}_*}$ = closed Weyl chamber; $K\backslash U/K = \Gamma_\Sigma\backslash\mathfrak{h}_{\mathfrak{p}_*}$ = closed polyhedron in $\mathfrak{h}_{\mathfrak{p}_*}$ determined by $D(U,K)$. In the course of describing all the spaces U/K associated with $(\mathfrak{u}, \theta)$ we determine the fundamental group of U, the unit lattice, and the center of the universal covering group $\tilde{U}$ in terms of the diagram $D(U)$. Here the singular set in U interferes, and a rigorous proof of the fact that it has no influence on the fundamental group of U requires some tools from dimension theory, collected in §12.

In §11 we prove that for U/K simply connected, all closed geodesics of minimal length are conjugate; so are all maximally curved totally geodesic submanifolds of the same dimension.

§ 1. The Contrast between the Compact Type and the Noncompact Type

Let $\mathfrak{u}$ be a compact semisimple Lie algebra and θ an involutive automorphism of $\mathfrak{u}$. Then θ extends uniquely to a (complex) involutive automorphism of $\mathfrak{g}$, the complexification of $\mathfrak{u}$. We denote this extension also by θ. Let $(\mathfrak{g}_0, s)$ denote the orthogonal symmetric Lie algebra which is dual to $(\mathfrak{u}, \theta)$. Then $\mathfrak{g}_0$ is a real form of $\mathfrak{g}$ and s is just the restriction of θ to $\mathfrak{g}_0$. As usual we adopt the notational conventions in §5, Chapter V. We have then the direct decompositions

$$\mathfrak{u} = \mathfrak{k}_0 + \mathfrak{p}_*, \qquad \mathfrak{g}_0 = \mathfrak{k}_0 + \mathfrak{p}_0$$

into eigenspaces for θ. We recall that a pair (U, K) is said to be associated with $(\mathfrak{u}, \theta)$ if U is a connected Lie group with Lie algebra $\mathfrak{u}$ and K is a

Lie subgroup of U with Lie algebra $\mathfrak{k}_0$. Such a pair is said to be of the compact type. This pair is said to be a Riemannian symmetric pair if K is closed, $\mathrm{Ad}_U(K)$ compact and if there exists an analytic involutive automorphism $\tilde{\theta}$ of U such that $(K_{\tilde{\theta}})_0 \subset K \subset K_{\tilde{\theta}}$. Such a $\tilde{\theta}$ is necessarily unique and $d\tilde{\theta} = \theta$. Finally, a Riemannian globally symmetric space M is said to be of the compact type if the pair $(I_0(M), H)$ is of the compact type, H being the isotropy subgroup of $I_0(M)$ at some point $o \in M$.

Proposition 1.1. *Let $(\mathfrak{u}, \theta)$ be an orthogonal symmetric Lie algebra of the compact type. Let (U, K) be an arbitrary pair associated with $(\mathfrak{u}, \theta)$. Then K is compact and $Ad_U(K)\,\mathfrak{p}_* \subset \mathfrak{p}_*$. The restriction of $-B$ to $\mathfrak{p}_* \times \mathfrak{p}_*$ gives rise to a U-invariant Riemannian structure on U/K, which turns U/K into a Riemannian locally symmetric space.*

Proof. Let K_0 denote the identity component of K and $N(K_0)$ the normalizer of K_0 in U, that is, the set of $u \in U$ such that $uK_0u^{-1} \subset K_0$ The group K_0 is a closed subgroup of U (Prop. 3.6, Chapter IV); hence $N(K_0)$ is closed in U. The group U is compact (Theorem 6.9, Chapter II), so $N(K_0)$ is compact. The Lie algebra of $N(K_0)$ is the normalizer $\mathfrak{n}(\mathfrak{k}_0)$ of $\mathfrak{k}_0$ in $\mathfrak{u}$, that is, the set of elements $X \in \mathfrak{u}$ such that $[\boldsymbol{R}X, \mathfrak{k}_0] \subset \mathfrak{k}_0$. If $X \in \mathfrak{n}(\mathfrak{k}_0) \cap \mathfrak{p}_*$, then $[\boldsymbol{R}X, \mathfrak{k}_0] \subset \mathfrak{k}_0 \cap \mathfrak{p}_* = \{0\}$. Using Corollary 1.7, Chapter V, it follows that $X = 0$ so $\mathfrak{n}(\mathfrak{k}_0) = \mathfrak{k}_0$. The group $N(K_0)$ has finitely many components and the same is true of K since $K \subset N(K_0)$. Hence K is compact. The group $\mathrm{Ad}_U(K)$ leaves $\mathfrak{k}_0$, and therefore its orthogonal complement, $\mathfrak{p}_*$, invariant. The local symmetry of U/K is clear from Prop. 3.6, Chapter IV (even if K is not connected).

The question is now: When is U/K Riemannian globally symmetric? The answer will be given in the present chapter (Theorem 9.1). It is more complicated than the answer to the analogous question for the noncompact type, where G/K is always globally symmetric (Theorem 1.1, Chapter VI). To begin with we establish a negative result which indicates what kind of complications the compact type presents.

Proposition 1.2. *Let $(\mathfrak{u}, \theta)$ be an orthogonal symmetric Lie algebra of the compact type. Let (U, K) be an arbitrary pair associated with $(\mathfrak{u}, \theta)$. Then*

(i) *The center of U does not in general belong to K.*

(ii) *Even if U/K is Riemannian globally symmetric, K is not necessarily connected.*

(iii) *Even if U/K is Riemannian globally symmetric, the automorphism θ does not necessarily correspond to an automorphism of U.*

Proof. An example for (i) is given by[†] $U = \boldsymbol{SU}(n)$ $(n \geqslant 3)$, $K = \boldsymbol{SO}(n)$ with the involutive automorphism $u \to \bar{u}$ (complex conjugation) of U. For (ii) we consider the two-dimensional real projective space $\boldsymbol{P}^2$, that is, $\boldsymbol{S}^2$ with all antipodal points identified. Then $\boldsymbol{P}^2 = U/K$ where $U = \boldsymbol{SO}(3)$ and K is the subgroup of U leaving a line l through 0 invariant. The group K is generated by the rotations around l and the reflection in a line through 0, perpendicular to l. Here U/K is Riemannian globally symmetric and K has two components. For (iii) let $\tilde{U} = \boldsymbol{SU}(2) \times \boldsymbol{SU}(2)$ and let $\tilde{\theta}$ denote the automorphism $(g_1, g_2) \to (g_2, g_1)$ of $\tilde{U}$. The subgroup $\tilde{K}$ of fixed points is isomorphic to $\boldsymbol{SU}(2)$. The center of $\boldsymbol{SU}(2)$ is a cyclic group of order 2; let z be the generator and let S denote the subgroup of the center $\tilde{Z}$ of $\tilde{U}$ consisting of the two elements (e, e), (e, z). Let $U = \tilde{U}/S$ and let $K = \pi(\tilde{K})$, π denoting the natural mapping of $\tilde{U}$ onto U. The pair (U, K) is associated with $(\mathfrak{u}, \theta)$ where $\mathfrak{u}$ is the Lie algebra of $\tilde{U}$ and θ is the automorphism of $\mathfrak{u}$ induced by $\tilde{\theta}$. The homomorphism Ad_U has a kernel consisting of two elements (since $\tilde{Z}$ has four elements). These elements are S and $(z, z)\, S$ both of which belong to K. Hence $U/K = \mathrm{Ad}_U(U)/\mathrm{Ad}_U(K)$ and this last space is globally symmetric since the automorphism θ induces an automorphism of $\mathrm{Int}\,(\mathfrak{u}) = \mathrm{Ad}_U\,(U)$. On the other hand, since $\tilde{\theta}(S) \not\subset S$, the following lemma shows that θ does not correspond to an automorphism of U.

Lemma 1.3.[§] *Let L be a connected Lie group and let (L^*, π) denote the universal covering group of L. Let Z^* denote the kernel of π. Let σ be any analytic automorphism of L^* and $d\sigma$ the corresponding automorphism of $\mathfrak{l}$, the Lie algebra of L. Then $d\sigma$ corresponds to an analytic automorphism of L if and only if $\sigma(Z^*) \subset Z^*$.*

Proof. If $\sigma(Z^*) \subset Z^*$, then σ induces the desired automorphism of $L^*/Z^* = L$. On the other hand, suppose λ is an automorphism of L such that $d\lambda = d\sigma$. Consider the mappings $\varphi = \pi \circ \sigma$ and $\psi = \lambda \circ \pi$ of L^* onto L. Then φ and ψ are homomorphism and $d\varphi = d\psi$. Consequently $\varphi = \psi$. Since Z^* is the kernel of π it follows that $\sigma(Z^*) \subset Z^*$ and the lemma is proved.

§ 2. The Weyl Group and the Restricted Roots

Let $(\mathfrak{u}, \theta)$ be an orthogonal symmetric Lie algebra of the compact type and let (U, K) be any pair associated with $(\mathfrak{u}, \theta)$. The notation

[†] For the notation $\boldsymbol{SU}(n)$ see Chapter X, §2.

[§] See Berger [2], p. 162.

of §1 will be preserved but for simplicity we shall now write Ad instead of Ad_U. For each $X \in \mathfrak{p}_*$, let T_X denote the restriction of $(\mathrm{ad}\, X)^2$ to $\mathfrak{p}_*$.

Let $\mathfrak{h}_{\mathfrak{p}_*}$ denote an arbitrary maximal abelian subspace of $\mathfrak{p}_*$. Then the space $\mathfrak{h}_{\mathfrak{p}_0} = i\mathfrak{h}_{\mathfrak{p}_*}$ is a maximal abelian subspace of $\mathfrak{p}_0$. Let $\mathfrak{h}_0$ be any maximal abelian subalgebra of $\mathfrak{g}_0$ containing $\mathfrak{h}_{\mathfrak{p}_0}$ and let $\mathfrak{h}$ denote the subalgebra of $\mathfrak{g}$ generated by $\mathfrak{h}_0$. Then $\mathfrak{h}$ is a Cartan subalgebra of $\mathfrak{g}$ (Lemma 3.2, Chapter VI), and as in §3, Chapter VI, we can form the subspaces $\mathfrak{h}_{\mathfrak{p}}$, $\mathfrak{h}_{\mathfrak{k}_0}$, $\mathfrak{h}_{\mathfrak{k}}$, and $\mathfrak{h}_R$. Let Δ denote the set of nonzero roots of $\mathfrak{g}$ with respect to $\mathfrak{h}$. Let $\Delta_{\mathfrak{p}}$ denote the set of roots in Δ which do not vanish identically on $\mathfrak{h}_{\mathfrak{p}}$. As in Chapter VI, §3, let Δ^+ denote the subset of Δ formed by the positive roots with respect to an ordering of Δ given by any compatible orderings in the dual spaces of $\mathfrak{h}_{\mathfrak{p}_0}$ and $\mathfrak{h}_R$, respectively. Let $P_+ = \Delta^+ \cap \Delta_{\mathfrak{p}}$. Finally, $\mathfrak{m}_0$ shall denote the centralizer of $\mathfrak{h}_{\mathfrak{p}_0}$ (or $\mathfrak{h}_{\mathfrak{p}_*}$) in $\mathfrak{k}_0$.

Let M and M', respectively, denote the centralizer and normalizer of $\mathfrak{h}_{\mathfrak{p}_*}$ in K. In other words,

$$M = \{k \in K : \mathrm{Ad}\,(k)\, H = H \text{ for each } H \in \mathfrak{h}_{\mathfrak{p}_*}\},$$

$$M' = \{k \in K : \mathrm{Ad}\,(k)\, \mathfrak{h}_{\mathfrak{p}_*} \subset \mathfrak{h}_{\mathfrak{p}_*}\}.$$

It is clear that M is a normal subgroup of M'.

Proposition 2.1. *The groups M and M' are compact and have the same Lie algebra, namely $\mathfrak{m}_0$.*

Proof. The groups M and M' are closed subgroups of K, hence compact. It is obvious that M has Lie algebra $\mathfrak{m}_0$. On the other hand, let Y belong to $\mathfrak{L}(M')$, the Lie algebra of M'. Then $[Y, H] \in \mathfrak{h}_{\mathfrak{p}_*}$ for each $H \in \mathfrak{h}_{\mathfrak{p}_*}$ so

$$B\,(\mathrm{ad}\, H(Y), \mathrm{ad}\, H(Y)) = -\, B((\mathrm{ad}\, H)^2\, Y, Y) = 0.$$

Hence $[H, Y] = 0$ for each $H \in \mathfrak{h}_{\mathfrak{p}_*}$ so $Y \in \mathfrak{m}_0$.

Definition. The factor group M'/M is called the *Weyl group* of the pair (U, K). It is denoted by $W(U, K)$ (or simply W).

Remark. It is clear from Prop. 2.1 that $W(U, K)$ is a finite group. The mapping $k \to \mathrm{Ad}\,(k)$ $(k \in M')$ induces an isomorphism of $W(U, K)$ into $\boldsymbol{GL}(\mathfrak{h}_{\mathfrak{p}_*})$. We can therefore regard $W(U, K)$ as a group of (complex) linear transformations of $\mathfrak{h}_{\mathfrak{p}}$. It will be shown later that for a fixed $\mathfrak{h}_{\mathfrak{p}_*}$, $W(U, K)$ only depends on $(\mathfrak{u}, \theta)$. On the other hand, it is clear from

Lemma 6.3, Chapter V, that different choices of $\mathfrak{h}_{\mathfrak{p}_*}$ lead to isomorphic Weyl groups.

Proposition 2.2. *Let $\mathfrak{a}$ be a subset of $\mathfrak{h}_{\mathfrak{p}_*}$ and suppose k is an element of K such that $Ad(k)\,\mathfrak{a} \subset \mathfrak{h}_{\mathfrak{p}_*}$. Then there exists an element $s \in W(U, K)$ such that $s \cdot A = Ad(k)\,A$ for each $A \in \mathfrak{a}$.*

Proof. The centralizer $Z_{\mathfrak{a}}$ of $\mathfrak{a}$ in U is a closed subgroup of U. Its Lie algebra is $\mathfrak{z}_{\mathfrak{a}}$, the centralizer of $\mathfrak{a}$ in $\mathfrak{u}$. Since $\mathfrak{z}_{\mathfrak{a}}$ is invariant under θ we have the direct decomposition

$$\mathfrak{z}_{\mathfrak{a}} = \mathfrak{z}_{\mathfrak{a}} \cap \mathfrak{k}_0 + \mathfrak{z}_{\mathfrak{a}} \cap \mathfrak{p}_*.$$

The spaces $\mathfrak{h}_{\mathfrak{p}_*}$ and $\mathrm{Ad}\,(k^{-1})\,\mathfrak{h}_{\mathfrak{p}_*}$ are maximal abelian subspaces of $\mathfrak{z}_{\mathfrak{a}} \cap \mathfrak{p}_*$. In view of Lemma 6.3, Chapter V, there exists an element $H \in \mathfrak{h}_{\mathfrak{p}_*}$ whose centralizer in $\mathfrak{p}_*$ coincides with $\mathfrak{h}_{\mathfrak{p}_*}$. Let X be any fixed element in $\mathfrak{z}_{\mathfrak{a}} \cap \mathfrak{p}_*$. The function $z \to B\,(H, \mathrm{Ad}\,(z)\,X)$ $(z \in Z_{\mathfrak{a}} \cap K)$ is real and attains its minimum, the group $Z_{\mathfrak{a}} \cap K$ being compact. If z_0 is a minimum point, we have

$$\left\{\frac{d}{dt} B(H, \mathrm{Ad}\,(\exp tT)\,\mathrm{Ad}\,(z_0)\,X)\right\}_{t=0} = 0$$

for each $T \in \mathfrak{z}_{\mathfrak{a}} \cap \mathfrak{k}_0$. It follows that

$$B(H, [T, \mathrm{Ad}\,(z_0)\,X]) = -\,B([H, \mathrm{Ad}\,(z_0)\,X], T) = 0 \qquad (1)$$

for each $T \in \mathfrak{z}_{\mathfrak{a}} \cap \mathfrak{k}_0$. Since $[H, \mathrm{Ad}\,(z_0)\,X] \in \mathfrak{z}_{\mathfrak{a}} \cap \mathfrak{k}_0$ we conclude from (1) that $[H, \mathrm{Ad}\,(z_0)\,X] = 0$, so, due to the choice of H, $\mathrm{Ad}\,(z_0)\,X \in \mathfrak{h}_{\mathfrak{p}_*}$. In particular, let $X = H'$ where H' is an element in $\mathrm{Ad}\,(k^{-1})\,\mathfrak{h}_{\mathfrak{p}_*}$ whose centralizer in $\mathfrak{p}_*$ is $\mathrm{Ad}\,(k^{-1})\,\mathfrak{h}_{\mathfrak{p}_*}$. Then from the above follows that $H' \in \mathrm{Ad}\,(z_0^{-1})\,\mathfrak{h}_{\mathfrak{p}_*}$ so

$$\mathrm{Ad}\,(z_0^{-1})\,\mathfrak{h}_{\mathfrak{p}_*} = \mathrm{Ad}\,(k^{-1})\,\mathfrak{h}_{\mathfrak{p}_*}.$$

Consequently $kz_0^{-1} \in M'$. Since $z_0 \in Z_{\mathfrak{a}} \cap K$, the restriction of $\mathrm{Ad}\,(kz_0^{-1})$ to $\mathfrak{h}_{\mathfrak{p}_*}$ is the desired element $s \in W(U, K)$.

The Killing form B is nondegenerate on $\mathfrak{h}_{\mathfrak{p}} \times \mathfrak{h}_{\mathfrak{p}}$. For each $\alpha \in \Delta_{\mathfrak{p}}$ there exists a unique vector $A_\alpha \in \mathfrak{h}_{\mathfrak{p}}$ such that $B(H, A_\alpha) = \alpha(H)$ for all $H \in \mathfrak{h}_{\mathfrak{p}}$. Since α is real on $\mathfrak{h}_{\mathfrak{p}_0}$ it follows that $A_\alpha \in \mathfrak{h}_{\mathfrak{p}_0}$.

Lemma 2.3. *For each $\alpha \in \Delta_{\mathfrak{p}}$ there exist nonzero vectors $Y_\alpha \in \mathfrak{p}_0$, $Z_\alpha \in \mathfrak{k}_0$ such that*

$$[H, Y_\alpha] = \alpha(H)\,Z_\alpha, \qquad [H, Z_\alpha] = \alpha(H)\,Y_\alpha$$

for all $H \in \mathfrak{h}_{\mathfrak{p}}$.

Proof. There exists a vector $X_\alpha \neq 0$ in $\mathfrak{g}$ such that $[H, X_\alpha] = \alpha(H) X_\alpha$ for all $H \in \mathfrak{h}$. Writing $X_\alpha = X_1 + iX_2$ where $X_1, X_2 \in \mathfrak{g}_0$, and noting that α is real on $\mathfrak{h}_{\mathfrak{p}_0}$, we obtain $[H, X_i] = \alpha(H) X_i$ for $H \in \mathfrak{h}_{\mathfrak{p}}$, $i = 1, 2$. At least one of the vectors X_i is nonzero and can be decomposed $Y_\alpha + Z_\alpha$ where $Y_\alpha \in \mathfrak{p}_0$, $Z_\alpha \in \mathfrak{k}_0$. Equating the components in $\mathfrak{p}_0$ and $\mathfrak{k}_0$, respectively, we get

$$[H, Y_\alpha] = \alpha(H) Z_\alpha$$

$$[H, Z_\alpha] = \alpha(H) Y_\alpha$$

for $H \in \mathfrak{h}_{\mathfrak{p}}$. At least one of the vectors Y_α, Z_α is $\neq 0$; it follows that both are $\neq 0$, since α does not vanish identically on $\mathfrak{h}_{\mathfrak{p}}$.

The form $-B$ induces a positive definite quadratic form on $\mathfrak{h}_{\mathfrak{p}_*}$. For each $\alpha \in \Delta_{\mathfrak{p}}$, let s_α denote the reflection of $\mathfrak{h}_{\mathfrak{p}_*}$ in the hyperplane $\alpha(H) = 0$ of $\mathfrak{h}_{\mathfrak{p}_*}$. Then s_α extends uniquely to a complex linear transformation of $\mathfrak{h}_{\mathfrak{p}}$ and

$$s_\alpha(H) = H - 2\frac{\alpha(H)}{\alpha(A_\alpha)} A_\alpha, \qquad H \in \mathfrak{h}_{\mathfrak{p}}.$$

Lemma 2.4. *Let $\alpha \in \Delta_{\mathfrak{p}}$. Then $s_\alpha \in W(U, K)$.*

Proof. Consider the vectors Y_α, Z_α from Lemma 2.3. We can assume, changing $Y_\alpha + Z_\alpha$ by a real factor if necessary, that $B(Z_\alpha, Z_\alpha) = -1$. Then, if $H \in \mathfrak{h}_{\mathfrak{p}}$,

$$B(H, [Z_\alpha, Y_\alpha]) = -B([H, Y_\alpha], Z_\alpha) = \alpha(H),$$

$$[H, [Z_\alpha, Y_\alpha]] = -[Z_\alpha, [Y_\alpha, H]] - [Y_\alpha, [H, Z_\alpha]] = 0.$$

The last relation implies that $[Z_\alpha, Y_\alpha] \in \mathfrak{h}_{\mathfrak{p}}$ and the first then shows that $[Z_\alpha, Y_\alpha] = A_\alpha$. It follows that

$$\mathrm{Ad}(\exp tZ_\alpha) A_\alpha = \sum_0^\infty \frac{1}{(2n)!}(t \operatorname{ad} Z_\alpha)^{2n}(A_\alpha) + \sum_0^\infty \frac{1}{(2n+1)!}(t \operatorname{ad} Z_\alpha)^{2n+1}(A_\alpha)$$

$$= \sum_0^\infty \frac{1}{(2n)!} t^{2n}(-\alpha(A_\alpha))^n A_\alpha + \sum_0^\infty \frac{1}{(2n+1)!} t^{2n+1}(-\alpha(A_\alpha))^{n+1} Y_\alpha.$$

Since $\alpha(A_\alpha) = B(A_\alpha, A_\alpha) > 0$, there exists a number $t_0 \in \boldsymbol{R}$ such that $t_0\sqrt{\alpha(A_\alpha)} = \pi$. Putting $k_0 = \exp t_0 Z_\alpha$ we obtain

$$\mathrm{Ad}(k_0) A_\alpha = -A_\alpha.$$

Moreover, since $[Z_\alpha, H] = 0$ if $\alpha(H) = 0$, the hyperplane $\alpha(H) = 0$ in $\mathfrak{h}_{\mathfrak{p}_*}$ is left pointwise fixed by $\text{Ad}(k_0)$. Hence s_α is the restriction of $\text{Ad}(k_0)$ to $\mathfrak{h}_{\mathfrak{p}_*}$. This proves the lemma.

Now we need some facts about toral subgroups (i.e., compact, abelian, connected subgroups) of compact Lie groups.

Theorem 2.5. *Let S be a compact connected Lie group and T a toral subgroup of S. Suppose a is an element in S which commutes with each member of T. Then there exists a torus $T' \subset S$ such that $T \subset T'$ and $a \in T'$.*

For the proof we need a lemma concerning monothetic groups. A topological group S is called *monothetic* if there exists an element $x \in S$ such that the sequence $e, x, x^2, \ldots,$ is dense in S. In this case, the element x is called a generator. Any torus is monothetic due to the classical theorem of Kronecker.

Lemma 2.6. *Let A be a compact abelian Lie group such that A/A_0 is cyclic, A_0 denoting the identity component of A. Then A is monothetic.*

Proof. The group A_0 is a torus, hence monothetic. Let a_0 be a generator for A_0 and let N denote the number of elements in A/A_0. Select a generator B for A/A_0 and an element b in the coset B. Then $b^N \in A_0$ and there exists an element $c \in A_0$ such that $b^N c^N = a_0$. Then bc is a generator of A.

In order to prove Theorem 2.5, let A denote the closed subgroup of S generated by T and a. The identity component A_0 of A is a torus containing T and the group $\bigcup_{n \in Z} A_0 a^n$ equals A. Since A is compact, some positive power of a lies in A_0. If N is the smallest such power of a, the group A/A_0 is cyclic of order N. By Lemma 2.6, A is monothetic. If b is a generator of A, let $\exp tX$ $(t \in \boldsymbol{R})$ be a one-parameter subgroup γ of S passing through b. The closure of γ in S is a torus containing a and T.

Corollary 2.7. *A maximal torus in a compact, connected Lie group is a maximal abelian subgroup.*

Corollary 2.8. *Let T be a toral subgroup of a compact connected Lie group S. The centralizer of T in S is connected.*

In fact, it is the union of the tori containing T.

Each root $\alpha \in \Delta_{\mathfrak{p}}$ defines a hyperplane $\alpha(H) = 0$ in the vector space $\mathfrak{h}_{\mathfrak{p}_*}$. These hyperplanes divide the space $\mathfrak{h}_{\mathfrak{p}_*}$ into finitely many connected components, called the *Weyl chambers*. These are open, convex subsets of $\mathfrak{h}_{\mathfrak{p}_*}$.

Lemma 2.9. *Let $H \in \mathfrak{h}_{\mathfrak{p}_*}$. Then the eigenvalues of T_H are:* (1) 0 *with multiplicity dim* $\mathfrak{h}_{\mathfrak{p}_*}$; (2) *the numbers* $\alpha(H)^2$ *as* α *runs through* P_+ *(with the right multiplicity).*

Proof. As proved in Chapter VI (Lemma 3.6), the space $\mathfrak{p}$ has a direct decomposition

$$\mathfrak{p} = \mathfrak{h}_{\mathfrak{p}} + \mathfrak{q} \qquad \text{where } \mathfrak{q} = \sum_{\alpha \in P_+} \boldsymbol{C}(X_\alpha - \theta X_\alpha).$$

Now $[H, X_\alpha] = \alpha(H) X_\alpha$, $[H, \theta X_\alpha] = -\alpha(H)\,\theta X_\alpha$, so

$$(\text{ad } H)^2 (X_\alpha - \theta X_\alpha) = \alpha(H)^2 (X_\alpha - \theta X_\alpha).$$

Let V be the eigenspace of $(\text{ad } H)^2$ in $\mathfrak{p}$ for the eigenvalue $\alpha(H)^2$. Since $\alpha(H)^2$ is real, we have $V = V \cap \mathfrak{p}_0 + V \cap \mathfrak{p}_*$ and the complex dimension of V equals the real dimension of $V \cap \mathfrak{p}_*$. The lemma now follows immediately. Lemma 2.3 gives another proof.

The maximal abelian subspace $\mathfrak{h}_{\mathfrak{p}}$ of $\mathfrak{p}$ can in general be extended to a Cartan subalgebra $\mathfrak{h}$ of $\mathfrak{g}$ in many different ways. However, the restrictions of the roots to $\mathfrak{h}_{\mathfrak{p}}$ do not depend on the choice of $\mathfrak{h}$, as Lemma 2.9 shows. More precisely, we have

Corollary 2.10. *Let λ be a real linear function on $\mathfrak{h}_{\mathfrak{p}_0}$. Then λ is the restriction of a root (of $\mathfrak{g}$ with respect to $\mathfrak{h}$) if and only if there exists a vector $X \neq 0$ in $\mathfrak{p}_0$ such that*

$$(\text{ad } H)^2 X = \lambda(H)^2 X \qquad \text{for } H \in \mathfrak{h}_{\mathfrak{p}_0}.$$

As an immediate consequence, we obtain

Corollary 2.11. *Let $k \in M'$ and for each real linear function λ on $\mathfrak{h}_{\mathfrak{p}_0}$, put $\lambda^k(H) = \lambda(Ad\,(k^{-1})\,H)$, $H \in \mathfrak{h}_{\mathfrak{p}_0}$. Then λ is the restriction of a root (of $\mathfrak{g}$ with respect to $\mathfrak{h}$) if and only if λ^k is the restriction of a root.*

Theorem 2.12. *Each $s \in W(U, K)$ permutes the Weyl chambers. The Weyl group is simply transitive on the set of Weyl chambers in $\mathfrak{h}_{\mathfrak{p}_*}$.*

Proof. Select $k \in M'$ such that s coincides with the restriction of Ad (k) to $\mathfrak{h}_{\mathfrak{p}_*}$. It follows from Cor. 2.11 that if some root in $\Delta_{\mathfrak{p}}$ vanishes at a point $H \in \mathfrak{h}_{\mathfrak{p}_*}$ then some root in $\Delta_{\mathfrak{p}}$ vanishes at Ad (k) H. Consequently, s permutes the Weyl chambers. Next we show that $W(U, K)$ is transitive. Let W' denote the subgroup of $W(U, K)$ generated by all s_α, $\alpha \in \Delta_{\mathfrak{p}}$. Let C_1 and C_2 be two arbitrary Weyl chambers and select $H_1 \in C_1$, $H_2 \in C_2$. If the segment $\overrightarrow{H_1 H_2}$ intersects a hyperplane $\alpha(H) = 0$, then it is clear that

$$| H_1 - H_2 | > | H_1 - s_\alpha \cdot H_2 |,$$

the norm in $\mathfrak{h}_{\mathfrak{p}_*}$ being denoted by $|\ |$. As s varies over the finite group W', the distance $|H_1 - s \cdot H_2|$ reaches a minimum, say for $s = s_0$. Then the segment from H_1 to $s_0 \cdot H_2$ intersects no hyperplane $\alpha(H) = 0$ and H_1 and $s_0 \cdot H_2$ lie in the same Weyl chamber. Hence $C_1 = s_0 C_2$ so the group W', and therefore $W(U, K)$ is transitive.

Suppose now that an element $s \in W(U, K)$ maps a chamber C into itself. Select any $H_0 \in C$ and let $H = N^{-1}(H_0 + sH_0 + \ldots + s^{N-1}H_0)$, where N is the order of s. Then $sH = H$ and, since C is convex, $H \in C$. In view of Cor. 3.7, Chapter VI, the centralizer $\mathfrak{z}_H$ of H in $\mathfrak{u}$ coincides with the centralizer of $\mathfrak{h}_{\mathfrak{p}_*}$ in $\mathfrak{u}$. Moreover, the centralizer Z_γ in U of the one-parameter subgroup $\gamma = \{\exp tH : t \in \boldsymbol{R}\}$ has Lie algebra $\mathfrak{z}_H$. The closure of γ in U is a torus, so by Cor. 2.8, Z_γ is connected. Select $k \in K$ such that s coincides with the restriction of $\operatorname{Ad}(k)$ to $\mathfrak{h}_{\mathfrak{p}_*}$. Then $\operatorname{Ad}(k)\, tH = tH$ for all $t \in \boldsymbol{R}$ so $k \in Z_\gamma$. Since Z_γ is generated by $\exp(\mathfrak{z}_H)$, it follows that the restriction of $\operatorname{Ad}(k)$ to $\mathfrak{h}_{\mathfrak{p}_*}$, that is, s, is the identity. This proves that $W(U, K)$ is simply transitive.

Corollary 2.13. *The Weyl group is generated by the reflections s_α, $\alpha \in \Delta_{\mathfrak{p}}$. Thus, for a fixed $\mathfrak{h}_{\mathfrak{p}_*}$, $W(U, K)$ depends only on $(\mathfrak{u}, \theta)$.*

In view of this corollary we shall often refer to $W(U, K)$ as the Weyl group of $(\mathfrak{u}, \theta)$ and denote it by $W(\mathfrak{u}, \theta)$.

Lemma 2.14. *Let $\mathfrak{a}$ be a subspace of $\mathfrak{h}_{\mathfrak{p}_*}$ and let $P_{\mathfrak{a}}$ denote the set of roots in P_+ which vanish identically on $\mathfrak{a}$. Let $\tilde{\mathfrak{a}}$ denote the subset of $\mathfrak{h}_{\mathfrak{p}_*}$ consisting of the points where all the roots in $P_{\mathfrak{a}}$ vanish. Then the centralizers $Z_{\mathfrak{a}}$ and $Z_{\tilde{\mathfrak{a}}}$ of $\mathfrak{a}$ and $\tilde{\mathfrak{a}}$ in U coincide, i.e.,*

$$Z_{\mathfrak{a}} = Z_{\tilde{\mathfrak{a}}}.$$

Proof. These centralizers are connected, due to Cor. 2.8. Therefore we only have to prove that their Lie algebras are the same. Each element $X \in \mathfrak{g}$ can be written

$$X = H_0 + \sum_{\alpha \in \Delta} a_\alpha X_\alpha \qquad (a_\alpha \in \boldsymbol{C}),$$

where $X_\alpha \in \mathfrak{g}^\alpha$, $H_0 \in \mathfrak{h}$. Then X commutes with each element in $\mathfrak{a}$ if and only if $a_\alpha \alpha(H) = 0$ for each $H \in \mathfrak{a}$ and each $\alpha \in \Delta$. Since α vanishes on $\mathfrak{a}$ if and only if it vanishes on $\tilde{\mathfrak{a}}$, it follows that $\mathfrak{a}$ and $\tilde{\mathfrak{a}}$ have the same centralizers in $\mathfrak{g}$ and also in $\mathfrak{u}$. This proves the lemma.

Theorem 2.15. *Let $\mathfrak{a}$ be a subspace of $\mathfrak{h}_{\mathfrak{p}_*}$ and let $W_{\mathfrak{a}}$ denote the group of elements in $W(\mathfrak{u}, \theta)$ which leave $\mathfrak{a}$ pointwise fixed. Then $W_{\mathfrak{a}}$ is generated by those reflections s_α $(\alpha \in \Delta_{\mathfrak{p}})$ for which α vanishes identically on $\mathfrak{a}$.*

Proof. Each element in $W_{\mathfrak{a}}$ leaves the set $\tilde{\mathfrak{a}}$ from Lemma 2.14 pointwise fixed. We can therefore assume, without restriction of generality, that $\mathfrak{a} = \tilde{\mathfrak{a}}$. Let $\mathfrak{z}_{\mathfrak{a}}$ and $Z_{\mathfrak{a}}$ denote the centralizers of $\mathfrak{a}$ in $\mathfrak{u}$ and U, respectively. Then $\mathfrak{z}_{\mathfrak{a}}$ is a compact Lie algebra invariant under θ so $\mathfrak{z}_{\mathfrak{a}} = \mathfrak{k}_0 \cap \mathfrak{z}_{\mathfrak{a}} + \mathfrak{p}_* \cap \mathfrak{z}_{\mathfrak{a}}$. Let $\mathfrak{c}$ denote the center of $\mathfrak{z}_{\mathfrak{a}}$. Then $\mathfrak{c} \cap \mathfrak{p}_* = \mathfrak{a}$. Let $\mathfrak{k}_1$ and $\mathfrak{p}_1$ denote the orthogonal complements of $\mathfrak{k}_0 \cap \mathfrak{c}$ in $\mathfrak{k}_0 \cap \mathfrak{z}_{\mathfrak{a}}$ and of $\mathfrak{a}$ in $\mathfrak{p}_* \cap \mathfrak{z}_{\mathfrak{a}}$. We put $\mathfrak{u}_1 = \mathfrak{k}_1 + \mathfrak{p}_1$ and let $\mathfrak{a}_1$ denote the orthogonal complement of $\mathfrak{a}$ in $\mathfrak{h}_{\mathfrak{p}_*}$. Since

$$\mathfrak{p}_1 = \mathfrak{a}_1 + \mathfrak{p}_* \cap \sum_{\alpha \in P_{\mathfrak{a}}} C(X_\alpha - \theta X_\alpha),$$

$\mathfrak{a}_1$ is a maximal abelian subspace of $\mathfrak{p}_1$. Also, the pair $(\mathfrak{u}_1, \theta)$ is an orthogonal symmetric Lie algebra of the compact type. Its Weyl group is generated by the reflections $s_{\bar{\alpha}}$ of $\mathfrak{a}_1$, if $\bar{\alpha}$ denotes the restriction of $\alpha \in P_{\mathfrak{a}}$ to $\mathfrak{a}_1$. If $s_{\bar{\alpha}}$ is extended to $\mathfrak{h}_{\mathfrak{p}_*}$ by defining it to be the identity mapping on $\mathfrak{a}$, then we obtain a member of $W_{\mathfrak{a}}$. It remains to be proved that the elements in $W_{\mathfrak{a}}$ thus obtained generate the whole of $W_{\mathfrak{a}}$.

Let $\tilde{Z}$ denote the universal covering group of $Z_{\mathfrak{a}}$. If $k \in Z_{\mathfrak{a}} \cap M'$ and $\tilde{k}$ is an element in $\tilde{Z}$ over k, then $\mathrm{Ad}\,(k)$ and $\mathrm{Ad}_{\tilde{Z}}\,(\tilde{k})$ agree on $\mathfrak{a}_1$. The group $\tilde{Z}$ decomposes according to the direct decomposition $\mathfrak{z}_{\mathfrak{a}} = \mathfrak{c} + \mathfrak{u}_1$

$$\tilde{Z} = C \times U_1,$$

where the groups C and U_1 have Lie algebras $\mathfrak{c}$ and $\mathfrak{u}_1$, respectively. If $\tilde{k}$ decomposes accordingly, $\tilde{k} = (c, k_1)$, then $\mathrm{Ad}_{\tilde{Z}}\,(\tilde{k})$ and $\mathrm{Ad}_{U_1}\,(k_1)$ agree on $\mathfrak{a}_1$. Since $k \in K \cap Z_{\mathfrak{a}}$, k_1 lies in a Lie subgroup of U_1 with Lie algebra $\mathfrak{k}_1$. Thus $\mathrm{Ad}_{U_1}\,(k_1)$ restricted to $\mathfrak{a}_1$ coincides with an element of the Weyl group of $(\mathfrak{u}_1, \theta)$. The same is therefore true of $\mathrm{Ad}\,(k)$ if $k \in Z_{\mathfrak{a}} \cap M'$. This proves the theorem.

For each linear form μ on $\mathfrak{h}_{\mathfrak{p}}$ let $A_\mu \in \mathfrak{h}_{\mathfrak{p}}$ be determined by $\mu(H) = B(A_\mu, H)$ for all $H \in \mathfrak{h}_{\mathfrak{p}}$ and put $\langle \mu, \lambda \rangle = B(A_\mu, A_\lambda)$ for any two such linear forms μ and λ. We now prove for the set Σ of restricted roots the analog of the integrality condition in Theorem 4.3, Chapter III.

Theorem 2.16. *For any* $\mu, \lambda \in \Sigma$,

$$2\frac{\langle \mu, \lambda \rangle}{\langle \mu, \mu \rangle} \in \boldsymbol{Z}.$$

Proof. Let $\lambda \neq \mu$. We may assume $\lambda - \mu \notin \Sigma$ because otherwise we replace λ by $\lambda - (p-1)\mu$ where p is the first integer $\geqslant 0$ such that $\lambda - p\mu \notin \Sigma$. Select nonzero vectors X_μ and X_λ in the restricted root spaces $\mathfrak{g}_{0,\mu}$ and $\mathfrak{g}_{0,\lambda}$, respectively. Then $[X_\mu, \theta X_\mu] \in \mathfrak{g}_{0,0} \cap \mathfrak{p}_0 = \mathfrak{h}_{\mathfrak{p}_0}$, and taking inner product with $H \in \mathfrak{h}_{\mathfrak{p}_0}$ we see that

$$[\theta X_\mu, X_\mu] = -B(X_\mu, \theta X_\mu)\, A_\mu.$$

Put $X = cX_\mu$ where $c > 0$ is determined by $-B(X, \theta X) = 2\langle\mu, \mu\rangle^{-1}$; put $Y = -\theta X$ and $H = 2\langle\mu, \mu\rangle^{-1} A_\mu$. Then

$$[H, X] = 2X, \qquad [H, Y] = -2Y, \qquad [X, Y] = H,$$

so $\boldsymbol{R}X + \boldsymbol{R}Y + \boldsymbol{R}H$ is a three-dimensional Lie algebra $\mathfrak{b}$ and $Z \to \operatorname{ad} Z$ is a representation of $\mathfrak{b}$ on $\mathfrak{g}$. Moreover, since $\lambda - \mu \notin \Sigma$,

$$[H, X_\lambda] = \lambda(H)\, X_\lambda, \qquad [Y, X_\lambda] = 0.$$

Put $e_n = (\operatorname{ad} X)^n(X_\lambda)$, $n \geqslant 0$, $e_{-1} = 0$. Then we obtain by induction and the Jacobi identity

$$[H, e_n] = (\lambda(H) + 2n)\, e_n,$$
$$[X, e_n] = e_{n+1},$$
$$[Y, e_n] = -n(\lambda(H) + n - 1)\, e_{n-1}.$$

If k is the last integer such that $e_k \neq 0$, then $[Y, e_{k+1}] = 0$; so $\lambda(H) + k = 0$ and the lemma follows. We also obtain the following consequence (compare with Theorem 4.3, Chapter III).

Corollary 2.17. *Suppose $\lambda, \mu \in \Sigma$ are proportional, $\mu = c\lambda$ $(c \in \boldsymbol{C})$. Then $c = \pm\frac{1}{2}, \pm 1, \pm 2$.*

As in Chapter VI, §3, let Σ^+ denote the set of positive restricted roots (that is, the restrictions of the members of P_+ to $\mathfrak{h}_\mathfrak{p}$). A $\lambda \in \Sigma^+$ is called *simple* if it cannot be written $\lambda = \alpha + \beta$ with $\alpha, \beta \in \Sigma^+$.

Lemma 2.18. *Let $\lambda, \mu \in \Sigma$.*

(i) *If $\langle\lambda, \mu\rangle > 0$, then $\lambda - \mu \in \Sigma$ (or $\lambda = \mu$).*
(ii) *If $\langle\lambda, \mu\rangle < 0$, then $\lambda + \mu \in \Sigma$ (or $\lambda = -\mu$).*
(iii) *If λ and μ are simple, then $\lambda - \mu \notin \Sigma$.*

Proof. If $\lambda - \mu \notin \Sigma$, the proof of Theorem 2.16 shows $\langle\lambda, \mu\rangle \leqslant 0$, proving (i) and (ii). Part (iii) is obvious.

Theorem 2.19. *Let* $\alpha_1, \ldots, \alpha_l$ *be the set of all simple restricted roots. Then* $l = \dim \mathfrak{h}_{\mathfrak{p}_0}$ *and each* $\beta \in \Sigma^+$ *has the form* $\beta = \sum_{i=1}^{l} n_i \alpha_i$ *where each* $n_i \in \boldsymbol{Z}^+$.

The proof is almost identical to that of Theorem 5.7, Chapter III; we just have to use $\langle \alpha_i, \alpha_j \rangle \leqslant 0$ for $i \neq j$, a property we just proved.

There is a simple connection between the Weyl chambers and the ordering of the restricted roots. A *Weyl chamber* in $\mathfrak{h}_{\mathfrak{p}_0}$ is by definition a component of the set $\mathfrak{h}'_{\mathfrak{p}_0}$ of elements in $\mathfrak{h}_{\mathfrak{p}_0}$ where all the members of Σ are $\neq 0$. These elements are said to be *regular*.

Lemma 2.20.

(i) *The set*

$$C^+ = \{H \in \mathfrak{h}_{\mathfrak{p}_0} : \alpha(H) > 0 \textit{ for all } \alpha \in \Sigma^+\}$$

is a Weyl chamber and its boundary is contained in the union of the hyperplanes $\alpha_1 = 0, \ldots, \alpha_l = 0$.

(ii)

$$\Sigma^+ = \{\alpha \in \Sigma : \alpha > 0 \textit{ on } C^+\}.$$

(iii) *If* $H = \sum_{i=1}^{l} a_i A_{\alpha_i}$ *lies in* C^+, *then* $a_i \geqslant 0$ *for each* i.

(iv) *If* $H, H' \in C^+$, *then* $B(H, H') > 0$ (*that is,* H *and* H' *form an acute angle*).

Proof. Because of Theorem 2.19, C^+ coincides with the set

$$\{H \in \mathfrak{h}_{\mathfrak{p}_0} : \alpha_1(H) > 0, \ldots, \alpha_l(H) > 0\},$$

whose boundary has the described property. But then C^+ is $\neq \emptyset$, so by its definition it is a maximal connected subset of $\mathfrak{h}'_{\mathfrak{p}_0}$, hence a Weyl chamber. This proves (i). Part (ii) is immediate from Theorem 2.19. For (iii) we put $e_i = A_{\alpha_i}$ $(1 \leqslant i \leqslant l)$ $e_{l+1} = -H$, $a_{l+1} = 1$. Then by Lemma 2.18 and the assumption $H \in C^+$ we have $B(e_i, e_j) \leqslant 0$ $(1 \leqslant i \neq j \leqslant l + 1)$ so

$$0 \leqslant B\left(\sum_1^{l+1} |a_i| e_i, \sum_1^{l+1} |a_i| e_i\right) \leqslant B\left(\sum_1^{l+1} a_i e_i, \sum_1^{l+1} a_i e_i\right).$$

Thus the relation $\sum_1^{l+1} a_i e_i = 0$ implies $\sum_1^{l+1} |a_i| e_i = 0$, whence $a_i = |a_i|$ $(1 \leqslant i \leqslant l)$, proving (iii). Part (iv) is now immediate since $B(H, H') = \sum_1^{l} a_i \alpha_i(H') > 0$.

Remark. Given a Weyl chamber C one can (by Theorem 2.12) order the dual of $\mathfrak{h}_{\mathfrak{p}_0}$ in such a way that the positive roots are the roots positive on C. The most canonical way of doing this is to use as a basis of the dual of $\mathfrak{h}_{\mathfrak{p}_0}$ the linear forms which form the walls of C.

An element $s \in W(\mathfrak{u}, \theta)$ acts on the dual of $\mathfrak{h}_{\mathfrak{p}_0}$ by $(s\lambda)(H) = \lambda(s^{-1}H)$. Thus $s\lambda = \lambda^s$ (§3, Chapter I). Also

$$s_\alpha \lambda = \lambda - \frac{2\langle \lambda, \alpha \rangle}{\langle \alpha, \alpha \rangle} \alpha, \qquad \alpha \in \Sigma, \quad \lambda \in (\mathfrak{h}_{\mathfrak{p}_0})^*.$$

Lemma 2.21. *Let α_i be a simple root. Then s_{α_i} permutes the elements in Σ^+ which are not proportional to α_i.*

Proof. If $\alpha \in \Sigma^+$, we have $\alpha = \sum_1^l n_j \alpha_j$ $(n_j \in \mathbf{Z}^+)$. Then

$$s_{\alpha_i} \alpha = \alpha - n\alpha_i = (n_i - n)\, \alpha_i + \sum_{j \neq i} n_j \alpha_j,$$

where $n = 2\langle \alpha, \alpha_i \rangle / \langle \alpha_i, \alpha_i \rangle$, which by Theorem 2.16 is an integer. Since $s_{\alpha_i} \alpha \in \Sigma$ the coefficients in this linear combination all have the same sign (Theorem 2.19). Thus $n_i \geqslant n$ unless α is a multiple of α_i.

Now we prove that the closure $\overline{C^+}$ is a fundamental domain for $W(\mathfrak{u}, \theta)$ acting on $\mathfrak{h}_{\mathfrak{p}_0}$.

Theorem 2.22. *Each orbit of $W(\mathfrak{u}, \theta)$ on $\mathfrak{h}_{\mathfrak{p}_0}$ intersects the closed chamber $\overline{C^+}$ in exactly one point H_0. The corresponding linear form $\lambda_0 : H \to B(H_0, H)$ on $\mathfrak{h}_{\mathfrak{p}_0}$ is maximal among its Weyl group transforms.*

Proof. By the proof of Theorem 2.12 each Weyl group orbit intersects $\overline{C^+}$. For the uniqueness let $H_1, H_2 \in C^+$. The cited proof showed that the distance $| H_1 - sH_2 |$ $(s \in W(\mathfrak{u}, \theta))$ is smallest if $s = e$. Consequently,

$$B(H_1, H_2) \geqslant B(sH_1, H_2),$$

and this relation remains valid for $H_1, H_2 \in \overline{C^+}$. If $s_0 H_1 \in \overline{C^+}$, we can in this relation exchange H_1 and $s_0 H_1$ and deduce $B(H_1 - s_0 H_1, H_2) = 0$, whence $s_0 H_1 = H_1$, proving the uniqueness.

If μ is maximal in the orbit $W(\mathfrak{u}, \theta) \cdot \lambda_0$, we have $\mu \geqslant s_\alpha \mu$ for all $\alpha \in \Sigma^+$, so $\langle \mu, \alpha \rangle \geqslant 0$ $(\alpha \in \Sigma^+)$, whence $A_\mu \in \overline{C^+}$. But then by the first part of the proof, $A_\mu = H_0$ so $\mu = \lambda_0$.

§ 3. Conjugate Points. Singular Points. The Diagram

Again, let $(\mathfrak{u}, \theta)$ be an orthogonal symmetric Lie algebra of the compact type and let (U, K) be any pair associated with $(\mathfrak{u}, \theta)$. The notation of the preceding section will be preserved.

The manifold U/K is a Riemannian locally symmetric space whose tangent space at o (the point K in U/K) is identified with $\mathfrak{p}_*$. Let $X \in \mathfrak{p}_*$. The formula for $d\,\mathrm{Exp}_X$ (Theorem 4.1, Chapter IV) is clearly valid here, even if K is not necessarily connected. By this formula, X is conjugate to o if and only if

$$\det\Big(\sum_0^\infty \frac{1}{(2n+1)!}(T_X)^n\Big) = 0. \tag{1}$$

According to Lemma 6.3, Chapter V, each $X \in \mathfrak{p}_*$ can be expressed $X = \mathrm{Ad}(k)H$ where $k \in K$, $H \in \mathfrak{h}_{\mathfrak{p}_*}$. Since $T_{\mathrm{Ad}(k)H} = \mathrm{Ad}(k)\circ T_H \circ \mathrm{Ad}(k^{-1})$, we obtain from Lemma 2.9,

$$\det\Big(\sum_0^\infty \frac{1}{(2n+1)!}(T_X)^n\Big) = \prod_{\alpha\in P_+} \frac{\sin \alpha(iH)}{\alpha(iH)}. \tag{2}$$

From this formula follows:

Proposition 3.1. *The point $X = Ad(k)H$ is conjugate to o if and only if $\alpha(H) \in \pi i(\boldsymbol{Z} - 0)$ for some $\alpha \in \Delta_{\mathfrak{p}}$.*

Consider now the coset space K/M, where M as before denotes the centralizer of $\mathfrak{h}_{\mathfrak{p}_*}$ in K. The mapping

$$\Phi : (kM, H) \to \mathrm{Exp}\,\mathrm{Ad}(k)H, \qquad k \in K,\ H \in \mathfrak{h}_{\mathfrak{p}_*},$$

is a differentiable mapping of $K/M \times \mathfrak{h}_{\mathfrak{p}_*}$ onto U/K. The mapping Φ can be decomposed $\Phi = \mathrm{Exp} \circ \beta$ where β is the mapping of $K/M \times \mathfrak{h}_{\mathfrak{p}_*}$ onto $\mathfrak{p}_*$ given by

$$\beta : (kM, H) \to \mathrm{Ad}(k)H, \qquad k \in K,\ H \in \mathfrak{h}_{\mathfrak{p}_*}.$$

As usual $\tau(x)$ ($x \in K$) denotes the mapping $kM \to xkM$ of K/M onto itself. As in Chapter VI, let $\mathfrak{l}_0$ denote the orthogonal complement of $\mathfrak{m}_0$ in $\mathfrak{k}_0$. According to Lemma 3.6, Chapter VI, the subspace $\mathfrak{l}$ of $\mathfrak{g}$ generated by $\mathfrak{l}_0$ is given by

$$\mathfrak{l} = \sum_{\alpha\in P_+} \boldsymbol{C}(X_\alpha + \theta X_\alpha). \tag{3}$$

The natural mapping of K onto K/M induces an isomorphism of $\mathfrak{l}_0$ onto the tangent space to K/M at $\{M\}$. We shall therefore denote this tangent space also by $\mathfrak{l}_0$. As usual (Remark in §2, No. 1, Chapter I), a finite-dimensional vector space will be identified with its tangent space at each point.

Let (k_0M, H_0) be an arbitrary point in $K/M \times \mathfrak{h}_{\mathfrak{p}_*}$. If L runs through $\mathfrak{l}_0$ and H runs through $\mathfrak{h}_{\mathfrak{p}_*}$, then $(d\tau(k_0)\cdot L, H)$ runs through the tangent space to $K/M \times \mathfrak{h}_{\mathfrak{p}_*}$ at (k_0M, H_0). Since

$$\beta(k_0(\exp tL)M, H_0) = \mathrm{Ad}(k_0)\,\mathrm{Ad}(\exp tL)H_0,$$

$$\beta(k_0, H_0 + tH) = \mathrm{Ad}(k_0)(H_0 + tH),$$

we find

$$d\beta_{(k_0M, H_0)}(d\tau(k_0)\, L, H) = \mathrm{Ad}\,(k_0)\,([L, H_0] + H). \tag{4}$$

Since

$$B([L, H_0], H) = B(L, [H_0, H]) = 0,$$

it is clear that the right-hand side of (4) vanishes if and only if $[L, H_0] = H = 0$. Using (3) we may write

$$L = \sum_{\alpha \in P_+} l_\alpha(X_\alpha + \theta X_\alpha), \qquad l_\alpha \in \boldsymbol{C},$$

so

$$[H_0, L] = \sum_{\alpha \in P_+} l_\alpha \alpha(H_0)\,(X_\alpha - \theta X_\alpha).$$

Consequently, the mapping β is regular at (k_0M, H_0) if and only if $\alpha(H_0) \neq 0$ for all $\alpha \in \Delta_{\mathfrak{p}}$. Combining this result with Prop. 3.1, we can state:

Proposition 3.2. *The mapping Φ is regular at the point (k_0M, H_0) if and only if $\alpha(iH_0)/\pi$ is not an integer for any $\alpha \in \Delta_{\mathfrak{p}}$.*

Definition. The set

$$\{H \in \mathfrak{h}_{\mathfrak{p}_*} : \alpha(H) \in \pi i \boldsymbol{Z} \quad \text{for some } \alpha \in \Delta_{\mathfrak{p}}\}$$

is called the *diagram* of the pair (U, K). It will be denoted by $D(U, K)$ or $D(\mathfrak{u}, \theta)$.

The diagram is therefore the union of finitely many families of equispaced hyperplanes. The complement $\mathfrak{h}_{\mathfrak{p}_*} - D(U, K)$ will be denoted by $(\mathfrak{h}_{\mathfrak{p}_*})_r$. It is obvious from Prop. 3.2 that $D(U, K)$ is invariant under each $s \in W(U, K)$.

Definition. The set $S_{U/K} = \Phi(K/M \times D(U, K))$ is called the *singular set* in U/K. The complement $U/K - S_{U/K}$ will be denoted $(U/K)_r$.

The topological dimension (see §12 in this chapter) of a subset S of a separable metric space will be denoted by dim S. This notation is permissible since the dimension of a separable C^∞ manifold coincides with its topological dimension (see Hurewicz-Wallman [1], Chapter IV).

For each $\alpha \in \Delta_{\mathfrak{p}}$ we put

$$\mathfrak{h}_\alpha = \{H \in \mathfrak{h}_{\mathfrak{p}_*} : \alpha(H) \in \pi i \boldsymbol{Z}\},$$

$$M_\alpha = \{k \in K : \mathrm{Exp}\,\mathrm{Ad}\,(k)\, H = \mathrm{Exp}\, H \text{ for all } H \in \mathfrak{h}_\alpha\}.$$

It is obvious that M_α is a closed subgroup of U containing M. We shall now prove the following stronger statement:

$$\dim M_\alpha > \dim M. \tag{5}$$

In fact, consider the vector $Z_\alpha \in \mathfrak{k}_0$ from Lemma 2.3. This lemma implies that

$$\operatorname{Ad}(\exp H) Z_\alpha = \cos \alpha(iH) Z_\alpha - i \sin \alpha(iH) Y_\alpha \qquad \text{for } H \in \mathfrak{h}_{\mathfrak{p}_*}.$$

This implies that

$$\exp \operatorname{Ad}(\exp tZ_\alpha) H = \exp H, \qquad \text{if } \cos \alpha(iH) = +1,$$

$$\exp \operatorname{Ad}(\exp tZ_\alpha) H = \exp H \exp(-2tZ_\alpha), \qquad \text{if } \cos \alpha(iH) = -1$$

for all $t \in \boldsymbol{R}$. In any case, we have

$$\operatorname{Exp} \operatorname{Ad}(\exp tZ_\alpha) H = \operatorname{Exp} H$$

for all $t \in \boldsymbol{R}$, if $H \in \mathfrak{h}_\alpha$. This means that $\exp tZ_\alpha \in M_\alpha$ for all t so Z_α belongs to the Lie algebra of M_α. Since $Z_\alpha \notin \mathfrak{m}_0$, relation (5) follows.

Theorem 3.3. *Let (U, K) be any pair associated with* $(\mathfrak{u}, \theta)$. *Then the singular set is closed and*

$$\dim S_{U/K} \leqslant \dim U/K - 2.$$

The same statement holds for the set of points in U/K which are conjugate to o.

Proof. *Let* $\alpha \in \Delta_{\mathfrak{p}}$. Consider the mapping Φ_α of $K/M_\alpha \times \mathfrak{h}_\alpha$ into U/K given by

$$\Phi_\alpha(kM_\alpha, H) = \operatorname{Exp} \operatorname{Ad}(k) H.$$

Then

$$\Phi_\alpha(K/M_\alpha \times \mathfrak{h}_\alpha) = \Phi(K/M \times \mathfrak{h}_\alpha). \tag{6}$$

Using Lemma 3.6, Chapter VI and relation (5) above we find

$$\dim K/M = \dim \mathfrak{p}_* - \dim \mathfrak{h}_{\mathfrak{p}_*},$$

$$\dim (K/M_\alpha \times \mathfrak{h}_\alpha) \leqslant \dim \mathfrak{p}_* - \dim \mathfrak{h}_{\mathfrak{p}_*} - 1 + (\dim \mathfrak{h}_{\mathfrak{p}_*} - 1)$$

$$= \dim \mathfrak{p}_* - 2.$$

Using (6) and Lemma 12.3 in the Appendix we conclude

$$\dim \Phi(K/M \times \mathfrak{h}_\alpha) \leqslant \dim U/K - 2.$$

Let $(\mathfrak{h}_{\mathfrak{p}_*})_e$ denote the set of $H \in \mathfrak{h}_{\mathfrak{p}_*}$ for which $\exp H = e$. As noted in the proof of Lemma 6.3, Chapter V, the subset $\exp \mathfrak{h}_{\mathfrak{p}_*}$ of U is compact; it follows that the factor space $\mathfrak{h}_{\mathfrak{p}_*}/(\mathfrak{h}_{\mathfrak{p}_*})_e$ is compact. Using this fact, it is easily seen that $\Phi(K/M \times \mathfrak{h}_\alpha)$ is closed in U/K. From the sum theorem in dimension theory (Appendix, Theorem 12.2), it follows that

$$\dim \Phi(K/M \times D(U, K)) \leqslant \dim U/K - 2.$$

Remark. For the example $S^2 = SO(3)/SO(2)$ the set of points conjugate to o consists of two points, namely, the antipodal point to o and o. Thus the inequality in Theorem 3.3 is the best possible. On the other hand, the equality sign does not in general hold in Theorem 3.3. This is seen from the example $S^3 = SO(4)/SO(3)$, the group of unit quaternions (Theorem 4.7).

§ 4. Applications to Compact Groups

As pointed out in §6, Chapter IV, a compact, connected Lie group is a Riemannian globally symmetric space when provided with a bi-invariant Riemannian structure.

Let U be a compact, connected semisimple Lie group. Let $\mathfrak{u}$ denote the Lie algebra of U. Let U^* denote the subgroup $\{(u, u) : u \in U\}$ of the product group $U \times U$. Then $(U \times U, U^*)$ is a Riemannian symmetric pair of the compact type associated with the orthogonal symmetric Lie algebra $(\mathfrak{u} \times \mathfrak{u}, d\sigma)$ where $d\sigma$ is the differential of the automorphism $\sigma : (u_1, u_2) \to (u_2, u_1)$ of $U \times U$. The coset space $U \times U/U^*$ is diffeomorphic to U under the mapping

$$(u_1, u_2)\, U^* \to u_1 u_2^{-1} \qquad (u_1, u_2 \in U).$$

Under this correspondence, the natural mapping of $U \times U$ onto $U \times U/U^*$ corresponds to the mapping

$$\pi : (u_1, u_2) \to u_1 u_2^{-1}$$

of $U \times U$ onto U, whose differential is

$$d\pi : (X, Y) \to X - Y, \qquad X, Y \in \mathfrak{u}.$$

The Lie algebra $\mathfrak{u} \times \mathfrak{u}$ decomposes into the eigenspaces of $d\sigma$ for the eigenvalues $+ 1$ and $- 1$:

$$\mathfrak{u} \times \mathfrak{u} = \mathfrak{u}^* + \mathfrak{v}^*,$$

where $\mathfrak{u}^*$ equals $\{(X, X) : X \in \mathfrak{u}\}$, the Lie algebra of U^*, and $\mathfrak{v}^* = \{(X, - X) : X \in \mathfrak{u}\}$.

Let T be a maximal torus in U and let $\mathfrak{t}_0$ denote the Lie algebra of T. Then the space

$$\mathfrak{t}^* = \{(H, -H) : H \in \mathfrak{t}_0\}$$

is a maximal abelian subspace of $\mathfrak{v}^*$. Let $\mathfrak{g}$ be the complexification of $\mathfrak{u}$ and let $\mathfrak{t}$ denote the subalgebra of $\mathfrak{g}$ generated by $\mathfrak{t}_0$. We shall now investigate the Weyl group of the symmetric pair $(U \times U, U^*)$ defined by means of the maximal abelian subspace $\mathfrak{t}^*$ of $\mathfrak{v}^*$. The spaces $\mathfrak{u}^*$, $\mathfrak{v}^*$, $\mathfrak{t}^*$, and $\mathfrak{t} \times \mathfrak{t}$, respectively, play the role of the spaces $\mathfrak{k}_0$, $\mathfrak{p}_*$, $\mathfrak{h}_{\mathfrak{p}_*}$, and $\mathfrak{h}$ from §2. In particular, $\mathfrak{t} \times \mathfrak{t}$ is a Cartan subalgebra of $\mathfrak{g} \times \mathfrak{g}$, the complexification of $\mathfrak{u} \times \mathfrak{u}$.

Let Δ^* denote the set of nonzero roots of $\mathfrak{g}$ with respect to $\mathfrak{t}$. Let $\alpha \in \Delta^*$. Then the linear functions α' and α'' on $\mathfrak{t} \times \mathfrak{t}$ given by

$$\alpha'(H_1, H_2) = \alpha(H_1),$$
$$\alpha''(H_1, H_2) = \alpha(H_2), \qquad H_1, H_2 \in \mathfrak{t},$$

are roots of $\mathfrak{g} \times \mathfrak{g}$ with respect to $\mathfrak{t} \times \mathfrak{t}$. In fact, if the vector $X_\alpha \in \mathfrak{g}$ satisfies $[H, X_\alpha] = \alpha(H) X_\alpha$ for all $H \in \mathfrak{t}$, then

$$[(H_1, H_2), (X_\alpha, 0)] = \alpha(H_1)(X_\alpha, 0), \quad [(H_1, H_2), (0, X_\alpha)] = \alpha(H_2)(0, X_\alpha)$$

for all $H_1, H_2 \in \mathfrak{t}$. By counting, it is clear that each nonzero root of $\mathfrak{g} \times \mathfrak{g}$ with respect to $\mathfrak{t} \times \mathfrak{t}$ arises in this manner from a member of Δ^*. The roots α' and α'' cannot vanish identically on $\mathfrak{t}^*$, moreover, their values on $\mathfrak{t}^*$ are purely imaginary.

Let $X \in \mathfrak{u}$. As in §2 we consider now the operator $T_{(X,-X)}$, which is the restriction of $(\operatorname{ad}(X, -X))^2$ to $\mathfrak{v}^*$. From Lemma 2.9 we know that if $H \in \mathfrak{t}_0$, the operator $T_{(H,-H)}$ has eigenvalues 0 (with multiplicity $\dim \mathfrak{t}_0$) and the numbers $\alpha(H)^2$ as α runs through Δ^*.

The manifold U has a bi-invariant Riemannian structure (for example, the one induced by the negative of the Killing form of $\mathfrak{u}$). The corresponding affine connection is always the same and $\operatorname{Exp} X = \exp X$ for $X \in \mathfrak{u}$. According to Theorem 6.4, Chapter V, each $X \in \mathfrak{u}$ can be written $X = \operatorname{Ad}(u) H$ where $u \in U$ and $H \in \mathfrak{t}_0$.

Proposition 4.1. *The point $X = \operatorname{Ad}(u) H$ is conjugate to e if and only if $\alpha(H) \in 2\pi i(\mathbf{Z} - 0)$ for some $\alpha \in \Delta^*$.*

Proof. From the formula for $d \exp_X$ (Theorem 1.7, Chapter II), it follows that X is conjugate to e if and only if

$$\det_{\mathfrak{u}}\left(\frac{1 - e^{-\operatorname{ad} X}}{\operatorname{ad} X}\right) = 0.$$

Since $\operatorname{ad} X = \operatorname{Ad}(u) \circ \operatorname{ad} H \circ \operatorname{Ad}(u^{-1})$, we have

$$\det_{\mathfrak{u}}\left(\frac{1 - e^{-\operatorname{ad} X}}{\operatorname{ad} X}\right) = \det_{\mathfrak{u}}\left(\frac{1 - e^{-\operatorname{ad} H}}{\operatorname{ad} H}\right).$$

The endomorphism $(1 - e^{-ad\,H})/\operatorname{ad} H$ of $\mathfrak{g}$ has obviously determinant

$$\prod_{\alpha \in \Delta^*} \frac{1 - e^{-\alpha(H)}}{\alpha(H)}. \tag{1}$$

The restriction to $\mathfrak{u}$ has the same determinant so the proposition follows immediately.

Remark. Proposition 4.1 above is of course a special case of Prop. 3.1. The reason for the appearance of the factor 2 in Prop. 4.1 is that $d\pi(X/2, -X/2) = X$ so X is conjugate to e if and only if

$$\det\left(\sum_0^\infty \frac{1}{(2n+1)!}\,(T_{(X/2,-X/2)})^n\right) = 0.$$

This determinant, however, equals

$$\prod_{\alpha \in \Delta^*} \frac{\sin \frac{1}{2}\alpha(iH)}{\frac{1}{2}\alpha(iH)} \tag{2}$$

and Prop. 4.1 follows again. Expressions (1) and (2) both give the determinant of $d\,\mathrm{Exp}_X$ (evaluated by orthonormal basis). It follows that these expressions are equal (as is easily seen anyway).

Consider now the diagram $D(U \times U, U^*) \subset \mathfrak{t}^*$. Let $D(U)$ denote the image of $D(U \times U, U^*)$ under $d\pi$. Since $d\pi(H, -H) = 2H$, $(H \in \mathfrak{t}_0)$ and $\alpha'(H, -H) = -\alpha''(H, -H) = \alpha(H)$, $(\alpha \in \Delta^*)$, it follows that

$$D(U) = \{H \in \mathfrak{t}_0 : \alpha(H) \in 2\pi i \boldsymbol{Z} \text{ for some } \alpha \in \Delta^*\}.$$

The set $D(U)$ is a union of a finite number of families of equispaced hyperplanes of $\mathfrak{t}_0$. It will be called the *diagram of* U. Under the mapping $d\pi : \mathfrak{t}^* \to \mathfrak{t}_0$ the Weyl group $W(U \times U, U^*)$ corresponds to a group $W(U)$ of endomorphisms of $\mathfrak{t}_0$. Since $W(U \times U, U^*)$ leaves $D(U \times U, U^*)$ invariant, it is clear that $W(U)$ leaves $D(U)$ invariant. Moreover, $W(U)$ is generated by the reflexions in the hyperplanes of $D(U)$ which pass through 0. On the other hand, the centralizer of $\mathfrak{t}^*$ in U^* is $T^* = \{(t, t) : t \in T\}$ and the normalizer of $\mathfrak{t}^*$ in U^* is $\{(n, n) : n \in N_T\}$ where N_T denotes the normalizer of T in U. It follows that the group N_T/T, considered as a group of linear transformations of $\mathfrak{t}_0$, coincides with $W(U)$. The *Weyl chambers* in $\mathfrak{t}_0$ are the components of the open subset of $\mathfrak{t}_0$ where all $\alpha \in \Delta^*$ are $\neq 0$. From Theorem 2.12 follows immediately that $W(U)$ is simply transitive on the set of Weyl chambers. Since $W(U)$ and $D(U)$ depend only on $\mathfrak{u}$, they will sometimes be denoted by $W(\mathfrak{u})$ and $D(\mathfrak{u})$, respectively.

In §3 we considered a mapping Φ of $K/M \times \mathfrak{h}_{\mathfrak{p}_*}$ onto U/K. In the present situation this is a mapping of $U^*/T^* \times \mathfrak{t}^*$ onto $U \times U/U^*$. We compare Φ with the mapping $\Psi : (uT, H) \to \exp \operatorname{Ad}(u) H$ which maps $(U/T \times \mathfrak{t}_0)$ onto U and consider the diagram

$$\begin{array}{ccc} U^*/T^* \times \mathfrak{t}^* & \xrightarrow{\Phi} & U \times U/U^* \\ f \big\downarrow & & \big\downarrow g \\ U/T \times \mathfrak{t}_0 & \xrightarrow{\Psi} & U \end{array} \qquad (3)$$

where the mappings f and g are given by

$$f : ((u, u)\, T^*, (H, -H)) \to (uT, 2H),$$

$$g : (u_1, u_2)\, U^* \to u_1 u_2^{-1}.$$

The diagram is then commutative. Since f and g are diffeomorphisms, we conclude from Prop. 3.2:

Proposition 4.2. Let $u_0 \in U$, $H_0 \in \mathfrak{t}_0$. *The mapping* $\Psi : (uT, H) \to$ *exp Ad* (u) H *which maps* $U/T \times \mathfrak{t}_0$ *onto* U *is regular at* $(u_0 T, H_0)$ *if and only if* $\alpha(iH_0)/2\pi$ *is not an integer for any* $\alpha \in \Delta^*$.

In analogy with the notations in §3 we make now the following definition.

Definition. The set $S = \Psi(U/T \times D(U))$ will be called the *singular set* in U and its elements will be called the *singular elements*. The complement $U - S$ will be denoted by U_r, and its elements will be called the *regular elements*. Finally $\mathfrak{t}_r$ shall denote the complement $\mathfrak{t}_0 - D(U)$. It is obvious from Prop. 4.4 below that $\Psi(U/T \times \mathfrak{t}_r) = U_r$.

Lemma 4.3. *Let* $H \in \mathfrak{t}_0$ *and put* $t = \exp H$. *Let* Z_t *denote the centralizer of* t *in* U. *Then*

$$\dim Z_t = \dim T + n,$$

where n *denotes the number of roots* $\alpha \in \Delta^*$ *for which* $\alpha(H) \in 2\pi i Z$.

Proof. The Cartan subalgebra $\mathfrak{t}$ of $\mathfrak{g}$ is invariant under the conjugation of $\mathfrak{g}$ with respect to $\mathfrak{u}$. Using Lemma 3.1, Chapter VI, we can for each $\alpha \in \Delta^*$ select a vector $X_\alpha \neq 0$ in $\mathfrak{g}$ such that $[H, X_\alpha] = \alpha(H) X_\alpha$ for all $H \in \mathfrak{t}$ and such that the vectors $E_\alpha = X_\alpha - X_{-\alpha}$, $F_\alpha = i(X_\alpha + X_{-\alpha})$ belong to $\mathfrak{u}$. Let Δ^* be ordered in some way and let $(\Delta^*)_+$ be the set of positive roots with respect to this ordering. Then E_α, F_α $(\alpha \in (\Delta^*)_+)$ is a basis of $\mathfrak{u}$ (mod $\mathfrak{t}_0$) and

$$[H, E_\alpha] = -i\alpha(H) F_\alpha, \qquad [H, F_\alpha] = i\alpha(H) E_\alpha \tag{4}$$

for all $H \in \mathfrak{t}_0$. It follows that

$$\begin{aligned} \operatorname{Ad}(t) E_\alpha &= \cos(i\alpha(H)) E_\alpha - \sin(i\alpha(H)) F_\alpha, \\ \operatorname{Ad}(t) F_\alpha &= \sin(i\alpha(H)) E_\alpha + \cos(i\alpha(H)) F_\alpha. \end{aligned} \tag{5}$$

The Lie algebra $\mathfrak{L}(Z_t)$ of Z_t is given by

$$\mathfrak{L}(Z_t) = \{X \in \mathfrak{u} : (\exp sX)\, t = t \exp sX \text{ for all } s \in \boldsymbol{R}\}.$$

Since $t \exp sX\, t^{-1} = \exp s \operatorname{Ad}(t) X$, we find

$$\mathfrak{L}(Z_t) = \{X \in \mathfrak{u} : \operatorname{Ad}(t) X = X\}.$$

Consequently, $\dim Z_t$ equals the dimension of the eigenspace of $\operatorname{Ad}(t)$ for the eigenvalue 1. From (5) we see that this eigenspace has dimension $\dim \mathfrak{t}_0$ + twice the number of $\alpha \in (\Delta^*)_+$ for which $\alpha(H) \in 2\pi i\boldsymbol{Z}$.

Proposition 4.4. *Let $x \in U$ and let Z_x denote the centralizer of x in U. Then x is singular (regular) if and only if $\dim Z_x > \dim T$ ($\dim Z_x = \dim T$).*

The element x can be written $x = utu^{-1}$ where $u \in U$, $t \in T$. Then $\dim Z_x = \dim Z_t$ and the proposition follows from Lemma 4.3.

For each $\alpha \in \Delta^*$, put

$$\begin{aligned} \mathfrak{t}_\alpha &= \{H \in \mathfrak{t}_0 : \alpha(H) \in 2\pi i\boldsymbol{Z}\}, \\ T_\alpha &= \{u \in U : \exp \operatorname{Ad}(u) H = \exp H \text{ for each } H \in \mathfrak{t}_\alpha\}. \end{aligned}$$

Then T_α is a closed subgroup of U containing T. Let $(\mathfrak{t}_\alpha)_0$ denote the hyperplane $\alpha(H) = 0$ in $\mathfrak{t}_0$.

Lemma 4.5. *The group T_α is connected and $\dim T_\alpha = \dim T + 2$.*

Proof. It follows from (5) that if $H \in \mathfrak{t}_\alpha$,

$$\operatorname{Ad}(\exp H)\, sE_\alpha = sE_\alpha, \qquad \operatorname{Ad}(\exp H)\, sF_\alpha = sF_\alpha \tag{6}$$

for all $s \in \boldsymbol{R}$. Now, (6) implies that

$$\exp \operatorname{Ad}(\exp sE_\alpha) H = \exp H, \qquad \exp \operatorname{Ad}(\exp sF_\alpha) H = \exp H$$

for all $H \in \mathfrak{t}_\alpha$ and all $s \in \boldsymbol{R}$. It follows that E_α and F_α lie in the Lie algebra of T_α so $\dim T_\alpha \geqslant \dim T + 2$. On the other hand, $T_\alpha \subset (T_\alpha)_0$ if $(T_\alpha)_0$ denotes the centralizer of $(\mathfrak{t}_\alpha)_0$ in U. Then $(T_\alpha)_0$ is the centralizer in U of the closure of $\exp (\mathfrak{t}_\alpha)_0$. By Cor. 2.8, $(T_\alpha)_0$ is connected. Furthermore, $(t_\alpha)_0$ contains an element H such that $\beta(H) \notin 2\pi i \boldsymbol{Z}$ for all $\beta \in \Delta^*$ different from $\pm \alpha$. By Lemma 4.3, $\dim (T_\alpha)_0 \leqslant \dim T + 2$ and Lemma 4.5 follows.

Let $(\mathfrak{t}_\alpha)_r$ denote the subset of $\mathfrak{t}_\alpha$ given by

$$(\mathfrak{t}_\alpha)_r = \{H \in \mathfrak{t}_\alpha : \beta(H) \notin 2\pi i \boldsymbol{Z} \text{ for } \beta \in \Delta^* - \{\alpha \cup -\alpha\}\}$$

and consider the mapping

$$\Psi_\alpha : (uT_\alpha, H) \to \exp \operatorname{Ad}(u) H$$

of $U/T_\alpha \times \mathfrak{t}_\alpha$ into U.

Lemma 4.6. *The mapping Ψ_α is regular on the subset $U/T_\alpha \times (\mathfrak{t}_\alpha)_r$ of $U/T_\alpha \times \mathfrak{t}_\alpha$.*

Proof. Consider the subspace $\mathfrak{u}_\alpha$ of $\mathfrak{u}$ spanned by all the vectors E_β, F_β as β varies through the positive roots in $\Delta^* - \{\alpha \cup -\alpha\}$. The natural mapping of U onto U/T_α has a differential which identifies $\mathfrak{u}_\alpha$ with the tangent space to U/T_α at $\{T_\alpha\}$. If $u \in U$, then as usual, $\tau(u)$ shall denote the mapping $xT_\alpha \to uxT_\alpha$ of U/T_α onto itself. The tangent space to the product $U/T_\alpha \times \mathfrak{t}_\alpha$ can be identified with the subspace $\mathfrak{u}_\alpha + (\mathfrak{t}_\alpha)_0$ of $\mathfrak{u}$, the subspaces $\mathfrak{u}_\alpha$ and $(\mathfrak{t}_\alpha)_0$ of $\mathfrak{u}$ being orthogonal. Now let $u_0 \in U$, $H_0 \in (\mathfrak{t}_\alpha)_r$, $X \in \mathfrak{u}_\alpha$. Then

$$\begin{aligned}\Psi_\alpha(u_0(\exp tX)\, T_\alpha, H_0) &= \exp(\operatorname{Ad}(u_0 \exp tX)\, H_0) \\ &= u_0 \exp(H_0 + t[X, H_0] + O(t^2))\, u_0^{-1}.\end{aligned}$$

Using Theorem 1.7, Chapter II, we obtain

$$(d\Psi_\alpha)_{(u_0T_\alpha, H_0)}\,(d\tau(u_0)\, X, 0) = dL_{u_0 \exp H_0 u_0^{-1}} \circ \operatorname{Ad}(u_0) \frac{1 - e^{-\operatorname{ad} H_0}}{\operatorname{ad} H_0}([X, H_0]). \tag{7}$$

Similarly, if $H \in (\mathfrak{t}_\alpha)_0$,

$$\Psi_\alpha(u_0T_\alpha, H_0 + tH) = \exp(\operatorname{Ad}(u_0)(H_0 + tH))$$

so

$$(d\Psi_\alpha)_{(u_0T_\alpha, H_0)}\,(0, H) = dL_{u_0 \exp H_0 u_0^{-1}} \circ \operatorname{Ad}(u_0)\, H. \tag{8}$$

Combining (7) and (8) we get

$$(d\Psi_\alpha)_{(u_0T_\alpha, H_0)}(d\tau(u_0)X, H) = dL_{u_0 \exp H_0 u_0^{-1}} \circ \mathrm{Ad}(u_0)\{(\mathrm{Ad}(\exp - H_0) - 1)X + H\}.$$

Since $(\mathrm{Ad}(\exp(-H_0)) - 1)X \in \mathfrak{u}_\alpha$, the right-hand side of this formula can vanish only if $H = 0$ and $\mathrm{Ad}(\exp H_0)X = X$. In view of (5), this would either require $X = 0$ or $\beta(H_0) \in 2\pi i \boldsymbol{Z}$ for some $\beta \in \Delta^* - \{\alpha \cup -\alpha\}$. But this last possibility is excluded by the assumption that $H_0 \in (\mathfrak{t}_\alpha)_r$. This proves the lemma.

Theorem 4.7. *Let S denote the singular set in U and let conj (U) denote the set of points in U which are conjugate to e. Then*

$$\dim S = \dim \mathrm{conj}(U) = \dim U - 3.$$

Proof. Let $\alpha \in \Delta^*$. It is obvious that

$$\Psi_\alpha(U/T_\alpha \times \mathfrak{t}_\alpha) = \Psi(U/T \times \mathfrak{t}_\alpha).$$

Using Lemmas 4.5 and 4.6 we find

$$\begin{aligned}\dim \Psi_\alpha(U/T_\alpha \times (\mathfrak{t}_\alpha)_r) &\geqslant \dim U/T_\alpha + \dim \mathfrak{t}_\alpha \\ &= \dim U - (\dim T + 2) + \dim T - 1 = \dim U - 3.\end{aligned}$$

Using Lemma 12.3 and Prop. 12.1 in the Appendix we have

$$\dim \Psi_\alpha(U/T_\alpha \times (\mathfrak{t}_\alpha)_r) \leqslant \dim \Psi_\alpha(U/T_\alpha \times \mathfrak{t}_\alpha) \leqslant \dim U - 3$$

so

$$\dim \Psi(U/T \times \mathfrak{t}_\alpha) = \dim U - 3.$$

Since

$$S = \bigcup_{\alpha \in \Delta^*} \Psi(U/T \times \mathfrak{t}_\alpha)$$

and since $\Psi(U/T \times \mathfrak{t}_\alpha)$ is closed in U, it follows from the sum theorem (Appendix, Theorem 12.2) that $\dim S = \dim U - 3$. The statement about conj(U) is proved in the same manner.

§ 5. Control over the Singular Set

We recall that an element $x \in U$ is regular or singular according to whether $\dim Z_x$ is equal or larger than $\dim T$. In view of Lemma 4.3 we can make the following definition.

Definition. An element $x \in U$ is called *singular of order* $n/2$ if $\dim Z_x = \dim T + n$ and $n > 0$.

Let k be an integer > 0, and let S_k denote the set of singular elements in U of order k. Then $S = \bigcup_{k>0} S_k$.

Consider now a fixed element $x_0 \in S_k$ and select $u \in U$, $H_0 \in \mathfrak{t}_0$ such that $x_0 = \exp \operatorname{Ad}(u) H_0$. Then there exist exactly k positive roots in Δ^*, say $\alpha_1, ..., \alpha_k$, for which $\alpha_j(H_0) \in 2\pi i \boldsymbol{Z}$ $(1 \leqslant j \leqslant k)$. Consider the group

$$T_{\alpha_1 \ldots \alpha_k} = \{u \in U : \exp \operatorname{Ad}(u)H = \exp H \text{ for each } H \in \mathfrak{t}_{\alpha_1} \cap ... \cap \mathfrak{t}_{\alpha_k}\}.$$

Lemma 5.1.

$$\dim T_{\alpha_1 \ldots \alpha_k} = \dim T + 2k.$$

Proof. We imitate the proof of Lemma 4.5. Just as there it can be verified that the vectors E_{α_j}, F_{α_j} $(1 \leqslant j \leqslant k)$ belong to the Lie algebra of $T_{\alpha_1 \ldots \alpha_k}$. Consequently,

$$\dim T_{\alpha_1 \ldots \alpha_k} \geqslant \dim T + 2k.$$

On the other hand,

$$\dim Z_{\exp H_0} = \dim T + 2k$$

according to Lemma 4.3. Since $T_{\alpha_1 \ldots \alpha_k} \subset Z_{\exp H_0}$, the lemma is proved.

Consider now the subset $(\mathfrak{t}_{\alpha_1 \ldots \alpha_k})_r$ of $\mathfrak{t}_{\alpha_1} \cap ... \cap \mathfrak{t}_{\alpha_k}$ consisting of all points H such that $\beta(H) \notin 2\pi i \boldsymbol{Z}$ unless β is among the roots $\pm \alpha_j$ $(1 \leqslant j \leqslant k)$. Then $(\mathfrak{t}_{\alpha_1 \ldots \alpha_k})_r$ is an open subset of $\mathfrak{t}_{\alpha_1} \cap ... \cap \mathfrak{t}_{\alpha_k}$ containing H_0. Consider the mapping

$$\Psi_{\alpha_1 \ldots \alpha_k} : (uT_{\alpha_1 \ldots \alpha_k}, H) \to \exp \operatorname{Ad}(u) H$$

of $(U/T_{\alpha_1 \ldots \alpha_k}) \times (\mathfrak{t}_{\alpha_1} \cap ... \cap \mathfrak{t}_{\alpha_k})$ into U.

Lemma 5.2. *The mapping* $\Psi_{\alpha_1 \ldots \alpha_k}$ *is regular on the subset*

$$(U/T_{\alpha_1 \ldots \alpha_k}) \times (\mathfrak{t}_{\alpha_1 \ldots \alpha_k})_r.$$

The proof is an immediate extension of that of Lemma 4.6 and can be omitted.

Definition. Let N be a subset of a manifold M; N is called a *quasi-submanifold* of M if there exists a connected manifold N^* and a regular differentiable mapping $f : N^* \to M$ such that $f(N^*) = N$.

If $\mathfrak{s}$ is a connected component of $(\mathfrak{t}_{\alpha_1 \ldots \alpha_k})_r$, then the image

$$\Psi(U/T \times \mathfrak{s}) = \Psi_{\alpha_1 \ldots \alpha_k}((U/T_{\alpha_1 \ldots \alpha_k}) \times \mathfrak{s})$$

is a quasisubmanifold of U due to Lemma 5.2.

Now let

$$\mathfrak{t}_e = \{H \in \mathfrak{t}_0 : \exp H = e\}.$$

The set $\mathfrak{t}_e$ is called the *unit lattice* in $\mathfrak{t}_0$. Then clearly $\alpha(H) \in 2\pi i \boldsymbol{Z}$ for all $H \in \mathfrak{t}_e$ and all $\alpha \in \Delta^*$. Therefore, if we consider $\mathfrak{t}_e$ as a group of translations of $\mathfrak{t}_0$, it leaves the diagram $D(U)$ invariant. Moreover, each transformation from $\mathfrak{t}_e$ maps $(\mathfrak{t}_{\alpha_1 \ldots \alpha_k})_r$ onto itself and therefore permutes the various components $\mathfrak{s}$ of $(\mathfrak{t}_{\alpha_1 \ldots \alpha_k})_r$.

Lemma 5.3. *There are only finitely many components of* $(\mathfrak{t}_{\alpha_1 \ldots \alpha_k})_r$ *which are incongruent modulo* $\mathfrak{t}_e$.

Proof. The factor space $\mathfrak{t}_0/\mathfrak{t}_e$ can be identified with T so the group $\mathfrak{t}_e$ has a bounded fundamental domain in $\mathfrak{t}_0$. Now on each component $\mathfrak{s}$ of $(\mathfrak{t}_{\alpha_1 \ldots \alpha_k})_r$ the roots $\alpha_1, \ldots, \alpha_k$ are constants and the other positive roots vary through an interval of $2\pi i$. It follows that the components $\mathfrak{s}$ are uniformly bounded. Consequently, there exists a closed ball $\mathfrak{b}$ in $\mathfrak{t}_0$ such that for each component $\mathfrak{s}$ of $(\mathfrak{t}_{\alpha_1 \ldots \alpha_k})_r$ there exists a vector $H \in \mathfrak{t}_e$ such that the translate of $\mathfrak{s}$ by $-H$, that is, $\mathfrak{s} - H$, lies in $\mathfrak{b}$. Since each root $\alpha \in \Delta^*$ is bounded on $\mathfrak{b}$ there can only be finitely many such sets $\mathfrak{s} - H$ in $\mathfrak{b}$. This proves the lemma.

The roots $\alpha_1, \ldots, \alpha_k$ were obtained by means of an arbitrary point $H_0 \in \mathfrak{t}_0$ which lies on exactly k hyperplanes in the diagram. As H_0 varies through all such points we get finitely many systems $(\alpha_1, \ldots, \alpha_k)$ of k positive roots. Let S_k^i $(i = 1, 2, \ldots)$ denote the images $\Psi(U/T \times \mathfrak{s})$ as $\mathfrak{s}$ varies through the components of $(\mathfrak{t}_{\alpha_1 \ldots \alpha_k})_r$ for the various systems $(\alpha_1, \ldots, \alpha_k)$.

Lemma 5.4. *The set of singular points of order k is a finite union*

$$S_k = \bigcup_i S_k^i,$$

where each S_k^i *is a quasisubmanifold of* U. *If* S_k^i *is given the relative topology of* U, *its boundary is contained in* $\bigcup_{p>k} S_p$.

Proof. The finiteness statement is a consequence of Lemma 5.3 since the components of $(\mathfrak{t}_{\alpha_1 \ldots \alpha_k})_r$ which are congruent mod $\mathfrak{t}_e$ give rise to the same set S_k^i. The last statement follows from the fact that the boundary points of $(\mathfrak{t}_{\alpha_1 \ldots \alpha_k})_r$ lie on more than k hyperplanes of the diagram.

Proposition 5.5. *Let $\gamma(t)$ and $\gamma'(t)$ $(0 \leqslant t \leqslant 1)$ be two continuous curves in U_r. Then γ is homotopic to γ' in U_r if and only if they are homotopic in U.*

Proof. We may assume that γ and γ' are homotopic in U. Let p and q, respectively, denote the beginning and end of γ. Let f denote the mapping of the unit square $\square$ $(0 \leqslant s \leqslant 1,\ 0 \leqslant t \leqslant 1)$ into U which sets up the assumed homotopy $\gamma \sim \gamma'$. In other words, $f(0, t) = \gamma(t)$, $f(1, t) = \gamma'(t)$ for $0 \leqslant t \leqslant 1$ and $f(s, 0) = p$, $f(s, 1) = q$ for $0 \leqslant s \leqslant 1$. We have to deform $f(\square)$ into U_r in such a way that the points p and q remain fixed. The deformations considered below are always understood to take place with p and q kept fixed.

Let $2m$ denote the number of elements in Δ^*. Then S_m is the center of U and S_p is empty if $p > m$. Since S_m is finite and $\dim U \geqslant 3$, we can, using Prop. 12.6 (Appendix), deform $f(\square)$ such that the resulting deformation, say $f_1(\square)$, satisfies

$$f_1(\square) \cap S_m = \emptyset.$$

Here f_1 denotes a new homotopy between γ and γ'. It follows that there exists a compact neighborhood N_1 of $f_1(\square)$ such that

$$N_1 \cap S_m = \emptyset.$$

Suppose now that we have found deformations $f_1(\square), \ldots, f_{m-k}(\square)$ of $f(\square)$ and compact neighborhoods $N_1, \ldots, N_{m-k}$ of $f_1(\square), \ldots, f_{m-k}(\square)$, respectively, such that

$$\begin{aligned} &N_1 \supset \ldots \supset N_{m-k}, \\ &N_1 \cap S_m = \ldots = N_{m-k} \cap S_{k+1} = \emptyset. \end{aligned} \tag{1}$$

We shall then show that there exists a deformation $f_{m-k+1}(\square)$ of $f_{m-k}(\square)$ and a compact neighborhood N_{m-k+1} of $f_{m-k+1}(\square)$ such that (1) holds with k replaced by $k-1$. Then the validity of (1) for $k = m-1$ implies its validity for $k = 0$, proving the proposition.

Starting now from (1), we consider the set $N_{m-k} \cap S_k^{\,i}$ where $S_k^{\,i}$ is one of the sets from Lemma 5.4. The last statement of that lemma, together with (1), implies that $N_{m-k} \cap S_k^{\,i}$ is compact. For a suitable system $(\alpha_1, \ldots, \alpha_k)$ of positive roots, the set $N_{m-k} \cap S_k^{\,i}$ is the image, under $\Psi_{\alpha_1 \ldots \alpha_k}$, of a compact subset of $U/T_{\alpha_1 \ldots \alpha_k} \times \mathfrak{s}$. This compact set can be covered by finitely many sets $U_j \times \mathfrak{b}_j$ $(j \in J)$, where the U_j are open subsets of $U/T_{\alpha_1 \ldots \alpha_k}$ and the $\mathfrak{b}_j$ are open balls whose closure is contained in $\mathfrak{s}$ such that for each $j \in J$, $\Psi_{\alpha_1 \ldots \alpha_k}$ maps some neighborhood of $\bar{U}_j \times \bar{\mathfrak{b}}_j$ in a one-to-one manner into S_k^i. We put now

$$\Psi_{\alpha_1 \ldots \alpha_k}(U_j \times \mathfrak{b}_j) = B_j \qquad (j \in J),$$

and turn B_j into a manifold diffeomorphic to $U_j \times \mathfrak{b}_j$. Then:

(a) Each B_j is a submanifold and a topological subspace of U.

(b) $B_j \subset S_k^i$ for each $j \in J$.

(c) $N_{m-k} \cap (\bigcup_{j \in J} B_j) = N_{m-k} \cap S_k^i$.

Consider now a fixed B_j. Using Props. 12.4 and 12.6 in the Appendix we can deform $f_{m-k}(\square)$ in the interior of N_{m-k} such that the resulting deformation, say $g_{m-k}(\square)$, is disjoint from the closure of B_j. Hence we can surround $g_{m-k}(\square)$ with a compact neighborhood $V_{m-k} \subset N_{m-k}$ disjoint from B_j. We treat the (finitely many) B_j ($j \in J$) successively in the same manner. The resulting deformation $G_{m-k}(\square)$ is then enclosed in a compact neighborhood W_{m-k} which is contained in N_{m-k} and is disjoint from $\bigcup_{j \in J} B_j$. It follows from (*c*) that $W_{m-k} \cap S_k^i = \emptyset$. The preceding process can be applied to the sets $S_k^1, S_k^2, \ldots$ successively. Since these are only finite in number, the result is a deformation $f_{m-k+1}(\square)$ enclosed in a compact neighborhood $N_{m-k+1} \subset N_{m-k}$ such that

$$N_{m-k+1} \cap S_k = \emptyset.$$

This concludes the proof.

§ 6. The Fundamental Group and the Center

Consider now the open set $\mathfrak{t}_r = \mathfrak{t}_0 - D(U)$. Let P_0 denote a component of $\mathfrak{t}_r$ whose closure $\bar{P}_0$ contains the origin. The polyhedron P_0 is an intersection of half-spaces in $\mathfrak{t}_0$, hence P_0 is an open, convex set. Since $\alpha(H) \in 2\pi i Z$ for all $\alpha \in \Delta^*$ and all $H \in \mathfrak{t}_e$, it is clear that each point in $\mathfrak{t}_e \cap \bar{P}_0$ is a vertex of P_0.

Theorem 6.1. *The number of points in $\mathfrak{t}_e \cap \bar{P}_0$ equals the order of the fundamental group $\pi_1(U)$ of U.*

The proof will require a few lemmas.

Lemma 6.2. *The coset space U/T is simply connected.*

Proof. Let $(\tilde{U}, \gamma)$ denote the simply connected covering group of U and put $\tilde{T} = \gamma^{-1}(T)$. Then $\tilde{U}$ is compact and the Lie algebra of $\tilde{T}$ is a maximal abelian subalgebra of the Lie algebra of $\tilde{U}$. Since the centralizer of a torus is connected, $\tilde{T}$ is a maximal torus in $\tilde{U}$. The space $\tilde{U}/\tilde{T}$ is simply connected and homeomorphic to U/T under the mapping $u\tilde{T} \to \gamma(u)\, T$, $(u \in \tilde{U})$.

Lemma 6.3. *Let ψ denote the restriction of Ψ to $U/T \times P_0$. Then $(U/T \times P_0, \psi)$ is the universal covering space of U_r.*

Proof. The connectedness of U_r is clear from Cor. 12.5 in the Appendix. Since P_0 is convex, it is simply connected. Hence $U/T \times P_0$ is simply connected. Let P_1 be an arbitrary component of $\mathfrak{t}_r$. If the images $\Psi(U/T \times P_0)$ and $\Psi(U/T \times P_1)$ are not disjoint there exist elements $H_0 \in P_0$, $H_1 \in P_1$, $x \in U$ such that $x \exp H_0 x^{-1} = \exp H_1$. It follows that the automorphism $u \to xux^{-1}$ $(u \in U)$ maps the centralizer $Z_{\exp H_0}$ onto the centralizer $Z_{\exp H_1}$. Owing to Lemma 4.3, T is the identity component of these centralizers. Consequently $xTx^{-1} = T$ and there exists an element $s \in W(U)$ which coincides with the restriction Ad (x) to $\mathfrak{t}_0$. Hence $\exp sH_0 = \exp H_1$, $sH_0 \in \mathfrak{t}_0$ and there exists a vector $A \in \mathfrak{t}_e$ such that $H_1 = sH_0 + A$. Since the groups $W(U)$ and $\mathfrak{t}_e$ leave the diagram invariant it follows that the transformation $H \to sH + A$ of $\mathfrak{t}_0$ maps P_0 onto P_1. Consequently $\Psi(U/T \times P_0) = \Psi(U/T \times P_1)$. The connectedness of U_r now implies that ψ maps $U/T \times P_0$ onto U_r.

Let $y \in U_r$ and suppose $\psi(u_1T, H_1) = \psi(u_2T, H_2) = y$. Then $u_1 \exp H_1 u_1^{-1} = u_2 \exp H_2 u_2^{-1}$, so by the above the element $x = u_2^{-1}u_1$ belongs to the normalizer N_T of T in U; if s denotes the corresponding Weyl group element, then for a suitable $A \in \mathfrak{t}_e$ the transformation $H \to sH + A$ maps P_0 onto itself and maps H_1 onto H_2. Thus the transformation $t \to xtx^{-1}$ leaves $\exp P_0$ invariant and maps $\exp H_1$ onto $\exp H_2$. Since exp is one-to-one on P_0, this shows that the preimage $\psi^{-1}(y)$ is in one-to-one correspondance with the subgroup of N_T/T which maps $\exp P_0$ into itself, in particular $\psi^{-1}(y)$ is finite and its cardinality independent of y. Since in addition ψ is a local homeomorphism (by Prop. 4.2), the lemma is proved.

Lemma 6.4. *The number of points in $\mathfrak{t}_e \cap \bar{P}_0$ equals the number of elements in $\pi_1(U_r)$, the fundamental group of U_r.*

Proof. Consider a neighborhood

$$V = \{X \in \mathfrak{u} : -B(X, X) < \rho^2\}$$

of 0 in $\mathfrak{u}$. We can select $\rho > 0$ so small that:

(a) $|\alpha(H)| < 2\pi$ *for* $H \in V \cap \mathfrak{t}_0$ *and* $\alpha \in \Delta^*$.

(b) exp *is one-to-one on* $2V$.

(c) $\exp(V \cap \mathfrak{t}_0) = (\exp V) \cap T$.

It is trivial to satisfy (a) and (b). For (c) one just has to make use of the fact that T is a topological subgroup of U (Lemma 2.5, Chapter II).

Fix an element $x \in U_r \cap (\exp V)$ and consider the inverse image

$$\psi^{-1}(x) = \{(u_1T, H_1), ..., (u_rT, H_r)\}.$$

For each i, $1 \leqslant i \leqslant r$, we have

$$\exp H_i \in u_i^{-1} (\exp V)\, u_i \subset \exp V.$$

From (c) and (b) it follows that there exists a *unique* vector $A_i \in \mathfrak{t}_e$ such that $H_i - A_i \in V$.

On the other hand, since

$$u_i \exp H_i\, u_i^{-1} = u_j \exp H_j\, u_j^{-1}, \qquad 1 \leqslant i, j \leqslant r,$$

it follows as in the last lemma, that $uTu^{-1} = T$ if $u = u_i^{-1}u_j$. Let s_{ij} denote the restriction of Ad (u) to $\mathfrak{t}_0$. Then there exists a vector $A_{ij} \in \mathfrak{t}_e$ such that $H_i = s_{ij}H_j + A_{ij}$. Since s_{ij} leaves $\mathfrak{t}_e$ invariant, there exists a vector $A^* \in \mathfrak{t}_e$ such that $s_{ij}A^* = A_i - A_{ij}$. Then $H_i - A_i = s_{ij}(H_j - A^*)$ so $H_j - A^* \in V$. By the uniqueness above, $A^* = A_j$ so

$$H_i - A_i = s_{ij}(H_j - A_j) \qquad (1 \leqslant i, j \leqslant r). \tag{1}$$

We shall now draw some consequences of this relation.

(d) *The points $H_1, ..., H_r$ are all different.*

In fact, if for example $H_1 = H_2$, then $A_1 = A_2$. But $W(U)$ is simply transitive on the set of Weyl chambers so (1) implies that $s_{12} = I$. Hence $u_1T = u_2T$, which is a contradiction.

(e) *The points $A_1, ..., A_r$ are all different.*

Suppose to the contrary that for example $A_1 = A_2$. The segment l joining H_1 and H_2 lies in P_0. It follows that the translated segment $l - A_1$ lies in $\mathfrak{t}_r$; in particular $H_1 - A_1$ and $H_2 - A_1$ lie in the same Weyl chamber. On the other hand, (1) implies $H_1 - A_1 = s_{12}(H_2 - A_1)$ so again by the simple transitivity $s_{12} = I$ which is a contradiction.

(f) *The points $A_1, ..., A_r$ belong to $\bar{P}_0$.*

In fact, if $A_i \notin \bar{P}_0$, then the interior of the segment from H_i to A_i intersects the boundary of P_0. Therefore, there exists a root $\alpha \in \Delta^*$ such that $|\alpha(H_i - A_i)| > 2\pi$. In view of (a), this contradicts $H_i - A_i \in V$.

(g) $\mathfrak{t}_e \cap \bar{P}_0$ *consists of precisely the points $A_1, ..., A_r$.*

Suppose to the contrary, that there were a point $A \in \mathfrak{t}_e \cap \bar{P}_0$ which does not occur among $A_1, \ldots, A_r$. Let C denote the Weyl chamber in $\mathfrak{t}_0$ with the property that P_0 is contained in the translated set $C + A$. There exists a unique element $s \in W(U)$ such that $s(H_1 - A_1) \in C$. Then $A + s(H_1 - A_1) \in C + A$. Moreover, due to (a), the open segment from 0 to $s(H_1 - A_1)$ lies in $\mathfrak{t}_r$; the same is true of the translated segment from A to $A + s(H_1 - A_1)$. Consequently the point

$$H_A = A + s(H_1 - A_1)$$

lies in P_0. Let u be an element in U such that $\mathrm{Ad}\,(u)$ and s^{-1} coincide on $\mathfrak{t}_0$; then

$$\begin{aligned}\psi(u_1 uT, H_A) &= u_1 u \exp\,(A + s(H_1 - A_1))\, u^{-1} u_1^{-1}\\ &= u_1 \exp\,(H_1 - A_1)\, u_1^{-1} = \psi(u_1 T, H_1) = x.\end{aligned}$$

On the other hand, $H_A \neq H_i$ for $i = 1, \ldots, r$ because the relation $H_A = H_i$ implies $|H_i - A| = |H_1 - A_1|$ which equals $|H_i - A_i|$ due to (1). But $|H_i - A| = |H_i - A_i|$ implies $A - A_i \in 2V$ which contradicts (b).

This finishes the proof of Lemma 6.4.

Now let u_0 be a point in U_r and γ a curve in U beginning and ending at u_0. Due to Prop. 12.4 in the Appendix, γ is homotopic to a curve $\gamma' \subset U_r$. Hence it follows from Prop. 5.5 that $\pi_1(U)$ and $\pi_1(U_r)$ are isomorphic. This concludes the proof of Theorem 6.1.

Let (M^*, π) be a simply connected covering space of a topological space M. A homeomorphism φ of M^* such that $\pi \circ \varphi = \pi$ is called a *covering transformation* of M^*. These homeomorphisms form a group which is isomorphic with the fundamental group $\pi_1(M)$ of M.

What are the covering transformations corresponding to the covering space $(U/T \times P_0, \psi)$ of U_r? In terms of the notation of Lemma 6.4, the transformation

$$(vT, H) \to (vu_1^{-1}u_i T, A_i + s_{i1}(H - A_1))$$

is a covering transformation of $U/T \times P_0$. In fact, due to (1) we have

$$A_i + s_{i1}(H - A_1) \in P_0 \qquad \text{for } H \in P_0$$

and

$$\psi(vu_1^{-1}u_i T, A_i + s_{i1}(H - A_1)) = \psi(vT, H)$$

for $H \in P_0$, $v \in U$. These covering transformations, $\pi_1(U)$ in number, are all different because the images $H_i = A_i + s_{i1}(H_1 - A_1)$ are all different as shown above. The group $\pi_1(U)$ is isomorphic with the group of transformations

$$\varphi_i : H \rightarrow A_i + s_{i1}(H - A_1)$$

of $\mathfrak{t}_0$, each of which maps P_0 into itself. The orbit of A_1 under this group is $\mathfrak{t}_e \cap \bar{P}_0$.

Lemma 6.5. *Let Z denote the center of U and let $\mathfrak{t}(\mathfrak{u})$ denote the set of points $H \in \mathfrak{t}_0$ for which exp $H \in Z$. Then*

$$\mathfrak{t}(\mathfrak{u}) = \{H \in \mathfrak{t}_0 : \alpha(H) \in 2\pi i \boldsymbol{Z} \text{ for each } \alpha \in \Delta^*\}.$$

Proof. It is obvious that $\mathfrak{t}(\mathfrak{u})$ is the set of $H \in \mathfrak{t}_0$ for which $\mathrm{Ad}(\exp H) = I$. The lemma now follows from relations (5), §4. The notation $\mathfrak{t}(\mathfrak{u})$ is to indicate that the set in question is the same for all groups that have Lie algebra $\mathfrak{u}$.

Corollary 6.6. *Let $\tilde{U}$ denote the universal covering group of U and let $\tilde{Z}$ denote the center of $\tilde{U}$. The number of points in $\mathfrak{t}(\mathfrak{u}) \cap \bar{P}_0$ (in geometric terms: the number of vertices of P_0 in $\mathfrak{t}(\mathfrak{u})$) equals the order of $\tilde{Z}$.*

Proof. This corollary results from using Theorem 6.1 on the group $\mathrm{Ad}(\tilde{U}) = \tilde{U}/\tilde{Z}$. For this group, the unit lattice $\mathfrak{t}_e$ coincides with $\mathfrak{t}(\mathfrak{u})$ and the fundamental group $\pi_1(\mathrm{Ad}(\tilde{U}))$ is isomorphic to $\tilde{Z}$.

Example. Let $U = \boldsymbol{SU}(n)$, the group of unitary $n \times n$ matrices of determinant one. The Lie algebra $\mathfrak{u} = \mathfrak{su}(n)$ consists of all $n \times n$ skew Hermitian matrices of trace 0, and the complexification $\mathfrak{g}$ is the Lie algebra $\mathfrak{sl}(n, \boldsymbol{C})$ of all $n \times n$ matrices of trace 0. The subset $\mathfrak{t}_0$ of $\mathfrak{u}$ consisting of all $n \times n$ purely imaginary diagonal matrices of trace 0 is a maximal abelian subalgebra of $\mathfrak{u}$. The subspace $\mathfrak{t}$ of $\mathfrak{g}$ generated by $\mathfrak{t}_0$ is a Cartan subalgebra of $\mathfrak{g}$ and consists of all diagonal matrices of trace 0. Let E_{ij} denote the matrix $(\delta_{ai}\delta_{bj})_{1 \leq a \leq n, 1 \leq b \leq n}$, and for each $H \in \mathfrak{t}$ let $e_i(H)$ denote the ith diagonal element in H. Then

$$[H, E_{ij}] = (e_i(H) - e_j(H))\, E_{ij}, \qquad H \in \mathfrak{t}, \tag{2}$$

so the linear function $\alpha_{ij} : H \rightarrow e_i(H) - e_j(H)$ is a root of $\mathfrak{g}$ with respect to $\mathfrak{t}$. In this manner we obtain $n(n-1)$ nonzero roots; on the other hand, the set Δ^* of all nonzero roots (of $\mathfrak{g}$ with respect to $\mathfrak{t}$) contains $\dim \mathfrak{g} - \dim \mathfrak{t} = n^2 - 1 - (n-1) = n^2 - n$ elements. Consequently

$$\Delta^* = \{\alpha_{ij} : 1 \leqslant i \neq j \leqslant n\}.$$

If $H, H' \in \mathfrak{t}$, then (2) implies that

$$B(H, H') = \operatorname{Tr}(\operatorname{ad} H \operatorname{ad} H') = \sum_{1 \le i,j \le n} (e_i(H) - e_j(H))(e_i(H') - e_j(H'))$$

$$= \sum_{i,j} e_i(HH') + e_j(HH') - \sum_{i,j} e_j(H)\, e_i(H') + e_i(H)\, e_j(H')$$

so

$$B(H, H') = 2n \operatorname{Tr}(HH'). \tag{3}$$

Now given a matrix $X \in \mathfrak{su}(n)$, there exists an element $u \in \boldsymbol{SU}(n)$ such that $uXu^{-1} \in \mathfrak{t}_0$. This well-known fact about matrices is a special case of Theorem 6.4, Chapter V. Since the mapping $X \to uXu^{-1}$ is an automorphism of $\mathfrak{su}(n)$ it follows that

$$B(X, Y) = 2n \operatorname{Tr}(XY)$$

for all $X, Y \in \mathfrak{su}(n)$, hence for all $X, Y \in \mathfrak{sl}(n, \boldsymbol{C})$. Let H_{ij} denote the vector in $\mathfrak{t}$ determined by

$$B(H_{ij}, H) = \alpha_{ij}(H), \qquad H \in \mathfrak{t}.$$

Then

$$H_{ij} = \frac{1}{2n}(E_{ii} - E_{jj}).$$

The case $n = 3$. Here $\dim \mathfrak{t}_0 = 2$ and the roots in Δ^* are given by $\pm\alpha_{12}$, $\pm\alpha_{13}$, $\pm\alpha_{23}$. The angle θ between the vectors H_{ij} and H_{kl} is given by

$$\cos\theta = \frac{B(H_{ij}, H_{kl})}{(B(H_{ij}, H_{ij})\, B(H_{kl}, H_{kl}))^{1/2}}.$$

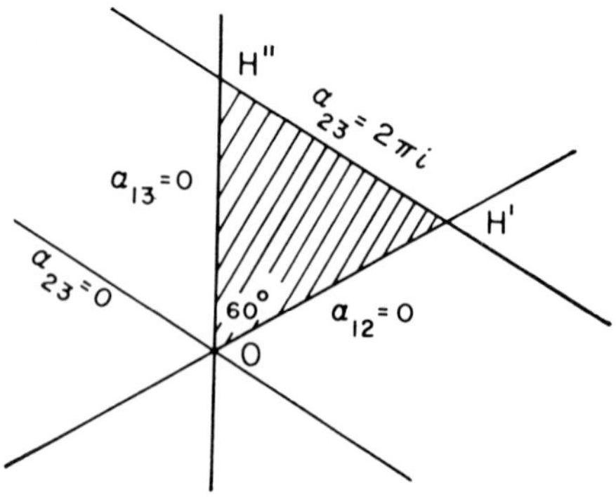

FIG. 2

The lines $\alpha_{12} = 0$, $\alpha_{13} = 0$, $\alpha_{23} = 0$ in $\mathfrak{t}_0$ are therefore situated as in Fig. 2. For the polyhedron P_0 we can select the triangle formed by the lines $\alpha_{12} = 0$, $\alpha_{13} = 0$ and $\alpha_{23} = 2\pi i$. The vectices of this triangle are the origin and the points

$$H' = i\frac{2\pi}{3}(E_{11} + E_{22} - 2E_{33})$$

$$H'' = i\frac{2\pi}{3}(-E_{11} + 2E_{22} - E_{33}).$$

If follows from Cor. 6.6 that the center $\tilde{Z}$ of $\tilde{U}$ has order 3 in this case. Moreover, since $\exp H' \neq e$ and $\exp H'' \neq e$, we see that $\mathfrak{t}_e \cap \bar{P}_0$ consists of the origin alone. In view of Theorem 6.1, the group $\boldsymbol{SU}(3)$ is simply connected. The same argument shows $\boldsymbol{SU}(n)$ simply connected $(n \geqslant 2)$.

Corollary 6.6 shows, that the order of $\tilde{Z}$ (in É. Cartan's terminology "indice de connexion") can be determined from the root pattern of $\mathfrak{u}$ as is done above in a very simple case. In Cartan's paper "La géométrie des groupes simples," *Annali di Mat.* **4** (1927), this method is used to determine the order of $\tilde{Z}$ corresponding to all simple compact Lie algebras. The result will be obtained in a different way in Chapter X.

A simple complement to Theorem 6.1 and Corollary 6.6 shows that the group $\pi_1(U)$ itself and not just its order, can be read off from the diagram and $\mathfrak{t}_e$.

Theorem 6.7. *Let $\tilde{\mathfrak{t}}_e$ denote the unit lattice for the group $\tilde{U}$. Considering $\tilde{\mathfrak{t}}_e$, $\mathfrak{t}_e$ and $\mathfrak{t}(\mathfrak{u})$ as groups of translations of $\mathfrak{t}_0$, the following isomorphisms hold:*

$$\pi_1(U) \approx \mathfrak{t}_e/\tilde{\mathfrak{t}}_e, \qquad \tilde{Z} \approx \mathfrak{t}(\mathfrak{u})/\tilde{\mathfrak{t}}_e.$$

Proof. Let $\tilde{T}$ denote the analytic subgroup of $\tilde{U}$ with Lie algebra $\mathfrak{t}_0$. Then the mapping $\exp : \mathfrak{u} \to \tilde{U}$ induces a homomorphism α of $\mathfrak{t}_0$ onto $\tilde{T}$. Then $\alpha(\mathfrak{t}(\mathfrak{u})) = \tilde{Z}$, $\tilde{\mathfrak{t}}_e = \alpha^{-1}(e)$ and $\alpha(\mathfrak{t}_e) = Z_1$ where Z_1 is the subgroup of $\tilde{Z}$ such that $U = \tilde{U}/Z_1$. Since $\pi_1(U) \approx Z_1$ the theorem follows immediately.

Corollary 6.8. *The diagram $D(\mathfrak{u})$ and the unit lattice $\mathfrak{t}_e$ determine the group U up to isomorphism.*

In fact, the Lie algebra $\mathfrak{u}$ is determined by $\mathfrak{t}_0$ and $D(\mathfrak{u})$ up to isomorphism (Theorem 5.4, Cor. 7.3, Chapter III). The group $\tilde{U}$ is determined by $\mathfrak{u}$ up to isomorphism and $U = \tilde{U}/\alpha(\mathfrak{t}_e)$ in the notation above.

§7. The Affine Weyl Group

Keeping the notation of §4–6, let $\mathfrak{t}_\Delta \subset \mathfrak{t}_0$ denote the set of integral linear combinations of the vectors

$$\frac{4\pi i}{\langle \alpha, \alpha \rangle} H_\alpha \qquad (\alpha \in \Delta^*).$$

Here the vector $H_\alpha \in \mathfrak{t}$ is determined by $B(H, H_\alpha) = \alpha(H)$ $(H \in \mathfrak{t})$. Let Γ_Δ denote the group of linear transformations of $\mathfrak{t}_0$ generated by the reflections in all the hyperplanes in the diagram $D(\mathfrak{u})$. This group is called the *affine Weyl group*. A component of $\mathfrak{t}_r$ will be called a *cell.*

Lemma 7.1. *Viewing $\mathfrak{t}_\Delta$ as a group of translations of $\mathfrak{t}_0$ we have*

$$\Gamma_\Delta = \mathfrak{t}_\Delta \cdot W(\mathfrak{u}) \qquad (\textit{semidirect product};\ \mathfrak{t}_\Delta\ \textit{normal}).$$

In particular, Γ_Δ leaves the diagram invariant.

Proof. Each element in Γ_Δ has the form $H \to A(H) + B$ where A is a linear transformation and B a vector. If this is given by a composition of reflections in certain planes, then $H \to A(H)$ is the corresponding composition of reflections in parallel planes through 0. Thus we have a homomorphism of Γ_Δ onto $W(\mathfrak{u})$, and the kernel consists of the translations in Γ_Δ. Those are given by the image $\Gamma_\Delta \cdot 0$, and it remains to prove that this equals $\mathfrak{t}_\Delta$. The reflection $\sigma_{\alpha,n}$ in the hyperplane $\alpha(H) = 2\pi in$ $(n \in \mathbf{Z})$ can be written

$$\sigma_{\alpha,n}(H) = s_\alpha(H) - s_\alpha\left(\frac{2\pi in}{\langle \alpha, \alpha \rangle} H_\alpha\right) + \frac{2\pi i}{\langle \alpha, \alpha \rangle} H_\alpha \cdot n$$

$$= H - \frac{2\alpha(H)}{\langle \alpha, \alpha \rangle} H_\alpha + \frac{4\pi in}{\langle \alpha, \alpha \rangle} H_\alpha;$$

so if $\beta \in \Delta^*$, $m \in \mathbf{Z}$,

$$\sigma_{\alpha,n}\sigma_{\beta,m}(0) = \sigma_{\alpha,n}\left(\frac{4\pi im}{\langle \beta, \beta \rangle} H_\beta\right) = m\frac{4\pi i}{\langle \beta, \beta \rangle} H_\beta + k\frac{4\pi i}{\langle \alpha, \alpha \rangle} H_\alpha,$$

where k is the integer $n - 2m\langle \alpha, \beta \rangle / \langle \beta, \beta \rangle$. An obvious induction shows that

$$\Gamma_\Delta \cdot 0 \subset \mathfrak{t}_\Delta.$$

The converse being trivial, the lemma is proved.

Proposition 7.2. *The group Γ_Δ is generated by the reflections in the walls of P_0.*

Proof. Let $\Gamma' \subset \Gamma_\Delta$ denote the subgroup generated by the reflections in the walls of P_0. Let Q be any cell and let $X_0 \in P_0$, $Y \in Q$. Let s_0 realize the minimum of the function $s \to |X_0 - sY|$ on Γ'. Then $Y_0 = s_0 Y \in P_0$ because otherwise the vector $X_0 Y_0$ intersects a wall of P_0 and $|X_0 Y'| < |X_0 Y_0|$ if Y' is the reflection of Y_0 in this wall. Since s_0 leaves the diagram invariant, $s_0 Q = P_0$.

Next let $\sigma_{\alpha,n}$ denote the reflection in the hyperplane $\pi_{\alpha,n}$: $\alpha(H) = 2\pi in$. This hyperplane bounds a cell in the diagram, say Q. By the above, $sQ = P_0$ for some $s \in \Gamma'$. Let σ be the reflection in the hyperplane $s\pi_{\alpha,n}$. Clearly $s\pi_{\alpha,n}$ bounds P_0 and $\sigma_{\alpha,n} = s^{-1}\sigma s$. Thus $\sigma_{\alpha,n} \in \Gamma'$ and the result is proved.

Now let $\pi_1, \ldots, \pi_n$ be the walls of P_0 and $s_1, \ldots, s_n$ the corresponding reflections. For each $\sigma \in \Gamma_\Delta$ let $l(\sigma)$ be the smallest number r such that σ is the product of r of the s_i. An expression $\sigma = s_{i_1} \ldots s_{i_r}$ is called *reduced* if $r = l(\sigma)$, and $l(\sigma)$ is called the *length function.*

Lemma 7.3. *Let $\sigma \in \Gamma_\Delta$. Then $l(\sigma)$ is the number of hyperplanes in the diagram separating P_0 and σP_0 (that is, the number of hyperplanes intersected by a segment $X_0 Y_0$ ($X_0 \in P_0$, $Y_0 \in \sigma P_0$)).*

Proof. Let π be a wall of P_0 and s the reflection in π. We now make two simple observations:

(a) If π' is a hyperplane of the diagram and $\pi' \neq \pi$, then P_0 and sP_0 are on the same side of π'.

(b) Assume P_0 and σP_0 are on the same side of π. Then (because of (a)) the hyperplanes of $D(\mathfrak{u})$ separating P_0 and $s\sigma P_0$ are precisely:

(i) π itself;

(ii) the hyperplanes $s\pi'$ where π' separates P_0 and σP_0.

Let $\sigma = s_{i_1} \ldots s_{i_r}$ be a reduced expression. We shall prove by induction on r that the hyperplanes of $D(\mathfrak{u})$ separating P_0 and σP_0 are precisely

$$s_{i_1} \ldots s_{i_{r-1}}(\pi_{i_r}), \quad s_{i_1} \ldots s_{i_{r-2}}(\pi_{i_{r-1}}), \quad \ldots, \quad s_{i_1}(\pi_{i_2}), \quad \pi_{i_1}$$

and that these are all different.

This is so, by (a), if $r = 1$. Let $\sigma' = s_{i_2} \ldots s_{i_r}$. This is a reduced expression and $\sigma = s_{i_1}\sigma'$. We shall prove that P_0 and $\sigma' P_0$ are on the same side of π_{i_1}, so using (b) the first part of our claim follows by the induction hypothesis. If P_0 and $\sigma' P_0$ were on different sides of π_{i_1}, then by the induction hypothesis

$$\pi_{i_1} = s_{i_2} \dots s_{i_m}(\pi_{i_{m+1}}) \qquad \text{for some} \quad m \geqslant 1. \tag{1}$$

Thus

$$s_{i_1} = (s_{i_2} \dots s_{i_m})\, s_{i_{m+1}} (s_{i_2} \dots s_{i_m})^{-1}$$

so

$$\dot{\sigma} = s_{i_2} \dots s_{i_m} s_{i_{m+2}} \dots s_{i_r},$$

contradicting the minimal property of r.

Finally, the induction hypothesis asserts that the hyperplanes

$$s_{i_2} \dots s_{i_{r-1}}(\pi_{i_r}), \quad \dots, \quad \pi_{i_2},$$

are all different; their images under s_{i_1} are of course different, and are also different from π_{i_1}. In fact, if $\pi_{i_1} = s_{i_1} \dots s_{i_m}(\pi_{i_{m+1}})$ $(m \geqslant 1)$, then (1) follows and again we have a contradiction.

Corollary 7.4. *The group Γ_Δ permutes the cells in $\mathfrak{t}_r$ simply transitively.*

We know from Lemma 7.1 that Γ_Δ permutes the cells, and the transitivity is contained in the proof of Prop. 7.2. The simple transitivity is obvious from Lemma 7.3.

Theorem 7.5. *Let $C_0 \subset \mathfrak{t}_0$ be any Weyl chamber.*

(i) *Each orbit of $W(\mathfrak{u})$ in $\mathfrak{t}_0$ intersects the closure $\bar{C}_0$ in exactly one point.*

(ii) *Each orbit of Γ_Δ in $\mathfrak{t}_0$ intersects the closure $\bar{P}_0$ in exactly one point.*

Proof. Part (i) is a special case of Theorem 2.22. For (ii) it suffices by Cor. 7.4 to prove that if $H, H' \in \bar{P}_0$, $\sigma \in \Gamma_\Delta$, and $H = \sigma H'$, then $H = H'$. We shall prove this by induction on the length r of σ. The case $r = 0$ being obvious let $\sigma = s_{i_1} s_{i_2} \dots s_{i_r}$ be a reduced expression. We write $\sigma = s_{i_1}\sigma'$, so $l(\sigma) > l(\sigma')$. We know from the proof of Lemma 7.3 that π_{i_1} separates P_0 and σP_0. Thus if S is the closed half-space with boundary π_{i_1} containing $\bar{P}_0$, we have $\bar{P}_0 \cap \sigma\bar{P}_0 \subset S \cap \sigma\bar{P}_0 \subset \pi_{i_1}$. Thus $H' \subset \pi_{i_1}$, so $H' = s_{i_1}H' = s_{i_1}\sigma H = \sigma' H$ and the induction assumption implies $H = H'$.

Remark. Comparing the proofs of Prop. 7.2 and Theorem 2.12, we see that $W(\mathfrak{u})$ is generated by the reflections in the walls of C_0 and that the length function has an obvious analog for $W(\mathfrak{u})$. Lemma 7.3 would have an analog for C_0, and the proof of (ii) above would give another proof of (i).

We shall now apply these Euclidean results about Γ_Δ to derive some global results for the group U and its unit lattice $\mathfrak{t}_e$.

Lemma 7.6. *We have*

$$\exp\left(\frac{4\pi i}{\langle\alpha, \alpha\rangle} H_\alpha\right) = e \qquad (\alpha \in \Delta^*),$$

that is, $\mathfrak{t}_\Delta \subset \mathfrak{t}_e$.

Proof. As shown in the proof of Lemma 4.5, T_α coincides with the centralizer in U of the hyperplane $\alpha(H) = 0$. In particular, T_α contains an element u such that $\mathrm{Ad}(u)$ restricted to $\mathfrak{t}_0$ is the reflection s_α in this hyperplane. But then the definition of T_α implies

$$\exp\left(\mathrm{Ad}(u)\frac{2\pi i}{\langle\alpha, \alpha\rangle} H_\alpha\right) = \exp\frac{2\pi i}{\langle\alpha, \alpha\rangle} H_\alpha,$$

and this proves the lemma.

Considering the unit lattice $\mathfrak{t}_e$ as a group of translations of $\mathfrak{t}_0$, let Γ_e denote the group generated by $W(\mathfrak{u})$ and $\mathfrak{t}_e$. Again Γ_e leaves the diagram invariant. Also $\Gamma_e = \mathfrak{t}_e \cdot W(\mathfrak{u})$, $\mathfrak{t}_e$ being a normal subgroup of Γ_e. Also Γ_Δ is a normal subgroup of Γ_e since $s_\alpha H - H \in \mathfrak{t}_\Delta$ for $H \in \mathfrak{t}_e$.

Theorem 7.7. *The subgroup Ω of Γ_e leaving P_0 invariant is isomorphic to $\mathfrak{t}_e/\mathfrak{t}_\Delta$, and its order is the cardinality of $\mathfrak{t}_e \cap \bar{P}_0$.*

Proof. Given $\gamma \in \Gamma_e$, there exists a unique $\sigma_\gamma \in \Gamma_\Delta$ such that $\gamma P_0 = \sigma_\gamma P_0$, that is, $\sigma_\gamma^{-1}\gamma \in \Omega$. Since $\Gamma_\Delta \subset \Gamma_e$ is a normal subgroup, the map $\gamma \to \sigma_\gamma^{-1}\gamma$ is a homomorphism of Γ_e into Ω. The image is Ω and the kernel is Γ_Δ, so $\Omega \approx \Gamma_e/\Gamma_\Delta = \mathfrak{t}_e/\mathfrak{t}_\Delta$.

Next, given $Y \in \mathfrak{t}_e \cap \bar{P}_0$, the cell $P_0 - Y$ has 0 in its closure; so for a unique $w \in W(\mathfrak{u})$, $P_0 - Y$ and wP_0 lie in the same Weyl chamber and contain 0 in their closures. Hence $P_0 - Y = wP_0$, so the mapping $X \to wX + Y$ belongs to Ω. Conversley, if a mapping $X \to w_1 X + Y_1$ ($w_1 \in W(\mathfrak{u})$, $Y_1 \in \mathfrak{t}_e$) belongs to Ω, then $w_1 \cdot 0 = Y_1 \in \mathfrak{t}_e \cap \bar{P}_0$. This gives the desired bijection of $\mathfrak{t}_e \cap \bar{P}_0$ onto Ω.

Considering Theorems 6.1 and 6.7, we conclude the following result.

Corollary 7.8. *We have $\mathfrak{t}_\Delta = \tilde{\mathfrak{t}}_e$, so*

$$\Omega \approx \pi_1(U) \qquad \textit{and} \qquad \tilde{Z} \approx \mathfrak{t}(\mathfrak{u})/\mathfrak{t}_\Delta.$$

In particular, for a compact, semisimple, simply connected Lie group the unit lattice is the lattice spanned by the vectors

$$\frac{4\pi i}{\langle \alpha, \alpha \rangle} H_\alpha \qquad (\alpha \in \Delta^*).$$

We can now state the conjugacy theorem (Theorem 6.4, Chapter V) in a sharper form.

Theorem 7.9. *Let U be a compact connected semisimple Lie group. Let $C_0 \subset \mathfrak{t}_0$ denote the Weyl chamber containing P_0 and $\bar{C}_0$ its closure.*

(i) *Each $X \in \mathfrak{u}$ is* $\mathrm{Ad}(U)$*-conjugate to a unique element $H \in \bar{C}_0$.*

(ii) *Assume U simply connected. Let $u \in U$. Then there exists a unique $H \in \bar{P}_0$ such that u is conjugate to* $\exp H$.

Proof. (i) By Theorem 2.12 and Chapter V, Theorem 6.4, X is $\mathrm{Ad}(U)$-conjugate to some $H \in \bar{C}_0$. If $H' \in \bar{C}_0$ is another such element, then by Prop. 2.2, $H' = sH$ for some $s \in W(\mathfrak{u})$, so by Theorem 7.5, $H = H'$.

For (ii) we first prove a global analog of Prop. 2.2. Recall that in a group $I(u)$ denotes the automorphism $g \to ugu^{-1}$.

Lemma 7.10. *Let $S \subset T$ be a subset and suppose $u \in U$ satisfies $I(u)(S) \subset T$. Then there exists an element $n \in N_T$ (the normalizer of T in U) such that*

$$I(n) \mid S = I(u) \mid S \qquad (\mid \text{ — restriction}).$$

Proof. The groups T and uTu^{-1} are maximal tori in the centralizer $Z_{uSu^{-1}}$. Hence there exists an element $v \in Z_{uSu^{-1}}$ such that $vTv^{-1} = uTu^{-1}$. Then $n = v^{-1}u$ has the desired property.

In order to conclude the proof of Theorem 7.9 we first observe from Lemma 6.3 that each $u \in U$ is conjugate to some $t \in \exp \bar{P}_0$. If $t' \in \exp \bar{P}_0$ is another such element, then by Lemma 7.10 $t' = ntn^{-1}$ for some $n \in N_T$. Writing $t' = \exp H'$, $t = \exp H$ $(H, H' \in \bar{P}_0)$ and w for the restriction of $\mathrm{Ad}(n)$ to $\mathfrak{t}_0$, we deduce $H' = wH + A$ where $A \in \mathfrak{t}_e$. In other words, H' and H are conjugate under Γ_e. By the simple connectedness, $\Gamma_\Delta = \Gamma_e$; so Theorem 7.5 implies $H = H'$ as desired.

8. Application to the Symmetric Space U/K

We return now to the situation in §1-§3 where $(\mathfrak{u}, \theta)$ is an arbitrary orthogonal symmetric Lie algebra of the compact type. To simplify the notation we now write $\mathfrak{a}_* = \mathfrak{h}_{\mathfrak{p}_*}$. We know that exp maps $\mathfrak{u}$ onto U and Exp maps $\mathfrak{p}_*$ onto U/K.

The unit lattice in $\mathfrak{a}_*$ is defined as the set

$$\mathfrak{a}_e = \{H \in \mathfrak{a}_* : \exp H = e\}.$$

If H belongs to this set then $\mathrm{Ad}(\exp H)$ is the identity mapping of $\mathfrak{u}$ so $\alpha(H) \in 2\pi i \boldsymbol{Z}$ for each root $\alpha \in \Delta$. Hence if we consider $\mathfrak{a}_e$ as a group of translations of $\mathfrak{a}_*$, each element in this group leaves the diagram $D(U, K)$ invariant and permutes the components of $\mathfrak{a}_r = \mathfrak{a}_* - D(U, K)$. The mapping

$$\Phi : (kM, H) \to \mathrm{Exp}\, \mathrm{Ad}(k)\, H$$

maps $K/M \times \mathfrak{a}_*$ onto U/K. We recall that the singular set in U/K is defined as the image

$$S_{U/K} = \Phi(U/K \times D(U, K))$$

and the regular set is defined as the complement

$$(U/K)_r = U/K - S_{U/K}.$$

Lemma 8.1. *Suppose the pair* (U, K) *associated with* $(\mathfrak{u}, \theta)$ *is a symmetric pair. Then the mapping* Φ *maps the subset* $K/M \times \mathfrak{a}_r$ *regularly onto* $(U/K)_r$.

Proof. Since the regularity is already established (Prop. 3.2), we only have to prove that if $H_s \in D(U, K)$ and $H_r \in \mathfrak{a}_r$, then the relation

$$\mathrm{Exp}\, \mathrm{Ad}(k_s)\, H_s = \mathrm{Exp}\, \mathrm{Ad}(k_r)\, H_r \tag{1}$$

is impossible for $k_r, k_s \in K$. Relation (1) implies for $k = k_r^{-1}k_s$ that

$$k \exp H_s k_1 = \exp H_r \tag{2}$$

for a suitable $k_1 \in K$. Applying the automorphism of U which corresponds to θ we obtain

$$k \exp 2H_s k^{-1} = \exp 2H_r.$$

The centralizers of the elements $\exp 2H_r$ and $\exp 2H_s$ in U are therefore isomorphic. This contradicts Lemma 4.3 because $\alpha(2H_s) \in 2\pi i\boldsymbol{Z}$ whenever $\alpha(2H_r) \in 2\pi i\boldsymbol{Z}$ but not conversely.

The lemma shows that Φ is a local homeomorphism of $K/M \times \mathfrak{a}_r$

onto $(U/K)_r$. It follows easily that if $p(t)$, $0 \leqslant t \leqslant 1$, is a continuous curve in $(U/K)_r$ and if the point $q_0 \in K/M \times \mathfrak{a}_r$ satisfies $\Phi(q_0) = p(0)$ then there exists a unique continuous curve $q(t)$, $0 \leqslant t \leqslant 1$, in $K/M \times \mathfrak{a}_r$ such that $\Phi(q(t)) = p(t)$ for all t and $q(0) = q_0$. In analogy with the terminology of covering spaces, $q(t)$ is called a *lift* of $p(t)$.

Theorem 8.2. *Let σ be an analytic involutive automorphism of a compact simply connected Lie group. Then the set of fixed points of σ is connected.*

Proof. A compact simply connected Lie group is necessarily semisimple (Prop. 6.6, Chapter II). In terms of the notation of Lemma 7.1, it suffices therefore to prove that U/K is simply connected if U is simply connected as we now assume.

Let Q_0 denote a component of $\mathfrak{a}_r$ whose closure contains the origin. Let (k^*M, H^*) be a fixed point in $K/M \times Q_0$ and put $p^* = \Phi(k^*M, H^*)$. Let $\gamma_0(t)$ $(0 \leqslant t \leqslant 1)$ be a continuous curve in U/K which begins and ends at the point $o = \{K\}$. Due to Prop. 12.4 in the Appendix, γ_0 is homotopic to a path $\gamma(t)$ $(0 \leqslant t \leqslant 1)$ which lies in $(U/K)_r$ except for the point $o = \gamma(0) = \gamma(1)$. Since $(U/K)_r$ is connected (Cor. 12.5), we may assume that $\gamma(\frac{1}{2}) = p^*$. Now let $0 < \epsilon < \frac{1}{2}$ and consider the path γ_ϵ given by $\gamma_\epsilon(t) = \gamma(t)$, $(\epsilon \leqslant t \leqslant 1 - \epsilon)$. Let $\Gamma_\epsilon(t)$ be the lift of γ_ϵ to $K/M \times \mathfrak{a}_r$ such that $\Gamma_\epsilon(\frac{1}{2}) = (k^*M, H^*)$. Put $\Gamma_\epsilon(t) = (k_tM, H_t)$ for $\epsilon \leqslant t \leqslant 1 - \epsilon$. Then, since Γ_ϵ is connected, $H_t \in Q_0$ for $\epsilon \leqslant t \leqslant 1 - \epsilon$. Let $0 < \epsilon_1 < \epsilon_2 < \frac{1}{2}$. By the uniqueness of the lift we have

$$\Gamma_{\epsilon_1}(t) = \Gamma_{\epsilon_2}(t) \qquad \text{for } \epsilon_2 \leqslant t \leqslant 1 - \epsilon_2.$$

Consequently we can define a continuous curve

$$\Gamma(t) = (k_tM, H_t), \qquad 0 < t < 1,$$

in $K/M \times Q_0$ such that $\Phi(\Gamma(t)) = \gamma(t)$. The set Q_0 is bounded since all roots are bounded on it. Let H_0 be any limit point of $\{H_t\}$ as $t \to 0$. Then there exists a sequence $t_n \to 0$ such that $k_{t_n} \to k_0 \in K$ and $H_{t_n} \to H_0$. It follows that

$$\gamma(t_n) = \operatorname{Exp} \operatorname{Ad}(k_{t_n}) H_{t_n} \to \operatorname{Exp}(\operatorname{Ad}(k_0)) H_0.$$

On the other hand, $\gamma(t_n) \to o$ so

$$\operatorname{Exp} \operatorname{Ad}(k_0) H_0 = o, \qquad \text{whence } \exp H_0 \in K.$$

Consequently $\exp 2H_0 = e$, i.e., $2H_0 \in \mathfrak{a}_e$.

We shall now apply Theorem 6.1 for $\mathfrak{t}_0 = \mathfrak{h}_{\mathfrak{k}_0} + \mathfrak{a}_*$. Since H_0 lies in the closure of Q_0 each root $\alpha \in \Delta^*$ has its values in the closed interval $[0, 2\pi i]$ on the segment joining 0 and $2H_0$. It follows that this segment is contained in a closed polyhedron $\bar{P}_0$, if P_0 is a suitably chosen component of $\mathfrak{t}_r$. Since $\alpha(2H_0) \in 2\pi i \boldsymbol{Z}$ for all $\alpha \in \Delta^*$, $2H_0$ is a vertex of $\bar{P}_0$. The group U being simply connected, Theorem 6.1 implies that $H_0 = 0$. Consequently $\lim_{t\to 0} H_t = 0$. In the same way we find that the limit $H_1 = \lim_{t\to 1} H_t$ exists and $H_1 = 0$. The curve H_t $(0 \leqslant t \leqslant 1)$ is therefore a closed curve in $\bar{Q}_0$. The mapping

$$\alpha(s, t) = sH_t, \qquad 0 \leqslant s \leqslant 1, \qquad 0 \leqslant t \leqslant 1,$$

is a homotopy of this curve and 0. The mapping β given by

$$\begin{aligned} \beta(s, t) &= \mathrm{Exp}(\mathrm{Ad}(k_t)\, sH_t), && 0 \leqslant s \leqslant 1, && 0 < t < 1, \\ \beta(s, 0) &= \beta(s, 1) = o, && 0 \leqslant s \leqslant 1, \end{aligned}$$

is then a homotopy of γ and o. This proves the theorem.

Remark. Theorem 8.2 does not hold if the assumption of simple connectedness is dropped. As an example let

$$s_0 = \begin{pmatrix} -1 & 0 & 0 \\ 0 & -1 & 0 \\ 0 & 0 & 1 \end{pmatrix}$$

and let σ denote the involutive automorphism $u \to s_0 u s_0$ of $\boldsymbol{SO}(3)$. The set K of fixed points consists of two components.

Now generalizing the situation in §7, let $\mathfrak{a}_\Sigma$ denote the lattice in $\mathfrak{a}_*$ spanned by the vectors (§2)

$$\frac{2\pi i}{\langle \mu, \mu \rangle} A_\mu \qquad (\mu \in \Sigma)$$

and Γ_Σ the group of linear transformations of $\mathfrak{a}_*$ generated by the reflections in all the hyperplanes in the diagram $D(\mathfrak{u}, \theta)$. Again the components of $\mathfrak{a}_r$ are called *cells* and Γ_Σ the *affine Weyl group*. Because of Theorem 2.16, we can imitate the proof of Lemma 7.1, whereas 7.2–7.5 require no change. We can therefore state the following result.

Theorem 8.3. *Let Q_0 be a component of $\mathfrak{a}_r$ whose closure contains the origin. Viewing $\mathfrak{a}_\Sigma$ as a group of translations of $\mathfrak{a}_*$, we have:*

(i) $\Gamma_\Sigma = \mathfrak{a}_\Sigma \cdot W(\mathfrak{u}, \theta)$ *(semidirect product).*

(ii) *The group Γ_Σ is generated by the reflections in the walls of Q_0.*

(iii) *The group* Γ_Σ *permutes the cells simply transitively.*

(iv) *Let* $C_* \subset \mathfrak{a}_*$ *be any Weyl chamber. Then each orbit of* $W(\mathfrak{u}, \theta)$ *in* $\mathfrak{a}_*$ *intersects the closure* $\bar{C}_*$ *in exactly one point.*

(v) *Each orbit of* Γ_Σ *in* $\mathfrak{a}_*$ *intersects the closure* $\bar{Q}_0$ *in exactly one point.*

We would now like to determine the unit lattic $\mathfrak{a}_e$ more explicitly, generalizing Cor. 7.8. For this we need some additional information about the restricted roots.

Lemma 8.4. *Let* $\alpha \in \Delta_\mathfrak{p}$. *Then its restriction* $\bar{\alpha}$ to $\mathfrak{h}_\mathfrak{p}$ satisfies

$$\langle\alpha, \alpha\rangle = m\langle\bar{\alpha}, \bar{\alpha}\rangle \qquad \textit{where} \quad m = 1, 2, \textit{or } 4.$$

If $m = 1$, *then* $H_\alpha = A_{\bar{\alpha}}$; *and if* $m = 4$, *then* $2\bar{\alpha} \in \Sigma$.

Proof. With the notation of §3, Chapter VI, we have $A_{\bar{\alpha}} = \frac{1}{2}(H_\alpha - H_{\alpha^\theta})$. If $\alpha = -\alpha^\theta$, then $m = 1$, so let us assume $\alpha + \alpha^\theta \neq 0$. If $\alpha + \alpha^\theta$ were a root, then $\mathfrak{g}^{\alpha+\alpha^\theta} \subset \mathfrak{k}$ whereas $[X_\alpha, \theta X_\alpha] \subset \mathfrak{p}$ for each $X_\alpha \in \mathfrak{g}^\alpha$. Because of Theorem 4.3, Chapter III, this is a contradiction; thus $\alpha + \alpha^\theta$ is not a root, so by the theorem quoted, $2\langle\alpha, \alpha^\theta\rangle/\langle\alpha, \alpha\rangle$ is an integer $\geqslant 0$, hence 0, 1, or 2. Since $\bar{\alpha} \neq 0$, the value 2 is excluded, so by $\langle\bar{\alpha}, \bar{\alpha}\rangle = \frac{1}{2}(\langle\alpha, \alpha\rangle - \langle\alpha, \alpha^\theta\rangle)$ we find $m = 2$ or 4. If $m = 4$, then $\langle\alpha, \alpha^\theta\rangle > 0$, so $\alpha - \alpha^\theta$ is a root and therefore $2\bar{\alpha} \in \Sigma$. This proves the lemma.

Theorem 8.5. *Let* U/K *be a symmetric space of the compact type,* U *simply connected, and* K *connected. Let* $\mathfrak{a}_K$ *denote the lattice*

$$\mathfrak{a}_K = \{H \in \mathfrak{a}_* : \exp H \in K\}.$$

Then

$$\mathfrak{a}_K = \tfrac{1}{2}\mathfrak{a}_e = \mathfrak{a}_\Sigma. \tag{3}$$

Proof. First we note that $\exp H \in K \Leftrightarrow \exp H = \exp(-H)$ by Theorem 8.2, proving the first relation in (3). Next, let $\mu \in \Sigma$ and select $\alpha \in \Delta_\mathfrak{p}$ such that $\bar{\alpha} = \mu$. Since $\alpha - \bar{\alpha} = \frac{1}{2}(\alpha + \alpha^\theta)$, we have $iH_\alpha - iA_\alpha \in \mathfrak{k}_0$ so Lemmas 7.6 and 8.4 imply

$$\exp\left(\frac{2\pi i}{\langle\mu, \mu\rangle} A_\mu\right) \in K, \tag{4}$$

in other words, $\mathfrak{a}_\Sigma \subset \mathfrak{a}_K$.

Considering $\mathfrak{a}_K$ as a group of translations of $\mathfrak{a}_*$, let Γ_K denote the group generated by $W(\mathfrak{u}, \theta)$ and $\mathfrak{a}_K$. Since $\alpha(H) \in \pi i\mathbf{Z}$ for $H \in \mathfrak{a}_K$ and $\alpha \in \Delta$,

Γ_K leaves $D(\mathfrak{u}, \theta)$ invariant and permutes the cells. Also $W(\mathfrak{u}, \theta)$ leaves $\mathfrak{a}_K$ invariant because of (4). Thus $\Gamma_K = \mathfrak{a}_K \cdot W(\mathfrak{u}, \theta)$, $\mathfrak{a}_K$ is a normal subgroup of Γ_K and Γ_Σ is a normal subgroup of Γ_K since $s_u H - H \in \mathfrak{a}_\Sigma$ for $H \in \mathfrak{a}_K$.

Since $\mathfrak{a}_K$ coincides with the set $\{H \in \mathfrak{a}_* : \operatorname{Exp} H = o\}$, the group Γ_K is the natural generalization of the group Γ_e. Repeating the argument of Theorem 7.7, we conclude that the group $\mathfrak{a}_K/\mathfrak{a}_\Sigma$ has order equal to the cardinality of $\mathfrak{a}_K \cap \bar{Q}_0$. But if $H_0 \in \mathfrak{a}_K \cap \bar{Q}_0$, then $2H_0 \in \mathfrak{a}_e$; so as in the proof of Theorem 8.2, $H_0 = 0$, so $\mathfrak{a}_K = \mathfrak{a}_\Sigma$, and

$$\Gamma_K = \Gamma_\Sigma. \tag{5}$$

This proves Theorem 8.5.

We can now prove a generalization of Theorem 7.9, sharpening Theorem 6.7, Chapter V.

Theorem 8.6. *With U/K as in Theorem 8.5 let $C_* \subset \mathfrak{a}_*$ be any Weyl chamber; $\bar{C}_*$ its closure. Let Q_0 be any component of $\mathfrak{a}_r$ whose closure contains 0.*

(i) *Each $X \in \mathfrak{p}_*$ is $\operatorname{Ad}_U(K)$-conjugate to a unique element $H \in \bar{C}_*$;*

(ii) $U = K \exp \bar{Q}_0 K$

and each $u \in U$ can be written $u = k_1 \exp H\, k_2$ $(k_1, k_2 \in K)$ with $H \in \bar{Q}_0$ unique.

Proof. Considering Theorem 8.3, the proof of (i) is the same as in Theorem 7.9. For part (ii) we need the following analog of Lemma 7.10, where, as before, M' denotes the normalizer of $\mathfrak{a}_*$ in K.

Lemma 8.7. *Let S be a subset of $A_* = \exp \mathfrak{a}_*$ and suppose $k \in K$ such that $kSk^{-1} \subset A_*$. Then there exists an element $m \in M'$ such that*

$$msm^{-1} = ksk^{-1} \qquad \text{for} \quad s \in S.$$

Proof. The centralizer $Z_{kSk^{-1}}$ of kSk^{-1} in U is invariant under the involutive automorphism θ, so its Lie algebra $\mathfrak{u}_1$ has a direct decomposition $\mathfrak{u}_1 = \mathfrak{k}_1 + \mathfrak{p}_1$ where $\mathfrak{k}_1 = \mathfrak{u}_1 \cap \mathfrak{k}_0$ and $\mathfrak{p}_1 = \mathfrak{u}_1 \cap \mathfrak{p}_*$. Let $P_1 = \exp \mathfrak{p}_1$. Then $A_* \subset P_1$, $kA_*k^{-1} \subset Z_{kSk^{-1}}$; but since $d\theta = -1$ on $\operatorname{Ad}(k)\,\mathfrak{a}_*$, we have $\operatorname{Ad}(k)\,\mathfrak{a}_* \subset \mathfrak{u}_1 \cap \mathfrak{p}_* = \mathfrak{p}_1$, so $kA_*k^{-1} \subset P_1$. The conjugacy arguments of §6, Chapter V, show that there exists an element $l \in \exp \mathfrak{k}_1$ such that $lA_*l^{-1} = kA_*k^{-1}$. Then the element $m = l^{-1}k$ has the desired property.

To conclude the proof of Theorem 8.6 let $u \in U$. Then $u = k_1 \exp Hk_2$ $(k_1, k_2 \in K, H \in \mathfrak{a}_*)$. Because of (5) and Theorem 8.3 (iii), we may assume

$H \in \bar{Q}_0$. If $u = l_1 \exp H' l_2$ ($l_1, l_2 \in K$, $H' \in \bar{Q}_0$) is another such expression, we obtain

$$u(\theta u)^{-1} = k_1 \exp(2H)\, k_1^{-1} = l_1 \exp(2H')\, l_1^{-1};$$

so by Lemma 8.7, $m \exp(2H)\, m^{-1} = \exp(2H')$ for some $m \in M'$. Thus by (3) and (5), H and H' are conjugate under the group Γ_Σ, so Theorem 8.3 (v) implies $H = H'$.

We conclude this section establishing some useful connections between the Weyl groups $W(\mathfrak{u})$ and $W(\mathfrak{u}, \theta)$.

Proposition 8.8. *With U/K as in Theorem* 8.5 *let* $P_* = \exp \mathfrak{p}_*$ *and M' the normalizer of $\mathfrak{a}_*$ in K. Then:*

(i) *A_* equals its centralizer in P_*.*

(ii) *$M'A_*$ equals the normalizer of $\mathfrak{a}_*$ in U.*

Proof. (i) Let $\mathfrak{b} \subset \mathfrak{a}_*$ be a subspace and suppose $p \in P_*$ commutes elementwise with the group $B = \exp \mathfrak{b}$. Let C' denote the subgroup of U generated by p and B and let C denote the closure of C' in U. Clearly C' consists of all products bp^n ($b \in B$, $n \in \mathbf{Z}$). Now $b = d^2$ ($d \in B$), $p = q^2$ ($q \in P_*$), so $bp^n = d^2p^n = dp^nd = dq^nq^nd = dq^n\theta((dq^n)^{-1})$. Since $P_* = \{u\theta(u^{-1}) : u \in U\}$ and is therefore closed in U, we deduce $C' \subset P_*$ and $C \subset P_*$. The identity component C_0 of C contains some power of p; if N is the smallest such power, then C/C_0 is cyclic of order N. By Lemma 2.6 there exists an element $c \in C$ such that the sequence e, $c, c^2, \ldots$ is dense in C. Select $X \in \mathfrak{p}_*$ such that $c = \exp X$. Then the closure of the one-parameter subgroup $\exp \mathbf{R}X$ is a torus, contained in P_*, containing both p and B. Taking $\mathfrak{b} = \mathfrak{a}_*$, (i) follows since $\mathfrak{a}_*$ is maximal abelian in $\mathfrak{p}_*$.

(ii) Let $\mathfrak{b} \subset \mathfrak{a}_*$ be a subspace and suppose $u \in U$ such that $\mathrm{Ad}(u)\mathfrak{b} \subset \mathfrak{a}_*$. Then $u = kp$ ($k \in K$, $p = \exp X$, $X \in \mathfrak{p}_*$), so $e^{\mathrm{ad}\,X}(\mathfrak{b}) \subset \mathfrak{p}_*$. Hence $\theta(e^{\mathrm{ad}\,X}(H)) = -e^{\mathrm{ad}\,X}(H)$ for $H \in \mathfrak{b}$, so $e^{2\,\mathrm{ad}\,X}H = H$ for $H \in \mathfrak{b}$. This means that p^2 commutes elementwise with $\exp \mathfrak{b}$. By the proof of (i) there exists a torus $T \subset P_*$ containing p^2 and $\exp \mathfrak{b}$. Let $\mathfrak{t}$ denote its Lie algebra, write $p^2 = \exp Z$ ($Z \in \mathfrak{t}$), and put $p_0 = \exp \frac{1}{2}Z$. Then $\theta(p_0 p^{-1}) = p_0^{-1}p = p_0 p^{-1}$, so by Theorem 8.2, $p = k_0 p_0$ where $k_0 \in K$. Thus $u = kk_0 p_0 = k'p_0$ ($k' \in K$) and

$$\mathrm{Ad}(u)H = \mathrm{Ad}(k')H \qquad (H \in \mathfrak{b}). \tag{6}$$

Taking $\mathfrak{b} = \mathfrak{a}_*$, we deduce from (i) that $p^2 \in \exp \mathfrak{b}$ and $T = A_*$, so $p_0 \in A_*$. Now (6) implies $k' \in M'$, so $u \in M'A_*$ as desired.

Corollary 8.9. *Let $W(\mathfrak{u})$ and $W(\mathfrak{u}, \theta)$ be the Weyl groups acting on the complex spaces $\mathfrak{h}$ and $\mathfrak{h}_{\mathfrak{p}} \subset \mathfrak{h}$, respectively. Then if two elements $H, H' \in \mathfrak{h}_{\mathfrak{p}}$ are conjugate under $W(\mathfrak{u})$, they are conjugate under $W(\mathfrak{u}, \theta)$.*

Let $u \in N_T$ (normalizer of T) such that $\mathrm{Ad}(u)\, H = H'$. Writing $H = H_1 + iH_2$, $H' = H_1' + iH_2'$ $(H_1, H_2, H_1', H_2' \in \mathfrak{a}_*)$, we obtain $\mathrm{Ad}(u)\, H_j = H_j'\, (j = 1, 2)$, so $\mathrm{Ad}(u)$ maps the subspace $\mathfrak{b} = \boldsymbol{R}H_1 + \boldsymbol{R}H_2$ into $\mathfrak{a}_*$. By (6) and by Proposition 2.2 there exists an $s \in W(\mathfrak{u}, \theta)$ such that $s = \mathrm{Ad}(u)$ on $\mathfrak{b}$. Then $sH = H'$ as we wanted to prove.

We consider now two interesting subgroups of the Weyl group $W(\mathfrak{u})$. Let W_θ denote the subgroup of $W(\mathfrak{u})$ leaving $\mathfrak{h}_{\mathfrak{p}}$ invariant and $W(\mathfrak{m}_0)$ the subgroup of $W(\mathfrak{u})$ leaving $\mathfrak{h}_{\mathfrak{p}}$ pointwise fixed. According to Theorem 2.15 for the group U, $W(\mathfrak{m}_0)$ is generated by the symmetries s_α of $\mathfrak{h}$ for $\alpha \in P_-$ (notation of §3, Chapter VI). The Lie algebra $\mathfrak{m}$ is a direct sum of an abelian Lie algebra and a semisimple Lie algebra; $\mathfrak{h}_{\mathfrak{k}} \subset \mathfrak{m}$ is a maximal abelian subalgebra, and the decomposition

$$\mathfrak{m} = \mathfrak{h}_{\mathfrak{k}} + \sum_{\alpha \in P_-} (\mathfrak{g}^\alpha + \mathfrak{g}^{-\alpha})$$

(cf. Lemma 3.6, Chapter VI) can be viewed as a "root space decomposition" of $\mathfrak{m}$. This justifies calling $W(\mathfrak{m}_0)$ the Weyl group of the Lie algebra $\mathfrak{m}_0$.

Proposition 8.10. *The restriction $s \to s \mid \mathfrak{h}_{\mathfrak{p}}$ is a homomorphism of W_θ onto $W(\mathfrak{u}, \theta)$ with kernel $W(\mathfrak{m}_0)$.*

Proof. Let $\sigma \in W_\theta$ and select a representative $u \in N_T$ for σ. Then u normalizes $\mathfrak{a}_*$, so by Proposition 8.8, $\sigma \mid \mathfrak{a}_* \in W(\mathfrak{u}, \theta)$. For the surjectivity let $s \in W(\mathfrak{u}, \theta)$ and $k \in M'$ be a representative for s. Then $\mathfrak{h}_{\mathfrak{k}_0}$ and $\mathrm{Ad}(k)\, \mathfrak{h}_{\mathfrak{k}_0}$ are two maximal abelian subalgebras of $\mathfrak{m}_0$, so by Theorem 6.4, Chapter V there exists an element $m \in M$ such that $\mathrm{Ad}(k)\, \mathfrak{h}_{\mathfrak{k}_0} = \mathrm{Ad}(m)\, \mathfrak{h}_{\mathfrak{k}_0}$. Then $m^{-1}k$ normalizes T and the corresponding element $\sigma \in W(\mathfrak{u})$ belongs to W_θ and $\sigma \mid \mathfrak{h}_{\mathfrak{p}} = s$.

§ 9. Classification of Locally Isometric Spaces[†]

Let $(\mathfrak{u}, \theta)$ be an orthogonal symmetric Lie algebra of the compact type. We recall that a Riemannian globally symmetric space M is said to be associated with $(\mathfrak{u}, \theta)$ if the Riemannian symmetric pair $(I_0(M), K)$ is associated with $(\mathfrak{u}, \theta)$, K being the isotropy subgroup of $I_0(M)$ at

[†] See also Exercises C, Chapter X and Exercise 10 in this chapter.

some point in M. We shall now return to the problem stated in §1, namely, to find all M associated with the given $(\mathfrak{u}, \theta)$. For this purpose we shall use Theorem 8.2.

Theorem 9.1. *Let* $(\mathfrak{u}, \theta)$ *be an orthogonal symmetric Lie algebra of the compact type and suppose that* $\mathfrak{k}_0$*, the fixed point set of* θ*, contains no ideal* $\neq \{0\}$ *in* $\mathfrak{u}$*. Let* $\tilde{U}$ *denote the simply connected Lie group with Lie algebra* $\mathfrak{u}$*, let* $\tilde{\theta}$ *denote the automorphism of* $\tilde{U}$ *such that* $d\tilde{\theta} = \theta$ *and let* $\tilde{K}$ *denote the set of fixed points of* $\tilde{\theta}$*. Let* $\tilde{Z}$ *denote the center of* $\tilde{U}$*.*

Let S *be any subgroup of* $\tilde{Z}$ *and put*

$$K_S = \{u \in \tilde{U} : u^{-1}\tilde{\theta}(u) \in S\}.$$

The Riemannian globally symmetric spaces M *associated with* $(\mathfrak{u}, \theta)$ *are exactly the spaces* $M = U/K$ *with any* U*-invariant metric, where*

$$U = \tilde{U}/S, \qquad K = K^*/S. \tag{1}$$

Here S *varies through all* $\tilde{\theta}$*-invariant subgroups of* $\tilde{Z}$ *and* K^* *varies through all* $\tilde{\theta}$*-invariant subgroups of* $\tilde{U}$ *such that* $\tilde{K}S \subset K^* \subset K_S$*.*

Proof. Consider first a pair (U, K) given by (1). Because of Lemma 1.3 $\tilde{\theta}$ induces an involutive automorphism σ of U such that $\varphi\tilde{\theta} = \sigma\varphi$ if $\varphi: \tilde{U} \to U$ denotes the natural projection. But then the group $\varphi(K_S)$, and therefore K, is left pointwise fixed by σ. This proves that the manifold U/K is globally symmetric. Furthermore, the set of elements in U which induce the identity mapping of U/K is a closed subgroup D of K whose Lie algebra is 0, being an ideal of $\mathfrak{u}$ contained in $\mathfrak{k}_0$. It follows that U/D is a semisimple subgroup of the isometry group $I(U/K)$, so by Theorem 4.1, Chapter V, $U/D = I_0(U/K)$. In particular, $I(U/K)$ has Lie algebra $\mathfrak{u}$, so the space U/K is associated with $(\mathfrak{u}, \theta)$.

On the other hand, suppose M is a Riemannian globally symmetric space associated with $(\mathfrak{u}, \theta)$. This means that there exists a point $o \in M$ such that the automorphism $\sigma : u \to s_o u s_o$ of $I_0(M)$ has differential $d\sigma = \theta$. If $U = I_0(M)$ and K denotes the isotropy subgroup of $I_0(M)$ at o, then the pair (U, K) is a Riemannian symmetric pair associated with $(\mathfrak{u}, \theta)$. There exists a subgroup S of $\tilde{Z}$ such that $U = \tilde{U}/S$. Let φ denote the natural mapping $\tilde{U} \to U$ and put $K^* = \varphi^{-1}(K)$. Then $K = K^*/S$, $\tilde{\theta}K^* \subset K^*$. Since $\sigma \circ \varphi = \varphi \circ \tilde{\theta}$, we find for $k \in K^*$,

$$\varphi(\tilde{\theta}k) = \sigma\varphi(k) = \varphi(k).$$

Consequently, $k^{-1}\tilde{\theta}(k) \in S$ so $K^* \subset K_S$. On the other hand, $\tilde{K} \subset K^*$ since $\tilde{K}$ is connected. This finishes the proof since S is $\tilde{\theta}$-invariant.

While Theorem 9.1 describes in terms of $\tilde{Z}$ the Riemannian globally symmetric spaces M associated with $(\mathfrak{u}, \theta)$ it is more delicate to decide which of these are actually different. For this problem see Exercises C, Chapter X. We can however conclude two illuminating corollaries of Theorem 9.1.

Corollary 9.2. *If $\tilde{Z}$ consists of the identity element alone,*[†] *then $\tilde{U}/\tilde{K}$ is the only Riemannian globally symmetric space M associated with* $(\mathfrak{u}, \theta)$.

The automorphism θ of $\mathfrak{u}$ induces an automorphism of $\mathrm{Int}(\mathfrak{u})$, also denoted θ; let $\mathrm{Int}(\mathfrak{u})_\theta$ denote the subgroup of fixed points. We call the globally symmetric space $\mathrm{Int}(\mathfrak{u})/\mathrm{Int}(\mathfrak{u})_\theta$ the *adjoint space* of $(\mathfrak{u}, \theta)$ because it generalizes the adjoint group.

Corollary 9.3. *The adjoint space satisfies*

$$\mathrm{Int}(\mathfrak{u})/\mathrm{Int}(\mathfrak{u})_\theta = \tilde{U}/K_{\tilde{Z}}.$$

All the globally symmetric spaces $U/K = \tilde{U}/K^*$ associated with $(\mathfrak{u}, \theta)$ *cover the adjoint space and are covered by $\tilde{U}/\tilde{K}$:*

$$\tilde{U}/\tilde{K} \to \tilde{U}/K^* \to \tilde{U}/K_{\tilde{Z}}.$$

Here K^* varies through the groups in Theorem 9.1.

§ 10. Geometry of U/K. Symmetric Spaces of Rank One

This chapter has, so far, dealt with group-theoretic properties of the compact symmetric space U/K. We shall now examine some of its geometric properties, and start with some remarks on closed geodesics in compact symmetric spaces.

Definition. Let $\gamma(t)$, $-\infty < t < \infty$ be a geodesic in a Riemannian manifold M. The geodesic is called *closed* if there exists a number $L > 0$ such that $\gamma(t + L) = \gamma(t)$ for all t. The geodesic is said to be *simply closed* if in addition $\gamma(t_1) \neq \gamma(t_2)$ for $0 < t_1 < t_2 \leqslant L$. If $|t|$ is the arc parameter, L is called the *length* of the simply closed geodesic.

Proposition 10.1. *Let M be a Riemannian globally symmetric space and $\gamma(t)$ $(-\infty < t < \infty)$ a geodesic in M which intersects itself. Then it is a simply closed geodesic.*

[†] This is the case for the exceptional algebras $\mathfrak{e}_8$, $\mathfrak{f}_4$, and $\mathfrak{g}_2$,

Proof. Let p_0 be some point where the geodesic intersects itself. We may assume that the parameter t is such that $\gamma(0) = \gamma(1) = p_0$ and $\gamma(t_1) \neq \gamma(t_2)$ for $0 < t_1 < t_2 \leqslant 1$. Using now the notation of Theorem 3.3, Chapter IV, we have $M = G/K$. There is a unique vector $X \in \mathfrak{p}$ such that $d\pi \cdot X = \dot{\gamma}(0)$. Then $\operatorname{Exp} d\pi X = \gamma(1) = p_0$ so $\exp X \in K$. If $Y \in M_{p_0}$, then by the theorem quoted, the vector $(d \exp X)_{p_0}(Y)$ is the parallel translate of Y along the curve segment $\gamma(t)$, $0 \leqslant t \leqslant 1$. Using (1) in Chapter IV, §3, we obtain

$$(d \exp X)_{p_0}(d\pi X) = d\pi(\operatorname{Ad}(\exp X)X) = d\pi X.$$

This shows that $\dot{\gamma}(0)$ is the parallel translate of $\dot{\gamma}(0)$ along $\gamma(t)$ $(0 \leqslant t \leqslant 1)$. Hence $\gamma(t+1) = \gamma(t)$ $(0 \leqslant t \leqslant 1)$ and the proposition follows.

Proposition 10.2. *Let M be a compact Riemannian globally symmetric space. Then M has a simply closed geodesic. If M is of rank one then all the geodesics in M are simply closed and have the same length.*

Proof. We follow again the notation of Theorem 3.3, Chapter IV. Then since M and the group K are compact, G is compact. Let $\mathfrak{a}$ be a maximal abelian subspace of $\mathfrak{p}$ and put $A = \exp \mathfrak{a}$. The closure $\bar{A}$ of A in G is a torus whose Lie algebra is contained in $\mathfrak{p}$. Using the maximality of $\mathfrak{a}$, it follows that $A = \bar{A}$. Being a torus, A contains a one-parameter subgroup $\exp tH$ $(t \in \boldsymbol{R}, H \in \mathfrak{a})$ which intersects itself. The geodesic $\pi(\exp tH)$ $(t \in \boldsymbol{R})$ in G/K is simply closed by Prop. 10.1. If M has rank one, then all geodesics (parametrized by arc length) are congruent under an isometry of M (Theorem 6.2, Chapter V). This proves the proposition.

Definition. Let M be a compact Riemannian manifold and p a point in M. The set of points in M of maximum distance from p will be called the *antipodal set* associated to p. It will be denoted by A_p.

If M is a compact Riemannian globally symmetric space of rank one and $p \in M$, then the isotropy subgroup of $I_0(M)$ at p acts transitively on A_p. In view of Prop. 4.4, Chapter II, A_p is a compact submanifold of M. It is obvious that $\dim A_p < \dim M$.

Theorem 10.3. *Let M be a compact Riemannian globally symmetric space of rank one. Let $2L$ denote the common length of the geodesics in M. Let p be any point in M. Then Exp_p is a diffeomorphism of the open ball $\| X \| < L$ in M_p onto the complement $M - A_p$.*

Proof. We first assert that A_p coincides with the set of midpoints of the geodesics $\gamma(s)$ $(0 \leqslant s \leqslant 2L)$, starting at p, s denoting the arc length measured from p. In fact, consider one of these geodesics γ. The point $\gamma(L)$ is clearly left fixed by the symmetry s_p. If Γ is a curve segment of length $d(p, \gamma(L))$ joining p and $\gamma(L)$ then $s_p \cdot \Gamma$ also has these properties. The curves $s_p \cdot \Gamma$ and Γ are geodesics with opposite tangent vectors at p. Using Prop. 10.1 we see that Γ followed by $s_p \cdot \Gamma$ (in opposite direction) forms a closed geodesic. Hence $L = d(p, \gamma(L))$, so γ realizes the shortest distance between any two of its points. Next, let $q \in A_p$. Then $d(p, q) \geqslant L$. We join p to q by a geodesic Γ' of shortest length. The closed geodesic in M tangent to Γ' at p must contain Γ'. Since this closed geodesic has length $2L$ we conclude that $d(p, q) \leqslant L$, hence $d(p, q) = L$. This proves the assertion above.

Next, let $\gamma_1(s)$ and $\gamma_2(s)$ $(0 \leqslant s \leqslant 2L)$ be two geodesics in M starting at p, s being arc length measured from p. Suppose they intersect at a point p' different from p and $\gamma_2(L)$. Consider the curve γ formed by the shortest part of γ_1 joining p and p' together with the shortest part of γ_2 joining p' and $\gamma_2(L)$. Then γ is a curve of length L joining p and $\gamma_2(L)$ and must be a geodesic due to Lemma 9.8, Chapter I. But this obviously implies that either $\gamma_1(s) \equiv \gamma_2(s)$ or $\gamma_1(s) \equiv \gamma_2(2L - s)$.

It has now been proved that Exp_p $(= \operatorname{Exp})$ is a one-to-one differentiable mapping of the open ball $\| X \| < L$ onto $M - A_p$. It remains to be proved that Exp is regular at $X \in M_p$ provided $0 < \| X \| < L$. For this purpose let Y be a tangent vector to M_p at X for which $(d \operatorname{Exp})_X (Y) = 0$. Let Q denote the Riemannian structure of M and consider the function $q(t) = Q_{\operatorname{Exp} tX}(d \operatorname{Exp}_{tX}(X), d \operatorname{Exp}_{tX}(Y))$ $(t \in \boldsymbol{R})$. Here Y is considered as a tangent vector to M_p at tX. If we decompose $Y = yX + Y_1$ where Y_1 is perpendicular to X we see from Lemma 9.7, Chapter I, that for small t,

$$q(t) = yQ_p(X, X).$$

Since $q(t)$ is an analytic function we conclude that it is a constant. But $q(1) = 0$ so $y = 0$ and Y is perpendicular to X.

Let K_p denote the isotropy subgroup of $G = I_0(M)$ at p. We may assume that $\dim M > 1$. Then $I_0(M)$ is semisimple and K_p acts transitively on each sphere $S_r(p)$ in M. In particular, the linear isotropy group K_p^* acts transitively on the sphere S around the origin in M_p through X. The group K_p^* is a compact linear group and is a Lie transformation group of S. In view of Prop. 4.3, Chapter II, there exists a vector Y_0 in the Lie algebra $\mathfrak{L}(K_p^*) = \mathfrak{L}(K_p)$ such that†

$$Y = \left\{ \frac{d}{dt} d\tau\,(\exp tY_0) \cdot X \right\}_{t=0}. \tag{1}$$

Then, if f is a differentiable function on M,

$$0 = (d\,\mathrm{Exp})_X(Y)f = \left\{ \frac{d}{dt} f(\mathrm{Exp}(d\tau\,(\exp tY_0) \cdot X) \right\}_{t=0}$$

$$= \left\{ \frac{d}{dt} f(\exp tY_0 \cdot \mathrm{Exp}\, X) \right\}_{t=0},$$

where we have used relations (1) and (3) in Chapter IV, §3. Let $s \in \boldsymbol{R}$ and let us use the last formula on the function $f^*(q) = f(\exp sY_0 \cdot q)$, $q \in M$. Then

$$0 = \left\{ \frac{d}{dt} f^*(\exp tY_0 \cdot \mathrm{Exp}\, X) \right\}_{t=0} = \left\{ \frac{d}{dt} f(\exp tY_0 \cdot \mathrm{Exp}\, X) \right\}_{t=s},$$

which shows that $f(\exp sY_0 \cdot \mathrm{Exp}\, X)$ is constant in s. Since f is arbitrary, this shows that the one-parameter subgroup $\exp sY_0$ ($s \in \boldsymbol{R}$) leaves the point $\mathrm{Exp}\, X$ fixed. The *unique* geodesic of shortest length joining p and $\mathrm{Exp}\, X$ is therefore left fixed by each $\exp sY_0$. Consequently, $d\tau(\exp sY_0) \cdot X = X$ for $s \in \boldsymbol{R}$ so $Y = 0$ by (1). Thus, Exp is regular and the theorem is proved.

For the remainder of this section we assume M is a compact Riemannian globally symmetric space of rank one and of dimension greater than one. Fix $o \in M$, write Exp for Exp_o, let U denote the compact semisimple group $I_0(M)$ and $\mathfrak{u}$ its Lie algebra. Let $\mathfrak{u} = \mathfrak{k} + \mathfrak{p}_*$ be the decomposition of $\mathfrak{u}$ into the eigenspaces of the involutive automorphism of $\mathfrak{u}$ which corresponds to the automorphism $u \to s_o u s_o$ of $I(M)$. Here $\mathfrak{k}$ is the Lie algebra of the isotropy subgroup of U at o. Changing the distance function d on M by a constant factor, we may, since $\mathfrak{u}$ is semisimple, assume that the differential of the mapping $u \to u \cdot o$ of $I(M)$ onto M gives an isometry of $\mathfrak{p}_*$ (with the metric of $-B$) onto the tangent space M_o.

Proposition 10.4. *For each $p \in M$, the antipodal manifold A_p, with the Riemannian structure induced by that of M, is a Riemannian globally symmetric space of rank one, and a totally geodesic submanifold of M.*

† Here exp is the exponential mapping for G and $\tau(x)$ is the mapping $gK_p \to xgK_p$ of G/K_p onto itself.

Proof. Let $q \in A_p$. Considering a geodesic in M through p and q, we see that p is fixed under the geodesic symmetry s_q; hence $s_q(A_p) = A_p$. If σ_q denotes the restriction of s_q to A_q, then σ_q is an involutive isometry of A_p with q as isolated fixed point. As the image of the sphere $S_L(0)$ in M_p under the continuous mapping Exp, A_p is also connected. Thus A_p is globally symmetric and σ_q is the geodesic symmetry with respect to q. Let $t \to \gamma(t)$ $(t \in \boldsymbol{R})$ be a geodesic in the Riemannian manifold A_p. We shall prove that γ is a geodesic in M. Consider the isometry $s_{\gamma(t)}s_{\gamma(0)}$ and a vector X in the tangent space $M_{\gamma(0)}$. Let $\tau_r : M_{\gamma(0)} \to M_{\gamma(r)}$ denote the parallel translation in M along the curve $\gamma(\rho)$ $(0 \leqslant \rho \leqslant r)$. Then the parallel field $\tau_r \cdot X$ $(0 \leqslant r \leqslant t)$ along the curve $r \to \gamma(r)$ $(0 \leqslant r \leqslant t)$ is mapped by $s_{\gamma(t)}$ onto a parallel field along the image curve $r \to s_{\gamma(t)}\gamma(r) = \sigma_{\gamma(t)}\gamma(r) = \gamma(2t - r)$ $(0 \leqslant r \leqslant t)$. Since $s_{\gamma(t)}\tau_t X = -\tau_t X$, we deduce that $s_{\gamma(t)}s_{\gamma(0)}X = -s_{\gamma(t)}X = \tau_{2t}X$. In particular, the parallel transport in M along γ maps tangent vectors to γ into tangent vectors to γ. Hence γ is a geodesic in M. Consequently, A_p is a totally geodesic submanifold of M, and by the definition of rank, A_p has rank one.

The argument above proves a more general result.

Corollary 10.5. *Let Q be a Riemannian globally symmetric space, N a connected submanifold of Q such that for each $n \in N$, N is invariant under the geodesic symmetry of Q with respect to n. Then N, with the Riemannian structure induced by that of Q, is a Riemannian globally symmetric space, totally geodesic in Q.*

Now select a vector $H \in \mathfrak{p}_*$ of length L, the diameter of M. Then $\mathfrak{a}_* = \boldsymbol{R}H$ is a maximal abelian subspace of $\mathfrak{p}_*$. Select a restricted root $\alpha \in \Sigma$ such that $\alpha(H) > 0$ and such that $\frac{1}{2}\alpha$ is the only other possible restricted root, positive on H (cf. Cor. 2.17). This means (cf. Lemma 3.6, Chapter VI) that the eigenvalues of $(\operatorname{ad} H)^2$ on $\mathfrak{u}$ are 0, $\alpha(H)^2$, and possibly $(\frac{1}{2}\alpha(H))^2$ with multiplicities $\dim \mathfrak{m} + 1$, $2m_\alpha$, and $2m_{\frac{1}{2}\alpha}$. Let

$$\mathfrak{u} = \mathfrak{u}_0 + \mathfrak{u}_\alpha + \mathfrak{u}_{\frac{1}{2}\alpha}, \qquad \mathfrak{k} = \mathfrak{k}_0 + \mathfrak{k}_\alpha + \mathfrak{k}_{\frac{1}{2}\alpha}, \qquad \mathfrak{p}_* = \mathfrak{a}_* + \mathfrak{p}_\alpha + \mathfrak{p}_{\frac{1}{2}\alpha}$$

be the corresponding decompositions of $\mathfrak{u}$, $\mathfrak{k}$, and $\mathfrak{p}_*$ into eigenspaces. We have the following commutation relations:

(i) $\mathfrak{k}_0 = \mathfrak{m}$, the centralizer of H in $\mathfrak{k}$;

(ii) $\operatorname{ad} H$ exchanges $\mathfrak{k}_\alpha$ and $\mathfrak{p}_\alpha$, $\mathfrak{k}_{\frac{1}{2}\alpha}$ and $\mathfrak{p}_{\frac{1}{2}\alpha}$;

(iii) $[\mathfrak{p}_\alpha, \mathfrak{p}_\alpha] \subset \mathfrak{k}_0$, $[\mathfrak{k}_\alpha, \mathfrak{p}_\alpha] \subset \mathfrak{a}_*$, $[\mathfrak{k}_0, \mathfrak{p}_\alpha] \subset \mathfrak{p}_\alpha$.

The first relation follows from $B([H, T], [H, T]) = -B(T, (\operatorname{ad} H)^2\, T)$ $(H \in \mathfrak{a}_*,\ T \in \mathfrak{k})$. Part (ii) is obvious. Part (iii) comes from observing that by Lemma 3.6, Chapter VI, $(\mathrm{ad}\, H)^2$ has eigenvalues 0, 0, and $\alpha(H)^2$ on the spaces $[\mathfrak{p}_\alpha, \mathfrak{p}_\alpha]$, $[\mathfrak{k}_\alpha, \mathfrak{p}_\alpha]$, and $[\mathfrak{k}_0, \mathfrak{p}_\alpha]$, respectively.

Proposition 10.6. *Let $S \subset K$ be the subgroup leaving the point $\mathrm{Exp}\, H \in M$ fixed and let $\mathfrak{s}$ denote its Lie algebra. Then:*

(i) $\mathfrak{s} = \mathfrak{k}_0 + \mathfrak{k}_\alpha$ *if H is conjugate to o,*

(ii) $\mathfrak{s} = \mathfrak{k}_0$ *if H is not conjugate*† *to o,*

(iii) *If $\frac{1}{2}\alpha$ is a restricted root, then H is conjugate to o.*

Proof. We first prove (iii). Consider the sphere in the tangent space M_o with center 0 and radius $2L$. The mapping Exp maps this sphere onto the point o, so the differential $(d\,\mathrm{Exp})_{2H} = 0$. But then Theorem 4.1, Chapter IV, implies

$$\sum_0^\infty \frac{T_{2H}^n}{(2n+1)!} = 0.$$

If $\frac{1}{2}\alpha$ is a restricted root, this implies $\sinh(\frac{1}{2}\alpha(2H)) = 0$, so $\alpha(H) \in \pi i \boldsymbol{Z}$ and (iii) follows from Prop. 3.1.

(i) and (ii). A vector $T \in \mathfrak{k}$ belongs to $\mathfrak{s}$ if and only if

$$\exp(-H)\exp(tT)\exp H \in K \quad \text{for all } t \in \boldsymbol{R}.$$

Thus $T \in \mathfrak{s} \Leftrightarrow \mathrm{Ad}(\exp H)\, T \in \mathfrak{k}$, which is equivalent to

$$\sinh(\mathrm{ad}\, H)T = 0. \tag{2}$$

In particular, $\mathfrak{s}$ is the sum of its intersections with $\mathfrak{k}_0$, $\mathfrak{k}_\alpha$, and $\mathfrak{k}_{\frac{1}{2}\alpha}$. If $T \neq 0$ in $\mathfrak{k}_\beta$ ($\beta = 0, \alpha, \frac{1}{2}\alpha$) condition (2) amounts to $\sinh \beta(H) = 0$ or equivalently, $\beta(H) \in \pi i \boldsymbol{Z}$.

Now, if H is not conjugate to o, then $\frac{1}{2}\alpha \notin \Sigma$ by (iii) and $\alpha(H) \notin \pi i \boldsymbol{Z}$ by Prop. 3.1, so $\mathfrak{s} = \mathfrak{k}_0$, proving (ii). For (i) suppose H conjugate to o. Whether or not $\frac{1}{2}\alpha$ is a restricted root, we have by the cited result $\alpha(H) \in \pi i \boldsymbol{Z}$, so $\mathfrak{k}_\alpha \subset \mathfrak{s}$. Finally $\mathfrak{s} \cap \mathfrak{k}_{\frac{1}{2}\alpha} = 0$ because otherwise $\frac{1}{2}\alpha(H) \in \pi i \boldsymbol{Z}$ which would imply that $\frac{1}{2}H$ is conjugate to o, contradicting Theorem 10.3 This concludes the proof.

Theorem 10.7. *Suppose H is conjugate to o, and recall the identification $M_o = \mathfrak{a}_* + \mathfrak{p}_\alpha + \mathfrak{p}_{\frac{1}{2}\alpha}$. Then:*

(i) $\mathrm{Exp}(\mathfrak{a}_* + \mathfrak{p}_\alpha)$, *with the Riemannian structure induced by that of M,*

† According to the list in Exercise G.2, Chapter X, this happens only for the real projective spaces.

is a sphere, totally geodesic in M, having o and $\operatorname{Exp} H$ *as antipodal points, and having curvature π^2/L^2.*

(ii) $\operatorname{Exp}(\mathfrak{p}_{\frac{1}{2}\alpha})$ *equals the antipodal set $A_{\operatorname{Exp} H}$, which is also totally geodesic in M.*

Proof. (i) Let S_0 denote the identity component of S. Since $[\mathfrak{s}, \mathfrak{a}_* + \mathfrak{p}_\alpha] \subset \mathfrak{a}_* + \mathfrak{p}_\alpha$, the orbit $\operatorname{Ad}_U(S_0)H$ lies in $\mathfrak{a}_* + \mathfrak{p}_\alpha$ and its tangent space is $[\mathfrak{s}, \boldsymbol{R}H] = \mathfrak{p}_\alpha$. Thus $\operatorname{Ad}(S_0)H$ is a sphere in $\mathfrak{a}_* + \mathfrak{p}_\alpha$ of radius L and center 0. Moreover, if $s \in S$, then the geodesic $t \to s \cdot \operatorname{Exp} tH = \operatorname{Exp}(\operatorname{Ad}(s)\, tH)$ passes through $\operatorname{Exp} H$. The commutation relations (i)-(iii) above show that $\mathfrak{a}_* + \mathfrak{p}_\alpha$ is a Lie triple system. Thus by Theorem 7.2, Chapter IV, and the subsequent remark, the manifold $M_\mathfrak{p} = \operatorname{Exp}(\mathfrak{a}_* + \mathfrak{p}_\alpha)$ is a globally symmetric space, totally geodesic in M and hence of rank one. If $Z \in \mathfrak{p}_\alpha$ is a unit vector, the curvature of $M_\mathfrak{p}$ along the plane section spanned by H and Z is by Chapter V, §3, given by

$$-L^{-2}B([H, Z], [H, Z]) = -L^{-2}\alpha(H)^2 = (\pi/L)^2.$$

Since $M_\mathfrak{p}$ has rank one, every plane section is congruent to one containing H; hence $M_\mathfrak{p}$ has constant curvature. Finally, $M_\mathfrak{p} - \{\operatorname{Exp} H\}$ is the diffeomorphic image of an open ball, hence simply connected. Since $\dim M_\mathfrak{p} > 1$, it follows that $M_\mathfrak{p}$ is simply connected too, hence a sphere (Exercise A.4, Chapter IV). This proves part (i).

For part (ii) we first note that the geodesics from o to $\operatorname{Exp} H$ intersect A_o under a right angle in $\operatorname{Exp} H$. In fact, the sphere $S_{tL}(o)$ in M equals $K \cdot \operatorname{Exp} tH = \operatorname{Exp} \operatorname{Ad}(K)\, tH$ $(t \leqslant 1)$ and by Lemma 9.7, Chapter I,

$$Q_{\operatorname{Exp} tH}(d \operatorname{Exp}_{tH}(tH), d \operatorname{Exp}_{tH}(\operatorname{Ad}(K)(tH))) = 0 \qquad (0 < t < 1),$$

so the statement follows letting $t \to 1$ because the tangent space $(A_o)_{\operatorname{Exp} H}$ equals $d \operatorname{Exp}_H(\operatorname{Ad}(K)\, H)$ (Chapter II, §4). Reversing o and $\operatorname{Exp} H$, it follows that the geodesics from $\operatorname{Exp} H$ to o intersect $A_{\operatorname{Exp} H}$ in o under a right angle. From part (i) we therefore deduce that $A_{\operatorname{Exp} H} = \operatorname{Exp}(\mathfrak{p}_{\frac{1}{2}\alpha})$ and the theorem is proved.

Proposition 10.8. *If H is not conjugate to o, then $\frac{1}{2}\alpha$ is not a restricted root and*

$$A_{\operatorname{Exp} H} = \operatorname{Exp}(\mathfrak{p}_\alpha).$$

Proof. The first statement was already proved. The second is a consequence of the following facts: (1) By Prop. 10.6, $\dim A_{\operatorname{Exp} H} = \dim A_o = \dim K - \dim S = \dim \mathfrak{k}_\alpha = \dim \mathfrak{p}_\alpha$; (2) the geodesic $t \to$

$\mathrm{Exp}(tH)$ is perpendicular to $A_{\mathrm{Exp}\,H}$ in o; (3) $A_{\mathrm{Exp}\,H}$ and $\mathrm{Exp}(\mathfrak{p}_\alpha)$ are both totally geodesic and connected.

We conclude this section exhibiting a kind of projective duality between points and antipodal manifolds.

Proposition 10.9. *Let $p, q \in M$. Then:*

(i) *$p \neq q$ implies $A_p \neq A_q$;*

(ii) *$p \in A_q$ if and only if $q \in A_p$.*

Proof. If $r \in A_p$, then the geodesics which meet A_p in r under a right angle all pass through a point r^* at distance L from r (Theorem 10.7 and Prop. 10.8); among these are the geodesics joining p and r. Thus $p = r^*$ and the result follows.

§ 11. Shortest Geodesics and Minimal Totally Geodesic Spheres

Let M be a compact *irreducible* (cf. Chapter VIII, §5) Riemannian globally symmetric space with the Riemannian structure induced by the negative of the Killing form of the Lie algebra of $I_0(M)$. Let κ be the maximum of the sectional curvatures of M whose values are then restricted to the interval $[0, \kappa]$ (cf. Chapter V, §3). By Theorem 6.2, Chapter V, the maximum-dimensional, flat, totally geodesic submanifolds of M are all conjugate under $I_0(M)$. This section is devoted to a proof of a counterpart of this result for the maximal curvature κ. Exercise 5 provides an illustration of the general concepts used.

Theorem 11.1. *The space M contains totally geodesic submanifolds of constant curvature κ. Any two such submanifolds of the same dimension are conjugate under $I_0(M)$. The maximum dimension of such submanifolds is $1 + m_{\bar\delta}$ where $m_{\bar\delta}$ is the multiplicity of the highest restricted root $\bar\delta$. Also $\kappa = \langle\bar\delta, \bar\delta\rangle$.*

During the proof of this theorem we establish also the following result.

Theorem 11.2. *Assume the space M above is simply connected. Then the closed geodesics in M of minimal length are permuted transitively by $I_0(M)$. The minimum length is $2\pi/\|\bar\delta\|$, $\|\ \|$ denoting the norm $\langle\ ,\ \rangle^{1/2}$.*

The assumption "simply connected" cannot be omitted as Exercise 6 shows.

We start with a few algebraic preliminaries. Let $\mathfrak{u}$ be a compact semisimple Lie algebra over $\boldsymbol{R}$, θ an involutive automorphism of $\mathfrak{u}$. Let $\mathfrak{u}^C$ denote the complexification of $\mathfrak{u}$, $\mathfrak{g}$ the real form of $\mathfrak{u}^C$ corresponding to $(\mathfrak{u}, \theta)$, that is, $\mathfrak{g} = \mathfrak{k} + \mathfrak{p}$, $\mathfrak{u} = \mathfrak{k} + \mathfrak{p}_*$ where $\mathfrak{k}$ and $\mathfrak{p}_*$ are the eigenspaces of θ for $+1$ and -1, respectively, and $\mathfrak{p} = i\mathfrak{p}_*$. Let $\mathfrak{a}_* \subset \mathfrak{p}_*$ be a maximal abelian subspace, put $\mathfrak{a} = i\mathfrak{a}_*$, and extend $\mathfrak{a}_*$ to a maximal abelian subalgebra $\mathfrak{t}$ of $\mathfrak{u}$. Then the complexification $\mathfrak{h} = \mathfrak{t}^C$ is a Cartan subalgebra of $\mathfrak{u}^C$. Let as before Δ denote the corresponding system of nonzero roots, $\Delta_\mathfrak{p}$ the set of roots which do not vanish identically on $\mathfrak{a}^C$ (the complexification of $\mathfrak{a}$ in $\mathfrak{u}^C$). Let Σ denote the corresponding set of restricted roots, and for each $\lambda \in \Sigma$ let m_λ denote its multiplicity. Putting $\mathfrak{t}_\mathfrak{k} = \mathfrak{k} \cap \mathfrak{t}$, $\mathfrak{t}_{\mathfrak{k}_*} = i\mathfrak{t}_\mathfrak{k}$, we know that each root in Δ takes real values on $\mathfrak{h}_{\boldsymbol{R}} = \mathfrak{a} + \mathfrak{t}_{\mathfrak{k}_*}$. Select compatible orderings in the dual spaces of $\mathfrak{a} + \mathfrak{t}_{\mathfrak{k}_*}$ and of $\mathfrak{a}$ and let Δ^+ and Σ^+ denote the sets of positive elements in Δ and Σ, respectively. If f is a function on $\mathfrak{t}^C$ its restriction to $\mathfrak{a}^C$ is denoted by $\bar{f}$.

For each linear form λ on $\mathfrak{a}^C$ put

$$\mathfrak{k}_\lambda = \{T \in \mathfrak{k} : (\operatorname{ad} H)^2\, T = \lambda(H)^2\, T \text{ for } H \in \mathfrak{a}_*\},$$

$$\mathfrak{p}_\lambda = \{X \in \mathfrak{p}_* : (\operatorname{ad} H)^2\, X = \lambda(H)^2\, X \text{ for } H \in \mathfrak{a}_*\}.$$

Then $\mathfrak{k}_\lambda = \mathfrak{k}_{-\lambda}$, $\mathfrak{p}_\lambda = \mathfrak{p}_{-\lambda}$, $\mathfrak{p}_0 = \mathfrak{a}_*$, and $\mathfrak{k}_0$ equals $\mathfrak{m}$, the centralizer of $\mathfrak{a}_*$ in $\mathfrak{k}$.

Lemma 11.3. *The following decompositions are direct*:

$$\mathfrak{k} = \mathfrak{m} + \sum_{\lambda \in \Sigma^+} \mathfrak{k}_\lambda, \qquad \mathfrak{p}_* = \mathfrak{a}_* + \sum_{\lambda \in \Sigma^+} \mathfrak{p}_\lambda.$$

In fact, the endomorphisms $(\operatorname{ad} H)^2$ $(H \in \mathfrak{a}_*)$ commute, are symmetric with respect to the Killing form B of $\mathfrak{u}^C$, leave $\mathfrak{k}$ and $\mathfrak{p}_*$ invariant, and have eigenvalues 0 and $\lambda(H)^2$ $(\lambda \in \Sigma^+)$; cf. §2.

Lemma 11.4. *Let* $\lambda, \mu \in \Sigma \cup \{0\}$. *Then*

$$[\mathfrak{k}_\lambda, \mathfrak{p}_\mu] \subset \mathfrak{p}_{\lambda+\mu} + \mathfrak{p}_{\lambda-\mu},$$

$$[\mathfrak{k}_\lambda, \mathfrak{k}_\mu] \subset \mathfrak{k}_{\lambda+\mu} + \mathfrak{k}_{\lambda-\mu},$$

$$[\mathfrak{p}_\lambda, \mathfrak{p}_\mu] \subset \mathfrak{k}_{\lambda+\mu} + \mathfrak{k}_{\lambda-\mu}.$$

Proof. For each $\alpha \in \Delta$ select $X_\alpha \neq 0$ in $\mathfrak{u}^C$ such that $[H, X_\alpha] = \alpha(H) X_\alpha$ for $H \in \mathfrak{t}^C$. Extending θ to an automorphism of $\mathfrak{u}^C$, also denoted θ, let

$$\mathfrak{u}^+(\alpha) = \boldsymbol{C}(X_\alpha + \theta X_\alpha), \qquad \mathfrak{u}^-(\alpha) = \boldsymbol{C}(X_\alpha - \theta X_\alpha).$$

Then by Lemma 3.6, Chapter VI, if $\lambda \in \Sigma$,

$$\mathfrak{k}_\lambda + i\mathfrak{k}_\lambda = \sum_{\bar{\alpha}=\lambda} \mathfrak{u}^+(\alpha), \qquad \mathfrak{p}_\lambda + i\mathfrak{p}_\lambda = \sum_{\bar{\alpha}=\lambda} \mathfrak{u}^-(\alpha). \tag{1}$$

It follows that

$$\mathfrak{k}_\lambda = \mathfrak{k} \cap \sum_{\bar{\alpha}=\lambda} \mathfrak{u}^+(\alpha), \qquad \mathfrak{p}_\lambda = \mathfrak{p}_* \cap \sum_{\bar{\alpha}=\lambda} \mathfrak{u}^-(\alpha) \tag{2}$$

and by the quoted lemma,

$$\mathfrak{k}_0 = \mathfrak{t}_\mathfrak{k} + \mathfrak{k} \cap \sum_{\bar{\alpha}=0} \boldsymbol{C}X_\alpha, \qquad \mathfrak{p}_0 = \mathfrak{a}_*. \tag{3}$$

Let $\lambda, \mu \in \Sigma^+$ and $\alpha, \beta \in \Delta$ such that $\bar{\alpha} = \lambda$, $\bar{\beta} = \mu$. Then

$$[\mathfrak{u}^+(\alpha), \mathfrak{u}^-(\beta)] \subset \mathfrak{u}^-(\alpha + \beta) + \mathfrak{u}^-(\alpha^\theta + \beta) \qquad \text{if} \quad \alpha^\theta + \beta \neq 0, \tag{4}$$

$$[\mathfrak{u}^+(\alpha), \mathfrak{u}^-(\beta)] \subset \mathfrak{u}^-(\alpha + \beta) + \boldsymbol{C}(H_\beta - \theta H_\beta) \qquad \text{if} \quad \alpha^\theta + \beta = 0, \tag{5}$$

so the first relation of the lemma follows from (2). The others follow similarly; and if λ or $\mu = 0$, the relations are clear from (3).

With A_λ as in §2 let $\mathfrak{a}_\lambda$ denote the subspace $\boldsymbol{R}iA_\lambda$ of $\mathfrak{a}_*$ ($\lambda \in \Sigma$). Then if $\alpha \in \Delta$ we have $H_\alpha - A_{\bar{\alpha}} \in \mathfrak{t}_{\mathfrak{k}_*}$ and $A_{\bar{\alpha}} = \frac{1}{2}(H_\alpha - H_{\alpha^\theta})$. Then we have by (3)-(5),

Lemma 11.5. *Let $\lambda \in \Sigma$. Then*

$$[\mathfrak{k}_\lambda, \mathfrak{p}_\lambda] \subset \mathfrak{p}_{2\lambda} + \mathfrak{a}_\lambda.$$

Let (cf. Lemma 2.20) $C \subset \mathfrak{h}_{\boldsymbol{R}}$ be the Weyl chamber where all $\alpha \in \Delta^+$ take positive values and let $C_\mathfrak{p} \subset \mathfrak{a}$ be the Weyl chamber where all $\lambda \in \Sigma^+$ take positive values. By the compatibility of the orderings, $C_\mathfrak{p} \subset \bar{C}$ (the closure of C).

Lemma 11.6. *Let $\delta \in \Delta^+$ denote the highest root. Then $\bar{\delta} \in \Sigma^+$ is the highest restricted root and $H_\delta \in \bar{C}$, $A_{\bar{\delta}} \in \bar{C}_\mathfrak{p}$.*

Proof. That $H_\delta \in \bar{C}$ is clear from Theorem 2.22. Since $\delta - \alpha \geqslant 0$ for all $\alpha \in \Delta$, we have by the compatibility of the orderings $\bar{\delta} - \lambda \geqslant 0$ for all $\lambda \in \Sigma$, so the remaining statements follow in the same way as the first.

Definition. The restricted root system Σ is called *irreducible* if there exists no partition $\Sigma = \Sigma_1 \cup \Sigma_2$ into disjoint, nonempty, orthogonal subsets Σ_1 and Σ_2.

We now relate this concept to the irreducibility of $(\mathfrak{u}, \theta)$ as defined in the following chapter, §5.

Lemma 11.7. *Assume $\mathfrak{k}$ contains no nonzero ideal in $\mathfrak{u}$. Then $(\mathfrak{u}, \theta)$ is irreducible if and only if Σ is irreducible.*

Proof. If Σ is not irreducible, let $\Sigma = \Sigma_1 \cup \Sigma_2$ be the associated partition. Then $\lambda \in \Sigma_1$, $\mu \in \Sigma_2$ implies $\lambda \pm \mu \notin \Sigma_1 \cup \Sigma_2$. Hence by Lemmas 11.3-11.5, $\sum_{\lambda \in \Sigma_1} (\mathfrak{k}_\lambda + \mathfrak{p}_\lambda)$ and $\sum_{\lambda \in \Sigma_2} (\mathfrak{k}_\lambda + \mathfrak{p}_\lambda)$ would generate ideals $\mathfrak{u}_1$ and $\mathfrak{u}_2$ in $\mathfrak{u}$ invariant under θ, and this shows $(\mathfrak{u}, \theta)$ not irreducible. Conversely, if $(\mathfrak{u}, \theta)$ is not irreducible, then Prop. 5.2, Chapter VIII, shows that Σ is not irreducible.

Lemma 11.8. *Suppose Σ is irreducible, let $\alpha_1, \ldots, \alpha_l$ be the simple restricted roots and $\delta = \sum_1^l d_i \alpha_i$ the highest restricted root where (by Theorem 2.19) $d_1, \ldots, d_l \in \boldsymbol{Z}^+$. Then:*

(i) *$d_i \geqslant 1$ for each i.*

(ii) *If $\alpha \in \Sigma^+$ and we write $\alpha = \sum_{i=1}^l a_i \alpha_i$ $(a_i \in \boldsymbol{Z}^+)$, then*

$$a_1 \leqslant d_1, \ldots, a_l \leqslant d_l.$$

Remark. Part (ii) shows that δ depends only on the Weyl chamber $C_\mathfrak{p}$ (which determines the subset $\Sigma^+ \subset \Sigma$) but not on the ordering of Σ chosen.

Proof. Let $\Sigma' = \{\alpha \in \Sigma : \frac{1}{2}\alpha \notin \Sigma\}$ be the set of "indivisible" roots. The irreducibility of Σ implies that the set $\{\alpha_1, \ldots, \alpha_l\}$ has no partition $S_1 \cup S_2$ into orthogonal, disjoint, nonempty subsets S_1 and S_2. In fact, let $\Sigma_i' = W(\mathfrak{u}, \theta) \cdot S_i (i = 1, 2)$. Since, by Theorem 2.12 and Lemma 2.20 each $\alpha \in \Sigma'$ is $W(\mathfrak{u}, \theta)$-conjugate to a simple root, we would have $\Sigma' = \Sigma_1' \cup \Sigma_2'$. Also $W(\mathfrak{u}, \theta)$ is generated by the reflections $s_{\alpha_1}, \ldots, s_{\alpha_l}$ (cf. remark following Theorem 7.5) and $s_\mu \lambda = \lambda$ if $\lambda \in S_1$, $\mu \in S_2$. The formula for s_{α_i} therefore shows that $W(\mathfrak{u}, \theta)\, S_1$ remains in the subspace spanned by S_1, so the union $\Sigma' = \Sigma_1' \cup \Sigma_2'$ is an orthogonal partition, contradicting the irreducibility of Σ.

Now let $\prec$ denote the following partial ordering of Σ: $\lambda \prec \mu$ means that $0 \neq \mu - \lambda = \sum_{i=1}^l c_i \alpha_i$ $(c_i \geqslant 0)$. Let $\mu = \sum_{i=1}^l m_i \alpha_i$ be a maximal element in this ordering. We shall prove $m_i > 0$ for each i. Otherwise we have a nontrivial partition $\{\alpha_1, \ldots, \alpha_l\} = S_1 \cup S_2$ where $m_i > 0$ for $\alpha_i \in S_1$, $m_i = 0$ for $\alpha_i \in S_2$. Then $\langle \mu, \alpha_i \rangle \leqslant 0$ for all $\alpha_i \in S_2$

(Lemma 2.18), whereas by the indecomposability shown above, some $\alpha_j \in S_2$ is not orthogonal to all members of S_1, and therefore satisfies $\langle \mu, \alpha_j \rangle < 0$. But then Lemma 2.18 implies $\mu + \alpha_j \in \Sigma$, contradicting the maximality of μ. Thus $\mu = \Sigma_1^l m_i \alpha_i$ where $m_i > 0$ for each i. The argument also proves $\langle \mu, \lambda \rangle \geqslant 0$ for each $\lambda \in \Sigma^+$ and therefore, by nondegeneracy, $\langle \mu, \lambda \rangle > 0$ for some $\lambda \in \Sigma^+$. If γ is another maximal root in the ordering $\prec$, then by the above $\gamma = \Sigma_1^l c_i \alpha_i$ where each $c_i > 0$ and $\langle \gamma, \alpha_i \rangle \geqslant 0$ for all i. Thus $\langle \gamma, \mu \rangle > 0$, so $\mu - \gamma \in \Sigma$, so either $\mu \succ \gamma$ or $\gamma \succ \mu$, which is impossible. Thus γ cannot exist. This uniqueness of μ shows that if $\alpha \in \Sigma^+$ is arbitrary and we write $\alpha = \Sigma_1^l a_i \alpha_i$, then $a_1 \leqslant m_1, ..., a_l \leqslant m_l$. Finally, by Lemma 11.6, $\langle \bar{\delta}, \mu \rangle > 0$, so again $\bar{\delta} - \mu$ is a restricted root α. But $\alpha > 0$ contradicts the maximality of μ for the ordering $\prec$, and $\alpha < 0$ contradicts $\bar{\delta}$ being the highest restricted root. Thus $\alpha = 0$, $\bar{\delta} = \mu$, and the lemma is proved.

Suppose now the Lie algebra $\mathfrak{u}$ is simple. Let U be simply connected Lie group with Lie algebra $\mathfrak{u}$, "extend" θ to an automorphism of U, and let K be the group of fixed points of θ. Then we know that U is compact and K connected. Let Exp denote the Exponential mapping of $\mathfrak{p}_*$ onto U/K. Let $|\ |$ denote the norm in $\mathfrak{h}_{\boldsymbol{R}}$ and its dual, $\|\ \|$ the norm in $\mathfrak{a}$ and its dual.

Proposition 11.9. *The shortest, periodic, one-parameter subgroups in a simple simply connected, compact Lie group U have length $4\pi/|\delta|$ and they are all conjugate in U.*

Proof. Let $C_* = iC \subset \mathfrak{t}$ and consider the polyhedron P_* given by

$$P_* = \left\{ H \in \mathfrak{t} : 0 < \frac{\alpha(H)}{2\pi i} < 1 \text{ for } \alpha \in \Delta^+ \right\}.$$

This is a cell in $\mathfrak{t}_r$. Since $\mathfrak{u}$ is simple and Δ therefore irreducible, we have by Lemma 11.8, $P_* = \{H \in C_* : (2\pi i)^{-1} \delta(H) < 1\}$. Since U is simply connected, we have by Theorem 6.1

$$\mathfrak{t}_e \cap \bar{P}_* = \{0\}. \tag{6}$$

From Lemma 7.6 we have for each $\alpha \in \Delta$

$$\exp 2H(\alpha) = e, \tag{7}$$

where $H(\alpha) = 2\pi i H_\alpha / |\alpha|^2$ is the projection of 0 onto the hyperplane $\alpha = 2\pi i$ in $\mathfrak{t}$. For the closed one-parameter subgroup $\tau : t \to \exp(2tH(\delta))$ let $t_0 > 0$ be the first value such that $\exp(2t_0 H(\delta)) = e$. Then $\delta(2t_0 H(\delta)) \in$

$2\pi i\boldsymbol{Z}$ (Lemma 6.5), so $2t_0 \in \boldsymbol{Z}$. But (6) implies $t_0 \neq \frac{1}{2}$, so (7) implies $t_0 = 1$ and the length of τ is $2 \mid H(\delta)\mid = 4\pi/\mid \delta \mid$. Now let $t \to \exp tX$ be any periodic one-parameter subgroup of U of length $\leqslant 2 \mid H(\delta)\mid$, the parameter being fixed such that $\exp tX \neq e$ for $0 < t < 1$ and $\exp X = e$. Then for some $u \in U$, $H_1 = \mathrm{Ad}(u)X$ lies in $\bar{C}_*$; and since $\exp H_1 = e$, $\delta(H_1) = n2\pi i$ $(n \in \boldsymbol{Z}^+)$. Since $H_1 \neq 0$, we have $n \neq 0$. Also $n \neq 1$ by (6). Thus H_1 belongs to the union of the hyperplanes $\delta = n2\pi i$ $(n \geqslant 2)$ and by our assumption on X, we have $\mid H_1 \mid \leqslant 2 \mid H(\delta)\mid$. But the point $2H(\delta)$ is the only point which minimizes the distance from 0 to the union of the hyperplanes $\delta = n2\pi i$ $(n \geqslant 2)$. Thus $H_1 = 2H(\delta)$ and the proposition is proved.

We can now prove Theorem 11.2. According to Theorems 5.3 and 5.4, Chapter VIII, there are two cases to consider, namely $M = U/K$ and $M = (U \times U)/U^*$, where U^* is the diagonal in $U \times U$. For the first case $M = U/K$ we first recall that if $H \in \mathfrak{a}_*$, then $\mathrm{Exp}\, H = o$ if and only if $\exp 2H = e$. Let $A(\bar{\alpha}) = \pi i A_{\bar{\alpha}}/\| \bar{\alpha} \|^2$ $(\alpha \in \Delta_{\mathfrak{p}})$. Then we have by (4), §8,

$$\mathrm{Exp}(2A(\bar{\alpha})) = o. \tag{8}$$

In particular $4A(\bar{\delta}) \in \mathfrak{t}_e$; and since $\delta(4A(\bar{\delta})) = 4\pi i$ and $4A(\bar{\delta}) \in \bar{C}_*$, we have by (6), $4tA(\bar{\delta}) \notin \mathfrak{t}_e$ for $0 < t < 1$. Consequently, the geodesic

$$t \to \mathrm{Exp}(2tA(\bar{\delta})) \tag{9}$$

is simply closed and has length $2 \| A(\bar{\delta})\| = 2\pi/\| \bar{\delta} \|$. Now let $t \to \mathrm{Exp}\, tX$ be a geodesic in M such that $\mathrm{Exp}\, X = o$, $\mathrm{Exp}\, t_0 X \neq o$ for $0 < t_0 < 1$. Then by Prop. 10.1 it is a simply closed geodesic of length $\| X \|$. Assume this length is $\leqslant 2\pi/\| \bar{\delta} \|$. Select $k \in K$ such that $H = \mathrm{Ad}_U(k)X \in \overline{C_{\mathfrak{p}_*}}$ where $C_{\mathfrak{p}_*} = iC_{\mathfrak{p}}$. Then $2H \in \mathfrak{t}_e$, so $\delta(H) = m\pi i$, $m \in \boldsymbol{Z}^+$. But $H \neq 0$, so $m \neq 0$ and $m \neq 1$ because of (6). But $\| H \| \leqslant 2\pi/\| \bar{\delta} \|$, whereas the point $2A(\bar{\delta})$ is the only point which minimizes the distance from 0 to the union of the hyperplanes $\bar{\delta} = m\pi i$ $(m \geqslant 2)$ in $\mathfrak{a}_*$. Thus $H = 2A(\bar{\delta})$ and Theorem 11.2 is proved for the space U/K.

Next we prove Theorem 11.2 for $M = (U \times U)/U^*$, with the Riemannian structure Q^* defined by the Killing form B^* of $\mathfrak{u} \times \mathfrak{u}$, the identification of M with U being made via the mapping $\tau : (u_1, u_2)\, U^* \to u_1 u_2^{-1}$. In the context of §4 the highest restricted root, say $\tilde{\delta}$, is given by $\tilde{\delta}(H, -H) = \delta(H)$ $(H \in \mathfrak{t})$; and since

$$B^*((X, -X), (Y, -Y)) = 2B(X, Y) \qquad (X, Y \in \mathfrak{u}),$$

we find

$$A_{\tilde{\delta}} = (\tfrac{1}{2}H_\delta, -\tfrac{1}{2}H_\delta), \qquad 2 \| \tilde{\delta} \|^2 = \mid \delta \mid^2. \tag{10}$$

Moreover, $d\tau_0(X, -X) = 2X$, so for each tangent vector Z to $(U \times U)/U^*$ we have

$$2Q^*(Z, Z) = Q(d\tau(Z), d\tau(Z)) \tag{11}$$

if Q is the Riemannian structure on U defined by $-B$. Thus in the Riemannian structure Q^* the one-parameter subgroups from Prop. 11.9 have length $2\pi\sqrt{2}/|\delta|$ which by (10) equals $2\pi/\|\bar{\delta}\|$. This and Prop. 11.9 concludes the proof of Theorem 11.2.

We note some simple consequences of the proof. First (7) and (8) imply the following result.

Corollary 11.10. *For* $\mathfrak{u}$ *simple we have*

$$|\alpha| \leqslant |\delta| \qquad \textit{and} \qquad \|\bar{\alpha}\| \leqslant \|\bar{\delta}\| \qquad \textit{for} \quad \alpha \in \Delta.$$

We can also deduce the following results.

Corollary 11.11. *Let* $\mathfrak{u}$ *be simple. Suppose* $\alpha \in \Delta^+$ *and* $\bar{\alpha} \neq \bar{\delta}$. *Then*

$$2\frac{\langle \bar{\alpha}, \bar{\delta} \rangle}{\langle \bar{\delta}, \bar{\delta} \rangle} = 0 \quad \textit{or} \quad 1.$$

In fact, $\delta(A(\bar{\delta})) = \pi i$ and by (8) $\alpha(A(\bar{\delta})) = \frac{1}{2}\pi i n$ where $n \in \mathbf{Z}^+$. But $\delta - \alpha$ is (by Lemma 11.8 for $\mathfrak{u}$ and Δ) a positive integral linear combination of the simple roots in Δ^+. Thus $(\delta - \alpha)(A(\bar{\delta})) = \frac{1}{2}\pi i m$ where $m \in \mathbf{Z}^+$. Thus $n = 0$, 1, or 2. But if $n = 2$, then $\langle \bar{\alpha}, \bar{\delta} \rangle = \|\bar{\delta}\|^2$, so Cor. 11.10 implies $\bar{\alpha} = \bar{\delta}$. This proves Cor. 11.11.

From Cor. 11.11 we can now deduce the fact that when the double of a restricted root is a restricted root, then it either is on the line $\mathbf{R}\bar{\delta}$ or is perpendicular to it.

Corollary 11.12. *Let* $\mathfrak{u}$ *be simple. Suppose* $\beta \in \Delta^+$ *and* $2\bar{\beta} \in \Sigma^+ - \{\bar{\delta}\}$. *Then* $\bar{\beta}$ *and* $\bar{\delta}$ *are orthogonal.*

In fact, select $\alpha \in \Delta^+$ such that $\bar{\alpha} = 2\bar{\beta}$. Since $2\bar{\beta} \in \Sigma^+ - \{\bar{\delta}\}$, we have also $\bar{\beta} \in \Sigma^+ - \{\bar{\delta}\}$. Thus Cor. 11.11 applies to both α and β, so

$$2\frac{\langle 2\bar{\beta}, \bar{\delta} \rangle}{\langle \bar{\delta}, \bar{\delta} \rangle} = 0 \quad \text{or} \quad 1, \qquad 2\frac{\langle \bar{\beta}, \bar{\delta} \rangle}{\langle \bar{\delta}, \bar{\delta} \rangle} = 0 \quad \text{or} \quad 1,$$

whence $\langle \bar{\beta}, \bar{\delta} \rangle = 0$.

Definition. Let M be a compact, irreducible simply connected Riemannian globally symmetric space. For $p \in M$ let A_p denote the set of midpoints of closed geodesics of minimal length passing through p; A_p is called the *midpoint locus* associated with p.

From Theorem 11.2 and Cor. 10.5 we have the following consequence.

Corollary 11.13. *For each $p \in M$ the midpoint locus A_p is a totally geodesic submanifold and is an orbit of the isotropy subgroup of $I_0(M)$ at p.*

For $\epsilon = 0, \frac{1}{2}, 1$ put

$$\mathfrak{k}(\epsilon) = \sum_{\lambda \neq 0, \langle \lambda, \bar{\delta} \rangle = \epsilon \langle \bar{\delta}, \bar{\delta} \rangle} \mathfrak{k}_\lambda, \qquad \mathfrak{p}(\epsilon) = \sum_{\lambda \neq 0, \langle \lambda, \bar{\delta} \rangle = \epsilon \langle \bar{\delta}, \bar{\delta} \rangle} \mathfrak{p}_\lambda.$$

Then by Cor. 11.11 we have $\mathfrak{k}(1) = \mathfrak{k}_{\bar{\delta}}$, $\mathfrak{p}(1) = \mathfrak{p}_{\bar{\delta}}$ and

$$\mathfrak{k} = \mathfrak{m} + \mathfrak{k}(0) + \mathfrak{k}(\tfrac{1}{2}) + \mathfrak{k}_{\bar{\delta}}, \qquad \mathfrak{p}_* = \mathfrak{a}_* + \mathfrak{p}(0) + \mathfrak{p}(\tfrac{1}{2}) + \mathfrak{p}_{\bar{\delta}}.$$

Theorem 11.14. *Let S denote the centralizer of* $\exp 2A(\bar{\delta})$ *in K. Then*:

(i) *The Lie algebra $\mathfrak{s}$ of S equals*

$$\mathfrak{s} = \mathfrak{m} + \mathfrak{k}(0) + \mathfrak{k}_{\bar{\delta}}.$$

(ii) *In U/K we have for the midpoint loci*

$$A_o = K/S, \qquad A_{\operatorname{Exp} A(\bar{\delta})} = \operatorname{Exp} \mathfrak{p}(\tfrac{1}{2}).$$

Proof. For (i) we note first that $T \in \mathfrak{s}$ if and only if $e^{\operatorname{ad}(2A(\bar{\delta}))}T = T$. Putting $c = \lambda(2A(\bar{\delta}))$ for $\lambda \in \Sigma^+$ we have

$$e^{\operatorname{ad}(2A(\bar{\delta}))}T = \cosh cT + \frac{\sinh c}{c}[2A(\bar{\delta}), T], \qquad T \in \mathfrak{k}_\lambda,$$

so (i) follows.

For (ii) we note that $k \in K$ commutes with $\exp 2A(\bar{\delta})$ if and only if $\exp(-A(\bar{\delta}))k \exp A(\bar{\delta})$ is fixed by θ, that is, belongs to K. But $\exp(-A(\bar{\delta}))k \exp A(\bar{\delta}) \in K$ if and only if k leaves the point $\operatorname{Exp} A(\bar{\delta})$ fixed. Thus $A_o = K/S$. Next we observe that since $\exp 2A(\bar{\delta}) \in K$, we have

$$\exp(-A(\bar{\delta})) \cdot A_o = \exp A(\bar{\delta}) A_o = A_{\operatorname{Exp} A(\bar{\delta})}$$

so by Cor. 11.13

$$A_{\operatorname{Exp} A(\bar{\delta})} = \exp(-A(\bar{\delta}))\, K \exp A(\bar{\delta}) \cdot o. \tag{12}$$

If $T \in \mathfrak{k}$, it is clear that

$$\operatorname{Ad}_U(\exp(-A(\bar{\delta})))T \equiv -\sinh(\operatorname{ad}(A(\bar{\delta})))T \qquad (\operatorname{mod} \mathfrak{k}),$$

so the curve

$$t \to \exp(-A(\bar{\delta})) \exp tT \exp A(\bar{\delta}) \cdot o$$

has tangent vector $-\sinh \operatorname{ad}(A(\bar{\delta}))T$ at $t = 0$. But using (2) we deduce

$$\sinh(\operatorname{ad}(A(\bar{\delta})))\mathfrak{k} = \mathfrak{p}(\tfrac{1}{2}),$$

which by (12) finishes the proof since we know in advance that $A_{\operatorname{Exp} A(\bar{\delta})}$ is totally geodesic in U/K.

After these preparations we can give a proof of Theorem 11.1. We first compute the maximal sectional curvature κ of U/K. For this it suffices to consider plane sections in $\mathfrak{p}_*$ spanned by orthonormal vectors $H \in C_{\mathfrak{p}_*}$ and $X \in \mathfrak{p}_*$. The corresponding curvature is by Chapter V, §3, given by

$$B((\operatorname{ad} H)^2 X, X). \tag{13}$$

Decomposing X according to Lemma 11.3

$$X = X_0 + \sum_{\lambda \in \Sigma^+} X_\lambda$$

we have

$$B((\operatorname{ad} H)^2 X, X) = \sum_{\lambda \in \Sigma^+} \lambda(H)^2 B(X_\lambda, X_\lambda) \leqslant -\delta(H)^2$$

because, by Lemma 11.8, $\delta - \lambda$ has positive values on $C_\mathfrak{p}$. This last inequality implies $\kappa \leqslant \| \bar{\delta} \|^2$. But choosing $X \in \mathfrak{p}_{\bar{\delta}}$, $H \in \mathfrak{a}_{\bar{\delta}}$, we see that the equality sign actually holds.

Now since $2\bar{\delta}$ is not a restricted root, it is easily seen from Lemmas 11.4-11.5 that the subspace $\mathfrak{a}_{\bar{\delta}} + \mathfrak{p}_{\bar{\delta}} \subset \mathfrak{p}_*$ is a Lie triple system. Also Cor. 11.11 implies that if $\lambda \in \Sigma^+$ is orthogonal to $\bar{\delta}$, then neither $\bar{\delta} + \lambda$ nor $\bar{\delta} - \lambda$ is a restricted root. Hence the Lie algebra $\mathfrak{s}$ satisfies

$$[\mathfrak{s}, \mathfrak{a}_{\bar{\delta}}] = \mathfrak{p}_{\bar{\delta}}, \qquad [\mathfrak{s}, \mathfrak{a}_{\bar{\delta}} + \mathfrak{p}_{\bar{\delta}}] = \mathfrak{a}_{\bar{\delta}} + \mathfrak{p}_{\bar{\delta}}.$$

Since $\mathfrak{a}_{\bar{\delta}}$ is a maximal abelian subspace of $\mathfrak{a}_{\bar{\delta}} + \mathfrak{p}_{\bar{\delta}}$, the totally geodesic subspace $M_{\bar{\delta}} = \operatorname{Exp}(\mathfrak{a}_{\bar{\delta}} + \mathfrak{p}_{\bar{\delta}})$ has rank one, and by the computation above, constant curvature $\| \bar{\delta} \|^2$. Let S_0 denote the identity component of S. Then the tangent space to the orbit $\operatorname{Ad}_U(S_0) \cdot A(\bar{\delta}) \subset \mathfrak{a}_{\bar{\delta}} + \mathfrak{p}_{\bar{\delta}}$ is $[\mathfrak{s}, A(\bar{\delta})] = \mathfrak{p}_{\bar{\delta}}$, so this orbit is the sphere in $\mathfrak{a}_{\bar{\delta}} + \mathfrak{p}_{\bar{\delta}}$ with center 0, passing through $A(\bar{\delta})$. Since all $s \in S_0$ leave $\operatorname{Exp} A(\bar{\delta})$ fixed, all geodesics in $M_{\bar{\delta}}$ through o pass through $o' = \operatorname{Exp} A(\bar{\delta})$. Hence $M_{\bar{\delta}} - o'$ is the diffeomorphic image of a ball (Theorem 10.3), so $M_{\bar{\delta}}$ is simply connected, hence a sphere.

Now the geodesic symmetry $s_{o'}$ of U/K leaves o fixed. If γ is a geodesic segment of minimum length joining o and o', then γ and $s_{o'}\gamma$ will by Prop. 10.1 form a simply closed geodesic, and by Theorem 11.2 it has length $2\pi/\| \bar{\delta} \|$ (since the length of γ by assumption is $\leqslant \| A(\bar{\delta})\|$). The closed geodesics of this length starting at o and passing through o' are permuted transitively by S, hence form finitely many spheres Σ_i of dimension $1 + m_{\bar{\delta}}$, namely the images of M_δ under S. Let σ_i denote the unit sphere in the tangent space $(\Sigma_i)_o$. Each σ_i is an orbit of $\mathrm{Ad}_U(S_0)$, so any two different σ_i are disjoint. Now let Σ be any totally geodesic sphere in U/K of curvature $\| \bar{\delta} \|^2$. Using an isometry $u \in U$, we may assume that Σ passes through o and o'. Then the unit sphere σ in the tangent space $(\Sigma)_o$ is contained in the union of the σ_i. But σ is connected (since $\dim \Sigma > 1$), so it is contained in a single σ_i. Hence there exists an element $s \in S$ such that $s \cdot \Sigma \subset M_\delta$.

Next we consider the subgroup $U_\delta = \{u \in U : uM_\delta \subset M_\delta\}$. Then $S_0 \subset U_\delta$ and the restriction mapping $\varphi : u \to u/M_\delta$ is a homomorphism of U_δ into the isometry group $I(M_\delta)$. If $X \in \mathfrak{a}_{\bar{\delta}} + \mathfrak{p}_{\bar{\delta}}$ (viewed in the Lie algebra of U_δ), then $\varphi(\exp tX)$ is a one-parameter group of transvections of M_δ along the geodesic $\mathrm{Exp}\, tX$; in particular, $d\varphi$ is one-to-one on $\mathfrak{a}_{\bar{\delta}} + \mathfrak{p}_{\bar{\delta}}$. Thus we identify $(M_\delta)_0 = \mathfrak{a}_{\bar{\delta}} + \mathfrak{p}_{\bar{\delta}}$ with a subspace of the Lie algebra $\mathfrak{L}(I(M_\delta))$ and have by Chapter V, §4,

$$\mathfrak{L}(I(M_\delta)) = [\mathfrak{a}_{\bar{\delta}} + \mathfrak{p}_{\bar{\delta}}, \mathfrak{a}_{\bar{\delta}} + \mathfrak{p}_{\bar{\delta}}] + (\mathfrak{a}_{\bar{\delta}} + \mathfrak{p}_{\bar{\delta}})$$

The first term on the right is a subalgebra $\mathfrak{k}' \subset \mathfrak{s} \subset \mathfrak{k}$, and $\mathrm{ad}_{\mathfrak{u}}(\mathfrak{k}')$ restricted to $\mathfrak{a}_{\bar{\delta}} + \mathfrak{p}_{\bar{\delta}}$ is the Lie algebra $\mathfrak{so}(\mathfrak{a}_{\bar{\delta}} + \mathfrak{p}_{\bar{\delta}})$ (cf. Exercise 8, Chapter V). Hence we have the inclusion

$$\boldsymbol{SO}((M_\delta)_0) \subset \mathrm{Ad}_U(S_0) \mid (M_\delta)_0,$$

so that any two subspaces of $(M_\delta)_0$ of the same dimension are conjugate under $\mathrm{Ad}_U(S_0)$. Thus any two totally geodesic spheres of the same dimension contained in M_δ are conjugate under a member of U; to finish the proof of Theorem 11.1 for the space U/K it remains to verify that any totally geodesic submanifold N of U/K of constant curvature $\| \bar{\delta} \|^2$ is a sphere. But if N were not a sphere, it is clear, passing to the universal covering of N, that N would contain a simply closed geodesic of length $< 2\pi/\| \bar{\delta} \|$ which is impossible.

If M is any compact Riemannian globally symmetric space such that $I_0(M)$ is simple, then Theorem 11.1 can be applied to the universal covering space of M and the theorem follows for M.

Next we prove Theorem 11.1 for the case $M = U$. Suppose U has the bi-invariant Riemannian structure given by $-B$ on $\mathfrak{u}$. Then if $X, Y \in \mathfrak{u}$ are orthonormal, the corresponding sectional curvature is

$$\tfrac{1}{4}B((\text{ad } X)^2 Y, Y)$$

(cf. Exercise A.6, Chapter II). The maximum is $\frac{1}{4} | \delta |^2$ as we see by taking X proportional to H_δ and Y in the corresponding root space. However, with the conventions of Theorem 11.1 we must view U as $(U \times U)/U^*$ whereby the Riemannian structure is multiplied by $\frac{1}{2}$ (cf. (11)). Accordingly all sectional curvatures are multiplied by 2 so the maximal curvature becomes $\kappa = 2 \cdot \frac{1}{4} | \delta |^2$, which, as we saw before, equals $\| \tilde{\delta} \|^2$. Moreover $m_{\tilde{\delta}} = 2$. The remaining statements of Theorem 11.1 are proved for U and any compact group covered by U just as for the space U/K. By Chapter X, §1, the simple compact Lie groups and the globally symmetric spaces covered by U/K exhaust all compact irreducible Riemannian globally symmetric spaces, so Theorem 11.1 is now completely proved.

§ 12. Appendix. Results from Dimension Theory

In this section we collect some results from dimension theory which have been used earlier in this chapter. The dimension concept is here the topological dimension of Brouwer, Menger, and Urysohn, defined for all separable metric spaces. This definition assigns to the empty set dimension -1 and by induction the dimension of an arbitrary separable metric space M is defined as the smallest integer n for which each point $p \in M$ has arbitrarily small neighborhoods with boundaries of dimension less than n. Whenever possible, we refer to the book of Hurewicz and Wallman [1] for proofs of the results below. *All spaces considered are assumed to be separable metric spaces.* An n-dimensional manifold has topological dimension n (Hurewicz and Wallman [1], p. 46).

Proposition 12.1. *If M is a subspace of N then*

$$\dim M \leqslant \dim N.$$

For the proof, see Hurewicz and Wallman [1], p. 26.

Theorem 12.2. *Suppose a space M is a countable union $M = \bigcup_n M_n$ of closed subspaces M_n. Then*

$$\dim M \leqslant \sup_n M_n.$$

For the proof, see Hurewicz and Wallman [1], p. 30.

Lemma 12.3.† *Let M and N be differentiable manifolds and f a differentiable mapping of M into N. Then*

$$\dim f(Q) \leqslant \dim Q$$

for each submanifold $Q \subset M$.

Proof. Let m and n denote the dimensions of M and N, respectively. Let $p \in M$ and (B, φ) a local chart around p. The set B is called an open ball if φ can be chosen such that $\varphi(B)$ is an open ball in $\boldsymbol{R}^m$ with center $\varphi(p)$. Since f is continuous there exists a countable family $B_1, B_2, \ldots$ of open balls in M such that $M = \bigcup_i B_i$ and for each $i, f(B_i)$ is contained in an open ball B_i' in N. Now $Q = \bigcup_i (Q \cap B_i)$ and $f(Q) = \bigcup_i (f(Q) \cap B_i')$ so due to Theorem 12.2 we may assume that $M = \boldsymbol{R}^m$ and $N = \boldsymbol{R}^n$ and that Q is a bounded subset in $\boldsymbol{R}^m$. If $|\ |$ denotes the norm in $\boldsymbol{R}^m$ (and $\boldsymbol{R}^n$) we have

$$|f(x) - f(y)| \leqslant c\,|x - y| \tag{1}$$

for all x, y in some cube containing Q, c being a constant. Let $q = \dim Q$. Then the $(q+1)$-dimensional Hausdorff measure of Q is 0 (Hurewicz and Wallman [1], p. 105). From (1) follows that the $(q+1)$-dimensional Hausdorff measure of $f(Q)$ is 0 and therefore $\dim f(Q) \leqslant q$.

Proposition 12.4.†† *Let M be a connected manifold and let S be a closed submanifold of M, $\dim S \leqslant \dim M - 2$. Let $\gamma(t)$, $0 \leqslant t \leqslant 1$, be a continuous curve in M. Then γ is homotopic to a continuous curve $\gamma'(t)$, $0 \leqslant t \leqslant 1$, such that $\gamma'(t) \in M - S$ for $0 < t < 1$.*

Proof. Let B be an open ball in M and let a, b be two points in $B - S$. We shall first show that a and b can be joined by means of a continuous curve, not intersecting S. Let B' be an open ball, "concentric" to B such that $a, b \in B'$ and $\bar{B}' \subset B$. The "central projection" with respect to b gives a differentiable mapping of B onto the boundary of B'. The image of $B \cap S$ under this mapping contains no open subset of the boundary of B' (Lemma 12.3). Therefore, if N_a is a neighborhood of a in B such that $N_a \cap S = \emptyset$, there exists a point $a' \in N_a$ such that the segment from b to a' is disjoint from S. Combining this with the segment aa' we obtain the desired path between a and b.

Suppose now that the end points $\gamma(0)$ and $\gamma(1)$ do not belong to S. Then there exists a positive number $\epsilon = 1/(2n)$ (n = integer) such that each segment $\gamma(t)$, $|t - t_0| \leqslant \epsilon$ is contained in an open ball $B(t_0)$

† Harish-Chandra [5], p. 615.

†† Compare Pontrjagin [1], p. 263, Teil 2.

($|t_0| \leqslant 1$). Let $0 < j < n$. Then the point $\gamma(2j\epsilon)$ belongs to the intersection $B((2j-1)\epsilon) \cap B((2j+1)\epsilon)$. We replace $\gamma(2j\epsilon)$ by another point $\gamma'(2j\epsilon)$ in this intersection, not belonging to S. Finally we put $\gamma'(0) = \gamma(0)$ and $\gamma'(1) = \gamma(1)$. From the first part of the proof follows that the points $y'(2j\epsilon)$ and $\gamma'((2j+2)\epsilon)$ can be connected by a path in $B((2j+1)\epsilon)$, not intersecting S. The desired path γ' is obtained by combining these small paths. Then γ is homotopic to γ', since each ball is simply connected.

Finally suppose that at least one of the points $\gamma(0)$, $\gamma(1)$ belongs to S. Suppose for example that $\gamma(0) \in S$, $\gamma(1) \notin S$. Then there exists a sequence $x_1, x_2, \ldots$ in $M - S$, converging to $\gamma(0)$. Select N so large that all x_n ($n \geqslant N$) and all $\gamma(t)$ ($0 \leqslant t \leqslant 1/N$) belong to a ball B around $\gamma(0)$. Combining the part of γ from $\gamma(1/N)$ to $\gamma(1)$ with a curve from x_N to $\gamma(1/N)$ we obtain a curve δ from x_N to $\gamma(1)$ whose end points do not lie in S. In view of the result proved there exists a curve δ' in $M - S$ homotopic to δ. If we combine δ' with a sequence of suitable paths in $B - S$ joining x_n and x_{n+1} ($n \geqslant N$), we obtain the desired path γ'.

Corollary 12.5. *Under the assumptions of Prop.* 12.4, *the set* $M - S$ *is connected.*

By Hurewicz–Wallman [1], p. 48, Cor. 12.5 holds if S is any closed subset of M, $\dim S \leqslant \dim M - 2$. Thus Prop. 12.4 holds for any such S.

Proposition 12.6. *Let M be a connected manifold and S a connected submanifold of M. We assume that $\dim S \leqslant \dim M - 3$ and that S is a topological subspace of M. Let $\gamma(t)$ and $\gamma'(t)$ $(0 \leqslant t \leqslant 1)$ be two continuous curves in the complement $M - S$. Then if γ and γ' are homotopic in M they are also homotopic in $M - S$.*

Proof. The homotopy $\gamma \sim \gamma'$ can be broken up into a sequence of homotopies

$$\gamma = \Gamma_0 \sim \Gamma_1 \sim \ldots \sim \Gamma_{n-1} \sim \Gamma_n = \gamma',$$

where, for each i, the curves $\Gamma_{i-1}(t)$ and $\Gamma_i(t)$ coincide except on a subinterval I_i of $0 \leqslant t \leqslant 1$ for which $\Gamma_{i-1}(I_i)$ and $\Gamma_i(I_i)$ lie in the same open ball (compare Seifert and Threlfall [1], §44). This means, roughly speaking, that every deformation is a finite sequence of small deformations. We can therefore assume that γ lies in an open ball V and that γ' reduces to a point. Since S is a topological subspace of M we may also assume (Prop. 3.2, Chapter I), that V and the coordinates $\{x_1, \ldots, x_m\}$ on V are such that $S \cap V$ is the submanifold of V given by $x_1 = x_2 = x_3 = 0$. Thus it can be assumed that $M = \boldsymbol{R}^m$ and that S is the subspace given by $x_1 = 0$, $x_2 = 0$, $x_3 = 0$. If $|\ |$ denotes the norm in M let $C(S)$ denote the set of points in M whose distance from S equals $|\gamma(0)|$. Then $C(S)$ is homeomorphic to $\boldsymbol{S}^2 \times \boldsymbol{R}^{m-3}$, in particular, $C(S)$ is simply connected. Now $M - S$ can be mapped onto $C(S)$ by "central projection" φ from S. This mapping is defined as follows: if $p \in M - S$ let $s(p)$ denote the unique point in S at shortest distance from p. The ray from $s(p)$ through p intersects $C(S)$ at a point which we call $\varphi(p)$. The image of γ under φ is homotopic in $M - S$ to γ and since $C(S)$ is simply connected, $\varphi \cdot \gamma$ is homotopic in $C(S) \subset M - S$ to the point $\gamma(0)$. This finishes the proof.

EXERCISES AND FURTHER RESULTS

1. Find the centralizer $\mathfrak{m}_0$ and the Weyl group $W(U, K)$ for the Riemannian symmetric pair (U, K) where $U = \boldsymbol{SU}(n)$, $K = \boldsymbol{SO}(n)$.

2. Let σ be an involutive automorphism of a compact connected Lie group U. Let H denote the set of fixed points. Let U be given any two-sided invariant Riemannian structure. The mapping

$$uH \to u\sigma(u^{-1})$$

is a diffeomorphism of U/H onto a closed totally geodesic submanifold of U. This submanifold is Riemannian globally symmetric with respect to the induced Riemannian structure (É. Cartan [16]).

3. Let M be a compact symmetric space of rank one, κ_1 and κ_2 the infimum and supremum, respectviely, of the sectional curvature. Show that $\kappa_1/\kappa_2 = 1$ or $\frac{1}{4}$.

4. Show, in analogy with Prop. 7.2, that $W(\mathfrak{u}, \theta)$ is generated by the reflections in the walls of any fixed Weyl chamber.

5. Let $0 < p \leqslant q$ be integers and $\mathfrak{u} = \mathfrak{su}(p + q)$ the Lie algebra of the simply connected group $U = \boldsymbol{SU}(p + q)$. With $I_{p,q}$ as in Chapter X, §2 let θ denote the involutive automorphism $\theta X = I_{p,q} X I_{p,q}$ of $\mathfrak{u}$. We now illustrate some of the concepts of this chapter by means of this example.

(i) Show that

$$\mathfrak{k} = \left\{ \begin{pmatrix} A & 0 \\ 0 & B \end{pmatrix} \middle| \begin{array}{l} A \in \mathfrak{u}(p),\ B \in \mathfrak{u}(q) \\ \mathrm{Tr}(A + B) = 0 \end{array} \right\}.$$

$$\mathfrak{p}_* = \left\{ \begin{pmatrix} 0 & Z \\ -{}^t\bar{Z} & 0 \end{pmatrix} \middle| Z \quad p \times q \text{ complex matrix} \right\}.$$

(ii) A maximal abelian subspace $\mathfrak{a}_* \subset \mathfrak{p}_*$ is given by the matrices

$$H_* = \left(\begin{array}{c|c|c} & \begin{matrix} h_1 & & \\ & \ddots & \\ & & h_p \end{matrix} & \\ \hline \begin{matrix} h_1 & & \\ & \ddots & \\ & & h_p \end{matrix} & & \\ \hline & & \\ & & \end{array}\right), \qquad h_j \in i\boldsymbol{R},$$

and a Cartan subalgebra $\mathfrak{h}$ of $\mathfrak{u}^C$ is given by the matrices

$$H = \left(\begin{array}{c|c|c} \begin{matrix} t_1 & & \\ & \ddots & \\ & & t_p \end{matrix} & \begin{matrix} h_1 & & \\ & \ddots & \\ & & h_p \end{matrix} & \\ \hline \begin{matrix} h_1 & & \\ & \ddots & \\ & & h_p \end{matrix} & \begin{matrix} t_1 & & \\ & \ddots & \\ & & t_p \end{matrix} & \\ \hline & & \begin{matrix} t_{2p+1} & & \\ & \ddots & \\ & & t_{p+q} \end{matrix} \end{array}\right), \qquad \begin{array}{l} h_j \in \boldsymbol{C}, \\ t_j \in \boldsymbol{C}, \\ \operatorname{Tr} H = 0. \end{array}$$

(iii) The set Δ of nonzero roots of $(\mathfrak{u}^C, \mathfrak{h})$ is given by

$$\Delta = \{\pm(e_i - e_j) : 1 \leqslant i < j \leqslant p + q\},$$

where

$$e_i(H) = \begin{cases} \frac{1}{2}(t_i + h_i), & 1 \leqslant i \leqslant p, \\ \frac{1}{2}(t_{i-p} - h_{i-p}), & p + 1 \leqslant i \leqslant 2p, \\ t_i, & 2p + 1 \leqslant i \leqslant p + q. \end{cases}$$

The set Σ of restricted roots is given by ($\pm$ signs read independently)

$$\Sigma = \{\pm f_i, \pm 2f_i, \pm f_i \pm f_j : 1 \leqslant i < j \leqslant p\}$$

with multiplicities $2(q-p)$, 1, and 2, respectively. Here

$$f_j(H_*) = \tfrac{1}{2}h_j, \qquad 1 \leqslant j \leqslant p.$$

(iv) The set

$$f_1 > f_2 > \ldots > f_p > 0,$$

is a Weyl chamber $C_\mathfrak{p} \subset \mathfrak{a}$ and

$$\alpha_1 = f_1 - f_2, \quad \alpha_2 = f_2 - f_3, \quad \ldots, \quad \alpha_{p-1} = f_{p-1} - f_p, \quad \alpha_p = f_p$$
$$(\alpha_p = 2f_p \text{ if } p = q)$$

is the corresponding set of simple restricted roots. The highest restricted root is

$$2f_1 = 2\alpha_1 + \ldots + 2\alpha_p \qquad (2f_1 = 2\alpha_1 + \ldots + 2\alpha_{p-1} + \alpha_p \text{ if } p = q).$$

(v) The shortest closed geodesic in U/K has length $2\pi(p+q)$.

6. Show that the adjoint group of $\boldsymbol{SU}(3)$ has two nonconjugate closed one-parameter subgroups of minimum length. (Thus the assumption of simple connectedness cannot be dropped in Theorem 11.2.)

7*. Let M be a compact irreducible Riemannian globally symmetric space. Assume M is not a real projective space. Then the maximum dimensional totally geodesic submanifolds of maximum sectional curvature κ (cf. Theorem 11.1) are spheres (Helgason [8]).

8. Show that the midpoint locus for $\boldsymbol{SU}(n)$ is given by

$$A_0 = \boldsymbol{SU}(n)/\boldsymbol{S}(\boldsymbol{U}_2 \times \boldsymbol{U}_{n-2})$$

in terms of the notation of Chapter X, §2.

9. With the notation of §8 suppose $H_1, H_2 \in \mathfrak{a}_*$ are $\mathrm{Ad}(U)$-conjugate. Show that they are conjugate under $W(\mathfrak{u}, \theta)$.

10. Let $(\mathfrak{u}, \theta)$ be an orthogonal symmetric Lie algebra of the compact type, $\mathfrak{k}$ the fixed point set of θ. Let (U, K) be any pair associated with $(\mathfrak{u}, \theta)$ (that is, U is a connected Lie group with Lie algebra $\mathfrak{u}$, and $K \subset U$ is a Lie subgroup with Lie algebra $\mathfrak{k}$). Assume K is connected. Then U/K is not just locally symmetric (Prop. 3.6, Chapter IV) but globally symmetric (Helgason [13], p. 275).

NOTES

The Weyl group $W(\mathfrak{u})$ has an interesting history. In his paper [3], Cartan determines the Galois group of the characteristic polynomial $p(\lambda) = \det(\lambda I - \text{ad } X)$; for example, if $\mathfrak{u}$ is the exceptional algebra $\mathfrak{e}_6$, then the Galois group is the direct product of Z_2 and the group of the 27 lines on a cubic surface. In [4] Cartan identifies the Galois group with Aut(Δ), the group of automorphisms of the root system Δ (cf. Chapter IX, §5) and for each simple $\mathfrak{u}$ determines the normal subgroup $W \subset \text{Aut}(\Delta)$ induced by Int($\mathfrak{u}$) and proves $\text{Aut}(\mathfrak{u})/\text{Int}(\mathfrak{u}) \approx \text{Aut}(\Delta)/W$ (Theorem 5.4, Chapter IX). At about the same time, Weyl ([1], Kap. III, §4) introduced $W(\mathfrak{u})$ as the group generated by the s_α $(\alpha \in \Delta)$ and used it in the determination of the characters of irreducible representations of U. The identity $W = W(\mathfrak{u})$ now being clear, Cartan (in [9]) determined: the fundamental domain of $W(\mathfrak{u})$ and of Γ_Δ (Theorem 7.5), the fundamental group $\pi_1(U)$ (Theorem 6.1), the classification of locally isomorphic compact simple Lie groups (Exercise C.1, Chapter X) the highest root (Theorem 3.28, Chapter X). In Cartan [10] these results are generalized to the group $W(\mathfrak{u}, \theta)$ and the compact, irreducible, globally symmetric spaces are classified (cf. Exercises C.2-4, Chapter X; also Takeuchi [1]). The connection between the Weyl groups and abstract transformation groups of $\boldsymbol{R}^n$ generated by reflections was determined by Coxeter [1] and Witt [2].

In recent years extensive literature has appeared on reflection groups, their geometric properties and invariant theory. As a sample we mention Chevalley [7], Steinberg [1], Iwahari and Matsumoto [1], Harish-Chandra [9, I, §3], Solomon [1], Bourbaki [2, Chapter IV-VI], Carter [1], Tits [5], Vinberg [4, 5, 6].

§2. Lemmas 2.3, 2.4 and Cor. 2.13 are given in Cartan [10], §7-§10. Theorem 2.5 goes back to Hopf [1]. Theorem 2.15 is used without proof by Harish-Chandra [9] (where it is attributed to Chevalley) and by Kostant [1]. Theorem 2.16 and Cor. 2.17 were known to Cartan ([10], §98); our proof is like that of Lemma 4 in Harish-Chandra [8], p. 196 based on the representation theory of $\mathfrak{sl}(2, \boldsymbol{C})$. An abstract generalization is given by Araki [1], Prop. 2.1.

§3. The results are mostly Cartan's ([10], §4-§6); see also Harish-Chandra [5], VI, §12.

§4-§6. The dimension of the singular set S ($\dim S = \dim U - 3$) is determined in Weyl [1], Kap. IV, §1. This equality was extensively used by H. Weyl and É. Cartan. Weyl used it to prove the conjugacy theorem (Theorem 6.4(iii), Chapter V) and the compactness of the universal covering group (Theorem 6.9, Chapter II). For these applications it would be sufficient to know that S is closed and has dimension $\leqslant \dim U - 2$ because only Prop. 12.4 is needed (see Pontrjagin [1], §64). Cartan, on the other hand, used the relation $\dim S = \dim U - 3$ to prove the more delicate Theorem 6.1, ([9], pp. 217-218), and Theorem 8.2, ([10], p. 430), which rely on the equality $\pi_1(U) = \pi_1(U - S)$. Here Cartan used Prop. 12.6, but did not enter into the difficulties which stem from the fact that S is not a manifold. That these difficulties are present can be seen from the fact (mentioned to the author by G. W. Whitehead) that Prop. 12.6 does not hold for a suitable 0-dimensional subset (Antoine's necklace) of $\boldsymbol{R}^3$. The reasoning actually gives $\pi_2(U) = 0$, (É. Cartan [20], cf. Borel [1]). It is also known that if U is simple then $\pi_3(U) = \boldsymbol{Z}$ (Bott [1]), and $\pi_4(U)$ can be read off from the diagram $D(U)$ and the unit lattice $\mathfrak{t}_e$ (Bott and Samelson [1]). The result is that $\pi_4(U)$ has two elements if the plane $\mu(H) = 2\pi i$ in $\mathfrak{t}_0$ contains no member of $\mathfrak{t}_e$, μ being the highest root with respect to a lexicographic ordering; otherwise $\pi_4(U) = 0$. In Bott and

Samelson [1], Theorem 8.2 is reduced to Theorem 6.1 in a different way. Bott has also (unpublished) extended Theorem 8.2 to all automorphisms, involutive or not. Proofs (based on Theorem 6.1) are given in Borel [8] and, with a generalization, in Raševski [3]. Bott has also used the behavior of the geodesics in the symmetric space U/K to prove a remarkable periodicity theorem for the stable homotopy of the classical groups [2].

§7-§8. The affine Weyl group goes back to Cartan [9]; see also Stiefel [2]. Our treatment, in particular the proof of Theorem 7.5, is based on Iwahori and Matsumoto [1]. The proof of Theorem 7.7 follows Loos [2], II. Lemma 8.4 is contained in Araki [1], and Theorem 8.5 probably in Cartan [10]. Proposition 8.8 is proved in the author's paper [11] and its consequence Cor. 8.9 was pointed out by S. Rallis. Proposition 8.10 was proved by Satake [3] and Harish-Chandra (unpublished). A generalization was given by Hirai [1]. An independent exposition of many results in §6-§7 was given by Dynkin and Oniščik [1]. See also Wallach [1], Chapter 4.

§10-§11. Apart from Theorem 10.3 (Cartan [10], p. 437) and Lemma 11.8 (Cartan [9], p. 257) the results of §10 and §11 on antipodal manifolds, shortest geodesics, minimal totally geodesic spheres and midpoint loci are from Helgason [8]. A generalization of Prop. 10.1 is given in Kostant [2], p. 260.

Compact symmetric spaces of rank one play a central role in the theory of Riemannian manifolds of positive curvature (see, e.g., Berger [3], Klingenberg [1], Rauch [4], or Cheeger and Ebin [1]).

CHAPTER VIII

HERMITIAN SYMMETRIC SPACES

A Hermitian symmetric space is a Riemannian globally symmetric space which has a complex structure invariant under each geodesic symmetry. Examples are provided by all simply connected two-dimensional Riemannian globally symmetric spaces. We shall mostly be concerned with Hermitian symmetric spaces of the compact type and the noncompact type. These are always simply connected and have the characteristic property that their isotropy groups are not semisimple and therefore have nondiscrete centers. In §7 it is shown that the Hermitian symmetric spaces G_0/K_0 of the noncompact type are exactly the bounded symmetric domains in the space of several complex variables. Moreover, the space G_0/K_0 can always be imbedded to the compact dual U/K_0 as an open subset. The simplest instance of this imbedding is the unit disk $|z| < 1$ situated in the extended complex plane.

The three first sections deal with some basic facts concerning complex manifolds. The main notions treated are Hermitian and Kählerian structures, Ricci curvature, and the Bergman kernel function.

§ 1. Almost Complex Manifolds

Definition. Let M be a C^∞ manifold. An *almost complex structure* on M is a tensor field J of type (1, 1) such that $J(JX) = -X$ for each vector field X on M.

An almost complex structure on M thus amounts to a rule which in a differentiable fashion assigns to each $p \in M$ an endomorphism $J_p : M_p \to M_p$ such that $(J_p)^2 = -I$ for each $p \in M$. An *almost complex manifold* is a pair (M, J) where M is a C^∞ manifold and J is an almost complex structure on M.

For reasons given in Example II below it is important to consider the mapping $S : \mathfrak{D}^1(M) \times \mathfrak{D}^1(M) \to \mathfrak{D}^1(M)$ given by

$$S(X, Y) = [X, Y] + J[JX, Y] + J[X, JY] - [JX, JY] \tag{1}$$

for $X, Y \in \mathfrak{D}^1(M)$. Using the relation

$$[fX, gY] = fg[X, Y] + f(Xg)\,Y - g(Yf)\,X$$

for $f, g \in C^\infty(M)$ it follows easily that $S(fX, gY) = fgS(X, Y)$. As customary, we identify S with the multilinear mapping $(\omega, X, Y) \to$

$\omega(S(X, Y))$ of $\mathfrak{D}_1 \times \mathfrak{D}^1 \times \mathfrak{D}^1$ into $C^\infty(M)$. Thus S is a tensor field of type (1, 2), called the *torsion tensor* of the almost complex structure J. Obviously S is skew symmetric, that is, $S(X, Y) = -S(Y, X)$. If $S = 0$, the almost complex structure is said to be *integrable.*

Example I. Let $M = \boldsymbol{R}^2$, considered as a manifold with local coordinates the ordinary Cartesian coordinates (x, y). For each $p \in M$ the endomorphism of M_p given by

$$J_p : a\left(\frac{\partial}{\partial x}\right)_p + b\left(\frac{\partial}{\partial y}\right)_p \to -b\left(\frac{\partial}{\partial x}\right)_p + a\left(\frac{\partial}{\partial y}\right)_p$$

for $a, b \in \boldsymbol{R}$, has square $-I$. The tensor field $p \to J_p$, $p \in M$ is an almost complex structure on M. This almost complex structure is integrable, since $S(\partial/\partial x, \partial/\partial y) = 0$.

Example II. Let M be a complex manifold of dimension m as defined in Chapter I, §1. There exists a covering $M = \bigcup_{\alpha \in A} U_\alpha$ of M by open subsets U_α each of which is homeomorphic to an open subset of $\boldsymbol{C}^m$ under a mapping φ_α such that for each pair $\alpha, \beta \in A$, $\varphi_\beta \circ \varphi_\alpha^{-1}$ is a holomorphic mapping† of $\varphi_\alpha(U_\alpha \cap U_\beta)$ onto $\varphi_\beta(U_\alpha \cap U_\beta)$. As remarked in Chapter I, §1, we can always assume that the system $(U_\alpha, \varphi_\alpha)_{\alpha \in A}$ is maximal with this property. In that case, the system is said to be a *complex structure* on the underlying topological space M.

Let $p \in M$ and let α be an index in A such that $p \in U_\alpha$. If $q \in U_\alpha$ then $\varphi_\alpha(q) = (z_1(q), \ldots, z_m(q))$ where each $z_j(q)$ is a complex number $x_j(q) + iy_j(q)$. The mapping

$$\psi_\alpha : q \to (x_1(q), y_1(q), \ldots, x_m(q), y_m(q)), \qquad q \in U_\alpha,$$

is a homeomorphism of U_α onto an open subset of $\boldsymbol{R}^{2m}$. The collection of open charts $(U_\alpha, \psi_\alpha)_{\alpha \in A}$ on M turns M into an analytic manifold whose analytic structure is said to be *underlying* the complex structure above. Thus a complex structure has a definite underlying analytic structure. On the other hand, it can happen that two different complex structures have the same underlying analytic structure.

The tangent space M_p of the analytic manifold M has a basis given by the vectors

$$\left(\frac{\partial}{\partial x_1}\right)_p, \left(\frac{\partial}{\partial y_1}\right)_p, \ldots, \left(\frac{\partial}{\partial x_m}\right)_p, \left(\frac{\partial}{\partial y_m}\right)_p.$$

† If O and O' are open subsets in $\mathbf{C}^m$ and $\mathbf{C}^n$, respectively, then a mapping $f: O \to O'$ is called *holomorphic* if the coordinates of $f(z_1, \ldots, z_m)$ are holomorphic functions of $z_1, \ldots, z_m$.

The endomorphism $J^\alpha : M_p \to M_p$ given by

$$J^\alpha\left(\frac{\partial}{\partial x_i}\right)_p = \left(\frac{\partial}{\partial y_i}\right)_p, \qquad J^\alpha\left(\frac{\partial}{\partial y_i}\right)_p = -\left(\frac{\partial}{\partial x_i}\right)_p$$

for $1 \leqslant i \leqslant m$ satisfies $(J^\alpha)^2 = -I$. Suppose now β is another index in A such that $p \in U_\beta$. For $q \in U_\beta$ we denote the complex coordinates of $\varphi_\beta(q)$ by $(w_1(q), ..., w_m(q))$ and put $w_j(q) = u_j(q) + iv_j(q)$ $(1 \leqslant j \leqslant m)$. If we consider $(u_1, v_1, ..., u_m, v_m)$ as local coordinates on the analytic manifold M, the vectors

$$\left(\frac{\partial}{\partial u_1}\right)_p, \left(\frac{\partial}{\partial v_1}\right)_p, ..., \left(\frac{\partial}{\partial u_m}\right)_p, \left(\frac{\partial}{\partial v_m}\right)_p$$

form a basis of the tangent space M_p. The endomorphism $J^\beta : M_p \to M_p$ given by

$$J^\beta\left(\frac{\partial}{\partial u_i}\right)_p = \left(\frac{\partial}{\partial v_i}\right)_p, \qquad J^\beta\left(\frac{\partial}{\partial v_i}\right)_p = -\left(\frac{\partial}{\partial u_i}\right)_p$$

for $1 \leqslant i \leqslant m$ satisfies $(J^\beta)^2 = -I$.

Lemma 1.1. *The endomorphisms J^α and J^β are identical.*

Proof. Since the mapping $\varphi_\beta \circ \varphi_\alpha^{-1}$ is a holomorphic mapping of $\varphi_\alpha(U_\alpha \cap U_\beta)$ onto $\varphi_\beta(U_\alpha \cap U_\beta)$, each function $w_i(z_1, ..., z_m)$, $1 \leqslant i \leqslant m$, is a holomorphic function in a neighborhood of $\varphi_\alpha(p)$. This being so, the corresponding real functions $u_i(x_1, y_1, ..., x_m, y_m)$, $v_i(x_1, y_1, ..., x_m, y_m)$ satisfy the Cauchy-Riemann equations

$$\frac{\partial u_i}{\partial x_j} - \frac{\partial v_i}{\partial y_j} = 0, \qquad \frac{\partial u_i}{\partial y_j} + \frac{\partial v_i}{\partial x_j} = 0, \qquad 1 \leqslant i, j \leqslant m.$$

It follows that

$$\frac{\partial}{\partial x_j} = \sum_i \left(\frac{\partial u_i}{\partial x_j}\frac{\partial}{\partial u_i} + \frac{\partial v_i}{\partial x_j}\frac{\partial}{\partial v_i}\right) = \sum_i \left(\frac{\partial u_i}{\partial x_j}\frac{\partial}{\partial u_i} - \frac{\partial u_i}{\partial y_j}\frac{\partial}{\partial v_i}\right),$$

$$\frac{\partial}{\partial y_j} = \sum_i \left(\frac{\partial u_i}{\partial y_j}\frac{\partial}{\partial u_i} + \frac{\partial v_i}{\partial y_j}\frac{\partial}{\partial v_i}\right) = \sum_i \left(\frac{\partial u_i}{\partial y_j}\frac{\partial}{\partial u_i} + \frac{\partial u_i}{\partial x_j}\frac{\partial}{\partial v_i}\right),$$

and consequently

$$J^\beta\left(\frac{\partial}{\partial x_j}\right)_p = \left(\frac{\partial}{\partial y_j}\right)_p, \qquad J^\beta\left(\frac{\partial}{\partial y_j}\right)_p = -\left(\frac{\partial}{\partial x_j}\right)_p,$$

which proves the lemma.

In view of this lemma the endomorphism $J_p = J^\alpha = J^\beta$ is independent of the choice of local coordinates around p. The tensor field $J : p \to J_p$ is an almost complex structure on M, which we call the *canonical* almost complex structure associated to the complex structure on M.

Let (M, J) and (M', J') be almost complex manifolds and Φ a differentiable mapping of M into M'; the mapping Φ is called *almost complex* if

$$d\Phi_p \circ J_p = J'_{\Phi(p)} \circ d\Phi_p \qquad \text{for } p \in M. \tag{2}$$

Suppose now M and M' are complex manifolds and J and J' their corresponding almost complex structures. A mapping of M into M' is called *holomorphic* if its expression in terms of complex local coordinates is given by holomorphic functions. It is obvious from the Cauchy-Riemann equations that a holomorphic mapping is almost complex. On the other hand, suppose a mapping $\Phi : M \to M'$ satisfies (2). Let $\{z_1, \ldots, z_m\}$ and $\{w_1, \ldots, w_n\}$ be complex local coordinates in a neighborhood of p in M and of $\Phi(p)$ in M'. Put $z_j = x_j + iy_j$ $(1 \leqslant j \leqslant m)$, $w_k = u_k + iv_k$ $(1 \leqslant k \leqslant n)$. Then

$$u_k = \varphi_k(x_1, y_1, \cdots, x_m, y_m),$$
$$v_k = \psi_k(x_1, y_1, \cdots, x_m, y_m),$$

where φ_k and ψ_k are differentiable functions $(1 \leqslant k \leqslant n)$. Condition (2) implies that

$$\frac{\partial \varphi_k}{\partial x_j} = \frac{\partial \psi_k}{\partial y_j}, \qquad \frac{\partial \varphi_k}{\partial y_j} = -\frac{\partial \psi_k}{\partial x_j},$$

which shows that w_k is a holomorphic function of each variable z_j, and therefore, by a classical theorem on holomorphic functions (see, e.g., Bochner and Martin [1], p. 33) w_k is a holomorphic function of $(z_1, \ldots, z_m)$. This shows that an almost complex mapping of a complex manifold into another is holomorphic.

Let M be a complex manifold and let J be the associated canonical almost complex structure. The tensor field J satisfies the integrability condition

$$S(X, Y) = 0, \qquad X, Y \in \mathfrak{D}^1(M), \tag{3}$$

where S is defined by (1). In fact, since S is $C^\infty(M)$-bilinear it suffices to check that (3) is satisfied in each coordinate neighborhood and there it obviously holds for the vector fields $\partial/\partial x_i$, $\partial/\partial y_j$. Thus the canonical almost complex structure associated with a complex structure is integrable. The converse is contained in the following theorem, first proved in full generality by Newlander and Nirenberg [1].

Theorem 1.2. *Let (M, J) be an almost complex manifold which satisfies the integrability condition* (3). *Then there exists a unique complex structure on M such that J is the associated almost complex structure.*

For the proof of this theorem which is too long to be given here, we refer to the cited article. However, we shall only use Theorem 1.2 in the case when M and J are assumed analytic.† Here much simpler proofs are available, see, e.g., Frölicher [1].

§ 2. Complex Tensor Fields. The Ricci Curvature

Let M be a C^∞ manifold. The set $C^\infty(M) + iC^\infty(M)$ of all complex-valued differentiable functions on M is an algebra over $\boldsymbol{C}$, denoted $\mathfrak{E}_0$. A *complex vector field* on M is, by definition, a derivation of the algebra $\mathfrak{E}_0$. Let $\mathfrak{E}^1$ denote the set of complex vector fields on M. Then $\mathfrak{E}^1$ is a module over $\mathfrak{E}_0$; also $\mathfrak{E}^1$ is closed under the bracket operation $[X, Y] = XY - YX$, $(X, Y \in \mathfrak{E}^1)$. If s is an integer, $s \geqslant 1$, we consider the $\mathfrak{E}_0$ module

$$\mathfrak{E}^1 \times \ldots \times \mathfrak{E}^1 \qquad (s \text{ times})$$

and let $\mathfrak{E}_s$ denote the $\mathfrak{E}_0$-module of all $\mathfrak{E}_0$-multilinear mappings of $\mathfrak{E}^1 \times \ldots \times \mathfrak{E}^1$ into $\mathfrak{E}_0$. The elements of $\mathfrak{E}_1$ are called *complex* 1-*forms* on M. It follows from Lemma 2.3, Chapter I, that $\mathfrak{E}^1$ and $\mathfrak{E}_1$ are dual modules. In analogy with Chapter I, the $\mathfrak{E}_0$-multilinear mappings of the module

$$\mathfrak{E}_1 \times \ldots \times \mathfrak{E}_1 \times \mathfrak{E}^1 \times \ldots \times \mathfrak{E}^1 \qquad (\mathfrak{E}_1 \ r \text{ times}, \ \mathfrak{E}^1 \ s \text{ times})$$

into $\mathfrak{E}_0$, are called *complex tensor fields*, contravariant of degree r, covariant of degree s. The set of these is denoted by $\mathfrak{E}^r_s$ (or $\mathfrak{E}^r_s(M)$).

The operation of conjugation in $\mathfrak{E}_0$ induces a similar operation in each $\mathfrak{E}^r_s$. If $Z \in \mathfrak{E}^1$, the complex vector field $\bar{Z}$ is defined by $\bar{Z}f = (Z\bar{f})^-$ for all $f \in \mathfrak{E}_0$. If $\omega \in \mathfrak{E}_1$, the complex 1-form $\bar{\omega}$ is defined by $\bar{\omega}(Z) = (\omega(\bar{Z}))^-$. Finally, if $\Omega \in \mathfrak{E}^r_s$, the tensor field $\bar{\Omega}$ is defined by

$$\bar{\Omega}(\omega_1, \ldots, \omega_r, Z_1, \ldots, Z_s) = (\Omega(\bar{\omega}_1, \ldots, \bar{\omega}_r, \bar{Z}_1, \ldots, \bar{Z}_s))^-$$

for $\omega_i \in \mathfrak{E}_1$, $Z_j \in \mathfrak{E}^1$. Each $X \in \mathfrak{D}^1$ can be regarded as a complex vector field on M by defining

$$XF = X\left(\frac{1}{2}(F + \bar{F})\right) + iX\left(\frac{1}{2i}(F - \bar{F})\right) \qquad (F \in \mathfrak{E}_0).$$

† Strictly speaking, Theorem 1.2 is not even necessary for our purposes. It will only be used to prove Prop. 4.2 of which an alternative proof is indicated in an exercise following Chapter VIII.

Similarly, $\mathfrak{D}_1$ can be regarded as a subset of $\mathfrak{E}_1$ and more generally, we shall regard the members of $\mathfrak{D}_s^r$ as complex tensor fields on M whenever this is called for by the context.

Let $p \in M$ and let M_p^C denote the complexification of the tangent space M_p. According to Chapter III, §6, M_p^C is a vector space over $\boldsymbol{C}$ consisting of all symbols $X + iY$ where $X, Y \in M_p$ with the vector space operations

$$(X_1 + iY_1) + (X_2 + iY_2) = (X_1 + X_2) + i(Y_1 + Y_2),$$

$$(a + ib)(X + iY) = aX - bY + i(bX + aY), \qquad a, b \in \boldsymbol{R}.$$

The elements of M_p^C are called *complex tangent vectors* at p. Each $X + iY \in M_p^C$ defines a complex linear function on $\mathfrak{E}_0$ given by

$$(X + iY)(f + ig) = Xf - Yg + i(Xg + Yf)$$

for $f, g \in C^\infty(M)$. Then

$$Z(FG) = F(p)\, ZG + G(p)\, ZF$$

for $Z \in M_p^C$ and $F, G \in \mathfrak{E}_0$. If Z is a complex vector field then the linear function $F \to (ZF)(p)$ on $\mathfrak{E}_0$ arises in this way from a complex tangent vector $Z_p \in M_p^C$. Thus, a complex vector field Z on M can be identified with a collection Z_p ($p \in M$) of complex tangent vectors to M varying differentiably with p. The elements of $\mathfrak{E}_s^r$ can be described similarly.

Suppose now J is an almost complex structure on M. For each $p \in M$, the endomorphism J_p can be extended uniquely to a complex linear mapping of M_p^C onto itself. The extension, also denoted by J_p, then satisfies $(J_p)^2 = -I$. Now, since $J \in \mathfrak{D}_1^1 \subset \mathfrak{E}_1^1$, JZ is a complex vector field for each $Z \in \mathfrak{E}^1$. It is clear that $(JZ)_p = J_p Z_p$ for each $p \in M$.

Definition. Let (M, J) be an almost complex manifold and let Z be a complex vector field on M. Then Z is said to be of type (1, 0) if $JZ = iZ$ and of type (0, 1) if $JZ = -iZ$.

Every complex vector field Z on an almost complex manifold can be written as a sum

$$Z = Z_{1,0} + Z_{0,1},$$

where $Z_{1,0}$ and $Z_{0,1}$ are complex vector fields of type (1, 0) and (0, 1), respectively. In fact, it suffices to put $Z_{1,0} = \frac{1}{2}(Z - iJZ)$, $Z_{0,1} = \frac{1}{2}(Z + iJZ)$. In Example I in §1, the vector fields $\partial/\partial x - i\partial/\partial y$ and $\partial/\partial x + i\partial/\partial y$ are of type (1, 0) and (0, 1), respectively. They are usually denoted by $2\partial/\partial z$ and $2\partial/\partial \bar{z}$ because, if $f(z) = F(x, y)$ is a holomorphic function, then

$$2\frac{\partial f(z)}{\partial z} = \frac{\partial F(x, y)}{\partial x} - i\frac{\partial F(x, y)}{\partial y}$$

due to the Cauchy-Riemann equations.

Definition. Let M be a connected manifold with almost complex structure J. A Riemannian structure g on M is said to be a *Hermitian structure* if

$$g(JX, JY) = g(X, Y) \qquad \text{for } X, Y \in \mathfrak{D}^1 \tag{1}$$

and a *Kählerian structure* if in addition

$$\nabla_X \cdot J = 0 \qquad \text{for } X \in \mathfrak{D}^1. \tag{2}$$

In other words, the Hermitian condition means that J_p is an isometry of M_p for each $p \in M$. The Kählerian condition means that in addition the tensor field J is invariant under parallelism.

Let g be any Riemannian structure on a connected manifold M. Let $X \to \nabla_X$, $(X \in \mathfrak{D}^1)$, be the corresponding Riemannian connection. Regarding now g as a complex tensor field, the covariant derivative ∇_Z can be defined by relation (2) in Chapter I, §9, for all $Z \in \mathfrak{C}^1$. Then

$$R(X, Y) = \nabla_X\nabla_Y - \nabla_Y\nabla_X - \nabla_{[X,Y]}, \qquad X, Y \in \mathfrak{C}^1,$$

because both sides are $\mathfrak{C}_0$-bilinear and coincide for $X, Y \in \mathfrak{D}^1$.

Lemma 2.1. *Let M be a connected manifold with almost complex structure J and Riemannian structure g.*

(i) *If g is Hermitian, then $g(X, Y) = 0$ if X and Y are both of type* (1, 0) (*or both of type* (0, 1)).

(ii) *If g is Kählerian and R denotes the curvature tensor, then* $R(X, Y) = 0$ *if X and Y are both of type* (1, 0) (*or both of type* (0, 1)).

Proof. Let X and Y be complex vector fields of type (1, 0). Then, if g is Hermitian,

$$g(X, Y) = g(JX, JY) = g(iX, iY) = -g(X, Y)$$

so $g(X, Y) = 0$. Now let Z, T be arbitrary complex vector fields on M. The Kählerian condition (2) implies $\nabla_Z(JX) = J\nabla_Z(X)$. It follows that $R(Z, T)X$ is of type (1, 0) as well as X. Hence by (i)

$$g(R(Z, T)X, Y) = 0. \tag{3}$$

The quadrilinear form $g(R(Z, T)\,U, V)$ on $\mathfrak{C}^1 \times \mathfrak{C}^1 \times \mathfrak{C}^1 \times \mathfrak{C}^1$ satisfies conditions (a), (b), (c) of Lemma 12.4, Chapter I. Owing to this lemma we have

$$g(R(U, V)\,Z, T) = g(R(Z, T)\,U, V)$$

and (3) implies $g(R\,(X, Y)\,Z, T) = 0$. Since Z and T are arbitrary and g_p is nondegenerate for each $p \in M$, the lemma follows.

Let M be a connected complex manifold of dimension m. Let p be a point in M and $\{z_1, ..., z_m\}$ local coordinates in an open neighborhood U of p. A complex-valued function f on M is said to be *holomorphic* at p if there exists a neighborhood of p where f is given by a convergent power series in the local coordinates $z_1 - z_1(p), ..., z_m - z_m(p)$. If f is holomorphic at each point of a set V, then f is said to be holomorphic on V. If we write $x_j = \frac{1}{2}\,(z_j + \bar{z}_j)$, $y_j = 1/(2i)\,(z_j - \bar{z}_j)$, then $\{x_1, y_1, ..., x_m, y_m\}$ is a coordinate system on the underlying analytic manifold U. The vector fields given by

$$\frac{\partial}{\partial z_j} = \frac{1}{2}\left(\frac{\partial}{\partial x_j} - i\,\frac{\partial}{\partial y_j}\right), \qquad \frac{\partial}{\partial \bar{z}_j} = \frac{1}{2}\left(\frac{\partial}{\partial x_j} + i\,\frac{\partial}{\partial y_j}\right)$$

$(1 \leqslant j \leqslant m)$ are complex vector fields on U of type (1, 0) and (0, 1), respectively. A function f which is holomorphic on U satisfies

$$\frac{\partial}{\partial \bar{z}_j} f = 0, \qquad \frac{\partial}{\partial z_j} \bar{f} = 0.$$

The differential forms

$$dz_j = dx_j + i\,dy_j, \qquad d\bar{z}_j = dx_j - i\,dy_j \qquad (1 \leqslant j \leqslant m)$$

are complex 1-forms on U. It is easily seen that

$$\overline{\left(\frac{\partial}{\partial z_j}\right)} = \frac{\partial}{\partial \bar{z}_j}, \qquad \overline{(dz_j)} = d\bar{z}_j \qquad (1 \leqslant j \leqslant m)$$

and

$$dz_i\left(\frac{\partial}{\partial z_j}\right) = d\bar{z}_i\left(\frac{\partial}{\partial \bar{z}_j}\right) = \delta^i{}_j,$$

$$dz_i\left(\frac{\partial}{\partial \bar{z}_j}\right) = d\bar{z}_i\left(\frac{\partial}{\partial z_j}\right) = 0$$

for $1 \leqslant i, j \leqslant m$. Let T be a complex tensor field on M of type (1, 2). The coefficients $T^k{}_{ij}$, $T^{k^*}{}_{ij^*}$, etc., are defined by

$$T\left(dz_k, \frac{\partial}{\partial z_i}, \frac{\partial}{\partial z_j}\right) = T^k{}_{ij}, \qquad T\left(d\bar{z}_k, \frac{\partial}{\partial z_i}, \frac{\partial}{\partial \bar{z}_j}\right) = T^{k^*}{}_{ij^*},$$

and similarly for tensor fields of other types. We also write for simplicity $Z_i = \partial/\partial z_i$, $Z_{j^*} = \partial/\partial \bar{z}_j$. If g is a Hermitian structure on M we have from Lemma 2.1

$$g_{ij} = g_{i^*j^*} = 0 \tag{4}$$

and if g is Kählerian, we have by the same lemma

$$R^{\alpha}{}_{\beta ij} = R^{\alpha}{}_{\beta i^*j^*} = 0, \tag{5}$$

where α, β are arbitrary indices, starred or not. If $X \to \nabla_X$ is an affine connection on M, the functions $\Gamma_{ij}{}^k$, $\Gamma_{i^*j}{}^k$, ..., $\Gamma_{i^*j^*}{}^{k^*}$ are defined by

$$\nabla_{Z_i}(Z_j) = \sum_k \Gamma_{ij}{}^k Z_k + \sum_k \Gamma_{ij}{}^{k^*} Z_{k^*}$$

and the similar equations for $\nabla_{Z_{i^*}}(Z_j)$, $\nabla_{Z_i}(Z_{j^*})$ and $\nabla_{Z_{i^*}}(Z_{j^*})$.

Lemma 2.2. *A Hermitian structure g on a connected complex manifold is Kählerian if and only if*

$$\Gamma_{jk}{}^{l^*} = \Gamma_{j^*k}{}^{l} = \Gamma_{jk^*}{}^{l^*} = \Gamma_{j^*k^*}{}^{l} = 0 \tag{6}$$

in each coordinate neighborhood.

Proof. If g is Kählerian, then $J\nabla_{Z_i}(Z_j) = \nabla_{Z_i}(JZ_j) = i\nabla_{Z_i}(Z_j)$ so $\Gamma_{jk}{}^{l^*} = 0$ and conversely. The other relations are proved similarly.

Let M be a manifold with an affine connection having curvature tensor R. Let $p \in M$ and $X, Y \in \mathfrak{D}^1(M)$. The mapping

$$L \to R_p(Y_p, L) \cdot X_p, \qquad L \in M_p,$$

is an endomorphism of M_p whose trace will be denoted by $r_p(X_p, Y_p)$. The tensor field r given by

$$(r(X, Y))(p) = r_p(X_p, Y_p)$$

is called the *Ricci curvature* of the affine connection.

Lemma 2.3. *On a Riemannian manifold M the Ricci curvature is a symmetric tensor, that is,*

$$r(X, Y) = r(Y, X), \qquad X, Y \in \mathfrak{D}^1(M).$$

Proof. Let $p \in M$ and let $X_1, \ldots, X_m$ be a basis of the vector fields on an open neighborhood U of p such that $g(X_i, X_j) = \delta^i{}_j$ on U, g being the Riemannian structure. Then

$$R(X_i, X_j) \cdot X_l = \sum_k R^k{}_{lij} X_k$$

so

$$r(X_l, X_i) = \sum_k R^k{}_{lik}.$$

On the other hand, if X, Y, S, T are any vector fields on M,

$$g(R(X, Y)\, S, T) = g(R(S, T)\, X, Y)$$

so

$$R^m{}_{lij} = R^j{}_{ilm}$$

and $r(X_l, X_i) = r(X_i, X_l)$ follows.

If M is a complex manifold, we consider r as a complex tensor field on M with coefficients r_{ij}, r_{i^*j}, r_{ij^*}, $r_{i^*j^*}$ defined as above.

We recall that a manifold M is said to be *orientable* if there exists a collection $(U_\alpha, \psi_\alpha)_{\alpha\in A}$ of local charts such that $\{U_\alpha\}_{\alpha\in A}$ is a covering of M and such that for any $\alpha, \beta \in A$, the mapping $\psi_\beta \circ \psi_\alpha^{-1}$ has strictly positive Jacobian determinant in its domain of definition $\psi_\alpha(U_\alpha \cap U_\beta)$. The manifold M is said to be *oriented* if such a collection $(U_\alpha, \psi_\alpha)_{\alpha\in A}$ has been chosen.

Let M be a complex manifold of dimension m. Let $(V_\alpha, \varphi_\alpha)_{\alpha\in A}$ be a collection of local charts covering M such that $\varphi_\beta \circ \varphi_\alpha^{-1}$ is holomorphic for each pair $\alpha, \beta \in A$. Let

$$\varphi_\alpha(p) = (z_1, \ldots, z_m), \qquad \varphi_\beta(q) = (w_1, \ldots, w_m)$$

for $p \in V_\alpha$, $q \in V_\beta$ and let $z_j = x_j + iy_j$, $w_j = u_j + iv_j$ $(1 \leqslant j \leqslant m)$. From the Cauchy-Riemann equations follows easily by induction

$$\frac{\partial(u_1, v_1, \ldots, u_m, v_m)}{\partial(x_1, y_1, \ldots, x_m, y_m)} = \left| \frac{\partial(w_1, \ldots, w_m)}{\partial(z_1, \ldots, z_m)} \right|^2. \tag{7}$$

If we define ψ_α by

$$\psi_\alpha : p \to (x_1, y_1, \ldots, x_m, y_m),$$

then M, with the local charts $(V_\alpha, \psi_\alpha)_{\alpha\in A}$, is an oriented manifold.

Let M be an oriented manifold with a Riemannian structure g. Let $\{x_1, \ldots, x_m\}$ be a coordinate system valid on an open subset U of M. Let

$$g_{ij} = g\left(\frac{\partial}{\partial x_i}, \frac{\partial}{\partial x_j}\right), \qquad \bar{g} = \det(g_{ij}).$$

Then $\bar{g} > 0$ and we can consider the m-form

$$\sqrt{\bar{g}}\, dx_1 \wedge \ldots \wedge dx_m \tag{8}$$

on U. If $\{y_1, \ldots, y_m\}$ is another coordinate system on U then it is easy to see that

$$\bar{g}(y_1, \ldots, y_m)^{1/2}\, dy_1 \wedge \ldots \wedge dy_m = \bar{g}(x_1, \ldots, x_m)^{1/2}\, dx_1 \wedge \ldots \wedge dx_m,$$

since the Jacobian determinant $(\partial y_j/\partial x_i)$ is positive. It follows that there exists an m-form ω on M, which on an arbitrary coordinate neighborhood has the expression (8). This form ω is called the *volume element* corresponding to the Riemannian structure g on the oriented manifold M.

Proposition 2.4. *Let ω be the volume element on an oriented Riemannian manifold M. Then*

$$\nabla_X \omega = 0$$

for each vector field X on M.

Proof. Let $p \in M$ and let $\{x_1, \ldots, x_m\}$ be a coordinate system valid in a neighborhood of p such that the tangent vectors $(\partial/\partial x_i)_p$ form an orthonormal basis of M_p. Let $X_1, \ldots, X_m$ be the vector fields on a normal neighborhood N_p of p adapted to this basis and let the forms $\omega^1, \ldots, \omega^m$ on N_p be determined by $\omega^i(X_j) = \delta^i{}_j$. Then

$$\omega = \omega^1 \wedge \ldots \wedge \omega^m$$

and since $\nabla_X \omega^i = 0$ at p, the relation $\nabla_X \omega = 0$ holds at p. The point p being arbitrary, the proposition follows.

Proposition 2.5. *Let M be a connected complex manifold and g a Riemannian structure on M. Let ω denote the corresponding volume element. In a local coordinate system $\{z_1, \ldots, z_m\}$, ω has an expression*

$$\omega = G\, dz_1 \wedge d\bar{z}_1 \wedge \ldots \wedge dz_m \wedge d\bar{z}_m,$$

where the function G is given by

$$(-2i)^m\, G = \bar{g}(x_1, y_1 \ldots, x_m, y_m)^{1/2}.$$

If g is Kählerian, the Ricci curvature satisfies

$$r_{ij^*} = Z_i Z_{j^*} \log |\, G \,|, \qquad r_{ij} = r_{i^*j^*} = 0.$$

Proof. The expression for ω is obvious since

$$dz_j \wedge d\bar{z}_j = (-2i)\, dx_j \wedge dy_j.$$

Now assume g Kählerian. The curvature tensor R satisfies

$$g(R(Z_i, Z_{j^*}) \cdot Z_l, Z_m) = g(R(Z_l, Z_m) \cdot Z_i, Z_{j^*}) \tag{9}$$

and this expression vanishes due to Lemma 2.1. Since g_p is nondegenerate for $p \in M$, it follows that

$$R(Z_i, Z_{j^*}) \cdot Z_l = \sum_k R^k{}_{lij^*} Z_k, \tag{10}$$

$$R(Z_i, Z_{j^*}) \cdot Z_{l^*} = \sum_k R^{k^*}{}_{l^*ij^*} Z_{k^*}. \tag{11}$$

Now the vector fields Z_i, Z_{j^*} make up a basis of the complex tangent space at each point. The trace of an endomorphism of a vector space is the same as the trace of the extension of the endomorphism to the complexified vector space. Therefore $r(Z_i, Z_{j^*})$ equals the trace of the complex endomorphism given by

$$Z_m \to R(Z_{j^*}, Z_m) \cdot Z_i,$$

$$Z_{m^*} \to R(Z_{j^*}, Z_{m^*}) \cdot Z_i.$$

It follows that

$$r_{ij^*} = \sum_m R^m{}_{ij^*m}. \tag{12}$$

Similarly, we find $r_{ij} = r_{i^*j^*} = 0$. Now, from Lemma 2.2 and the fact that covariant differentiation commutes with contractions, it follows that

$$\nabla_{Z_l}(dz_i) = -\sum_j \Gamma_{lj}{}^i\, dz_j, \qquad \nabla_{Z_l}(d\bar{z}_i) = 0.$$

Combining these equations with

$$\nabla_{Z_l}(\omega) = 0$$

one finds that

$$Z_l G = G \sum_i \Gamma_{li}{}^i. \tag{13}$$

On the other hand, we have from Lemma 2.2

$$R(Z_{j^*}, Z_k) \cdot Z_l = (\nabla_{Z_{j^*}} \nabla_{Z_k} - \nabla_{Z_k} \nabla_{Z_{j^*}}) \cdot Z_l$$

$$= \nabla_{Z_{j^*}} \Big(\sum_s \Gamma_{kl}{}^s Z_s\Big) = \sum_s (Z_{j^*} \Gamma_{kl}{}^s)\, Z_s.$$

Comparing with (10) we find

$$R^s{}_{lj^*k} = Z_{j^*} \Gamma_{kl}{}^s. \tag{14}$$

Since $\Gamma_{kl}^{s} = \Gamma_{lk}^{s}$ we find from (12)-(14)

$$r_{ij^*} = Z_{j^*}\left(\frac{1}{G} Z_i\, G\right),$$

and since G is a constant multiple of $|\, G\,|$ the desired expression for r_{ij*} follows.

§ 3. Bounded Domains. The Kernel Function

In this chapter, a *bounded domain* shall mean a bounded, open connected subset of the product $\boldsymbol{C}^N$, N being an integer > 0.

Let D be a bounded domain in $\boldsymbol{C}^N$. Let $L^2(D)$ denote the Hilbert space of complex functions on D for which $\int_D |\, f\,|^2\, d\mu < \infty$, the measure $d\mu$ denoting the Lebesgue measure on $\boldsymbol{R}^{2N}$. The inner product on $L^2(D)$ is

$$(f, g) = \int_D f(\zeta)\,\overline{g(\zeta)}\; d\mu(\zeta),$$

and as usual the norm is defined by $\|\, f\,\| = (f, f)^{1/2}$. Functions which coincide except on a set of measure 0 are considered as the same member of $L^2(D)$. Let $\mathfrak{H}(D)$ denote the set of functions in $L^2(D)$ which are holomorphic in D.

Proposition 3.1. *Let A be a compact subset of D. Then there exists a number N_A such that*

$$|f(z)| \leqslant N_A \|\, f\,\|$$

for all $z \in A$ and all $f \in \mathfrak{H}(D)$.

Proof. Let $\zeta = (\zeta_1, ..., \zeta_N) \in A$ and let $C(\zeta, \epsilon)$ be any polycylinder $|\, z_1 - \zeta_1\,| < \epsilon_1, ..., |\, z_N - \zeta_N\,| < \epsilon_N$ contained in D. Then the power series expansion for f,

$$f(z_1, ..., z_N) = \sum_{r_i \geqslant 0} a_{r_1 \ldots r_N} (z_1 - \zeta_1)^{r_1} \ldots (z_N - \zeta_N)^{r_N}$$

is absolutely convergent in $C(\zeta, \epsilon)$ (cf. Bochner and Martin [1], p. 33). Now the terms in the series are mutually orthogonal with respect to the inner product

$$(g, h)_\epsilon = \int_{C(\zeta, \epsilon)} g(z)\, \overline{h(z)}\; d\mu(z).$$

Consequently

$$\int_D |f(z)|^2 \, d\mu(z) \geqslant \int_{C(\zeta,\varepsilon)} |f(z)|^2 \, d\mu(z) \geqslant \int_{C(\zeta,\varepsilon)} |a_{0\ldots 0}|^2 \, d\mu(z), \tag{1}$$

so

$$\mu(C(\zeta, \epsilon))^{1/2} \, |f(\zeta)| \leqslant \|f\|. \tag{2}$$

Since A is compact, there exists a number $\epsilon > 0$ such that for each $(\zeta_1, \ldots, \zeta_N) \in A$, the polycylinder $|z_i - \zeta_i| < \epsilon$ $(1 \leqslant i \leqslant N)$ belongs to D. If the volume of this polycylinder is denoted by $(1/N_A)^2$, Prop. 3.1 follows from (2).

Corollary 3.2. *The set $\mathfrak{H}(D)$ is a closed linear subspace of $L^2(D)$, hence a Hilbert space with the inner product $\int_D f(z)\,(g(z))^- \, d\mu(z)$.*

In fact, let (f_n) be a sequence in $\mathfrak{H}(D)$ which converges to an element $f \in L^2(D)$. Then by Prop. 3.1

$$|f_n(\zeta) - f_m(\zeta)| \leqslant N_A \, \|f_m - f_n\| \tag{3}$$

for all $\zeta \in A$. It follows that there exists a function g on D such that $f_n \to g$ uniformly on each compact subset of D. Hence g is holomorphic on D. By (3) we have for $\zeta \in A$

$$|f_n(\zeta) - g(\zeta)| \leqslant N_A \, \|f_n - f\|. \tag{4}$$

Given A, there exists an integer K such that the right-hand side of (4) is $\leqslant 1$ for $n \geqslant K$ and such that $\|f_K\| \leqslant \|f\| + 1$. Then

$$\Big[\int_A |g(\zeta)|^2 \, d\mu\Big]^{1/2} \leqslant \Big[\int_A |f_K(\zeta) - g(\zeta)|^2 \, d\mu\Big]^{1/2} + \Big[\int_A |f_K(\zeta)|^2 \, d\mu\Big]^{1/2}$$

$$\leqslant \mu(D)^{1/2} + \|f\| + 1.$$

Hence $g \in \mathfrak{H}(D)$. Finally, since

$$\lim \int_A |f_n(z) - f(z)|^2 \, d\mu = 0, \qquad \lim \int_A |f_n(z) - g(z)|^2 \, d\mu = 0$$

it follows that $f = g$ almost everywhere.

Theorem 3.3. *Let $\varphi_0, \varphi_1, \ldots$ be any orthonormal basis of the Hilbert space $\mathfrak{H}(D)$. Then the series*

$$\sum_0^\infty \varphi_n(z) \, \overline{\varphi_n(\zeta)}$$

converges uniformly on each compact subset of $D \times D$. *The sum, denoted* $K(z, \bar{\zeta})$, *is independent of the choice of orthonormal basis and*

$$F(z) = \int_D K(z, \bar{\zeta})\, F(\zeta)\, d\mu(\zeta)$$

for each $F \in \mathfrak{H}(D)$.

Proof. Let $z \in D$. In view of Prop. 3.1, the linear functional $F \to F(z)$ on $\mathfrak{H}(D)$ is continuous. Now every continuous linear functional on a Hilbert space is representable as the inner product with some fixed vector in the space. Hence there exists a function $K_z \in \mathfrak{H}(D)$ such that

$$F(z) = \int_D F(\zeta)\, \overline{K_z(\zeta)}\, d\mu(\zeta) \qquad (F \in \mathfrak{H}(D)). \tag{5}$$

We now put $K(z, \bar{\zeta}) = \overline{K_z(\zeta)}$ for $z, \zeta \in D$. The vector K_z can be expressed by means of the basis $\varphi_0, \varphi_1, \ldots,$

$$K_z(\zeta) = \sum_0^\infty a_n \varphi_n(\zeta), \tag{6}$$

where the series converges in the L_2-norm and

$$a_n = \int_D K_z(\zeta)\, \overline{\varphi_n(\zeta)}\, d\mu(\zeta) \qquad (n = 0, 1, \ldots).$$

In view of Prop. 3.1, the series (6) converges uniformly on each compact subset of D. From (5) we have $a_n = \overline{\varphi_n(z)}$ and therefore

$$K(z, \bar{\zeta}) = \sum_0^\infty \varphi_n(z)\, \overline{\varphi_n(\zeta)}. \tag{7}$$

This shows that $\overline{K(\zeta, \bar{z})} = K(z, \bar{\zeta})$; consequently, for a given ζ, the function $z \to K(z, \bar{\zeta})$ belongs to $\mathfrak{H}(D)$. In order to prove that the series (7) converges uniformly on each compact subset of $D \times D$, it suffices, due to the inequality

$$2\, |\, \varphi_n(z)\, \overline{\varphi_n(\zeta)}\, | \leqslant |\, \varphi_n(z)\, |^2 + |\, \varphi_n(\zeta)\, |^2,$$

to prove that the series $\Sigma_0^\infty\, |\, \varphi_n(z)\, |^2$ converges to $K(z, \bar{z})$, uniformly on each compact subset A of D. Let $\epsilon > 0$. There exists a number $\delta > 0$ such that for each $(\zeta_1, \ldots, \zeta_N) \in A$ the closed polycylinder $|\, z_1 - \zeta_1\, | \leqslant \delta, \ldots, |\, z_N - \zeta_N\, | \leqslant \delta$ belongs to D. Let A_δ denote the union of these polycylinders. Then by (1) there exists a constant M_δ such that

$$|f(z)|^2 \leqslant M_\delta \int_{A_\delta} |f(\zeta)|^2 \, d\mu$$

for all $z \in A$ and all $f \in \mathfrak{H}(D)$. Since

$$\sum_0^\infty \int_{A_\delta} |\varphi_n(\zeta)|^2 \, d\mu = \int_{A_\delta} K(\zeta, \bar{\zeta}) \, d\mu < \infty,$$

there exists an integer P such that

$$\sum_{P+1}^\infty |\varphi_n(z)|^2 \leqslant M_\delta \sum_{P+1}^\infty \int_{A_\delta} |\varphi_n(\zeta)|^2 \, d\mu < \epsilon$$

for all $z \in A$. This concludes the proof.

Definition. The function K is called the *kernel function* for D.

Let $(z_1, \ldots, z_N)$ denote the components of $z \in \boldsymbol{C}^N$ and consider the complex tensor field H on D given by

$$H = \sum_{1 \leqslant i,j \leqslant N} Z_i Z_{j^*} \log K(z, \bar{z}) \, dz_i \otimes \overline{dz_j}.$$

Here $dz_i \otimes \overline{dz_j}$ denotes the complex tensor field

$$(X, Y) \rightarrow dz_i(X) \, \overline{dz_j}(Y), \qquad X, Y \in \mathfrak{E}^1(D),$$

which is covariant of degree 2. Let g denote the real part of the restriction of H to $\mathfrak{D}^1(D) \times \mathfrak{D}^1(D)$.

Proposition 3.4. *The tensor field g is a Riemannian structure on D which is Kählerian.*

Proof. If f is a holomorphic function on D, then

$$Z_i \bar{f} = Z_{j^*} f = 0, \qquad 1 \leqslant i \leqslant N,$$

due to the Cauchy-Riemann equations. We use this on the series

$$K(z, \bar{\zeta}) = \sum_0^\infty \varphi_n(z) \, \overline{\varphi_n(\zeta)},$$

which by Theorem 3.3 can be differentiated term by term. We obtain

$$Z_{j^*} \log K(z, \zeta) = \frac{1}{K} \sum_n \varphi_n(z) \, (Z_{j^*}\overline{\varphi_n}) \, (\zeta)$$

and writing $\log K$ for $\log K(z, \bar{z})$ we obtain, since

$$\left\{ \frac{\partial}{\partial z_i} \frac{\partial}{\partial \bar{\zeta}_j} \log K(z, \bar{\zeta}) \right\}_{\zeta = z} = \frac{\partial}{\partial z_i} \frac{\partial}{\partial \bar{z}_j} \log K(z, \bar{z}),$$

$$Z_i Z_{j^*} \log K = \frac{1}{K^2} \left\{ \sum_n \varphi_n \overline{\varphi_n} \sum_m (Z_i \varphi_m) \, (Z_{j^*}\overline{\varphi_m}) - \sum_m \overline{\varphi_m} (Z_i \varphi_m) \sum_n \varphi_n (Z_{j^*}\overline{\varphi_n}) \right\},$$

so

$$Z_i Z_{j^*} \log K = \frac{1}{K^2} \sum_{n > m} (\varphi_n(Z_i\varphi_m) - \varphi_m(Z_i\varphi_n)) \, (\varphi_n(Z_j\varphi_m) - \varphi_m(Z_j\varphi_n))^-, \tag{8}$$

since $Z_j f = Z_{j^*}\bar{f}$. Let $X, Y \in \mathfrak{E}^1(D)$ and put $\xi_i = dz_i(X)$, $\eta_j = d\bar{z}_j(Y)$. Then, by (8),

$$H(X, \bar{X}) = \sum_{i,j} (Z_i Z_{j^*} \log K) \, \xi_i \bar{\xi}_j \geqslant 0, \tag{9}$$

$$H(X, Y) = \sum_{i,j} (Z_i Z_{j^*} \log K) \, \xi_i \eta_j = \overline{H(\bar{Y}, \bar{X})}. \tag{10}$$

For $X, Y \in \mathfrak{D}^1(D)$ this implies that $g(X, X) \geqslant 0$ and $2g(X, Y) = H(X, Y) + \overline{H(X, Y)} = H(X, Y) + H(Y, X)$ so $g(X, Y) = g(Y, X)$. Suppose now that $g(X, X) = 0$ at some point $p \in D$. Then, by (8), we have at the point p

$$\sum \xi_i(\varphi_n(Z_i\varphi_m) - \varphi_m(Z_i\varphi_n)) = 0 \tag{11}$$

for all n, m. Now, since D is bounded, the functions $1, z_1, \ldots, z_N$ all belong to $\mathfrak{H}(D)$. We can choose the system $\varphi_0, \varphi_1, \ldots$ in such a way that the functions $\varphi_0, \varphi_1, \ldots, \varphi_N$ are obtained from $1, z_1, \ldots, z_N$ by the usual orthonormalization process. Then the matrix (b_{ij}) given by

$$b_{ij} = Z_i \varphi_j \qquad (1 \leqslant i, j \leqslant N)$$

is an upper triangular matrix whose diagonal elements are constants $\neq 0$. Hence $\det (b_{ij}) \neq 0$. On the other hand, (11) implies, φ_0 being constant, that

$$\sum \xi_i (Z_i \varphi_j) = 0, \qquad 1 \leqslant j \leqslant N,$$

at the point p. It follows that all ξ_i vanish at p, so $X_p = 0$. It has now been shown that g is a Riemannian structure on D. The relations $dz_i(JX) = i\, dz_i(X)$ and $d\bar{z}_j(JY) = -i\, d\bar{z}_j(Y)$ imply that $g(JX, JY) = g(X, Y)$ so g is Hermitian. Now suppose g is extended to a complex tensor field. Then we have

$$2g(X, Y) = H(X, Y) + H(Y, X), \qquad X, Y \in \mathfrak{E}^1(D),$$

because both sides of this equation are complex tensor fields which coincide for $X, Y \in \mathfrak{D}^1(D)$. It follows from (9) and (10) that

$$g_{ij} = g_{i^*j^*} = 0, \qquad g_{ij^*} = \tfrac{1}{2} Z_i Z_{j^*} \log K. \tag{12}$$

If α, β, γ, δ are any indices, starred or not, we have from formula (2), Chapter I, §9,

$$2 \sum_{\delta} g_{\alpha\delta} \Gamma_{\beta\gamma}{}^{\delta} = Z_\beta g_{\alpha\gamma} + Z_\gamma g_{\alpha\beta} - Z_\alpha g_{\gamma\beta}. \tag{13}$$

Using the fact that g is nondegenerate one derives from (12) and (13)

$$\Gamma_{jk}{}^{l^*} = \Gamma_{j^*k}{}^{l} = \Gamma_{jk^*}{}^{l^*} = \Gamma_{j^*k^*}{}^{l} = 0. \tag{14}$$

By Lemma 2.2, g is Kählerian.

Definition. The metric induced by the Riemannian structure g is called the *Bergman metric* on D.

Let φ be a holomorphic diffeomorphism of a bounded domain $D \subset C^N$ onto a bounded domain $D' \subset C^N$; expressed in coordinates, we have

$$\varphi(z_1, \ldots, z_N) = (w_1(z_1, \ldots, z_N), \ldots, w_N(z_1, \ldots, z_N)).$$

Then the Jacobian determinant

$$J_\varphi = \frac{\partial(w_1, \ldots, w_N)}{\partial(z_1, \ldots, z_N)}$$

is a holomorphic function on D. For the real coordinates given by $z_j = x_j + iy_j$, $w_j = u_j + iv_j$ $(1 \leqslant j \leqslant N)$, we have

$$|\, J_\varphi \,|^2 = \frac{\partial(u_1, v_1, \ldots, u_N, v_N)}{\partial(x_1, y_1, \ldots, x_N, y_N)} \tag{15}$$

as noted earlier. Let μ and μ' denote the Euclidean measures on D and D', respectively; then

$$\mu'(\varphi(M)) = \int_M |\, J_\varphi \,|^2 \, d\mu$$

for each Borel subset M of D. Consequently, the mapping $f \to (f \circ \varphi) J_\varphi$ is an isometry of $\mathfrak{H}(D')$ onto $\mathfrak{H}(D)$. It follows that the kernel functions K and K' are related by

$$K(z, \bar{z}) = K'(\varphi(z), \overline{\varphi(z)}) \mid J_\varphi \mid^2 \qquad (z \in D). \tag{16}$$

Proposition 3.5. *Let D and D' be bounded domains in $\mathbf{C}^N$ and let g and g' denote the Riemannian structures on D and D' induced by the kernel functions. Then each holomorphic diffeomorphism φ of D onto D' is an isometry.*

Proof. Using the notation above, we have

$$dw_j = du_j + idv_j, \; d\bar{w}_j = du_j - idv_j$$

Furthermore,

$$\varphi^*(du_j) = \sum_k \left(\frac{\partial u_j}{\partial x_k} dx_k + \frac{\partial u_j}{\partial y_k} dy_k \right)$$

and similarly for $\varphi^*(dv_j)$. Since

$$dw_j(d\varphi \cdot X) = \varphi^*(du_j)(X) + i\varphi^*(dv_j)(X), \qquad X \in \mathfrak{D}^1(D),$$

it follows from the Cauchy-Riemann equations that

$$dw_j(d\varphi \cdot X) = \sum_k \frac{\partial w_j}{\partial z_k} dz_k(X),$$

$$d\bar{w}_j(d\varphi \cdot X) = \sum_k \left(\frac{\partial w_j}{\partial z_k} \right)^- \overline{dz_k}(X).$$

Hence

$$g'(d\varphi \cdot X, d\varphi \cdot X) = \sum_{i,j} \frac{\partial^2 \log K'}{\partial w_i \partial \overline{w_j}} dw_i(d\varphi X) \overline{dw_j}(d\varphi X)$$

$$= \sum_{k,l} Z_k Z_{l^*} \log K'(\varphi(z), \overline{\varphi(z)}) \, dz_k(X) \, \overline{dz_l}(X)$$

$$= \sum_{k,l} Z_k Z_{l^*} \log K(z, \bar{z}) \, dz_k(X) \, \overline{dz_l}(X) = g(X, X),$$

where we have used (16) and the relation

$$\frac{\partial^2 \log |f|^2}{\partial z_k \partial \overline{z_l}} = 0,$$

which is valid for an arbitrary holomorphic function f in D without zeros. In fact, we have in a neighborhood of each point in D

$$\log |f|^2 = \log f + \log \bar{f} = \log f + (\log f)^-$$

so

$$Z_k Z_{l^*} \log |f|^2 = Z_k Z_{l^*} \log f + Z_{l^*} Z_k (\log f)^- = 0.$$

Let M be a complex manifold and let φ and ψ be holomorphic diffeomorphisms of M onto M. Then the diffeomorphisms $\varphi \circ \psi$ and φ^{-1} are almost complex, hence holomorphic. Consequently, the set of holomorphic diffeomorphisms of M onto itself forms a group, denoted $H(M)$, the group operation being composition of mappings. If $H(M)$ is transitive on M, M is said to be *homogeneous*.

Let D be a bounded domain with Riemannian structure given by the kernel function. According to Prop. 3.5 we have

$$H(D) \subset I(D).$$

If, as usual, $I(D)$ is taken with the compact open topology (Chapter IV, §2), it is clear that $H(D)$ is a closed subgroup of $I(D)$.

Proposition 3.6. *Let D be a bounded domain with Riemannian structure g given by the kernel function K. If D is homogeneous, then*

$$K(z, \bar{z}) = c\bar{g}(x_1, y_1, \ldots, x_N, y_N)^{1/2} \qquad (c = \text{constant}) \tag{17}$$

and

$$r = 2g,$$

r being the Ricci curvature.

Proof. Let $\varphi \in H(D)$. Then (16) implies

$$K(z, \bar{z}) = K(\varphi(z), \overline{\varphi(z)}) \, | J_\varphi |^2.$$

Furthermore, relation (15) shows that

$$\bar{g}(x_1, y_1, \ldots, x_N, y_N) = \bar{g}(u_1, v_1, \ldots, u_N, v_N) \, | J_\varphi |^4.$$

Formula (17) now follows from the homogeneity assumption; the formula $r = 2g$ follows from (12) and Prop. 2.5.

§ 4. Hermitian Symmetric Spaces of the Compact Type and the Noncompact Type

Let M be a connected complex manifold with a Hermitian structure. The set of holomorphic isometries of M forms a group, denoted $A(M)$, the group operation being composition of mappings. We have obviously

$$A(M) = H(M) \cap I(M).$$

Definition. Let M be a connected complex manifold with a Hermitian structure; M is said to be a *Hermitian symmetric space* if each point $p \in M$ is an isolated fixed point of an involutive holomorphic isometry s_p of M.

A Hermitian symmetric space M is of course a Riemannian symmetric space of even dimension. Hence the group $I(M)$ has a Lie group structure compatible with the compact open topology (Chapter IV, Lemma 3.2) and is a Lie transformation group of M. The group $A(M)$ is a closed subgroup of $I(M)$ and is therefore also a Lie transformation group of M. It is transitive on M since it contains all the symmetries. The identity component $A_0(M)$ of $A(M)$ is also transitive on M (Chapter II, Prop. 4.3(b)). Let $o \in M$ and let K be the subgroup of $G = A_0(M)$ leaving o fixed. With the automorphism $g \to s_o g s_o$ of G, the pair (G, K) is a Riemannian symmetric pair and M is diffeomorphic to G/K.

Let $X \in M_o$ and let $\gamma_X(t)$ $(t \in \boldsymbol{R})$ denote the geodesic in M having tangent vector X for $t = 0$. Let s_t denote the geodesic symmetry (extended to M) with respect to the point $\gamma_X(t)$. We have seen (Chapter IV, §3) that if $T_t = s_{t/2} s_o$ then $t \to T_t$ is a one-parameter subgroup of G, $T_t \cdot o = \gamma_X(t)$ and $(dT_t)_o$ is the parallel translation along γ_X. Since the elements of G by definition leave the complex structure of M invariant, the same is true of the parallel translation $(dT_t)_o$. This proves

Proposition 4.1. *The Hermitian structure of a Hermitian symmetric space is Kählerian.*

Next we consider the problem of constructing a complex structure on a given coset space. Let G be a connected Lie group, H a closed subgroup of G. Suppose J is an almost complex structure on the coset space $M = G/H$, invariant under the action of G. Let o denote the point $\{H\}$ in G/H. Then J_o is an endomorphism of the tangent space M_o satisfying the following conditions:

(i) $J_o^2 = -I$.

(ii) J_o commutes with each element in the linear isotropy group H^*.

On the other hand, if J_o is an endomorphism of M_o satisfying (i) and (ii), then the coset space $M = G/H$ has a unique almost complex structure which coincides with J_o at o and is invariant under the action of G.

Proposition 4.2. *Let (G, K) be a Riemannian symmetric pair. Let π denote the natural mapping of G onto $M = G/K$ and put $o = \pi(e)$. Let Q be any G-invariant Riemannian structure on M. Suppose A is an endomorphism of the tangent space M_o such that:*

(a) $A^2 = -I$.

(b) $Q_o(AX, AY) = Q_o(X, Y)$ for $X, Y \in M_o$.

(c) *A commutes with each element of the linear isotropy group K^*.*

Then M has a unique G-invariant almost complex structure J such that $J_o = A$. The structure Q is Hermitian, J is integrable, and with the corresponding† *complex structure, M is a Hermitian symmetric space.*

Proof. The existence and uniqueness of J is already mentioned above. That Q is Hermitian is clear from (b) since Q and J are G-invariant. We shall now verify that J is invariant under the symmetry s_o (and therefore under each symmetry s_p, $p \in M$). Let σ be any involutive analytic automorphism of G such that $(K_\sigma)_0 \subset K \subset K_\sigma$. Then, according to Prop. 3.4, Chapter IV, $s_o \circ \pi = \pi \circ \sigma$. Let $p \in M$ and $Z \in M_p$. Select $g \in G$ such that $\tau(g) \cdot o = p$ and put $Z_0 = d\tau(g^{-1}) Z$. Then using the invariance of J under G and the relations $s_o \circ \tau(g) = \tau(\sigma(g)) \circ s_o$, $ds_o J_o = J_o\, ds_o$ it follows that

$$ds_o(J_p Z) = ds_o d\tau(g)\, J_o Z_0 = d\tau(\sigma(g)) \circ J_o(ds_o Z_0)$$

$$= J_{s_o \cdot p}(ds_o d\tau(g)\, Z_o) = J_{s_o \cdot p}(ds_o Z);$$

hence J is invariant under s_o. Next we verify that J satisfies the integrability condition

$$[X, Y] + J[JX, Y] + J[X, JY] - [JX, JY] = 0 \tag{1}$$

for arbitrary vector fields X, Y on M. Owing to the homogeneity of M it suffices to verify (1) at the point o. Moreover, since the left-hand side of (1) is $C^\infty(M)$-bilinear, we can assume that the vector fields X, Y are (in a neighborhood of o) adapted to their values X_o, Y_o at o. Since J is invariant under G, in particular under parallelism, it follows that the vector fields JX, JY are adapted to their values at o. But since the torsion is 0 we have

$$[U, V]_o = (\nabla_U(V))_o - (\nabla_V(U))_o = 0$$

† See Theorem 1.2.

for any vector fields U, V adapted to their values at o, so (1) follows. The complex structure on M corresponding to J (Theorem 1.2) is, due to its uniqueness, invariant under each s_p, so M is a Hermitian symmetric space.

Remark. The proof above uses the deep Theorem 1.2 which is unproved in this book. In an exercise following this chapter we outline a direct proof of Prop. 4.2 (under a mild restriction), which does not make use of Theorem 1.2.

The example $\boldsymbol{C}^2$ shows that sometimes the groups $A_0(M)$ and $I_0(M)$ are different. This however, is somewhat exceptional as the following lemma shows.

Lemma 4.3. *Let M be a Hermitian symmetric space. Then $I_0(M)$ is semisimple if and only if $A_0(M)$ is semisimple. In this case $A_0(M) = I_0(M)$.*

Proof. Let $G = I_0(M)$ and let $\mathfrak{g}$ denote the Lie algebra of G. Let s denote the automorphism of $\mathfrak{g}$ which corresponds to the automorphism $g \to s_o g s_o$ of G and let $\mathfrak{p}$ denote the set of vectors $X \in \mathfrak{g}$ such that $sX = -X$. Since $A(M)$ contains the symmetries with respect to all points in M, it is clear that $A(M)$ contains all one-parameter subgroups $\exp tX$, $X \in \mathfrak{p}$. Thus the Lie algebra of $A(M)$ contains $\mathfrak{p}$ and $[\mathfrak{p}, \mathfrak{p}]$. If $I(M)$ is semisimple, then $[\mathfrak{p}, \mathfrak{p}] + \mathfrak{p} = \mathfrak{g}$ so $A_0(M) = I_0(M)$. On the other hand, if $A_0(M)$ is semisimple, then $A_0(M) = I_0(M)$ by Theorem 4.1, Chapter V.

Definition. Let M be a Hermitian symmetric space; M is said to be of the *compact type* or the *noncompact type* according to the type of the Riemannian symmetric pair $(A_0(M), K)$, K being the isotropy subgroup of $A_0(M)$ at some point $o \in M$.

Proposition 4.4. *Let M be a simply connected Hermitian symmetric space. Then M is a product*

$$M = M_o \times M_- \times M_+,$$

where all the factors are simply connected Hermitian symmetric spaces and $M_0 = \boldsymbol{C} \times \ldots \times \boldsymbol{C}$, M_- and M_+ are of the compact type and noncompact type, respectively.

Proof. Let $G = A_0(M)$, let o be a point in M and let K denote the isotropy subgroup of G at o. Let $(\tilde{G}, \varphi)$ be the universal covering group of G and let $\tilde{K}$ denote the identity component of $\varphi^{-1}(K)$. Then, if ψ denotes the mapping $g\tilde{K} \to \varphi(g) K$ of $\tilde{G}/\tilde{K}$ onto G/K, the pair $(\tilde{G}/\tilde{K}, \psi)$ is a covering space of G/K (Lemma 13.4, Chapter I); consequently $M = \tilde{G}/\tilde{K}$. Let $\mathfrak{g}$ denote the Lie algebra of G and let s denote the automorphism of $\mathfrak{g}$ which corresponds to the automorphism $g \to s_o g s_o$ of G. Then the pair $(\mathfrak{g}, s)$ is an effective orthogonal symmetric Lie algebra. We can decompose $\mathfrak{g}$ and $\mathfrak{k}$, the Lie algebra of $\tilde{K}$, according to Theorem 1.1, Chapter V,

$$\mathfrak{g} = \mathfrak{g}_0 + \mathfrak{g}_- + \mathfrak{g}_+, \qquad \mathfrak{k} = \mathfrak{k}_0 + \mathfrak{k}_- + \mathfrak{k}_+.$$

The groups $\tilde{G}$ and $\tilde{K}$ decompose accordingly

$$\tilde{G} = G_0 \times G_- \times G_+, \qquad \tilde{K} = K_0 \times K_- \times K_+$$

and the spaces $M_0 = G_0/K_0$, $M_- = G_-/K_-$, $M_+ = G_+/K_+$ are simply connected Riemannian globally symmetric spaces whose product is M. Let $\mathfrak{p}$, $\mathfrak{p}_0$, $\mathfrak{p}_-$, $\mathfrak{p}_+$ denote the eigenspaces for the eigenvalue -1 of s. As usual, these can be identified with tangent spaces to M, M_0, M_-, M_+ and

$$\mathfrak{p} = \mathfrak{p}_0 + \mathfrak{p}_- + \mathfrak{p}_+. \tag{2}$$

Let J and Q denote the almost complex structure and Riemannian structure, respectively, on M. Since $\mathfrak{p}$ is identified with the tangent space to M at o, Q_o is a bilinear form on $\mathfrak{p} \times \mathfrak{p}$ and J_o is an endomorphism of $\mathfrak{p}$. Let $Y_0 \in \mathfrak{p}_0$. Then $J_o Y_0$ can be decomposed according to (2),

$$J_o Y_0 = X_0 + X_- + X_+. \tag{3}$$

Let Ad denote the adjoint representation of $\tilde{G}$. Then

$$\mathrm{Ad}\,(k)\, Z_0 = Z_0 \qquad \text{for } k \in K_- \times K_+, \qquad Z_0 \in \mathfrak{p}_0, \tag{4}$$

by Lemma 1.5, Chapter V. Since $\mathrm{Ad}\,(k)$ and J_o commute, (3) and (4) imply

$$\mathrm{Ad}\,(k)\,(X_- + X_+) = X_- + X_+ \qquad \text{for } k \in K_- \times K_+.$$

This last relation, however, holds for all $k \in \tilde{K}$ since for $k_0 \in K_0$, $\mathrm{Ad}\,(k_0)$ keeps every vector in $\mathfrak{p}_- + \mathfrak{p}_+$ fixed. From Cor. 1.7, Chapter V, we deduce therefore that $X_- + X_+ = 0$ so $J_o\mathfrak{p}_0 \subset \mathfrak{p}_0$. Since J_o leaves Q_o invariant, it is clear that $J_o(\mathfrak{p}_- + \mathfrak{p}_+) \subset \mathfrak{p}_- + \mathfrak{p}_+$. Repeating the argument above, we find that $\mathfrak{p}_-$ and $\mathfrak{p}_+$ are invariant under J_o. Now Prop. 4.2 implies that M_- and M_+ are Hermitian symmetric.

Theorem 4.5. *Let M be a Hermitian symmetric space for which $A_0(M)$ is semisimple. Let $o \in M$ and let K denote the isotropy subgroup of $A_0(M)$*

at o. The corresponding linear isotropy group and its Lie algebra are denoted by K^* and $\mathfrak{k}^*$, respectively. Then:

(i) *The complex structure J_o of M_o belongs to the center $\mathfrak{c}$ of $\mathfrak{k}^*$.*

(ii) *The symmetry s_o is contained in the identity component of the center Z_K of K.*

Proof. Let Q and R, respectively, denote the Riemannian structure and curvature tensor of M. Then, according to Theorem 4.1, Chapter V, the Lie algebra $\mathfrak{k}^*$ consists of those endomorphisms of M_o which, when extended to the mixed tensor algebra over M_o as derivations commuting with contractions, annihilate Q_o and R_o. Now, if $X, Y \in M_o$ we have by (1) and (2), Chapter IV, §5,

$$(J_o \cdot Q_o)(X, Y) = -Q_o(X, J_oY) - Q_o(J_oX, Y), \tag{5}$$

$$(J_o \cdot R_o)(X, Y) = [J_o, R_o(X, Y)] - R_o(J_oX, Y) - R_o(X, J_oY). \tag{6}$$

The right-hand side of (5) vanishes since Q is Hermitian. The first term on the right-hand side of (6) vanishes since J_o commutes elementwise with K^* and $R_o(X, Y) \in \mathfrak{k}^*$. Finally, considering R as a complex tensor field we have by Lemma 2.1,

$$R_o(X - iJ_oX, Y - iJ_oY) = 0. \tag{7}$$

Considering the imaginary part in (7) we find that the right-hand side of (6) vanishes. Hence $J_o \in \mathfrak{k}^*$ and therefore $J_o \in \mathfrak{c}$. Identifying the Lie algebras $\mathfrak{k}$ and $\mathfrak{k}^*$ we have

$$\exp(tJ_o) \in Z_K \qquad \text{for each } t \in \mathbf{R}. \tag{8}$$

But on the space M_0 ($= \mathfrak{p}$) we have

$$e^{\pi J_o} = -I,$$

so $\exp(\pi J_o) = s_o$. This finishes the proof.

A compact semisimple Lie group U is a Riemannian globally symmetric space in each two-sided invariant Riemannian structure. However, this can never make U Hermitian symmetric as Theorem 4.5 shows.

Theorem 4.6. *Let M be a Hermitian symmetric space of the compact type or the noncompact type. Then M is simply connected.*

Proof. Since every Riemannian globally symmetric space of the noncompact type is simply connected we can assume that M is of the compact type. Let $U = I_0(M)$ ($= A_0(M)$); in the notation of Theorem 4.5 we have $M = U/K$. The Lie algebra $\mathfrak{u}$ of U decomposes $\mathfrak{u} = \mathfrak{k}_0 + \mathfrak{p}_*$

where $\mathfrak{k}_0$ is the Lie algebra of K and $\mathfrak{p}_*$ is the orthogonal complement of $\mathfrak{k}_0$ in $\mathfrak{u}$ with respect to the Killing form of $\mathfrak{u}$. Since we can consider $\mathfrak{k}_0$ as the Lie algebra of the linear isotropy group, corresponding to K, we have by Prop. 4.5, $J_o \in \mathfrak{k}_0$. Let S denote the closure in K of the one-parameter subgroup $\exp tJ_o$, ($t \in \boldsymbol{R}$). Since J_o annihilates no vector in $\mathfrak{p}_*$, it follows that $\mathfrak{k}_0$ is the centralizer of J_o in $\mathfrak{u}$. The centralizer Z_S of S in U therefore has Lie algebra $\mathfrak{k}_0$. From (8) we conclude $K \subset Z_S$. Being the centralizer of a torus, Z_S is connected so K is connected.

Let $(\tilde{U}, \varphi)$ be the universal covering group of U. The mapping $\sigma : u \to s_o u s_o$ is an automorphism of U. Let $\tilde{\sigma}$ denote the automorphism of $\tilde{U}$ such that $d\tilde{\sigma} = d\sigma$. Let $\tilde{K}$ denote the set of fixed points of $\tilde{\sigma}$. By Theorem 7.2, Chapter VII, the group $\tilde{K}$ is connected. Hence $\varphi(\tilde{K}) = K$ and there exists by Theorem 4.5 an element $z \in \tilde{K}$ such that $\varphi(z) = s_0$. The automorphism $\Sigma : u \to zuz^{-1}$ of $\tilde{U}$ satisfies $\varphi \circ \Sigma = \sigma \circ \varphi$ so $d\Sigma = d\sigma$. It follows that $\tilde{\sigma} = \Sigma$ and $\tilde{K}$ is the centralizer of z in $\tilde{U}$. In particular $\tilde{K}$ contains the center of $\tilde{U}$ so $\tilde{K} = \varphi^{-1}(K)$. Consequently, $U/K = \tilde{U}/\tilde{K}$ which is simply connected.

Remark. It will be proved later in the chapter that the class of symmetric bounded domains coincides with the class of Hermitian symmetric spaces of the noncompact type.

Example. It is possible for two Riemannian globally symmetric spaces M_1 and M_2 to be associated with the same orthogonal symmetric Lie algebra such that M_1 is Hermitian symmetric while M_2 is not. As an example take $M_1 = \boldsymbol{S}^2$ (two-dimensional sphere) and $M_2 = \boldsymbol{P}^2$ (two-dimensional projective space). Both are Riemannian globally symmetric (Chapter VII, Prop. 1.2) and $\boldsymbol{S}^2$ is Hermitian symmetric. However, $\boldsymbol{P}^2$ is not Hermitian symmetric since it is not even orientable.

§ 5. Irreducible Orthogonal Symmetric Lie Algebras

It is now convenient to make a further decomposition of the orthogonal symmetric Lie algebras of compact type and noncompact type.

Definition. Let $(\mathfrak{l}, s)$ be an orthogonal symmetric Lie algebra, $\mathfrak{u}$ and $\mathfrak{e}$ the eigenspaces of s for the eigenvalues $+1$ and -1, respectively; $(\mathfrak{l}, s)$ is said to be *irreducible* if the two following conditions are satisfied:

(i) $\mathfrak{l}$ is semisimple and $\mathfrak{u}$ contains no ideal $\neq \{0\}$ of $\mathfrak{l}$.

(ii) The algebra $\mathrm{ad}_{\mathfrak{l}}(\mathfrak{u})$ acts irreducibly on $\mathfrak{e}$.

Let (L, U) be a pair associated with $(\mathfrak{l}, s)$; then (L, U) is said to be irreducible if $(\mathfrak{l}, s)$ is irreducible. A Riemannian globally symmetric space M is called irreducible if the pair $(I_0(M), K)$ is irreducible, K being the isotropy subgroup of $I_0(M)$ at some point in M.

Let (L, U) be an irreducible Riemannian symmetric pair. Then all L-invariant Riemannian structures on L/U coincide except for a constant factor. In fact, $\mathrm{Ad}_L(U)$ is a compact linear group acting irreducibly on $\mathfrak{e}$ and the endomorphism $b : \mathfrak{e} \to \mathfrak{e}$ (from the proof of Lemma 1.2, Chapter V) commutes with each element of $\mathrm{Ad}_L(U)$. Hence b can only have one eigenvalue so the forms $Q(X, X)$ and $B(X, X)$ in the cited lemma are proportional. Thus L/U has an essentially unique L-invariant Riemannian structure. We can therefore always assume that this Riemannian structure is induced by $\pm B$ where B is the Killing form of $\mathfrak{l}$.

It is obvious that $(\mathfrak{l}, s)$ is irreducible if and only if the dual $(\mathfrak{l}^*, s^*)$ is irreducible.

The condition (ii) above can be described in different terms.

Proposition 5.1. *In the notations above suppose that the condition (i) is satisfied. Then (ii) holds if and only if* $\mathfrak{u}$ *is a maximal proper subalgebra of* $\mathfrak{l}$.

Proof. Assume first that (ii) holds. If $\mathfrak{u}$ were not maximal, there would exist a subalgebra $\mathfrak{u}^*$ of $\mathfrak{l}$ satisfying the *proper* inclusions $\mathfrak{u} \subset \mathfrak{u}^* \subset \mathfrak{l}$. Put $\mathfrak{e}^* = \mathfrak{u}^* \cap \mathfrak{e}$. Then $[\mathfrak{u}, \mathfrak{e}^*] \subset \mathfrak{u}^* \cap \mathfrak{e} = \mathfrak{e}^*$ so, due to the irreducibility, $\mathfrak{e}^* = \{0\}$ or $\mathfrak{e}^* = \mathfrak{e}$. Now the identity $\mathfrak{e}^* = \mathfrak{e}$ implies $\mathfrak{u}^* = \mathfrak{l}$ which is impossible. The identity $\mathfrak{e}^* = \{0\}$ is also impossible because if $Z \in \mathfrak{u}^*$, $Z \notin \mathfrak{u}$, then $Z = T + X$, where $T \in \mathfrak{u}$, $X \in \mathfrak{e}$, $X \neq 0$. It follows that $X = Z - T \in \mathfrak{u}^* \cap \mathfrak{e} = \mathfrak{e}^*$, which is a contradiction. The converse is trivial because if $\mathfrak{e}'$ were a proper invariant subspace of $\mathfrak{e}$, then $\mathfrak{u} + \mathfrak{e}'$ would be a proper subalgebra of $\mathfrak{l}$, properly containing $\mathfrak{u}$.

Proposition 5.2. *Let* $(\mathfrak{l}, s)$ *be an orthogonal symmetric Lie algebra. Let* $\mathfrak{l} = \mathfrak{u} + \mathfrak{e}$ *be the decomposition of* $\mathfrak{l}$ *into the eigenspaces of* s *for the eigenvalue* $+1$ *and* -1, *respectively. Assume* $\mathfrak{l}$ *is semisimple and that* $\mathfrak{u}$ *contains no ideal* $\neq \{0\}$ *of* $\mathfrak{l}$. *Then there exists ideals* $\mathfrak{l}_i$ *in* $\mathfrak{l}$ *such that:*

(a) $\mathfrak{l} = \sum_i \mathfrak{l}_i$ *(direct sum).*

(b) *The ideals* $\mathfrak{l}_i$ *are mutually orthogonal with respect to the Killing form* B *of* $\mathfrak{l}$ *and they are invariant under* s.

(c) *Denoting by* s_i *the restriction of* s *to* $\mathfrak{l}_i$, *each* $(\mathfrak{l}_i, s_i)$ *is an irreducible orthogonal symmetric Lie algebra.*

Proof. The proof proceeds along the same lines as that of Theorem 1.1, Chapter V. Let Q and b be as in Lemma 1.2, Chapter V. Then b is an endomorphism of $\mathfrak{e}$ which is symmetric with respect to Q, i.e.,

$$Q(bX, Y) = Q(X, bY), \qquad X, Y \in \mathfrak{e}.$$

Let

$$\mathfrak{e} = \sum_j \mathfrak{f}_j$$

be the decomposition of $\mathfrak{e}$ into the eigenspaces of b. The spaces $\mathfrak{f}_j$ are mutually orthogonal with respect to Q and B. Each $\mathfrak{f}_j$ is invariant under $\text{ad}_{\mathfrak{l}}(\mathfrak{u})$ and can be decomposed into irreducible subspaces which are mutually orthogonal with respect to Q and B. Thus we get a direct decomposition

$$\mathfrak{e} = \sum_i \mathfrak{e}_i,$$

orthogonal with respect to Q and B, where the spaces $\mathfrak{e}_i$ are invariant and irreducible under $\text{ad}_{\mathfrak{l}}(\mathfrak{u})$. We put $\mathfrak{u}_i = [\mathfrak{e}_i, \mathfrak{e}_i]$ and $\mathfrak{l}_i = \mathfrak{u}_i + \mathfrak{e}_i$. It can be proved just as in Chapter V, §1, that the spaces $\mathfrak{l}_i$ have the properties of Prop. 5.2.

The next theorems give an important description of the irreducible orthogonal symmetric Lie algebras.

Theorem 5.3. *The irreducible orthogonal symmetric Lie algebras of the compact type are:*

I. $(\mathfrak{l}, s)$ *where* $\mathfrak{l}$ *is a compact simple Lie algebra and* s *any involutive automorphism of* $\mathfrak{l}$.

II. $(\mathfrak{l}, s)$ *where the compact algebra* $\mathfrak{l}$ *is the direct sum* $\mathfrak{l} = \mathfrak{l}_1 + \mathfrak{l}_2$ *of simple ideals which are interchanged by an involutive automorphism* s *of* $\mathfrak{l}$.

Theorem 5.4. *The irreducible, orthogonal symmetric Lie algebras of the noncompact type are*

III. $(\mathfrak{l}, s)$ *where* $\mathfrak{l}$ *is a simple, noncompact Lie algebra over* $\boldsymbol{R}$, *the complexification* $\mathfrak{l}^C$ *is a simple Lie algebra over* $\boldsymbol{C}$ *and* s *is an involutive automorphism of* $\mathfrak{l}$ *such that the fixed points form a compactly imbedded subalgebra.*

IV. $(\mathfrak{l}, s)$ *where* $\mathfrak{l} = \mathfrak{g}^R$, $\mathfrak{g}$ *being a simple Lie algebra over* $\boldsymbol{C}$. *Here* s *is the conjugation of* $\mathfrak{l}$ *with respect to a maximal compactly imbedded subalgebra.*

Furthermore, if $(\mathfrak{l}^*, s^*)$ *denotes the dual of* $(\mathfrak{l}, s)$,

$$(\mathfrak{l}, s) \textit{ is of type } \text{III} \Leftrightarrow (\mathfrak{l}^*, s^*) \textit{ is of type } \text{I},$$

$$(\mathfrak{l}, s) \textit{ is of type } \text{IV} \Leftrightarrow (\mathfrak{l}^*, s^*) \textit{ is of type } \text{II}.$$

Proof of Theorem 5.3. It is obvious from Prop. 5.2 that each $(\mathfrak{l}, s)$ of type I is irreducible. Next, let $(\mathfrak{l}, s)$ be of type II. Then according to

Cor. 6.3, Chapter II, the only ideals in $\mathfrak{l}$ are $\{0\}$, $\mathfrak{l}_1$, $\mathfrak{l}_2$, and $\mathfrak{l}$. Again, Prop. 5.2 shows that $(\mathfrak{l}, s)$ is irreducible.

On the other hand, suppose $(\mathfrak{l}, s)$ is irreducible and $\mathfrak{l} = \mathfrak{u} + \mathfrak{e}$ compact. Let

$$\mathfrak{l} = \mathfrak{a}_1 + \dots + \mathfrak{a}_n$$

be the decomposition into the simple ideals of $\mathfrak{l}$ (Cor. 6.3, Chapter II). Then s permutes the ideals $\mathfrak{a}_i$. If $s\mathfrak{a}_i = \mathfrak{a}_i$, put $\mathfrak{l}_i = \mathfrak{a}_i$. If $s\mathfrak{a}_i \neq \mathfrak{a}_i$, put $\mathfrak{l}_i = \mathfrak{a}_i + s\mathfrak{a}_i$. We have then a direct decomposition

$$\mathfrak{l} = \sum_i \mathfrak{l}_i,$$

where each $\mathfrak{l}_i$ is an ideal in $\mathfrak{l}$, invariant under s, and can therefore be decomposed into eigenspaces, $\mathfrak{l}_i = \mathfrak{u}_i + \mathfrak{e}_i$. Since $\mathfrak{u} = \sum_i \mathfrak{u}_i$, the irreducibility of $(\mathfrak{l}, s)$ implies that all $\mathfrak{e}_i$ vanish except one, say $\mathfrak{e}_1$. But then condition (i) for irreducibility shows that $\mathfrak{u}_i = \{0\}$ for $i \neq 1$. Thus $\mathfrak{l}_i = \{0\}$ for $i \neq 1$, and this proves the theorem.

Proof of Theorem 5.4. Let $(\mathfrak{l}, s)$ be an orthogonal symmetric Lie algebra and $(\mathfrak{l}^*, s^*)$ its dual. Since irreducibility is preserved under the duality, it suffices to prove the last two statements of the theorem. Suppose first that $(\mathfrak{l}^*, s^*)$ is of type I. Then the decomposition $\mathfrak{l} = \mathfrak{u} + \mathfrak{e}$ into eigenspaces of s is a Cartan decomposition of $\mathfrak{l}$. If $\mathfrak{l}$ were not simple we would have $\mathfrak{l} = \mathfrak{a}_1 + \mathfrak{a}_2$ where $\mathfrak{a}_1$ and $\mathfrak{a}_2$ are nonzero ideals. Let $\mathfrak{a}_1 = \mathfrak{k}_1 + \mathfrak{p}_1$, $\mathfrak{a}_2 = \mathfrak{k}_2 + \mathfrak{p}_2$ be Cartan decompositions of $\mathfrak{a}_1$ and $\mathfrak{a}_2$. (If $\mathfrak{a}_1$ or $\mathfrak{a}_2$ is compact, then $\mathfrak{p}_1$ or $\mathfrak{p}_2$ is $\{0\}$.) Since the Cartan decompositions $\mathfrak{l} = \mathfrak{u} + \mathfrak{e}$ and $\mathfrak{l} = (\mathfrak{k}_1 + \mathfrak{k}_2) + (\mathfrak{p}_1 + \mathfrak{p}_2)$ are conjugate we may assume $\mathfrak{k}_1 + \mathfrak{k}_2 = \mathfrak{u}$, $\mathfrak{p}_1 + \mathfrak{p}_2 = \mathfrak{e}$. But then $(\mathfrak{k}_1 + i\mathfrak{p}_1) + (\mathfrak{k}_2 + i\mathfrak{p}_2)$ is a decomposition of $\mathfrak{l}^*$ into nonzero ideals. Hence $\mathfrak{l}$ must be simple. The complex algebra $\mathfrak{l}^C$ is also simple because otherwise $\mathfrak{l}^C$ is a direct sum $\mathfrak{l}^C = \mathfrak{n}_1 + \mathfrak{n}_2$ where $\mathfrak{n}_1$, $\mathfrak{n}_2$ are nonzero ideals. These being semisimple, let $\mathfrak{k}_1$, $\mathfrak{k}_2$ be compact real forms of $\mathfrak{n}_1$ and $\mathfrak{n}_2$, respectively. Then $\mathfrak{l}^*$ is isomorphic to $\mathfrak{k}_1 + \mathfrak{k}_2$ which is not simple. Thus $(\mathfrak{l}, s)$ is of type III. On the other hand, let $(\mathfrak{l}, s)$ be of type III. Then $(\mathfrak{l}, s)$ and therefore $(\mathfrak{l}^*, s^*)$ is irreducible. If $(\mathfrak{l}^*, s^*)$ were of type II, the complexification $(\mathfrak{l}^*)^C = \mathfrak{l}^C$ would not be simple. Hence $(\mathfrak{l}^*, s^*)$ is of type I. Next, suppose $(\mathfrak{l}^*, s^*)$ is of type II. Then $\mathfrak{l}^* = \mathfrak{l}_1^* + \mathfrak{l}_2^*$ where $\mathfrak{l}_1^*$ and $\mathfrak{l}_2^*$ are simple ideals interchanged by s^*. Then we know from Theorem 2.4, Chapter V, and the subsequent remark that $\mathfrak{l}$ has a complex structure J, and if $\mathfrak{l} = \mathfrak{k} + J\mathfrak{k}$ is a Cartan decomposition of $\mathfrak{l}$, then $\mathfrak{k}$ and $\mathfrak{l}_1^*$ are isomorphic. Hence $\mathfrak{k}$ is simple and $\mathfrak{l}$ is simple (as a Lie algebra over $\boldsymbol{C}$). Thus $(\mathfrak{l}, s)$ is of type IV. Reversing these arguments and using Theorem 2.4,

Chapter V, again we find that if $(\mathfrak{l}, s)$ is of type IV then $(\mathfrak{l}^*, s^*)$ is of type II.

Proposition 5.5. *Let M be a simply connected Riemannian globally symmetric space of the compact type or the noncompact type. Then M is a product*

$$M = M_1 \times \ldots \times M_r,$$

where the factors M_i are irreducible. If M is Hermitian, then each M_i is Hermitian.

Proof. As in the proof of Prop. 4.2, Chapter V, let $G = I_0(M)$ and let K denote the isotropy subgroup of G at some point o in M. If $(\tilde{G}, \varphi)$ is the universal covering group of G and $\tilde{K}$ is the identity component of $\varphi^{-1}(K)$, then $M = \tilde{G}/\tilde{K}$. Using Prop. 5.2 we get a decomposition

$$\tilde{G} = G_1 \times \ldots \times G_r,$$

$$\tilde{K} = K_1 \times \ldots \times K_r, \tag{1}$$

where each pair (G_i, K_i) is irreducible. If we put $M_i = G_i/K_i$ we have

$$M = M_1 \times \ldots \times M_r.$$

Moreover, G_i is semisimple and the Lie algebra of K_i contains no ideal $\neq \{0\}$ of the Lie algebra of G_i. In view of Theorem 4.1, Chapter V, G_i and $I(M_i)$ have the same Lie algebra. Thus M_i is irreducible. Finally suppose M is Hermitian symmetric and let J denote the corresponding almost complex structure on M. As proved in §4, the groups $A_0(M)$ and $I_0(M)$ coincide and J_o lies in the center of the Lie algebra of K. According to (1), J_o is decomposed $J_o = J_1 \times \ldots \times J_r$ where each J_i is an endomorphism of square $-I$ of the tangent space to M_i at $\{K_i\}$ and J_i lies in the center of the Lie algebra of K_i. Since each K_i is connected, the group $\mathrm{Ad}_{G_i}(K_i)$ commutes elementwise with J_i. Proposition 4.2 now shows that M_i is Hermitian.

§ 6. Irreducible Hermitian Symmetric Spaces

Theorem 6.1.

(i) *The noncompact irreducible Hermitian symmetric spaces are exactly the manifolds G/K where G is a connected noncompact simple Lie group with center $\{e\}$ and K has nondiscrete center and is a maximal compact subgroup of G.*

(ii) *The compact irreducible Hermitian symmetric spaces are exactly the manifolds U/K where U is a connected compact simple Lie group with center $\{e\}$ and K has nondiscrete center and is a maximal connected proper subgroup of U.*

Proof. Let M be an irreducible Hermitian symmetric space. Then $M = \tilde{G}/\tilde{K}$ where $\tilde{G}$ is the simply connected covering group of $I_0(M)$ and $\tilde{K}$ is connected and contains the center of $\tilde{G}$, (Theorem 1.1, Chapter VI, and Theorem 4.6). Hence $M = \mathrm{Ad}(\tilde{G})/\mathrm{Ad}(\tilde{K})$, where $\mathrm{Ad} = \mathrm{Ad}_{\tilde{G}}$. This representation of M has the properties stated in (i) and (ii) as a glance at Theorems 4.5, 5.3, and 5.4 and Prop. 5.1 shows.

On the other hand, suppose U and K have the properties in (ii). Then the center of K contains an element j of order 4. Let $s = j^2$, and let Z_s denote the centralizer of s in U. Since $s \neq e$, we have $Z_s \neq U$ so K coincides with the identity component of Z_s. The automorphism $\sigma : u \to sus^{-1}$ $(u \in U)$ turns (U, K) into a Riemannian symmetric pair. Let $\mathfrak{p}_*$ denote the eigenspace for the eigenvalue -1 of the automorphism $d\sigma$, and let J denote the restriction of $\mathrm{Ad}_U(j)$ to $\mathfrak{p}_*$. Then $J^2 = -I$ so J gives rise to a U-invariant almost complex structure on U/K and, according to Prop. 4.2, U/K is Hermitian symmetric. The statement (i) now follows by use of duality.

Proposition 6.2. *The center Z_K of the group K in Theorem* 6.1 (i) *and* (ii) *is analytically isomorphic to the circle group.*

Consider for example case (i). The action of $\mathrm{Ad}_G(K)$ on $\mathfrak{p}_0$ is irreducible and so is the action of $\mathrm{Ad}_G(K)$ on the complex vector space $\tilde{\mathfrak{p}}_0$ (Chapter III, §6), an action which is $\boldsymbol{C}$-linear since $\mathrm{Ad}_G(K)$ commutes elementwise with J_0. For $c \in Z_K$ let c^* denote the restriction $\mathrm{Ad}_G(c) \mid \tilde{\mathfrak{p}}_0$. From Schur's lemma (see e.g. Chevalley [2], p. 183) c^* is a scalar multiple of the identity. Since Z_K is nondiscrete, the image Z_K^* is the whole circle $\boldsymbol{S}^1$. But if $\mathrm{Ad}_G(c) \mid \mathfrak{p}_0 = 1$, then since $\mathrm{Ad}_G(c) \mid \mathfrak{k}_0 = 1$ we have $c \in \mathrm{center}(G) = \{e\}$. Thus Z_K is isomorphic to $\boldsymbol{S}^1$.

§ 7. Bounded Symmetric Domains

Definition. A bounded domain D is called *symmetric* if each $p \in D$ is an isolated fixed point of an involutive holomorphic diffeomorphism of D onto itself.

Theorem 7.1.

(i) *Each bounded symmetric domain D is, when equipped with the Bergman metric, a Hermitian symmetric space of the noncompact type.*

In particular, a bounded symmetric domain is necessarily simply connected.

(ii) *Let M be a Hermitian symmetric space of the noncompact type. Then there exists a bounded symmetric domain D and a holomorphic diffeomorphism of M onto D.*

Proof of (i). Let D be a bounded symmetric domain. Let Q be the Riemannian structure on D corresponding to the Bergman metric. Prop. 3.5 shows that D is a Hermitian symmetric space. Let o be a fixed point in D, let σ denote the automorphism $g \to s_o g s_o$ of $A_0(D)$ and let $\mathfrak{l}$ denote the Lie algebra of $A_0(D)$. If we put $s = d\sigma$, then $(\mathfrak{l}, s)$ is an effective orthogonal symmetric Lie algebra. From Theorem 1.1. Chapter V we have the direct decompositions

$$\mathfrak{l} = \mathfrak{u} + \mathfrak{e}, \qquad \mathfrak{e} = \mathfrak{e}_0 + \mathfrak{e}_- + \mathfrak{e}_+,$$

and in order to prove (i) above, it suffices to show that $\mathfrak{e}_0 = \mathfrak{e}_- = \{0\}$. Let $X \in \mathfrak{e}$. As before, let T_X denote the restriction of $(\text{ad}_{\mathfrak{l}}\, X)^2$ to $\mathfrak{e}$. Then the curvature tensor R of D satisfies

$$R_o(X, Y)X = T_X Y, \qquad X, Y \in \mathfrak{e}, \tag{1}$$

so by Prop. 3.6

$$2Q_o(X, X) = \text{Trace}\,(T_X). \tag{2}$$

Now $[\mathfrak{e}_0, \mathfrak{e}] = \{0\}$ by Lemma 1.3, Chapter V, so (2) implies that $\mathfrak{e}_0 = \{0\}$. Next, suppose $X \in \mathfrak{e}_-$. Then $T_X \mathfrak{e}_+ = \{0\}$, $T_X \mathfrak{e}_- \subset \mathfrak{e}_-$ and

$$Q_o(T_X Y, Y) = Q_o(R_o(X, Y)X, Y) \leqslant 0$$

for $Y \in \mathfrak{e}_-$, since the curvature along two-dimensional subspaces of $\mathfrak{e}_-$ is $\geqslant 0$. Thus (2) implies that $X = 0$, so $\mathfrak{e}_- = \{0\}$. This proves (i).

Proof of Theorem 7.1 (ii) (Algebraic part). In view of Theorem 4.6 and Prop. 5.5 it can be assumed that M is irreducible. Then by Theorem 6.1, the group $I_0(M)$ is simple. Let $\mathfrak{g}_0$ denote its Lie algebra, let θ denote the involutive automorphism of $\mathfrak{g}_0$ which arises from the symmetry with respect to some point in M and let $\mathfrak{g}_0 = \mathfrak{k}_0 + \mathfrak{p}_0$ be the decomposition of $\mathfrak{g}_0$ into eigenspaces of θ for the eigenvalues $+1$ and -1, respectively. Let $\mathfrak{c}_0$ be the center of $\mathfrak{k}_0$ and let $\mathfrak{h}_0$ be some maximal abelian subalgebra of $\mathfrak{k}_0$. Then $\mathfrak{c}_0 \subset \mathfrak{h}_0$ and $\mathfrak{h}_0$ is a maximal abelian subalgebra of $\mathfrak{g}_0$. In fact, the centralizer of $\mathfrak{c}_0$ in $\mathfrak{g}_0$ contains $\mathfrak{k}_0$ but differs from $\mathfrak{g}_0$ so by Prop. 5.1, it must coincide with $\mathfrak{k}_0$.

This maximality of $\mathfrak{h}_0$ carries with it important relationship between the Cartan decomposition $\mathfrak{g}_0 = \mathfrak{k}_0 + \mathfrak{p}_0$ and the root space decomposition of the complexification $\mathfrak{g}$ of $\mathfrak{g}_0$. Let $\mathfrak{c}$, $\mathfrak{h}$, $\mathfrak{k}$, $\mathfrak{p}$ be the subspaces of $\mathfrak{g}$ spanned by $\mathfrak{c}_0$, $\mathfrak{h}_0$, $\mathfrak{k}_0$, $\mathfrak{p}_0$. Then $\mathfrak{u} = \mathfrak{k}_0 + i\mathfrak{p}_0$ is a compact real form of $\mathfrak{g}$.

Let σ and τ denote the conjugations of $\mathfrak{g}$ with respect to $\mathfrak{g}_0$ and $\mathfrak{u}$, respectively, and let B denote the Killing form of $\mathfrak{g}$. The Hermitian form B_τ on $\mathfrak{g} \times \mathfrak{g}$ given by $B_\tau(X, Y) = - B(X, \tau Y)$ is strictly positive definite and

$$B_\tau([Z, X], Y) + B_\tau(X, [Z, Y]) = 0$$

for $Z \in \mathfrak{u}$, $X, Y \in \mathfrak{g}$. It follows that the endomorphism $\operatorname{ad} H$ of $\mathfrak{g}$ is semisimple for each $H \in \mathfrak{h}_0 \cup i\mathfrak{h}_0$. Since all $\operatorname{ad} H (H \in \mathfrak{h})$ commute, they are semisimple endomorphisms of $\mathfrak{g}$ so $\mathfrak{h}$ is a Cartan subalgebra of $\mathfrak{g}$. Let Δ denote the set of nonzero roots of $\mathfrak{g}$ with respect to $\mathfrak{h}$. Let $\alpha \in \Delta$. Since $[\mathfrak{h}, \mathfrak{k}] \subset \mathfrak{k}$ and $[\mathfrak{h}, \mathfrak{p}] \subset \mathfrak{p}$, it is clear that either $\mathfrak{g}^\alpha \subset \mathfrak{k}$ or $\mathfrak{g}^\alpha \subset \mathfrak{p}$. In the first case the root α is called *compact*, in the second case *noncompact* and we have the direct decompositions

$$\mathfrak{k} = \mathfrak{h} + \sum_\alpha \mathfrak{g}^\alpha, \qquad \mathfrak{p} = \sum_\beta \mathfrak{g}^\beta, \tag{3}$$

where α runs over all the compact roots, and β runs over all the noncompact roots. Let $\Delta_\mathfrak{c}$ denote the set of roots in Δ which do not vanish identically on $\mathfrak{c}$. In view of Lemma 3.1, Chapter VI, each root α is real valued on $i\mathfrak{h}_0$. We introduce compatible orderings in the duals of the real vector spaces $i\mathfrak{h}_0$ and $i\mathfrak{c}_0$. This gives an ordering of Δ which will be used in the rest of the proof. Let Δ^+ denote the set of positive roots in Δ, put $Q_+ = \Delta^+ \cap \Delta_\mathfrak{c}$ and

$$\mathfrak{p}_+ = \sum_{\beta \in Q_+} \mathfrak{g}^\beta, \qquad \mathfrak{p}_- = \sum_{-\beta \in Q_+} \mathfrak{g}^\beta.$$

Proposition 7.2. *The spaces $\mathfrak{p}_+$ and $\mathfrak{p}_-$ are abelian subalgebras of $\mathfrak{g}$ and*

$$[\mathfrak{k}, \mathfrak{p}_-] \subset \mathfrak{p}_-, \qquad [\mathfrak{k}, \mathfrak{p}_+] \subset \mathfrak{p}_+, \qquad \mathfrak{p} = \mathfrak{p}_- + \mathfrak{p}_+.$$

Proof. Let $\alpha \in \Delta$ be compact. Then $[\mathfrak{c}, \mathfrak{g}^\alpha] = \{0\}$ so α vanishes identically on $\mathfrak{c}$. Consequently $\mathfrak{p}_- + \mathfrak{p}_+ \subset \mathfrak{p}$. Using in addition the compatibility of the orderings we have $[\mathfrak{g}^\alpha, \mathfrak{p}_+] \subset \mathfrak{p}_+$, $[\mathfrak{g}^\alpha, \mathfrak{p}_-] \subset \mathfrak{p}_-$. The relations $[\mathfrak{h}, \mathfrak{p}_-] \subset \mathfrak{p}_-$, $[\mathfrak{h}, \mathfrak{p}_+] \subset \mathfrak{p}_+$ being obvious, we derive $[\mathfrak{k}, \mathfrak{p}_-] \subset \mathfrak{p}_-$, $[\mathfrak{k}, \mathfrak{p}_+] \subset \mathfrak{p}_+$ from (3).

Next, let $\beta, \gamma \in Q_+$. Then $[\mathfrak{g}^\beta, \mathfrak{g}^\gamma] \subset \mathfrak{g}^{\beta+\gamma}$ and if $\beta + \gamma$ is a root, then $\beta + \gamma \in Q_+$. But on the other hand $[\mathfrak{p}_+, \mathfrak{p}_+] \subset \mathfrak{k}$, so $[\mathfrak{p}_+, \mathfrak{p}_+] = \{0\}$. Also $\mathfrak{p}_-$ is abelian because $\tau \cdot \mathfrak{g}^\delta = \mathfrak{g}^{-\delta}$ for any $\delta \in \Delta$ (Lemma 3.1, Chapter VI).

Finally, in order to prove $\mathfrak{p} = \mathfrak{p}_- + \mathfrak{p}_+$, let $\mathfrak{q}$ denote the orthogonal complement of $\mathfrak{p}_- + \mathfrak{p}_+$ in $\mathfrak{p}$ with respect to B_τ and put

$$\mathfrak{g}_+ = \mathfrak{p}_+ + \mathfrak{p}_- + [\mathfrak{p}_+, \mathfrak{p}_-].$$

We shall prove that $\mathfrak{g}_+$ is an ideal in $\mathfrak{g}$; for this purpose it suffices to prove that

$$[\mathfrak{p}_+, \mathfrak{q}] = [\mathfrak{p}_-, \mathfrak{q}] = \{0\}. \tag{4}$$

Let $T \in \mathfrak{k}$, $X \in \mathfrak{p}_+$, $Y \in \mathfrak{q}$. Since $\tau \cdot T \in \mathfrak{k}$, and $\tau \cdot [X, \tau \cdot T] \in \mathfrak{p}_-$ we have

$$B_\tau([X, Y], T) = -B([X, Y], \tau \cdot T) = -B_\tau(Y, \tau \cdot [X, \tau \cdot T]) = 0$$

so $[\mathfrak{p}_+, \mathfrak{q}] = \{0\}$ and similarly $[\mathfrak{p}_-, \mathfrak{q}] = \{0\}$. Now, the orthogonal symmetric Lie algebra $(\mathfrak{g}_0, \theta)$ is of type III so by Theorem 5.4, $\mathfrak{g}$ is simple. We have therefore either $\mathfrak{g}_+ = \{0\}$ or $\mathfrak{g}_+ = \mathfrak{g}$. The first case implies that all the roots in Δ vanish identically on $\mathfrak{c}$ which is impossible. Thus $\mathfrak{g}_+ = \mathfrak{g}$, so $\mathfrak{p} = \mathfrak{p}_- + \mathfrak{p}_+$ and the proposition is proved.

Corollary 7.3. *A root $\alpha \in \Delta$ is compact if and only if it vanishes identically on $\mathfrak{c}$.*

In Chapters VI and VII much use was made of maximal abelian subspaces $\mathfrak{a}_0$ of $\mathfrak{p}_0$. Whereas all of these are conjugate under the linear isotropy group it is possible in the special situation here to select $\mathfrak{a}_0$ with particular reference to Δ. For each $\alpha \in \Delta$ we select a nonzero vector $X_\alpha \in \mathfrak{g}^\alpha$. Two roots $\alpha, \beta \in \Delta$ are called *strongly orthogonal* if $\alpha \pm \beta \notin \Delta$. Let $s = \dim \mathfrak{a}_0$.

Proposition 7.4. *There exists a subset $\gamma_1, \ldots, \gamma_s$ of Q_+ consisting of strongly orthogonal roots. Thus the subspace:*

$$\mathfrak{a} = \sum_{i=1}^{s} C(X_{\gamma_i} + X_{-\gamma_i}) \tag{5}$$

is a maximal abelian subspace of $\mathfrak{p}$.

The proof requires some preparation. If Q is any subset of Q_+, let

$$\mathfrak{p}_Q = \sum_{\gamma \in Q} (\mathfrak{g}^\gamma + \mathfrak{g}^{-\gamma}).$$

Let β be the lowest root in Q and let $Q(\beta)$ denote the set of all $\gamma \in Q$ such that $\gamma \neq \beta$ and neither $\gamma + \beta$ nor $\gamma - \beta$ is a root. Then the centralizer of $\mathfrak{g}^{-\beta} + \mathfrak{g}^\beta$ in $\mathfrak{p}_Q$ coincides with $\mathfrak{p}_{Q(\beta)}$.

Lemma 7.5. *The centralizer of $X_\beta + X_{-\beta}$ in $\mathfrak{p}_Q$ is $C(X_\beta + X_{-\beta}) + \mathfrak{p}_{Q(\beta)}$.*

Proof. Let $X \in \mathfrak{p}_Q$ and let Q' denote the complement of $\{\beta\}$ in Q.

Then

$$X = c_\beta X_\beta + c_{-\beta} X_{-\beta} + \sum_{\gamma \in Q'} (c_\gamma X_\gamma + c_{-\gamma} X_{-\gamma}),$$

where the coefficients are complex numbers. Now $\mathfrak{g} = \mathfrak{h} + \Sigma_{\alpha \in \Delta} \mathfrak{g}^\alpha$ and the component of $[X, X_\beta + X_{-\beta}]$ in $\mathfrak{h}$ is $(c_\beta - c_{-\beta}) [X_\beta, X_{-\beta}]$. Suppose X and $X_\beta + X_{-\beta}$ commute. We have then $c_\beta = c_{-\beta}$ and the vector

$$Y = \sum_{\gamma \in Q'} (c_\gamma X_\gamma + c_{-\gamma} X_{-\gamma})$$

commutes with $X_\beta + X_{-\beta}$. Since $\mathfrak{p}_+$ and $\mathfrak{p}_-$ are abelian we obtain

$$[Y, X_\beta + X_{-\beta}] = \sum_{\gamma \in Q'} (c_\gamma [X_\gamma, X_{-\beta}] + c_{-\gamma} [X_{-\gamma}, X_\beta]) = 0.$$

Here

$$c_\gamma [X_\gamma, X_{-\beta}] = c_{-\gamma} [X_{-\gamma}, X_\beta] = 0 \tag{6}$$

for each $\gamma \in Q'$. Otherwise, suppose, for example, $c_\gamma [X_\gamma, X_{-\beta}] \neq 0$. Then $\mathfrak{g}^{\gamma-\beta}$ and $\mathfrak{g}^{\beta-\gamma}$ are $\neq \{0\}$ and there exists a $\delta \in Q'$ such that

$$c_\gamma [X_\gamma, X_{-\beta}] + c_{-\delta} [X_{-\delta}, X_\beta] = 0.$$

This implies that $\alpha = \gamma - \beta = -\delta + \beta$ is a root $\neq 0$ but the relations $\gamma = \alpha + \beta$, $\delta = \beta - \alpha$ contradict the fact that β is the lowest root in Q. It follows from (6) that $Y \in \mathfrak{p}_{Q(\beta)}$ and the lemma is proved.

Proof of Proposition 7.4. We define a sequence of spaces $\mathfrak{p} = \mathfrak{p}_1 \supset \mathfrak{p}_2 \supset \dots \supset \mathfrak{p}_{s+1} = \{0\}$, each of which has the form $\mathfrak{p}_i = \mathfrak{p}_{Q_i}$, $(Q_i \subset Q_+)$, as follows: Let $Q_1 = Q_+$ and let γ_1 be the lowest positive root in Q_1. Let $\mathfrak{p}_2$ denote the centralizer of $\mathfrak{g}^{-\gamma_1} + \mathfrak{g}^{\gamma_1}$ in $\mathfrak{p}_1 = \mathfrak{p}_{Q_1}$; then $\mathfrak{p}_2 = \mathfrak{p}_{Q_2}$ where $Q_2 = Q_1(\gamma_1)$. Denoting by γ_2 the lowest positive root in Q_2, let $\mathfrak{p}_3$ denote the centralizer of $\mathfrak{g}^{-\gamma_2} + \mathfrak{g}^{\gamma_2}$ in $\mathfrak{p}_2$ etc. Then the roots $\gamma_1, \dots, \gamma_s$ are all different and form the desired subset of Q_+. In fact, it is clear from the construction that the space $\mathfrak{a}$ in (5) is abelian. Moreover, suppose $X \in \mathfrak{p}$ commutes with each element in $\mathfrak{a}$. We wish to prove $X \in \mathfrak{a}$. Suppose this were false. Then there exists an integer r $(1 \leqslant r \leqslant s)$, such that $X \in \mathfrak{p}_r + \mathfrak{a}$ but $X \notin \mathfrak{p}_{r+1} + \mathfrak{a}$. Let $X = Y + Z$ $(Y \in \mathfrak{p}_r, Z \in \mathfrak{a})$. Since X and Z commute with $X_{\gamma_r} + X_{-\gamma_r}$ the same is true of Y. Thus Lemma 7.5 implies that

$$Y = c(X_{\gamma_r} + X_{-\gamma_r}) + Y_1,$$

where $Y_1 \in \mathfrak{p}_{r+1}$ and $c \in \boldsymbol{C}$. Now, $Z_1 = Z + c(X_{\gamma_r} + X_{-\gamma_r})$ lies in $\mathfrak{a}$

and therefore

$$X = Y_1 + Z_1 \in \mathfrak{p}_{r+1} + \mathfrak{a},$$

which contradicts the definition of r. This proves Prop. 7.4.

Corollary 7.6. *In accordance with Lemma 3.1, Chapter VI, let the vectors $X_\alpha \in \mathfrak{g}^\alpha$ be chosen such that for each $\alpha \in \Delta$*

$$(X_\alpha - X_{-\alpha}), \qquad i(X_\alpha + X_{-\alpha}) \in \mathfrak{u},$$
$$[X_\alpha, X_{-\alpha}] = (2/\alpha(H_\alpha))\, H_\alpha.$$

Then the space

$$\mathfrak{a}_0 = \sum_{i=1}^{s} \boldsymbol{R}(X_{\gamma_i} + X_{-\gamma_i})$$

equals $\mathfrak{a} \cap \mathfrak{p}_0$ and is therefore a maximal abelian subspace of $\mathfrak{p}_0$.

In fact, owing to the choice of X_α we have $X_{\gamma_i} + X_{-\gamma_i} \in i\mathfrak{u} \cap \mathfrak{p} = \mathfrak{p}_0$ so $\mathfrak{a}_0 \subset \mathfrak{a} \cap \mathfrak{p}_0$. On the other hand, if

$$X = \sum_{i=1}^{s} c_i(X_{\gamma_i} + X_{-\gamma_i}) \in \mathfrak{p}_0 \qquad (c_i \in \boldsymbol{C}),$$

then $\tau \cdot X = -X$ so $c_i \in \boldsymbol{R}$.

Lemma 7.7. *Let $\mathfrak{l}$ be the three-dimensional Lie algebra over $\boldsymbol{C}$ given by the vector space $\boldsymbol{C}H + \boldsymbol{C}X + \boldsymbol{C}Y$ with the bracket defined by*

$$[X, Y] = H, \qquad [H, X] = 2X, \qquad [H, Y] = -2Y.$$

If L is any Lie group with Lie algebra $\mathfrak{l}^R$, then

$$\exp t(X + Y) = \exp(\tanh t)Y \exp(\log(\cosh t))H \exp(\tanh t)X \tag{7}$$

for $t \in \boldsymbol{R}$.

Proof. Consider the group $\boldsymbol{SL}(2, \boldsymbol{C})$ of all complex 2×2 matrices of determinant 1. The Lie algebra $\mathfrak{sl}(2, \boldsymbol{C})$ of this group consists of all complex 2×2 matrices of trace 0. It is isomorphic to $\mathfrak{l}$ under the mapping

$$\begin{pmatrix} 1 & 0 \\ 0 & -1 \end{pmatrix} \to H, \qquad \begin{pmatrix} 0 & 1 \\ 0 & 0 \end{pmatrix} \to X, \qquad \begin{pmatrix} 0 & 0 \\ 1 & 0 \end{pmatrix} \to Y.$$

The group $\boldsymbol{SL}(2, \boldsymbol{C})$ contains $\boldsymbol{SU}(2)$ as a maximal compact subgroup.

This group consists of all matrices

$$\begin{pmatrix} \alpha & \beta \\ -\bar{\beta} & \bar{\alpha} \end{pmatrix}$$

of determinant 1. Thus $\boldsymbol{SU}(2)$ is homeomorphic to the three dimensional sphere $\boldsymbol{S}^3$, in particular $\boldsymbol{SU}(2)$ is simply connected. In view of Theorem 2.2, Chapter VI, the group $\boldsymbol{SL}(2, \boldsymbol{C})$ is also simply connected. It suffices therefore to prove (7) for the group $L = \boldsymbol{SL}(2, \boldsymbol{C})$. This can be done by a direct computation, which is left to the reader.

Proof of Theorem 7.1 (ii) (Geometric part). The Lie algebra $\mathfrak{g}$ is the (vector space) direct sum of the Lie algebras $\mathfrak{k}$, $\mathfrak{p}_-$, and $\mathfrak{p}_+$. We shall now study the corresponding global situation.

Let G denote the simply connected Lie group with Lie algebra $\mathfrak{g}^R$. Let U, K, P_-, P_+, A^* denote the analytic subgroups of G corresponding to the subalgebras $\mathfrak{u}$, $\mathfrak{k}$, $\mathfrak{p}_-$, $\mathfrak{p}_+$, and $i\mathfrak{h}_0$, respectively, considered as real subalgebras of $\mathfrak{g}^R$. As in Chapter VI, §6, let

$$\mathfrak{n}_+ = \sum_{\alpha \in \Delta^+} \mathfrak{g}^\alpha, \qquad \mathfrak{n}_- = \sum_{-\alpha \in \Delta^+} \mathfrak{g}^\alpha,$$

considered as real subalgebras of $\mathfrak{g}^R$ and let N_+, N_-, G_0, K_0 denote the analytic subgroups of G corresponding to $\mathfrak{n}_+$, $\mathfrak{n}_-$, $\mathfrak{g}_0$, and $\mathfrak{k}_0$, respectively. Let exp denote the usual exponential mapping of $\mathfrak{g}^R$ into G, and let ad and Ad denote the adjoint representations of $\mathfrak{g}^R$ and G, respectively. Let θ, σ, τ denote the automorphisms of G which correspond to the automorphisms θ, σ, τ of $\mathfrak{g}^R$.

Lemma 7.8. *The mapping exp induces a diffeomorphism of* $\mathfrak{p}_-$ *onto* P_- *and of* $\mathfrak{p}_+$ *onto* P_+.

Proof. According to Lemma 3.5, Chapter VI, there exists a basis of $\mathfrak{g}$ with respect to which the matrices expressing ad $(\mathfrak{n}_+)$ are lower triangular with zeros in the diagonal. In view of Cor. 4.4 and Lemma 4.5, Chapter VI, the mapping ad $X \to e^{\text{ad} X} = \text{Ad}(\exp X)$ is a diffeomorphism of ad $(\mathfrak{n}_+)$ onto Ad (N_+). Since ad is one-to-one and Ad is one-to-one on N_+, the mapping $\exp : \mathfrak{n}_+ \to N_+$ is a diffeomorphism of $\mathfrak{n}_+$ onto N_+. Using the fact that $\mathfrak{p}_+ \subset \mathfrak{n}_+$ and $\tau \cdot \mathfrak{p}_+ = \mathfrak{p}_-$, the lemma follows.

Lemma 7.9. *The mapping* $(q, k, p) \to qkp$ *is diffeomorphism of* $P_- \times K \times P_+$ *onto an open submanifold of* G, *containing* G_0.

Proof. We prove first that $P_-K \cap P_+ = \{e\}$. Suppose to the contrary that $y \in P_-K \cap P_+$, $y \neq e$. Select $Y \in \mathfrak{p}_+$ such that $\exp Y = y$. Since $[\mathfrak{k}, \mathfrak{p}_-] \subset \mathfrak{p}_-$ we have $\operatorname{Ad}(y)\,\mathfrak{p}_- \subset \mathfrak{p}_-$. Writing $Y = \sum_{\alpha \in Q_+} c_\alpha X_\alpha$ ($c_\alpha \in \boldsymbol{C}$), let β denote the lowest root in Q_+ such that $c_\beta \neq 0$. Then $[Y, X_{-\beta}] \equiv c_\beta[X_\beta, X_{-\beta}] \pmod{\mathfrak{n}_+}$ and it follows that

$$\operatorname{Ad}(y)\,X_{-\beta} \equiv X_{-\beta} + c_\beta[X_\beta, X_{-\beta}] \qquad \pmod{\mathfrak{n}_+}.$$

Reading this relation $\bmod\,(\mathfrak{n}_- + \mathfrak{n}_+)$, we find that $\operatorname{Ad}(y)\,X_{-\beta} \notin \mathfrak{p}_-$ which contradicts $\operatorname{Ad}(y)\,\mathfrak{p}_- \subset \mathfrak{p}_-$. This shows that $P_-K \cap P_+ = \{e\}$. Applying the mapping $x \to \tau(x^{-1})$ $(x \in G)$, it follows that $P_- \cap KP_+ = \{e\}$. In order to prove that the mapping in the lemma is one-to-one suppose $q_1k_1p_1 = q_2k_2p_2$. This implies

$$(k_2^{-1}q_2^{-1}q_1k_2)\,k_2^{-1}k_1 = p_2p_1^{-1},$$

$$q_1^{-1}q_2 = k_1k_2^{-1}(k_2p_1p_2^{-1}k_2^{-1}),$$

which shows that $p_1 = p_2$, $q_1 = q_2$, $k_1 = k_2$. The regularity of the mapping follows by using Lemma 5.2, Chapter VI, twice, first on the algebra $\mathfrak{p}_- + \mathfrak{k}$ and then on the algebra $(\mathfrak{p}_- + \mathfrak{k}) + \mathfrak{p}_+$. The image is therefore a submanifold of G of dimension

$$\dim_{\boldsymbol{R}} \mathfrak{p}_- + \dim_{\boldsymbol{R}} \mathfrak{k} + \dim_{\boldsymbol{R}} \mathfrak{p}_+ = \dim \mathfrak{g}^{\boldsymbol{R}}$$

and is therefore an open submanifold of G. Finally, we know from Theorem 1.1, Chapter VI, that $G_0 = P_0K_0 = K_0P_0$ where $P_0 = \exp \mathfrak{p}_0$. Let $X \in \mathfrak{p}_0$ and $p = \exp \frac{1}{2}X$. From Theorem 6.3, Chapter VI,

$$p = uan, \qquad u \in U,\ a \in A^*,\ n \in N_+$$

and applying τ

$$\tau(p) = p^{-1} = ua^{-1}\tau(n),$$

so

$$\exp X = p^2 = \tau(n^{-1})\,a^2n \in N_-A^*N_+.$$

Moreover, $\mathfrak{n}_+ + \mathfrak{h} \subset \mathfrak{p}_+ + \mathfrak{k}$, $\mathfrak{n}_- \subset \mathfrak{p}_- + \mathfrak{k}$ so

$$N_-A^*N_+ \subset P_-KP_+,$$

and the lemma follows.

Remark. In general $P_-KP_+ \neq G$ as can be seen by considering the example $G = \boldsymbol{SL}(2, \boldsymbol{C})$, $G_0 = \boldsymbol{SL}(2, \boldsymbol{R})$, $K_0 = \boldsymbol{SO}(2)$. Here $K = \boldsymbol{SO}(2,\boldsymbol{C})$, the group of complex orthogonal 2×2 matrices of determinant 1.

As a result of Theorem 1.1, Chapter VI, each complex orthogonal matrix can be written uniquely as $\alpha e^{i\beta}$ where α is a real orthogonal matrix and β is a real skew symmetric matrix. Hence $K = K_0 \times \boldsymbol{R}$ topologically. Thus P_-KP_+ has fundamental group $\boldsymbol{Z}$ whereas G is simply connected.

Lemma 7.10. *The set G_0KP_+ is open in P_-KP_+ and $G_0 \cap KP_+ = K_0$.*

Proof. Suppose $p \in P_0$ has the form $p = kp_+$ where $k \in K$ and $p_+ \in P_+$. Applying the automorphism $\theta = \sigma\tau$ we get $p^{-1} = k(p_+)^{-1}$ so $p^2 = (p_+)^2$. Applying τ we get $p^{-2} \in \tau(P_+) \subset P_-$, so $p_+^2 \in P_-$. Hence $p = k = p_+ = e$. Since $G_0 = K_0P_0$ this shows that $G_0 \cap KP_+ = K_0$.

Consider now the group KP_+. The group P_+ is closed in N_+, hence closed in G. The group K is closed in G since it is the identity component of the set of fixed points of θ. Let (k_np_n) be a sequence in KP_+ which converges in G. Applying θ we see that the sequence (p_n^2) and therefore the sequences (k_n) and (p_n) are convergent. Thus KP_+ is closed in G and due to Lemma 7.9, its Lie algebra is $\mathfrak{k} + \mathfrak{p}_+$. Consider the mapping $\psi : (g, x) \to gx$ of $G_0 \times (KP_+)$ into G. Let $Y \in \mathfrak{g}_0$, $Z \in \mathfrak{k} + \mathfrak{p}_+$. Then

$$\psi(g \exp tY, x) = gx \exp (t \operatorname{Ad}(x^{-1})\, Y),$$

$$\psi(g, x \exp tZ) = gx \exp tZ$$

and consequently

$$d\psi_{(g,x)}(dL_gY, dL_xZ) = dL_{gx}(\operatorname{Ad}(x^{-1})\, Y + Z).$$

It follows that

$$\begin{aligned} d\psi(\{G_0 \times (KP_+)\}_{(g,x)}) &= dL_{gx} \circ \operatorname{Ad}(x^{-1})\, (\mathfrak{g}_0 + \operatorname{Ad}(x)\, (\mathfrak{k} + \mathfrak{p}_+)) \\ &= dL_{gx} \circ \operatorname{Ad}(x^{-1})\, (\mathfrak{g}_0 + \mathfrak{k} + \mathfrak{p}_+), \end{aligned}$$

and this image under $d\psi$ covers the whole tangent space G_{gx} due to the fact that

$$\mathfrak{g} = \mathfrak{p}_0 + \mathfrak{k} + \mathfrak{p}_+ = \mathfrak{g}_0 + \mathfrak{k} + \mathfrak{p}_+.$$

The lemma now follows from Lemma 7.9.

Lemma 7.11. *Let $Z \in \mathfrak{a}_0$,* i.e.,

$$Z = \sum_{i=1}^{s} t_i(X_{\gamma_i} + X_{-\gamma_i}), \qquad t_i \in \boldsymbol{R}. \tag{8}$$

Then

$$\exp Z = \exp Y \exp H \exp X,$$

where

$$Y = \sum_{i=1}^{s} (\tanh t_i)\, X_{-\gamma_i}, \qquad X = \sum_{i=1}^{s} (\tanh t_i)\, X_{\gamma_i},$$

$$H = \sum_{i=1}^{s} \log(\cosh t_i)\, [X_{\gamma_i}, X_{-\gamma_i}].$$

This follows from Lemma 7.7, combined with the fact that X_{γ_i} and $X_{-\gamma_j}$ commute if $i \neq j$.

The complex vector space $\mathfrak{g}$ becomes a finite-dimensional Hilbert space under the inner product B_τ. Let $\| X \| = B_\tau(X, X)^{1/2}$ for $X \in \mathfrak{g}$. According to Lemma 7.8, exp induces a one-to-one mapping of $\mathfrak{p}_-$ onto P_-. Let log denote the inverse mapping. For $x \in G_0$, let $\zeta(x)$ denote the unique element in P_- such that $x \in \zeta(x)\, KP_+$ (Lemma 7.9).

Lemma 7.12. *The norm $\| \log \zeta(x) \|$ is bounded as x varies through G_0.*

Proof. Since $[\mathfrak{k}, \mathfrak{p}_-] \subset \mathfrak{p}_-$, we have $\zeta(kxk') = k\zeta(x)\, k^{-1}$ for $x \in G_0$, $k, k' \in K_0$. From Lemma 6.3, Chapter V, and $G_0 = K_0P_0$, it follows that $G_0 = K_0A_0K_0$ where A_0 is the analytic subgroup of G_0 with Lie algebra $\mathfrak{a}_0$. Writing an arbitrary $x \in G_0$ as $x = kak'$ $(k, k' \in K_0,\ a \in A_0)$, we get $\zeta(x) = k\zeta(a)\, k^{-1}$ and

$$\| \log \zeta(x) \| = \| \mathrm{Ad}(k) \log \zeta(a) \| = \| \log \zeta(a) \|.$$

Now $a = \exp Z$ where Z has the form (8), and from Lemma 7.11 follows that

$$\log \zeta(a) = \sum_{i=1}^{s} (\tanh t_i)\, X_{-\gamma_i},$$

and since $| \tanh t | \leqslant 1$ for $t \in \boldsymbol{R}$,

$$\| \log \zeta(x) \| \leqslant \sum_{i=1}^{s} \| X_{-\gamma_i} \|$$

for all $x \in G_0$, which proves the lemma.

Completion of the proof. The Hermitian manifold M is diffeomorphic to G_0/K_0 (Theorem 1.1, Chapter VI) and the complex structure on M corresponds to an endomorphism J_0 on $\mathfrak{p}_0$ which commutes with all $\mathrm{Ad}(k)$ $(k \in K_0)$ and satisfies $J_0^2 = -I$.

For any coset space X/Y let $\tau(x)$ as usual denote the mapping $\xi Y \rightarrow x\xi Y$ of X/Y onto itself.

As remarked earlier, the group KP_+ is closed in G and the coset space G/KP_+ contains P_-KP_+/KP_+ as an open subset which in turn contains G_0KP_+/KP_+ as an open subset. Let these imbeddings be

$$\begin{array}{ccccc} G_0KP_+/KP_+ & \xrightarrow{I_1} & P_-KP_+/KP_+ & \xrightarrow{I_2} & G/KP_+ \\ \downarrow{\scriptstyle \psi_1} & & \downarrow{\scriptstyle \psi_2} & & \\ G_0/K_0 & \xrightarrow{\psi_0} & P_- & \xrightarrow{\log} & \mathfrak{p}_- \end{array}$$

denoted by I_2 and I_1, respectively (see diagram). In this diagram ψ_0, ψ_1, and ψ_2 denote the mappings

$$\psi_1 : gkpKP_+ \to gK_0 \qquad (g \in G_0,\, k \in K,\, p \in P_+),$$

$$\psi_2 : qkpKP_+ \to q \qquad (q \in P_-,\, k \in K,\, p \in P_+),$$

$$\psi_0 = \psi_2 \circ I_1 \circ \psi_1^{-1}.$$

The mapping ψ_1 is a diffeomorphism of G_0KP_+/KP_+ onto G_0/K_0 (Lemma 7.10). The mapping ψ_2 is a diffeomorphism of P_-KP_+/KP_+ onto P_- (Lemma 7.9). Thus it follows from Lemma 7.10 that ψ_0 is a diffeomorphism of G_0/K_0 onto an open subset of P_-. Combining this with Lemma 7.12 it is clear that the mapping $\psi = \log \circ \psi_0$ is a diffeomorphism of G_0/K_0 onto a bounded domain D in the complex vector space $\mathfrak{p}_-$. Moreover $\psi(xK_0) = \log \zeta(x)$, so for $k \in K_0$, $\psi \circ \tau(k) = \mathrm{Ad}(k) \circ \psi$ and $d\psi(\mathrm{Ad}(k)\, X) = \mathrm{Ad}(k)\, d\psi(X)$ for $X \in \mathfrak{p}_0$. Let J^* denote the endomorphism of $\mathfrak{p}_-$ defined by

$$d\psi(J_0X) = J^*\, d\psi(X) \qquad (X \in \mathfrak{p}_0). \tag{9}$$

Then J^* commutes with all $\mathrm{Ad}(k)$, $(k \in K_0)$. Since $\mathrm{Ad}(K_0)$ is irreducible on $\mathfrak{p}_-$, Schur's lemma implies $J^* = cI$ $(c \in \boldsymbol{C})$, so since $(J^*)^2 = -I$, $c = \pm i$. Replacing J_0 by $-J_0$ if necessary, we have $J^* = iI$. The mapping $I_2 \circ \psi_2^{-1} \circ \psi_0 = I_2 \circ I_1 \circ \psi_1^{-1}$ is just the mapping $\varphi : gK_0 \to gKP_+$ of G_0/K_0 into G/KP_+, so it commutes with the action of G_0. The groups G, K, and P_+ are *complex Lie groups*, that is, they have a complex structure in which the group operations are holomorphic, and the coset space G/KP_+ has a complex structure invariant under the action of G. Combining (9), the G_0-commutation of $I_2 \circ \psi_2^{-1} \circ \psi_0$, and the fact that $I_2 \circ \psi_2^{-1}$ is holomorphic, we conclude that $I_2 \circ \psi_2^{-1} \circ \psi_0$, and therefore ψ_0 and ψ, are holomorphic. Thus Theorem 7.1 is proved.

Corollary 7.13. *The complex structure* J_0 *is given by* $J_0 = \mathrm{ad}_{\mathfrak{g}_0}(H_0)$ *where* $H_0 \in \mathfrak{c}_0$ *is determined by*

$$\alpha(H_0) = -i \qquad \textit{for} \quad \alpha \in Q_+.$$

Moreover, the mapping $\psi : G_0/K_0 \to \mathfrak{p}_-$ *has differential* $d\psi : \mathfrak{p}_0 \to \mathfrak{p}_-$ *given by*

$$d\psi(X) = \tfrac{1}{2}(X - iJ_0X) \qquad (X \in \mathfrak{p}_0).$$

In fact, with the decomposition $X = X_- + X_+$ $(X_\pm \in \mathfrak{p}_\pm)$ we have $d\psi(X) = X_-$. By Theorem 4.5, $J_0 = \mathrm{ad}\, H_0$ for some $H_0 \in \mathfrak{c}_0$. But then the relation $d\psi(J_0X) = i\, d\psi(X)$ (that is, (9)) implies

$$[H_0, X_-] + [H_0, X_+] = [H_0, X] = J_0X \equiv iX_- \qquad (\mathrm{mod}\ \mathfrak{p}_+)$$

so $\alpha(H_0) = -i$ $(\alpha \in Q_+)$. Thus $J_0X = iX_- - iX_+$ so $X - iJ_0X = 2X_- = 2\, d\psi(X)$.

The complex structure on $M = G_0/K_0$ which is determined by the endomorphism $J_0 : \mathfrak{p}_0 \to \mathfrak{p}_0$ gives rise to a U-invariant complex structure on U/K_0 which at the tangent space $(U/K_0)_o = \mathfrak{p}_* = i\mathfrak{p}_0$ is the map $iX \to iJ_0X$ $(X \in \mathfrak{p}_0)$.

Proposition 7.14. *In the notation above, the mapping* $f : uK_0 \to uKP_+$ *is a holomorphic diffeomorphism of* U/K_0 *onto* G/KP_+. *Thus the compact Hermitian symmetric space* U/K_0 *contains the dual* G_0/K_0 *as an open submanifold.*

Proof. Let $u \in U \cap KP_+$. It is clear that $u^{-1}\theta(u) \in P_+$. Applying τ we find that $\theta(u) = u$. Since U is simply connected, it follows from Theorem 7.2, Chapter VII, that $u \in K_0$. Thus $U \cap KP_+ = K_0$ and consequently the mapping f is one-to-one. Since f is regular and $\dim U/K_0 = \dim G/KP_+$, the image $f(U/K_0)$ is an open submanifold of G/KP_+. Being compact this submanifold must coincide with G/KP_+. The complex structures on the tangent spaces $(U/K)_o$ and $(G/KP_+)_o = (G_0/K_0)_o$ correspond under $(df)_o$, in fact, if $X \in \mathfrak{p}_0$, then $df_o(iX) = i\, d\psi(X)$, so

$$df_o(iJ_0(X)) = i\, d\psi(J_0X) = -d\psi(X) = i(df_o(iX)).$$

Moreover, f commutes with the action of U, hence is almost complex, hence holomorphic (§1).

Corollary 7.15. *Each mapping* $\tau(g)$, $(g \in G_0)$ *of* G_0/K_0 *extends to a holomorphic diffeomorphism of* U/K_0.

Example. We can now illustrate the general concepts by means of the group $\boldsymbol{SL}(2, \boldsymbol{C})$. For this example we add a superscript $*$ to the general notation. Thus

$$G_0^* = \boldsymbol{SU}(1,1) = \left\{ \begin{pmatrix} a & b \\ \bar{b} & \bar{a} \end{pmatrix} : |a|^2 - |b|^2 = 1 \right\},$$

$$K_0^* = \boldsymbol{SO}(2) = \left\{ \begin{pmatrix} a & 0 \\ 0 & \bar{a} \end{pmatrix} : |a|^2 = 1 \right\}.$$

Here

$$G^* = \boldsymbol{SL}(2, \boldsymbol{C}), \qquad K^* = \boldsymbol{SO}(2, \boldsymbol{C}), \qquad U^* = \boldsymbol{SU}(2)$$

and

$$\mathfrak{p}^* = \left\{ \begin{pmatrix} 0 & b \\ c & 0 \end{pmatrix} : b, c \in \boldsymbol{C} \right\}.$$

The group G^* acts on the *Riemann sphere* $\boldsymbol{C} \cup \{\infty\}$ by the maps

$$z \to \frac{az + b}{cz + d}, \qquad \begin{pmatrix} a & b \\ c & d \end{pmatrix} \in G^*$$

and the isotropy subgroup at the origin O is

$$K^*P_+^* = \left\{ \begin{pmatrix} a & 0 \\ c & d \end{pmatrix} : \text{ad} = 1, c \in \boldsymbol{C} \right\}.$$

The imbedding $\varphi : G_0^*/K_0^* \to G^*/K^*P_+^*$ is the imbedding of the orbit $G_0^* \cdot O$ (which is the open unit disk) into the Riemann sphere. The mapping $f : U^*/K_0^* \to G^*/K^*P_+^*$ is the mapping

$$\begin{pmatrix} a & b \\ -\bar{b} & \bar{a} \end{pmatrix} K_0^* \to b\bar{a}^{-1}$$

and $P_-^* = \boldsymbol{C}$ is dense in $G^*/K^*P_+^*$.

Returning to the general case, let o and o_c denote the identity cosets in G_0/K_0 and G/KP_+, respectively, and let $\xi(Y) = \exp Y \cdot o_c$ $(Y \in \mathfrak{p}_-)$. Denoting by M^* the Hermitian symmetric space dual to M, we have the diagram,

$$\begin{array}{ccc} \mathfrak{p}_- & \xrightarrow{\xi} & G/KP_+ = U/K_0 = M^* \\ | & & | \\ D & \xrightarrow{\xi} & G_0 \cdot o_c = G_0/K_0 = M \end{array} \qquad (10)$$

and will now describe D and ξ more explicitly. Let Γ denote the set $\{\gamma_1, \ldots, \gamma_s\}$ of strongly orthogonal roots from Prop. 7.4.

Theorem 7.16. *In the diagram* (10) *the mapping*

$$\xi : Y \to \exp Y \cdot o_c$$

is a holomorphic diffeomorphism of $\mathfrak{p}_-$ *onto an open dense subset of* M^*.

The only statement not already proved is that $\xi(\mathfrak{p}_-)$ is dense in M^*. As in Chapter VII let $\mathfrak{a}_* = i\mathfrak{a}_0$, $A_* = \exp \mathfrak{a}_*$ so that since $U = K_0 A_* K_0$, we have $M^* = K_0 A_* \cdot o_c$. But by Lemma 7.11 and the fact that the group operations in G are holomorphic, we have for $t_\gamma \in \mathbf{R}$ and $\cos t_\gamma \neq 0$ ($\gamma \in \Gamma$),

$$a = \exp\Big(\sum_{\gamma\in\Gamma} it_\gamma(X_\gamma + X_{-\gamma})\Big) \quad \Rightarrow \quad a \cdot o_c = \exp\Big(\sum_{\gamma\in\Gamma} i \tan t_\gamma X_{-\gamma}\Big) \cdot o_c.$$

The mapping $d\psi : \mathfrak{p}_0 \to \mathfrak{p}_-$ commutes with the action of $\mathrm{Ad}_G(K_0)$ and maps $\mathfrak{a}_0$ onto the space

$$\mathfrak{a}_- = \sum_{\gamma\in\Gamma} \mathbf{R} X_{-\gamma}.$$

Using Prop. 6.2, we therefore deduce,

$$\mathfrak{p}_- = \bigcup_{k\in K_0} \mathrm{Ad}(k)\, \mathfrak{a}_- = \bigcup_{k\in K_0} \mathrm{Ad}(k)(i\mathfrak{a}_-).$$

Thus the image $\xi(\mathfrak{p}_-)$ fills up M^* except possibly for the set of points $ka \cdot o_c$ where $k \in K_0$ and a has some of the "coordinates" t_γ satisfying $\cos t_\gamma = 0$. This proves the theorem.

We conclude this section with two direct corollaries of Lemma 7.11 which give a somewhat better picture of the domain $D \subset \mathfrak{p}_-$.

Corollary 7.17. *The cube* $\square = \{\sum_{\gamma\in\Gamma} x_\gamma X_{-\gamma} : |x_\gamma| < 1\}$ *in* $\mathfrak{a}_-$ *generates the bounded domain* $D \subset \mathfrak{p}_-$ *in the sense that*

$$D = \mathrm{Ad}_G(K_0)(\square).$$

In fact $G_0 = K_0 A_0 K_0$ where $A_0 = \exp \mathfrak{a}_0$ and by Lemma 7.11

$$A_0 \cdot o_c = A_0 \cdot \xi(0) = \xi(\square)$$

so

$$\xi(D) = G_0 \cdot o_c = K_0 A_0 \cdot o_c = K_0 \cdot \xi(\square).$$

Hence

$$D = \mathrm{Ad}_G(K_0)(\square).$$

Finally, we show that the geodesics in D starting at 0 have very special limit points on the boundary of D, in fact these limit points constitute the $\mathrm{Ad}_G(K_0)$-orbits of the vertices of $\square$.

Corollary 7.18. *Let $k \in K_0$, $H \in \mathfrak{a}_0$, and*

$$\gamma(t) = \xi^{-1}(k \exp tH \cdot o_c) \qquad (t \in \boldsymbol{R}).$$

Then

$$\lim_{t\to+\infty} \gamma(t) = \mathrm{Ad}_G(k) \sum_{\gamma\in\Gamma} \epsilon_\gamma X_{-\gamma},$$

where $\epsilon_\gamma = 0$ *or* ± 1.

In fact suppose $H = \sum_{\gamma\in\Gamma} t_\gamma(X_\gamma + X_{-\gamma})$. Then

$$k \exp tH \cdot o_c = \xi\left(\mathrm{Ad}_G(k) \sum_{\gamma\in\Gamma} \tanh(tt_\gamma)\, X_{-\gamma}\right)$$

so the result is immediate.

Remark. The proof shows that all geodesics $\gamma(t) = \xi^{-1}(\exp tH \cdot o_c)$ with H in the "octant" $t_\gamma > 0$ $(\gamma \in \Gamma)$ converge to the same point $\sum_{\gamma\in\Gamma} X_{-\gamma}$.

EXERCISES AND FURTHER RESULTS†

A. Complex Structures

1. Let g be a Hermitian structure on an arbitrary complex manifold (M, J). Show that the tensor field ω given by

$$\omega(X, Y) = g(X, JY), \qquad X, Y \in \mathfrak{D}^1(M),$$

is a 2-form on M. Show that the following conditions are equivalent:

(i) g is Kählerian.

(ii) $\nabla_X(\omega) = 0$ for $X \in \mathfrak{D}^1(M)$.

(iii) $d\omega = 0$.

2. Let G be a Lie group with Lie algebra $\mathfrak{g}$. Suppose the Lie algebra $\mathfrak{g}$ has a complex structure. Show that G has a complex structure in which

† See also Exercises D, Chapter X.

it is a complex Lie group. (Hint: The complex strucutre on $\mathfrak{g}$ induces a left invariant almost complex structure on G. As a result of Theorem 1.7 Chapter II, exp is an almost complex mapping of $\mathfrak{g}$ into G.)

3. Show that the Riemannian structure corresponding to the kernel function for the unit disk $|z| < 1$ turns the disk into a Riemannian manifold of constant negative curvature.

B. Bounded Symmetric Domains

1. Let D be a bounded symmetric domain. Representing D as $H(D)/K$ ($H(D)$ as in §3, K compact), D acquires a natural metric in two different ways: Firstly from the kernel function for D and secondly from the Killing form B of the Lie algebra of $H(D)$. Show that these two metrics coincide. (In view of (2), §7, it suffices to prove $B(X, X) = 2\,\text{Trace}\,(T_X)$ for $X \in \mathfrak{e}$.)

2. With the notation as in §7 let $\gamma \in \Gamma$ and consider the three-dimensional simple algebra

$$\mathfrak{g}(\gamma) = \boldsymbol{C}H_\gamma + \mathfrak{g}^\gamma + \mathfrak{g}^{-\gamma} \subset \mathfrak{g}$$

and the real forms

$$\mathfrak{g}_0(\gamma) = \mathfrak{g}_0 \cap \mathfrak{g}(\gamma), \qquad \mathfrak{u}(\gamma) = \mathfrak{u} \cap \mathfrak{g}(\gamma).$$

Let $G(\gamma) \subset G$, $G_0(\gamma) \subset G_0$, and $U(\gamma) \subset U$ be the corresponding analytic subgroups. Prove that the isomorphism $\sigma : \mathfrak{sl}(2, \boldsymbol{C}) \to \mathfrak{g}(\gamma)$ which sends

$$\boldsymbol{C}\begin{pmatrix} 1 & 0 \\ 0 & -1 \end{pmatrix} \to \boldsymbol{C}H_\gamma, \qquad \boldsymbol{C}\begin{pmatrix} 0 & 1 \\ 0 & 0 \end{pmatrix} \to \mathfrak{g}^{-\gamma}, \qquad \boldsymbol{C}\begin{pmatrix} 0 & 0 \\ 1 & 0 \end{pmatrix} \to \mathfrak{g}^\gamma$$

induces a holomorphic diffeomorphism of:

(i) Riemann sphere onto $G(\gamma) \cdot o_c = U(\gamma) \cdot o_c$;
(ii) complex plane onto $\exp(\mathfrak{g}^{-\gamma}) \cdot o_c$;
(iii) unit disk onto $G_0(\gamma) \cdot o_c$.

3. With the notation of §7 consider the action of G_0 on G/KP_+. Prove that the subgroup $A_0 = \exp(\mathfrak{a}_0) \subset G_0$ leaves the subset $\xi(\mathfrak{a}_-)$ invariant (Korányi-Wolf [1]).

Hint: Use the matrix identity

$$\exp\begin{pmatrix} 0 & y \\ y & 0 \end{pmatrix} \exp\begin{pmatrix} 0 & 0 \\ x & 0 \end{pmatrix} = \exp\begin{pmatrix} 0 & 0 \\ x_1 & 0 \end{pmatrix} \exp\begin{pmatrix} s & 0 \\ 0 & -s \end{pmatrix} \exp\begin{pmatrix} 0 & y_1 \\ 0 & 0 \end{pmatrix}$$

with

$$e^s = x \sinh y + \cosh y, \qquad x_1 = e^{-s}(x \cosh y + \sinh y), \qquad y_1 = e^{-s} \sinh y,$$

and strong orthogonality (as in Lemma 7.11) to prove

$$a \cdot \xi(X) = \xi\Big(\sum_{\gamma\in\Gamma} \frac{x_\gamma \cosh y_\gamma + \sinh y_\gamma}{x_\gamma \sinh y_\gamma + \cosh y_\gamma} X_{-\gamma}\Big)$$

if

$$a = \exp\Big(\sum_{\gamma\in\Gamma} y_\gamma(X_\gamma + X_{-\gamma})\Big), \qquad X = \sum_{\gamma\in\Gamma} x_\gamma X_{-\gamma}.$$

4. Let $M = G_0/K_0$ be an irreducible bounded symmetric domain, $I(M)$ its group of isometries, and $H \subset I(M)$ the subgroup fixing the origin $o \in M$. Prove that if $h \in H$, then the complex structure J_o on $\mathfrak{p}_0$ satisfies

$$\mathrm{Ad}_{I(M)}(h) \circ J_o = \pm J_o \circ \mathrm{Ad}_{I(M)}(h).$$

Deduce that each isometry of M is either holomorphic or antiholomorphic.

C. Siegel's Generalized Upper Half-Plane

Let I denote the $n \times n$ unit matrix and let

$$J = \begin{pmatrix} 0 & I \\ -I & 0 \end{pmatrix}, \qquad z = \frac{1}{\sqrt{2}}\begin{pmatrix} I & I \\ -I & I \end{pmatrix}.$$

Let $\boldsymbol{Sp}(n, \boldsymbol{R})$ denote the group of all real $2n \times 2n$ matrices g satisfying

$${}^t g J g = J.$$

1. Show that the group $G = \boldsymbol{Sp}(n, \boldsymbol{R})$ is semisimple, and that the group $K = \boldsymbol{Sp}(n, \boldsymbol{R}) \cap \boldsymbol{SO}(2n)$ is a maximal compact subgroup.

2. Let $\mathfrak{g}_0 = \mathfrak{k}_0 + \mathfrak{p}_0$ be a Cartan decomposition of the Lie algebra $\mathfrak{g}_0$ of G, $\mathfrak{k}_0$ denoting the Lie algebra of K. Then $\mathrm{Ad}_G(J)$ restricted to $\mathfrak{p}_0$ is -1; also $z^2 = +J$ and z lies in the center of K.

3. The restriction of $\mathrm{Ad}_G(z)$ to $\mathfrak{p}_0$ gives rise to an invariant complex structure on the space

$$M = \boldsymbol{Sp}(n, \boldsymbol{R})/(\boldsymbol{SO}(2n) \cap \boldsymbol{Sp}(n, \boldsymbol{R}))$$

turning M into a Hermitian symmetric space of the noncompact type.

4. The mapping $p \to g\sigma(g^{-1})$ from Exercise A.5, Chapter VI, is a diffeomorphism of M onto the submanifold S of $\boldsymbol{Sp}(n, \boldsymbol{R})$ formed by the matrices in $\boldsymbol{Sp}(n, \boldsymbol{R})$ which are symmetric and strictly positive definite.

5. In the complex manifold of $n \times n$ complex symmetric matrices consider the open submanifold $\mathscr{S}_n$ of complex symmetric matrices $Z = X + iY$ where X and Y are real and Y is strictly positive definite. Each $g \in \boldsymbol{Sp}(n, \boldsymbol{R})$ gives rise to a holomorphic diffeomorphism

$$T_g : Z \to (AZ + B)(CZ + D)^{-1}$$

of $\mathscr{S}_n$, the matrix g being written

$$g = \begin{pmatrix} A & B \\ C & D \end{pmatrix}.$$

Prove that the mapping $gK \to T_g \cdot (iI)$ is a well-defined holomorphic diffeomorphism of M onto $\mathscr{S}_n$.

6. Prove that the mapping

$$Z \to (I + iZ)(I - iZ)^{-1}$$

is a holomorphic diffeomorphism of $\mathscr{S}_n$ onto the bounded domain in $\boldsymbol{C}^{\frac{1}{2}n(n+1)}$ consisting of all complex symmetric $n \times n$ matrices W for which $I - \overline{W}W$ is strictly positive definite (the generalized unit disk).

D. An Alternative Proof of Prop. 4.2

Let the assumptions be as in Prop. 4.2 but suppose in addition that the identity component K_0 of K leaves no $X \neq 0$ in M_o fixed. Show through the following steps† that $M = G/K$ is a Hermitian symmetric space.

1. Let $\mathfrak{g}_0 = \mathfrak{k}_0 + \mathfrak{p}_0$ be the decomposition of the Lie algebra of G into eigenspaces of $d\sigma$. Complexify $\mathfrak{g}_0, \mathfrak{k}_0, \mathfrak{p}_0$ to $\mathfrak{g}$, $\mathfrak{k}$, and $\mathfrak{p}$, respectively. Then $\mathfrak{g} = \mathfrak{k} + \mathfrak{p}$ and J_o extends to an endomorphism of $\mathfrak{p}$ of square -1. Let $\mathfrak{p} = \mathfrak{p}_+ + \mathfrak{p}_-$ be the decomposition of $\mathfrak{p}$ into the eigenspaces of J_o for the eigenvalues $+i$ and $-i$, respectively. As in Chapter IV, §5, show that there exists a simply connected Lie group G^c with Lie algebra $\mathfrak{g}$.

2. Let L denote the analytic subgroup of G^c with Lie algebra $\mathfrak{k} + \mathfrak{p}_-$. Then L is closed in G^c (consider the normalizer of $\mathfrak{k} + \mathfrak{p}_-$ in G^c).

3. The Lie groups L and G^c being complex (Exercise A.2 above), show that G^c/L has a complex structure invariant under the action of G^c.

4. The identity mapping $\mathfrak{g}_0 \to \mathfrak{g}$ induces a mapping $g \to \varphi(g)$ of a neighborhood of e in G into G^c. Let ψ denote the induced mapping

† This approach follows a suggestion of I. Singer, cf. Frölicher [1], §20.

$gK \to \varphi(g)L$ of a neighborhood of $\{K\}$ in G/K into G^c/L. Show that ψ is regular at $\{K\}$ and thus is a diffeomorphism of a neighborhood of $\{K\}$ in G/K onto a neighborhood of $\{L\}$ in G^c/L.

5. Show that

$$d\psi_K(J_oX) = i\, d\psi_K(X), \qquad X \in \mathfrak{p}_0,$$

$$\tau(g) \circ \psi = \psi \circ \tau(g)$$

for all g in a suitable neighborhood of e in G. Deduce that the mapping ψ is almost complex and therefore G/K has a complex structure corresponding to J.

NOTES

§1-§3. The torsion tensor S was introduced by Eckmann and Frölicher, see Frölicher [1]. The equivalence of "complex structure" and "integrable almost complex structure" was first proved by Newlander and Nirenberg [1]. The analytic case had been settled by several authors, see Libermann [1]. In [1] Kähler first studied the class of Hermitian structures named after him. Lemma 2.1 and Prop. 2.5 are proved in Bochner [1]. The kernel function was introduced by Bergman [1] and Bochner [3].

§5. The decomposition of a symmetric space into irreducible ones is due to É. Cartan [16]. Theorem 5.3 is also given there.

§4, §6, §7. Theorems 4.5, 4.6, and 6.1 are proved in Borel and Lichnerowicz [1]. The theory of bounded symmetric domains was developed by É. Cartan [19]. He proved Theorem 7.1 (i), namely, that a bounded symmetric domain is a symmetric space of the noncompact type. His proof uses the Liouville theorem that a bounded holomorphic function $f(z)$ in $|z| < \infty$ is constant; the proof differs from the one given here. The second part of Theorem 7.1 states that the bounded symmetric domains exhaust the class of Hermitian symmetric spaces of the noncompact type. This fact was verified in É. Cartan [19] by means of an explicit construction for the irreducible Hermitian symmetric spaces for which $A_0(M)$ is a classical group. This leaves out two exceptional Hermitian symmetric spaces for which Cartan stated the result without proof ([19], p. 151). The first *a priori* proof was given by Harish-Chandra [5], p. 591; this proof is reproduced here in the text. In [19] Cartan raised the question whether a bounded homogeneous domain is necessarily symmetric. He answered it affirmatively for dimensions 1, 2 and 3; for dimension 3 he did not publish the proof, considering its length, for the time being, out of proportion to the interest of the result. This situation has now changed since Pyatetzki-Shapiro [1, 2] answered the question negatively for dimensions 4 and 5. The imbedding in Proposition 7.14 was shown by Borel [2, 3]. Theorem 7.16 and Cor. 7.17 are from Korányi and Wolf [1] pp. 268, 269, 286, and Cor. 7.18 was mentioned to the author by Korányi. For further studies of Hermitian symmetric spaces see, e.g., Chow [1], Wolf [3], Wolf and Korányi [1], Koecher [3], and a generalization by Shapiro [1].

CHAPTER IX

STRUCTURE OF SEMISIMPLE LIE GROUPS

This chapter, which deals with noncompact semisimple Lie groups, is a continuation of Chapter VI, but now we take the theory of the Weyl group and of the restricted roots from Chapter VII into account. This leads to a sharpening of the Cartan decomposition from Chapter V, to a conjugacy of all Iwasawa decompositions, and to the Bruhat decomposition. In §7 we relate the Iwasawa decomposition to the Jordan decomposition of a matrix. In §2 and §3 we imbed a family of rank-one spaces into the given space G/K, and in a rank-one space we imbed a complex two-dimensional ball. These imbeddings are very useful in analysis on G/K. Sections 4 and 6 deal with some simple facts concerning general Cartan subalgebras and multiplicities, while §5 establishes the existence and uniqueness of the normal form of a semisimple Lie algebra over $\boldsymbol{C}$. We also give a criterion for θ to be an inner automorphism of $\mathfrak{g}$ and for the geodesic symmetry to lie in the identity component of the isometry group.

§ 1. Cartan, Iwasawa, and Bruhat Decompositions

Since we are now primarily concerned with noncompact semisimple Lie groups, we will simplify a bit the notational conventions made in §5, Chapter V.

Let $\mathfrak{g}$ be an arbitrary semisimple Lie algebra over $\boldsymbol{R}$, B its Killing form, θ any Cartan involution of $\mathfrak{g}$, and $\mathfrak{g} = \mathfrak{k} + \mathfrak{p}$ the corresponding Cartan decomposition of $\mathfrak{g}$ ($\mathfrak{k}$ the fixed point set of θ). Let $\mathfrak{a} \subset \mathfrak{p}$ be any maximal abelian subspace (all such subspaces have the same dimension), and let $\mathfrak{m}$ denote the centralizer of $\mathfrak{a}$ in $\mathfrak{k}$. For each λ in the dual space $\mathfrak{a}^*$ of $\mathfrak{a}$ let

$$\mathfrak{g}_\lambda = \{X \in \mathfrak{g} : [H, X] = \lambda(H)X \text{ for } H \in \mathfrak{a}\}.$$

Then λ is called a *root* of $(\mathfrak{g}, \mathfrak{a})$ (or a *restricted root*) if $\lambda \neq 0$ and $\mathfrak{g}_\lambda \neq 0$. The simultaneous diagonalization of the $\mathrm{ad}_{\mathfrak{g}}(\mathfrak{a})$ gives the decomposition

$$\mathfrak{g} = \mathfrak{g}_0 + \sum_{\lambda \in \Sigma} \mathfrak{g}_\lambda, \qquad \mathfrak{g}_0 = \mathfrak{a} + \mathfrak{m}, \tag{1}$$

where Σ is the set of restricted roots. The spaces $\mathfrak{g}_\lambda$ are called *root subspaces*. A point $H \in \mathfrak{a}$ is called *regular* if $\lambda(H) \neq 0$ for all $\lambda \in \Sigma$, otherwise *singular*. The subset $\mathfrak{a}' \subset \mathfrak{a}$ of regular elements consists of the complement of finitely many hyperplanes, and its components are called Weyl chambers. Fix a Weyl chamber $\mathfrak{a}^+$ and call a root λ *positive* ($\lambda > 0$) if λ has positive values on $\mathfrak{a}^+$. A root $\lambda > 0$ is called *simple* if it is not the sum of two positive roots. Then $\mathfrak{a}^+$ is given by

$$\mathfrak{a}^+ = \{H \in \mathfrak{a} : \alpha_1(H) > 0, ..., \alpha_l(H) > 0\}, \tag{2}$$

where $\{\alpha_1, ..., \alpha_l\}$ is the set of simple roots. Let the dual space $\mathfrak{a}^*$ be ordered lexicographically with respect to this basis, $\{\alpha_1, ..., \alpha_l\}$.

Let G be any connected Lie group with Lie algebra $\mathfrak{g}$ and $K \subset G$ any Lie subgroup with Lie algebra $\mathfrak{k}$. We know from Chapter VI that K is connected and closed and that $\mathrm{Ad}_G(K)$ is compact. Let M and M', respectively, denote the centralizer and normalizer of $\mathfrak{a}$ in K. The factor group M'/M is the Weyl group $W = W(\mathfrak{g}, \theta)$ (which in Chapter VII was denoted by $W(\mathfrak{u}, \theta)$). Let $A = \exp \mathfrak{a}$, $A^+ = \exp \mathfrak{a}^+$, and $\overline{A^+}$ the closure of A^+ in G. The dimension $\dim \mathfrak{a}$ is called the *real rank* of $\mathfrak{g}$ and of G. We have now the following refinement of Theorem 6.7, Chapter V, which we refer to as the *Cartan decomposition* of G.

Theorem 1.1. *Let G be any connected semisimple Lie group with Lie algebra $\mathfrak{g}$. Then*

$$G = K\overline{A^+}K,$$

that is, each $g \in G$ can be written $g = k_1ak_2$ where $k_1, k_2 \in K$ and $a \in \overline{A^+}$. Moreover, $a = a^+(g)$ is unique.

Because of Theorem 6.7, Chapter V, and Theorem 2.12, Chapter VII, only the uniqueness remains to be proved. But suppose $g = k \exp Hk' = \exp H'$ with H and H' in the closed Weyl chamber $\overline{\mathfrak{a}^+}$. The automorphism θ "extends" to an automorphism of G, also denoted θ, and we derive $k \exp(-H)k' = \exp(-H')$. Eliminating k', we get $\exp(\mathrm{Ad}(k)(-2H)) = \exp(-2H')$; and since $\mathrm{Ad}(k)H \in \mathfrak{p}$ and exp is one-to-one on $\mathfrak{p}$, we find $\mathrm{Ad}(k)H = H'$. Because of Prop. 2.2 Chapter VII, H and H' are conjugate under the Weyl group $W(\mathfrak{g}, \theta)$, so $H = H'$ by Theorem 2.22, Chapter VII. This proves the theorem.

We write $a^+(g) = \exp A^+(g)$ where $A^+(g) \in \overline{\mathfrak{a}^+}$.

As a consequence we get a kind of a *polar coordinate decomposition* of the symmetric space $X = G/K$. Let $A' = \exp \mathfrak{a}'$, $G' = KA'K$ and $X' = G' \cdot o$ where o is the identity coset.

Corollary 1.2. *We have $X = K\overline{A^+} \cdot o$ and*

$$X' = (K/M) \times (A^+ \cdot o)$$

in the sense that $(kM, a) \to ka \cdot o$ is a diffeomorphism of $(K/M) \times A^+$ onto X'.

Since $\mathfrak{a}' = W(\mathfrak{g}, \theta) \cdot \mathfrak{a}^+$, the relations $X = K\overline{A^+} \cdot o$ and $X' = KA^+ \cdot o$ are obvious. If $k_1 \exp H_1 \cdot o = k_2 \exp H_2 \cdot o$ $(H_1, \ H_2 \in \mathfrak{a}^+)$, then $k = k_2^{-1}k_1$ satisfies $k \exp H_1 k' = \exp H_2$ for $k' \in K$. Applying θ, we get as before $\mathrm{Ad}(k)H_1 = H_2$, so considering the centralizers of H_1 and H_2 in $\mathfrak{p}$ we deduce $k \in M'$; by Theorem 2.12, Chapter VII, $k \in M$ as desired.

Next we restate the Iwasawa decomposition in our present notation. Let Σ^+ denote the set of positive elements in Σ, $\mathfrak{n}$ the subalgebra

$$\mathfrak{n} = \sum_{\lambda \in \Sigma^+} \mathfrak{g}_\lambda$$

of $\mathfrak{g}$ and N the corresponding analytic subgroup of G.

Theorem 1.3. *Let G be any connected semisimple Lie group with Lie algebra $\mathfrak{g}$. Then*

$$\mathfrak{g} = \mathfrak{k} + \mathfrak{a} + \mathfrak{n} \qquad \textit{(direct vector space sum)}$$

$$G = KAN,$$

that is, the mapping $(k, a, n) \to kan$ is a diffeomorphism of $K \times A \times N$ onto G.

If $g \in G$ we write the decomposition

$$g = k(g) \exp H(g)\, n(g), \qquad k(g) \in K, \quad H(g) \in \mathfrak{a}, \quad n(g) \in N.$$

Since $\mathrm{Ad}_G(m)$ $(m \in M)$ leaves $\mathfrak{a}$ pointwise fixed, it maps each root subspace $\mathfrak{g}_\lambda$ into itself. Hence $B = MAN$ is a group, and in fact a closed subgroup of G. We shall now give the explicit decomposition of G into the double cosets of B, the Bruhat decomposition of G. For s in the Weyl group W fix a representative $m_s \in M'$.

Theorem 1.4. *Let G be any noncompact semisimple Lie group. Then*

$$G = \bigcup_{s \in W} Bm_sB \qquad \textit{(disjoint union).}$$

We begin with a lemma about N.

Lemma 1.5. *Let $H \in \mathfrak{a}'$. Then the mapping $\varphi : n \to Ad(n)H - H$ is a diffeomorphism of N onto $\mathfrak{n}$.*

Proof. Let $X \in \mathfrak{n}$. Then $\mathrm{Ad}(\exp X)H - H = e^{\mathrm{ad}\, X}H - H \in \mathfrak{n}$ since $[\mathfrak{a}, \mathfrak{n}] \subset \mathfrak{n}$. Since exp maps $\mathfrak{n}$ onto N (Cor. 4.4, Chapter VI), it is clear that $\varphi(N) \subset \mathfrak{n}$. Next we prove that φ is one-to-one. Suppose $n_1, n_2 \in N$ such that $\mathrm{Ad}(n_1)H - H = \mathrm{Ad}(n_2)H - H$. Then if $X \in \mathfrak{n}$ is chosen such that $\exp X = n_2^{-1}n_1$, we have $e^{\operatorname{ad} X}H = H$. But then Lemma 4.5, Chapter VI, implies $\operatorname{ad} X(H) = 0$ or equivalently, $\operatorname{ad} H(X) = 0$, so since H is regular, $X = 0$ and $n_2 = n_1$.

Next we prove $\varphi(N) = \mathfrak{n}$. If this were not the case, let $Z \in \mathfrak{n} - \varphi(N)$. Then $Z = \sum_{\lambda \in \Sigma^+} c_\lambda Z_\lambda$ ($c_\lambda \in \boldsymbol{R}$, $Z_\lambda \neq 0$ in $\mathfrak{g}_\lambda$). Let β be the lowest root in Σ^+ for which $c_\beta \neq 0$. We can assume Z chosen in $\mathfrak{n} - \varphi(N)$ such that this β is as high as possible. Since $\operatorname{ad} H$ is nonsingular on $\mathfrak{n}$, there exists an element $Z_1 \in \mathfrak{n}$ such that $[H, Z_1] = Z$. Putting $n_1 = \exp Z_1$, we have

$$\operatorname{Ad}(n_1)(H + Z) - H \equiv [Z_1, H] + Z = 0 \qquad \Big(\operatorname{mod} \sum_{\lambda > \beta} \mathfrak{g}_\lambda\Big)$$

so by the choice of Z there exists an element $n' \in N$ such that

$$\operatorname{Ad}(n')H - H = \operatorname{Ad}(n_1)(H + Z) - H.$$

Then $\operatorname{Ad}(n_1^{-1}n')H - H = Z$. This contradiction shows that $\varphi(N) = \mathfrak{n}$.

Finally, to show that φ is regular we compute its differential at an arbitrary point $n \in N$. Each tangent vector in N_n has the form $dL_n(X)$ ($X \in \mathfrak{n}$). Since

$$\begin{aligned}\varphi(n \exp tX) &= \operatorname{Ad}(n)\, e^{\operatorname{ad} tX}H - H \\ &= \operatorname{Ad}(n)H - H + \operatorname{Ad}(n)\Big(t[X, H] + \frac{t^2}{2}[X, [X, H]] + \dots\Big),\end{aligned}$$

it follows that

$$d\varphi_n(dL_n(X)) = -\operatorname{Ad}(n) \operatorname{ad} H(X).$$

Thus φ is regular and the lemma is proved.

Now let $\mathfrak{b}$ denote the Lie algebra of B which by (1) is given by $\mathfrak{b} = \sum_{\lambda \geqslant 0} \mathfrak{g}_\lambda$. Let $\mathfrak{g}^C$ denote the complexification of $\mathfrak{g}$ and if $\mathfrak{c} \subset \mathfrak{g}$ is any subalgebra let $\mathfrak{c}^C$ denote the complex subalgebra of $\mathfrak{g}^C$, generated by $\mathfrak{c}$.

Lemma 1.6.

(i) $\mathfrak{n}$ *is the set of* $Z \in \mathfrak{b}$ *such that* $\operatorname{ad}_{\mathfrak{g}}(Z)$ *is nilpotent.*

(ii) $\mathfrak{a} + \mathfrak{n}$ *is the set of elements* $Z \in \mathfrak{b}$ *such that* $\operatorname{ad}_{\mathfrak{g}^C}(Z)$ *has all its eigenvalues real.*

Proof. Let $Z \in \mathfrak{b}$ be written $Z = T + H + X$, $T \in \mathfrak{m}$, $H \in \mathfrak{a}$, $X \in \mathfrak{n}$. Extend $\boldsymbol{R}T$ to a maximal abelian subalgebra $\mathfrak{t}$ of $\mathfrak{m}$. Then $\mathfrak{h} = \mathfrak{t} + \mathfrak{a}$ is a maximal abelian subalgebra of $\mathfrak{g}$ and $\mathfrak{h}^C$ is a Cartan subalgebra of $\mathfrak{g}^C$ (Lemma 3.2, Chapter VI). Let Δ denote the corresponding system of roots, let $\mathfrak{h}_R \subset \mathfrak{h}^C$ be the subset where all $\alpha \in \Delta$ are real and consider an ordering of the dual of $\mathfrak{h}_R$ compatible with our ordering of $\mathfrak{a}^*$. Then we know from §3, Chapter VI, that $X \in \sum_{\alpha>0} (\mathfrak{g}^C)^\alpha$. Expressing $\mathrm{ad}_{\mathfrak{g}^C}(Z)$ in matrix form corresponding to the decomposition

$$\mathfrak{g}^C = \mathfrak{h}^C + \sum_{\alpha \in \Delta} (\mathfrak{g}^C)^\alpha, \tag{3}$$

we obtain a triangular matrix with diagonal elements $\alpha(T + H)$ $(\alpha \in \Delta)$ and 0. Since $\alpha(H)$ is real and $\alpha(T)$ purely imaginary, the lemma follows.

Lemma 1.7. *For each $x \in G$ put $\mathfrak{b}_x = \mathfrak{b} \cap \mathrm{Ad}(x)\mathfrak{b}$. Then*

$$\mathfrak{b} = \mathfrak{b}_x + \mathfrak{n} \qquad (x \in G).$$

Proof. The inclusion $\supset$ being obvious, it suffices to prove

$$\dim(\mathfrak{b}_x + \mathfrak{n}) = \dim \mathfrak{b}.$$

By Theorem 1.3 we may assume $x = k \in K$. Let $\perp$ denote orthogonal complementation in $\mathfrak{g}$ with respect to the positive definite form $B_\theta(X, Y) = -B(X, \theta Y)$. Then since $\theta\mathfrak{g}_\lambda = \mathfrak{g}_{-\lambda}$, we have $\mathfrak{b}^\perp = \theta\mathfrak{n}$ and

$$(\mathfrak{b} + \mathrm{Ad}(k)\mathfrak{b})^\perp = \theta\mathfrak{n} \cap \mathrm{Ad}(k)\, \theta\mathfrak{n} = \theta(\mathfrak{n} \cap \mathrm{Ad}(k)\mathfrak{n}). \tag{4}$$

But using Lemma 1.6 (i) we have

$$\mathfrak{n} \cap \mathrm{Ad}(k)\mathfrak{n} = \mathfrak{n} \cap \mathfrak{b}_k,$$

so using (4)

$$\begin{aligned} 2 \dim \mathfrak{b} - \dim \mathfrak{b}_k &= \dim(\mathfrak{b} + \mathrm{Ad}(k)\mathfrak{b}) \\ &= \dim \mathfrak{g} - \dim(\mathfrak{n} \cap \mathfrak{b}_k) \\ &= \dim \mathfrak{g} + \dim(\mathfrak{b}_k + \mathfrak{n}) - \dim \mathfrak{n} - \dim \mathfrak{b}_k. \end{aligned}$$

This proves $\dim(\mathfrak{b}_k + \mathfrak{n}) = \dim \mathfrak{b}$ and the lemma.

We can now prove Theorem 1.4. Let $x \in G$. Pick $H \in \mathfrak{a}'$. By Lemma 1.7, there exists an $X \in \mathfrak{n}$ such that $H + X \in \mathfrak{b}_x$ and therefore by Lemma 1.5 an $n_1 \in N$ such that $\mathrm{Ad}(n_1)H \in \mathfrak{b}_x$. Thus the element $Z = \mathrm{Ad}(x^{-1}n_1)H \in \mathfrak{b}$. The eigenvalues of ad Z being the same as those of ad H, we have by Lemma 1.6, $Z = H' + X'$ $(H' \in \mathfrak{a},\ X' \in \mathfrak{n})$. But ad H and ad H' have the same eigenvalues, so $H' \in \mathfrak{a}'$. Hence by Lemma 1.5 there exists an $n_2 \in N$ such that $H' = \mathrm{Ad}(n_2)Z = \mathrm{Ad}(n_2x^{-1}n_1)H$. But $\mathfrak{m} + \mathfrak{a}$ is the centralizer of H and of H' in $\mathfrak{g}$, so we conclude $\mathrm{Ad}(n_2x^{-1}n_1)(\mathfrak{m} + \mathfrak{a}) =$

$\mathfrak{m} + \mathfrak{a}$. But $\mathfrak{a}$ is the set of elements $A \in \mathfrak{m} + \mathfrak{a}$ such that $\mathrm{ad}_{\mathfrak{g}^c}(A)$ has all eigenvalues real. Thus

$$\mathrm{Ad}(n_2 x^{-1} n_1)\mathfrak{a} = \mathfrak{a},$$

so $n_2 x^{-1} n_1 \in M'$. This proves the theorem except for disjointness. So suppose $s \neq t$ in W such that

$$m_s b_2 = b_1 m_t \qquad (b_1, b_2 \in B). \tag{5}$$

Select $H \in \mathrm{Ad}(m_t^{-1})\mathfrak{a}^+$ and apply both sides of (5) to it. Then if $H_t = \mathrm{Ad}(m_t)H$,

$$\mathrm{Ad}(m_s)\,\mathrm{Ad}(b_2)H = \mathrm{Ad}(b_1)\,H_t \tag{6}$$

and $\mathrm{Ad}(b_2)H = H + X_2$, $\mathrm{Ad}(b_1)H_t = H_t + X_1$ where $X_1, X_2 \in \mathfrak{n}$. Since $\mathrm{Ad}(m_s)$ permutes the spaces $\mathfrak{g}_\lambda$ $(\lambda \in \Sigma)$, we conclude from (6), $\mathrm{Ad}(m_s)H = H_t$. Thus $\mathrm{Ad}(m_s m_t^{-1})H_t = H_t$, so $m_s m_t^{-1} \in M$ and $s = t$. This contradiction concludes the proof.

For a better understanding of the decomposition in Theorem 1.4 we write it in the form

$$G = \bigcup_{s \in W} MA(NN^s)\, m_s,$$

where $N^s = m_s N m_s^{-1}$; here we used the fact that m_s normalizes MA and that MA normalizes N. The Lie algebra $\mathfrak{n}^s$ of N^s is given by

$$\mathrm{Ad}(m_s) \sum_{\alpha > 0} \mathfrak{g}_\alpha = \sum_{\alpha^{s^{-1}} > 0} \mathfrak{g}_\alpha, \qquad \text{where} \quad \alpha^{s^{-1}}(H) = \alpha(sH).$$

Thus the group $N \cap N^s$ has Lie algebra

$$\sum_{\alpha > 0,\, \alpha^{s^{-1}} > 0} \mathfrak{g}^\alpha$$

which is 0 exactly if s is the Weyl group element s^* which maps $\mathfrak{a}^+$ into $-\mathfrak{a}^+$. Then $N^{s^*} = \theta N$ which we also denote by $\bar{N}$. Consider now the mapping $\pi : G \to G/MAN$ given by $\pi(g) = g \cdot o$, where o is the identity coset. The tangent space $(G/MAN)_o = d\pi(\mathfrak{g}) = \mathfrak{g}/(\mathfrak{m} + \mathfrak{a} + \mathfrak{n})$, so the tangent space $(\bar{N} \cdot o)_o$ fills up all of $d\pi(\mathfrak{g})$, in other words the orbit $\bar{N} \cdot o$ is an open submanifold of G/MAN. The inverse image $\pi^{-1}(\bar{N} \cdot o)$ equals $\bar{N}MAN$ which is therefore an open submanifold of G. Thus $Bm_{s^*}B$ is an open submanifold of G. If $s \neq s^*$, then $\mathfrak{n} \cap \mathfrak{n}^s \neq \{0\}$ so the orbit $N^s \cdot o$ is a submanifold of G/MAN of lower dimension; since π is a submersion, we deduce from Theorem 15.5, Chapter I, that the inverse image $\pi^{-1}(N^s \cdot o)$ which equals N^sMAN is also a submanifold of G. Summarizing, we have the following consequence.

Corollary 1.8. *In the decomposition*

$$G = \bigcup_{s \in W} Bm_sB = \bigcup_{s \in W} MANN^sm_s$$

the term $Bm_{s*}B$ is an open submanifold of G, and the other terms are submanifolds of lower dimension.

The mapping $kM \to kMAN$ is a bijection of K/M onto G/MAN which is regular at the origin, hence everywhere, so is a diffeomorphism. As a consequence of Exercise B.2, Chapter VI, the components of an element $g \in \bar{N}MAN$ are uniquely determined by g. We write

$$g = \bar{n}(g)\, m(g) \exp B(g)\, n_B(g)$$

where $\bar{n}(g) \in \bar{N}$, $m(g) \in M$, $B(g) \in \mathfrak{a}$, and $n_B(g) \in N$.

Corollary 1.9. *The mapping $\bar{n} \to k(\bar{n})M$ is a diffeomorphism of $\bar{N}$ onto an open submanifold of K/M whose complement consists of finitely many disjoint manifolds of lower dimension.*

We conclude this section with two useful bracket relations. For $\lambda \in \mathfrak{a}^*$ let $A_\lambda \in \mathfrak{a}$ be determined by $B(H, A_\lambda) = \lambda(H)$ $(H \in \mathfrak{a})$. If $\alpha \in \Sigma$ and $X_\alpha \in \mathfrak{g}_\alpha$, $X_{-\alpha} \in \mathfrak{g}_{-\alpha}$, then $[X_\alpha, \theta X_\alpha] \in \mathfrak{p} \cap \mathfrak{g}_0 = \mathfrak{a}$, $[X_\alpha, X_{-\alpha}] \in \mathfrak{a} + \mathfrak{m}$; so we deduce

$$[X_\alpha, \theta X_\alpha] = B(X_\alpha, \theta X_\alpha)\, A_\alpha, \tag{7}$$

$$[X_\alpha, X_{-\alpha}] - B(X_\alpha, X_{-\alpha})\, A_\alpha \in \mathfrak{m}. \tag{8}$$

§ 2. The Rank-One Reduction

In this section we construct for the symmetric space G/K a family of rank-one spaces G^α/K^α totally geodesic in G/K. Let $\mathfrak{g}$ be a semisimple Lie algebra over $\mathbf{R}$, and θ, $\mathfrak{k}$, $\mathfrak{p}$, $\mathfrak{a}$, $\mathfrak{n}$, $\mathfrak{m}$, Σ, Σ^+ as in §1. Recalling Cor. 2.17, Chapter VII, a root $\alpha \in \Sigma$ is called *indivisible* if $c\alpha \in \Sigma \Rightarrow c = \pm 1, \pm 2$. Let Σ_0 denote the set of indivisible roots and put $\Sigma_0^+ = \Sigma_0 \cap \Sigma^+$.

Proposition 2.1. *Let $\alpha \in \Sigma_0^+$ and let $\mathfrak{g}^\alpha$ denote the subalgebra of $\mathfrak{g}$ generated by $\mathfrak{g}_\alpha$ and $\mathfrak{g}_{-\alpha}$. Then $\mathfrak{g}^\alpha$ is semisimple, has a Cartan decomposition*

$$\mathfrak{g}^\alpha = \mathfrak{k}^\alpha + \mathfrak{p}^\alpha, \qquad \textit{where} \quad \mathfrak{k}^\alpha = \mathfrak{k} \cap \mathfrak{g}^\alpha, \quad \mathfrak{p}^\alpha = \mathfrak{p} \cap \mathfrak{g}^\alpha, \tag{1}$$

and $\mathbf{R}A_\alpha$ is a maximal abelian subspace $\mathfrak{a}^\alpha$ of $\mathfrak{p}^\alpha$ (so $\mathfrak{g}^\alpha$ has real rank one).

We start with a general lemma about Cartan decompositions (cf. Exercise A.8 (ii), Chapter VI).

Lemma 2.2. *Let* $\mathfrak{s} \subset \mathfrak{g}$ *be a semisimple subalgebra invariant under the Cartan involution* θ *of* $\mathfrak{g}$. *Then the restriction* $\theta \mid \mathfrak{s}$ *is a Cartan involution of* $\mathfrak{s}$.

Proof. Consider the adjoint group $\text{Int}(\mathfrak{g})$ with Lie algebra $\text{ad}_{\mathfrak{g}}(\mathfrak{g})$ and the analytic subgroups, K^*, $S \subset \text{Int}(\mathfrak{g})$ with Lie algebra $\text{ad}_{\mathfrak{g}}(\mathfrak{k})$ and $\text{ad}_{\mathfrak{g}}(\mathfrak{s})$, respectively. Then K^* is compact, S is a closed subgroup (Exercise B.7, Chapter III), $K^* \cap S$ is compact and so is the group of restrictions $(K^* \cap S) \mid \mathfrak{s}$. Thus $\mathfrak{k} \cap \mathfrak{s}$ is a compactly imbedded subalgebra of $\mathfrak{s}$. It is also maximal with this property because otherwise there would exist $X \neq 0$ in $\mathfrak{p} \cap \mathfrak{s}$ such that the powers $e^{n\, \text{ad}_{\mathfrak{s}}(X)}$ lie in a compact matrix group. This contradicts the fact that $\text{ad}_{\mathfrak{s}}(X)$ is symmetric with respect to the positive definite quadratic form $B_\theta \mid \mathfrak{s}$ and thus has real eigenvalues, not all 0. Since $\mathfrak{k} \cap \mathfrak{s}$ and $\mathfrak{p} \cap \mathfrak{s}$ are orthogonal with respect to the Killing form of $\mathfrak{s}$, this proves the lemma.

Turning now to the proof of Prop. 2.1, let $\mathfrak{g}^C$ denote the complexification of $\mathfrak{g}$ and extend $\mathfrak{a}$ to a Cartan subalgebra $\mathfrak{h}^C$ of $\mathfrak{g}^C$ as in Chapter VI. Then

$$\mathfrak{g}_\alpha = \Big(\sum_{\bar{\beta}=\alpha} (\mathfrak{g}^C)^\beta\Big) \cap \mathfrak{g}, \tag{2}$$

where β is a root of $(\mathfrak{g}^C, \mathfrak{h}^C)$ and $\bar{\beta}$ denotes the restriction $\beta \mid \mathfrak{a}$. The subalgebra $\tilde{\mathfrak{g}}$ of $\mathfrak{g}^C$ generated by $\sum_{\bar{\beta}=\pm\alpha} (\mathfrak{g}^C)^\beta$ is semisimple (Theorem 4.3 (iv), Chapter III, and Exercise B.1, Chapter III) and so is its real form $\mathfrak{g}^\alpha = \tilde{\mathfrak{g}} \cap \mathfrak{g}$. Hence by Lemma 2.2 decomposition (1) is a Cartan decomposition.

Next we note that $[\mathfrak{g}_\alpha, \mathfrak{g}_\alpha] = \mathfrak{g}_{2\alpha}$. In fact, if not, let $X \neq 0$ in $\mathfrak{g}_{2\alpha}$ satisfy

$$B(\theta X, [\mathfrak{g}_\alpha, \mathfrak{g}_\alpha]) = 0.$$

Thus the element $X_{-2\alpha} = \theta X \in \mathfrak{g}_{-2\alpha}$ satisfies

$$B([X_{-2\alpha}, \mathfrak{g}_\alpha], \theta\mathfrak{g}_{-\alpha}) = 0,$$

so $[X_{-2\alpha}, \mathfrak{g}_\alpha] = 0$. Combining this with $[X_{-2\alpha}, \mathfrak{g}_{-\alpha}] = 0$, we get by the Jacobi identity for all $X_\alpha \in \mathfrak{g}_\alpha$

$$[[X_\alpha, \theta X_\alpha], X_{-2\alpha}] = 0$$

and now (7), §1 gives a contradiction. Using the Jacobi identity we therefore have $[\mathfrak{g}_{2\alpha}, \mathfrak{g}_{-2\alpha}] \subset [\mathfrak{g}_\alpha, \mathfrak{g}_{-\alpha}]$ so, putting

$$\mathfrak{m}^\alpha = [\mathfrak{g}_\alpha, \mathfrak{g}_{-\alpha}] \cap \mathfrak{m}, \qquad \mathfrak{g}_0^\alpha = \boldsymbol{R}A_\alpha + \mathfrak{m}^\alpha$$

we conclude from (8), §1 the decomposition

$$\mathfrak{g}^{\alpha} = \mathfrak{g}_{-2\alpha} + \mathfrak{g}_{-\alpha} + \mathfrak{g}_0^{\alpha} + \mathfrak{g}_{\alpha} + \mathfrak{g}_{2\alpha}. \tag{3}$$

This shows $\boldsymbol{R}A_{\alpha}$ maximal abelian in $\mathfrak{p}^{\alpha}$, so Prop. 2.1 is proved.

We note that (3) is the root space decomposition (1), §1 for the Lie algebra $\mathfrak{g}^{\alpha}$. The half-line $\mathfrak{a}_+^{\alpha} = \{tA_{\alpha} : t > 0\}$ is a Weyl chamber in $\mathfrak{a}^{\alpha}$, and the corresponding algebra $\mathfrak{n}^{\alpha}$ equals $\mathfrak{g}_{\alpha} + \mathfrak{g}_{2\alpha}$. Let G^{α}, K^{α}, A^{α}, and N^{α} denote the analytic subgroups of G with Lie algebras $\mathfrak{g}^{\alpha}$, $\mathfrak{k}^{\alpha}$, $\mathfrak{a}^{\alpha}$, and $\mathfrak{n}^{\alpha}$, respectively, and let M^{α} denote the centralizer of A^{α} in K^{α}. The following lemma shows that the three decomposition of §1 can be made compatible for G^{α} and G.

Lemma 2.3. *With the notation above,*

$$K^{\alpha} = G^{\alpha} \cap K, \qquad A^{\alpha} = G^{\alpha} \cap A,$$

$$N^{\alpha} = G^{\alpha} \cap N, \qquad M^{\alpha} = G^{\alpha} \cap M.$$

Proof. The first relation follows from Theorem 1.1, Chapter VI, and the two next relations follow from $G^{\alpha} = K^{\alpha}A^{\alpha}N^{\alpha}$ and $A^{\alpha} \subset G^{\alpha} \cap A$, $N^{\alpha} \subset G^{\alpha} \cap N$. Finally, $G^{\alpha} \cap M \subset M^{\alpha}$ is obvious; also $M^{\alpha} \subset G^{\alpha} \cap M$ follows from the fact that if $m \in M^{\alpha}$, then not only do we have $m \exp tA_{\alpha} m^{-1} = \exp tA_{\alpha}$ $(t \in \boldsymbol{R})$, but in addition, if $H \in \mathfrak{a}$, $\alpha(H) = 0$, then by (3), G^{α} commutes elementwise with $\exp H$. Thus $M^{\alpha} \subset M$.

Corollary 2.4. *The mapping $gK^{\alpha} \to gK$ imbeds G^{α}/K^{α} in G/K as a totally geodesic submanifold.*

This is immediate from Lemma 2.3 and Theorem 7.2, Chapter IV.

§ 3. The *SU*(2, 1) Reduction

The three decompositions in §1 are basic tools in analysis on the group G. We shall now prove a result which for G of real rank one reduces their explicit computation to matrix calculations in the group $SU(2, 1)$.

Theorem 3.1. *Let $\mathfrak{g}$ be a semisimple Lie algebra of real rank one, θ a Cartan involution of $\mathfrak{g}$, and $\mathfrak{g} = \mathfrak{k} + \mathfrak{p}$ the corresponding Cartan decomposition. Let $\mathfrak{a} \subset \mathfrak{p}$ be a maximal abelian subspace and assume α and 2α are roots of $(\mathfrak{g}, \mathfrak{a})$. Select $X_{\alpha} \in \mathfrak{g}_{\alpha}$, $X_{2\alpha} \in \mathfrak{g}_{2\alpha}$, both $\neq 0$. Then the subalgebra $\mathfrak{g}^* \subset \mathfrak{g}$ generated by X_{α}, $X_{2\alpha}$, θX_{α}, $\theta X_{2\alpha}$ is isomorphic to $\mathfrak{su}(2, 1)$, the Lie algebra of $SU(2, 1)$.*

This is a consequence of a few lemmas. We put $Y_{\alpha} = [\theta X_{\alpha}, X_{2\alpha}]$.

Lemma 3.2. $[X_\alpha, Y_\alpha] = cX_{2\alpha}$ *where* $c = 2\alpha(A_\alpha)\, B(X_\alpha, \theta X_\alpha)$

Proof. By the Jacobi identity

$$[X_\alpha, [\theta X_\alpha, X_{2\alpha}]] + [\theta X_\alpha, [X_{2\alpha}, X_\alpha]] + [X_{2\alpha}, [X_\alpha, \theta X_\alpha]] = 0.$$

The middle term is 0 (since $3\alpha \notin \Sigma$), so the lemma follows from (7), §1.

Lemma 3.3. *The vector* $[X_\alpha, \theta Y_\alpha]$ is $\neq 0$ *and lies in* $\mathfrak{m}$.

Proof. We have $[X_\alpha, \theta Y_\alpha] \in \mathfrak{g}_0 = \mathfrak{m} + \mathfrak{a}$; also if $H \in \mathfrak{a}$,

$$B(H, [X_\alpha, \theta Y_\alpha]) = B([H, X_\alpha], \theta Y_\alpha) = \alpha(H)\, B(X_\alpha, [X_\alpha, \theta X_{2\alpha}]) = 0.$$

Thus $[X_\alpha, \theta Y_\alpha] \in \mathfrak{m}$. To see that this vector is $\neq 0$ consider the vector $[[X_\alpha, \theta Y_\alpha], X_\alpha]$. By the first part of the proof it equals

$$\begin{aligned}[[\theta X_\alpha, Y_\alpha], X_\alpha] &= -[[Y_\alpha, X_\alpha], \theta X_\alpha] - [[X_\alpha, \theta X_\alpha], Y_\alpha] \\ &= c[X_{2\alpha}, \theta X_\alpha] - B(X_\alpha, \theta X_\alpha)\, \alpha(A_\alpha)\, Y_\alpha,\end{aligned}$$

so by Lemma 3.2

$$[[X_\alpha, \theta Y_\alpha], X_\alpha] = -3\alpha(A_\alpha)\, B(X_\alpha, \theta X_\alpha)\, Y_\alpha \neq 0. \tag{1}$$

Lemma 3.4. *The linear transformation* $ad_{\mathfrak{g}}([X_\alpha, \theta Y_\alpha])$ *leaves the plane* $\boldsymbol{R}X_\alpha + \boldsymbol{R}Y_\alpha$ *invariant.*

Proof. We have

$$B(Y_\alpha, \theta Y_\alpha) = B(Y_\alpha, [X_\alpha, \theta X_{2\alpha}]) = -B([X_\alpha, Y_\alpha], \theta X_{2\alpha}),$$

so

$$B(Y_\alpha, \theta Y_\alpha) = -2\alpha(A_\alpha)\, B(X_\alpha, \theta X_\alpha)\, B(X_{2\alpha}, \theta X_{2\alpha}). \tag{2}$$

Hence

$$\begin{aligned}[[X_\alpha, \theta Y_\alpha], Y_\alpha] &= -[[\theta Y_\alpha, Y_\alpha], X_\alpha] - [[Y_\alpha, X_\alpha], \theta Y_\alpha] \\ &= B(Y_\alpha, \theta Y_\alpha)\, \alpha(A_\alpha)\, X_\alpha + c[X_{2\alpha}, [X_\alpha, \theta X_{2\alpha}]] \\ &= B(Y_\alpha, \theta Y_\alpha)\, \alpha(A_\alpha)\, X_\alpha - c\alpha(A_{2\alpha})\, B(X_{2\alpha}, \theta X_{2\alpha})\, X_\alpha,\end{aligned}$$

so

$$[X_\alpha, \theta Y_\alpha], Y_\alpha] = -6\alpha(A_\alpha)^2\, B(X_\alpha, \theta X_\alpha)\, B(X_{2\alpha}, \theta X_{2\alpha})\, X_\alpha, \tag{3}$$

which together with (1) proves the lemma.

Lemma 3.5.

$$[[X_\alpha, \theta Y_\alpha], X_{2\alpha}] = 0.$$

Proof. Putting $T = [X_\alpha, \theta Y_\alpha]$, we have seen above that $[T, X_\alpha] \in \boldsymbol{R}Y_\alpha$, $[T, Y_\alpha] \subset \boldsymbol{R}X_\alpha$, so $[T, [X_\alpha, Y_\alpha]] = 0$. Now use Lemma 3.2.

Lemma 3.6.

$$[Y_\alpha, \theta X_{2\alpha}] = 2\alpha(A_\alpha)\, B(X_{2\alpha}, \theta X_{2\alpha})\, \theta X_\alpha.$$

In fact,

$$\begin{aligned}[Y_\alpha, \theta X_{2\alpha}] &= [[\theta X_\alpha, X_{2\alpha}], \theta X_{2\alpha}] \\ &= -[[X_{2\alpha}, \theta X_{2\alpha}], \theta X_\alpha] - [[\theta X_{2\alpha}, \theta X_\alpha], X_{2\alpha}] \\ &= \alpha(A_{2\alpha})\, B(X_{2\alpha}, \theta X_{2\alpha})\, \theta X_\alpha.\end{aligned}$$

These lemmas show that

$$\begin{aligned}\mathfrak{g}^* = (\boldsymbol{R}[X_\alpha, \theta Y_\alpha] + \boldsymbol{R}A_\alpha) + (\boldsymbol{R}X_\alpha + \boldsymbol{R}Y_\alpha) \\ + \boldsymbol{R}X_{2\alpha} + (\boldsymbol{R}\theta X_\alpha + \boldsymbol{R}\theta Y_\alpha) + \boldsymbol{R}\theta X_{2\alpha},\end{aligned}$$

and that this is a decomposition into the eigenspaces of ad A_α.

It will now be convenient to give all general Lie algebra concepts connected with $\mathfrak{g}^*$ the superscript *. (Dual spaces will not be considered, so no confusion with our customary use of * should result.)

Although we do not yet know that $\mathfrak{g}^*$ is semisimple, we have for $H \in \boldsymbol{R}A_\alpha$

$$B^*(H, H) = \operatorname{Tr}(\operatorname{ad}_{\mathfrak{g}^*}(H)\operatorname{ad}_{\mathfrak{g}^*}(H)) = 4\alpha(H)^2 + 2(2\alpha(H))^2 = 12\alpha(H)^2,$$

so let us define

$$A^*_\alpha = \frac{1}{12\langle\alpha, \alpha\rangle} A_\alpha, \qquad X^*_\alpha = c_\alpha X_\alpha,$$

$$X^*_{2\alpha} = c_{2\alpha} X_{2\alpha}, \qquad Y^*_\alpha = [\theta X^*_\alpha, X^*_{2\alpha}],$$

where the constants c_α and $c_{2\alpha}$ are determined by

$$c_\alpha^2 \alpha(A_\alpha)\, B(X_\alpha, \theta X_\alpha) = -2, \qquad c_{2\alpha}^2 \alpha(A_\alpha)\, B(X_{2\alpha}, \theta X_{2\alpha}) = -2. \tag{4}$$

Then Lemmas 3.3-3.5 hold for the starred vectors in $\mathfrak{g}^*$, and we find the relations

$$[X^*_\alpha, Y^*_\alpha] = -4X^*_{2\alpha}, \qquad [X^*_\alpha, [X^*_\alpha, \theta Y^*_\alpha]] = -6Y^*_\alpha$$

$$[Y^*_\alpha, \theta X^*_{2\alpha}] = -4\theta X^*_\alpha, \qquad [Y^*_\alpha, \theta Y^*_\alpha] = -96A^*_\alpha$$

$$[[X^*_\alpha, \theta Y^*_\alpha], Y^*_\alpha] = -24X^*_\alpha$$

from Lemma 3.2, (1), Lemma 3.6, (2) and (3), respectively.

Next we consider the Lie algebra $\mathfrak{su}(2, 1)$ and add the subscript 0 to all general Lie algebra concepts connected with it. Since $\mathfrak{su}(2, 1)$ is a real form of $\mathfrak{sl}(3, \boldsymbol{C})$, its Killing form is by (3), §8, Chapter III, and Lemma 6.1, Chapter III, given by

$$B_0(X, Y) = 6\,\mathrm{Tr}(XY)$$

and the automorphism $X \to I_{2,1}XI_{2,1}$ of $\mathfrak{su}(2, 1)$ where

$$I_{2,1} = \begin{pmatrix} -1 & 0 & 0 \\ 0 & -1 & 0 \\ 0 & 0 & 1 \end{pmatrix}$$

is a Cartan involution θ_0 of $\mathfrak{su}(2, 1)$ (cf. Chapter X). Here $\mathfrak{k}_0$ consists of the matrices

$$\left(\begin{array}{c|c} A & \begin{matrix} 0 \\ 0 \end{matrix} \\ \hline 0 \quad 0 & -\mathrm{Tr}(A) \end{array}\right) \qquad \text{(A skew-hermitian)}$$

and $\mathfrak{p}_0$ of the matrices

$$\left(\begin{array}{cc|c} 0 & 0 & z_1 \\ 0 & 0 & z_2 \\ \hline \bar{z}_1 & \bar{z}_2 & 0 \end{array}\right) \qquad (z_1, z_2 \in \boldsymbol{C}).$$

For $\mathfrak{a}_0$ and $\mathfrak{a}_0^+$ we can take $\mathfrak{a}_0 = \boldsymbol{R}H_0$, $\mathfrak{a}_0^+ = \{tH_0 : t > 0\}$ where

$$H_0 = \begin{pmatrix} 0 & 0 & 1 \\ 0 & 0 & 0 \\ 1 & 0 & 0 \end{pmatrix}.$$

Then K_0 and M_0 are defined by the convention above. The restricted roots are α_0, $2\alpha_0$, $-\alpha_0$, $-2\alpha_0$ where

$$\alpha_0(H_0) = 1, \qquad \text{so} \qquad A_{\alpha_0} = H_0/12.$$

Also $\rho_0 = 2\alpha_0$ and $\mathfrak{m}_0 = \boldsymbol{R}T_0$ where

$$T_0 = \begin{pmatrix} i & 0 & 0 \\ 0 & -2i & 0 \\ 0 & 0 & i \end{pmatrix}.$$

The root spaces $\mathfrak{g}_{\alpha_0}$ and $\mathfrak{g}_{-\alpha_0}$ are given by the set of matrices

$$\begin{pmatrix} 0 & z & 0 \\ -\bar{z} & 0 & \bar{z} \\ 0 & z & 0 \end{pmatrix} \text{ for } \mathfrak{g}_{\alpha_0}, \qquad \begin{pmatrix} 0 & z & 0 \\ -\bar{z} & 0 & -\bar{z} \\ 0 & -z & 0 \end{pmatrix} \text{ for } \mathfrak{g}_{-\alpha_0},$$

where $z \in \boldsymbol{C}$. The root spaces $\mathfrak{g}_{2\alpha_0}$ and $\mathfrak{g}_{-2\alpha_0}$ are given by the set of matrices

$$\begin{pmatrix} it & 0 & -it \\ 0 & 0 & 0 \\ it & 0 & -it \end{pmatrix} \text{ for } \mathfrak{g}_{2\alpha_0}, \qquad \begin{pmatrix} it & 0 & it \\ 0 & 0 & 0 \\ -it & 0 & -it \end{pmatrix} \text{ for } \mathfrak{g}_{-2\alpha_0},$$

where $t \in \boldsymbol{R}$. Consider now the linear mapping $\tau : \mathfrak{g}^* \to \mathfrak{su}(2, 1)$ given by

$$X_\alpha^* \to \begin{pmatrix} 0 & 1 & 0 \\ -1 & 0 & 1 \\ 0 & 1 & 0 \end{pmatrix}, \qquad X_{2\alpha}^* \to \begin{pmatrix} i & 0 & -i \\ 0 & 0 & 0 \\ i & 0 & -i \end{pmatrix},$$

$$\theta X_\alpha^* \to \begin{pmatrix} 0 & 1 & 0 \\ -1 & 0 & -1 \\ 0 & -1 & 0 \end{pmatrix}, \qquad \theta X_{2\alpha}^* \to \begin{pmatrix} i & 0 & i \\ 0 & 0 & 0 \\ -i & 0 & -i \end{pmatrix},$$

$$Y_\alpha^* \to -2 \begin{pmatrix} 0 & i & 0 \\ i & 0 & -i \\ 0 & i & 0 \end{pmatrix}, \qquad \theta Y_\alpha^* \to -2 \begin{pmatrix} 0 & i & 0 \\ i & 0 & i \\ 0 & -i & 0 \end{pmatrix},$$

$$[X_\alpha^*, \theta Y_\alpha^*] \to -4 \begin{pmatrix} i & 0 & 0 \\ 0 & -2i & 0 \\ 0 & 0 & i \end{pmatrix}, \qquad A_\alpha^* \to \frac{1}{12} \begin{pmatrix} 0 & 0 & 1 \\ 0 & 0 & 0 \\ 1 & 0 & 0 \end{pmatrix}.$$

The commutation relations in $\mathfrak{g}^*$ show quickly that τ is an isomorphism, whereby Prop. 3.1 is proved. Note that the vector A_α^* satisfies

$$B^*(H, A_\alpha^*) = \alpha(H), \qquad H \in \mathfrak{a}.$$

Now let G be any connected Lie group with Lie algebra $\mathfrak{g}$ and let K, A, N, $\bar{N}$, G^*, K^*, A^*, N^*, and $\bar{N}^*$ be the analytic subgroups corresponding to $\mathfrak{k}$, $\mathfrak{a}$, $\mathfrak{n}$, $\bar{\mathfrak{n}} = \theta\mathfrak{n}$, $\mathfrak{g}^*$, $\mathfrak{k}^*$, $\mathfrak{a}^*$, $\mathfrak{n}^*$, and $\bar{\mathfrak{n}}^*$, respectively. Here $\mathfrak{n}^* = \boldsymbol{R}X_\alpha + \boldsymbol{R}Y_\alpha$ and $\bar{\mathfrak{n}}^* = \theta\mathfrak{n}^*$. As usual, let M be the centralizer of A in K and M^* the centralizer of A^* in K^*. The following lemma ensures compatibility of the three decompositions in §1 for the groups G^* and G.

Lemma 3.7. *With the notation above*

$$K^* = G^* \cap K, \qquad A^* = G^* \cap A,$$

$$N^* = G^* \cap N, \qquad M^* = G^* \cap M.$$

The first three follow as in the proof of Lemma 2.3. Since $\mathfrak{m}^* = \boldsymbol{R}[X_\alpha, \theta Y_\alpha]$, it is clear that $\mathfrak{m}^* = \mathfrak{g}^* \cap \mathfrak{m}$ and of course $G^* \cap M \subset M^*$. On the other hand, the symmetric space G^*/K^* has rank one, so K^*/M^* is a 3-sphere (Lemma 6.3, Chapter V), hence simply connected, so M^* is connected. Thus $M^* \subset G^* \cap M$ and the lemma is proved.

We recall now the Cartan, Iwasawa, and Bruhat decompositions

$$g = k_1 \exp A^+(g)\, k_2, \qquad g = k(g) \exp H(g)\, n(g),$$
$$g = \bar{n}(g)\, m(g) \exp B(g)\, n_B(g), \tag{5}$$

where $A^+(g) \in \overline{\mathfrak{a}^+}$ and $H(g), B(g) \in \mathfrak{a}$. The importance of Theorem 3.1 is that if g lies in a group G^*, then, by Lemma 3.7, all the components of g above can be found by matrix computation within the elementary group $\boldsymbol{SU}(2, 1)$.

Select $m^* \in K$ such that $\mathrm{Ad}_G(m^*)$ interchanges the Weyl chambers $\mathfrak{a}^+$ and $-\mathfrak{a}^+$. If $\bar{n} \neq e$, then $m^*\bar{n} \notin m^*MAN$, so $m^*\bar{n} \in \bar{N}MAN$.

Theorem 3.8. *Let G be a semisimple Lie group of real rank one and put $|Z|^2 = -B(Z, \theta Z)$ for $Z \in \mathfrak{g}$. If $\bar{n} \in \bar{N}$ and we write $\bar{n} = \exp(X + Y)$ $(X \in \mathfrak{g}_{-\alpha}, Y \in \mathfrak{g}_{-2\alpha})$, then*

$$2 \cosh 2\alpha(A^+(\bar{n})) = 1 + 2c\,|X|^2 + (1 + c\,|X|^2)^2 + 4c\,|Y|^2,$$

$$e^{\rho(H(\bar{n}))} = [(1 + c\,|X|^2)^2 + 4c\,|Y|^2]^{\frac{1}{4}(m_\alpha + 2m_{2\alpha})},$$

$$e^{\rho(B(m^*\bar{n}))} = [c^2\,|X|^4 + 4c\,|Y|^2]^{\frac{1}{4}(m_\alpha + 2m_{2\alpha})} \qquad (\bar{n} \neq e).$$

Here m_α and $m_{2\alpha}$ are the multiplicities of the roots α and 2α and $c^{-1} = 4(m_\alpha + 4m_{2\alpha})$.

We begin with the case $\boldsymbol{SU}(2, 1)$ and will follow the notational conventions (subscript 0) made for this group above. Let m^* be chosen as the element

$$m^* = \begin{pmatrix} -1 & 0 & 0 \\ 0 & -1 & 0 \\ 0 & 0 & 1 \end{pmatrix}.$$

Lemma 3.9. *In the group $G_0 = \boldsymbol{SU}(2, 1)$ let $\bar{n} \neq e$ in $\bar{N}_0$ and write (according to the formula for $\mathfrak{g}_{-\alpha_0}$ and $\mathfrak{g}_{-2\alpha_0}$)*

$$\bar{n} = \exp \begin{pmatrix} it & z & it \\ -\bar{z} & 0 & -\bar{z} \\ -it & -z & -it \end{pmatrix}, \qquad \bar{n}(m^*\bar{n}) = \exp \begin{pmatrix} iu & w & iu \\ -\bar{w} & 0 & -\bar{w} \\ -iu & -w & -iu \end{pmatrix}.$$

Then

$$\cosh \rho_0(A^+(\bar{n})) = \tfrac{1}{2} + | z |^2 + \tfrac{1}{2}[(1 + | z |^2)^2 + 4t^2],$$

$$e^{\rho_0(H(\bar{n}))} = (1 + | z |^2)^2 + 4t^2, \qquad e^{\rho_0(B(m^*\bar{n}))} = | z |^4 + 4t^2.$$

Furthermore

$$w = -\frac{z}{| z |^2 + 2it}, \qquad u = \frac{-t}{| z |^4 + 4t^2},$$

and writing

$$n(\bar{n}) = \exp \begin{pmatrix} i\tau & \zeta & -i\tau \\ -\bar{\zeta} & 0 & \bar{\zeta} \\ i\tau & \zeta & -i\tau \end{pmatrix},$$

we have

$$\zeta = \frac{-z}{| z |^2 - 2it + 1}, \qquad \tau = \frac{-t}{(1 + | z |^2)^2 + 4t^2}.$$

Proof. We write out (5) for $g = m^*\bar{n}$,

$$m(m^*\bar{n}) = \begin{pmatrix} e^{i\alpha} & 0 & 0 \\ 0 & e^{-2i\alpha} & 0 \\ 0 & 0 & e^{i\alpha} \end{pmatrix}, \qquad \exp B(m^*\bar{n}) = \begin{pmatrix} \cosh r & 0 & \sinh r \\ 0 & 1 & 0 \\ \sinh r & 0 & \cosh r \end{pmatrix}$$

$$n_B(m^*\bar{n}) = \exp \begin{pmatrix} iu' & w' & -iu' \\ -\bar{w}' & 0 & \bar{w}' \\ iu' & w' & -iu' \end{pmatrix}.$$

We have also by exponentiation

$$\bar{n} = \begin{pmatrix} 1 + it - \tfrac{1}{2} | z |^2 & z & it - \tfrac{1}{2} | z |^2 \\ -\bar{z} & 1 & -\bar{z} \\ -it + \tfrac{1}{2} | z |^2 & -z & 1 - it + \tfrac{1}{2} | z |^2 \end{pmatrix}.$$

Now the last equation in (5) represents nine equations in the variables u, w, α, r, u', and w', which can be solved in a routine fashion. Denoting the matrix $m^*\bar{n}$ by (a_{ij}), we find

$$\tfrac{1}{2}(a_{21} + a_{23}) = \bar{z} = -\bar{w}e^{i\alpha}e^r, \tag{6}$$

$$\tfrac{1}{2}(a_{21} - a_{23}) = 0 = -\bar{w}'e^{-2i\alpha} + \bar{w}e^{i\alpha}e^r(| w' |^2 - 2iu'), \tag{7}$$

$$\tfrac{1}{2}(a_{12} + a_{32}) = -z = w'e^{i\alpha}e^r, \tag{8}$$

$$\tfrac{1}{2}(a_{12} - a_{32}) = 0 = we^{-2i\alpha} - w'e^{i\alpha}e^r(| w |^2 - 2iu). \tag{9}$$

From (6) and (8) we derive

$$\bar{w}w'e^{2i\alpha} = |\,z\,|^2\,e^{-2r} = w\bar{w}'e^{-2i\alpha}, \tag{10}$$

$$|\,w\,|^2 = |\,w'\,|^2 = |\,z\,|^2\,e^{-2r}, \tag{11}$$

and invoking (9),

$$e^{i\alpha}e^{r} = (|\,w\,|^2 - 2iu)^{-1}. \tag{12}$$

But (5) can also be written

$$m^*\bar{n}(m^*\bar{n}) = \bar{n}m(m^*\bar{n})^{-1}\exp(-B(m^*\bar{n}))\,n' \qquad (n' \in N),$$

so (6) and (12) give by symmetry

$$e^{-i\alpha}e^{-r} = (|\,z\,|^2 - 2it)^{-1}, \qquad \bar{w} = -\bar{z}e^{-i\alpha}e^{-r}. \tag{13}$$

Now (12) and (13) give the formulas for w and u. Secondly, $\rho_0 = \frac{1}{2}(2\alpha_0 + 2\alpha_0) = 2\alpha_0$ so

$$e^{\rho_0(B(m^*\bar{n}))} = e^{2r} = |\,z\,|^4 + 4t^2.$$

Next we have $\bar{n} = k(\bar{n})\exp H(\bar{n})\,n(\bar{n})$ which implies

$$\theta(\bar{n})^{-1}\,\bar{n} = \theta(n(\bar{n}))^{-1}\exp 2H(\bar{n})\,n(\bar{n}). \tag{14}$$

Let $s = \alpha_0(H(\bar{n}))$ so $H(\bar{n}) = sH_0$. Denoting the matrix (14) by (b_{ij}), we obtain

$$b_{22} = 1 + 2\,|\,z\,|^2 = 1 + 2\,|\,\zeta\,|^2\,e^{2s}, \tag{15}$$

$$b_{21} = 2it\bar{z} - \bar{z}\,|\,z\,|^2 = -\bar{\zeta} + \bar{\zeta}e^{2s}(1 + 2i\tau - |\,\zeta\,|^2), \tag{16}$$

$$b_{32} = -2z - 2itz - z\,|\,z\,|^2 = \zeta + \zeta e^{2s}(1 + 2i\tau + |\,\zeta\,|^2). \tag{17}$$

We conjugate (17) and add to (16); this gives

$$-\bar{z}(1 - 2it + |\,z\,|^2) = \bar{\zeta}e^{2s}. \tag{18}$$

Conjugating (16) and subtracting from (17) gives

$$-z = \zeta + \zeta e^{2s}(2i\tau + |\,\zeta\,|^2). \tag{19}$$

From (18) and (15) we deduce the desired formula for $e^{\rho_0(H(\bar{n}))}$. Then (18) and (19) give the formulas for ζ and τ.

Finally, $\bar{n} = k_1 \exp A^+(\bar{n})\, k_2$, so $(b_{ij}) = \theta(\bar{n})^{-1}\bar{n} = k_2^{-1} \exp(2A^+(\bar{n})k_2$. Letting $v = \alpha_0(A^+(\bar{n}))$ so $A^+(\bar{n}) = vH_0$, we get

$$\operatorname{Trace}(b_{ij}) = 2 \cosh 2v + 1 = 2 \cosh(\rho_0(A^+(\bar{n}))) + 1.$$

But

$$b_{11} = 1 + \tfrac{1}{2}(|\, z\,|^4 + 4t^2), \qquad b_{33} = 1 + 2\,|\, z\,|^2 + \tfrac{1}{2}(|\, z\,|^4 + 4t^2);$$

so using (15), the formula for $A^+(\bar{n})$ follows.

We turn now to the proof of Theorem 3.8. Assume first both $\mathfrak{g}_{-\alpha}$ and $\mathfrak{g}_{-2\alpha}$ are $\neq 0$; take $X = \theta X_\alpha$, $Y = \theta X_{2\alpha}$ in Theorem 3.1 and construct $\mathfrak{g}^*$, etc. For the Killing forms we have

$$B(H, H) = \operatorname{Tr}(\operatorname{ad} H \operatorname{ad} H) = 2m_\alpha \alpha(H)^2 + 2m_{2\alpha}(2\alpha(H))^2 = 2(m_\alpha + 4m_{2\alpha})\, \alpha(H)^2$$
$$B^*(H, H) = 12\alpha(H)^2$$

so $A_\alpha^* = \frac{1}{6}(m_\alpha + 4m_{2\alpha})\, A_\alpha$ and by (7), §1,

$$B(X_\alpha, \theta X_\alpha) = \tfrac{1}{6}(m_\alpha + 4m_{2\alpha})\, B^*(X_\alpha, \theta X_\alpha), \tag{20}$$

$$B(X_{2\alpha}, \theta X_{2\alpha}) = \tfrac{1}{6}(m_\alpha + 4m_{2\alpha})\, B^*(X_{2\alpha}, \theta X_{2\alpha}). \tag{21}$$

Let $\bar{n} = \exp(X + Y)$ and note that by Lemma 3.7, $H(\bar{n})$ is the same in $\mathfrak{g}^*$ and in $\mathfrak{g}$. We have

$$e^{\rho(H(\bar{n}))} = (e^{\rho^*(H(\bar{n}))})^{\frac{1}{4}(m_\alpha + 2m_{2\alpha})}. \tag{22}$$

Under the isomorphism $\tau : \mathfrak{g}^* \to \mathfrak{su}(2, 1)$ let $X + Y$ correspond to

$$\begin{pmatrix} it & z & it \\ -\bar{z} & 0 & -\bar{z} \\ -it & -z & -it \end{pmatrix}.$$

Then by the formula for B_0

$$B^*(X, \theta X) = |\, z\,|^2\, B^*(\theta X_\alpha^*, X_\alpha^*) = -24\,|\, z\,|^2,$$

$$B^*(Y, \theta Y) = t^2 B^*(\theta X_{2\alpha}^*, X_{2\alpha}^*) = -24t^2,$$

so by (20) and (21)

$$|\, X\,|^2 = 4(m_\alpha + 4m_{2\alpha})\,|\, z\,|^2, \qquad |\, Y\,|^2 = 4(m_\alpha + 4m_{2\alpha})\, t^2.$$

But by Lemma 3.9, $e^{\rho^*(H(\bar{n}))} = (1 + |\, z\,|^2)^2 + 4t^2$, so (22) gives the formula for $e^{\rho(H(\bar{n}))}$.

Next we note that $B(m^*\bar{n})$ is independent of the choice of m^*. In particular we take m^* in the group K^*. Then $B(m^*\bar{n})$ is the same in $\mathfrak{g}^*$ and in $\mathfrak{g}$, so the formula for it follows from Lemma 3.9. The formula for $A^+(\bar{n})$ is obtained in the same way.

This proves Theorem 3.8 for the case $2\alpha \in \Sigma$. If $\mathfrak{g}_{2\alpha} = 0$, the proof is much easier by just considering the subalgebra (isomorphic to $\mathfrak{su}(1, 1)$) generated by X_α and θX_α.

§ 4. Cartan Subalgebras

The structure theory of our real semisimple Lie algebra $\mathfrak{g}$ has been founded on the study of the family $\mathrm{ad}_{\mathfrak{g}}(\mathfrak{a})$ in analogy with the study in Chapter III of the family $\mathrm{ad}_{\mathfrak{g}^C}(\mathfrak{h}^C)$, $\mathfrak{h}^C$ being a Cartan subalgebra of the complex semisimple Lie algebra $\mathfrak{g}^C$.

Definition. Let $\mathfrak{g}$ be a semisimple Lie algebra over $\boldsymbol{R}$ and $\mathfrak{g}^C$ its complexification. A subalgebra $\mathfrak{h} \subset \mathfrak{g}$ is called a *Cartan subalgebra* if its complexification $\mathfrak{h}^C = \mathfrak{h} + i\mathfrak{h}$ in $\mathfrak{g}^C$ is a Cartan subalgebra of $\mathfrak{g}^C$.

Considering now the definition in §3, Chapter III, a Cartan subalgebra $\mathfrak{h} \subset \mathfrak{g}$ is characterized by the two conditions: (1) $\mathfrak{h}$ is a maximal abelian subalgebra of $\mathfrak{g}$; (2) for each $H \in \mathfrak{h}$, $\mathrm{ad}_{\mathfrak{g}}(H)$ is semisimple. (We recall that an endomorphism of a real vector space V is called *semisimple* if its extension to the complexification V^C is in a suitable basis of V^C expressed by means of a diagonal matrix.) We are now going to relate these Cartan subalgebras to the Cartan involution θ of $\mathfrak{g}$ and the associated Cartan decomposition $\mathfrak{g} = \mathfrak{k} + \mathfrak{p}$. We saw already in Chapter VI, §3 that any maximal abelian subalgebra of $\mathfrak{g}$ containing $\mathfrak{a}$ is a Cartan subalgebra.

It is clear that all Cartan subalgebras of $\mathfrak{g}$ have the same dimension. This is called the *rank* of $\mathfrak{g}$ (and of G). We note also that if $\mathfrak{g}$ has a complex structure J and $\mathfrak{h} \subset \mathfrak{g}$ is a Cartan subalgebra, then the complex subalgebra $\tilde{\mathfrak{h}} \subset \tilde{\mathfrak{g}}$ (§6, Chapter III) is a Cartan subalgebra in the sense of Chapter III, §3.

Lemma 4.1. *Let $\mathfrak{h} \subset \mathfrak{g}$ be a Cartan subalgebra and $\mathfrak{h}^C \subset \mathfrak{g}^C$ the complexifications. Then there exists a compact real form $\mathfrak{v}$ of $\mathfrak{g}^C$ such that:*

(i) $\sigma\mathfrak{v} \subset \mathfrak{v}$ ($\sigma =$ *conjugation of* $\mathfrak{g}^C$ *with respect to* $\mathfrak{g}$).

(ii) $\mathfrak{h}^C \cap \mathfrak{v}$ *is maximal abelian in* $\mathfrak{v}$.

Proof. Let $\mathfrak{g}_k$ be the compact real form of $\mathfrak{g}^C$ constructed in the proof of Theorem 6.3, Chapter III, and let τ denote the conjugation of $\mathfrak{g}^C$ with respect to $\mathfrak{g}_k$. According to Theorem 7.1, Chapter III, and its proof, the automorphism $\varphi = (\sigma\tau)^2)^{\frac{1}{4}}$ of $\mathfrak{g}^C$ maps $\mathfrak{g}_k$ into a compact real form $\mathfrak{v}$ satisfying (i). Since σ and τ both leave $\mathfrak{h}^C$ invariant, so does φ; thus (ii) follows from the fact that $\mathfrak{h}^C \cap \mathfrak{g}_k$ is maximal abelian in $\mathfrak{g}_k$ (cf. (2), §6, Chapter III).

Corollary 4.2. *Each Cartan subalgebra $\mathfrak{h} \subset \mathfrak{g}$ is conjugate under Int($\mathfrak{g}$) to one which is θ-invariant.*

Property (i) in Lemma 4.1 means that σ commutes with the conjugation η of $\mathfrak{g}^C$ with respect to $\mathfrak{v}$. By (ii) $\mathfrak{h}^C = \mathfrak{h}^C \cap \mathfrak{v} + \mathfrak{h}^C \cap (i\mathfrak{v})$, so $\mathfrak{h}^C$ is η-invariant as well as σ-invariant. Thus $\mathfrak{h}$ is invariant under the Cartan involution $\eta | \mathfrak{g}$ of $\mathfrak{g}$, so the corollary follows from Theorem 7.2, Chapter III.

We shall now prove that $\mathfrak{g}$ has at most finitely many nonconjugate Cartan subalgebras.

Lemma 4.3. *Let $\mathfrak{h}_1, \mathfrak{h}_2 \subset \mathfrak{g}$ be two Cartan subalgebras such that $\mathfrak{h}_i \cap \mathfrak{p} \subset \mathfrak{a}$, $\theta\mathfrak{h}_i \subset \mathfrak{h}_i$ $(i = 1, 2)$. Let*

$$\Sigma_i = \{\alpha \in \Sigma : \alpha(\mathfrak{h}_i \cap \mathfrak{p}) \equiv 0\} \qquad (i = 1, 2).$$

Then if $\Sigma_1 = \Sigma_2$, there exists a $k \in K$ such that $\mathfrak{h}_2 = Ad(k)\mathfrak{h}_1$.

Proof. The zeros in $\mathfrak{h}_1 \cap \mathfrak{p}$ of the members of $\Sigma - \Sigma_1$ constitute finitely many hyperplanes, so we can select $H_1 \in \mathfrak{h}_1 \cap \mathfrak{p}$ such that $\alpha(H_1) \neq 0$ for all $\alpha \in \Sigma - \Sigma_1$. Let $\mathfrak{a}_1 \subset \mathfrak{a}$ be the common nullspace of the members of Σ_1. Of course $\mathfrak{a}_1 \supset \mathfrak{h}_1 \cap \mathfrak{p}$. We shall now prove $\mathfrak{a}_1 = \mathfrak{h}_1 \cap \mathfrak{p}$. Let $\mathfrak{m}_1$ denote the centralizer of H_1 in $\mathfrak{k}$. Of course $\mathfrak{m}_1 \subset \mathfrak{g}_0 + \sum_{\alpha \in \Sigma_1} \mathfrak{g}_\alpha$, so $[\mathfrak{m}_1, \mathfrak{a}_1] = 0$. But $\mathfrak{h}_1 \cap \mathfrak{k} \subset \mathfrak{m}_1$, so $\mathfrak{a}_1 + \mathfrak{h}_1 \cap \mathfrak{k}$ is abelian. Since

$$\mathfrak{a}_1 + \mathfrak{h}_1 \cap \mathfrak{k} \supset \mathfrak{h}_1 \cap \mathfrak{p} + \mathfrak{h}_1 \cap \mathfrak{k} = \mathfrak{h}_1,$$

the maximality of $\mathfrak{h}_1$ implies $\mathfrak{a}_1 = \mathfrak{h}_1 \cap \mathfrak{p}$.

It follows that if $\Sigma_1 = \Sigma_2$, then $\mathfrak{h}_1 \cap \mathfrak{p} = \mathfrak{h}_2 \cap \mathfrak{p}$. The centralizer M_1 of H_1 in $\text{Ad}_G(K)$ is compact and has Lie algebra $\mathfrak{m}_1$. Also $\mathfrak{h}_1 \cap \mathfrak{k}$ is a maximal abelian subalgebra of $\mathfrak{m}_1$; since $\mathfrak{h}_2 \cap \mathfrak{k} \subset \mathfrak{m}_1$, we conclude by dimensionality that $\mathfrak{h}_2 \cap \mathfrak{k} = \text{Ad}(k)(\mathfrak{h}_1 \cap \mathfrak{k})$ for some k in the identity component of M_1. But then $\text{Ad}(k)(\mathfrak{h}_1 \cap \mathfrak{p}) = \mathfrak{h}_1 \cap \mathfrak{p}$, so $\mathfrak{h}_2 = \text{Ad}(k)\mathfrak{h}_1$.

Corollary 4.4. *A semisimple Lie algebra $\mathfrak{g}$ over **R** has at most finitely many nonconjugate Cartan subalgebras.*

In fact any such subalgebra is conjugate to a Cartan subalgebra $\mathfrak{h}$ satisfying $\theta\mathfrak{h} \subset \mathfrak{h}$ and $\mathfrak{h} \cap \mathfrak{p} \subset \mathfrak{a}$ (Cor. 4.2 and Lemma 6.3, Chapter V). Since Σ is finite, the corollary follows from Lemma 4.3.

Proposition 4.5. *Let $\mathfrak{h}_1, \mathfrak{h}_2 \subset \mathfrak{g}$ be two Cartan subalgebras invariant under θ. Then they are conjugate under Int($\mathfrak{g}$) if and only if their "vector parts" $\mathfrak{h}_1 \cap \mathfrak{p}$ and $\mathfrak{h}_2 \cap \mathfrak{p}$ are conjugate.*

Proof. Suppose $\mathfrak{h}_1 \cap \mathfrak{p}$ and $\mathfrak{h}_2 \cap \mathfrak{p}$ are conjugate. Using suitable elements from $\mathrm{Ad}_G(K)$ (which necessarily commute with θ), we may assume that the vector parts $\mathfrak{a}_1 = \mathfrak{h}_1 \cap \mathfrak{p}$ and $\mathfrak{a}_2 = \mathfrak{h}_2 \cap \mathfrak{p}$ both lie in $\mathfrak{a}$. Now if $g\mathfrak{a}_1 = \mathfrak{a}_2$, $(g \in \mathrm{Int}(\mathfrak{g}))$, we write $g = k \exp X$ $(k \in \mathrm{Ad}_G(K)$, $X \in \mathfrak{p})$ and deduce $\mathrm{Ad}(\exp X)\mathfrak{a}_1 \subset \mathfrak{p}$. Applying θ, we derive

$$\mathrm{Ad}(\exp 2X) \mid \mathfrak{a}_1 \equiv I.$$

But ad X is semisimple with real eigenvalues, so we conclude $[X, \mathfrak{a}_1] = 0$ and consequently $k\mathfrak{a}_1 = \mathfrak{a}_2$. Thus we may assume $\mathfrak{h}_1 \cap \mathfrak{p} = \mathfrak{h}_2 \cap \mathfrak{p}$. By Lemma 4.3, $\mathfrak{h}_1$ and $\mathfrak{h}_2$ are conjugate.

On the other hand, suppose $\mathfrak{h}_2 = \mathrm{Ad}(g)\mathfrak{h}_1$ and $\theta\mathfrak{h}_i \subset \mathfrak{h}_i$ $(i = 1, 2)$. Since

$$\mathfrak{h}_i \cap \mathfrak{p} = \{H \in \mathfrak{h}_i : \text{ad } H \text{ has real eigenvalues}\}$$

it follows that $\mathfrak{h}_2 \cap \mathfrak{p} = \mathrm{Ad}(g)(\mathfrak{h}_1 \cap \mathfrak{p})$. Q.E.D.

We recall that $X \in \mathfrak{g}$ is called *semisimple* if $\mathrm{ad}_{\mathfrak{g}}(X)$ is semisimple. These elements can now be characterized.

Proposition 4.6. *Let $\mathfrak{g}$ be a semisimple Lie algebra over **R**, $\mathfrak{g}_s$ the set of semisimple elements in $\mathfrak{g}$. Then*

$$\mathfrak{g}_s = \bigcup_{\mathfrak{h} \in CS} \mathfrak{h}$$

CS denoting the set of Cartan subalgebras of $\mathfrak{g}$.

Proof. We have to prove that each semisimple element H in $\mathfrak{g}$ lies in a Cartan subalgebra. Considering the centralizer $\mathfrak{z}_H$ of H in the complexification $\mathfrak{g}^C$ and putting $(\mathfrak{g}^C)^H = [\mathfrak{g}^C, H]$, we have the vector space isomorphism $\mathfrak{g}^C/\mathfrak{z}_H \cong (\mathfrak{g}^C)^H$. But by the semisimplicity of H, ad H and $(\mathrm{ad}\, H)^2$ have the same null space, so $\mathfrak{g}^C = \mathfrak{z}_H \oplus (\mathfrak{g}^C)^H$. The mapping $\tau : \mathfrak{z}_H \times (\mathfrak{g}^C)^H \to \mathfrak{g}^C$ given by

$$\tau(X, Y) = e^{\mathrm{ad}\, Y}(X + H) \tag{1}$$

has differential at (0, 0) given by $d\tau_{(0,0)}(X, Y) = X + [Y, H]$. Thus $d\tau_{(0,0)}$ is one-to-one, so the image $\tau(\mathfrak{z}_H \times (\mathfrak{g}^C)^H)$ contains a neighborhood N of 0 in $\mathfrak{g}^C$. Let a_l denote the polynomial from Exercise B.2, Chapter III, whose zeros (in $\mathfrak{g}^C$) consist of the nonregular elements. Clearly N will contain a regular element and so will $\mathfrak{z}_H$ by (1) and the invariance of a_l.

But then a_l is not identically 0 on $\mathfrak{z}_H \cap \mathfrak{g}$, so $\mathfrak{z}_H \cap \mathfrak{g}$ contains an element Z which is regular in $\mathfrak{g}^C$. But then $\mathfrak{z}_Z$ is a Cartan subalgebra of $\mathfrak{g}^C$, and $Z \in \mathfrak{g}$ implies $\mathfrak{z}_Z = \mathfrak{z}_Z \cap \mathfrak{g} + i(\mathfrak{z}_Z \cap \mathfrak{g})$. Thus $\mathfrak{z}_Z \cap \mathfrak{g}$ is a Cartan subalgebra of $\mathfrak{g}$ containing H. This proves the proposition.

§ 5. Automorphisms

We saw in Chapter V, §2 how noncompact real semisimple Lie algebras $\mathfrak{g}$ correspond to pairs $(\mathfrak{u}, s)$ where $\mathfrak{u}$ is a compact semisimple Lie algebra and s an involutive automorphism of $\mathfrak{u}$. In this section we consider the group Aut($\mathfrak{u}$) of all automorphisms of $\mathfrak{u}$ which is a closed subgroup of the group $\boldsymbol{GL}(\mathfrak{u})$ of all invertible endomorphisms of $\mathfrak{u}$. Each member of Aut($\mathfrak{u}$) leaves the Killing form of $\mathfrak{u}$ invariant. It follows that Aut($\mathfrak{u}$) is compact. The adjoint group Int($\mathfrak{u}$) is the identity component of Aut($\mathfrak{u}$). Let $\mathfrak{t}$ be a maximal abelian subalgebra of $\mathfrak{u}$ and let $\mathfrak{t}^C$ denote the subalgebra generated by $\mathfrak{t}$ in the complexification $\mathfrak{u}^C$ of $\mathfrak{u}$. Then $\mathfrak{t}^C$ is a Cartan subalgebra and we can use the results of Chapter III. An endomorphism φ of $\mathfrak{t}$ will be called an automorphism of Δ if it maps the set of vectors $iH_\alpha (\alpha \in \Delta)$ onto itself. Each element in the Weyl group $W = W(\mathfrak{u})$ of $\mathfrak{u}$ is an automorphism of Δ. Let Aut(Δ) denote the group of all automorphisms of Δ.

Definition. Let $\mathfrak{u}$ be a compact, semisimple Lie algebra, $\mathfrak{t}$ a maximal abelian subalgebra, $\mathfrak{u}^C$ the complexification of $\mathfrak{u}$ and $\mathfrak{t}^C$ the subalgebra of $\mathfrak{u}^C$ generated by $\mathfrak{t}$. A *Weyl basis* of $\mathfrak{u}^C$ mod $\mathfrak{t}^C$ with respect to $\mathfrak{u}$ is a basis $\{X_\alpha : \alpha \in \Delta\}$ of $\mathfrak{u}^C$ mod $\mathfrak{t}^C$ with the following properties.

(i) $X_\alpha \in (\mathfrak{u}^C)^\alpha$ and $[X_\alpha, X_{-\alpha}] = H_\alpha$ for each $\alpha \in \Delta$.

(ii) $[X_\alpha, X_\beta] = N_{\alpha,\beta} X_{\alpha+\beta}$ if $\alpha, \beta, \alpha + \beta \in \Delta$ where the constants $N_{\alpha,\beta}$ satisfy $N_{\alpha,\beta} = -N_{-\alpha,-\beta}$.

(iii) $X_\alpha - X_{-\alpha} \in \mathfrak{u}$, $i(X_\alpha + X_{-\alpha}) \in \mathfrak{u}$ for each $\alpha \in \Delta$.

The existence of such a basis is clear from Cor. 7.3, Chapter III and the proof of Theorem 6.3, Chapter III. Let τ denote the conjugation of $\mathfrak{u}^C$ with respect to $\mathfrak{u}$. Then $\tau(X_\alpha) = -X_{-\alpha} (\alpha \in \Delta)$.

Theorem 5.1. *Let A be an automorphism of $\mathfrak{u}$ leaving $\mathfrak{t}$ invariant. Then the restriction of A to $\mathfrak{t}$ is an automorphism of Δ. On the other hand, each automorphism of Δ can be extended to an automorphism of $\mathfrak{u}$.*

Proof. It is clear that A extends uniquely to an automorphism of $\mathfrak{u}^C$. Denoting this extension by A, we have $A\mathfrak{t}^C \subset \mathfrak{t}^C$. For each $\alpha \in \Delta$ the linear function α^A on $\mathfrak{t}^C$ given by $\alpha^A(H) = \alpha(A^{-1}H)$ is a root. Since $A \cdot H_\alpha = H_{\alpha^A}$, the restriction of A to $\mathfrak{t}$ lies in Aut(Δ). On the other hand, suppose $\varphi \in \text{Aut}(\Delta)$. The extension of φ to an endomorphism of $\mathfrak{t}^C$ will also be denoted φ. Let $\alpha' \in \Delta$ be defined by $H_{\alpha'} = \varphi H_\alpha$. Then we have from Theorem 4.3, Chapter III,

$$\frac{\beta'(H_{\alpha'})}{\alpha'(H_{\alpha'})} = \frac{\beta(H_\alpha)}{\alpha(H_\alpha)}, \qquad \alpha, \beta \in \Delta. \tag{1}$$

As shown in the proof of Theorem 5.4, Chapter III, relation (1) implies that

$$B(\varphi H, \varphi H') = B(H, H') \qquad \text{for } H, H' \in \mathfrak{t}^C. \tag{2}$$

From (2) it is easily seen that ${}^t\varphi \cdot \alpha' = \alpha (\alpha \in \Delta)$ in the sense of Theorem 5.4, Chapter III. Hence φ can be extended to an automorphism A of $\mathfrak{u}^C$. We shall now replace A by an automorphism which leaves $\mathfrak{u}$ invariant and coincides with A on $\mathfrak{t}^C$. Let $\{X_\alpha : \alpha \in \Delta\}$ be a Weyl basis of $\mathfrak{u}^C$ mod $\mathfrak{t}^C$ with respect to $\mathfrak{u}$. For each $\alpha \in \Delta$, let a_α be determined by $AX_\alpha = a_\alpha X_{\alpha'}$. Since $[X_\alpha, X_{-\alpha}] = H_\alpha$ it follows that

$$a_\alpha a_{-\alpha} = 1. \tag{3}$$

The numbers $N_{\alpha,\beta}$ determined by $[X_\alpha, X_\beta] = N_{\alpha,\beta} X_{\alpha+\beta}$ satisfy (Theorem 5.5), Chapter III)

$$N_{\alpha,\beta}^2 = \tfrac{1}{2}\alpha(H_\alpha)\, q(1 - p)$$

and this number is determined by the root pattern. It follows that $N_{\alpha,\beta} = \pm N_{\alpha',\beta'}$ so

$$a_\alpha a_\beta = \pm a_{\alpha+\beta} \qquad \text{if } \alpha, \beta, \alpha + \beta \in \Delta. \tag{4}$$

Let $H' \in \mathfrak{t}^C$. Then the automorphism $B = e^{\text{ad}(H')}$ of $\mathfrak{u}^C$ leaves $\mathfrak{t}^C$ pointwise fixed and

$$BX_\alpha = e^{\alpha(H')} X_\alpha, \qquad \alpha \in \Delta.$$

Let $\alpha_1, \ldots, \alpha_r$ be the system of simple roots. Since $a_\alpha \neq 0$, it is clear that $H' \in \mathfrak{t}^C$ can be chosen such that

$$a_{\alpha_i} = e^{-\alpha_i(H')}, \qquad 1 \leqslant i \leqslant r.$$

Then it follows from (4) by induction that

$$a_\alpha = \pm \prod (a_{\alpha_i})^{n_i} = \pm e^{-\alpha(H')}$$

if $\alpha = \sum_1^r n_i \alpha_i$. This implies by (3) that

$$ABX_\alpha = \epsilon_\alpha X_{\alpha'} \quad (\epsilon_\alpha = \epsilon_{-\alpha} = \pm 1) \qquad \text{and} \qquad ABX_{\alpha_i} = X_{\alpha_i'} \tag{5}$$

for $1 \leqslant i \leqslant r$. Since $\mathfrak{u}$ is spanned by $i\mathfrak{t}$ and $X_\alpha - X_{-\alpha}$, $i(X_\alpha + X_{-\alpha})$ $(\alpha \in \Delta)$, it is clear that AB leaves $\mathfrak{u}$ invariant and therefore gives the desired extension of φ.

Corollary 5.2. *Let A be an automorphism of $\mathfrak{u}$ leaving $\mathfrak{t}$ invariant. Then the extension of A to $\mathfrak{u}^C$ satisfies*

$$AX_\alpha = a_\alpha X_{\alpha^A},$$

where $a_\alpha a_{-\alpha} = 1$ and $| a_\alpha | = 1$.

In fact, $a_\alpha a_{-\alpha} = 1$ as before, but now we have in addition $A\tau = \tau A$. Since $\tau X_\alpha = -X_{-\alpha}$, this implies $a_{-\alpha} = \bar{a}_\alpha$ so $| a_\alpha | = 1$.

Proposition 5.3. *An automorphism A of $\mathfrak{u}$ leaves $\mathfrak{t}$ pointwise fixed if and only if it has the form*

$$A = e^{\operatorname{ad} H}$$

for a suitable element $H \in \mathfrak{t}$.

Proof. Let the extension of A to $\mathfrak{u}^C$ also be denoted by A. In the notation above we have $\alpha = \alpha'$. Hence (4) and (5) take the form $a_\alpha a_\beta = a_{\alpha+\beta}$ and $ABX_\alpha = +X_\alpha$ so AB is the identity. Thus $A = e^{\operatorname{ad} H}$ for some $H \in \mathfrak{t}^C$. The eigenvalues of A are 1 and $e^{\alpha(H)} (\alpha \in \Delta)$. Since the powers of A form a bounded set, $\alpha(H)$ must be purely imaginary for each $\alpha \in \Delta$. Hence $H \in \mathfrak{t}$ and the proposition is proved.

Theorem 5.4. *The factor group $Aut(\mathfrak{u})/Int(\mathfrak{u})$ is isomorphic to $Aut(\Delta)/W$, W denoting the Weyl group, $Aut(\Delta)$ denoting the group of all automorphisms of Δ.*

Proof. Let $E \in \operatorname{Aut}(\mathfrak{u})$. Then $E\mathfrak{t}$ is a maximal abelian subalgebra of $\mathfrak{u}$ so there exists a $B_1 \in \operatorname{Int}(\mathfrak{u})$ such that $B_1 E\mathfrak{t} = \mathfrak{t}$. Consequently, each element in $\operatorname{Aut}(\mathfrak{u})/\operatorname{Int}(\mathfrak{u})$ contains an automorphism leaving $\mathfrak{t}$ invariant. On the other hand, W consists of the automorphisms of Δ induced by members of $\operatorname{Int}(\mathfrak{u})$ leaving $\mathfrak{t}$ invariant. Since W is generated by the reflections in the planes $\alpha(H) = 0$, it is easy to see that W is a normal subgroup of $\operatorname{Aut}(\Delta)$. We obtain now a well-defined mapping S of $\operatorname{Aut}(\mathfrak{u})/\operatorname{Int}(\mathfrak{u})$ into $\operatorname{Aut}(\Delta)/W$ as follows: In each class of $\operatorname{Aut}(\mathfrak{u})$ mod $\operatorname{Int}(\mathfrak{u})$ select an element A leaving $\mathfrak{t}$ invariant; let $A_\mathfrak{t}$ denote the restriction of A to $\mathfrak{t}$ and let

$$S : A \rightarrow A_\mathfrak{t} W.$$

Since W is a normal subgroup of $\text{Aut}(\Delta)$, S is a homomorphism. It is clear from the above that S is one-to-one and Theorem 5.1 shows that S is onto.

Corollary 5.5. *The factor group* $Aut(\mathfrak{u})/Int(\mathfrak{u})$ *is isomorphic to the group* $W_C(\Delta)$ *of automorphisms of* Δ *leaving a given Weyl chamber* $C \subset \mathfrak{t}$ *invariant.*

In fact, each automorphism of Δ permutes the Weyl chambers; thus if $t \in \text{Aut}(\Delta)$, there exists by Theorem 2.12, Chapter VII, a unique $s = s(t) \in W$ such that $s^{-1}t$ maps C into itself. Thus we get a mapping $\psi : t \to s^{-1}t$ of $\text{Aut}(\Delta)$ onto $W_C(\Delta)$. This mapping is a homomorphism: Let $t_1, t_2 \in \text{Aut}(\Delta)$, $s, s_1, s_2 \in W$ such that

$$s_1^{-1}t_1, \qquad s_2^{-1}t_2, \qquad s^{-1}t_1t_2 \in W_C(\Delta);$$

and since W is normal in $\text{Aut}(\Delta)$, select $s^* \in W$ such that $s^*t_1t_2 = s_1^{-1}t_1s_2^{-1}t_2$. But then sC and $(s^*)^{-1}C$ are the same Weyl chamber t_1t_2C, so $s^* = s^{-1}$, proving that ψ is a homomorphism. The kernel of ψ is clearly W, so $W_C(\Delta)$ is isomorphic to the factor group $\text{Aut}(\Delta)/W$. Now apply Theorem 5.4.

The *rank* of a compact Lie algebra $\mathfrak{u}$ is defined as the dimension r of any maximal abelian subalgebra. This is, by definition, also the rank of the complexification $\mathfrak{u}^C$.

Theorem 5.6. *Let* θ *be an involutive automorphism of a compact semisimple Lie algebra* $\mathfrak{u}$. *Let* $\mathfrak{k}$ *denote the set of fixed points of* θ. *Then* $\theta \in Int(\mathfrak{u})$ *if and only if rank* $\mathfrak{u}$ = *rank* $\mathfrak{k}$.

Proof. Let U be any connected Lie group with Lie algebra $\mathfrak{u}$. Suppose first $\theta \in \text{Int}(\mathfrak{u})$. Then $\theta = \text{Ad}(u)$ for some $u \in U$. Now u lies in a maximal torus T of U and θ leaves the Lie algebra $\mathfrak{t}$ of T pointwise fixed. Hence $\mathfrak{t} \subset \mathfrak{k}$ so rank $\mathfrak{k}$ = rank $\mathfrak{u}$. On the other hand, if $\mathfrak{k}$ and $\mathfrak{u}$ have the same rank, there exists a subalgebra $\mathfrak{t} \subset \mathfrak{k}$ which is maximal abelian in $\mathfrak{u}$. Proposition 5.3 shows at once that $\theta \in \text{Int}(\mathfrak{u})$.

Theorem 5.7. *Let* $(\mathfrak{g}, \theta)$ *be an orthogonal symmetric Lie algebra of the noncompact type and let* $\mathfrak{k}$ *denote the set of fixed points of* θ. *Then* $\theta \in Int(\mathfrak{g})$ *if and only if* $\mathfrak{k}$ *contains a maximal abelian subalgebra of* $\mathfrak{g}$.

Let as usual $\mathfrak{g} = \mathfrak{k} + \mathfrak{p}$ and let $\mathfrak{u}$ denote the subspace $\mathfrak{k} + i\mathfrak{p}$ of the complexification $\mathfrak{g}^C$ of $\mathfrak{g}$. Then θ extends to an involutive automorphism of $\mathfrak{g}^C$, also denoted θ, leaving the compact real form $\mathfrak{u}$ invariant. Suppose $\mathfrak{t}$ is a maximal abelian subalgebra of $\mathfrak{g}$ contained in $\mathfrak{k}$. Then $\mathfrak{t}$ is also maximal abelian in $\mathfrak{u}$ so $\theta = e^{\text{ad}\, H}$ where $H \in \mathfrak{t}$. On the other hand, suppose $\theta \in \mathrm{Int}(\mathfrak{g})$. Since $\theta \in \mathrm{Aut}(\mathfrak{u})$ we see that θ lies in the Lie subgroup $\mathrm{Int}(\mathfrak{g}) \cap \mathrm{Aut}(\mathfrak{u})$ of $\mathrm{Int}(\mathfrak{g})$ which has Lie algebra $\mathfrak{g} \cap \mathfrak{u} = \mathfrak{k}$. It follows from Theorem 1.1, Chapter VI, that $\mathrm{Int}(\mathfrak{g}) \cap \mathrm{Aut}(\mathfrak{u}) = \mathrm{Int}(\mathfrak{g}) \cap \mathrm{Int}(\mathfrak{u})$. Hence $\theta \in \mathrm{Int}(\mathfrak{u})$ and Theorem 5.7 follows from Theorem 5.6.

Corollary 5.8. *Let $M = I_0(M)/K$ be a Riemannian globally symmetric space of the noncompact type, K being the isotropy subgroup of $I_0(M)$ at some point $o \in M$. Let $I_0(M)$ and K have Lie algebras $\mathfrak{g}$ and $\mathfrak{k}$, respectively. Then the geodesic symmetry s_0 belongs to $I_0(M)$ if and only if $\mathfrak{k}$ contains a maximal abelian subalgebra of $\mathfrak{g}$.*

In fact, if $s_0 \in I_0(M)$ and if Ad denotes the adjoint representation of $I_0(M)$ then $\theta = \mathrm{Ad}(s_0) \in \mathrm{Int}(\mathfrak{g})$. On the other hand, if $\theta \in \mathrm{Int}(\mathfrak{g})$, then $\theta \in \mathrm{Int}(\mathfrak{g}) \cap \mathrm{Aut}(\mathfrak{u}) = \mathrm{Ad}(K)$ by Theorem 1.1, Chapter VI. Let $s \in K$ such that $\mathrm{Ad}(s) = \theta$. Then $(ds)_o = -I$ so $s = s_o$.

Let θ_1 and θ_2 be two involutive automorphisms of $\mathfrak{u}$ leaving a maximal abelian subalgebra $\mathfrak{t}$ invariant, such that θ_1 and θ_2 are identical on $\mathfrak{t}$. In general θ_1 and θ_2 are not conjugate† in $\mathrm{Aut}(\mathfrak{u})$. However, we have

Theorem 5.9. *Let θ_1 and θ_2 be two involutive automorphisms of $\mathfrak{u}$ such that $\theta_1(H) = \theta_2(H) = -H$ for§ $H \in \mathfrak{t}$. Then there exists a $\sigma \in \mathrm{Int}(\mathfrak{u})$ such that $\theta_2 = \sigma\theta_1\sigma^{-1}$.*

Proof. Let $\{X_\alpha : \alpha \in \Delta\}$ be a Weyl basis of $\mathfrak{u}^C$ mod $\mathfrak{t}^C$ with respect to $\mathfrak{u}$. Then for each $\alpha \in \Delta$

$$\theta_1 X_\alpha = a_\alpha X_{-\alpha}, \tag{6}$$

where the number a_α satisfies

$$a_\alpha a_{-\alpha} = 1, \qquad | a_\alpha | = 1, \qquad a_\alpha a_\beta = -a_{\alpha+\beta},$$

if $\alpha, \beta, \alpha + \beta \in \Delta$. Suppose now Δ ordered in some way; let Δ^+ denote the set of positive roots, $\alpha_1, \ldots, \alpha_r$ the simple roots. There exists a vector $H_1 \in \mathfrak{t}^C$ such that

$$a_{\alpha_j} = -e^{\alpha_j(H_1)} \qquad (1 \leqslant j \leqslant r), \tag{7}$$

and since each a_α has modulus 1, we have $H_1 \in \mathfrak{t}$. Now $(-a_\alpha)(-a_\beta) = (-a_{\alpha+\beta})$ so we obtain from (7) by induction

$$-a_\alpha = e^{\alpha(H_1)}, \qquad \alpha \in \Delta.$$

† An example is given by the spaces *A III* in É. Cartan's list of Riemannian globally symmetric spaces.

§ Such automorphisms exist as a result of Theorem 5.1.

Thus we have for all $\alpha \in \Delta$

$$\theta_1 X_\alpha = -e^{\alpha(H_1)} X_{-\alpha}, \qquad \theta_2 X_\alpha = -e^{\alpha(H_2)} X_{-\alpha}, \tag{8}$$

where H_1 and H_2 are certain fixed vectors in $\mathfrak{t}$. Consider now the automorphism

$$\sigma = e^{\frac{1}{2}\operatorname{ad}(H_1 - H_2)}$$

of $\mathfrak{u}^C$. This automorphism leaves $\mathfrak{u}$ invariant and keeps $\mathfrak{t}$ pointwise fixed. Since

$$\theta_2 \sigma X_\alpha = \theta_2 e^{\frac{1}{2}\alpha(H_1 - H_2)} X_\alpha = -e^{\frac{1}{2}\alpha(H_1 + H_2)} X_{-\alpha},$$

$$\sigma \theta_1 X_\alpha = -e^{\alpha(H_1)} \sigma X_{-\alpha} = -e^{\frac{1}{2}\alpha(H_1 + H_2)} X_{-\alpha},$$

the automorphism σ has the required properties.

Definition. Let $\mathfrak{l}$ be a semisimple Lie algebra over $\boldsymbol{C}$. A real form $\mathfrak{g}$ of $\mathfrak{l}$ is called *normal* if in each Cartan decomposition $\mathfrak{g} = \mathfrak{k} + \mathfrak{p}$ the space $\mathfrak{p}$ contains a maximal abelian subalgebra of $\mathfrak{g}$.

Theorem 5.10. *Each semisimple Lie algebra $\mathfrak{g}^C$ over $\boldsymbol{C}$ has a normal real form and this is unique up to isomorphism.*

Proof. Let $\mathfrak{u}_0$ be a compact real form of $\mathfrak{g}^C$, $\mathfrak{t}_0$ a maximal abelian subalgebra of $\mathfrak{u}_0$, and $\mathfrak{t}_0^C$ the subspace of $\mathfrak{g}^C$ generated by $\mathfrak{t}_0$. Let $\{X_\alpha : \alpha \in \Delta\}$ be a Weyl basis of $\mathfrak{g}^C$ mod $\mathfrak{t}_0^C$ with respect to $\mathfrak{u}_0$. Then the subspace

$$\mathfrak{g}_0 = \sum_{\alpha \in \Delta} \boldsymbol{R} H_\alpha + \sum_{\alpha \in \Delta} \boldsymbol{R} X_\alpha$$

is a real form of $\mathfrak{g}^C$. The conjugation τ of $\mathfrak{g}^C$ with respect to $\mathfrak{u}_0$ leaves $\mathfrak{g}_0$ invariant and if $\mathfrak{k}_0 = \mathfrak{g}_0 \cap \mathfrak{u}_0$, $\mathfrak{p}_0 = \mathfrak{g}_0 \cap (i\mathfrak{u}_0)$, then

$$\mathfrak{g}_0 = \mathfrak{k}_0 + \mathfrak{p}_0$$

is a Cartan decomposition of $\mathfrak{g}_0$. Moreover, $\mathfrak{p}_0$ contains the subspace $\sum_{\alpha \in \Delta} \boldsymbol{R} H_\alpha$ which is a maximal abelian subalgebra of $\mathfrak{g}_0$. Hence $\mathfrak{g}_0$ is a normal real form. On the other hand, let $\mathfrak{g}_1$ be another normal real form of $\mathfrak{g}^C$, and $\mathfrak{g}_1 = \mathfrak{k}_1 + \mathfrak{p}_1$ any Cartan decomposition of $\mathfrak{g}_1$. Then $\mathfrak{u}_1 = \mathfrak{k}_1 + i\mathfrak{p}_1$ is a compact real form of $\mathfrak{g}^C$. The mapping $\theta_i : T + X \to T - X (T \in \mathfrak{k}_i, X \in i\mathfrak{p}_i)$ is an involutive automorphism of $\mathfrak{u}_i (i = 0, 1)$. There exists a maximal abelian subalgebra $\mathfrak{t}_i$ of $\mathfrak{u}_i$ such that $\theta_i(H) = -H$ for $H \in \mathfrak{t}_i$ $(i = 0, 1)$. Here $\mathfrak{t}_0$ is the same as that above. Owing to previous conjugacy theorems, there exists an automorphism A of $\mathfrak{g}^C$ mapping $\mathfrak{u}_1$ onto $\mathfrak{u}_0$ such that $A\mathfrak{t}_1 = \mathfrak{t}_0$. According to Theorem 5.9, the automorphisms $A\theta_1 A^{-1}$ and θ_0 of $\mathfrak{u}_0$ are conjugate within $\operatorname{Int}(\mathfrak{u}_0)$. If $\mathfrak{g}_A$

denotes the real form of $\mathfrak{g}^C$ which corresponds to the involution $A\theta_1 A^{-1}$ of $\mathfrak{u}_0$, then Prop. 2.2, Chapter V, shows that $\mathfrak{g}_A$ and $\mathfrak{g}_0$ are isomorphic. On the other hand, A induces (by restriction) an isomorphism of $\mathfrak{g}_1$ onto $\mathfrak{g}_A$ and consequently $\mathfrak{g}_0$ and $\mathfrak{g}_1$ are isomorphic.

Let $M = I_0(M)/K$ be an irreducible Riemannian globally symmetric space of the noncompact type. The list of such spaces given in the next chapter shows that the Lie group K by itself does not determine M, not even locally. Nevertheless, the next theorem shows that the linear isotropy group K^* determines the curvature tensor of M at $\{K\}$ and therefore (M being simply connected) determines the space M. It is known that for an irreducible Riemannian globally symmetric space the holonomy group and the linear isotropy group have the same identity component. Therefore, the theorem below shows that problem 2 stated in the introduction to Chapter IV has an affirmative solution (after decomposition into irreducible factors).

Theorem 5.11. *Let $\mathfrak{g}_1 = \mathfrak{k}_1 + \mathfrak{p}_1$, $\mathfrak{g}_2 = \mathfrak{k}_2 + \mathfrak{p}_2$ be Cartan decompositions of two semisimple Lie algebras $\mathfrak{g}_1$ and $\mathfrak{g}_2$ over* **R**. *Assume that $\mathrm{ad}_{\mathfrak{g}_i}(\mathfrak{k}_i)$ acts irreducibly on $\mathfrak{p}_i$ $(i = 1, 2)$. Let φ be a one-to-one linear mapping of $\mathfrak{g}_1$ onto $\mathfrak{g}_2$ such that:*

(i) $\varphi(\mathfrak{p}_1) = \mathfrak{p}_2$.

(ii) *The restriction of φ to $\mathfrak{k}_1$ is an isomorphism of $\mathfrak{k}_1$ onto $\mathfrak{k}_2$.*

(iii) $\varphi([T, X]) = [\varphi(T), \varphi(X)]$ *for* $T \in \mathfrak{k}_1$, $X \in \mathfrak{p}_1$.

Then $\mathfrak{g}_1$ and $\mathfrak{g}_2$ are isomorphic.

Proof. Let B_1, B_2, Q_1, and Q_2 denote the Killing forms of $\mathfrak{g}_1$, $\mathfrak{g}_2$, $\mathfrak{k}_1$, and $\mathfrak{k}_2$, respectively.

Then

$$B_i(T, T) = Q_i(T, T) + \mathrm{Tr}_{\mathfrak{p}_i}(\mathrm{ad}_{\mathfrak{g}_i}(T)\,\mathrm{ad}_{\mathfrak{g}_i}(T)) \qquad (T \in \mathfrak{k}_i)$$

for $i = 1, 2$. Consider the bilinear form Q on $\mathfrak{p}_1 \times \mathfrak{p}_1$ given by

$$Q(X, Y) = B_2(\varphi(X), \varphi(Y)), \qquad X, Y \in \mathfrak{p}_1.$$

By (iii) Q is invariant under the action of $\mathfrak{k}_1$ on $\mathfrak{p}_1$, that is,

$$Q([T, X], Y) + Q(X, [T, Y]) = 0, \qquad X, Y \in \mathfrak{p}_1, \qquad T \in \mathfrak{k}_1.$$

Since the action of $\mathfrak{k}_1$ on $\mathfrak{p}_1$ is irreducible, Q is proportional to the restriction of B_1 to $\mathfrak{p}_1 \times \mathfrak{p}_1$; hence

$$B_2(\varphi(X), \varphi(Y)) = dB_1(X, Y), \qquad X, Y \in \mathfrak{p}_1, \tag{9}$$

where d is a constant, $d > 0$. On the other hand, let us compare

$$B_2([\varphi(X), \varphi(Y)], \varphi(T)) \qquad \text{and} \qquad B_2(\varphi([X, Y]), \varphi(T))$$

for $X, Y \in \mathfrak{p}_1$, $T \in \mathfrak{k}_1$. We have

$$\begin{aligned} B_2(\varphi([X, Y]), \varphi(T)) &= Q_2(\varphi([X, Y], \varphi(T)) + \mathrm{Tr}_{\mathfrak{p}_2}(\mathrm{ad}(\varphi([X, Y)]) \,\mathrm{ad}(\varphi(T))) \\ &= Q_1([X, Y], T) + \mathrm{Tr}_{\mathfrak{p}_1}(\mathrm{ad}([X, Y]) \,\mathrm{ad}(T)), \end{aligned}$$

where we have used (ii) and the relation

$$\mathrm{ad}\, T(X) = (\varphi^{-1} \circ \mathrm{ad}(\varphi(T)) \circ \varphi)(X).$$

This proves that

$$B_2(\varphi([X, Y]), \varphi(T)) = B_1([X, Y], T). \tag{10}$$

On the other hand, using (iii) and (9) we have

$$\begin{aligned} B_2([\varphi(X), \varphi(Y)], \varphi(T)) &= B_2(\varphi(X), [\varphi(Y), \varphi(T)]) = B_2(\varphi(X), \varphi([Y, T]) \\ &= dB_1(X, [Y, T]) = dB_1([X, Y], T) \end{aligned}$$

so by (10)

$$B_2([\varphi(X), \varphi(Y)], \varphi(T)) = dB_2(\varphi([X, Y]), \varphi(T)). \tag{11}$$

Since B_2 is strictly negative definite on $\mathfrak{k}_2$, (11) implies

$$\varphi([X, Y]) = d^{-1}[\varphi(X), \varphi(Y)], \qquad X, Y \in \mathfrak{p}_1.$$

The desired isomorphism of $\mathfrak{g}_1$ onto $\mathfrak{g}_2$ is now obtained by defining

$$\psi(T) = \varphi(T) \quad (T \in \mathfrak{k}_1), \qquad \psi(X) = d^{-1/2}\varphi(X) \quad (X \in \mathfrak{p}_1).$$

Remark. We shall prove later (Chapter X, §6) with considerably more effort that two real forms of a simple Lie algebra over $\boldsymbol{C}$ are isomorphic if their maximal compactly imbedded subalgebras are isomorphic.

§6. The Multiplicities

The multiplicities of the restricted roots are important invariants of the noncompact Lie algebra $\mathfrak{g}$ (or the associated symmetric space G/K). In fact, Cartan's classification shows that the triple $(\mathfrak{a}, \Sigma, \boldsymbol{m})$, where $\boldsymbol{m}$ is the multiplicity function, determines $\mathfrak{g}$ up to isomorphism (cf. Chapter X, Exercise F9).

Theorem 6.1. *The following properties of the symmetric space G/K are equivalent*:

(i) *All Cartan subalgebras of* $\mathfrak{g}$ *are conjugate under* $Int(\mathfrak{g})$.

(ii) G/K *has split rank, that is,*

$$\operatorname{rank}(G) = \operatorname{rank}(K) + \operatorname{rank}(G/K).$$

(iii) *All restricted roots have even multiplicity.*

Let $\mathfrak{a} \subset \mathfrak{p}$ be maximal abelian and $\mathfrak{h} \subset \mathfrak{g}$ a Cartan subalgebra containing $\mathfrak{a}$. Let Δ be the set of roots of $(\mathfrak{g}^C, \mathfrak{h}^C)$ and Σ the set of restrictions $\bar{\alpha}$ as $\alpha \in \Delta$. The mapping $H \to -\theta H$ of $\mathfrak{h}^C \to \mathfrak{h}^C$ induces a permutation of Δ, also denoted by $-\theta$. We first prove a simple lemma.

Lemma 6.2. *Let* $\psi \in \Sigma$ *and* $\Delta_\psi = \{\alpha \in \Delta : \bar{\alpha} = \psi\}$. *Then* Δ_ψ *contains an element* α *such that* $\alpha^\theta = -\alpha$ *if and only if the multiplicity* m_ψ *is odd.*

In fact, the map $-\theta$ permutes the elements of Δ_ψ. Identifying an element of Σ with a linear function on $\mathfrak{h}^C$ vanishing on $\mathfrak{h} \cap \mathfrak{k}$, we have $\bar{\alpha} = \frac{1}{2}(\alpha - \alpha^\theta)$. If $\alpha \in \Delta_\psi$, then $-\alpha^\theta = \alpha$ if and only if $\alpha = \psi$; so $-\theta$ permutes the set $\Delta_\psi - (\psi)$ without fixed points. Thus $\psi \in \Delta_\psi$ if and only if the cardinality of Δ_ψ is odd.

Passing now to the proof of Theorem 6.1, assume (i). Extend $\mathfrak{h} \cap \mathfrak{k}$ to a maximal abelian subalgebra $\mathfrak{t}$ of $\mathfrak{k}$ and extend this to a maximal abelian subalgebra $\tilde{\mathfrak{t}}$ of $\mathfrak{g}$. If $H \in \mathfrak{t}$, $Z \in \tilde{\mathfrak{t}}$, then $Z + \theta Z \in \mathfrak{k}$ and

$$[H, Z + \theta Z] = 0 + \theta[\theta H, Z] = \theta[H, Z] = 0, \qquad \text{so} \qquad Z + \theta Z \in \mathfrak{t},$$

whence $\theta\tilde{\mathfrak{t}} \subset \tilde{\mathfrak{t}}$. Therefore

$$\tilde{\mathfrak{t}} = \tilde{\mathfrak{t}} \cap \mathfrak{k} + \tilde{\mathfrak{t}} \cap \mathfrak{p} = \mathfrak{t} + \tilde{\mathfrak{t}} \cap \mathfrak{p}.$$

The elements of $\operatorname{ad}_\mathfrak{g}(\mathfrak{t})$ and of $\operatorname{ad}_\mathfrak{g}(\tilde{\mathfrak{t}} \cap \mathfrak{p})$ are semisimple; since they all commute, the elements of $\operatorname{ad}_\mathfrak{g}(\tilde{\mathfrak{t}})$ are semisimple, so $\tilde{\mathfrak{t}}$ is a Cartan subalgebra. But assuming (i), $\mathfrak{h}$ and $\tilde{\mathfrak{t}}$ are conjugate, so by Prop. 4.5, $\mathfrak{t} = \mathfrak{h} \cap \mathfrak{k}$. This proves (ii).

Now assume (ii). If (iii) were false, select $\psi \in \Sigma$ such that m_ψ is odd. Then by Lemma 6.2 there exists an $\alpha \in \Delta$ such that $\alpha(\mathfrak{h} \cap \mathfrak{k}) = 0$ and $\bar{\alpha} = \psi$. Select $X_\alpha \neq 0$ in $\mathfrak{g}^C$ such that $[H, X_\alpha] = \alpha(H)X_\alpha$ $(H \in \mathfrak{h})$. Writing $X_\alpha = Z_1 + iZ_2$ $(Z_1, Z_2 \in \mathfrak{g})$, we deduce $[H, Z_i] = \alpha(H)Z_i$ $(i = 1, 2)$ since $\alpha \mid \mathfrak{h}$ is real. Assuming, say, $Z_1 \neq 0$, we write $Z_1 = T + X$ $T \in \mathfrak{k}$, $X \in \mathfrak{p}$. Then both T and X are $\neq 0$ and $[H, T] = 0$ for $\mathfrak{h} \cap \mathfrak{k}$. This contradicts (ii).

Finally assume (iii). For each $\alpha \in \Delta$ let $H_\alpha \in \mathfrak{h}^C$ as in Theorem 4.2, Chapter III. Then by Lemma 6.2, $H_\alpha \notin \mathfrak{a}$ for all $\alpha \in \Delta$ which do not vanish identically on $\mathfrak{a}$, hence for all $\alpha \in \Delta$. With X_α as above we have

$$\mathfrak{k}^C = (\mathfrak{h} \cap \mathfrak{k})^C + \sum_{\alpha \in \Delta} C(X_\alpha + \theta X_\alpha).$$

But then $H_\alpha \notin \mathfrak{a}$ $(\alpha \in \Delta)$ implies that $(\mathfrak{h} \cap \mathfrak{k})^C$ is maximal abelian in $\mathfrak{k}^C$, so $\mathfrak{h} \cap \mathfrak{k}$ is maximal abelian in $\mathfrak{k}$. If (i) were false, there would by Prop. 4.5 exist a Cartan subalgebra $\mathfrak{h}_1 \subset \mathfrak{g}$ such that $\theta\mathfrak{h}_1 \subset \mathfrak{h}_1$ and $\mathfrak{h}_1 \cap \mathfrak{p} \subset \mathfrak{a}$ (proper inclusion). But then $\dim(\mathfrak{h}_1 \cap \mathfrak{k}) > \dim(\mathfrak{h} \cap \mathfrak{k})$ which is a contradiction. This proves the theorem.

Remark. Property (ii) can also be stated: $\mathfrak{h} \cap \mathfrak{k}$ is maximal abelian in $\mathfrak{k}$ for any θ-invariant Cartan subalgebra $\mathfrak{h}$ of $\mathfrak{g}$.

Proposition 6.3. *A semisimple Lie algebra $\mathfrak{g}$ which is a normal real form of its complexification has all its restricted roots of multiplicity* 1. *The converse holds if $\mathfrak{k}$ contains no ideal $\neq\{0\}$ of $\mathfrak{g}$.*

Proof. If $\mathfrak{g}$ is a normal real form, the complexification $\mathfrak{a}^C$ is a Cartan subalgebra of $\mathfrak{g}^C$ and $(\mathfrak{g}^C)^\alpha \cap \mathfrak{g} = \mathfrak{g}_\alpha$ $(\alpha \in \Sigma)$, so each multiplicity is 1. On the other hand, suppose $m_\lambda = 1$ for each $\lambda \in \Sigma$. Then $\dim \mathfrak{p}_\lambda = 1$ for each $\lambda \in \Sigma^+$ (§11, Chapter VII). Now $[\mathfrak{p}, \mathfrak{p}] = \mathfrak{k}$ since the orthogonal complement of $[\mathfrak{p}, \mathfrak{p}]$ in $\mathfrak{k}$ would be an ideal in $\mathfrak{g}$. By Lemma 11.3, Chapter VII, this implies $\mathfrak{k}_0 = \sum_{\lambda\in\Sigma^+} [\mathfrak{p}_\lambda, \mathfrak{p}_\lambda] = 0$, that is, $\mathfrak{m} = 0$, so $\mathfrak{a}$ is maximal abelian in $\mathfrak{g}$.

§7. Jordan Decompositions

We recall that an element $X \in \mathfrak{gl}(n, R)$ is called *semisimple* if as an element in $\mathfrak{gl}(n, C)$ it is conjugate to a diagonal matrix. We call such an X *real semisimple* if the eigenvalues are all real. A matrix u is called *unipotent* if $u - 1$ is nilpotent. A semisimple matrix $g \in GL(n, R)$ is called *elliptic* (respectively, *hyperbolic*) if all its (complex) eigenvalues have modulus 1 (respectively, are >0).

Lemma 7.1. *Each $g \in GL(n, R)$ can be uniquely written*

$$g = ehu \tag{1}$$

where $e, h, u \in GL(n, R)$ are elliptic, hyperbolic, and unipotent, respectively, and all three commute.

Proof. By (1), §1, Chapter III, we have the *additive Jordan decomposition* $g = s + n_1$ (s semisimple, n_1 nilpotent, $sn_1 = n_1 s$). Putting $n = s^{-1}n_1$, $u = 1 + s^{-1}n_1$ we have the *multiplicative Jordan decomposition* $g = su$, where u is unipotent and $su = us$. Uniqueness is clear. Now s is conjugate to a diagonal matrix so can be written $s = eh$ where e is elliptic, h hyperbolic, and $eh = he$. Here e and h are uniquely determined because the decomposition of a nonsingular matrix into a product of a unitary and a positive definite matrix is unique. Since

$$ehu = g = ugu^{-1} = ueu^{-1}uhu^{-1}u,$$

the uniqueness of s, u, e, and h implies $ue = eu$, $uh = hu$. Finally, since g is fixed under complex conjugation, the uniqueness of e, h, u in (1) implies that e, h, u all belong to $\boldsymbol{GL}(n, \boldsymbol{R})$ and the lemma is proved.

The decomposition in Lemma 7.1 is called the *complete multiplicative Jordan decomposition.*

We now characterize elliptic, hyperbolic, and unipotent elements in terms of the Iwasawa decomposition.

Theorem 7.2. *Let* $\mathfrak{g}$ *be a semisimple Lie algebra over* $\boldsymbol{R}$ *and* G *the adjoint group* Int($\mathfrak{g}$). *Let* $G = KAN$ *be any Iwasawa decomposition of* G. *Then*:

(i) $g \in G$ *is elliptic if and only if it is conjugate to an element in* K.

(ii) $g \in G$ *is hyperbolic if and only if it is conjugate to an element in* A.

(iii) $g \in G$ *is unipotent if and only if it is conjugate to an element in* N.

We identify the Lie algebra of G with $\mathfrak{g}$ and start with a lemma.

Lemma 7.3. *Let* G *be as in Theorem* 7.2.

(i) *Each hyperbolic element in* G *lies on a one-parameter subgroup of* G.

(ii) *Each unipotent element in* G *lies on a one-parameter subgroup of* G.

Proof. (i) Let θ be a Cartan involution on $\mathfrak{g}$. The transpose $g \to {}^t g$ with respect to the positive definite bilinear form B_θ on $\mathfrak{g}$ leaves Aut($\mathfrak{g}$) invariant; in fact ${}^t g = \theta g^{-1} \theta$ as is easily seen by direct computation. Also Aut($\mathfrak{g}$) is a pseudoalgebraic subgroup of $\boldsymbol{GL}(\mathfrak{g})$ (as defined in Chapter X); in fact the condition $x \in$ Aut($\mathfrak{g}$) is expressed in terms of a basis of $\mathfrak{g}$ by a system $\{P\}$ of second degree polynomials P in the matrix entries of x with coefficients formed by the structure constants of $\mathfrak{g}$. Let $h \in G$ be hyperbolic. Then $h = e^H$ where H is real semisimple. The proof of Lemma 2.3, Chapter X, shows that $e^{tH} \in$ Aut($\mathfrak{g}$) for all $t \in \boldsymbol{R}$. Since G is the identity component of Aut($\mathfrak{g}$), this means $H \in \mathfrak{g}$, proving (i).

Part (ii) is proved in a similar fashion. For this, let $u \in G$ be a unipotent element. By Lemma 4.5, Chapter VI, we have $u = e^{\log u}$ where

$$\log u = N - \tfrac{1}{2}N^2 + \tfrac{1}{3}N^3 - \dots \qquad (N = u - I).$$

Then

$$e^{t \log u} = I + t(N - \tfrac{1}{2}N^2 + \dots) + \tfrac{1}{2}t^2(N - \tfrac{1}{2}N^2 + \dots)^2 + \dots$$

where by the nilpotency of N, these series are finite, so each matrix entry $(e^{t \log u})_{ij}$ is a polynomial $q_{ij}(t)$. Let $P(x_{ij})$ be one of the polynomials from the system $\{P\}$ above definining $\text{Aut}(\mathfrak{g})$. Then

$$u^n \in \text{Aut}(\mathfrak{g}), \quad n \in \boldsymbol{Z} \qquad \Rightarrow \qquad P(q_{ij}(t)) = 0 \quad (t \in \boldsymbol{Z});$$

so since $P(q_{ij}(t))$ is a polynomial in t, it must vanish identically in t. This means $e^{t \log u} \in \text{Aut}(\mathfrak{g})$, so as before, $\log u \in \mathfrak{g}$, proving (ii).

Remark. The elliptic elements of course lie on one-parameter subgroups as well (cf. Prop. 6.10. Chapter II).

Proof of Theorem 7.2. (i) By the compactness of K each $k \in K$ is is elliptic. Conversely, each elliptic element lies in a compact subgroup of G, so by Theorem 2.1, Chapter VI, it is conjugate to an element of K.

For (ii) we note first that the root space decomposition ((1), §1) shows that each element in A is hyperbolic. On the other hand, suppose $h \in G$ hyperbolic. By Lemma 7.3, $h = e^H$ $(H \in \mathfrak{g})$, and by Prop. 4.6, H lies in a Cartan subalgebra of $\mathfrak{g}$. Having real eigenvalues, H lies in a vector part of a Cartan subalgebra, so by Cor. 4.2 (and Lemma 6.3, Chapter V) H is conjugate to an element of $\mathfrak{a}$.

For part (iii) we need the following result which provides useful information about the nilpotent elements.

Theorem 7.4. *Let $\mathfrak{g}$ be semisimple and suppose $X \neq 0$ in $\mathfrak{g}$ such that* $\text{ad}_{\mathfrak{g}}(X)$ *is nilpotent. Then there exist elements H, $Y \in \mathfrak{g}$ such that*

$$[H, X] = 2X, \qquad [H, Y] = -2Y, \qquad [X, Y] = H. \tag{2}$$

(Hence the subalgebra $\mathfrak{l} = \boldsymbol{R}H + \boldsymbol{R}X + \boldsymbol{R}Y$ is isomorphic to $\mathfrak{sl}(2, \boldsymbol{R})$ *under the linear map given by*

$$H \to \begin{pmatrix} 1 & 0 \\ 0 & -1 \end{pmatrix}, \qquad X \to \begin{pmatrix} 0 & 1 \\ 0 & 0 \end{pmatrix}, \qquad Y \to \begin{pmatrix} 0 & 0 \\ 1 & 0 \end{pmatrix}.)$$

This is based on two simple lemmas.

Lemma 7.5. *Let A and B be linear transformations of a finite-dimensional real vector space V. Assume A nilpotent and $[A, [A, B]] = 0$. Then $[A, B]$ and AB are both nilpotent.*

Proof. Putting $C = [A, B]$, we have $[A, C] = 0$, so for $p \in Z^+$,

$$[A, BC^p] = [A, B]\, C^p = C^{p+1}.$$

Thus Trace $(C^p) = 0$ for $p \geqslant 1$, so using (1), §1, Chapter III, we see that C is nilpotent. Now since $D \to [B, D]$ is a derivation and since $[[B, A], A] = 0$, we have $[B, A^p] = p[B, A]A^{p-1}$ $(p \geqslant 1)$. Let $\lambda \in C$ be an eigenvalue of AB on the complexification V^C and $x \neq 0$ a corresponding eigenvector, $ABx = \lambda x$. Let r be the smallest integer such that $A^r x = 0$. Then

$$\lambda A^{r-1}x = A^{r-1}ABx = BA^r x - [B, A^r]\, x = -r[B, A]\, A^{r-1}x.$$

Since $[B, A]$ is nilpotent and $A^{r-1}x \neq 0$, we conclude $\lambda = 0$. Thus AB has all eigenvalues 0, hence is nilpotent.

Lemma 7.6. *Let $H, X \in \mathfrak{g}$ be nonzero elements such that*

$$[H, X] = 2X, \qquad H \in [X, \mathfrak{g}].$$

Then there exists an element $Y \in \mathfrak{g}$ such that (2) *is satisfied.*

Proof. Let $\mathfrak{g}'$ denote the solvable Lie algebra $RH + RX$ and consider the representation $Z \to \operatorname{ad}(Z)$ of $\mathfrak{g}'$ on the complexification $\mathfrak{g}^C$. Using Cor. 2.3, Chapter III, we see that $2\operatorname{ad}_{\mathfrak{g}}(X) = [\operatorname{ad} H, \operatorname{ad} X]$ is nilpotent. Its kernel, say $\mathfrak{c}$, satisfies $[H, \mathfrak{c}] \subset \mathfrak{c}$. By assumption there exists a $Z \in \mathfrak{g}$ such that $H = [Z, X]$. By induction, we have

$$[\operatorname{ad} Z, (\operatorname{ad} X)^n] = n(\operatorname{ad} H - n + 1)(\operatorname{ad} X)^{n-1}.$$

Thus if $\mathfrak{g}_n = (\operatorname{ad} X)^n \mathfrak{g}$ $(n \geqslant 0)$, we have for $T \in \mathfrak{g}_{n-1}$

$$n(\operatorname{ad} H - n + 1)\, T \in \operatorname{ad} Z \operatorname{ad} X(T) + \mathfrak{g}_n,$$

whence by $[H, \mathfrak{c}] \subset \mathfrak{c}$

$$(\operatorname{ad} H - n + 1)(\mathfrak{c} \cap \mathfrak{g}_{n-1}) \subset \mathfrak{c} \cap \mathfrak{g}_n. \tag{3}$$

Suppose $v \in \mathfrak{c}$ is an eigenvector of ad H, ad $H(v) = cv$ $(c \in R)$. Since $\mathfrak{g}_n = 0$ for n sufficiently large, there exists a $k \geqslant 1$ such that $v \in \mathfrak{g}_{k-1}$, $v \notin \mathfrak{g}_k$. Then (3) implies $c = k - 1$. In particular, ad $H + 2$ is nonsingular on $\mathfrak{c}$. But by the Jacobi identity $[X, [H, Z] + 2Z] = 0$, that is, $[H, Z] + 2Z \in \mathfrak{c}$, so there exists a $Z' \in \mathfrak{c}$ satisfying $(\operatorname{ad} H + 2)(Z') = [H, Z] + 2Z$. Then the element $Y = Z' - Z$ has the desired properties and the lemma is proved.

Turning now to the proof of Theorem 7.4, let Z belong to the kernel $\mathfrak{d}$ of $(\text{ad } X)^2$. Then Lemma 7.5 implies $\text{ad } X \circ \text{ad } Z$ nilpotent, so in particular $\text{Tr}(\text{ad } X \text{ ad } Z) = 0$, that is, $B(X, \mathfrak{d}) = 0$. But since

$$B(\text{ad } X)^2 Y, Y') = B(Y, (\text{ad } X)^2 Y')$$

for all $Y, Y' \in \mathfrak{g}$ and since B is nondegenerate, the image $(\text{ad } X)^2\mathfrak{g}$ is the set of elements orthogonal to $\mathfrak{d}$. Thus $X = (\text{ad } X)^2(Y')$ for a suitable $Y' \in \mathfrak{g}$. Then the element $H = -2[X, Y']$ satisfies $[H, X] = 2X$, so the theorem follows from Lemma 7.6.

We can now finish the proof of Theorem 7.2. Let $u \in G$ be a unipotent element. By Lemma 7.3, $u = e^X$ where $X \in \mathfrak{g}$ is nilpotent. By Theorem 7.4, X can be imbedded in a three-dimensional algebra $\mathfrak{l}$ such that (2) is satisfied. Considering the isomorphism of $\mathfrak{l}$ with $\mathfrak{sl}(2, \mathbf{R})$, we see that the linear map θ_0 determined by $(H, X, Y) \to (-H, -Y, -X)$ is a Cartan involution of $\mathfrak{l}$. We extend this to a Cartan involution θ of $\mathfrak{g}$ (Exercise A.8 (ii), Chapter VI). Since $\theta H = -H$, Lemma 1.2, Chapter VI, shows that H is real semisimple. Thus, by (ii), we can conjugate H and X such that H lands in the closed Weyl chamber $\bar{\mathfrak{a}}^+$. But the relation $[H, X] = 2X$ and (1), §1 then shows that X lands in $\mathfrak{n}$. Thus $u = e^X$ is conjugate to an element in N and Theorem 7.2 is proved.

EXERCISES AND FURTHER RESULTS

A. The Decompositions

1. Write out the Bruhat decomposition for $G = \mathbf{SL}(2, \mathbf{R})$, $G = \mathbf{SL}(2, \mathbf{C})$.

2. By direct matrix computation prove the following case of Theorem 1.4 for $G = \mathbf{SL}(k, \mathbf{R})$ with N and $\bar{N}$, respectively, the groups of unipotent uppertriangular (resp. lower-triangular) matrices and MA the group of diagonal matrices of determinant 1.

(i) Each $g \in \mathbf{SL}(k, \mathbf{R})$ can be written in the form

$$g = n_1 s n_2,$$

where $n_1, n_2 \in N$ and s is a matrix which in each row and in each column has only one nonzero entry (Gelfand and Naimark [1], §18)

(ii) Let $g = (g_{lm})$ and $\Delta_i(g) = \det((g_{lm})_{1 \leqslant l,m \leqslant i})$. Then (cf. Godement [3]),

$$\bar{N}MAN = \{g \in G : \Delta_i(g) \neq 0 \text{ for } 1 \leqslant i \leqslant k\},$$

and the diagonal matrix $d(g) = m(g) \exp B(g)$ has entries

$$d(g)_{ii} = \Delta_i(g)/\Delta_{i-1}(g).$$

(iii) Prove the relation ${}^t gg = {}^t n \exp 2H(g)n$ (${}^t g$ = transpose, $n \in N$) from which $H(g)$ is computable.

3. With the notation of §1 suppose G^C is a connected Lie group with Lie algebra $\mathfrak{g}^C$ (the complexification of $\mathfrak{g}$) and assume G taken as the analytic subgroup of G^C with Lie algebra $\mathfrak{g}$. Then (cf. Satake [3])

$$M = M^0(\exp(i\mathfrak{a}) \cap K)$$

if M^0 denotes the identity component of M. (Hint: Use Theorem 2.5, Chapter VII, on the compact analytic subgroup $U \in G^C$ with Lie algebra $\mathfrak{u} = \mathfrak{k} + i\mathfrak{p}$.)

4. Show that if $H_1, H_2 \in \mathfrak{a}$ are Ad(G)-conjugate, then they are $W(\mathfrak{g}, \theta)$-conjugate (compare Exercise 9, Chapter VII).

5. Deduce from Theorem 7.2, Chapter III, Lemma 6.3, Chapter V, and Theorem 2.12, Chapter VII that all Iwasawa decompositions of a connected semisimple Lie group are conjugate.

6.* Let $\mathfrak{g}$ be a semisimple Lie algebra over $\boldsymbol{R}$. For $X \in \mathfrak{g}$ and $g \in \mathrm{Int}(\mathfrak{g})$ let $\mathrm{ad}\, X = S + N$, $g = su$, $g = ehu$, respectively, be the additive, multiplicative and complete multiplicative Jordan decompositions. Then

(i) $S, N \in \mathrm{ad}(\mathfrak{g})$.

(ii) $e, h, u \in \mathrm{Int}(\mathfrak{g})$

(cf. Chevalley [6], Vol. II, §14, Mostow [5], Varadarajan [1], Chapter 3.)

7.* For $a \in A$ let $C(a) \subset \mathfrak{a}$ denote the convex hull of the points $s(\log a)$ $(s \in W)$. Then

$$C(a) = \{H(ak) : k \in K\}$$

and as a consequence, $G = KNK$ (cf. Kostant [8]).

B. The Rank-One Case

1. If $\mathfrak{g}$ is noncompact, semisimple of real rank 1, and not isomorphic to $\mathfrak{sl}(2, \boldsymbol{R})$, then M is connected (cf. the proof of Lemma 3.7).

2. For an indivisible positive root α let $\mathfrak{g}^\alpha \subset \mathfrak{g}$ denote the semisimple subalgebra constructed in Prop. 2.1. Prove that:

(i) $\mathfrak{g}^\alpha$ is simple.

(ii) If $\mathfrak{g}$ has complex structure, $\mathfrak{g}^\alpha \approx \mathfrak{sl}(2, \boldsymbol{C})^{\boldsymbol{R}}$.

3. With the notation of Theorem 3.8 show that the components $A^+(g)$, $H(g)$, and $B(g)$ of the Cartan, Iwasawa, and Bruhat decompositions satisfy the relations:

(i) $2e^{2\alpha(H(\bar{n}))} - e^{2\alpha(B(m^*\bar{n}))} = 2\cosh 2\alpha(A^+(\bar{n})) \quad (\bar{n} \neq e)$.

(ii) Given $\bar{n} \in \bar{N}$, let $\bar{n}_0 \in \bar{N}$ denote the unique element such that

$$m^*k(\bar{n}_0)\,M = k(\bar{n})^{-1}\,m^*M.$$

Then

$$A^+(\bar{n}_0) = A^+(\bar{n}), \qquad H(\bar{n}_0) = H(\bar{n}).$$

(cf. Helgason [1], p. 465).

4. With the notation of §3 show that:

(i) The mapping $gK^* \to gK$ imbeds G^*/K^* in G/K as a totally geodesic submanifold.

(ii) G^*/K^* is the ball $z_1\bar{z}_1 + z_2\bar{z}_2 < 1$ with the action of $\boldsymbol{SU}(2, 1)$ given by

$$Z = \begin{pmatrix} z_1 \\ z_2 \end{pmatrix} \to g \cdot Z = (AZ + B)(CZ + D)^{-1}$$

if

$$g = \begin{pmatrix} A & B \\ C & D \end{pmatrix} \in \boldsymbol{SU}(2, 1).$$

C. Cartan Subalgebras

1. In $\mathfrak{sl}(2, \boldsymbol{R})$ consider the subalgebras

$$\boldsymbol{R}\begin{pmatrix} 0 & 1 \\ 0 & 0 \end{pmatrix}, \qquad \boldsymbol{R}\begin{pmatrix} 1 & 0 \\ 0 & -1 \end{pmatrix}, \qquad \boldsymbol{R}\begin{pmatrix} 0 & 1 \\ -1 & 0 \end{pmatrix}.$$

Show that they are maximal abelian in $\mathfrak{sl}(2, \boldsymbol{R})$, the first is not a Cartan subalgebra, and the last two are nonconjugate Cartan subalgebras.

2. Let G be a connected semisimple Lie group whose Lie algebra $\mathfrak{g}$ has all Cartan subalgebras conjugate. Then K is semisimple (hence compact).

3. Let $\mathfrak{g}$ be a semisimple Lie algebra over $\boldsymbol{C}$ and $\mathfrak{h}$ a Cartan subalgebra. An automorphism σ of $\mathfrak{g}$ leaves $\mathfrak{h}$ pointwise fixed if and only if it has the the form $\sigma = e^{\operatorname{ad} H}$ for some $H \in \mathfrak{h}$.

NOTES

§1. For Theorems 1.2, 1.3, see Notes to Chapter VI-VII. The Bruhat decomposition (Theorem 1.4) was proved by Gelfand and Naimark [1], §18 for $\boldsymbol{SL}(n, \boldsymbol{C})$, for $\boldsymbol{SO}(n, \boldsymbol{C})$, $\boldsymbol{Sp}(n, \boldsymbol{C})$ by Bruhat (cf. [1], p. 187). It was generalized in Chevalley's paper [5] (to "Chevalley groups") and by Harish-Chandra [6] to all semisimple Lie groups. We have followed Harish-Chandra's proof and the customary terminology "Bruhat decomposition."

§2. The rank-one reduction was used by Gindikin and Karpelevič [1] to prove a product formula for Harish-Chandra's **c**-function [9], I. It is also used in Araki's classification method [1].

§3. This section is from Helgason [9], Chapter III, §1. Another computation of the components $H(\bar{n})$, $B(m^*\bar{n})$ is given in Schiffmann [1], p. 24.

§4. The proof of finiteness (Lemma 4.3 and Cor. 4.4) and Prop. 4.5 follows Harish-Chandra [7], §2. These results were also found by Kostant [1] and A. Borel. A relationship between the conjugacy classes of Cartan subalgebras of $\mathfrak{g}$ and the connected components of the subset of regular semisimple elements in $\mathfrak{g}$ is established in Rothschild [1]. A classification of the Cartan subalgebras was carried out by Kostant (unpublished, cf. [1]) and Sugiura [1].

§5-§6. This section is mostly based on Gantmacher [1]. As mentioned in the Notes to Chapter VII, Theorem 5.4 goes back to Cartan [4], p. 366. In Cartan [10] the number of components of $I(M)$ is determined for each noncompact irreducible M and Cor. 5.8 is verified in each case. The existence and uniqueness of the normal real form (Theorem 5.10) is also established by a case-by-case verification in É. Cartan [2]. This verification is particularly cumbersome for the exceptional Lie algebras. The present proof, as well as that of Theorem 5.9, is apparently new. The proof of Theorem 5.11 was worked out jointly with M. Berger. It has also been proved by Kostant, see Simons [1]. Theorem 6.1 is proved in Araki [2].

§7. For Lemma 7.1 and Theorem 7.2 see Mostow [5], [6], §2 and Kostant [8]. Theorem 7.4 was stated by Morozov [1] and given a complete proof by Jacobson [2]. We have followed the presentation in Kostant [4], §3, and Bourbaki [2], Chapter VIII, §11.

CHAPTER X

THE CLASSIFICATION OF SIMPLE LIE ALGEBRAS AND OF SYMMETRIC SPACES

In his papers from the years 1926 and 1927 É. Cartan accomplished a complete classification of irreducible Riemannian globally symmetric spaces. Locally, the question amounts to a classification of all simple Lie algebras over $\boldsymbol{R}$, a problem which É. Cartan had solved already in 1914; this classification is a significant refinement of the Killing-Cartan classification of simple Lie algebras over $\boldsymbol{C}$. Both classifications are carried out in this chapter, and some special properties of the corresponding spaces developed.

In §1 the classification is reduced to Lie algebra problems. In §2 we describe in some detail the symmetric spaces associated with the classical groups. Section 3 deals with the theory of abstract root systems and their classification by means of Dynkin diagrams. As an application, the centers of the simple simply connected compact Lie groups are determined. In §4 the construction and classification of the simple Lie algebras over $\boldsymbol{C}$ is given and their finite-order automorphisms are described in §5. Section 6 is a synthesis of the previous ones, giving the symmetric space classification.

§ 1. Reduction of the Problem

We recall that two orthogonal symmetric Lie algebras $(\mathfrak{l}_1, s_1)$ and $(\mathfrak{l}_2, s_2)$ are called isomorphic if there exists an isomorphism φ of $\mathfrak{l}_1$ onto $\mathfrak{l}_2$ such that $\varphi \circ s_1 = s_2 \circ \varphi$.

The next lemma shows that the classification of simply connected, irreducible Riemannian globally symmetric spaces up to isometry is equivalent to the classification of irreducible orthogonal symmetric Lie algebras up to isomorphism. Here it is assumed as usual that the Riemannian structure is that induced by the Killing form.

Lemma 1.1.

(i) *Let M_1 and M_2 be two irreducible Riemannian globally symmetric spaces and Φ an isometry of M_1 onto M_2. Let $p_1 \in M_1$, $p_2 \in M_2$ such that $\Phi(p_1) = p_2$. Let σ_i denote the automorphism of $I_0(M_i)$ given by $\sigma_i(g) = s_{p_i} \circ g \circ s_{p_i}$ $(i = 1, 2)$. Let s_i denote the corresponding automorphism of the Lie algebra $\mathfrak{l}_i$ of $I_0(M_i)$. Then the orthogonal symmetric*

Lie algebras $(\mathfrak{l}_1, s_1)$ *and* $(\mathfrak{l}_2, s_2)$ *are isomorphic under the differential of the isomorphism* $g \to \Phi \circ g \circ \Phi^{-1}$ *of* $I_0(M_1)$ *onto* $I_0(M_2)$.

(ii) *Let* $(\mathfrak{l}_1, s_1)$ *and* $(\mathfrak{l}_2, s_2)$ *be two irreducible orthogonal symmetric Lie algebras. Let* (L_1, U_1) *and* (L_2, U_2) *be the corresponding Riemannian symmetric pairs,* L_1 *and* L_2 *simply connected,* U_1 *and* U_2 *connected. Let* φ *be an isomorphism of* $(\mathfrak{l}_1, s_1)$ *onto* $(\mathfrak{l}_2, s_2)$. *Then there exists an isometry* Φ *of* L_1/U_1 *onto* L_2/U_2 *such that* φ *is the differential of the isomorphism* $g \to \Phi \circ g \circ \Phi^{-1}$ *of* $I_0(L_1/U_1)$ *onto* $I_0(L_2/U_2)$.

The proof, which is quite canonical, can be omitted.

We have seen that given an irreducible orthogonal symmetric Lie algebra of the noncompact type, there is a associated with it exactly one Riemannian globally symmetric space and this space is simply connected. Owing to the duality for symmetric spaces (Chapter V, §1), it suffices therefore to classify the irreducible compact Riemannian symmetric spaces.

Definition. Let $(\mathfrak{l}, s)$ be an irreducible orthogonal symmetric Lie algebra and let M be a Riemannian globally symmetric space associated with $(\mathfrak{l}, s)$. The space M is said to be of type i ($i =$ I, II, III, IV) if $(\mathfrak{l}, s)$ is of type i in the notation of Theorems 5.3 and 5.4 in Chapter VIII.

As mentioned above it suffices to consider the types I and II. Let us first consider type II.

Proposition 1.2. *The Riemannian globally symmetric spaces of type II are precisely the compact, connected simple Lie groups provided with a Riemannian structure invariant under left and right translations.*

Proof. It is clear from §6, Chapter IV, that a compact, connected, simple Lie group with a bi-invariant Riemannian structure is a Riemannian globally symmetric space of type II.

On the other hand, let $(\mathfrak{l}, s)$ be an orthogonal symmetric Lie algebra of type II. Then $\mathfrak{l} = \mathfrak{l}_1 + \mathfrak{l}_2$ (direct sum) where the ideals $\mathfrak{l}_1$ and $\mathfrak{l}_2$ are interchanged by s. Let $\mathfrak{l}_0$ be a Lie algebra isomorphic to both $\mathfrak{l}_1$ and $\mathfrak{l}_2$ and let I_i denote the isomorphism of $\mathfrak{l}_i$ onto $\mathfrak{l}_0$, $(i = 1, 2)$. Then the mapping

$$I_0 : X + Y \to (I_1X, I_2Y), \qquad X \in \mathfrak{l}_1,\ Y \in \mathfrak{l}_2,$$

is an isomorphism of $\mathfrak{l}$ onto the product algebra $\bar{\mathfrak{l}} = \mathfrak{l}_0 \times \mathfrak{l}_0$. Consider the automorphisms $\bar{s}$ and σ of $\bar{\mathfrak{l}}$ given by

$$\bar{s}(I_1X, I_2Y) = (I_2Y, I_1X),$$

$$\sigma(I_1X, I_2Y) = (I_1X, I_1sY)$$

for $X \in \mathfrak{l}_1$, $Y \in \mathfrak{l}_2$. Then $(\mathfrak{l}, s)$ and $(\bar{\mathfrak{l}}, \bar{s})$ are isomorphic under the mapping $\sigma \circ I_0 : \mathfrak{l} \to \bar{\mathfrak{l}}$. Let (L, H) and $(\bar{L}, \bar{H})$ be corresponding Riemannian symmetric pairs, L and $\bar{L}$ simply connected, H and $\bar{H}$ connected. Then $\bar{L}$ is the product $L_0 \times L_0$ where L_0 is a simply connected Lie group with Lie algebra $\mathfrak{l}_0$ and $\bar{H} = \{(x, x) : x \in L_0\}$. Let Q and $\bar{Q}$ be the Riemannian structures on L/H and $\bar{L}/\bar{H}$, respectively, and let $\psi : L/H \to \bar{L}/\bar{H}$ be the isometry from Lemma 1.1, induced by the isomorphism $\sigma \circ I_0 : \mathfrak{l} \to \bar{\mathfrak{l}}$. Now $\bar{L}/\bar{H}$ is a group $\bar{G}$ with the multiplication

$$(x_1, x_2)\,\bar{H} \cdot (y_1, y_2)\,\bar{H} = (x_1 x_2^{-1} y_1 y_2^{-1}, e)\,\bar{H}$$

(§6, Chapter IV). Note that $\bar{H}$ is *not* a normal subgroup of $\bar{L}$. The Riemannian structure $\bar{Q}$ is invariant under left and right translations on $\bar{G}$. The mapping ψ turns L/H into a group G isomorphic to $\bar{G}$ and Q is invariant under left and right translations on G.

In order to conclude the proof of Prop. 1.2 we need a simple lemma.

Lemma 1.3. *Let N be a subgroup of L such that H is a normal subgroup of N. The the product in L/H satisfies*

$$(xH)(nH) = xnH, \qquad x \in L,\ n \in N.$$

In particular, the factor group N/H is a subgroup of L/H (note that H is not normal in L).

Proof. Let $\bar{N}$ denote the subgroup of $\bar{L}$ which corresponds to N under the isomorphism $\sigma \circ I_0 : \mathfrak{l} \to \bar{\mathfrak{l}}$. Then $\bar{H}$ is a normal subgroup of $\bar{N}$ and it suffices to prove the lemma for $\bar{N}$, $\bar{L}$, and $\bar{H}$ instead of N, L, and H. Consider two arbitrary elements $(n_1, n_2)\bar{H} \in \bar{N}/\bar{H}$ and $(x_1, x_2) \in \bar{L}/\bar{H}$. Since $(n_1, n_2)\ (x, x)\ (n_1, n_2)^{-1} \in \bar{H}$ for each $x \in L_0$, it follows that $n_1^{-1}n_2$ and (therefore) $n_1 n_2^{-1}$ belong to the center of L_0. Hence the product in $\bar{G} = \bar{L}/\bar{H}$ is

$$\begin{aligned}(x_1, x_2)\,\bar{H}(n_1, n_2)\,\bar{H} &= (x_1 x_2^{-1} n_1 n_2^{-1}, e)\,\bar{H} = (x_1 n_1 n_2^{-1} x_2^{-1}, e)\,\bar{H} \\ &= (x_1 n_1, x_2 n_2)\,\bar{H}\end{aligned}$$

and the lemma is proved.

Turning now to Prop. 1.2, let M be an arbitrary Riemannian globally symmetric space associated with $(\mathfrak{l}, s)$. Then there exists a symmetric pair (L_1, H_1) associated with $(\mathfrak{l}, s)$ such that $L_1 = I_0(M)$ and $M = L_1/H_1$. Let π denote the homomorphism of L onto L_1 such that $d\pi$ is the identity mapping $\mathfrak{l} \to \mathfrak{l}$. Then $\pi(H) \subset H_1$. Let φ denote the mapping $xH \to \pi(x)H_1$ of the group $G = L/H$ onto the manifold $M = L_1/H_1$ (see diagram). Then (G, φ) is a covering manifold of M (Lemma 13.4,

Chapter I), and if $o = \varphi(e)$, the geodesic symmetries s_e and s_0 of G and M, respectively, are related by $\varphi \circ s_e = s_o \circ \varphi$. Consider the closed subset $\Gamma = \varphi^{-1}(o)$ of G. We shall prove that Γ is a normal subgroup of G. The set $\tilde{H} = \pi^{-1}(H_1)$ is a closed subgroup of L; its identity component is H and $\Gamma = \tilde{H}/H$ (as subsets of G). Since H is a normal subgroup of $\tilde{H}$, Lemma 1.3 shows that $\varphi(g\gamma) = \varphi(g)$ for $g \in G$, $\gamma \in \Gamma$. Thus Γ is a subgroup of G and we can define a mapping $\beta : g\Gamma \to \varphi(g)$ of the coset space G/Γ onto M.

$$\begin{array}{ccc} G = L/H & & \\ \varphi \downarrow & & \\ M & \xleftarrow{\beta} & G/\Gamma \\ s_0 \downarrow & & \downarrow \eta \\ M & \xleftarrow{\beta} & G/\Gamma \end{array}$$

If $\varphi(x_1H) = \varphi(x_2H)$, then $x_2^{-1}x_1 = h \in \tilde{H}$ so $(x_2H)\,(hH) = x_1H$. This shows that β is one-to-one. Finally consider the mapping $\eta : G/\Gamma \to G/\Gamma$ which corresponds to s_o under β. Then since $s_e(g) = g^{-1}$ and $\varphi \circ s_e = s_o \circ \varphi$, we find that $\eta(g\Gamma) = g^{-1}\Gamma$, $(g \in G)$. This requires that $(g\gamma)^{-1}\Gamma = g^{-1}\Gamma$ for each $g \in G$, $\gamma \in \Gamma$, so Γ is a normal subgroup of G and we can turn $M = \beta(G/\Gamma)$ into a group by requiring β to be an isomorphism. Finally since β is an isometry, the metric on M is invariant under left and right translations. This finishes the proof of Prop. 1.2.

We turn now to the type I. In view of Lemma 1.1, Theorem 5.3, Chapter VIII, and Theorem 9.1, Chapter VII, the classification problem for type I reduces to the following three problems:

A. *Find all compact simple Lie algebras, isomorphic Lie algebras not distinguished.*

B. *For each compact simple Lie algebra* $\mathfrak{u}$, *find all involutive automorphisms of* $\mathfrak{u}$, *not distinguishing automorphisms which are conjugate within the group* $Aut(\mathfrak{u})$.

C. *Find the centers of all compact, simple, simply connected Lie groups.*

Now every complex semisimple Lie algebra $\mathfrak{g}$ has a compact real form $\mathfrak{u}$ which is unique up to an inner automorphism (Cor. 7.3, Chapter III). It is clear that $\mathfrak{g}$ is simple if and only if $\mathfrak{u}$ is simple. Problem A is therefore equivalent to

A′. *Find all simple Lie algebras over* **C**, *isomorphic Lie algebras not distinguished.*

If $\mathfrak{u}$ runs through all compact real forms of a complex semisimple Lie algebra $\mathfrak{g}$, and s runs through all involutive automorphisms of $\mathfrak{u}$, then $\mathfrak{g}_0$, in the dual $(\mathfrak{g}_0, s^*)$ to $(\mathfrak{u}, s)$, runs through all noncompact real forms of $\mathfrak{g}$. We have also seen (Prop. 2.2, Chapter V) that conjugate automorphisms s correspond to isomorphic real forms $\mathfrak{g}_0$. Hence, problem B is equivalent to:

B′. *For each simple Lie algebra* $\mathfrak{g}$ *over* $\boldsymbol{C}$, *find all noncompact real forms of* $\mathfrak{g}$ *up to isomorphism.*

Another problem equivalent to problem *B* is the following:

B″. *For each simple Lie algebra* $\mathfrak{g}$ *over* $\boldsymbol{C}$ *find all involutive automorphisms of* $\mathfrak{g}$, *not distinguishing automorphisms which are conjugate within* Aut($\mathfrak{g}$).

To see the equivalence, let $\mathfrak{u}$ be a compact real form of $\mathfrak{g}$. Let Inv($\mathfrak{u}$) denote the set of involutive automorphisms of $\mathfrak{u}$ and Inv($\mathfrak{u}$)/Aut($\mathfrak{u}$) the set of conjugacy classes in Aut($\mathfrak{u}$) of the elements in Inv($\mathfrak{u}$). We define Inv($\mathfrak{g}$)/Aut($\mathfrak{g}$) similarly. Each $s \in \mathrm{Inv}(\mathfrak{u})$ extends uniquely to $s^c \in \mathrm{Inv}(\mathfrak{g})$; and if s_1, s_2 are conjugate within Aut($\mathfrak{u}$), then s_1^c and s_2^c are conjugate within Aut($\mathfrak{g}$). The following converse gives the equivalence of problems B and B″.

Proposition 1.4. *The mapping*

$$\tau : \mathrm{Inv}(\mathfrak{u})/\mathrm{Aut}(\mathfrak{u}) \rightarrow \mathrm{Inv}(\mathfrak{g})/\mathrm{Aut}(\mathfrak{g})$$

induced by $s \rightarrow s^c$ *is a bijection.*

Proof. If $\sigma \in \mathrm{Inv}(\mathfrak{g})$, then Chapter III, Exercise B.4 implies that σ leaves invariant a compact real form, which by Chapter III, Cor. 7.3, we can write $\varphi\mathfrak{u}$ where $\varphi \in \mathrm{Aut}(\mathfrak{g})$. If s denotes the restriction $\sigma \mid \varphi\mathfrak{u}$, then $\varphi^{-1}s\varphi \in \mathrm{Inv}(\mathfrak{u})$ and $(\varphi^{-1}s\varphi)^c = \varphi^{-1}s^c\varphi = \varphi^{-1}\sigma\varphi$. Hence σ is conjugate to a $\sigma' \in \mathrm{Inv}(\mathfrak{g})$ leaving $\mathfrak{u}$ invariant; thus τ is surjective. Finally, to prove τ is one-to-one suppose $\sigma_1, \sigma_2 \in \mathrm{Inv}(\mathfrak{g})$ are conjugate in Aut($\mathfrak{g}$), $\sigma_2 = g\sigma_1 g^{-1}$, and leave $\mathfrak{u}$ invariant. To see that the restrictions $s_1 = \sigma_1 \mid \mathfrak{u}$ and $s_2 = \sigma_2 \mid \mathfrak{u}$ are conjugate in Aut($\mathfrak{u}$) we write the compact real form $g\mathfrak{u}$ as $g_0\mathfrak{u}$, where $g_0 \in \mathrm{Int}(\mathfrak{g})$ (Cor. 7.3, Chapter III). Using Theorem 1.1, Chapter VI, on g_0, we have $g = pu$ where $p \in \exp(J\mathfrak{u})$, $u \in \tilde{U}$. Here J is the complex structure of $\mathfrak{g}$ and $\tilde{U}$ the normalizer of $\mathfrak{u}$ in Aut($\mathfrak{g}^{\boldsymbol{R}}$). Thus $pu\sigma_1u^{-1}p^{-1} = \sigma_2$ and $\sigma_1, \sigma_2 \in \tilde{U}$. If θ is the Cartan involution of $\mathfrak{g}^{\boldsymbol{R}}$ with respect to $\mathfrak{u}$, we apply the corresponding automorphism of Aut($\mathfrak{g}^{\boldsymbol{R}}$) to this last equation, deriving $p^{-1}u\sigma_1u^{-1}p = \sigma_2$; hence p^2 commutes with $u\sigma_1u^{-1}$. But since exp is one-to-one on $J\mathfrak{u}$, this implies that p itself commutes with $u\sigma_1u^{-1}$, so $u\sigma_1u^{-1} = \sigma_2$, and s_1, s_2 are conjugate within Aut($\mathfrak{u}$) as stated.

The solutions to problems A, B, and C will be given later in the chapter. We conclude this section with a description of the simple Lie algebras in the spirit of Theorems 5.3 and 5.4, Chapter VIII.

Proposition 1.5. *Let* $\mathfrak{g}$ *be a simple Lie algebra over* $\boldsymbol{C}$. *Then* $\mathfrak{g}^{\boldsymbol{R}}$ *is simple and* $(\mathfrak{g}^{\boldsymbol{R}})^{\boldsymbol{C}}$ *is not simple. The simple Lie algebras over* $\boldsymbol{R}$ *fall into two disjoint classes*:

A. The simple Lie algebras over $\boldsymbol{C}$, *considered as real Lie algebras.*

B. The real forms of simple Lie algebras over $\boldsymbol{C}$.

A real simple Lie algebra $\mathfrak{l}$ *belongs to class A if* $\mathfrak{l}^{\boldsymbol{C}}$ *is not simple, and to class B if* $\mathfrak{l}^{\boldsymbol{C}}$ *is simple. In the first case* $\mathfrak{l}^{\boldsymbol{C}}$ *is the direct sum of two simple isomorphic ideals.*

Proof. Let $\mathfrak{a} \subset \mathfrak{g}^{\boldsymbol{R}}$ be an ideal. Then if J is the complex structure of $\mathfrak{g}^{\boldsymbol{R}}$,

$$\mathfrak{a} \supset [\mathfrak{a}, \mathfrak{g}^{\boldsymbol{R}}] \supset [\mathfrak{a}, J\mathfrak{a}] = J[\mathfrak{a}, \mathfrak{a}] = J\mathfrak{a},$$

so $\mathfrak{a}$ is invariant under J and thus an ideal in the complex algebra $\mathfrak{g}$. But $\mathfrak{g}$ is simple, so $\mathfrak{a} = 0$ or $\mathfrak{g}$; thus $\mathfrak{g}^{\boldsymbol{R}}$ is simple. If θ is a Cartan involution of $\mathfrak{g}^{\boldsymbol{R}}$, the dual to $(\mathfrak{g}^{\boldsymbol{R}}, \theta)$ is of type II (Theorem 5.4, Chapter VIII), so $(\mathfrak{g}^{\boldsymbol{R}})^{\boldsymbol{C}}$ is the direct sum of two simple isomorphic ideals. Thus A and B are disjoint classes of simple Lie algebras.

Finally, let $\mathfrak{l}$ be a real simple Lie algebra, $\mathfrak{l}^{\boldsymbol{C}} = \sum_{i=1}^{n} \mathfrak{l}_i$ the decomposition of $\mathfrak{l}^{\boldsymbol{C}}$ into its simple ideals, σ the conjugation of $\mathfrak{l}^{\boldsymbol{C}}$ with respect to $\mathfrak{l}$. Since $\sigma(aZ) = \bar{a}\sigma(Z)(a \in \boldsymbol{C}, Z \in \mathfrak{l}^{\boldsymbol{C}})$ each $\sigma\mathfrak{l}_i$ is a simple ideal in $\mathfrak{l}^{\boldsymbol{C}}$ and if σ_i is any conjugation of $\mathfrak{l}_i$ then $Z \to \sigma\sigma_i Z$ is an isomorphism of $\mathfrak{l}_i$ onto $\sigma\mathfrak{l}_i$. Since σ must permute the $\mathfrak{l}_i$ we may number them such that

$$\mathfrak{l}^{\boldsymbol{C}} = \left\{\sum_{j=1}^{k} \mathfrak{l}_j \oplus \sigma\mathfrak{l}_j\right\} \oplus \sum_{2k+1}^{n} \mathfrak{l}_i\,,$$

where $\sigma\mathfrak{l}_i = \mathfrak{l}_i$ $(2k + 1 \leqslant i \leqslant n)$. Superscript σ denoting fixed points of σ we get

$$\mathfrak{l} = \sum_{j=1}^{k} (\mathfrak{l}_j \oplus \sigma\mathfrak{l}_j)^{\sigma} \oplus \sum_{2k+1}^{n} \mathfrak{l}_i^{\sigma}$$

and each term is an ideal in $\mathfrak{l}$. By simplicity, either $k = 0$ $(n = 1)$ or $k = 1$ $(n = 2)$. In the first case $\mathfrak{l}$ belongs to class A; in the second case $\mathfrak{l}^{\boldsymbol{C}} = \mathfrak{l}_1 \oplus \sigma\mathfrak{l}_1$ and the mapping $X \to X + \sigma X$ is an isomorphism of $\mathfrak{l}_1^{\boldsymbol{R}}$ onto $\mathfrak{l}$. Thus $\mathfrak{l}$ belongs to class B.

§2. The Classical Groups and Their Cartan Involutions

1. Some Matrix Groups and Their Lie Algebras

In order to describe the real and complex classical groups, we adopt the following (mostly standard) notation. Let $(x_1, ..., x_n)$ and $(z_1, ..., z_n)$ be variable points in $\mathbf{R}^n$ and $\mathbf{C}^n$, respectively. A matrix $A = (a_{ij})_{1 \leqslant i,j \leqslant n}$ operates on $\mathbf{C}^n$ by the rule

$$\begin{pmatrix} z_1 \\ \cdot \\ \cdot \\ \cdot \\ z_n \end{pmatrix} \to \begin{pmatrix} a_{11} & \dots & a_{1n} \\ \cdot & & \cdot \\ \cdot & & \cdot \\ \cdot & & \cdot \\ a_{n1} & & a_{nn} \end{pmatrix} \cdot \begin{pmatrix} z_1 \\ \cdot \\ \cdot \\ \cdot \\ z_n \end{pmatrix}.$$

As before, E_{ij} denotes the matrix $(\delta_{ai}\delta_{bj})_{1 \leqslant a,b \leqslant n}$. The transpose and conjugate of a matrix A are denoted by ${}^tA$ and $\bar{A}$, respectively; A is called skew symmetric if $A + {}^tA = 0$, Hermitian if ${}^tA = \bar{A}$, skew Hermitian if ${}^tA + \bar{A} = 0$.

If I_n denotes the unit matrix of order n, we put

$$I_{p,q} = \begin{pmatrix} -I_p & 0 \\ 0 & I_q \end{pmatrix}, \qquad J_n = \begin{pmatrix} 0 & I_n \\ -I_n & 0 \end{pmatrix},$$

$$K_{p,q} = \begin{pmatrix} -I_p & 0 & 0 & 0 \\ 0 & I_q & 0 & 0 \\ 0 & 0 & -I_p & 0 \\ 0 & 0 & 0 & I_q \end{pmatrix}.$$

The multiplicative group of complex numbers of modulus 1 will be denoted by $\mathbf{T}$.

$\mathbf{GL}(n, \mathbf{C})$, $(\mathbf{GL}(n, \mathbf{R}))$: The group of complex (real) $n \times n$ matrices of determinant $\neq 0$.

$\mathbf{SL}(n, \mathbf{C})$, $(\mathbf{SL}(n, \mathbf{R}))$: The group of complex (real) $n \times n$ matrices of determinant 1.

$\mathbf{U}(p, q)$: The group of matrices g in $\mathbf{GL}(p + q, \mathbf{C})$ which leave invariant the Hermitian form

$$-z_1\bar{z}_1 - \dots - z_p\bar{z}_p + z_{p+1}\bar{z}_{p+1} + \dots + z_{p+q}\bar{z}_{p+q}, \qquad \text{i.e.,} \quad {}^tgI_{p,q}\bar{g} = I_{p,q}.$$

We put $\mathbf{U}(n) = \mathbf{U}(0, n) = \mathbf{U}(n, 0)$ and $\mathbf{SU}(p, q) = \mathbf{U}(p, q) \cap \mathbf{SL}(p + q, \mathbf{C})$, $\mathbf{SU}(n) = \mathbf{U}(n) \cap \mathbf{SL}(n, \mathbf{C})$. Moreover, let $\mathbf{S}(\mathbf{U}_p \times \mathbf{U}_q)$ denote the set of matrices

$$\begin{pmatrix} g_1 & 0 \\ 0 & g_2 \end{pmatrix},$$

where $g_1 \in \mathbf{U}(p)$, $g_2 \in \mathbf{U}(q)$ and $\det g_1 \det g_2 = 1$.

$SU^*(2n)$: The group of matrices in $SL(2n, C)$ which commute with the transformation ψ of C^{2n} given by

$$(z_1, \ldots, z_n, z_{n+1}, \ldots, z_{2n}) \to (\bar{z}_{n+1}, \ldots, \bar{z}_{2n}, -\bar{z}_1, \ldots, -\bar{z}_n).$$

$SO(n, C)$: The group of matrices g in $SL(n, C)$ which leave invariant the quadratic form

$$z_1^2 + \ldots + z_n^2, \qquad \text{i.e.,} \quad {}^t g g = I_n.$$

$SO(p, q)$: The group of matrices g in $SL(p + q, R)$ which leave invariant the quadratic form

$$-x_1^2 - \ldots - x_p^2 + x_{p+1}^2 + \ldots + x_{p+q}^2, \qquad \text{i.e.,} \quad {}^t g I_{p,q} g = I_{p,q}.$$

We put $SO(n) = SO(0, n) = SO(n, 0)$.

$SO^*(2n)$: The group of matrices in $SO(2n, C)$ which leave invariant the skew Hermitian form

$$-z_1\bar{z}_{n+1} + z_{n+1}\bar{z}_1 - z_2\bar{z}_{n+2} + z_{n+2}\bar{z}_2 - \ldots - z_n\bar{z}_{2n} + z_{2n}\bar{z}_n.$$

Thus $g \in SO^*(2n) \Leftrightarrow {}^t g J_n \bar{g} = J_n, \ {}^t g g = I_{2n}$.

$Sp(n, C)$: The group of matrices g in $GL(2n, C)$ which leave invariant the exterior form

$$z_1 \wedge z_{n+1} + z_2 \wedge z_{n+2} + \ldots + z_n \wedge z_{2n}, \qquad \text{i.e.,} \quad {}^t g J_n g = J_n.$$

$Sp(n, R)$: The group of matrices g in $GL(2n, R)$ which leave invariant the exterior form

$$x_1 \wedge x_{n+1} + x_2 \wedge x_{n+2} + \ldots + x_n \wedge x_{2n}, \qquad \text{i.e.,} \quad {}^t g J_n g = J_n.$$

$Sp(p, q)$: The group of matrices g in $Sp(p + q, C)$ which leave invariant the Hermitian form

$${}^t Z K_{p,q} \bar{Z}, \qquad \text{i.e.,} \quad {}^t g K_{p,q} \bar{g} = K_{p,q}.$$

We put $Sp(n) = Sp(0, n) = Sp(n, 0)$. It is clear that $Sp(n) = Sp(n, C) \cap U(2n)$.

The groups listed above are all topological Lie subgroups of a general linear group. The Lie algebra of the general linear group $GL(n, C)$ can (as in Chapter II, §1) be identified with the Lie algebra $\mathfrak{gl}(n, C)$ of all complex $n \times n$ matrices, the bracket operation being $[A, B] = AB - BA$. The Lie algebra for each of the groups above is then canonically identified with a subalgebra of $\mathfrak{gl}(n, C)$, considered as a real Lie algebra. These Lie algebras will be denoted by the corresponding small German letters, $\mathfrak{sl}(n, R)$, $\mathfrak{su}(p, q)$, etc.

Now, if G is a Lie group with Lie algebra $\mathfrak{g}$, then the Lie algebra $\mathfrak{h}$ of a topological Lie subgroup H of G is given by

$$\mathfrak{h} = \{X \in \mathfrak{g} : \exp tX \in H \text{ for } t \in \boldsymbol{R}\}. \tag{1}$$

Using this fact (Chapter II, §2) we can describe the Lie algebras of the groups above more explicitly. Since the computation is fairly similar for all the groups we shall give the details only in the cases $\boldsymbol{SU}^*(2n)$ and $\boldsymbol{Sp}(n, \boldsymbol{C})$. Case $\boldsymbol{SO}(p, q)$ was done in Chapter V, §2.

$\mathfrak{gl}(n, \boldsymbol{C})$, $(\mathfrak{gl}(n, \boldsymbol{R}))$: {all $n \times n$ complex (real) matrices},

$\mathfrak{sl}(n, \boldsymbol{C})$, $(\mathfrak{sl}(n, \boldsymbol{R}))$: {all $n \times n$ complex (real) matrices of trace 0},

$$\mathfrak{u}(p, q) : \left\{\begin{pmatrix} Z_1 & Z_2 \\ {}^t\bar{Z}_2 & Z_3 \end{pmatrix} \,\middle|\, \begin{array}{l} Z_1, Z_3 \text{ skew Hermitian of order } p \text{ and } q, \\ \text{respectively, } Z_2 \text{ arbitrary} \end{array}\right\},$$

$$\mathfrak{su}(p, q) : \left\{\begin{pmatrix} Z_1 & Z_2 \\ {}^t\bar{Z}_2 & Z_3 \end{pmatrix} \,\middle|\, \begin{array}{l} Z_1, Z_3 \text{ skew Hermitian, of order } p \text{ and } q, \\ \text{respectively, } \operatorname{Tr} Z_1 + \operatorname{Tr} Z_3 = 0,\ Z_2 \text{ arbitrary} \end{array}\right\},$$

$$\mathfrak{su}^*(2n) : \left\{\begin{pmatrix} Z_1 & Z_2 \\ -\bar{Z}_2 & \bar{Z}_1 \end{pmatrix} \,\middle|\, \begin{array}{l} Z_1, Z_2\ n \times n \text{ complex matrices} \\ \operatorname{Tr} Z_1 + \operatorname{Tr} \bar{Z}_1 = 0 \end{array}\right\},$$

$\mathfrak{so}(n, \boldsymbol{C})$: {all $n \times n$ skew symmetric complex matrices},

$$\mathfrak{so}(p, q) : \left\{\begin{pmatrix} X_1 & X_2 \\ {}^tX_2 & X_3 \end{pmatrix} \,\middle|\, \begin{array}{l} \text{All } X_i \text{ real, } X_1, X_3 \text{ skew symmetric of order} \\ p \text{ and } q, \text{ respectively, } X_2 \text{ arbitrary} \end{array}\right\},$$

$$\mathfrak{so}^*(2n) : \left\{\begin{pmatrix} Z_1 & Z_2 \\ -\bar{Z}_2 & \bar{Z}_1 \end{pmatrix} \,\middle|\, \begin{array}{l} Z_1, Z_2\ n \times n \text{ complex matrices,} \\ Z_1 \text{ skew, } Z_2 \text{ Hermitian} \end{array}\right\},$$

$$\mathfrak{sp}(n, \boldsymbol{C}) : \left\{\begin{pmatrix} Z_1 & Z_2 \\ Z_3 & -{}^tZ_1 \end{pmatrix} \,\middle|\, \begin{array}{l} Z_i \text{ complex } n \times n \text{ matrices,} \\ Z_2 \text{ and } Z_3 \text{ symmetric} \end{array}\right\},$$

$$\mathfrak{sp}(n, \boldsymbol{R}) : \left\{\begin{pmatrix} X_1 & X_2 \\ X_3 & -{}^tX_1 \end{pmatrix} \,\middle|\, \begin{array}{l} X_1, X_2, X_3 \text{ real } n \times n \text{ matrices,} \\ X_2, X_3 \text{ symmetric} \end{array}\right\},$$

$$\mathfrak{sp}(p, q) : \left\{\begin{pmatrix} Z_{11} & Z_{12} & Z_{13} & Z_{14} \\ {}^t\bar{Z}_{12} & Z_{22} & {}^tZ_{14} & Z_{24} \\ -\bar{Z}_{13} & \bar{Z}_{14} & \bar{Z}_{11} & -\bar{Z}_{12} \\ {}^t\bar{Z}_{14} & -\bar{Z}_{24} & -{}^tZ_{12} & \bar{Z}_{22} \end{pmatrix} \,\middle|\, \begin{array}{l} Z_{ij} \text{ complex matrix; } Z_{11} \text{ and } Z_{13} \text{ of} \\ \text{order } p,\ Z_{12} \text{ and } Z_{14}\ p \times q \text{ matrices,} \\ Z_{11} \text{ and } Z_{22} \text{ are skew Hermitian,} \\ Z_{13} \text{ and } Z_{24} \text{ are symmetric} \end{array}\right\}.$$

Proof for $\boldsymbol{SU}^*(2n)$. By the definition of this group, we have $g \in \boldsymbol{SU}^*(2n)$ if and only if $g\psi = \psi g$ and $\det g = 1$. This shows that $A \in \mathfrak{su}^*(2n)$ if and only if $A\psi = \psi A$ and $\operatorname{Tr} A = 0$. Writing A in the form

$$A = \begin{pmatrix} A_1 & A_2 \\ A_3 & A_4 \end{pmatrix},$$

where A_i are $n \times n$ complex matrices we see that if U and V are $n \times 1$ matrices, then

$$A\psi\begin{pmatrix}U\\V\end{pmatrix} = \begin{pmatrix}A_1 & A_2\\A_3 & A_4\end{pmatrix}\begin{pmatrix}\bar{V}\\-\bar{U}\end{pmatrix} = \begin{pmatrix}A_1\bar{V} - A_2\bar{U}\\A_3\bar{V} - A_4\bar{U}\end{pmatrix},$$

$$\psi A\begin{pmatrix}U\\V\end{pmatrix} = \psi\begin{pmatrix}A_1U + A_2V\\A_3U + A_4V\end{pmatrix} = \begin{pmatrix}\bar{A}_3\bar{U} + \bar{A}_4\bar{V}\\-\bar{A}_1\bar{U} - \bar{A}_2\bar{V}\end{pmatrix}.$$

It follows that $\bar{A}_3 = -A_2$, $A_1 = \bar{A}_4$ as desired.

Proof for $\boldsymbol{Sp}(n, \boldsymbol{C})$. Writing symbolically

$$2(z_1 \wedge z_{n+1} + \dots + z_n \wedge z_{2n}) = (z_1, \dots, z_{2n}) \wedge J_n{}^t(z_1, \dots, z_{2n})$$

it is clear that $g \in \boldsymbol{Sp}(n, \boldsymbol{C})$ if and only if

$${}^tgJ_ng = J_n.$$

Using this for $g = \exp tZ$ $(t \in \boldsymbol{R})$, we find since $A \exp ZA^{-1} = \exp(AZA^{-1})$, ${}^t(\exp Z) = \exp {}^tZ$,

$$\exp t(J_n^{-1}\,{}^tZJ_n) = \exp(-tZ) \qquad (t \in \boldsymbol{R}),$$

so $Z \in \mathfrak{sp}(n, \boldsymbol{C})$ if an only if

$${}^tZJ_n + J_nZ = 0. \tag{2}$$

Writing Z in the form

$$Z = \begin{pmatrix}Z_1 & Z_2\\Z_3 & Z_4\end{pmatrix},$$

where Z_i is a complex $n \times n$ matrix, condition (2) is equivalent to ${}^tZ_1 + Z_4 = 0$, $Z_2 = {}^tZ_2$, $Z_3 = {}^tZ_3$.

2. Connectivity Properties

Having described the Lie algebras, we shall now discuss the connectivity of the groups defined.

Lemma 2.1. *Let $\approx$ denote topological isomorphism, and $\sim$ a homeomorphism. We then have*

(a) $\boldsymbol{SO}(2n) \cap \boldsymbol{Sp}(n) \approx \boldsymbol{U}(n)$.

(b) $\boldsymbol{Sp}(p, q) \cap \boldsymbol{U}(2p + 2q) \approx \boldsymbol{Sp}(p) \times \boldsymbol{Sp}(q)$.

(c) $\boldsymbol{Sp}(n, \boldsymbol{R}) \cap \boldsymbol{U}(2n) \approx \boldsymbol{U}(n)$.

(d) $\boldsymbol{SO}^*(2n) \cap \boldsymbol{U}(2n) \approx \boldsymbol{U}(n)$.

(e) $\boldsymbol{SU}(p, q) \cap \boldsymbol{U}(p+q) = \boldsymbol{S}(\boldsymbol{U}_p \times \boldsymbol{U}_q) \sim \boldsymbol{SU}(p) \times \boldsymbol{T} \times \boldsymbol{SU}(q)$.

(f) $\boldsymbol{SU}^*(2n) \cap \boldsymbol{U}(2n) = \boldsymbol{Sp}(n)$.

Proof. (a) Each $g \in \boldsymbol{Sp}(n)$ has determinant 1 so $g \in \boldsymbol{SO}(2n) \cap \boldsymbol{Sp}(n)$ is equivalent to ${}^tgg = I_{2n}$, ${}^tgJ_ng = J_n$, ${}^tg\bar{g} = I_{2n}$. Writing

$$g = \begin{pmatrix} A & B \\ C & D \end{pmatrix}$$

these last relations amount to g real, $A = D$, $B = -C$, $A^tB - B^tA = 0$, $A^tA + B^tB = I_n$. But the last two formulas express simply $A + iB \in \boldsymbol{U}(n)$. For part (b), let

$$V = \{g \in \boldsymbol{GL}(2p+2q, \boldsymbol{C}) : {}^tgK_{p,q}\bar{g} = K_{p,q}\}.$$

Then

$$g \in \boldsymbol{U}(2p+2q) \cap V \quad \Leftrightarrow \quad {}^tg\bar{g} = I_{2p+2q}, \qquad {}^tgK_{p,q}\bar{g} = K_{p,q}.$$

But the last two relations are equivalent to

$$g = \begin{pmatrix} X_{11} & 0 & X_{13} & 0 \\ 0 & X_{22} & 0 & X_{24} \\ X_{31} & 0 & X_{33} & 0 \\ 0 & X_{42} & 0 & X_{44} \end{pmatrix} \qquad \text{where} \qquad \begin{matrix} \begin{pmatrix} X_{11} & X_{13} \\ X_{31} & X_{33} \end{pmatrix} \in \boldsymbol{U}(2p) \\ \begin{pmatrix} X_{22} & X_{24} \\ X_{42} & X_{44} \end{pmatrix} \in \boldsymbol{U}(2q). \end{matrix} \tag{3}$$

By definition

$$\boldsymbol{Sp}(p, q) = \boldsymbol{Sp}(p+q, \boldsymbol{C}) \cap V$$

so

$$\boldsymbol{Sp}(p, q) \cap \boldsymbol{U}(2p+2q) = \boldsymbol{Sp}(p+q, \boldsymbol{C}) \cap \boldsymbol{U}(2p+2q) \cap V.$$

Thus, g in (3) belongs to $\boldsymbol{Sp}(p, q) \cap \boldsymbol{U}(2p+2q)$ if and only if ${}^tgJ_{p+q}g = J_{p+q}$ or equivalently

$$\begin{pmatrix} X_{11} & X_{13} \\ X_{31} & X_{33} \end{pmatrix} \in \boldsymbol{U}(2p) \cap \boldsymbol{Sp}(p, \boldsymbol{C}) = \boldsymbol{Sp}(p)$$

and

$$\begin{pmatrix} X_{22} & X_{24} \\ X_{42} & X_{44} \end{pmatrix} \in U(2q) \cap \boldsymbol{Sp}(q, \boldsymbol{C}) = \boldsymbol{Sp}(q).$$

This proves (b). For (c) we only have to note that

$$\boldsymbol{Sp}(n, \boldsymbol{R}) \cap \boldsymbol{U}(2n) = \boldsymbol{Sp}(n) \cap \boldsymbol{SO}(2n),$$

which by (a) is isomorphic to $U(n)$. Part (d) is also easy; in fact, $g \in SO^*(2n)$ by definition if and only if ${}^t g g = I_{2n}$ and ${}^t g J_n \bar{g} = J_n$.
Thus

$$SO^*(2n) \cap U(2n) = SO(2n) \cap Sp(n, C) = SO(2n) \cap Sp(n) \approx U(n).$$

Part (e). We have

$$g \in SU(p, q) \cap U(p+q) \quad \Leftrightarrow \quad g = \begin{pmatrix} g_1 & 0 \\ 0 & g_2 \end{pmatrix},$$

where $g_1 \in U(p)$, $g_2 \in U(q)$ and $\det g_1 \det g_2 = 1$. Such a matrix can be written

$$\begin{pmatrix} g_1 & 0 \\ 0 & g_2 \end{pmatrix} = \begin{pmatrix} \det g_1 & 0 & 0 & 0 \\ 0 & 1 & & \\ \vdots & & \ddots & \\ 0 & & 1 & 0 \\ 0 & & 0 & \det g_2 \end{pmatrix} \cdot \begin{pmatrix} \gamma_1 & 0 \\ 0 & \gamma_2 \end{pmatrix},$$

where $\gamma_1 \in SU(p)$, $\gamma_2 \in SU(q)$. We have therefore a mapping

$$g \to (\gamma_1, \det g_1, \gamma_2)$$

of $SU(p, q) \cap U(p+q)$ into $SU(p) \times T \times SU(q)$. This mapping is not in general a homomorphism but it is continuous, one-to-one and onto; hence $SU(p, q) \cap U(p+q)$ is homeomorphic to $SU(p) \times T \times SU(q)$. Finally, $g \in SU^*(2n)$ if and only if $\bar{g} J_n = J_n g$ and $\det g = 1$. Hence $g \in SU^*(2n) \cap U(2n)$ if and only if $\bar{g} J_n = J_n g$, ${}^t g \bar{g} = I_{2n}$, $\det g = 1$. However, these conditions are equivalent to ${}^t g J_n g = J_n$, ${}^t g \bar{g} = I_{2n}$ or $g \in Sp(n)$. This finishes the proof of the lemma.

The following lemma is well known, see, e.g., Chevalley [2].

Lemma 2.2.

(a) *The groups* $GL(n, C)$, $SL(n, C)$, $SL(n, R)$, $SO(n, C)$, $SO(n)$, $SU(n)$, $U(n)$, $Sp(n, C)$, $Sp(n)$ *are all connected.*

(b) *The group* $GL(n, R)$ *has two connected components.*

In order to determine the connectivity of the remaining groups we need another lemma.

Definition. Let G be a subgroup of the general linear group $GL(n, C)$. Let $z_{ij}(\sigma)$ $(1 \leqslant i, j \leqslant n)$ denote the matrix elements of an arbitrary $\sigma \in GL(n, C)$, and let $x_{ij}(\sigma)$ and $y_{ij}(\sigma)$ be the real and imaginary part of $z_{ij}(\sigma)$. The group G is called a *pseudoalgebraic* subgroup of $GL(n, C)$

if there exists a set of polynomials P_β in $2n^2$ arguments such that $\sigma \in G$ if and only if $P_\beta(\dots x_{ij}(\sigma), y_{ij}(\sigma), \dots) = 0$ for all P_β.

A pseudoalgebraic subgroup of $\boldsymbol{GL}(n, \boldsymbol{C})$ is a closed subgroup, hence a topological Lie subgroup.

Lemma 2.3.† *Let G be a pseudoalgebraic subgroup of $\boldsymbol{GL}(n, \boldsymbol{C})$ such that the condition $g \in G$ implies ${}^t\bar{g} \in G$. Then there exists an integer $d \geq 0$ such that G is homeomorphic to the topological product of $G \cap \boldsymbol{U}(n)$ and $\boldsymbol{R}^d$.*

Proof. We first remark that if an exponential polynomial $Q(t) = \sum_{j=1}^{n} c_j e^{b_j t}$ $(b_j \in \boldsymbol{R},\ c_j \in \boldsymbol{C})$ vanishes whenever t is an integer then $Q(t) = 0$ for all $t \in \boldsymbol{R}$. Let $\mathfrak{h}(n)$ denote the vector space of all Hermitian $n \times n$ matrices. Then exp maps $\mathfrak{h}(n)$ homeomorphically onto the space $\boldsymbol{P}(n)$ of all positive definite Hermitian $n \times n$ matrices (see Chevalley [2], Prop. 5, §IV, Chapter I). Let $H \in \mathfrak{h}(n)$. We shall prove

$$\text{If } \exp H \in G \cap \boldsymbol{P}(n), \text{ then } \exp tH \in G \cap \boldsymbol{P}(n) \text{ for } t \in \boldsymbol{R}. \tag{4}$$

There exists a matrix $u \in \boldsymbol{U}(n)$ such that uHu^{-1} is a diagonal matrix. Since the group uGu^{-1} is pseudoalgebraic as well as G, we may assume that H in (4) is a diagonal matrix. Let $h_1, \dots, h_n$ be the (real) diagonal elements of H. The condition $\exp H \in G \cap \boldsymbol{P}(n)$ means that the numbers $e^{h_1}, \dots, e^{h_n}$ satisfy a certain set of algebraic equations. Since $\exp kH \in G \cap \boldsymbol{P}(n)$ for each integer k, the numbers $e^{kh_1}, \dots, e^{kh_n}$ also satisfy these algebraic equations and by the remark above the same is the case if k is any real number. This proves (4).

Each $g \in \boldsymbol{GL}(n, \boldsymbol{C})$ can be decomposed uniquely $g = up$ where $u \in \boldsymbol{U}(n)$, $p \in \boldsymbol{P}(n)$. Here u and p depend continuously on g. If $g \in G$, then ${}^t\bar{g}g = p^2 \in G \cap \boldsymbol{P}(n)$ so by (4) $p \in G \cap \boldsymbol{P}(n)$ and $u \in G \cap \boldsymbol{U}(n)$. The mapping $g \to (u, p)$ is a one-to-one mapping of G onto the product $(G \cap \boldsymbol{U}(n)) \times (G \cap \boldsymbol{P}(n))$ and since G carries the relative topology of $\boldsymbol{GL}(n, \boldsymbol{C})$, this mapping is a homeomorphism.

The Lie algebra $\mathfrak{gl}(n, \boldsymbol{C})$ is a direct sum

$$\mathfrak{gl}(n, \boldsymbol{C}) = \mathfrak{u}(n) + \mathfrak{h}(n).$$

Since the Lie algebra $\mathfrak{g}$ of G is invariant under the involutive automorphism $X \to -{}^t\bar{X}$ of $\mathfrak{gl}(n, \boldsymbol{C})$ we have

$$\mathfrak{g} = \mathfrak{g} \cap \mathfrak{u}(n) + \mathfrak{g} \cap \mathfrak{h}(n).$$

It is obvious that $\exp(\mathfrak{g} \cap \mathfrak{h}(n)) \subset G \cap \boldsymbol{P}(n)$. On the other hand, each $p \in G \cap \boldsymbol{P}(n)$ can be written uniquely $p = \exp H$ where $H \in \mathfrak{h}(n)$; by

† Compare Chevalley [2], p. 201.

(4), $H \in \mathfrak{h}(n) \cap \mathfrak{g}$, so exp induces a homeomorphism of $\mathfrak{g} \cap \mathfrak{h}(n)$ onto $G \cap \boldsymbol{P}(n)$. This proves the lemma.

Lemma 2.4.

(a) *The groups* $\boldsymbol{SU}(p, q)$, $\boldsymbol{SU}^*(2n)$, $\boldsymbol{SO}^*(2n)$, $\boldsymbol{Sp}(n, \boldsymbol{R})$, *and* $\boldsymbol{Sp}(p, q)$ *are all connected.*

(b) *The group* $\boldsymbol{SO}(p, q)$ $(0 < p < p + q)$ *has two connected components.*

Proof. All these groups are pseudoalgebraic subgroups of the corresponding general linear group and have the property that $g \in G \Rightarrow {}^t\bar{g} \in G$. Part (a) is therefore an immediate consequence of Lemma 2.3 and Lemma 2.1. For (b) we consider the intersection $\boldsymbol{SO}(p, q) \cap \boldsymbol{U}(p + q) = \boldsymbol{SO}(p, q) \cap \boldsymbol{SO}(p + q)$. This consists of all matrices of the form

$$\begin{pmatrix} A & 0 \\ 0 & B \end{pmatrix},$$

where A and B are orthogonal matrices of order p and q respectively satisfying $\det A \det B = 1$. It follows again from Lemma 2.3 that $\boldsymbol{SO}(p, q)$ has two components.

3. The Involutive Automorphisms of the Classical Compact Lie Algebras

Let $\mathfrak{u}$ be a compact simple Lie algebra, θ an involutive automorphism of $\mathfrak{u}$; let $\mathfrak{u} = \mathfrak{k}_0 + \mathfrak{p}_*$ be the decomposition of $\mathfrak{u}$ into eigenspaces of θ and let $\mathfrak{g}_0 = \mathfrak{k}_0 + \mathfrak{p}_0$ (where $\mathfrak{p}_0 = i\mathfrak{p}_*$). Then $\mathfrak{g}_0$ is a real form of the complexification $\mathfrak{g} = \mathfrak{u}^C$. We list below the "classical" $\mathfrak{u}$, that is, $\mathfrak{su}(n)$, $\mathfrak{so}(n)$, and $\mathfrak{sp}(n)$ and for each give various θ; later these will be shown to exhaust all possibilities for θ up to conjugacy. Then $\mathfrak{g}_0$ runs through all noncompact real forms of $\mathfrak{g}$ up to isomorphism. The simply connected Riemannian globally symmetric spaces corresponding to $(\mathfrak{u}, \theta)$ and $\mathfrak{g}_0$ are also listed (for $\mathfrak{u}$ classical). As earlier, $\mathfrak{h}_{\mathfrak{p}_*}$ and $\mathfrak{h}_{\mathfrak{p}_0}$ denote maximal abelian subspaces of $\mathfrak{p}_*$ and $\mathfrak{p}_0$, respectively.

Type A I $\quad \mathfrak{u} = \mathfrak{su}(n)$; $\theta(X) = \bar{X}$.

Here $\mathfrak{k}_0 = \mathfrak{so}(n)$ and $\mathfrak{p}_*$ consists of all symmetric purely imaginary $n \times n$ matrices of trace 0. Thus $\mathfrak{g}_0 = \mathfrak{k}_0 + \mathfrak{p}_0 = \mathfrak{sl}(n, \boldsymbol{R})$. The corresponding simply connected symmetric spaces are

$$\boldsymbol{SL}(n, \boldsymbol{R})/\boldsymbol{SO}(n), \qquad \boldsymbol{SU}(n)/\boldsymbol{SO}(n) \qquad (n > 1).$$

The diagonal matrices in $\mathfrak{p}_*$ form a maximal abelian subspace. Hence the *rank* is $n - 1$. Since $\mathfrak{g} = \mathfrak{a}_{n-1}$, the algebra $\mathfrak{g}_0$ is a *normal* real form of $\mathfrak{g}$.

Type A II $\mathfrak{u} = \mathfrak{su}(2n)$; $\theta(X) = J_n \bar{X} J_n^{-1}$.
Here $\mathfrak{k}_0 = \mathfrak{sp}(n)$ and

$$\mathfrak{p}_* = \left\{ \begin{pmatrix} Z_1 & Z_2 \\ \bar{Z}_2 & -\bar{Z}_1 \end{pmatrix} \,\middle|\, Z_1 \in \mathfrak{su}(n),\ Z_2 = \mathfrak{so}(n, \boldsymbol{C}) \right\}.$$

Hence $\mathfrak{g}_0 = \mathfrak{k}_0 + \mathfrak{p}_0 = \mathfrak{su}^*(2n)$. The corresponding simply connected symmetric spaces are

$$\boldsymbol{SU}^*(2n)/\boldsymbol{Sp}(n), \qquad \boldsymbol{SU}(2n)/\boldsymbol{Sp}(n) \qquad (n > 1).$$

The diagonal matrices in $\mathfrak{p}_*$ form a maximal abelian subspace of $\mathfrak{p}_*$. Hence the *rank* is $n - 1$.

Type A III $\mathfrak{u} = \mathfrak{su}(p + q)$; $\theta(X) = I_{p,q} X I_{p,q}$.
Here

$$\mathfrak{k}_0 = \left\{ \begin{pmatrix} A & 0 \\ 0 & B \end{pmatrix} \,\middle|\, \begin{matrix} A \in \mathfrak{u}(p),\ B \in \mathfrak{u}(q) \\ \operatorname{Tr}(A + B) = 0 \end{matrix} \right\},$$

$$\mathfrak{p}_* = \left\{ \begin{pmatrix} 0 & Z \\ -{}^t\bar{Z} & 0 \end{pmatrix} \,\middle|\, Z \quad p \times q \text{ complex matrix} \right\}.$$

The decomposition

$$\begin{pmatrix} A & 0 \\ 0 & B \end{pmatrix}$$

$$= \begin{pmatrix} A - \frac{1}{p}(\operatorname{Tr} A) I_p & 0 \\ 0 & 0 \end{pmatrix} + \begin{pmatrix} \frac{1}{p}(\operatorname{Tr} A) I_p & 0 \\ 0 & \frac{1}{q}(\operatorname{Tr} B) I_q \end{pmatrix} + \begin{pmatrix} 0 & 0 \\ 0 & B - \frac{1}{q}(\operatorname{Tr} B) I_q \end{pmatrix}$$

shows that $\mathfrak{k}_0$ is isomorphic to the product

$$\mathfrak{su}(p) \times \mathfrak{c}_0 \times \mathfrak{su}(q),$$

where $\mathfrak{c}_0$ is the center of $\mathfrak{k}_0$. Also $\mathfrak{g}_0 = \mathfrak{k}_0 + \mathfrak{p}_0 = \mathfrak{su}(p, q)$. The corresponding simply connected symmetric spaces are

$$\boldsymbol{SU}(p, q)/\boldsymbol{S}(\boldsymbol{U}_p \times \boldsymbol{U}_q), \qquad \boldsymbol{SU}(p + q)/\boldsymbol{S}(\boldsymbol{U}_p \times \boldsymbol{U}_q) \qquad (p \geqslant 1, q \geqslant 1, p \geqslant q).$$

A maximal abelian subspace of $\mathfrak{p}_*$ is given by

$$\mathfrak{h}_{\mathfrak{p}_*} = \sum_{i=1}^{q} \boldsymbol{R}(E_{i\ p+i} - E_{p+i\ i}). \tag{5}$$

Consequently, the *rank* is q. The spaces are *Hermitian symmetric*. For $q = 1$, these spaces are the so-called *Hermitian hyperbolic space* and the *complex projective space*.

Type BD I $\mathfrak{u} = \mathfrak{so}(p+q)$; $\theta(X) = I_{p,q} X I_{p,q}$ $(p \geqslant q)$.
Here

$$\mathfrak{k}_0 = \left\{ \begin{pmatrix} A & 0 \\ 0 & B \end{pmatrix} \,\middle|\, A \in \mathfrak{so}(p),\ B \in \mathfrak{so}(q) \right\},$$

$$\mathfrak{p}_* = \left\{ \begin{pmatrix} 0 & X \\ -{}^tX & 0 \end{pmatrix} \,\middle|\, X \text{ real } p \times q \text{ matrix} \right\}.$$

As shown in Chapter V, §2, the mapping

$$\begin{pmatrix} A & iX \\ -i\,{}^tX & B \end{pmatrix} \to \begin{pmatrix} A & X \\ {}^tX & B \end{pmatrix}$$

is an isomorphism of $\mathfrak{g}_0 = \mathfrak{k}_0 + \mathfrak{p}_0$ onto $\mathfrak{so}(p, q)$. The simply connected symmetric spaces associated with $\mathfrak{so}(p, q)$ and $(\mathfrak{u}, \theta)$ are

$$\boldsymbol{SO}_0(p,q)/\boldsymbol{SO}(p) \times \boldsymbol{SO}(q), \qquad \boldsymbol{SO}(p+q)/\boldsymbol{SO}(p) \times \boldsymbol{SO}(q) \quad \begin{pmatrix} p > 1, q \geqslant 1 \\ p+q \neq 4, p \geqslant q \end{pmatrix}.$$

Here $\boldsymbol{SO}_0(p, q)$ denotes the identity component of $\boldsymbol{SO}(p, q)$. The compact space is the manifold of oriented p-planes of $(p+q)$-space, which is known (see, e.g., Steenrod [1], p. 134) to be simply connected. A maximal abelian subspace of $\mathfrak{p}_*$ is again given by (5), so the *rank* is q. If $p+q$ is even then $\mathfrak{g}_0$ is a *normal* real form of $\mathfrak{g}$ if and only if $p = q$. If $p + q$ is odd then $\mathfrak{g}_0$ is a *normal* real form of $\mathfrak{g}$ if and only if $p = q + 1$.

For $q = 1$, the spaces are the *real hyperbolic space* and the *sphere.* These are the simply connected Riemannian manifolds of constant sectional curvature $\neq 0$ and dimension $\neq 3$. Those of dimension 3 are $\boldsymbol{SL}(2, \boldsymbol{C})/\boldsymbol{SU}(2)$ and $\boldsymbol{SU}(2)$, i.e., $\mathfrak{a}_n$ for $n = 1$.

If $q = 2$, then $\mathfrak{k}_0$ has nonzero center and the spaces are *Hermitian symmetric.*

Type D III $\mathfrak{u} = \mathfrak{so}(2n)$; $\theta(X) = J_n X J_n^{-1}$.

Here $\mathfrak{k}_0 = \mathfrak{so}(2n) \cap \mathfrak{sp}(n)$ which by Lemma 2.1 is isomorphic to $\mathfrak{u}(n)$. Moreover,

$$\mathfrak{p}_* = \left\{ \begin{pmatrix} X_1 & X_2 \\ X_2 & -X_1 \end{pmatrix} \,\middle|\, X_1, X_2 \in \mathfrak{so}(n) \right\}.$$

Hence $\mathfrak{g}_0 = \mathfrak{k}_0 + \mathfrak{p}_0 = \mathfrak{so}^*(2n)$. The symmetric spaces are

$$\boldsymbol{SO}^*(2n)/\boldsymbol{U}(n), \qquad \boldsymbol{SO}(2n)/\boldsymbol{U}(n) \qquad (n > 2).$$

Here the imbedding of $\boldsymbol{U}(n)$ into $\boldsymbol{SO}(2n)$, (and $\boldsymbol{SO}^*(2n)$), is given by the mapping

$$A + iB \to \begin{Bmatrix} A & B \\ -B & A \end{Bmatrix}, \tag{6}$$

where $A + iB \in U(n)$, A, B real. The spaces are *Hermitian symmetric* since $\mathfrak{k}_0$ has nonzero center. In view of Theorem 4.6, Chapter VIII, they are simply connected. A maximal abelian subspace of $\mathfrak{p}_*$ is spanned by the matrices

$$(E_{12} - E_{21}) - (E_{n+1\ n+2} - E_{n+2\ n+1}), \quad (E_{23} - E_{32}) - (E_{n+2\ n+3} - E_{n+3\ n+2}), \ldots.$$

Consequently, the *rank* is $[n/2]$.

Type C I $\mathfrak{u} = \mathfrak{sp}(n)$; $\theta(X) = \bar{X}\ (= J_n X J_n^{-1})$.
Here $\mathfrak{k}_0 = \mathfrak{sp}(n) \cap \mathfrak{so}(2n)$ which is isomorphic to $\mathfrak{u}(n)$.

$$\mathfrak{p}_* = \left\{ \begin{pmatrix} Z_1 & Z_2 \\ Z_2 & -Z_1 \end{pmatrix} \middle| \begin{array}{l} Z_1 \in \mathfrak{u}(n), \text{ purely imaginary} \\ Z_2 \text{ symmetric, purely imaginary} \end{array} \right\}.$$

Hence $\mathfrak{g}_0 = \mathfrak{k}_0 + \mathfrak{p}_0 = \mathfrak{sp}(n, \boldsymbol{R})$. The corresponding simply connected symmetric spaces are

$$\boldsymbol{Sp}(n, \boldsymbol{R})/\boldsymbol{U}(n), \qquad \boldsymbol{Sp}(n)/\boldsymbol{U}(n) \qquad (n \geqslant 1).$$

Here the imbedding of $\boldsymbol{U}(n)$ into $\boldsymbol{Sp}(n)$ (and $\boldsymbol{Sp}(n, \boldsymbol{R})$) is given by (6). The diagonal matrices in $\mathfrak{p}_*$ form a maximal abelian subspace. Thus the spaces have *rank n* and $\mathfrak{g}_0$ is a *normal* real form of $\mathfrak{g}$. The spaces are *Hermitian symmetric.*

Type C II $\mathfrak{u} = \mathfrak{sp}(p + q)$; $\theta(X) = K_{p,q} X K_{p,q}$.
Here

$$\mathfrak{k}_0 = \left\{ \begin{pmatrix} X_{11} & 0 & X_{13} & 0 \\ 0 & X_{22} & 0 & X_{24} \\ -\bar{X}_{13} & 0 & \bar{X}_{11} & 0 \\ 0 & -\bar{X}_{24} & 0 & \bar{X}_{22} \end{pmatrix} \middle| \begin{array}{l} X_{11} \in \mathfrak{u}(p), X_{22} \in \mathfrak{u}(q) \\ X_{13}\ p \times p \text{ symmetric} \\ X_{24}\ q \times q \text{ symmetric} \end{array} \right\},$$

$$\mathfrak{p}_* = \left\{ \begin{pmatrix} 0 & Y_{12} & 0 & Y_{14} \\ -{}^t\bar{Y}_{12} & 0 & {}^tY_{14} & 0 \\ 0 & -\bar{Y}_{14} & 0 & \bar{Y}_{12} \\ -{}^t\bar{Y}_{14} & 0 & -{}^tY_{12} & 0 \end{pmatrix} \middle| \begin{array}{l} Y_{12} \text{ and } Y_{14} \text{ arbitrary} \\ \text{complex } p \times q \text{ matrices} \end{array} \right\}.$$

It is clear that $\mathfrak{k}_0$ is isomorphic to the direct product $\mathfrak{sp}(p) \times \mathfrak{sp}(q)$. Moreover, $\mathfrak{g}_0 = \mathfrak{k}_0 + \mathfrak{p}_0 = \mathfrak{sp}(p, q)$. The corresponding simply connected symmetric spaces are

$$\boldsymbol{Sp}(p, q)/\boldsymbol{Sp}(p) \times \boldsymbol{Sp}(q), \qquad \boldsymbol{Sp}(p + q)/\boldsymbol{Sp}(p) \times \boldsymbol{Sp}(q) \qquad (p \geqslant q \geqslant 1).$$

Here the imbedding of $\boldsymbol{Sp}(p) \times \boldsymbol{Sp}(q)$ into $\boldsymbol{Sp}(p+q)$ (and $\boldsymbol{Sp}(p, q)$) is given by the mapping

$$\left(\begin{pmatrix} A_1 & B_1 \\ C_1 & D_1 \end{pmatrix}, \begin{pmatrix} A_2 & B_2 \\ C_2 & D_2 \end{pmatrix}\right) \to \begin{pmatrix} A_1 & 0 & B_1 & 0 \\ 0 & A_2 & 0 & B_2 \\ C_1 & 0 & D_1 & 0 \\ 0 & C_2 & 0 & D_2 \end{pmatrix}$$

A maximal abelian subspace of $\mathfrak{p}_*$ is obtained by taking $Y_{14} = 0$ and letting Y_{12} run through the space $\boldsymbol{R}E_{11} + \boldsymbol{R}E_{22} + \cdots + \boldsymbol{R}E_{qq}$. Consequently, the *rank* is q. For $q = 1$, the spaces are the so-called *quaternian hyperbolic spaces* and the *quaternian projective spaces*.

This will be shown to exhaust all involutive automorphisms of the compact classical simple Lie algebras. The restriction on the indices is made in order that the algebras should be simple, the spaces of dimension >0, and the condition $p \geqslant q$ is required in order to avoid repetition within the same class.

§ 3. Root Systems

1. Generalities

Let V be a finite-dimensional vector space over $\boldsymbol{R}$ and $\alpha \in V$, $\alpha \neq 0$. A *reflection along* α is any linear transformations of V satisfying the two conditions:

(i) $s\alpha = -\alpha$;

(ii) The fixed points of s constitute a hyperplane in V.

A reflection s along α is determined by its fixed point set and satisfies $s\beta = \beta + c_\beta \alpha (c_\beta \in \boldsymbol{R})$. It has order 2.

Lemma 3.1. *Let $R \subset V$ be a finite subset which generates V. Let $\alpha \neq 0$ in V. Then there exists at most one reflection along α leaving R invariant.*

Proof. If s and s' are two such reflections, the linear transformation $\sigma = ss'$ satisfies: (a) $\sigma(R) = R$; (b) $\sigma\alpha = \alpha$; (c) σ induces the identity map of $V/\boldsymbol{R}\alpha$ (since s and s' both do). Thus there is a linear function f on V such that

$$\sigma(x) = x + f(x)\,\alpha, \qquad x \in V$$

and by (ii), $f(\alpha) = 0$. By iteration we get $\sigma^n x = x + nf(x)\alpha$ $(n \in \boldsymbol{Z}^+)$. But R is finite, so (a) implies that σ has finite order; hence $f \equiv 0$.

Definition. Let V be a real finite-dimensional vector space and $R \subset V$ a finite set of nonzero vectors; R is called a *root system* in V (and its members called *roots*) if:

(i) R generates V.

(ii) For each $\alpha \in R$ there exists a reflection s_α along α leaving R invariant (by Lemma 3.1, s_α is unique).

(iii) For all $\alpha, \beta \in R$ the number $a_{\beta,\alpha}$ determined by

$$s_\alpha \beta = \beta - a_{\beta,\alpha}\alpha$$

is an integer, that is, $a_{\beta,\alpha} \in \boldsymbol{Z}$.

Remark. If $\alpha, \beta \in R$ are proportional, $\beta = m\alpha$ ($m \in \boldsymbol{R}$), then

$$m = \pm\tfrac{1}{2}, \quad \pm 1, \quad \pm 2. \tag{1}$$

In fact, the numbers $a_{\alpha,m\alpha} = 2/m$ and $a_{m\alpha,\alpha} = 2m$ are both integers. A root system R is said to be *reduced* if $\alpha, \beta \in R$, $\beta = m\alpha$ implies $m = \pm 1$. A root $\alpha \in R$ is called *indivisible* if $\frac{1}{2}\alpha \notin R$, and *unmultipliable* if $2\alpha \notin R$.

Examples. (i) The set $\Delta = \Delta(\mathfrak{g}, \mathfrak{h})$ of roots of a semisimple Lie algebra $\mathfrak{g}$ over $\boldsymbol{C}$ with respect to a Cartan subalgebra $\mathfrak{h}$ is a reduced root system (Chapter III, Theorem 4.3 and Exercise C.1).

(ii) The set Σ of restricted roots is a root system which in general is not reduced (Chapter VII, Theorem 2.16 and Exercise 5).

Lemma 3.2. *Let R be a root system in V and put*

$$R' = \{\alpha \in R : \alpha/2 \notin R\}, \qquad R'' = \{\alpha \in R : 2\alpha \notin R\}.$$

Then R' and R'' are reduced root systems in V.

It is obvious that R' and R'' are root systems in V. If $\alpha, 2\alpha \in R$, then $\alpha/2, 4\alpha \notin R$, so $\alpha \in R'$, $2\alpha \notin R'$ and $\alpha \notin R''$, $2\alpha \in R''$. In particular, R' and R'' are reduced.

Given a root system R in V let $\mathrm{Aut}(R)$ denote the (finite) group of linear transformations of V leaving R invariant. The subgroup $W(R)$ (or simply W) of $\mathrm{Aut}(R)$ generated by the reflections s_α ($\alpha \in R$) is called the *Weyl group* of R.

Lemma 3.3. *Let $\langle\ ,\ \rangle$ denote any positive definite scalar product on V, invariant under* $\mathrm{Aut}(R)$ *(such $\langle\ ,\ \rangle$ exist). Then*

$$a_{\beta,\alpha} = 2\frac{\langle \beta, \alpha \rangle}{\langle \alpha, \alpha \rangle}, \qquad \alpha, \beta \in R.$$

In fact, by the s_α-invariance of $\langle\ ,\ \rangle$,

$$\langle\alpha, s_\alpha\beta + \beta\rangle = \langle -\alpha, \beta + s_\alpha\beta\rangle$$

so $\langle\alpha, s_\alpha\beta + \beta\rangle = 0$, proving the lemma.

The scalar product $\langle\ ,\ \rangle$ will be fixed for the time being. Because of Lemma 3.3, the sign of $\langle\alpha, \beta\rangle$ (and, in particular, "orthogonality" of $\alpha, \beta \in R$) is independent of the choice of $\langle\ ,\ \rangle$.

Lemma 3.4. *Let $\alpha, \beta \in R$ be linearly independent. Then:*

(i) $0 \leqslant a_{\alpha,\beta}a_{\beta,\alpha} \leqslant 3$;

(ii) $a_{\alpha,\beta} > 0 \Rightarrow \alpha - \beta \in R$;

(iii) $a_{\alpha,\beta} < 0 \Rightarrow \alpha + \beta \in R$.

Proof. Part (i) is obvious from Schwarz' inequality. If $a_{\alpha,\beta} > 0$, then by (i) at least one of the integers $a_{\alpha,\beta}$, $a_{\beta,\alpha}$ equals 1. In the first case $s_\beta\alpha = \alpha - a_{\alpha,\beta}\beta = \alpha - \beta \in R$; in the second case $s_\alpha\beta = \beta - a_{\beta,\alpha}\alpha = \beta - \alpha \in R$, so $\alpha - \beta \in R$. This proves (ii); part (iii) follows by replacing β by $-\beta$.

We can now deduce an analog of Theorem 4.3(i), Chapter III, for the present abstract situation.

Corollary 3.5. *Let α and β be nonproportional roots. The set of roots of the form $\beta + n\alpha$ (n = integer) is an uninterupted string*

$$\beta + n\alpha, \qquad p \leqslant n \leqslant q.$$

Moreover,

$$-a_{\beta,\alpha} = p + q.$$

In fact, let $p \leqslant q$ be the extremal values of n with $\beta + n\alpha \in R$. If there were a gap, that is, an interval $r < n < s$ such that

$$\beta + r\alpha \in R, \qquad \beta + n\alpha \notin R \quad (r < n < s), \qquad \beta + s\alpha \in R,$$

then Lemma 3.4 implies

$$\langle\alpha, \beta + r\alpha\rangle \geqslant 0, \qquad \langle\alpha, \beta + s\alpha\rangle \leqslant 0,$$

contradicting $r < s$.

Finally, $s_\alpha(\beta + n\alpha) = \beta - (a_{\beta,\alpha} + n)\alpha$, so s_α maps the string onto itself. Thus the map $n \to -n - a_{\beta,\alpha}$ is a mapping of the integer interval $[p, q]$ onto itself. It must map p onto q, that is, $-p - a_{\beta,\alpha} = q$, as desired.

Definition. Let R be a root system in V. A subset $B \subset R$ is called a *basis of R* if:

(i) B is a basis of V.

(ii) Each $\beta \in R$ can be written

$$\beta = \sum_{\alpha \in B} n_\alpha \alpha,$$

where the n_α are integers of the same sign.

Theorem 3.6.

(i) *Each root system has a basis.*

(ii) *Any two bases are conjugate under a unique Weyl group element.*

(iii) $a_{\beta,\alpha} \leqslant 0$ *for any two different elements* α, β *in the same basis.*

Proof. An element $\gamma \in V$ is called *regular* if $\langle \alpha, \gamma \rangle \neq 0$ for all $\alpha \in R$. For $\gamma \in V$ regular, let $R^+(\gamma) = \{\alpha \in R : \langle \alpha, \gamma \rangle > 0\}$ and call a root $\alpha \in R^+(\gamma)$ *simple* if it cannot be written as a sum of two members of $R^+(\gamma)$. If α, β are simple, then $\alpha - \beta \notin R$, so by Lemma 3.4, $\langle \alpha, \beta \rangle \leqslant 0$. Now the proof of Theorem 5.7, Chapter III, shows that the simple roots in $R^+(\gamma)$ form a basis, say $B(\gamma)$.

Suppose $B' = \{\alpha_1, \ldots, \alpha_l\}$ is another basis. Then the element $\gamma' = \sum_{i=1}^{l} \gamma_i$, where $\langle \gamma_i, \alpha_j \rangle = \delta_{ij}$ satisfies $\langle \alpha_i, \gamma' \rangle > 0$ for all i. Moreover, B' is the set $B(\gamma')$ of simple roots in $R^+(\gamma')$.

For $\alpha \in R$ let π_α denote the hyperplane $\{\gamma : \langle \alpha, \gamma \rangle = 0\}$ in V. The components of $V - \bigcup_{\alpha \in R} \pi_\alpha$ are called *Weyl chambers*. Each regular element γ lies in a unique Weyl chamber $C(\gamma)$. The equality $C(\gamma) = C(\gamma')$ amounts to γ and γ' being on the same side of each π_α, that is, $R^+(\gamma) = R^+(\gamma')$ or equivalently $B(\gamma) = B(\gamma')$. Since $W(R)$ acts simply transitively on the set of Weyl chambers (cf. proof of Theorem 2.12, Chapter VII), the theorem is proved.

A root system R is called *irreducible* if it cannot be decomposed into two disjoint nonempty orthogonal subsets.

Proposition 3.7. *A root system R in V decomposes uniquely as the union of irreducible root systems* R_i *(in subspaces* $V_i \subset V$*) and* $V = \sum_i V_i$ *(orthogonal direct sum).*

Proof. If R is not irreducible, $R = R_1 \cup R_2$ where R_1 and R_2 are nonempty, disjoint, and orthogonal. If V_i is the span of R_i, then $V = V_1 \oplus V_2$ and R_i is a root system in V_i. Now the result follows by iteration.

2. Reduced Root Systems

Let R be a reduced root system in V and $B = \{\alpha_1, \ldots, \alpha_l\}$ any basis of R. The matrix $(a_{ij})_{1 \leqslant i,j \leqslant l}$ where $a_{ij} = a_{\alpha_i, \alpha_j}$ is called the *Cartan matrix* of R. From Lemma 3.3 it is clear that the Cartan matrix is nonsingular. The next result shows that the Cartan matrix of R determines R.

Proposition 3.8. *Let $R' \subset V'$ be a reduced root system with basis $B' = \{\alpha_1', \ldots, \alpha_l'\}$, and put $a_{ij}' = a_{\alpha_i', \alpha_j'}$.* If $a_{ij} = a_{ij}'$ $(1 \leqslant i, j \leqslant l)$, *then the mapping $\alpha_i \to \alpha_i'$ of B onto B' extends uniquely to a linear bijection $\varphi : V \to V'$ mapping R onto R'.*

Moreover,

$$a_{\varphi(\alpha), \varphi(\beta)} = a_{\alpha, \beta} \qquad \textit{for all} \quad \alpha, \beta \in R. \tag{2}$$

We need a simple lemma on the passage from B to R.

Lemma 3.9. *$W(R)$ is generated by the simple reflections s_α $(\alpha \in B)$, and $W(R)(B) = R$.*

Proof. Let W' denote the subgroup of $W(R)$ generated by the simple reflections. These are the reflections in the walls of the chamber $\{\gamma : a_{\gamma, \alpha} > 0$ for $\alpha \in B\}$, so the proof of Theorem 2.12, Chapter VII, shows that W' permutes the set of Weyl chambers (and therefore the set of bases) transitively.

Now if $\alpha \in R$, we can select $\gamma \in \pi_\alpha$ such that $\gamma \notin \pi_\beta$ for $\beta \in R$, $\beta \notin \mathbf{R}\alpha$. Choosing γ' close to γ, we can ensure $\langle \alpha, \gamma' \rangle = \epsilon > 0$ and $|\langle \beta, \gamma' \rangle| > \epsilon$ for $\beta \neq \pm\alpha$. Then α belongs to the base $B(\gamma')$, so by the above, $\alpha = \sigma(\alpha_i)$ for some $\sigma \in W'$, and some i. But then $s_\alpha = \sigma s_{\alpha_i} \sigma^{-1}$, proving $W(R) = W'$ and the lemma.

Turning now to Prop. 3.8, let $\varphi : V \to V'$ be the unique linear bijection sending α_i to α_i' $(1 \leqslant i \leqslant l)$. The assumption $a_{ij} = a_{ij}'$ then implies

$$s_{\alpha_i'} \circ \varphi = \varphi \circ s_{\alpha_i}; \tag{3}$$

using Lemma 3.9, we deduce that $\sigma \to \varphi \circ \sigma \circ \varphi^{-1}$ is an isomorphism of $W(R)$ onto $W(R')$. But then

$$\varphi(R) = \varphi(W(R)B) = W(R')(\varphi(B)) = R'$$

and $s_{\varphi(\alpha)} \circ \varphi = \varphi \circ s_\alpha$ $(\alpha \in R)$ proving (2) and Prop. 3.8.

While Prop. 3.8 reduces the classification of R to that of the Cartan matrices, it is of interest to have results showing how R is in a more constructive fashion determined by (a_{ij}). Note that R is a disjoint union $R = R^+ \cup (-R^+)$ where

$$R^+ = \left\{\alpha \in R : \alpha = \sum_{i=1}^{l} n_i\alpha_i,\ n_i \in Z^+\right\}.$$

Lemma 3.10. (i) *Let* $\alpha \in R^+ - B$. *Then for some* $\beta \in B$, $\alpha - \beta$ *is a root.*

(ii) *Each* $\beta \in R^+$ *can be written*

$$\beta = \beta_1 + \ldots + \beta_k$$

where all β_j *belong to* B (*not necessarily distinct*) *such that each partial sum* $\beta_1 + \ldots + \beta_i$ *is a root.*

Proof. (i) If $\alpha - \beta \notin R$ for all $\beta \in B$, then by Lemma 3.4 $\langle\alpha, \beta\rangle \leqslant 0$ for all $\beta \in B$. It follows that $\langle\alpha, \alpha\rangle \leqslant 0$, which is a contradiction. Part (ii) follows from (i) by iteration.

Construction of R from B and (a_{ij})

Given $\alpha \in R^+$, $\alpha = \sum_i n_i\alpha_i$, the sum $\sum_i n_i$ is called the *height* of α. The roots of height one are the simple roots. Since $\alpha_i - \alpha_j \notin R$, the α_i-series through α_j is

$$\alpha_j, \quad \alpha_j + \alpha_i, \quad \ldots, \quad \alpha_j + q\alpha_i \qquad (q = -a_{ji}) \tag{4}$$

This process gives in particular all roots $\alpha = \alpha_k + \alpha_m$ of height 2 as well as a_{α,α_j}. Then the α_j-series through α

$$\alpha + p\alpha_j, \quad \ldots, \quad \alpha + q\alpha_j$$

is determined because $p = -1$ or 0 depending on whether $j \in \{k, m\}$ or not, so q is determined by $p + q = -a_{\alpha,\alpha_j}$. Lemma 3.10 guarantees that all $\beta \in R^+$ are obtained by repetition of this process.

Example. Suppose the Cartan matrix is

$$\begin{pmatrix} 2 & -1 \\ -2 & 2 \end{pmatrix}$$

that is,

$$a_{12} = \frac{2\langle\alpha_1, \alpha_2\rangle}{\langle\alpha_2, \alpha_2\rangle} = -1, \qquad a_{21} = \frac{2\langle\alpha_2, \alpha_1\rangle}{\langle\alpha_1, \alpha_1\rangle} = -2.$$

The strings (4) take the form

$$\alpha_1, \quad \alpha_1 + \alpha_2,$$

$$\alpha_2, \quad \alpha_2 + \alpha_1, \quad \alpha_2 + 2\alpha_1.$$

The only $\alpha \in R^+$ of height 2 is $\alpha_1 + \alpha_2$; and since $\alpha_1 + 2\alpha_2$ is not a root, the only positive root of height 3 is $\alpha_2 + 2\alpha_1$. Since $2\alpha_1 + 2\alpha_2 = 2(\alpha_1 + \alpha_2)$ is not a root, there are no roots of height 4, so by Lemma 3.10 all the four positive roots have been enumerated.

Lemma 3.11. *Let* $\rho = \frac{1}{2}\sum_{\alpha \in R^+} \alpha$ *and let* $\alpha_i \in B$. *Then*:

(i) $s_{\alpha_i}(R^+ - \{\alpha_i\}) = R^+ - \{\alpha_i\}$.

(ii) $2\langle \rho, \alpha_i \rangle / \langle \alpha_i, \alpha_i \rangle = 1$.

Proof. If $\alpha \in R^+$ is written $\alpha = \sum_j n_j \alpha_j$ ($n_j \in \mathbf{Z}^+$), we have

$$s_{\alpha_i}\alpha = \alpha - a_{\alpha,\alpha_i}\alpha_i = (n_i - a_{\alpha,\alpha_i})\,\alpha_i + \sum_{j \neq i} n_j \alpha_j.$$

All the coefficients on the right must have the same sign; this sign is negative only if $n_j = 0$ for $j \neq i$, that is, if $\alpha = \alpha_i$. This proves (i) and implies $s_{\alpha_i}\rho = \rho - \alpha_i$, from which (ii) follows.

3. Classification of Reduced Root Systems. Coxeter Graphs and Dynkin Diagrams

Let R be a reduced root system in V with basis $B = \{\alpha_1, ..., \alpha_l\}$, R^+ the corresponding set of positive roots. From Lemma 3.4 we know that if $\alpha, \beta \in R^+$ are distinct, the integer $a_{\alpha,\beta}a_{\beta,\alpha}$ equals 0, 1, 2, or 3. The *Coxeter graph* of R is the graph in V with l vertices $P_1, ..., P_l$, the ith joined to the jth by $a_{ij}a_{ji}$ nonintersecting lines ($a_{ij} = a_{\alpha_i,\alpha_j}$).

The root system R is irreducible if and only if B cannot be decomposed into two nonempty disjoint orthogonal subsets (cf. the proof of Lemma 11.8, Chapter VII). This is equivalent to the (set-theoretic) connectivity of the Coxeter graph.

Suppose now R is irreducible. Then the Weyl group $W(R)$ acts irreducibly on V; in fact otherwise V is the orthogonal direct sum $V = V_1 + V_2$ of nonzero subspaces, both invariant under each s_α ($\alpha \in R$). Considering the (-1)-eigenspace of s_α, that is $\mathbf{R}\alpha$, we see that either $\alpha \in V_1$ or $\alpha \in V_2$, contrary to the irreducibility of R. Because of this irreducibility of $W(R)$, the invariant inner product $\langle\ ,\ \rangle$ is unique up to a constant factor. Giving the ith vertex P_i in the Coxeter graph a weight proportional to $\langle \alpha_i, \alpha_i \rangle$, the resulting figure is called the *Dynkin diagram*. The Dynkin diagram determines the Cartan matrix, hence determines R.

We now write down the Dynkin diagrams for the classical complex Lie algebras, using the description of the roots from Chapter III, §8, and Exercises B.5 and B.6. The scalar product is now given by the Killing form.

The algebra $\mathfrak{a}_l$ $(l \geqslant 1)$. Here

$$R = \{e_i - e_j : 1 \leqslant i \neq j \leqslant l + 1\}$$

and the vector space V spanned by R is the hyperplane

$$\left\{ \sum_{i=1}^{l+1} x_i e_i : \sum_i x_i = 0 \right\}$$

in the vector space $\sum_{i=1}^{l+1} \boldsymbol{R} e_i$. The roots

$$\alpha_i = e_i - e_{i+1} \qquad (1 \leqslant i \leqslant l)$$

form a basis B of R. Using (5), §8, Chapter III, we get

$$\langle \alpha_i, \alpha_i \rangle = (l + 1)^{-1} \qquad (1 \leqslant i \leqslant l)$$

and the Cartan matrix is

$$\mathfrak{a}_l : \begin{pmatrix} 2 & -1 & 0 & & & \cdots & & 0 \\ -1 & 2 & -1 & 0 & & \cdots & & 0 \\ 0 & -1 & 2 & -1 & 0 & \cdots & & 0 \\ \cdot & \cdot & \cdot & \cdot & & & & \\ & & & & & & 2 & -1 \\ 0 & 0 & 0 & 0 & & \cdots & -1 & 2 \end{pmatrix}.$$

The Coxeter graph is connected, hence $\mathfrak{a}_l$ is simple and the Dynkin diagram is

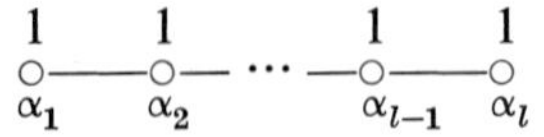

The algebra $\mathfrak{b}_l$ $(l \geqslant 1)$. Here

$$R = \{\pm e_i\,(1 \leqslant i \leqslant l),\ \pm e_i \pm e_j\,(1 \leqslant i \neq j \leqslant l,\ \pm \text{ independent})\}$$

in the vector space $V = \sum_{i=1}^{l} \boldsymbol{R} e_i$. The roots

$$\alpha_i = e_i - e_{i+1} \quad (1 \leqslant i \leqslant l-1), \qquad \alpha_l = e_l$$

form a basis B of R; in fact

$$\begin{aligned} e_i &= \alpha_i + \dots + \alpha_l, & 1 \leqslant i \leqslant l, \\ e_i - e_j &= \alpha_i + \dots + \alpha_{j-1}, & 1 \leqslant i < j \leqslant l. \end{aligned}$$

Using Chapter III, (15), §8 and the root description in Exercise B.6, we find

$$\langle \alpha_i, \alpha_i \rangle = \begin{cases} (2l-1)^{-1}, & 1 \leqslant i \leqslant l-1 \\ \frac{1}{2}(2l-1)^{-1}, & i = l \end{cases}$$

and the Cartan matrix is

$$\mathfrak{b}_l : \begin{pmatrix} 2 & -1 & 0 & & \dots & & & 0 \\ -1 & 2 & -1 & 0 & \dots & & & 0 \\ 0 & -1 & 2 & -1 & \dots & & & 0 \\ \cdot & \cdot & \cdot & & & & & \\ 0 & 0 & 0 & & \dots & -1 & 2 & -2 \\ 0 & 0 & 0 & & & 0 & -1 & 2 \end{pmatrix}.$$

The Coxeter graph is connected, hence $\mathfrak{b}_l$ is simple and the Dynkin diagram is

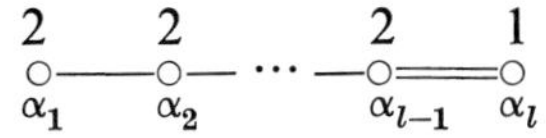

Note that for the case $l = 2$ the graph is ○═○.

The algebra $\mathfrak{c}_l$ $(l \geqslant 1)$. Here

$$R = \{\pm 2e_i\,(1 \leqslant i \leqslant l), \pm e_i \pm e_j\,(1 \leqslant i \neq j \leqslant l, \pm \text{ independent})\}$$

in the vector space $V = \sum_{i=1}^{l} \boldsymbol{R} e_i$. The roots

$$\alpha_i = e_i - e_{i+1} \quad (1 \leqslant i \leqslant l-1), \qquad \alpha_l = 2e_l$$

form a basis B of R; in fact

$$\begin{aligned} 2e_i &= 2\alpha_i + \dots + 2\alpha_{l-1} + \alpha_l, & 1 \leqslant i \leqslant l-1, \\ e_i - e_j &= \alpha_i + \dots + \alpha_{j-1}, & 1 \leqslant i < j \leqslant l. \end{aligned}$$

Using (22) §8, Chapter III, we find

$$\langle \alpha_i, \alpha_i \rangle = \begin{cases} \frac{1}{2}(l+1)^{-1}, & 1 \leqslant i \leqslant l-1, \\ (l+1)^{-1}, & i = l, \end{cases}$$

and the Cartan matrix is

$$\mathfrak{c}_n : \begin{pmatrix} 2 & -1 & 0 & & \cdots & & & 0 \\ -1 & 2 & -1 & & \cdots & & & 0 \\ 0 & -1 & 2 & -1 & \cdots & & & 0 \\ \cdot & \cdot & \cdot & & & & & \\ 0 & 0 & 0 & & \cdots & -1 & 2 & -1 \\ 0 & 0 & 0 & & & 0 & -2 & 2 \end{pmatrix}.$$

The Coxeter graph is connected, hence $\mathfrak{c}_l$ is simple and the Dynkin diagram is

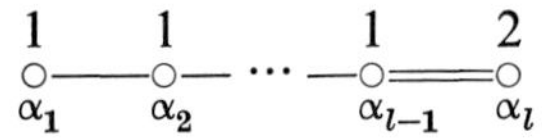

For $l = 2$ the graph is o=o.

The algebra $\mathfrak{d}_l$ ($l \geqslant 2$). Here

$$R = \{\pm e_i \pm e_j \; (1 \leqslant i \neq j \leqslant l, \pm \text{ independent})\}$$

in the vector space $V = \sum_{i=1}^{l} \boldsymbol{R} e_i$. The roots

$$\alpha_i = e_i - e_{i+1} \quad (1 \leqslant i \leqslant l-1), \qquad \alpha_l = e_{l-l} + e_l$$

form a basis B of R; in fact

$$e_i - e_j = \alpha_i + \ldots + \alpha_{j-1}, \qquad 1 \leqslant i < j \leqslant l,$$
$$e_i + e_j = (\alpha_i + \ldots + \alpha_{l-2}) + (\alpha_j + \ldots + \alpha_l), \qquad 1 \leqslant i < j \leqslant l.$$

Using Chapter III, (11), §8 and the root description in Exercise B.5, we find

$$\langle \alpha_i, \alpha_i \rangle = \tfrac{1}{2}(l-1)^{-1}, \qquad 1 \leqslant i \leqslant l,$$

and the Cartan matrix is

$$\mathfrak{d}_l : \begin{pmatrix} 2 & -1 & 0 & \cdots & & & & 0 \\ -1 & 2 & -1 & \cdots & & & & 0 \\ 0 & -1 & 2 & \cdots & & & & 0 \\ \cdot & \cdot & \cdot & & & & & \\ 0 & 0 & 0 & \cdots & 2 & -1 & 0 & 0 \\ 0 & 0 & 0 & & -1 & 2 & -1 & -1 \\ 0 & 0 & 0 & & 0 & -1 & 2 & 0 \\ 0 & 0 & 0 & \cdots & 0 & -1 & 0 & 2 \end{pmatrix}.$$

The Coxeter graph is connected except for $l = 2$. Thus $\mathfrak{d}_l$ is simple ($l \geqslant 3$) and $\mathfrak{d}_2$ is a direct sum of two simple Lie algebras (cf. Example II, §2, Chapter V). The Dynkin diagram is

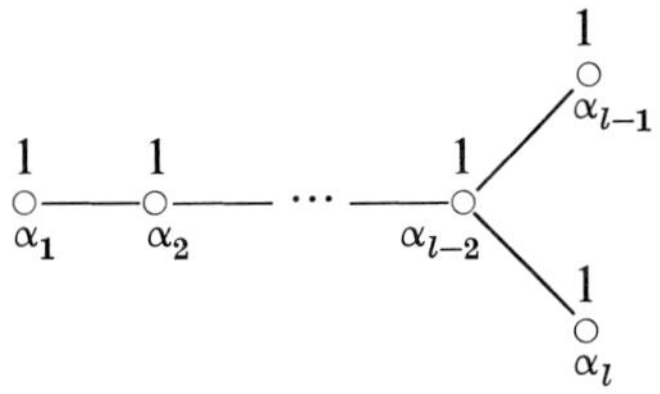

For $l = 3$ this is

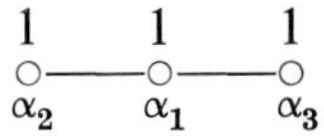

For $l = 2$ the Coxeter graph is 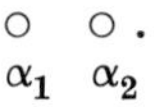 .

Looking now at the Cartan matrices (or equivalently, the Dynkin diagrams), we can conclude the following result from Prop. 3.8 and Theorem 5.4, Chapter III.

Theorem 3.12. *The following are the only isomorphisms which hold between the complex Lie algebras* $\mathfrak{a}_l$, $\mathfrak{b}_l$, $\mathfrak{c}_l$ ($l \geqslant 1$) *and* $\mathfrak{d}_l$ ($l \geqslant 2$):

$$\mathfrak{a}_1 \approx \mathfrak{b}_1 \approx \mathfrak{c}_1, \qquad \mathfrak{b}_2 \approx \mathfrak{c}_2, \qquad \mathfrak{a}_3 \approx \mathfrak{d}_3, \qquad \mathfrak{d}_2 = \mathfrak{a}_1 \times \mathfrak{a}_1.$$

We now proceed to determine all possibilities for Dynkin diagrams. Let V be a real vector space with positive definite scalar product $\langle\ ,\ \rangle$ and associated norm $|\ |$. A *diagram* is a subset $B \subset V$ with the following properties:

(i) The elements of B (called *vertices*) are linearly independent.

(ii) If $\alpha, \beta \in B$, $\alpha \neq \beta$, then

$$a_{\alpha,\beta} = 2\frac{\langle\alpha, \beta\rangle}{\langle\beta, \beta\rangle} \in -\boldsymbol{Z}^+$$

(so by Schwarz' inequality, $0 \leqslant a_{\alpha,\beta}a_{\beta,\alpha} \leqslant 3$).

(iii) Each vertex $\alpha \in B$ is given weight proportional to $\langle\alpha, \alpha\rangle$ and two distinct vertices $\alpha, \beta \in B$ are joined by $a_{\alpha,\beta}a_{\beta,\alpha}$ lines.

The collection of lines in (iii) is called the *graph* of B. In the respective cases $a_{\alpha,\beta}a_{\beta,\alpha} = 0$, 1, 2, or 3, the pair $\{\alpha, \beta\}$ is said to be *unconnected*, a simple, double, or a triple *link*. A subset of a diagram is also a diagram (called a *subdiagram*). A diagram is *connected* if for each pair $\alpha, \beta \in B$ there exist $\alpha_0, \ldots, \alpha_k \in B$ such that $\alpha_0 = \alpha$, $\alpha_k = \beta$ and each pair (α_i, α_{i+1}) is a link. A subset $\{\alpha_1, \ldots, \alpha_k\} \subset B$ is called a *cycle* if $k > 1$ and the pairs $\{\alpha_1, \alpha_2\}$, $\{\alpha_2, \alpha_3\}$, ..., $\{\alpha_{k-1}, \alpha_k\}$, $\{\alpha_k, \alpha_1\}$ are all links. A subdiagram $\{\alpha_1, \ldots, \alpha_k\} \subset B$ is called a *chain* if its only links are $\{\alpha_i, \alpha_{i+1}\}$ $(1 \leqslant i \leqslant k-1)$.

Lemma 3.13. *Let l denote the number of elements in a diagram B. The number of links in B is $\leqslant l-1$.*

Proof. If $B = \{\alpha_1, \ldots, \alpha_l\}$ put $\alpha = \sum_{i=1}^{l} \epsilon_i$, where $\epsilon_i = |\alpha_i|^{-1} \alpha_i$. Then by (i) $\alpha \neq 0$ and

$$0 < \langle \alpha, \alpha \rangle = l + 2 \sum_{i<j} \langle \epsilon_i, \epsilon_j \rangle. \tag{5}$$

If $\{\alpha_i, \alpha_j\}$ is a link, then $a_{\alpha_i,\alpha_j} a_{\alpha_j,\alpha_i} = 4\langle \epsilon_i, \epsilon_j \rangle^2 = 1, 2,$ or 3 so $2\langle \epsilon_i, \epsilon_j \rangle \leqslant -1$. Now (5) gives the lemma.

Corollary 3.14. *B contains no cycles.*

In fact, a cycle would be a subdiagram with at least as many links as vertices, contrary to Lemma 3.13.

Lemma 3.15. *The number of lines originating at a given vertex is at most* 3.

Let $\alpha \in B$, $\beta_1, \ldots, \beta_k \in B$ such that each $\{\alpha, \beta_i\}$ is a link and all β_i distinct. By Cor. 3.14, $\langle \beta_i, \beta_j \rangle = 0$ for $i \neq j$. Let α' be the orthogonal projection of α on the span of $\{\beta_1, \ldots, \beta_k\}$ and put $\beta_0 = \alpha - \alpha'$. Then all β_i $(0 \leqslant i \leqslant k)$ are orthogonal and $\langle \alpha, \beta_0 \rangle \neq 0$. Writing $\alpha = \sum_{i=0}^{k} c_i \beta_i$, we derive $2c_i = a_{\alpha,\beta_i}$ and $2 = \sum_{i=0}^{k} c_i a_{\beta_i,\alpha}$. Hence

$$\sum_{i=1}^{k} a_{\alpha,\beta_i} a_{\beta_i,\alpha} = \sum_{i=0}^{k} a_{\alpha,\beta_i} a_{\beta_i,\alpha} - a_{\alpha,\beta_0} a_{\beta_0,\alpha} < 4,$$

proving the lemma.

Corollary 3.16. *If a connected subdiagram B' contains a triple link, then* $B' = \overset{1}{\circ}\!\equiv\!\overset{3}{\circ}$. *(This diagram is denoted* $\mathfrak{g}_2$.*)*

Clear from Lemma 3.15.

Lemma 3.17. *Let $C = \{\alpha_1, ..., \alpha_k\}$ be a chain all of whose links are simple and let $\alpha = \alpha_1 + ... + \alpha_k$. Then:*

(i) $\langle\alpha, \alpha\rangle = \langle\alpha_i, \alpha_i\rangle$ $(1 \leqslant i \leqslant k)$.

(ii) *The set $(B - C) \cup \{\alpha\}$ is a diagram and its graph is obtained from the graph of B by replacing C by the vertex α and joining each $\beta \in B - C$ to α by p lines if p is the number of lines joining β to a vertex α_i of C (by Cor. 3.14 there is at most one such vertex).*

Proof. By assumption,

$$a_{\alpha_i,\alpha_j} = \begin{cases} 2 & \text{if } i = j, \\ -1 & \text{if } |i - j| = 1, \\ 0 & \text{otherwise.} \end{cases} \tag{6}$$

This implies (i). For (ii) we observe $\langle\beta, \alpha\rangle = \langle\beta, \alpha_i\rangle$ and $a_{\beta,\alpha}a_{\alpha,\beta} = a_{\beta,\alpha_i}a_{\alpha_i,\beta} = p$.

Lemma 3.18. *B contains no chain with more than one double link.*

Proof. In fact such a chain would contain a subdiagram $\{\alpha_0, ..., \alpha_{k+1}\}$ whose graph has the form

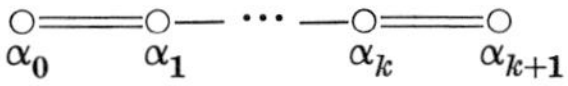

where $(\alpha_1, ..., \alpha_k)$ is a chain all of whose links are simple. Using Lemma 3.17 on this last chain, we get a contradiction to Lemma 3.15.

Corollary 3.19. *Let B be a connected diagram.*

(i) *If B contains a double link $\{\alpha, \beta\}$, it cannot contain a vertex $\gamma \neq \alpha, \beta$ from which three lines originate (a triple vertex)*

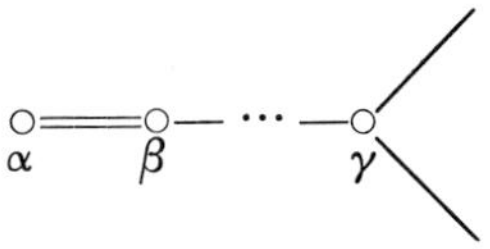

(ii) *B cannot contain two triple vertices.*

The proof is the same as that of Lemma 3.18.

We have now proved that the graph of a connected diagram is one of the following four types:

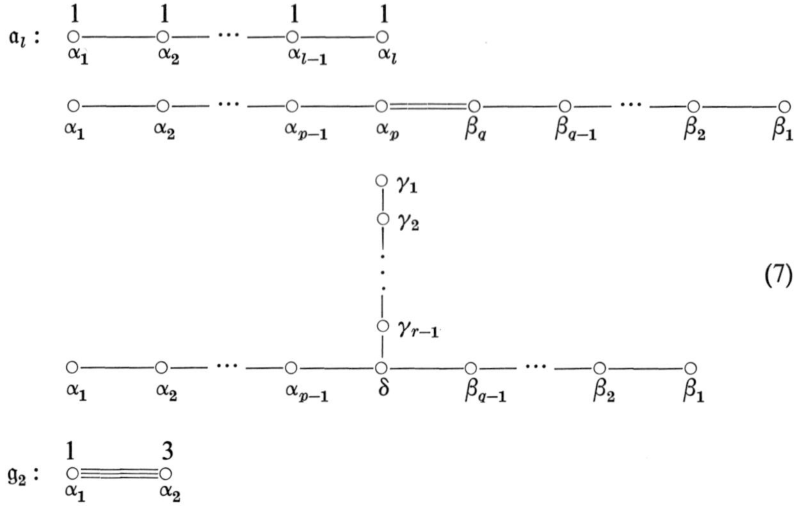

(7)

It remains to determine the possibilities for p, q, and r and the weights for the second and third type. For the second type put $\alpha = \sum_{i=1}^{p} i\alpha_i$, $\beta = \sum_{j=1}^{q} j\beta_j$. Then $\langle \alpha, \beta \rangle = pq\langle \alpha_p, \beta_q \rangle$. Also (6) implies

$$\langle \alpha, \alpha \rangle = \tfrac{1}{2}p(p+1)\langle \alpha_p, \alpha_p \rangle, \qquad \langle \beta, \beta \rangle = \tfrac{1}{2}q(q+1)\langle \beta_q, \beta_q \rangle. \tag{8}$$

Furthermore,

$$| \alpha_1 | = \ldots = | \alpha_p |, \qquad | \beta_1 | = \ldots = | \beta_q |,$$

and $\langle \alpha_p, \alpha_p \rangle / \langle \beta_q, \beta_q \rangle = \frac{1}{2}$ or 2; so we can assume $\langle \alpha_p, \alpha_p \rangle = 2\langle \beta_q, \beta_q \rangle$. But then, since

$$a_{\alpha_p, \beta_q} a_{\beta_q, \alpha_p} = 2,$$

it follows that $\langle \alpha_p, \beta_q \rangle = -\langle \beta_q, \beta_q \rangle$. Using Schwarz' inequality on the linearly independent vectors α, β (cf. (i)) we obtain $2pq < (p+1)(q+1)$, that is, $(p-1)(q-1) < 2$. Since p and q are integers $\geqslant 1$, we get the possibilities

$$\min(p, q) = 1, \qquad \text{or} \qquad p = q = 2.$$

This proves

Lemma 3.20. *A diagram of the second type in* (7) *is one of the following*:

$\mathfrak{b}_l$:
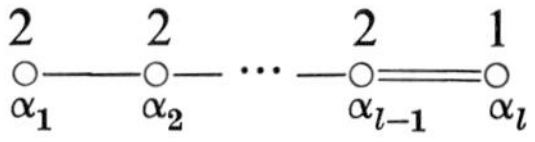

$\mathfrak{c}_l$:
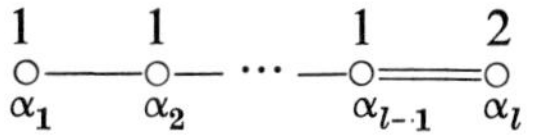

$\mathfrak{f}_4$:
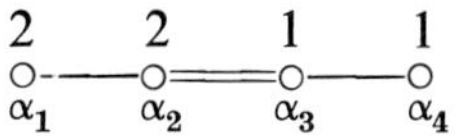

Passing now to the third type in (7), we put

$$\alpha = \sum_{i=1}^{p-1} i\alpha_i, \qquad \beta = \sum_{j=1}^{q-1} j\beta_j, \qquad \gamma = \sum_{k=1}^{r-1} k\gamma_k.$$

Since all links occurring are simple, all vertices have the same weight, say

$$c = |\alpha_i|^2 = |\beta_j|^2 = |\gamma_k|^2 = |\delta|^2.$$

The vectors α, β, γ are mutually orthogonal and δ is not spanned by them. Thus, by the proof of Lemma 3.15, the angles θ_p, θ_q, and θ_r between δ and α, β, and γ, respectively, satisfy

$$\cos^2\theta_p + \cos^2\theta_q + \cos^2\theta_r < 1. \tag{9}$$

But (cf. (8)), $\langle\alpha, \alpha\rangle = \frac{1}{2}p(p-1)c$ and

$$\langle\alpha, \delta\rangle = (p-1)\langle\alpha_{p-1}, \delta\rangle = -\tfrac{1}{2}(p-1)\,c,$$

so

$$\cos^2\theta_p = \frac{1}{2}\left(1 - \frac{1}{p}\right).$$

Now (9) implies the crucial formula

$$\frac{1}{p} + \frac{1}{q} + \frac{1}{r} > 1. \tag{10}$$

Here we can take $p \geqslant q \geqslant r \geqslant 2$ because if one of them equals 1, we have the first type in (7). Then $p^{-1} \leqslant q^{-1} \leqslant r^{-1}$, so by (10) $3r^{-1} > 1$.

Hence $r = 2$, and (10) implies $p^{-1} + q^{-1} > \frac{1}{2}$. This gives quickly the possibilities

$$r = q = 2, \qquad p \geqslant 2, \text{ arbitrary}$$

$$r = 2, \quad q = 3, \qquad p = 3, 4, 5.$$

Thus we have proved the following result.

Theorem 3.21. *Each connected diagram of rank l is one of the following*:

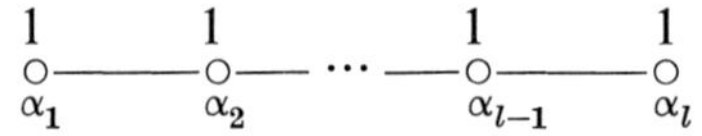

$\mathfrak{a}_l \quad (l \geqslant 1)$

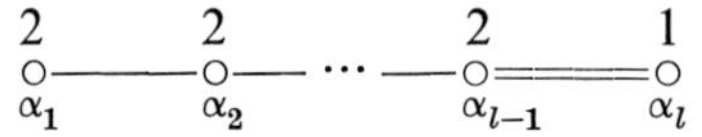

$\mathfrak{b}_l \quad (l \geqslant 2)$

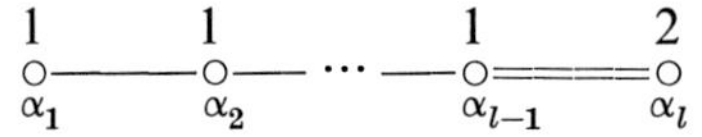

$\mathfrak{c}_l \quad (l \geqslant 3)$

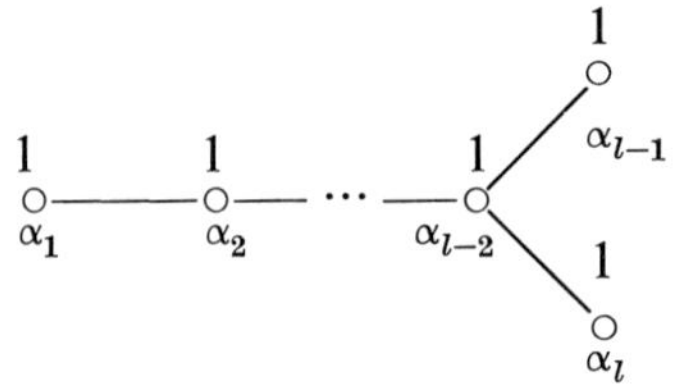

$\mathfrak{d}_l \quad (l \geqslant 4)$

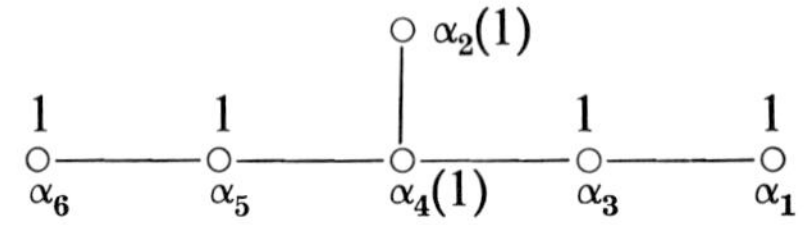

$\mathfrak{e}_6 \quad (l = 6)$

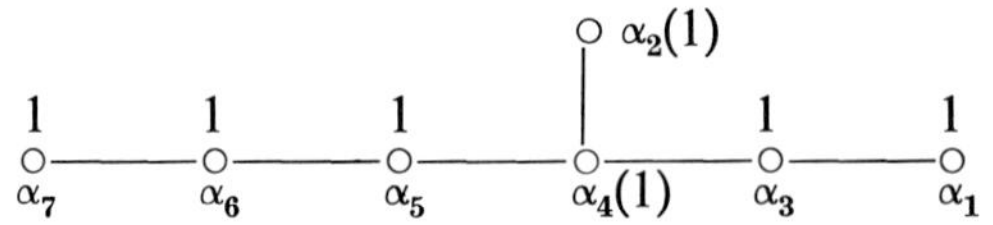

$\mathfrak{e}_7 \quad (l = 7)$

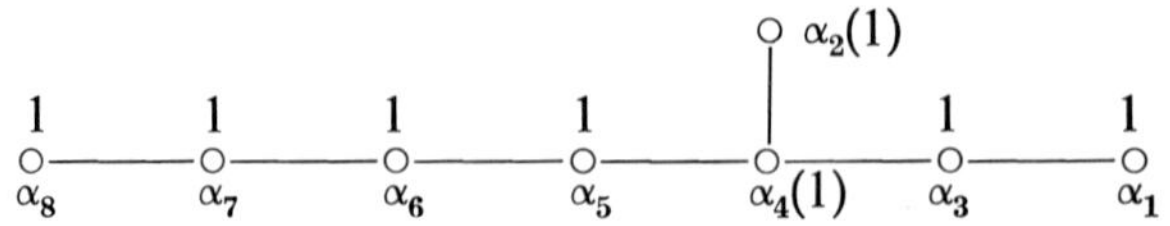

$\mathfrak{e}_8 \quad (l = 8)$

$\mathfrak{f}_4 \quad (l = 4)$

$\mathfrak{g}_2 \quad (l = 2)$.

We shall now verify that all these possibilities do indeed occur as Dynkin diagrams of reduced root systems. This will complete the classification of irreducible reduced root systems. This verification is already done for the "classical" diagrams $\mathfrak{a}_l$, $\mathfrak{b}_l$, $\mathfrak{c}_l$, and $\mathfrak{d}_l$, so it remains to consider the "exceptional" ones, $\mathfrak{e}_6$, $\mathfrak{e}_7$, $\mathfrak{e}_8$, $\mathfrak{f}_4$, and $\mathfrak{g}_2$.

Consider the Euclidean space $E = \boldsymbol{R}^n$ with its standard basis $(e_1, ..., e_n)$ and scalar product $\langle\ ,\ \rangle$. We consider subgroups V_{n-1}, L_0, L_1, L_2, L_3 of E defined as follows:

$$V_{n-1} = \left\{x = \sum_1^n x_i e_i : \sum_i x_i = 0\right\},$$

$$L_0 = \boldsymbol{Z}e_1 + ... + \boldsymbol{Z}e_n,$$

$$L_1 = \left\{x = \sum_{i=1}^n x_i e_i \in L_0 : \sum_i x_i \text{ even}\right\},$$

$$L_2 = L_0 + \boldsymbol{Z}\left\{\tfrac{1}{2}\sum_1^n e_i\right\},$$

$$L_3 = L_1 + \boldsymbol{Z}\left\{\tfrac{1}{2}\sum_1^n e_i\right\}.$$

Proposition 3.22.

(i) *The root system for* $\mathfrak{d}_l$ *is given by the sphere*

$$\{\alpha \in L_0 \subset \boldsymbol{R}^l : \langle\alpha, \alpha\rangle = 2\}.$$

(ii) *The root system for* $\mathfrak{b}_l$ *is given by*

$$\{\alpha \in L_0 \subset \boldsymbol{R}^l : \langle\alpha, \alpha\rangle = 1 \text{ or } 2\}.$$

(iii) *The mapping* $\alpha \to 2\alpha/\langle\alpha, \alpha\rangle$ (*which always maps a root system onto a root system*) *maps* $\mathfrak{b}_l$ *onto* $\mathfrak{c}_l$.

(iv) *The root system for* $\mathfrak{a}_l$ *is given by*

$$\{\alpha \in L_0 \subset \boldsymbol{R}^{l+1} : \langle\alpha, \alpha\rangle = 2\} \cap V_l.$$

The proof is immediate from the description of the classical root systems already given. We shall now construct the exceptional root systems in a similar spirit.

1. The root system $\mathfrak{g}_2$. Let

$$R = \{\alpha \in L_0 \subset \boldsymbol{R}^3 : \langle\alpha, \alpha\rangle = 2 \text{ or } 6\} \cap V_2.$$

The elements in R are then given by

$$\pm(e_2 - e_3), \quad \pm(e_3 - e_1), \quad \pm(e_1 - e_2)$$
$$\pm(2e_1 - e_2 - e_3), \quad \pm(2e_2 - e_1 - e_3), \quad \pm(2e_3 - e_1 - e_2).$$

The system R is indeed a root system (in V_2); in fact a direct verification gives the integrality condition $a_{\beta,\alpha} \in \boldsymbol{Z}$ for $\alpha, \beta \in R$, so L_0 is invariant under each s_α ($\alpha \in R$) and so is V_2 and $\langle \ , \ \rangle$.

The roots

$$\alpha_1 = e_1 - e_2, \qquad \alpha_2 = -2e_1 + e_2 + e_3$$

form a basis B of R and

$$\langle \alpha_1, \alpha_1 \rangle = 2, \qquad \langle \alpha_2, \alpha_2 \rangle = 6, \qquad a_{\alpha_1,\alpha_2} a_{\alpha_2,\alpha_1} = 3.$$

Thus the Dynkin diagram of R is $\mathfrak{g}_2$, and R has cardinality 12.

2. The root system $\mathfrak{f}_4$. Let

$$R = \{\alpha \in L_2 \subset \boldsymbol{R}^4 : \langle \alpha, \alpha \rangle = 1 \text{ or } 2\}.$$

The elements of R are then given by

$$\pm e_i, \qquad \pm e_i \pm e_j \quad (i < j), \qquad \tfrac{1}{2}(\pm e_1 \pm e_2 \pm e_3 \pm e_4),$$

where the $\pm$ signs are taken independently. To see this observe that a vector $\alpha \in R$ can only have coordinates $0, \pm 1, \pm \frac{1}{2}$ (not $\pm \frac{3}{2}$). To see that R is a root system it suffices to verify $a_{\beta,\alpha} \in \boldsymbol{Z}$ for $\alpha, \beta \in R$ (because then L_2 is invariant under each s_α ($\alpha \in R$)). This verification is trivial. For a basis of R we can take the roots

$$\alpha_1 = e_2 - e_3, \qquad \alpha_2 = e_3 - e_4, \qquad \alpha_3 = e_4, \qquad \alpha_4 = \tfrac{1}{2}(e_1 - e_2 - e_3 - e_4),$$

and the Dynkin diagram is $\mathfrak{f}_4$. The cardinality of R is $2 \cdot 4 + \binom{4}{2} \cdot 4 + 2^4 = 48$.

3. The root system $\mathfrak{e}_8$. Let

$$R = \{\alpha \in L_3 \subset \boldsymbol{R}^8 : \langle \alpha, \alpha \rangle = 2\}.$$

Then R consists of the vectors

$$\pm e_i \pm e_j \quad (i < j), \qquad \tfrac{1}{2} \sum_{i=1}^{8} (-1)^{\nu(i)} e_i \quad \Big(\sum_{1}^{8} \nu(i) \text{ even}\Big) \tag{11}$$

where the $\pm$ signs are taken independently. In fact, these vectors clearly belong to R; conversely, if $\alpha \in R$, its coordinates have values 0, $\pm\frac{1}{2}$, ± 1, and the form (11) is easily verified. Since $\langle \alpha, \beta \rangle \in \boldsymbol{Z}$ for all $\alpha, \beta \in R$, the integrality condition $a_{\beta,\alpha} \in \boldsymbol{Z}$ holds, so R is a root system. The roots

$$\alpha_1 = \tfrac{1}{2}(e_1 + e_8) - \tfrac{1}{2}(e_2 + e_3 + e_4 + e_5 + e_6 + e_7)$$

$$\alpha_2 = e_1 + e_2, \qquad \alpha_3 = e_2 - e_1, \qquad \alpha_4 = e_3 - e_2, \qquad \alpha_5 = e_4 - e_3$$

$$\alpha_6 = e_5 - e_4, \qquad \alpha_7 = e_6 - e_5, \qquad \alpha_8 = e_7 - e_6$$

form a basis of R. Instead of verifying this directly one can proceed as follows: The vector $\rho = (0, 1, 2, 3, 4, 5, 6, 23) \in \boldsymbol{R}^8$ is regular and $R^+(\rho)$ consists of the roots

$$\pm e_i \pm e_j \quad (i < j), \qquad \tfrac{1}{2}\left(e_8 + \sum_{i=1}^{7} (-1)^{\nu(i)}\, e_i\right) \quad \left(\sum_{i=1}^{7} \nu(i) \text{ even}\right).$$

We have $\langle \alpha, \rho \rangle \in \boldsymbol{Z}^+$ for each $\alpha \in R^+(\gamma)$ and $\langle \alpha_i, \rho \rangle = 1$ for $1 \leqslant i \leqslant 8$. It follows that the $\alpha_1, ..., \alpha_8$ are simple roots in $R^+(\gamma)$, and being also linearly independent will form a basis of R (cf. proof of Theorem 3.6). Finally, we have

$$\langle \alpha_4, \alpha_5 \rangle = \langle \alpha_5, \alpha_6 \rangle = \langle \alpha_6, \alpha_7 \rangle = \langle \alpha_7, \alpha_8 \rangle = \langle \alpha_4, \alpha_2 \rangle = \langle \alpha_4, \alpha_3 \rangle = \langle \alpha_3, \alpha_1 \rangle = -1$$

and $\langle \alpha_i, \alpha_j \rangle = 0$ for the other (i, j). The Dynkin diagram is therefore of type $\mathfrak{e}_8$. The cardinality of R is $\binom{8}{2} \cdot 4 + 2^7 = 240$.

4. **The root system** $\mathfrak{e}_7$. Let F^7 be the subspace of $\boldsymbol{R}^8$ spanned by the simple roots $\alpha_1, \alpha_2, ..., \alpha_7$ in the construction of the root system R_8 of type $\mathfrak{e}_8$. Let $R = R_8 \cap F^7$. Then R is a root system in F^7 with basis $\alpha_1, ..., \alpha_7$, and the Dynkin diagram is of type $\mathfrak{e}_7$. More specifically, since F^7 has normal $e_7 + e_8$, R is given by

$$\pm e_i \pm e_j \quad (1 \leqslant i < j \leqslant 6), \qquad \pm(e_7 - e_8),$$

$$\pm\tfrac{1}{2}\left(e_7 - e_8 + \sum_{1}^{6} (-1)^{\nu(i)}\, e_i\right) \quad \left(\sum_{i=1}^{6} \nu(i) \text{ odd}\right).$$

The cardinality of R is $\binom{6}{2} \cdot 4 + 2 + 2^6 = 126$.

5. **The root system** $\mathfrak{e}_6$. Let F^6 be the subspace of $\boldsymbol{R}^8$ spanned by the simple roots $\alpha_1, ..., \alpha_6$ in the construction of the root system R_8 of $\mathfrak{e}_8$. Let $R = R_8 \cap F^6$. Then R is a root system in F^6 with basis $(\alpha_1, ..., \alpha_6)$,

and the Dynkin diagram is of type $\mathfrak{e}_6$. Since F^6 is the set of $x \in F^7$ perpendicular to $e_6 + e_8$, R consists of the vectors,

$$\pm e_i \pm e_j \quad (1 \leqslant i < j \leqslant 5),$$

$$\pm \tfrac{1}{2}\left(e_8 - e_7 - e_6 + \sum_{i=1}^{5} (-1)^{\nu(i)} e_i\right) \quad \left(\sum_{1}^{5} \nu(i) \text{ even}\right).$$

The cardinality of R is $\binom{5}{2} \cdot 4 + 2^4 \cdot 2 = 72$.

Since an irreducible root system is uniquely determined by the Dynkin diagram, we have proved the following result.

Theorem 3.23. *The root systems* $\mathfrak{a}_l$ $(l \geqslant 1)$, $\mathfrak{b}_l$ $(l \geqslant 2)$, $\mathfrak{c}_l$ $(l \geqslant 3)$, $\mathfrak{d}_l$ $(l \geqslant 4)$, $\mathfrak{e}_6$, $\mathfrak{e}_7$, $\mathfrak{e}_8$, $\mathfrak{f}_4$, *and* $\mathfrak{g}_2$ *exhaust all irreducible reduced root systems.*

For completeness we write down the Cartan matrices for these root systems. As remarked earlier, the Dynkin diagram determines the Cartan matrix uniquely. For $\mathfrak{a}_l$, $\mathfrak{b}_l$, $\mathfrak{c}_l$, and $\mathfrak{d}_l$ the Cartan matrices were already given, so we just have to consider the exceptional ones. From the description in Theorem 3.21 we have the Cartan matrices

$$\mathfrak{g}_2 : \begin{pmatrix} 2 & -3 \\ -1 & 2 \end{pmatrix} \qquad \mathfrak{f}_4 : \begin{pmatrix} 2 & -1 & 0 & 0 \\ -1 & 2 & -2 & 0 \\ 0 & -1 & 2 & -1 \\ 0 & 0 & -1 & 2 \end{pmatrix}$$

$$\mathfrak{e}_8 : \begin{pmatrix} 2 & 0 & -1 & 0 & 0 & 0 & 0 & 0 \\ 0 & 2 & 0 & -1 & 0 & 0 & 0 & 0 \\ -1 & 0 & 2 & -1 & 0 & 0 & 0 & 0 \\ 0 & -1 & -1 & 2 & -1 & 0 & 0 & 0 \\ 0 & 0 & 0 & -1 & 2 & -1 & 0 & 0 \\ 0 & 0 & 0 & 0 & -1 & 2 & -1 & 0 \\ 0 & 0 & 0 & 0 & 0 & -1 & 2 & -1 \\ 0 & 0 & 0 & 0 & 0 & 0 & -1 & 2 \end{pmatrix}$$

$\mathfrak{e}_7$: Remove last row and last column from $\mathfrak{e}_8$.

$\mathfrak{e}_6$: Remove the last two rows and last two columns from $\mathfrak{e}_8$.

4. The Nonreduced Root Systems

It is now easy to determine all the irreducible nonreduced root systems. Let R be one such and as in §3, no. 1 let $R' \neq R$ be the set of indivisible roots in R; then by Lemma 3.2, R' is a reduced root system. Let B be a basis of R' (and R).

Lemma 3.24. *$2B \cap R$ consists of one element.*

Proof. Since each $\alpha \in R'$ is Weyl group conjugate to an element in B (Theorem 3.6), the assumption $R' \neq R$ implies $2B \cap R \neq \emptyset$. Let $\alpha \in B$, $2\alpha \in R$ and let $\beta \in B - \{\alpha\}$ be arbitrary such that $\langle \alpha, \beta \rangle \neq 0$. Then $a_{\beta,\alpha} = 2a_{\beta,2\alpha}$ and $0 < a_{\beta,\alpha}a_{\alpha,\beta} \leqslant 3$. Hence $\{\alpha, \beta\}$ is a double link in the Dynkin diagram and $a_{\beta,\alpha} = -2$, $a_{\alpha,\beta} = -1$, so $2\langle \alpha, \alpha \rangle = \langle \beta, \beta \rangle$. Thus α has a double link with any $\beta \in B$ to which it is connected and $\langle \alpha, \alpha \rangle = \frac{1}{2}\langle \beta, \beta \rangle$. It follows from Theorem 3.21 that R' is of type $\mathfrak{b}_l$ ($l \geqslant 1$). Now Lemma 3.18, and the fact that $2\beta \in R$ would imply $a_{\alpha,\beta} = 2a_{\alpha,2\beta} \in 2\boldsymbol{Z}$, shows that $2B \cap R$ consists of just one element.

Now we know that there exists an orthonormal basis $(e_1, ..., e_l)$ of the span of R such that

$$R' = \{\pm e_i \, (1 \leqslant i \leqslant l), \pm e_i \pm e_j \, (1 \leqslant i < j \leqslant l)\}$$

and $B = \{\alpha_1, ..., \alpha_l\}$ where

$$\alpha_i = e_i - e_{i+1} \quad (1 \leqslant i \leqslant l-1), \qquad \alpha_l = e_l.$$

Then, as proved above, $2\alpha_l = 2e_l \in R$. But the reflection $s_{e_i - e_l}$ sends e_l to e_i, so $2e_i \in R$ for $1 \leqslant i \leqslant l$. On the other hand, the inclusion $2(\pm e_i \pm e_j) \in R$ would violate the integrality condition $a_{\alpha,\beta} \in \boldsymbol{Z}$ for $\alpha = e_i$, $\beta = 2(\pm e_i \pm e_j)$.

Theorem 3.25. *The irreducible nonreduced root systems are precisely the root systems*

$$(\mathfrak{bc})_l = \{\pm e_i \pm e_j : 1 \leqslant i < j \leqslant l, \pm e_i \, (1 \leqslant i \leqslant l), \pm 2e_i \, (1 \leqslant i \leqslant l)\}$$

for $l \geqslant 1$, and a basis is given by

$$\alpha_i = e_i - e_{i+1} \quad (1 \leqslant i \leqslant l-1), \qquad \alpha_l = e_l.$$

In fact the above discussion shows that there are no other possibilities. On the other hand, $(\mathfrak{bc})_l$ is a root system (direct verification or Exercise 5, Chapter VII).

5. The Highest Root

Proposition 3.26. *Let R be an irreducible root system, $B = (\alpha_1, ..., \alpha_l)$ any basis of R. Then there exists a unique root $\delta = \sum_1^l d_i\alpha_i$ in R such that for any root $\alpha = \sum_1^l a_i\alpha_i$ in R we have $a_1 \leqslant d_1, ..., a_l \leqslant d_l$.*

In fact, the proof of Lemma 11.8, Chapter VII, is valid in the present abstract context because of Lemma 3.9. The root δ is called the *highest root* (for B) or *the maximal root*. The following lemma is useful for determining the highest root.

Lemma 3.27. *Let R be an irreducible root system in V. Let $B = (\alpha_1, ..., \alpha_l)$ be a basis of R and $(f_1, ..., f_l)$ any basis of V such that each α_i is positive for the lexicographic ordering of V given by $(f_1, ..., f_l)$. Then the highest root for this lexicographic ordering coincides with the highest root (for B).*

Proof. In fact if $\alpha \in R$, then $\delta - \alpha = \sum_1^l b_i\alpha_i$ where $b_i \geqslant 0$ for each i, so $\delta - \alpha$ is positive for the lexicographic ordering.

Theorem 3.28. *The irreducible root systems R have the highest roots given as follows:*

$\mathfrak{a}_l$: $\delta = e_1 - e_{l+1} = \alpha_1 + ... + \alpha_l$

$\mathfrak{b}_l$: $\delta = e_1 + e_2 = \alpha_1 + 2\alpha_2 + ... + 2\alpha_l$

$\mathfrak{c}_l$: $\delta = 2e_1 = 2\alpha_1 + 2\alpha_2 + ... + 2\alpha_{l-1} + \alpha_l$

$\mathfrak{d}_l$: $\delta = e_1 + e_2 = \alpha_1 + 2\alpha_2 + ... + 2\alpha_{l-2} + \alpha_{l-1} + \alpha_l$

$\mathfrak{e}_6$: $\delta = \frac{1}{2}(e_1 + e_2 + e_3 + e_4 + e_5 - e_6 - e_7 + e_8)$
$\quad = \alpha_1 + 2\alpha_2 + 2\alpha_3 + 3\alpha_4 + 2\alpha_5 + \alpha_6$

$\mathfrak{e}_7$: $\delta = e_8 - e_7 = 2\alpha_1 + 2\alpha_2 + 3\alpha_3 + 4\alpha_4 + 3\alpha_5 + 2\alpha_6 + \alpha_7$

$\mathfrak{e}_8$: $\delta = e_7 + e_8 = 2\alpha_1 + 3\alpha_2 + 4\alpha_3 + 6\alpha_4 + 5\alpha_5 + 4\alpha_6 + 3\alpha_7 + 2\alpha_8$

$\mathfrak{f}_4$: $\delta = e_1 + e_2 = 2\alpha_1 + 3\alpha_2 + 4\alpha_3 + 2\alpha_4$

$\mathfrak{g}_2$: $\delta = -e_1 - e_2 + 2e_3 = 3\alpha_1 + 2\alpha_2$

$(\mathfrak{bc})_l$: $\delta = 2e_1 = 2\alpha_1 + 2\alpha_2 + ... + 2\alpha_l$.

Proof. For $\mathfrak{a}_l$, $\mathfrak{b}_l$, $\mathfrak{c}_l$, $\mathfrak{d}_l$, $(\mathfrak{bc})_l$ this is obvious. For the others we make use of Lemma 3.27.

For $\mathfrak{g}_2$ we use the basis

$$f_1 = e_3 - e_1, \qquad f_2 = e_1 - e_2$$

of the space spanned by R. Then $\alpha_1 = f_2$, $\alpha_2 = f_1 - f_2$, and these roots are positive for the lexicographic ordering given by (f_1, f_2). But $2e_3 - e_1 - e_2$ is clearly the highest of the roots with respect to this lexicographic ordering; thus by Lemma 3.27 it coincides with δ.

For $\mathfrak{f}_4$ we use the basis

$$f_1 = e_1, \qquad f_2 = e_2, \qquad f_3 = e_3, \qquad f_4 = e_4.$$

The elements α_1, α_2, α_3, α_4 are positive for the corresponding lexicographic ordering so $\delta = e_1 + e_2$ follows from Lemma 3.27.

For $\mathfrak{e}_8$ we use the basis $(e_8, e_7, ..., e_1)$. Then each α_i is positive for the corresponding lexicographic ordering, and for this ordering the highest root is clearly $e_8 + e_7$. Hence $\delta = e_7 + e_8$.

For $\mathfrak{e}_7$ we first note that the space F^7 spanned by the roots is the hyperplane in $\boldsymbol{R}^8$ with normal $e_7 + e_8$. We can take

$$e_8 - e_7, \quad e_6, \quad e_5, \quad e_4, \quad e_3, \quad e_2, \quad e_1$$

for a basis of F^7. The roots $\alpha_1, ..., \alpha_7$ are positive for the corresponding lexicographic ordering, and for this ordering the highest root is $e_8 - e_7$. Hence $\delta = e_8 - e_7$.

Finally, for $\mathfrak{e}_6$ we use the basis

$$e_8 - e_7 - e_6, \quad e_5, \quad e_4, \quad e_3, \quad e_2, \quad e_1$$

of F^6. The roots $\alpha_1, ..., \alpha_6$ are positive in the corresponding lexicographic ordering, and for this ordering the highest root is half the sum of the basis elements.

We now rewrite the Dynkin diagrams omitting the weights of the vertices but instead putting an arrow pointing toward the shorter root in case two roots of unequal length are joined. Now we furnish each vertex with the coefficient it has in the formula for the highest root.

Dynkin Diagrams with the Coefficients of the Highest Root

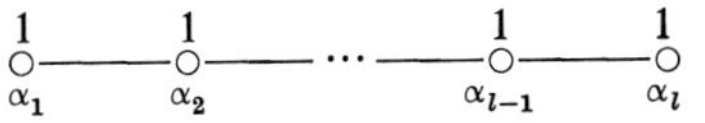

$\mathfrak{a}_l \ l \geqslant 1$

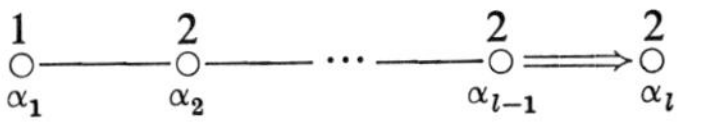

$\mathfrak{b}_l \ (l \geqslant 2)$

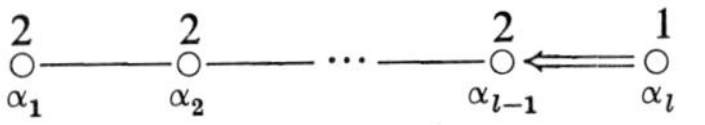

$\mathfrak{c}_l \ (l \geqslant 3)$

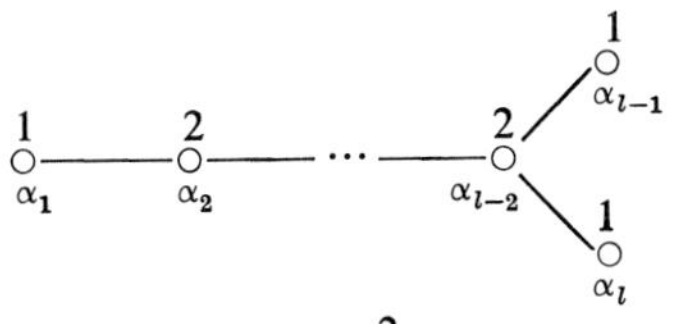

$\mathfrak{d}_l \ (l \geqslant 4)$

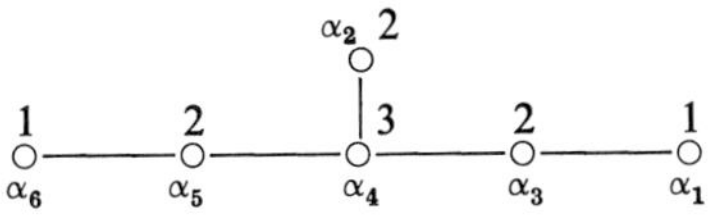

$\mathfrak{e}_6$

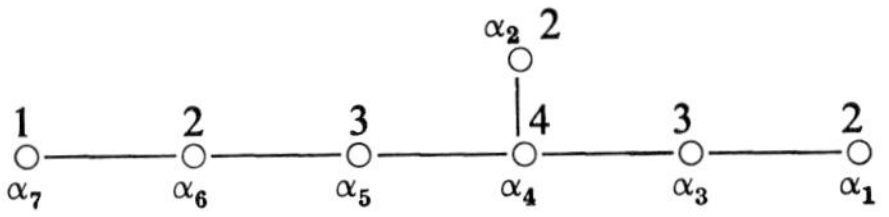

$\mathfrak{e}_7$

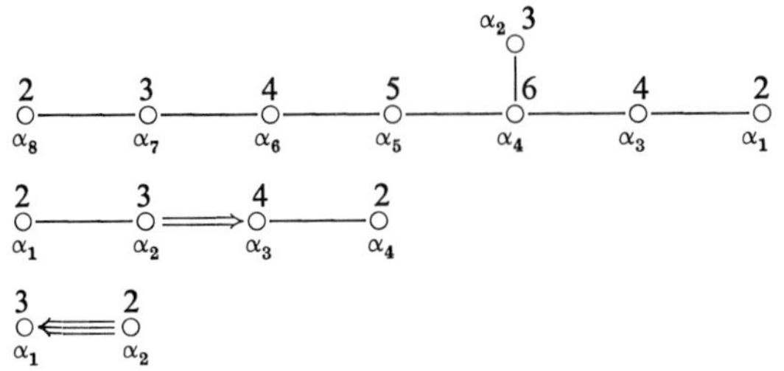

$$
\begin{array}{c}
\overset{\alpha_2}{\underset{3}{\circ}} \\
\overset{2}{\underset{\alpha_8}{\circ}} - \overset{3}{\underset{\alpha_7}{\circ}} - \overset{4}{\underset{\alpha_6}{\circ}} - \overset{5}{\underset{\alpha_5}{\circ}} - \overset{6}{\underset{\alpha_4}{\circ}} - \overset{4}{\underset{\alpha_3}{\circ}} - \overset{2}{\underset{\alpha_1}{\circ}}
\end{array}
\qquad \mathfrak{e}_8
$$

$$
\overset{2}{\underset{\alpha_1}{\circ}} - \overset{3}{\underset{\alpha_2}{\circ}} \Rightarrow \overset{4}{\underset{\alpha_3}{\circ}} - \overset{2}{\underset{\alpha_4}{\circ}}
\qquad \mathfrak{f}_4
$$

$$
\overset{3}{\underset{\alpha_1}{\circ}} \Lleftarrow \overset{2}{\underset{\alpha_2}{\circ}}
\qquad \mathfrak{g}_2.
$$

An *automorphism* of the Dynkin diagram is a permutation σ of the vertices preserving both the weights and the multiplicity of the links. Equivalently, an automorphism of the Dynkin diagram is a permutation σ of the vertices preserving the graph (including the arrows).

6. Outer Automorphisms and the Covering Index

Let R be an irreducible, reduced root system in V and $\langle \, , \, \rangle$ any positive definite scalar product on V invariant under $\mathrm{Aut}(R)$ (unique up to a constant factor). The Weyl group $W(R)$ is a normal subgroup of $\mathrm{Aut}(R)$.

Theorem 3.29. *The factor group* $\mathrm{Aut}(R)/W(R)$ *is isomorphic to the group of automorphisms of the Dynkin diagram. It is given as follows:*

(i)   *For* $\mathfrak{b}_l, \mathfrak{c}_l, \mathfrak{e}_7, \mathfrak{e}_8, \mathfrak{f}_4, \mathfrak{g}_2$: $\mathbf{Z}_1$

(ii) *For* $\mathfrak{a}_l, \mathfrak{d}_l \, (l > 4), \mathfrak{e}_6$: $\mathbf{Z}_2,$

(iii) *For* $\mathfrak{d}_4$: $\mathfrak{S}_3.$

where $\mathbf{Z}_m$ *is the cyclic group of order* m *and* $\mathfrak{S}_3$ *is the permutation group on three letters.*

In fact, since $W(R)$ permutes the bases of R simply transitively, $\mathrm{Aut}(R)/W(R)$ is (cf. proof of Cor. 5.5, Chapter IX) isomorphic to the subgroup of $\sigma \in \mathrm{Aut}(R)$ leaving a basis of R (or equivalently, a Weyl chamber) invariant. But this amounts to σ permuting the vertices of the Dynkin diagram, preserving both the weights of the vertices as well as the multiplicity of the links. The diagrams for $\mathfrak{b}_l, \mathfrak{c}_l, \mathfrak{e}_7, \mathfrak{e}_8, \mathfrak{f}_4$, and $\mathfrak{g}_2$ admit no such permutation, so part (i) follows. For $\mathfrak{a}_l$ and $\mathfrak{e}_6$ we have a symmetry of the diagram with respect to a vertical axis, and for $\mathfrak{d}_l$ a symmetry with respect to a horizontal axis. This gives (ii). If $l = 4$, $\mathfrak{d}_l$ has additional symmetries, resulting in (iii).

Now let $T(R)$ be the subgroup of V generated by the vectors in R and let

$$\tilde{T}(R) = \left\{x \in V : 2\frac{\langle x, \alpha\rangle}{\langle \alpha, \alpha\rangle} \in \boldsymbol{Z} \text{ for } \alpha \in R\right\}.$$

We shall now relate $T(R) \subset \tilde{T}(R)$ to some concepts from Chapter VII. Let $\mathfrak{u}$ be a simple compact Lie algebra, $\mathfrak{t}_0$ a maximal abelian subalgebra. Let $\mathfrak{g}$ and $\mathfrak{t}$ denote the respective complexifications and Δ^* the set of nonzero roots of $\mathfrak{g}$ with respect to $\mathfrak{t}$. As in §§6, 7, Chapter VII, consider the lattices

$$\mathfrak{t}(\mathfrak{u}) = \{H \in \mathfrak{t}_0 : \alpha(H) \in 2\pi i \boldsymbol{Z} \text{ for all } \alpha \in \Delta^*\}$$

$$\mathfrak{t}_\Delta = \left\{\text{integral linear combinations of } \frac{4\pi i}{\langle \alpha, \alpha\rangle} H_\alpha : \alpha \in \Delta^*\right\},$$

where of course $\mathfrak{t}_\Delta \subset \mathfrak{t}(\mathfrak{u})$. We recall (Cor. 7.8, Chapter VII) the isomorphism

$$\tilde{Z} \approx \mathfrak{t}(\mathfrak{u})/\mathfrak{t}_\Delta, \tag{12}$$

where $\tilde{Z}$ denotes the center of the simply connected Lie group $\tilde{U}$ with Lie algebra $\mathfrak{u}$. Because of the isomorphism π_1 (Int($\mathfrak{u}$)) $\approx \tilde{Z}$, we call the order of $\tilde{Z}$ *the covering index* (É. Cartan's terminology in [9] is *indice de connexion*).

Lemma 3.30. *The pairing*

$$(x + T(\Delta^*), H + \mathfrak{t}_\Delta) \to e^{x(H)}$$

turns the groups $\tilde{T}(\Delta^*)/T(\Delta^*)$ *and* $\mathfrak{t}(\mathfrak{u})/\mathfrak{t}_\Delta$ *into character groups of each other*; *in particular, they are isomorphic.*

This lemma is contained in the general duality theory for locally compact abelian groups, which we summarize below, following Weil [1], §28. (The case dealt with in the lemma could be verified much more directly.)

The locally compact abelian groups come in pairs G and $\hat{G}$, each being the character group of the other. If $(x, \hat{x})$ is the function on $G \times \hat{G}$ which defines the characters of G and of $\hat{G}$, there is a one-to-one correspondance $g \to \gamma$ between the closed subgroups $g \subset G$ and the closed subgroups $\gamma \subset \hat{G}$ satisfying

$$\gamma = \{\hat{x} \in \hat{G} : (x, \hat{x}) = 1 \text{ for } x \in g\},$$

$$g = \{x \in G : (x, \hat{x}) = 1 \text{ for } \hat{x} \in \gamma\}.$$

If $g \supset g'$, we have $\gamma \subset \gamma'$ and the groups g/g' and γ'/γ are character groups of each other; in particular

$$g = (\hat{G}/\gamma)^{\wedge}, \qquad \gamma = (G/g)^{\wedge}.$$

Lemma 3.30 follows by taking

$$G = \sum_{\alpha \in \Delta^*} \boldsymbol{R}\alpha, \qquad \hat{G} = \mathfrak{t}_0$$

$$g = \tilde{T}(\Delta^*), \qquad \gamma' = \mathfrak{t}(\mathfrak{u})$$

$$g' = T(\Delta^*), \qquad \gamma = \mathfrak{t}_\Delta.$$

Now let R be any irreducible, reduced root system. Because of Lemma 3.30, the order of the factor group $\tilde{T}(R)/T(R)$ is called *the covering index.*

Proposition 3.31. *The covering index equals the determinant of the Cartan matrix, that is,*

$$|\, \tilde{T}(R)/T(R)| = \det((a_{ij})_{1 \leqslant i,j \leqslant l}).$$

Proof. Let $B = (\alpha_1, \ldots, \alpha_l)$ be any basis of R. Then $T(R) = \Sigma_{i=1}^l \boldsymbol{Z}\alpha_i$. Put $\alpha^\vee = 2\alpha/\langle\alpha, \alpha\rangle$ for $\alpha \in R$. The lattice $\Sigma_{i=1}^l \boldsymbol{Z}\alpha_i^\vee$ is invariant under each s_{α_k}, hence under $W(R)$ (Lemma 3.9). Since $W(R)(B) = R$, we conclude that

$$\tilde{T}(R) = \{x \in V : \langle x, \alpha_i^\vee\rangle \in \boldsymbol{Z},\ 1 \leqslant i \leqslant l\}.$$

Let $\omega_1, \ldots, \omega_l$ be the "dual basis" given by $\langle \omega_j, \alpha_i^\vee\rangle = \delta_{ij}$. Then

$$\tilde{T}(R) = \sum_{j=1}^{l} \boldsymbol{Z}\omega_j \qquad \text{and} \qquad \alpha_i = \sum_{j=1}^{l} a_{ij}\omega_j.$$

Thus the lattice $T(R)$ is the image of $\tilde{T}(R)$ under the linear transformation given by the Cartan matrix. Now the result follows from volume considerations since the determinant is always found to be positive.

Theorem 3.32. *The factor group $\tilde{T}(R)/T(R)$ is given as follows:*

(i) *For* $\mathfrak{g}_2, \mathfrak{f}_4, \mathfrak{e}_8$: $\boldsymbol{Z}_1$,
(ii) *For* $\mathfrak{b}_l, \mathfrak{c}_l, \mathfrak{e}_7$: $\boldsymbol{Z}_2$,
(iii) *For* $\mathfrak{e}_6$: $\boldsymbol{Z}_3$,
(iv) *For* $\mathfrak{d}_{2k+1}$: $\boldsymbol{Z}_4$,
(v) *For* $\mathfrak{d}_{2k}$: $\boldsymbol{Z}_2 \times \boldsymbol{Z}_2$,
(vi) *For* $\mathfrak{a}_l$: $\boldsymbol{Z}_{l+1}$.

Proof. For (i), (ii), and (iii) this is clear from Prop. 3.31 by computation of the determinant. For the others we write down the groups $T(R)$ and $\tilde{T}(R)$.

For $\mathfrak{a}_l$, $T(R)$ consists of the integral linear combinations $\sum_1^{l+1} n_i e_i$ ($\sum_i n_i = 0$), and $\tilde{T}(R)$ consists of the vectors

$$\left\{x = \sum_{i=1}^{l+1} x_i e_i : \sum_i x_i = 0, \sum_i n_i x_i \in \boldsymbol{Z} \text{ whenever } \sum_{i=1}^{l+1} n_i = 0\right\}.$$

Thus

$$\tilde{T}(R) = T(R) + \boldsymbol{Z}\left(e_1 - \frac{1}{l+1}\sum_{i=1}^{l+1} e_i\right),$$

so (vi) follows.

For $\mathfrak{d}_l$ we note that $T(R)$ consists of the lattice of integral linear combinations $\sum_{i=1}^{l} n_i e_i$ with $\sum_i n_i \in 2\boldsymbol{Z}$ and that $\tilde{T}(R)$ consists of the vectors

$$\left\{x = \sum_1^l x_i e_i : \sum_i n_i x_i \in \boldsymbol{Z} \text{ whenever } \sum_i n_i \in 2\boldsymbol{Z}\right\}.$$

Thus

$$\tilde{T}(R) = \sum_{i=1}^{l} \boldsymbol{Z} e_i + \boldsymbol{Z}\left(\tfrac{1}{2}\sum_{i=1}^{l} e_i\right),$$

so $\tilde{T}(R)/T(R)$ has order at most 4. If l is odd, the element $(\frac{1}{2}\sum_1^l e_i) + T(R)$ has order 4 in $\tilde{T}(R)/T(R)$; however if l is even, $\tilde{T}(R)/T(R)$ is generated by the two second order elements

$$\left(\tfrac{1}{2}\sum_1^l e_i\right) + T(R) \qquad \text{and} \qquad \left(\tfrac{1}{2}\sum_1^l e_i\right) - e_1 + T(R).$$

This proves the theorem.

§ 4. The Classification of Simple Lie Algebras over C

We have seen that to each simple Lie algebra $\mathfrak{g}$ over $\boldsymbol{C}$ is associated an irreducible, reduced root system R which determines $\mathfrak{g}$ up to isomorphism. We shall now prove that to each such R corresponds a $\mathfrak{g}$; thus Theorem 3.23 gives all simple Lie algebras over $\boldsymbol{C}$. In order to motivate the proof we construct for a given semisimple Lie algebra $\mathfrak{g}$ over $\boldsymbol{C}$ a specific system of generators.

Let $\mathfrak{h} \subset \mathfrak{g}$ be a Cartan subalgebra, R the set of nonzero roots of $\mathfrak{g}$ with respect to $\mathfrak{h}$, and $B = \{\alpha_1, ..., \alpha_l\}$ any basis of R. Now we denote the Killing form of $\mathfrak{g}$ by $\langle\ ,\ \rangle$; it induces a bilinear form (also denoted $\langle\ ,\ \rangle$) on the dual space of $\mathfrak{h}$. Let $H_i \in \mathfrak{h}$ be determined by

$$H_i = \frac{2}{\langle \alpha_i, \alpha_i \rangle} H_{\alpha_i} \qquad (1 \leqslant i \leqslant l)$$

where H_{α_i} is given by Theorem 4.2, Chapter III. Fix elements X_i, Y_i in the one-dimensional spaces $\mathfrak{g}^{\alpha_i}$, $\mathfrak{g}^{-\alpha_i}$ such that $[X_i, Y_i] = H_i\,(1 \leqslant i \leqslant l)$. As in §3, we write for simplicity $a_{ij} = a_{\alpha_i, \alpha_j}$ for the entries in the Cartan matrix. Since $\alpha_i - \alpha_j \notin R$, the α_i-series containing α_j is given by

$$\alpha_j, \quad \alpha_j + \alpha_i, \quad ..., \quad \alpha_j + q\alpha_i \qquad (-q = a_{ji}).$$

Proposition 4.1. *The semisimple Lie algebra* $\mathfrak{g}$ *is generated by the above vectors*

$$X_i, \quad Y_i, \quad H_i \qquad (1 \leqslant i \leqslant l)$$

and these generators satisfy the following relations:

(i) $[H_i, H_j] = 0$,

(ii) $[X_i, Y_i] = H_i$, $[X_i, Y_j] = 0$ if $i \neq j$,

(iii) $[H_i, X_j] = a_{ji}X_j$, $[H_i, Y_j] = -a_{ji}Y_j$,

(iv) $(\operatorname{ad} X_i)^{1-a_{ji}}(X_j) = 0 \quad (i \neq j)$.

(v) $(\operatorname{ad} Y_i)^{1-a_{ji}}(Y_j) = 0 \quad (i \neq j)$.

The relations being clear from the above remarks, we just have to verify that X_i, Y_i, H_i do indeed generate $\mathfrak{g}$. This however is obvious from Lemma 3.10 and Theorem 4.3 (iv), Chapter III. The vectors X_i, Y_i, H_i $(1 \leqslant i \leqslant l)$ are called the *canonical generators*.

We shall now prove a converse to this result. Let R be any reduced root system, $B = \{\beta_1, ..., \beta_l\}$ any basis of R, and $(a_{ij})_{1 \leqslant i,j \leqslant l}$ the associated Cartan matrix $(a_{ij} = a_{\beta_i, \beta_j})$. We shall now construct a semisimple Lie algebra and a basis for its root system whose Cartan matrix is (a_{ij}). Roughly speaking, this Lie algebra will be the one defined by $3l$ generators and the relations (i)–(v).

We consider first the *free Lie algebra* $\hat{\mathfrak{l}}$ on $3l$ generators ξ_i, η_i, ζ_i $(1 \leqslant i \leqslant l)$. This is defined as follows. The tensor algebra $T(E)$, where E is the vector space over $\boldsymbol{C}$ having ξ_i, η_i, ζ_i $(1 \leqslant i \leqslant l)$ as basis, is a Lie algebra with bracket $[a, b] = a \otimes b - b \otimes a$ and $\hat{\mathfrak{l}}$ is defined as the Lie subalgebra generated by E. Let $\hat{\mathfrak{k}}$ denote the ideal in $\hat{\mathfrak{l}}$ generated by the elements

$$[\zeta_i, \zeta_j], \quad [\xi_i, \eta_j] - \delta_{ij}\zeta_i, \quad [\zeta_i, \xi_j] - a_{ji}\xi_j, \quad [\zeta_i, \eta_j] + a_{ji}\eta_j.$$

Put $\mathfrak{l} = \hat{\mathfrak{l}}/\hat{\mathfrak{k}}$ and let x_i, y_i, h_i denote the images in $\mathfrak{l}$ of the generators ξ_i, η_i, ζ_i, respectively. Then x_i, y_i, and h_i satisfy the relations (i)–(iii).

The next result describes the structure of $\mathfrak{l}$, which in general is infinite-dimensional.

Theorem 4.2. *The elements x_i, y_i, h_i $(1 \leqslant i \leqslant l)$ are linearly independent and the Lie algebra $\mathfrak{l}$ is given by*

$$\mathfrak{l} = X + \mathfrak{h} + Y \qquad (\textit{direct vector space sum})$$

where X, $\mathfrak{h}$, and Y are the subalgebras generated by $(x_i)_{1 \leqslant i \leqslant l}$, $(h_i)_{1 \leqslant i \leqslant l}$, and $(y_i)_{1 \leqslant i \leqslant l}$, respectively. Moreover, $\mathfrak{h}$ is abelian and has dimension l.

Remark. For later proposes we remark that although the matrix (a_{ij}), as a Cartan matrix for a root system, is nonsingular, the proof we give of Theorem 4.2 is valid for an arbitrary matrix (a_{ij}).

For the proof we construct a family of representations of $\hat{\mathfrak{l}}$ on the tensor algebra T over a complex vector space with basis $t_1, \ldots, t_l$, viewing T as a vector space with basis

$$1, \qquad t_i \quad (1 \leqslant i \leqslant l), \qquad t_{i_1} \otimes t_{i_2} \quad (1 \leqslant i_1, i_2 \leqslant l), \quad \ldots.$$

For simplicity we write $t_{i_1} \ldots t_{i_r}$ instead of $t_{i_1} \otimes \ldots \otimes t_{i_r}$ (which for $r = 0$ is to mean 1). For $\lambda_1, \ldots, \lambda_l \in \boldsymbol{C}$ arbitrary consider the representation of $\hat{\mathfrak{l}}$ on T where the action on the basis of T is given by

$$\eta_j \cdot 1 = t_j, \qquad \eta_j \cdot t_{i_1} \ldots t_{i_r} = t_j t_{i_1} \ldots t_{i_r},$$

$$\zeta_j \cdot 1 = \lambda_j, \qquad \zeta_j \cdot t_{i_1} \ldots t_{i_r} = (\lambda_j - a_{i_1 j} - \ldots - a_{i_r j})\, t_{i_1} \ldots t_{i_r},$$

$$\xi_j \cdot 1 = 0, \qquad \xi_j \cdot t_{i_1} \ldots t_{i_r} = t_{i_1}(\xi_j \cdot t_{i_2} \ldots t_{i_r}) + \delta_{i_1 j}\zeta_j \cdot t_{i_2} \ldots t_{i_r}.$$

This in fact gives a representation of the Lie algebra $T(E)$ on T, and, by restriction, a homomorphism $\hat{\varphi} : \hat{\mathfrak{l}} \to \mathfrak{gl}(T)$.

Lemma 4.3. *The kernel $\mathfrak{k}_{\hat{\varphi}}$ of $\hat{\varphi}$ contains $\hat{\mathfrak{k}}$.*

Proof. The specified basis of T diagonalizes the action of ζ_j, so clearly $[\zeta_i, \zeta_j] \in \mathfrak{k}_{\hat{\varphi}}$. Next we consider the action of η_j. We have

$$([\zeta_i, \eta_j] + a_{ji}\eta_j) \cdot 1 = \zeta_i t_j - \eta_j \lambda_i + a_{ji} t_j = (\lambda_i - a_{ji})\, t_j - \lambda_i t_j + a_{ji} t_j = 0$$

and

$$([\zeta_i, \eta_j] + a_{ji}\eta_j) \cdot t_{i_1} \dots t_{i_r}$$
$$= (\zeta_i + a_{ji}) \cdot t_j t_{i_1} \dots t_{i_r} - (\lambda_i - a_{i_1 i} - \dots - a_{i_r i})\, t_j t_{i_1} \dots t_{i_r} = 0,$$

so $[\zeta_i, \eta_j] + a_{ji}\eta_j \in \mathfrak{k}_\Phi$. Next $[\xi_i, \eta_j] - \delta_{ij}\zeta_i \in \mathfrak{k}_\Phi$ because

$$([\xi_i, \eta_j] - \delta_{ij}\zeta_i) \cdot 1 = \delta_{ji}\lambda_i - \delta_{ij}\lambda_i = 0$$

$$([\xi_i, \eta_j] - \delta_{ij}\zeta_i) \cdot t_{i_1} \dots t_{i_r} = t_j(\xi_i \cdot t_{i_1} \dots t_{i_r}) + \delta_{ij}\zeta_i \cdot t_{i_1} \dots t_{i_r}$$
$$- t_j\xi_i \cdot t_{i_1} \dots t_{i_r} - \delta_{ij}\zeta_i \cdot t_{i_1} \dots t_{i_r} = 0.$$

Finally, we must show $[\zeta_i, \xi_j] - a_{ji}\xi_j \in \mathfrak{k}_\Phi$. This amounts to proving the identity

$$\zeta_i \cdot \xi_j \cdot t_{i_1} \dots t_{i_r} = (\lambda_i + a_{ji} - a_{i_1 i} - \dots - a_{i_r i})\, \xi_j \cdot t_{i_1} \dots t_{i_r}. \tag{1}$$

For this we observe that the left multiplication $t \to t_k t$ on $T\,(1 \leqslant k \leqslant l)$ satisfies

$$\zeta_j \cdot t_k t - t_k \zeta_j \cdot t = -a_{kj} t_k t. \tag{2}$$

The relation (1) is obvious for $r = 1$; and assuming it for products of length $r - 1$, we have, using (2),

$$\zeta_i \cdot \xi_j \cdot t_{i_1} \dots t_{i_r}$$
$$= \zeta_i \cdot t_{i_1}(\xi_j \cdot t_{i_2} \dots t_{i_r}) + \zeta_i \delta_{i_1 j}\zeta_j \cdot t_{i_2} \dots t_{i_r}$$
$$= t_{i_1}\zeta_i\xi_j \cdot t_{i_2} \dots t_{i_r} - a_{i_1 i} t_{i_1}\xi_j \cdot t_{i_2} \dots t_{i_r} + \delta_{i_1 j}\zeta_i\zeta_j \cdot t_{i_2} \dots t_{i_r}$$
$$= (\lambda_i + a_{ji} - a_{i_1 i} - \dots - a_{i_r i})\, t_{i_1}(\xi_j \cdot t_{i_2} \dots t_{i_r})$$
$$+ (\lambda_i - a_{i_2 i} - \dots - a_{i_r i})\, \delta_{i_1 j}\zeta_j \cdot t_{i_2} \dots t_{i_r},$$

which reduces to the right-hand side of (1) (both for $i_1 = j$ and for $i_1 \neq j$). This proves the lemma.

Lemma 4.4. *The elements x_i, y_i, h_i $(1 \leqslant i \leqslant l)$ are linearly independent.*

It suffices to prove that the vector space E spanned by ξ_i, η_i, ζ_i $(1 \leqslant i \leqslant l)$ satisfies

$$E \cap \hat{\mathfrak{k}} = 0. \tag{3}$$

But if $\sum_i (a_i\xi_i + b_i\eta_i + c_i\zeta_i) \in \hat{\mathfrak{k}}$ $(a_i, b_i, c_i \in \boldsymbol{C})$, then by Lemma 4.3

$$\sum_i \hat{\varphi}(a_i\xi_i + b_i\eta_i + c_i\zeta_i) = 0.$$

Applying the left-hand side to $1 \in T$, we obtain

$$\sum_i (b_i t_i + c_i \lambda_i) = 0.$$

Since the λ_i are arbitrary, we deduce $b_i = c_i = 0$ for all i. Thus $\sum_i a_i \hat{\varphi}(\xi_i) = 0$, and applying this to $t_j \in T$, we obtain $a_j = 0$. This proves the lemma.

The elements $h_1, ..., h_l$ commute and since they are linearly independent, we can define the linear functions α_j' on $\mathfrak{h}$ by $\alpha_j'(h_i) = a_{ji}$ $(1 \leqslant i, j \leqslant l)$.

Lemma 4.5.

(i) *The algebra X is spanned by the elements*

$$[x_{i_1}, ..., x_{i_r}] = \text{ad}\, x_{i_1} \ldots \text{ad}\, x_{i_{r-1}}(x_{i_r}) \qquad (\text{ad} = \text{ad}_{\mathfrak{l}}) \tag{4}$$

$(1 \leqslant i_1, ..., i_r \leqslant l)$ *which satisfy for* $h \in \mathfrak{h}$

$$\text{ad}\, h[x_{i_1}, ..., x_{i_r}] = (\alpha_{i_1}'(h) + \ldots + \alpha_{i_r}'(h))[x_{i_1}, ..., x_{i_r}]. \tag{5}$$

(ii) *The algebra Y is spanned by the elements*

$$[y_{i_1}, ..., y_{i_r}] = \text{ad}\, y_{i_1} \ldots \text{ad}\, y_{i_{r-1}}(y_{i_r}) \tag{6}$$

$(1 \leqslant i_1, ..., i_r \leqslant l)$ *which satisfy for* $h \in \mathfrak{h}$

$$\text{ad}\, h[y_{i_1}, ..., y_{i_r}] = -(\alpha_{i_1}'(h) + \ldots + \alpha_{i_r}'(h))[y_{i_1}, ..., y_{i_r}]. \tag{7}$$

Proof. Let $X' \subset X$ be the span of the elements (4), $[x_{i_1}]$ meaning x_{i_1}. Define Y' similarly and let $\mathfrak{l}' = X' + \mathfrak{h} + Y'$. We now prove by the method of Lemma 4.4 that this sum is direct. Suppose $x' + h + y' = 0$, so that corresponding representatives in $\hat{\mathfrak{l}}$,

$$\xi' = \sum a_{i_1 \ldots i_r}[\xi_{i_1}, ..., \xi_{i_r}], \qquad \eta' = \sum b_{j_1 \ldots j_s}[\eta_{j_1}, ..., \eta_{j_s}]$$

satisfy $\xi' + \zeta + \eta' \in \hat{\mathfrak{k}}$ and therefore, by Lemma 4.3, $\hat{\varphi}(\xi') + \hat{\varphi}(\zeta) + \hat{\varphi}(\eta') = 0$. Applying the left-hand side to $1 \in T$, we obtain

$$\hat{\varphi}(\zeta) \cdot 1 + \sum b_{j_1 \ldots j_s}[t_{j_1}, ..., t_{j_s}] = 0.$$

But $\hat{\varphi}$ is arbitrary, so we deduce $\zeta = 0$ and then the resulting relation in the tensor algebra T implies the same relation in $T(E)$, that is, $\eta' = 0$. Hence $y' = 0$ and $x' = 0$.

The following steps now show that $\mathfrak{l}' \subset \mathfrak{l}$ is a subalgebra:

(a) Formulas (5) and (7) follow from (iii) by induction.

(b) $[x_i, X'] = \operatorname{ad} x_i(X') \subset X'$ and $\operatorname{ad}[x_{i_1}, ..., x_{i_r}]$ is a polynomial in the $\operatorname{ad} x_i$, so X' is a subalgebra of $\mathfrak{l}$, hence $X' = X$. Similarly, $Y' = Y$, so Lemma 4.5 is proved.

(c) The relations

$$\operatorname{ad} y_j[x_{i_1}, ..., x_{i_r}] \subset X' \qquad (r \geqslant 2)$$

$$\operatorname{ad} x_j[y_{i_1}, ..., y_{i_r}] \subset Y' \qquad (r \geqslant 2)$$

follow from (ii) and (iii) by induction.

Since the subalgebra $\mathfrak{l}'$ contains the generators x_i, y_i, h_i of $\mathfrak{l}$, we have $\mathfrak{l}' = \mathfrak{l}$ and Theorem 4.2 is proved.

So far, relations (iv)–(v) have not been taken into account. But now they motivate the construction of a factor algebra of $\mathfrak{l}$ which will be semisimple and will have Cartan matrix (a_{ij}). For $i \neq j$ we put

$$x_{ij} = (\operatorname{ad} x_i)^{1-a_{ji}}(x_j), \qquad y_{ij} = (\operatorname{ad} y_i)^{1-a_{ji}}(y_j)$$

remembering that $a_{ji} \leqslant 0$.

Lemma 4.6. *The elements x_{ij} and y_{ij} satisfy*

$$\operatorname{ad} x_k(y_{ij}) = \operatorname{ad} y_k(x_{ij}) = 0 \qquad (1 \leqslant k \leqslant l, i \neq j).$$

Proof. Suppose first $k \neq i$. Then by (ii), $\operatorname{ad} x_k$ and $\operatorname{ad} y_i$ commute. Thus

$$\operatorname{ad} x_k(y_{ij}) = (\operatorname{ad} y_i)^{1-a_{ji}} \operatorname{ad} x_k(y_j), \tag{8}$$

which vanishes for $k \neq j$. If $k = j$, expression (8) equals $(\operatorname{ad} y_i)^{1-a_{ji}}(h_j)$, which by (iii) vanishes both if $a_{ji} = 0$ and if $a_{ji} \neq 0$ (remembering $i \neq j$).

Finally, suppose $k = i$ and consider the algebra $\boldsymbol{C}x_i + \boldsymbol{C}y_i + \boldsymbol{C}h_i$ which by (ii)–(iii) has the bracket relations

$$[x_i, y_i] = h_i, \qquad [h_i, x_i] = 2x_i, \qquad [h_i, y_i] = -2y_i.$$

The vector y_j satisfies (since $i \neq j$) the relation

$$[h_i, y_j] = -a_{ji}y_j, \qquad [x_i, y_j] = 0.$$

Put $e_n = (\text{ad } y_i)^n(y_j)$, $n \geqslant 0$, $e_{-1} = 0$. Then we get by induction and the Jacobi identity (cf. the proof of Theorem 2.16, Chapter VII)

$$[h_i, e_n] = (-a_{ji} - 2n)\, e_n$$
$$[y_i, e_n] = e_{n+1}$$
$$[x_i, e_n] = n(-a_{ji} - n + 1)\, e_{n-1}.$$

This last expression is 0 for $n = 1 - a_{ji}$, so the lemma is proved.

Now define

$\mathfrak{k}$: The ideal in $\mathfrak{l}$ generated by all x_{ij}, y_{ij} $(i \neq j)$.
X_0: The ideal in X generated by all x_{ij} $(i \neq j)$.
Y_0: The ideal in Y generated by all y_{ij} $(i \neq j)$.

Lemma 4.7. *X_0 and Y_0 are ideals in $\mathfrak{l}$ and $\mathfrak{k} = X_0 + Y_0$.*

Proof. Clearly Y_0 is spanned by elements of the form

$$\text{ad } y_{i_1} \ldots \text{ad } y_{i_r}(y_{ij}).$$

Hence $\text{ad } \mathfrak{h}(Y_0) \subset Y_0$. Since $\text{ad } x_k \text{ ad } y_j = \text{ad } y_j \text{ ad } x_k + \text{ad } h$ $(h \in \mathfrak{h})$, Lemma 4.6 implies $\text{ad } x_k(Y_0) \subset Y_0$. Thus $\text{ad } \mathfrak{l}(Y_0) \subset Y_0$ and similarly $\text{ad } \mathfrak{l}(X_0) \subset X_0$. Since $\mathfrak{k} \supset X_0 + Y_0$ and since $X_0 + Y_0$ is an ideal in $\mathfrak{l}$, $\mathfrak{k} = X_0 + Y_0$.

From Lemma 4.7 and the directness in Theorem 4.2 we deduce the crucial relation

$$\mathfrak{h} \cap \mathfrak{k} = 0. \tag{9}$$

Let X_i, Y_i, and H_i denote the images of x_i, y_i, and h_i, respectively, under the canonical mapping $\pi : \mathfrak{l} \to \mathfrak{g}$, $\mathfrak{g}$ denoting the factor algebra $\mathfrak{l}/\mathfrak{k}$.

Lemma 4.8. *The elements X_i, Y_i, H_i $(1 \leqslant i \leqslant l)$ in $\mathfrak{g}$ are linearly independent. They satisfy relations* (i)–(v).

By (9) the H_i are linearly independent. The matrix (a_{ij}) being nonsingular, the α_j' are linearly independent and we can select $h_0 \in \mathfrak{h}$ such that the numbers $\pm\alpha_j'(h_0)$ $(1 \leqslant j \leqslant l)$ are all different and $\neq 0$. Then a vector $\sum c_i H_i$ and X_j, Y_j $(1 \leqslant j \leqslant l)$ are eigenvectors of $\text{ad } h_0$ with distinct eigenvalues so are linearly independent. This proves the lemma.

We are now going to prove that the Lie algebra $\mathfrak{g}$ gives the desired realization of the root system R. We put

$$\mathfrak{h}_{\boldsymbol{R}} = \sum_{i=1}^{l} \boldsymbol{R} H_i, \qquad \mathfrak{n}^+ = \pi(X), \qquad \mathfrak{n}^- = \pi(Y)$$

and write $\mathfrak{h}$ for $\pi(\mathfrak{h})$ because of (9). By Theorem 4.2,

$$\mathfrak{g} = \mathfrak{n}^+ + \mathfrak{h} + \mathfrak{n}^-. \tag{10}$$

Defining $\alpha_j \in \mathfrak{h}_{\boldsymbol{R}}^{\wedge}$ (the dual space of $\mathfrak{h}_{\boldsymbol{R}}$) by $\alpha_j(H_i) = a_{ji}$ $(1 \leqslant i, j \leqslant l)$, we have from (5) and (7)

$$\mathfrak{n}^+ = \sum_{\lambda > 0} \mathfrak{g}^\lambda, \qquad \mathfrak{h} = \mathfrak{g}^0, \qquad \mathfrak{n}^- = \sum_{-\lambda > 0} \tag{11}$$

where for each $\lambda \in \mathfrak{h}_{\boldsymbol{R}}^{\wedge}$ we put

$$\mathfrak{g}^\lambda = \{Z \in \mathfrak{g} : \operatorname{ad} H(Z) = \lambda(H)Z \text{ for } H \in \mathfrak{h}_{\boldsymbol{R}}\}$$

and the notation $\lambda > 0$ means that $\lambda \neq 0$ and is a positive integral linear combination of α_i $(1 \leqslant i \leqslant l)$. Because of (11), the sum in (10) is direct. By (5) and (7) each $\mathfrak{g}^\lambda$ $(\lambda \neq 0)$ is finite-dimensional and by (10) is 0 if neither $\lambda > 0$ nor $-\lambda > 0$.

Lemma 4.9. *The endomorphisms* ad X_i *and* ad Y_i $(1 \leqslant i \leqslant l)$ *of* $\mathfrak{g}$ *are locally nilpotent (that is, each* $Z \in \mathfrak{g}$ *is annihilated by some power of them).*

Proof. For a fixed i let $\mathfrak{m}_i$ denote the space of $Z \in \mathfrak{g}$ annihilated by some power of ad X_i. By the proof of Lemma 3.2, Chapter III, $\mathfrak{m}_i$ is a subalgebra of $\mathfrak{g}$. But relations (ii)–(iv) imply that $\mathfrak{m}_i$ contains the generators X_j, Y_j, H_j $(1 \leqslant j \leqslant l)$, so $\mathfrak{m}_i = \mathfrak{g}$ as desired.

This lemma implies that, although we have not yet proved that $\dim \mathfrak{g} < \infty$, $e^{\operatorname{ad} X_i}$ is a well-defined automorphism of $\mathfrak{g}$.

The linear mapping of $\sum_{i=1}^{l} \boldsymbol{R}\beta_i$ onto $\sum_{i=1}^{l} \boldsymbol{R}\alpha_i = \mathfrak{h}_{\boldsymbol{R}}^{\wedge}$ given by $\beta_i \to \alpha_i$ sends R into a root system (which we denote by Δ) in $\mathfrak{h}_{\boldsymbol{R}}^{\wedge}$ with basis $\{\alpha_1, \ldots, \alpha_l\}$ and corresponding Cartan matrix $(a_{ij})_{1 \leqslant i,j \leqslant l}$. The reflection s_{α_i} $(1 \leqslant i \leqslant l)$ is then given by

$$s_{\alpha_i}\lambda = \lambda - \lambda(H_i)\,\alpha_i, \qquad \lambda \in \mathfrak{h}_{\boldsymbol{R}}^{\wedge}. \tag{12}$$

These reflections generate the Weyl group $W(\Delta)$ of Δ.

Lemma 4.10. *The automorphism*

$$\theta_i = e^{\operatorname{ad} X_i} e^{-\operatorname{ad} Y_i} e^{\operatorname{ad} X_i}$$

leaves $\mathfrak{h}_{\boldsymbol{R}}$ *invariant and* s_{α_i} *coincides with the transpose of the restriction* $\theta_i \mid \mathfrak{h}_{\boldsymbol{R}}$.

Proof. We have by (i)–(iii)

$$\begin{aligned}\theta_i(H_j) &= e^{\operatorname{ad} X_i} e^{-\operatorname{ad} Y_i}(H_j - a_{ij}X_i) = e^{\operatorname{ad} X_i}(H_j - a_{ij}X_i - a_{ij}H_i) \\ &= H_j - a_{ij}X_i - a_{ij}H_i - a_{ij}X_i + 2a_{ij}X_i\end{aligned}$$

so

$$\theta_i(H_j) = H_j - a_{ij}H_i,$$

proving the first statement. Also by (12)

$$\lambda(\theta_i(H_j)) = \lambda(H_j) - \lambda(H_i)\, a_{ij} = s_{\alpha_i}(\lambda)(H_j),$$

proving the second statement.

Corollary 4.11. *Suppose* λ, $\mu \in \mathfrak{h}^\wedge$ *are conjugate under* $W(\Delta)$. *Then there exists an automorphism of* $\mathfrak{g}$ *mapping* $\mathfrak{g}^\lambda$ *onto* $\mathfrak{g}^\mu$.

In fact we may assume $\mu = s_{\alpha_i}\lambda$ and then $\theta_i^{-1}\mathfrak{g}^\lambda \subset \mathfrak{g}^\mu$ is immediate. But $s_{\alpha_i} = s_{\alpha_i}^{-1}$, so θ_i actually interchanges $\mathfrak{g}^\lambda$ and $\mathfrak{g}^\mu$.

Lemma 4.12. *Suppose* $\lambda \neq 0$ *in* $\hat{\mathfrak{h}}_{\boldsymbol{R}}$ *such that* $c\lambda \notin \Delta$ *for all* $c \in \boldsymbol{R}$. *Then there exists a* $\sigma \in W(\Delta)$ *such that neither* $\sigma\lambda > 0$ *nor* $-\sigma\lambda > 0$.

Proof. Since Δ is finite, we can select $H_0 \in \mathfrak{h}_{\boldsymbol{R}}$ such that $\alpha(H_0) \neq 0$ for all $\alpha \in \Delta$ but $\lambda(H_0) = 0$. By Theorem 3.6 there exists a $\sigma \in W(\Delta)$ which, identified with $({}^t\sigma)^{-1}$, satisfies $\alpha_i(\sigma H_0) > 0$ $(1 \leqslant i \leqslant l)$. Writing $\sigma\lambda = \sum_{i=1}^{l} a_i\alpha_i$, we have

$$0 = \lambda(H_0) = \sigma\lambda(\sigma H_0) = \sum_{i=1}^{l} a_i\alpha_i(\sigma H_0),$$

so the lemma follows.

Lemma 4.13. *Let* $\lambda \neq 0$ *in* $\hat{\mathfrak{h}}_{\boldsymbol{R}}$. *If* $\lambda \in \Delta$, *then* $\dim \mathfrak{g}^\lambda = 1$. *If* $\lambda \notin \Delta$, *then* $\mathfrak{g}^\lambda = 0$.

Proof. We have $\dim \mathfrak{g}^\lambda = 1$ if $\lambda = \alpha_i$ by (5), hence for all $\lambda \in \Delta$ by Cor. 4.11 because λ is $W(\Delta)$-conjugate to some α_i.

We know from Lemma 4.5 that $\mathfrak{n}^+$ is spanned by the elements $[X_{i_1}, ..., X_{i_r}]$. Consequently, if $c \neq 0, \pm 1$, then $\mathfrak{g}^{c\alpha} = 0$ for $\alpha = \alpha_i$, hence by Cor. 4.11 for $\alpha \in \Delta$. Hence if $\lambda \notin \Delta$ but $c\lambda = \alpha \in \Delta$ for some $c \in \boldsymbol{R}$, then $\mathfrak{g}^\lambda = \mathfrak{g}^{\alpha/c} = 0$ since $c^{-1} \neq 0, \pm 1$.

Finally, if λ is not proportional to a root, then by Cor. 4.11 and Lemma 4.12, $\mathfrak{g}^\lambda = 0$.

Lemma 4.14. *The Lie algebra* $\mathfrak{g}$ *is finite-dimensional, semisimple, and*

$$\mathfrak{g} = \mathfrak{h} + \sum_{\lambda \in \Delta} \mathfrak{g}^\lambda \qquad \textit{(direct sum)}.$$

By Lemma 4.13 and (10) we just have to prove the semisimplicity, that is, that the Killing form $\langle\ ,\ \rangle$ is nondegenerate. Let $\mathfrak{z} = \{Z \in \mathfrak{g} : \langle Z, \mathfrak{g}\rangle = 0\}$. Then since ad $\mathfrak{h}$ leaves the ideal $\mathfrak{z}$ invariant,

$$\mathfrak{z} = \mathfrak{z} \cap \mathfrak{h} + \sum_{\lambda \in \Delta} \mathfrak{z} \cap \mathfrak{g}^\lambda. \tag{13}$$

But $\langle \mathfrak{z} \cap \mathfrak{h}, \mathfrak{h}_R \rangle = 0$ implies

$$\sum_{\lambda \in \Delta} \lambda(H)\, \lambda(H_R) = 0, \qquad H \in \mathfrak{z} \cap \mathfrak{h}, \quad H_R \in \mathfrak{h}_R.$$

Writing $H = H_1 + iH_2$ $(H_1, H_2 \in \mathfrak{h}_R)$, we deduce $\lambda(H) = 0$ for $H \in \mathfrak{z} \cap \mathfrak{h}$, $\lambda \in \Delta$, so $\mathfrak{z} \cap \mathfrak{h} = 0$. If $\mathfrak{z} \cap \mathfrak{g}^{\lambda} \neq 0$, then $[\mathfrak{g}^{-\lambda}, \mathfrak{z} \cap \mathfrak{g}^{\lambda}]$ is a nonzero subspace of $\mathfrak{z} \cap \mathfrak{h}$ ((ii) and Cor. 4.11) which is impossible. Thus, by (13), $\mathfrak{z} = 0$, so $\mathfrak{g}$ is semisimple.

Summarizing, we have proved the following result.

Theorem 4.15. *Let R be a reduced root system in a vector space V, $B = \{\beta_1, ..., \beta_l\}$ any basis, and $(a_{ij})_{1 \leqslant i,j \leqslant l}$ the associated Cartan matrix. Let $\mathfrak{g}$ be the complex Lie algebra generated by $3l$ elements X_i, Y_i, H_i $(1 \leqslant i \leqslant l)$ subject to the relations* (i)–(v). *Then $\mathfrak{g}$ is finite-dimensional, semisimple, has $\mathfrak{h} = \sum_{i=1}^{l} CH_i$ as a Cartan subalgebra, and the linear mapping τ of V onto the dual $\hat{\mathfrak{h}}_R$ defined by $\tau(\beta_i)(H_j) = a_{ij}$ gives a bijection of R onto the set $\Delta(\mathfrak{g}, \mathfrak{h})$ of roots of $\mathfrak{g}$ with respect to $\mathfrak{h}$.*

§ 5. Automorphisms of Finite Order of Semisimple Lie Algebras†

Let $\mathfrak{g}$ be a Lie algebra over $\boldsymbol{C}$. If A is an abelian group, an *A-gradation* of $\mathfrak{g}$ is, by definition, a direct decomposition

$$\mathfrak{g} = \bigoplus_{i \in A} \mathfrak{g}_i \qquad \text{such that} \qquad [\mathfrak{g}_i, \mathfrak{g}_j] \subset \mathfrak{g}_{i+j}. \tag{1}$$

For example, if $\mathfrak{l} = \mathfrak{u} + \mathfrak{e}$ is a Cartan decomposition of a semisimple Lie algebra (Chapter V), then the decomposition $\mathfrak{l}^{\boldsymbol{C}} = \mathfrak{u}^{\boldsymbol{C}} + \mathfrak{e}^{\boldsymbol{C}}$ is a $\boldsymbol{Z}_2 = \boldsymbol{Z}/2\boldsymbol{Z}$-gradation of the complexification $\mathfrak{l}^{\boldsymbol{C}}$.

An ideal $\mathfrak{s} \subset \mathfrak{g}$ is called an *A-graded ideal* if $\mathfrak{s} = \bigoplus_{i \in A} \mathfrak{s} \cap \mathfrak{g}_i$. It is clear that in (1) $\mathfrak{g}_0$ is a subalgebra of $\mathfrak{g}$; writing $X(Y_i) = [X, Y_i]$ for $X \in \mathfrak{g}_0$, $Y_i \in \mathfrak{g}_i$, we have a representation of $\mathfrak{g}_0$ on $\mathfrak{g}_i$.

Now suppose $\mathfrak{g}$ has finite dimension and let σ be an automorphism of $\mathfrak{g}$ satisfying $\sigma^m = e$. Fix a primitive mth root ϵ of unity. Then each eigenvalue of σ has the form ϵ^i $(i \in \boldsymbol{Z}_m = \boldsymbol{Z}/m\boldsymbol{Z})$ and since σ is necessarily semisimple (it leaves invariant a positive definite Hermitian form), we have the direct decomposition

$$\mathfrak{g} = \bigoplus_{i \in \boldsymbol{Z}_m} \mathfrak{g}_i, \tag{2a}$$

† This section is an edition of written and oral expositions by Victor Kač to the author, covering the results of the announcement Kač [1].

where $\mathfrak{g}_i$ is the eigenspace of σ for the eigenvalue ϵ^i. Clearly (2a) is a $\boldsymbol{Z}_m$-gradation of $\mathfrak{g}$. Conversely, if a $\boldsymbol{Z}_m$-gradation (2a) of $\mathfrak{g}$ is given, the linear transformation of $\mathfrak{g}$ given by multiplying the vectors in $\mathfrak{g}_i$ by ϵ^i is an automorphism σ of $\mathfrak{g}$ satisfying $\sigma^m = e$.

Let $\boldsymbol{C}[x, x^{-1}]$ denote the algebra of Laurent polynomials in the indeterminate x (that is, all finite sums $\sum_{j\in\boldsymbol{Z}} c_j x^j$, $c_i \in \boldsymbol{C}$). Considering $\mathfrak{g}$ and $\boldsymbol{C}[x, x^{-1}]$ as vector spaces over $\boldsymbol{C}$, we can form the tensor product

$$\boldsymbol{C}[x, x^{-1}] \otimes \mathfrak{g} = \bigoplus_{j\in\boldsymbol{Z}} x^j\mathfrak{g}. \tag{2b}$$

With the bracket operation $[x^jY, x^kZ] = x^{j+k}[Y, Z]$ (2b) becomes a $\boldsymbol{Z}$-graded Lie algebra over $\boldsymbol{C}$ of infinite dimension. With the automorphism σ of $\mathfrak{g}$ and $\mathfrak{g}_i$ as in (2a) we associate a subalgebra $L(\mathfrak{g}, \sigma)$ of (2b):

$$L(\mathfrak{g}, \sigma) = \bigoplus_{j\in\boldsymbol{Z}} x^j\mathfrak{g}_{j \bmod m}.$$

This will be called a *covering Lie algebra* of $\mathfrak{g}$ and the homomorphism $\varphi : L(\mathfrak{g}, \sigma) \to \mathfrak{g}$ defined by $\varphi(x^kY) = Y$ $(Y \in \mathfrak{g}_{k \bmod m})$ will be called a *covering homomorphism*. The algebra (2b) is $L(\mathfrak{g}, e)$ (e = identity automorphism). By studying the $\boldsymbol{Z}$-graded algebras $L(\mathfrak{g}, \sigma)$ ($\mathfrak{g}$ simple) we shall obtain a description of all $\boldsymbol{Z}_m$-gradations of simple Lie algebras. The plan is to develop the weight theory for $L(\mathfrak{g}, \sigma)$, that is, the analog of the root theory for $\mathfrak{g}$. Thereby we establish an isomorphism $L(\mathfrak{g}, \sigma) \approx L(\mathfrak{g}, \nu)$, where ν is an automorphism of a very special type, namely induced by an automorphism of the Dynkin diagram. This results in an explicit description of σ in terms of ν (Theorem 5.15).

Assume now $\mathfrak{g}$ is a semisimple Lie algebra over $\boldsymbol{C}$ and σ an automorphism of $\mathfrak{g}$ of order m. Let B denote the Killing form of $\mathfrak{g}$.

Lemma 5.1.

(i) $B(\mathfrak{g}_i, \mathfrak{g}_j) = 0$ *for* $i, j \in \boldsymbol{Z}_m$, $i + j \neq 0$.

(ii) *For each* $X \neq 0$ *in* $\mathfrak{g}_i$ *there exists a* $Y \in \mathfrak{g}_{-i}$ *such that* $B(X, Y) \neq 0$. *In particular, the restriction of* B *to* $\mathfrak{g}_0 \times \mathfrak{g}_0$ *is nondegenerate.*

Proof. Let $X \in \mathfrak{g}_i$, $Y \in \mathfrak{g}_j$. By the invariance of B,

$$B(X, Y) = B(\sigma X, \sigma Y) = \epsilon^{i+j}B(X, Y).$$

This gives (i); part (ii) follows from (i) since B is nondegenerate.

Because of Exercise A.8, Chapter VI, we have the following result.

Lemma 5.2. *There exists a compact real form* $\mathfrak{u}$ *of* $\mathfrak{g}$ *invariant under* σ.

As a consequence, (2a) gives a $\boldsymbol{Z}_m$-gradation $\mathfrak{u} = \bigoplus_i \mathfrak{u}_i$, where $\mathfrak{u}_i = \mathfrak{u} \cap \mathfrak{g}_i$ and $\mathfrak{g}_i = (\mathfrak{u}_i)^{\boldsymbol{C}}$. Being the fixed point set of σ on the compact semisimple Lie algebra $\mathfrak{u}$, the subalgebra $\mathfrak{u}_0$ is compact. By Prop. 6.6, Chapter II, we get the direct decompositions into semisimple and abelian ideals,

$$\mathfrak{u}_0 = [\mathfrak{u}_0, \mathfrak{u}_0] \oplus \mathfrak{z}_{\mathfrak{u}_0}, \qquad \mathfrak{g}_0 = [\mathfrak{g}_0, \mathfrak{g}_0] \oplus \mathfrak{z}_{\mathfrak{g}_0} \tag{3}$$

the latter terms denoting the centers. Taking a maximal abelian subalgebra $\mathfrak{t}_0'$ of $[\mathfrak{u}_0, \mathfrak{u}_0]$, the algebra $\mathfrak{t}_0 = \mathfrak{t}_0' + \mathfrak{z}_{\mathfrak{u}_0}$ is maximal abelian in $\mathfrak{u}_0$ and its complexification $\mathfrak{h} = \mathfrak{t}_0^C$ is maximal abelian in $\mathfrak{g}_0$.

Lemma 5.3. *The centralizer* $\mathfrak{z}(\mathfrak{h})$ *of* $\mathfrak{h}$ *in* $\mathfrak{g}$ *is a Cartan subalgebra of* $\mathfrak{g}$. *In particular*, $\mathfrak{g}_0 \neq 0$.

Proof. Clearly $\mathfrak{z}(\mathfrak{h})$ is the complexification of the centralizer $\mathfrak{z}(\mathfrak{t}_0)$ of $\mathfrak{t}_0$ in $\mathfrak{u}$. Thus we have to prove only that $\mathfrak{z}(\mathfrak{h})$ is abelian. For this we use the analogs of (3) for $\mathfrak{z}(\mathfrak{t}_0)$ and $\mathfrak{z}(\mathfrak{h})$; thus $\mathfrak{z}(\mathfrak{h}) = \tilde{\mathfrak{h}} \oplus \mathfrak{s}$ where $\tilde{\mathfrak{h}}$ is abelian and contains $\mathfrak{h}$ and $\mathfrak{s} = [\mathfrak{z}(\mathfrak{h}), \mathfrak{z}(\mathfrak{h})]$ is semisimple. Now σ leaves $\mathfrak{h}$, $\mathfrak{z}(\mathfrak{h})$, and $\mathfrak{s}$ invariant; and since $\mathfrak{h}$ is maximal abelian in $\mathfrak{g}_0$, the action of σ on $\mathfrak{s}$ has no nonzero fixed point. From this we shall draw the desired conclusion $\mathfrak{s} = 0$.

Consider the Z_m-gradation $\mathfrak{s} = \bigoplus_i \mathfrak{s}_i$ induced by (2a). Numbering the elements of Z_m by the corresponding integers in the set $N_m = \{0, 1, ..., m-1\}$ and defining $\mathfrak{s}_u = \mathfrak{s}_v$ if $v \in N_m$, $u \equiv v \pmod m$, we shall prove $\mathfrak{s}_k = 0$ by induction on k.

We know $\mathfrak{s}_0 = 0$; let $k > 0$ and $X \in \mathfrak{s}_k$. Then $(\operatorname{ad} X)^r(\mathfrak{s}_i) \subset \mathfrak{s}_{kr+i}$ $(i \in N_m)$. Select $r \in Z^+$ such that $k(r-1) < m - i \leqslant kr$. Then $kr + i = m + t$ $(0 \leqslant t < k)$, so by the inductive assumption, $\mathfrak{s}_{kr+i} = \mathfrak{s}_t = 0$. Thus $\operatorname{ad} X \mid \mathfrak{s}$ is nilpotent. Similarly, $\operatorname{ad} Y \mid \mathfrak{s}$ is nilpotent if $Y \in \mathfrak{s}_{-k}$. But $[\mathfrak{s}_k, \mathfrak{s}_{-k}] \subset \mathfrak{s}_0 = 0$, so $\operatorname{ad} X$ and $\operatorname{ad} Y$ commute on $\mathfrak{s}$ and by the nilpotency, $\operatorname{Tr}_{\mathfrak{s}}(\operatorname{ad} X \operatorname{ad} Y) = 0$. Now Lemma 5.1 (ii) implies $\mathfrak{s}_k = 0$, so the lemma is proved.

Let $\alpha \in \mathfrak{h}^\wedge$ (dual of $\mathfrak{h}$) and $i \in Z_m$. A pair $\bar{\alpha} = (\alpha, i)$ will be called a *root* (of $\mathfrak{g}$ with respect to $\mathfrak{h}$) if the joint eigenspace

$$\mathfrak{g}^{\bar{\alpha}} = \{X \in \mathfrak{g}_i : [H, X] = \alpha(H)X \text{ for } H \in \mathfrak{h}\}$$

is $\neq 0$. We add pairs by $(\alpha, i) + (\alpha', i') = (\alpha + \alpha', i + i')$. Note that $[\mathfrak{g}^{\bar{\alpha}}, \mathfrak{g}^{\bar{\beta}}] \subset \mathfrak{g}^{\bar{\alpha}+\bar{\beta}}$. Let $\bar{\Delta}$ denote the set of nonzero roots and $\bar{\Delta}^0$ the set of roots of the form $(0, i)$, $i \in Z_m$. Then we have the direct decompositions

$$\mathfrak{g} = \mathfrak{h} + \sum_{\bar{\alpha} \in \bar{\Delta}} \mathfrak{g}^{\bar{\alpha}}, \qquad \mathfrak{h} = \mathfrak{g}^{(0,0)}, \tag{4}$$

$$\mathfrak{z}(\mathfrak{h}) = \sum_{\bar{\alpha} \in \bar{\Delta}^0} \mathfrak{g}^{\bar{\alpha}}, \tag{5}$$

and

$$\mathfrak{g} = \mathfrak{z}(\mathfrak{h}) + \sum_{\bar{\alpha} \in \bar{\Delta} - \bar{\Delta}^0} \mathfrak{g}^{\bar{\alpha}}. \tag{6}$$

Using Lemmas 5.1 and 5.3 and the proofs from §4, Chapter III, we obtain the following analogs to Theorems 4.2–4.4 of Chapter III.

Lemma 5.4.

(i) *dim* $\mathfrak{g}^{\bar{\alpha}} = 1$ *for each* $\bar{\alpha} \in \bar{\Delta} - \bar{\Delta}^0$.

(ii) *The restriction of B to* $\mathfrak{h} \times \mathfrak{h}$ *is nondegenerate. For each* $\beta \in \mathfrak{h}^\wedge$ *there exists a unique element* $H_\beta \in \mathfrak{h}$ *such that*

$$B(H, H_\beta) = \beta(H) \qquad (H \in \mathfrak{h}).$$

We put $\langle \beta, \gamma \rangle = B(H_\beta, H_\gamma)$.

(iii) *If* $\bar{\alpha} \in \bar{\Delta} - \bar{\Delta}^0$, *then* $-\bar{\alpha} \in \bar{\Delta} - \bar{\Delta}^0$ *and*

$$[\mathfrak{g}^{\bar{\alpha}}, \mathfrak{g}^{-\bar{\alpha}}] = CH_\alpha, \qquad \alpha(H_\alpha) \neq 0.$$

Lemma 5.5. *Let* $\bar{\beta}$ *be a root and* $\bar{\alpha} \in \bar{\Delta} - \bar{\Delta}^0$.

(i) *The set of all roots of the form* $\bar{\beta} + n\bar{\alpha}$ $(n \in \boldsymbol{Z})$ *has the form* $\bar{\beta} + n\bar{\alpha}$ $(p \leqslant n \leqslant q)$. *Also*

$$-2\,\frac{\beta(H_\alpha)}{\alpha(H_\alpha)} = p + q.$$

(ii) *The only roots proportional to* $\bar{\alpha}$ *are* $-\bar{\alpha}$, 0, *and* $\bar{\alpha}$.

(iii) *If* $\bar{\beta}, \bar{\alpha} + \bar{\beta} \in \bar{\Delta}$, *then there exist* $X \in \mathfrak{g}^{\bar{\alpha}}$, $Y \in \mathfrak{g}^{\bar{\beta}}$ *for which* $[X, Y] \neq 0$. *In particular,*

$$[\mathfrak{g}^{\bar{\alpha}}, \mathfrak{g}^{\bar{\beta}}] = \mathfrak{g}^{\bar{\alpha}+\bar{\beta}} \qquad \textit{if} \quad \bar{\alpha} + \bar{\beta} \notin \bar{\Delta}^0.$$

Lemma 5.6. *Let* $\mathfrak{h}_{\boldsymbol{R}} = \sum_{\bar{\alpha} \in \bar{\Delta} - \bar{\Delta}^0} \boldsymbol{R} H_\alpha$. *Then:*

(i) *B is real and strictly positive definite on* $\mathfrak{h}_{\boldsymbol{R}} \times \mathfrak{h}_{\boldsymbol{R}}$.

(ii) $\mathfrak{h} = \mathfrak{h}_{\boldsymbol{R}} + i\mathfrak{h}_{\boldsymbol{R}}$ *(direct sum)*.

We consider now the covering $\boldsymbol{Z}$-graded Lie algebra $L(\mathfrak{g}, \sigma) = \bigoplus_{j \in \boldsymbol{Z}} L_j$ where $L_j = x^j \mathfrak{g}_{j \bmod m}$. We identify L_0 with $\mathfrak{g}_0$ and give definitions for $L(\mathfrak{g}, \sigma)$ analogous to those for $\mathfrak{g}$. For $\alpha \in \mathfrak{h}^\wedge$, $j \in \boldsymbol{Z}$, a pair $\tilde{\alpha} = (\alpha, j)$ is called a *root* (of $L(\mathfrak{g}, \sigma)$ with respect to $\mathfrak{h}$) if the space

$$L^{\tilde{\alpha}} = \{X \in L_j : [H, X] = \alpha(H)X \text{ for } H \in \mathfrak{h}\}$$

is $\neq 0$. We put $(\alpha, j) + (\alpha', j') = (\alpha + \alpha', j + j')$. Let $\tilde{\Delta}$ denote the set of all nonzero roots, $\tilde{\Delta}^0$ the set of roots of the form $(0, j)$, $j \in \boldsymbol{Z}$. We have $[L^{\tilde{\alpha}}, L^{\tilde{\beta}}] \subset L^{\tilde{\alpha}+\tilde{\beta}}$ and the direct decomposition

$$L(\mathfrak{g}, \sigma) = \mathfrak{h} + \sum_{\tilde{\alpha} \in \tilde{\Delta}} L^{\tilde{\alpha}} \tag{7}$$

because, for each $H \in \mathfrak{h}$, the restriction $(\text{ad } H) \mid L_j$ is semisimple (Lemma 5.3). Note that if (α, j) is a root and $j \equiv j' \pmod m$, then (α, j') is a root; moreover, the map

$$\tilde{\alpha} = (\alpha, j) \rightarrow (\alpha, j \bmod m) = \bar{\alpha}$$

maps the set of roots of $L(\mathfrak{g}, \sigma)$ with respect to $\mathfrak{h}$ onto the set of roots of $\mathfrak{g}$ with respect to $\mathfrak{h}$ and $L^{\tilde{\alpha}} = x^j \mathfrak{g}^{\bar{\alpha}}$. This connection between $\bar{\Delta}$, $\bar{\Delta}^0$ and $\tilde{\Delta}$, $\tilde{\Delta}^0$ gives the following lemmas for $L(\mathfrak{g}, \sigma)$ and $\tilde{\Delta}$.

Lemma 5.4′.

(i) *dim* $L^{\tilde{\alpha}} = 1$ *for each* $\tilde{\alpha} \in \tilde{\Delta} - \tilde{\Delta}^0$.

(ii) *If* $\tilde{\alpha} \in \tilde{\Delta} - \tilde{\Delta}^0$, *then* $-\tilde{\alpha} \in \tilde{\Delta} - \tilde{\Delta}^0$ *and*

$$[L^{\tilde{\alpha}}, L^{-\tilde{\alpha}}] = CH_{\alpha}.$$

This is immediate from Lemma 5.4 and the above remarks.

Lemma 5.5′. *Let* $\tilde{\beta}$ *be a root and* $\tilde{\alpha} \in \tilde{\Delta} - \tilde{\Delta}^0$.

(i) *The set of all roots of the form* $\tilde{\beta} + n\tilde{\alpha}$ $(n \in \boldsymbol{Z})$ *has the form* $\tilde{\beta} + n\tilde{\alpha}$ $(p \leqslant n \leqslant q)$. *Also*

$$-2 \frac{\beta(H_\alpha)}{\alpha(H_\alpha)} = p + q.$$

Moreover, if $e_{\tilde{\alpha}} \neq 0$ *in* $L^{\tilde{\alpha}}$,

$$(\text{ad } e_{\tilde{\alpha}})^{q-p}(L^{\tilde{\beta}+p\tilde{\alpha}}) \neq 0.$$

(ii) *The only roots proportional to* $\tilde{\alpha}$ *are* $-\tilde{\alpha}$, 0, $\tilde{\alpha}$.

(iii) *If* $\tilde{\beta}$, $\tilde{\alpha} + \tilde{\beta} \in \tilde{\Delta}$, *then there exist* $e_{\tilde{\alpha}} \in L^{\tilde{\alpha}}$, $e_{\tilde{\beta}} \in L^{\tilde{\beta}}$ *for which* $[e_{\tilde{\alpha}}, e_{\tilde{\beta}}] \neq 0$. *In particular,*

$$[L^{\tilde{\alpha}}, L^{\tilde{\beta}}] = L^{\tilde{\beta}+\tilde{\alpha}} \qquad \textit{if} \quad \tilde{\alpha} + \tilde{\beta} \notin \tilde{\Delta}^0.$$

Proof. Part (iii) is immediate from Lemma 5.5 (iii) above. Part (ii) and the first statement in (i) follow just like the analogous statements in Theorem 4.3, Chapter III. Now put $h = 2\langle\alpha, \alpha\rangle^{-1} H_\alpha$ and select $e_{\tilde{\alpha}} \in L^{\tilde{\alpha}}$, $e_{-\tilde{\alpha}} \in L^{-\tilde{\alpha}}$ such that

$$[e_{\tilde{\alpha}}, e_{-\tilde{\alpha}}] = h, \qquad [h, e_{\tilde{\alpha}}] = 2e_{\tilde{\alpha}}, \qquad [h, e_{-\tilde{\alpha}}] = -2e_{-\tilde{\alpha}}.$$

and consider the adjoint representation of $Ce_{\tilde{\alpha}} + Ce_{-\tilde{\alpha}} + Ch$ on $\bigoplus_{p \leqslant n \leqslant q} L^{\tilde{\beta}+n\tilde{\alpha}}$. Fix $Z \neq 0$ in $L^{\tilde{\beta}+p\tilde{\alpha}}$ and put $e_n = (\text{ad } e_{\tilde{\alpha}})^n(Z)$, $n \geqslant 0$, $e_{-1} = 0$. Since $[h, Z] = (p - q)Z$ and $[e_{-\tilde{\alpha}}, Z] = 0$, the proof of Theorem 2.16, Chapter VII, gives

$$[e_{-\tilde{\alpha}}, e_n] = -n(p - q + n - 1)\, e_{n-1}.$$

If k is the last integer such that $e_k \neq 0$, then we deduce $k = q - p$, proving (i).

Now choose a basis of the root system Δ_0 of $[L_0, L_0]$ with respect to $\mathfrak{h} \cap [L_0, L_0]$ and let Δ_0^+ denote the corresponding set of positive roots. Defining each $\alpha \in \Delta_0$ to be $\equiv 0$ on the center of L_0, it can be considered a linear function on $\mathfrak{h}$. We identify α with the root $(\alpha, 0)$ and observe that all roots of the form $(\beta, 0)$ are obtained in this way. The set

$$\tilde{\Delta}^+ = \Delta_0^+ \cup \{(\alpha, j) \in \tilde{\Delta} : j > 0\}$$

will be called the set of positive roots in $\tilde{\Delta}$. A root $\tilde{\alpha} \in \tilde{\Delta}^+$ will be called *simple* if it is not a sum of two members of $\tilde{\Delta}^+$. Let $\tilde{\Pi} = \{(\alpha_0, s_0), (\alpha_1, s_1), \ldots\}$ be the set of simple roots and put $\Pi = \{\alpha_0, \alpha_1, \ldots\}$. Part (iv) in the lemma below shows that all the α_i are different; in particular Π and $\tilde{\Pi}$ are finite sets. Let N be their cardinality.

Lemma 5.7.

(i) *Each* $\tilde{\alpha} \in \tilde{\Delta}$ *can be written in the form* $\tilde{\alpha} = \pm \sum_i k_i \tilde{\alpha}_i$ *where* $k_i \in \boldsymbol{Z}^+$, $\tilde{\alpha}_i \in \tilde{\Pi}$.

(ii) $\tilde{\Pi} \subset \tilde{\Delta} - \tilde{\Delta}^0$.

(iii) *The system* Π *is a linearly dependent system of vectors which spans* $\mathfrak{h}^\wedge$.

(iv) *For* $i \neq j$ *we have*

$$a_{ij} = 2\frac{\langle \alpha_i, \alpha_j \rangle}{\langle \alpha_j, \alpha_j \rangle} \in -\boldsymbol{Z}^+;$$

in particular, $\alpha_i \neq \alpha_j$.

(v) *If* $\tilde{\alpha} \in \tilde{\Delta}^+$ *is not simple, then* $\tilde{\alpha} - \tilde{\alpha}_i \in \tilde{\Delta}$ *for some* $\tilde{\alpha}_i \in \tilde{\Pi}$.

Proof. (i) If $\tilde{\alpha} \in \tilde{\Delta}^+$ is not simple, then $\tilde{\alpha} = \tilde{\beta} + \tilde{\gamma}$ $(\tilde{\beta}, \tilde{\gamma} \in \tilde{\Delta}^+)$. Using this on $\tilde{\beta}$, $\tilde{\gamma}$, etc., we derive (i).

For (v) we first prove (for $\tilde{\alpha}$ not simple):

$$\text{If } \tilde{\alpha} \in \tilde{\Delta}^+ - \tilde{\Delta}^0, \text{ then } \tilde{\alpha} - \tilde{\beta} \in \tilde{\Delta} \text{ for some } \tilde{\beta} \in \tilde{\Pi} - \tilde{\Delta}^0.$$

In fact, otherwise Lemma 5.5′ (i) implies $\langle \alpha, \beta \rangle \leqslant 0$ for all such $\tilde{\beta}$; since $\langle \alpha, \beta \rangle = 0$ for $\tilde{\beta} \in \tilde{\Delta}^0$, part (i) and Lemma 5.6(i) gives the contradiction $\alpha = 0$. This being proved, Lemma 5.5′ (iii) and Lemma 5.4′ (i) imply because of (7):

$$L(\mathfrak{g}, \sigma) \text{ is generated by } L^{\tilde{\beta}}, L^{-\tilde{\beta}}, L^{\tilde{\gamma}} \; (\tilde{\beta} \in \tilde{\Pi} - \tilde{\Delta}^0, \tilde{\gamma} \in \tilde{\Delta}^0).$$

Secondly:

$$\text{If } \tilde{\alpha} \in \tilde{\Delta}^0, \text{ then } \tilde{\alpha} - \tilde{\beta} \in \tilde{\Delta} \text{ for some } \tilde{\beta} \in \tilde{\Pi} - \tilde{\Delta}^0.$$

In fact, otherwise Lemma 5.5′ (i) implies $\tilde{\alpha} + \tilde{\beta} \notin \tilde{\Delta}$ for all $\tilde{\beta} \in \tilde{\Pi} - \tilde{\Delta}^0$ ($p = q = 0$). Thus $[L^{\tilde{\alpha}}, L^{\pm\tilde{\beta}}] = 0$; and since $[L^{\tilde{\alpha}}, L^{\tilde{\gamma}}] = 0$ for $\tilde{\gamma} \in \tilde{\Delta}^0$, we have by the above $[L^{\tilde{\alpha}}, L(\mathfrak{g}, \sigma)] = 0$, which amounts to $[\mathfrak{g}^{\tilde{\alpha}}, \mathfrak{g}] = 0$, contradicting the semisimplicity of $\mathfrak{g}$. This proves (v).

For part (ii) suppose $\tilde{\alpha} \in \tilde{\Pi} \cap \tilde{\Delta}^0$. By simplicity, $\tilde{\alpha} - \tilde{\beta} \notin \tilde{\Delta} \cup \{0\}$ for all $\tilde{\beta} \in \tilde{\Pi} - \tilde{\Delta}^0$, so by Lemma 5.5′ (i), $\tilde{\alpha} + \tilde{\beta} \notin \tilde{\Delta} \cup \{0\}$ for all such $\tilde{\beta}$. Again we get the contradiction $[L^{\tilde{\alpha}}, L(\mathfrak{g}, \sigma)] = 0$; this proves (ii).

Since $(0, m) \in \tilde{\Delta}^0 \cap \tilde{\Delta}^+$, part (i) implies that the vectors $\alpha_0, \alpha_1, \ldots$ are linearly dependent and by (ii) they are all nonzero. They span $\mathfrak{h}^\wedge$ because of (i) and Lemma 5.6 (ii). This proves (iii). Finally, (iv) follows from Lemma 5.5′ (i).

For $0 \leqslant i \leqslant N - 1$ we put now $h_i = 2\langle\alpha_i, \alpha_i\rangle^{-1} H_{\alpha_i}$ and, by virtue of Lemma 5.4′ (ii), choose vectors $e_i \in L^{\tilde{\alpha}_i}, f_i \in L^{-\tilde{\alpha}_i}$ such that $[e_i, f_i] = h_i$. Then we have the relations

$$[h_i, h_j] = 0, \qquad [e_i, f_j] = \delta_{ij}h_i, \qquad [h_i, e_j] = a_{ji}e_j, \qquad [h_i, f_j] = -a_{ji}f_j. \tag{8}$$

The matrix $A = (a_{ij})_{0 \leqslant i,j \leqslant N-1}$ is called the *Cartan matrix* of the Lie algebra $L(\mathfrak{g}, \sigma)$. If M denotes the abelian group generated by $\tilde{\alpha}_0, \ldots, \tilde{\alpha}_{N-1}$, the decomposition (7) is an M-gradation of $L(\mathfrak{g}, \sigma)$,

$$L(\mathfrak{g}, \sigma) = \bigoplus_{\tilde{\alpha} \in M} L^{\tilde{\alpha}},$$

where $L^0 = \mathfrak{h}$, $L^{\tilde{\alpha}} = 0$ if $\tilde{\alpha} \notin \tilde{\Delta}$.

Lemma 5.8.

(i) *The elements* e_i, f_i, h_i $(0 \leqslant i \leqslant N - 1)$ *form a system of generators of* $L(\mathfrak{g}, \sigma)$.

(ii) *The M-graded Lie algebra* $L(\mathfrak{g}, \sigma)$ *has no nonzero M-graded ideals I* (*that is,* $I = \bigoplus_{\tilde{\alpha} \in M} (I \cap L^{\tilde{\alpha}})$) *for which* $I \cap \sum_i Ce_i = 0$.

(iii) *If σ is an indecomposable automorphism* (*that is,* $\mathfrak{g}$ *cannot be decomposed into a direct sum of σ-invariant ideals*), *then Π is indecomposable* (*as a union of two disjoint orthogonal nonempty subsets*).

Proof. (i) Let L' be the subalgebra of $L(\mathfrak{g}, \sigma)$ generated by e_i, f_i, h_i ($i = 0, \ldots, N - 1$). We claim

$$L^{\tilde{\alpha}} \subset L' \qquad \text{for } \tilde{\alpha} \in \tilde{\Delta} - \tilde{\Delta}^0.$$

It suffices to consider $\tilde{\alpha} \in \tilde{\Delta}^+ - \tilde{\Delta}^0$. By Lemma 5.7 (v), if $\tilde{\alpha}$ is not simple, $\tilde{\alpha} - \tilde{\alpha}_i \in \tilde{\Delta}^+$ for some i. If $\tilde{\alpha} - \tilde{\alpha}_i \notin \tilde{\Delta}^0$, then we write $L^{\tilde{\alpha}} = [e_i, L^{\tilde{\alpha}-\tilde{\alpha}_i}]$ by Lemma 5.5′ (iii); if $\tilde{\alpha} - \tilde{\alpha}_i \in \tilde{\Delta}^0$, then $p + q = -2$ in Lemma 5.5′ (i), so $(\tilde{\alpha} - \tilde{\alpha}_i) - \tilde{\alpha}_i \in \tilde{\Delta}^+ - \tilde{\Delta}^0$ and by Lemma 5.5′ (i)

$$L^{\tilde{\alpha}} = [e_i, [e_i, L^{\tilde{\alpha}-2\tilde{\alpha}_i}]].$$

By iteration, we find that $L^{\tilde{\alpha}}$ is generated by the e_i; this proves $L^{\tilde{\alpha}} \subset L'$ ($\tilde{\alpha} \in \tilde{\Delta} - \tilde{\Delta}^0$), so of course $[L^{\tilde{\alpha}}, L'] \subset L'$. Also if $\tilde{\alpha} \in \tilde{\Delta}^0$, we have by (7) and the commutativity of $\sum_{\tilde{\gamma} \in \tilde{\Delta}^0} L^{\tilde{\gamma}}$ (Lemma 5.3) that

$$[L^{\tilde{\alpha}}, L'] \subset 0 + \Big[L^{\tilde{\alpha}}, \sum_{\tilde{\beta} \in \tilde{\Delta} - \tilde{\Delta}^0} L^{\tilde{\beta}}\Big] \subset \sum_{\tilde{\beta} \in \tilde{\Delta} - \tilde{\Delta}^0} L^{\tilde{\beta}} \subset L'.$$

Thus L' is an ideal in $L(\mathfrak{g}, \sigma)$; and since each element in L' is a linear combination of multiple brackets of e_i, f_j, we have $L' = \bigoplus_{\tilde{\alpha} \in M} (L^{\tilde{\alpha}} \cap L')$, that is, L' is a graded ideal.

Now we define an invariant nondegenerate form $\tilde{B}$ on $L(\mathfrak{g}, \sigma) \times L(\mathfrak{g}, \sigma)$ by putting

$$\tilde{B}(x^k Y, x^l Z) = B(Y, Z), \qquad Y, Z \in \mathfrak{g}.$$

Let L'' denote the orthogonal complement of L' in $L(\mathfrak{g}, \sigma)$ with respect to $\tilde{B}$. Then $L'' \subset \sum_{\tilde{\alpha} \in \tilde{\Delta}^0} L^{\tilde{\alpha}}$. Hence if $Z \in L(\mathfrak{g}, \sigma)$ and we write according to (7)

$$Z = Z^0 + Z', \qquad Z^0 \in \sum_{\tilde{\alpha} \in \tilde{\Delta}^0} L^{\tilde{\alpha}}, \qquad Z' \in \sum_{\tilde{\alpha} \in \tilde{\Delta} - \tilde{\Delta}^0} L^{\tilde{\alpha}},$$

we have

$$[L'', Z] = [L'', Z'] \subset \sum_{\tilde{\alpha} \in \tilde{\Delta} - \tilde{\Delta}^0} L^{\tilde{\alpha}},$$

so, L'' being an ideal, $[L'', Z] = 0$, whence $L'' \subset$ center $(L(\mathfrak{g}, \sigma))$. But center $(\mathfrak{g}) = 0$, therefore $L(\mathfrak{g}, \sigma)$ has center 0, so $L'' = 0$; since L' is a graded ideal, this proves (i).

For (ii) suppose I were a nonzero M-graded ideal of $L(\mathfrak{g}, \sigma)$ such that $I \cap \sum Ce_i = 0$. Being M-graded, I will for some $\tilde{\alpha} \in \tilde{\Delta}$ contain a nonzero vector $e_{\tilde{\alpha}} \in L^{\tilde{\alpha}}$. If $\tilde{\alpha} \in \tilde{\Delta}^0$, then for some $i \in \{0, ..., N-1\}$ either $[e_{\tilde{\alpha}}, e_i] \neq 0$ or $[e_{\tilde{\alpha}}, f_i] \neq 0$ because otherwise part (i) would imply that $e_{\tilde{\alpha}}$ is in the center of $L(\mathfrak{g}, \sigma)$. Thus it suffices to consider the case $I \ni e_{\tilde{\alpha}} \neq 0$, where $\tilde{\alpha} \in \tilde{\Delta} - \tilde{\Delta}^0$. But then by Lemma 5.4′ (ii), $CH_\alpha \subset [L^{\tilde{\alpha}}, L^{-\tilde{\alpha}}] \subset I$. Since $\alpha_i(H_\alpha) \neq 0$ for some i, we obtain $L^{\tilde{\alpha}_i} = [L^{\tilde{\alpha}_i}, H_\alpha] \subset I$, which is a contradiction.

(iii) Let $\bar{e}_i, \bar{f}_i, \bar{h}_i$ be the images in $\mathfrak{g}$ of the elements e_i, f_i, h_i ($0 \leqslant i \leqslant N-1$) under the covering mapping of $L(\mathfrak{g}, \sigma)$ onto $\mathfrak{g}$. Suppose Π were decomposable: $\Pi = \Pi_1 \cup \Pi_2$. We obtain a corresponding decomposition $\mathfrak{h} = \mathfrak{h}_1 \oplus \mathfrak{h}_2$ where

$$\mathfrak{h}_s = \sum_{\alpha_i \in \Pi_s} Ch_i \qquad (s = 1, 2).$$

If $\alpha_i \in \Pi_1$, $\alpha_j \in \Pi_2$, then $\alpha_i(\bar{h}_j) = 0$, so by Lemma 5.5′ (i), $[\bar{e}_i, \bar{e}_j] = [\bar{f}_i, \bar{f}_j] = 0$. It follows that if $\mathfrak{g}^s$ ($s = 1, 2$) denotes the subalgebra of $\mathfrak{g}$

generated by those $\bar{e}_i, \bar{f}_i, \bar{h}_i$ for which $\alpha_i \in \Pi_s$, then by (i) we obtain a decomposition of $\mathfrak{g}$ into a direct sum of the σ-invariant ideals $\mathfrak{g}^1$ and $\mathfrak{g}^2$. This proves the lemma.

From now on, σ denotes an indecomposable automorphism of $\mathfrak{g}$ of finite order. Let $E = \sum_0^{N-1} \boldsymbol{R}\alpha_i$, put $\Pi = \{\alpha_0, ..., \alpha_{N-1}\}$ and let $n = \dim E$. The inner product $\langle\ ,\ \rangle$ is, by Lemma 5.6, positive definite on E, and we have the following properties:

(Π_1) $a_{ij} = 2\langle\alpha_i, \alpha_j\rangle/\langle\alpha_j, \alpha_j\rangle \in -\boldsymbol{Z}^+$ for $i \neq j$.

(Π_2) Π is an indecomposable system of vectors.

(Π_3) Π is a linearly dependent system of vectors generating E. In particular, $\det(a_{ij}) = 0$.

We shall classify all such systems of vectors.

Lemma 5.9.

(i) *Every proper subsystem of Π is a linearly independent system of vectors. Moreover, $N = n + 1$.*

(ii) *The system $\tilde{\Pi}$ is independent over $\boldsymbol{Z}$, that is, if $\sum_0^{N-1} c_i\tilde{\alpha}_i = 0$ $(c_i \in \boldsymbol{Z})$, then $c_i \equiv 0$.*

Proof. (i) If the first statement were false, there would be a relation

$$\sum_{\alpha\in\Pi'} a_\alpha\alpha - \sum_{\beta\in\Pi''} a_\beta\beta = 0,$$

where $\Pi' \cup \Pi'' \subset \Pi$ (proper inclusion), $\Pi' \neq \emptyset$, $\Pi' \cap \Pi'' = \emptyset$, and $a_\alpha > 0$, $a_\beta > 0$. Then by (Π_1)

$$\left\langle\sum a_\alpha\alpha, \sum a_\alpha\alpha\right\rangle = \left\langle\sum a_\alpha\alpha, \sum a_\beta\beta\right\rangle \leqslant 0,$$

so $\sum a_\alpha\alpha = 0$. But if $\beta \in \Pi - \Pi'$, we have $\langle\beta, \alpha\rangle = 0$ $(\alpha \in \Pi')$ because otherwise $0 = \langle\sum a_\alpha\alpha, \beta\rangle = \sum_\alpha a_\alpha\langle\alpha, \beta\rangle < 0$. Thus the decomposition $\Pi = \Pi' \cup (\Pi - \Pi')$ contradicts (Π_2). The statement $N = n + 1$ now follows from (Π_3).

For part (ii) we observe that if m is the order of σ, then $(0, m) \in \tilde{\Delta}^0 \cap \tilde{\Delta}^+$, so $(0, m) = \sum_0^{N-1} a_i\tilde{\alpha}_i$ where $a_i \in \boldsymbol{Z}^+$. But by part (i) the relations $0 = \sum_0^{N-1} a_i\alpha_i$ and $\sum_0^{N-1} c_i\alpha_i = 0$ must be proportional, that is, $a_i = cc_i$ $(c \in \boldsymbol{R})$. But then $m = 0$ which is a contradiction.

Lemma 5.10. *Let for the Lie algebras $L(\mathfrak{g}, \sigma)$ and $L(\mathfrak{g}', \sigma')$ the maximal abelian subalgebras $\mathfrak{h} \subset \mathfrak{g}_0$, $\mathfrak{h}' \subset \mathfrak{g}_0'$ and corresponding sets of simple roots*

$$\tilde{\Pi} = (\tilde{\alpha}_0, ..., \tilde{\alpha}_n), \qquad \tilde{\Pi}' = (\tilde{\alpha}_0', ..., \tilde{\alpha}_{n'}')$$

be given. Assume $n = n'$ *and that under the mapping* $\tilde{\alpha}_i \to \tilde{\alpha}_i'$ $(0 \leqslant i \leqslant n)$, *which by Lemma* 5.9 (ii) *gives a bijection* $\tau : M \to M'$, *the Cartan matrices* A *and* A' *coincide. Then*:

(i) *There exists an isomorphism*

$$\tilde{\psi} : L(\mathfrak{g}', \sigma') \to L(\mathfrak{g}, \sigma),$$

under which the M*-gradation and the* M'*-gradation correspond, that is,* $\tilde{\psi}((L')^{\tau(\tilde{\alpha})}) = L^{\tilde{\alpha}}$, *where* $L' = L(\mathfrak{g}', \sigma')$.

(ii) *Let* $\tilde{\psi}$ *be any isomorphism satisfying* (i). *If* $\mathfrak{g}$ *and* $\mathfrak{g}'$ *are simple, there exists an isomorphism* $\psi : \mathfrak{g}' \to \mathfrak{g}$ *and a constant* $c \in \boldsymbol{C} - (0)$ *for which the following diagram is commutative*:

$$\begin{array}{ccccc} L(\mathfrak{g}', \sigma') & \xrightarrow{\tilde{\psi}} & L(\mathfrak{g}, \sigma) & \xrightarrow{\mu_c} & L(\mathfrak{g}, \sigma) \\ \varphi' \downarrow & & & \swarrow \varphi & \\ \mathfrak{g}' & & \xrightarrow[\psi]{} & & \mathfrak{g} \end{array}$$

where φ *and* φ' *are the covering homomorphisms and* μ_c *is the automorphism of* $L(\mathfrak{g}, \sigma)$ *which corresponds to changing* x *to* cx.

Proof. (i) We consider the Lie algebra $\tilde{L}(A)$ with generators $\tilde{e}_i, \tilde{f}_i, \tilde{h}_i$ $(0 \leqslant i \leqslant n)$ and defining relations (8). (The algebra $\tilde{L}(A)$ is the analog of $\mathfrak{l}$ in §4; it is the free Lie algebra $\hat{L}(A)$ on $3(n+1)$ generators $\hat{e}_i, \hat{f}_i, \hat{h}_i$ modulo the ideal defined by relations (8).) By Lemma 5.9, each $\tilde{\alpha} \in M$ can be written $\tilde{\alpha} = \sum_0^n k_i \tilde{\alpha}_i$ where the $k_i \in \boldsymbol{Z}$ are unique. We write $\tilde{\alpha} > 0$ if $\tilde{\alpha} \neq 0$ and $k_i \geqslant 0$ for all i. If $\tilde{\alpha} > 0$ (respectively $-\tilde{\alpha} > 0$), we denote by $\tilde{L}^{\tilde{\alpha}}$ the linear span of all commutators

$$[\tilde{e}_{i_1}, \ldots, \tilde{e}_{i_r}] = \operatorname{ad} \tilde{e}_{i_1} \ldots \operatorname{ad} \tilde{e}_{i_{r-1}}(\tilde{e}_{i_r})$$

for which $\tilde{\alpha}_{i_1} + \ldots + \tilde{\alpha}_{i_r} = \tilde{\alpha}$ (respectively, all commutators

$$[\tilde{f}_{i_1}, \ldots, \tilde{f}_{i_r}] = \operatorname{ad} \tilde{f}_{i_1} \ldots \operatorname{ad} \tilde{f}_{i_{r-1}}(\tilde{f}_{i_r})$$

for which $\tilde{\alpha}_{i_1} + \ldots + \tilde{\alpha}_{i_r} = -\tilde{\alpha}$). For all other $\tilde{\alpha} \neq 0$ in M we put $\tilde{L}^{\tilde{\alpha}} = 0$ and we define $\tilde{L}^0 = \sum_0^n \boldsymbol{C}\tilde{h}_i$. By Theorem 4.2, the subsequent remark, and Lemma 4.5, we have the direct decomposition $\tilde{L}(A) = \tilde{L}^+ + \tilde{L}^0 + \tilde{L}^-$, where

$$\tilde{L}^+ = \sum_{\tilde{\alpha} > 0} \tilde{L}^{\tilde{\alpha}}, \qquad \tilde{L}^- = \sum_{-\tilde{\alpha} > 0} \tilde{L}^{\tilde{\alpha}}.$$

These sums are direct; suppose for example we have $\sum_{\tilde{\alpha}>0} a_{\tilde{\alpha}} e^{\tilde{\alpha}} = 0$ ($a_\alpha \in \boldsymbol{C}$) where

$$e^{\tilde{\alpha}} = \sum_{\tilde{\alpha}_{i_1}+\ldots+\tilde{\alpha}_{i_r}=\tilde{\alpha}} c_{(i)}[\tilde{e}_{i_1}, \ldots, \tilde{e}_{i_r}] \in \tilde{L}^{\tilde{\alpha}}, \qquad c_{(i)} \in \boldsymbol{C}.$$

Then by the method of proof of Lemma 4.5, (where $y' = 0$ implied $\eta' = 0$) we derive the relation

$$\sum_{\tilde{\alpha}>0} a_{\tilde{\alpha}} \hat{e}^{\tilde{\alpha}} = 0$$

for the representatives

$$\hat{e}^{\tilde{\alpha}} = \sum_{(i)} c_{(i)}[\hat{e}_{i_1}, \ldots, \hat{e}_{i_r}] \in \hat{L}(A).$$

But if $\hat{e}^{\tilde{\alpha}}$ is written as a sum of different monomials in the tensor algebra with nonzero coefficients, then for $\tilde{\alpha} \neq \tilde{\beta}$, $\hat{e}^{\tilde{\beta}}$ and $\hat{e}^{\tilde{\beta}}$ have no terms in common. Hence $a_{\tilde{\alpha}} = 0$ for all $\tilde{\alpha}$, so we have the direct decomposition

$$\tilde{L}(A) = \bigoplus_{\tilde{\alpha} \in M} \tilde{L}^{\tilde{\alpha}}.$$

This turns $\tilde{L}(A)$ into an M-graded Lie algebra. In fact $[\tilde{L}^{\tilde{\alpha}}, \tilde{L}^{\tilde{\beta}}] \subset \tilde{L}^{\tilde{\alpha}+\tilde{\beta}}$ as a consequence of the relations $[\tilde{L}^{\tilde{\alpha}}, \tilde{e}_i] \subset \tilde{L}^{\tilde{\alpha}+\tilde{\alpha}_i}$ and $[\tilde{L}^{\tilde{\alpha}}, \tilde{f}_i] \subset \tilde{L}^{\tilde{\alpha}-\tilde{\alpha}_i}$, which follow from (8) by induction.

The algebra $\tilde{L}(A)$ contains a unique maximal M-graded ideal $I(A)$ with the property $I(A) \cap \sum_i \boldsymbol{C}\tilde{e}_i = 0$, namely the subspace of $\tilde{L}(A)$ spanned by all M-graded ideals J_γ with this property. In fact, $\sum_\gamma J_\gamma$ is an ideal and using the fact that it, and each J_γ, is M-graded, one derives readily that $(\sum_\gamma J_\gamma) \cap \sum \boldsymbol{C}\tilde{e}_i = 0$.

Taking brackets with the f_j, it is clear that the elements e_i ($0 \leqslant i \leqslant n$) in $L(\mathfrak{g}, \sigma)$ are linearly independent. The homomorphism of $\tilde{L}(A)$ onto $L(\mathfrak{g}, \sigma)$ given by $\tilde{h}_i \to h_i$, $\tilde{e}_i \to e_i$, $\tilde{f}_i \to f_i$ has a kernel which is an M-graded ideal which, by the linear independence of the e_i, has 0 intersection with $\sum_i \boldsymbol{C}\tilde{e}_i$, and by Lemma 5.8 (ii) it is maximal with this property. Thus $L(\mathfrak{g}, \sigma) = \tilde{L}(A)/I(A)$, proving (i).

(ii) We recall that the *centroid* of a Lie algebra $\mathfrak{l}$ is the set of linear transformations A of $\mathfrak{l}$ which commute with each ad X ($X \in \mathfrak{l}$). It is well known (cf. Jacobson [1], p. 290) that if $\mathfrak{l} = [\mathfrak{l}, \mathfrak{l}]$, then the centroid is commutative. In fact,

$$A_1A_2[X_1, X_2] = A_1[X_1, A_2X_2] = -A_1[A_2X_2, X_1] = -[A_2X_2, A_1X_1]$$
$$A_2A_1[X_1, X_2] = -A_2A_1[X_2, X_1] = -A_2[X_2, A_1X_1] = [A_1X_1, A_2X_2],$$

so $A_1A_2 - A_2A_1$ annihilates $[\mathfrak{l}, \mathfrak{l}] = \mathfrak{l}$.

We now fix an isomorphism $\tilde{\psi}$ satisfying (i). The orders of σ and σ' being m and m', respectively, consider the linear transformations A of $L = L(\mathfrak{g}, \sigma)$, A' of $L' = L(\mathfrak{g}', \sigma')$ given by

$$Ag = x^m g, \qquad A'g = x^{m'} g.$$

Clearly, A and A' belong to the respective centroids. By the definition of the covering map the kernel of φ' equals $I' = \{g - A'g : g \in L'\}$. We put $\bar{I} = \tilde{\psi}(I')$. Then $\bar{I} = \{g - \bar{A}g : g \in L\}$ where $\bar{A} = \tilde{\psi}A'\tilde{\psi}^{-1}$, which belongs to the centroid of $L(g, \sigma)$. Now $A'(L'^{\tilde{\alpha}'}) \subset (L')^{\tilde{\alpha}'+\tilde{\gamma}'}$ where $\tilde{\gamma}' = (0, m') \in (\tilde{\Delta}')^0 \cap (\tilde{\Delta}')^+$. It follows that $\bar{A}(L^{\tilde{\alpha}}) \subset L^{\tilde{\alpha}+\tilde{\gamma}}$ where $\tilde{\gamma} = (\gamma, t) \in \tilde{\Delta}^+$ satisfies $\tau(\tilde{\gamma}) = \tilde{\gamma}'$. Now $\mathfrak{g}$ is simple, so by Lemma 5.8, Π is indecomposable. Thus, given any two indices $0 \leqslant i \leqslant j \leqslant n$, there exist $i = k_1, k_2, \ldots, k_r = j$ such that all $a_{k_p k_q} \neq 0$. Hence it follows from (8) that $[e_i, L] = L$. Therefore, by the above, A and $\bar{A}$ commute. Hence $\bar{A}$ leaves the kernel $I = (1 - A)\, L(\mathfrak{g}, \sigma)$ of φ invariant and induces on $\mathfrak{g} = L(\mathfrak{g}, \sigma)/I$ an element $\bar{a}$ from the centroid of $\mathfrak{g}$. But $\mathfrak{g}$ is a simple Lie algebra over C, so by Schur's lemma, $\bar{a}$ is a scalar. Choosing a basis vector $e_{\tilde{\beta}}$ in each $L^{\tilde{\beta}}$ ($\tilde{\beta} \in \tilde{\Delta} - \tilde{\Delta}^0$), we have for $\tilde{\alpha} \in \tilde{\Delta} - \tilde{\Delta}^0$, $\tilde{\alpha} = (\alpha, j)$,

$$\bar{A}e_{\tilde{\alpha}} = \bar{a}e_{\tilde{\alpha}} + (1 - x^m)\Big(l_{\tilde{\alpha}} + \sum_{\tilde{\beta} \in \tilde{\Delta} - \tilde{\Delta}^0} c_{\tilde{\beta},\tilde{\alpha}} e_{\tilde{\beta}}\Big),$$

where the c are constants and $l_{\tilde{\alpha}} \in \sum_{\tilde{\gamma} \in \tilde{\Delta}^0} L^{\tilde{\gamma}}$. If $\alpha + \gamma = 0$, this implies

$$\bar{a}e_{\tilde{\alpha}} + (1 - x^m) \sum_{\tilde{\beta} \in \tilde{\Delta} - \tilde{\Delta}^0} c_{\tilde{\beta},\tilde{\alpha}} e_{\tilde{\beta}} = 0.$$

But this equation is impossible because the sum is finite and $x^m e_{\tilde{\beta}} \neq 0$. Thus $\alpha + \gamma \neq 0$, and we have $\bar{A}e_{\tilde{\alpha}} = ce_{\tilde{\alpha}+\tilde{\gamma}}$ ($c \neq 0$), so the equation above implies $l_\alpha = 0$ and

$$ce_{\tilde{\alpha}+\tilde{\gamma}} = \bar{a}e_{\tilde{\alpha}} + (1 - x^m) \sum_{\tilde{\beta} \in \tilde{\Delta} - \tilde{\Delta}^0} c_{\tilde{\beta},\tilde{\alpha}} e_{\tilde{\beta}}.$$

This equation implies $\tilde{\gamma} = (0, dm)$, $d \in Z^+$, so $\bar{A}$ and A^d are proportional on each $L^{\tilde{\alpha}}$. In particular $(\bar{A} - aA^d)e_1 = 0$ for some $a \in C$, whence $(\bar{A} - aA^d)(L) = (\bar{A} - aA^d)([L, e_1]) = 0$ since $\bar{A}$ and $\bar{A}^d$ belong to the centroid of L. Thus $aA^d = \bar{A} = \tilde{\psi}A'\tilde{\psi}^{-1}$. Applying this to the isomorphism $\tilde{\psi}^{-1}$, we get $a'(A')^{d'} = \tilde{\psi}^{-1}A\tilde{\psi}$ where $d' \in Z^+$ and $a' \in C$. Hence $A^{dd'} = a_0 A$ ($a_0 \in C$) which implies $d = d' = 1$, whence $\bar{A}g = ax^m g$, $\bar{I} = (1 - ax^m)\, L(\mathfrak{g}, \sigma)$, and (iii) follows by putting $c = a^{-1/m}$.

In analogy with the procedure in §3 we associate with each matrix $A = (a_{ij})$ a diagram as follows. We take $n + 1$ vertices, we join the ith and jth vertices by $a_{ij}a_{ji}$ lines; and if $| a_{ij} | < | a_{ji} |$, these lines have an arrow, pointing toward the ith vertex (the shorter root).

Lemma 5.11. *Any diagram $S(A)$ belongs to one of the Tables 1–3 below. The numerical labels on the vertices are coefficients of the linear dependence between the corresponding rows of the matrix A.*

Moreover, $S(A)$ determines the Cartan matrix A.

Proof. From Lemma 5.9 and Theorem 3.21 we have:

(a) Every proper subdiagram of $S(A)$ is the disconnected union of Dynkin diagrams of type $\mathfrak{a}_l$, $\mathfrak{b}_l$, $\mathfrak{c}_l$, $\mathfrak{d}_l$, $\mathfrak{e}_6$, $\mathfrak{e}_7$, $\mathfrak{e}_8$, $\mathfrak{f}_4$, $\mathfrak{g}_2$.

We also have here:

(b) $S(A)$ is connected;

(c) $\det(A) = 0$.

Every subdiagram of $S(A)$ consisting of l vertices satisfies

$$\sum_{i<j} (a_{ij}a_{ji})^{\frac{1}{2}} \leqslant l. \qquad (9)$$

(This is (5), §3, except that now α could be 0.) Taking into account (a) and Cor. 3.14, we have:

(d) The only cycle which could occur in a diagram is a "simple" cycle of N vertices ($\mathfrak{a}_n^{(1)}$ in Table 1). If $S(A) \neq \mathfrak{a}_n^{(1)}$, then $S(A)$ does not contain any cycles.

Inequality (9) now implies:

(e) If $S(A)$ contains a triple link, then all other links are simple ($3^{\frac{1}{2}} + 2^{\frac{1}{2}} > 3$).

(f) $S(A)$ cannot contain more than two double links ($3 \cdot 2^{\frac{1}{2}} > 4$).

Property (c) implies:

(g) $S(A)$ is not a Dynkin diagram from (a).

(h) For $N = 2$, $S(A)$ is either $\mathfrak{a}_1^{(1)}$ or $\mathfrak{a}_2^{(2)}$ in the tables (quadruple link because of (g)).

We shall now prove that if $S(A)$ satisfies (a) (b), (d)–(h), then $S(A)$ belongs to Tables 1–3. Because of (d) we may assume $S(A)$ contains no cycles. We claim now that there is a vertex $\rho \in S(A)$, for which $S(A) - \{\rho\}$ is a connected diagram (and thus from the list in (a)). To see this we fix a point on $S(A)$ and move along $S(A)$ by the following rule: each time we move to a vertex where we have not been before. If $\{\alpha_r, \alpha_{r+1}\}$ is the last link on our path, then the "dead end" α_{r+1} can serve as ρ. In fact, any vertex in $S(A) - \{\rho\}$ linked to ρ must have been visited before, in other words, it is connected to α_r within $S(A) - \{\rho\}$. Hence

Tables of Diagrams $S(A)$

TABLE 1

$\mathfrak{a}_n^{(1)}$ $(n > 1)$

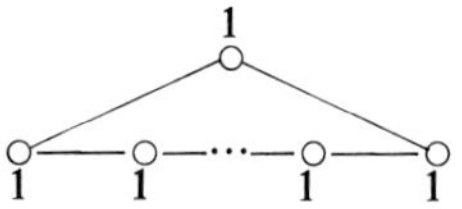

$\mathfrak{a}_1^{(1)}$

$\mathfrak{b}_n^{(1)}$ $(n > 2)$

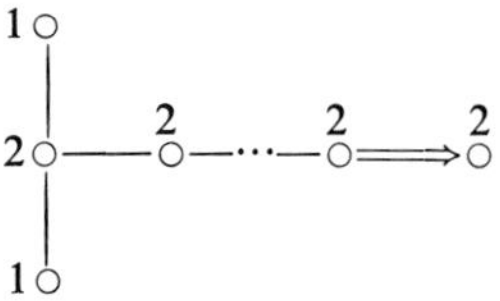

$\mathfrak{c}_n^{(1)}$ $(n > 1)$

$\mathfrak{d}_n^{(1)}$ $(n > 3)$

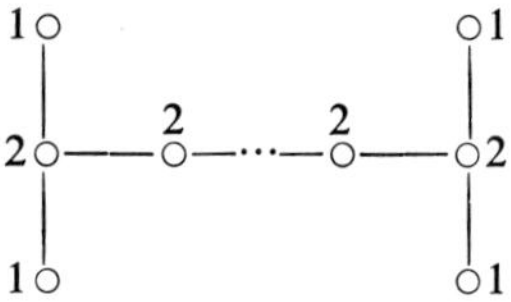

$\mathfrak{e}_6^{(1)}$

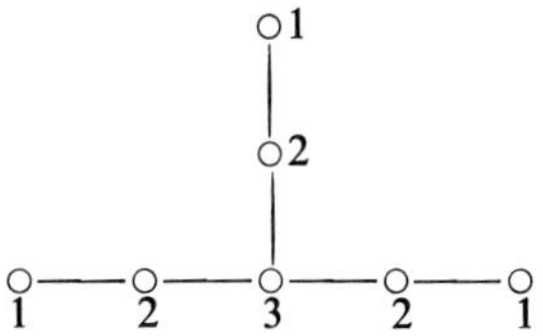

$\mathfrak{e}_7^{(1)}$

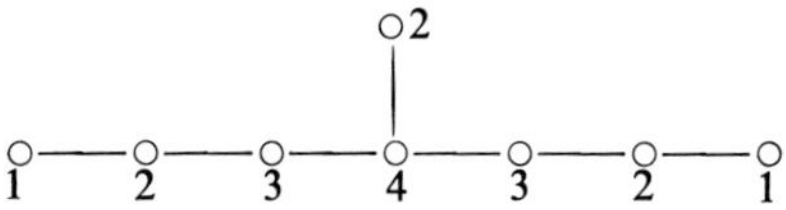

$\mathfrak{e}_8^{(1)}$

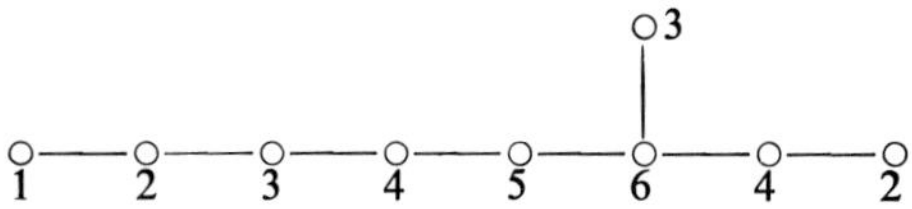

$\mathfrak{f}_4^{(1)}$

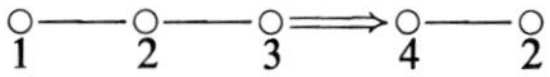

$\mathfrak{g}_2^{(1)}$

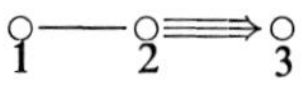

TABLE 2

$\mathfrak{a}_{2n}^{(2)}$ $(n > 1)$

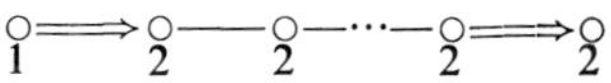

$\mathfrak{a}_2^{(2)}$

$\mathfrak{d}_{n+1}^{(2)}$ $(n > 1)$

$\mathfrak{a}_{2n-1}^{(2)}$ $(n > 2)$

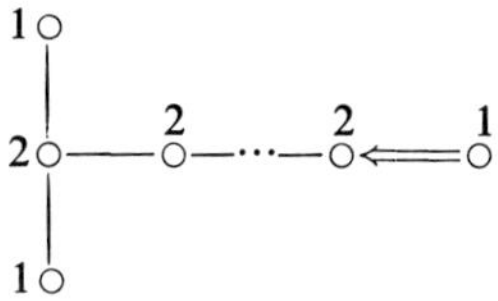

$\mathfrak{e}_6^{(2)}$

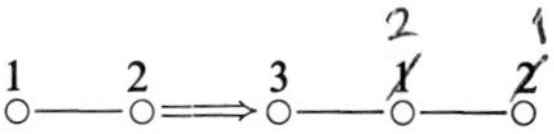

TABLE 3

$\mathfrak{d}_4^{(3)}$

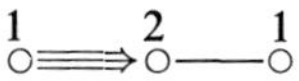

For $\mathfrak{a}_n^{(1)}$, $\mathfrak{a}_{2n}^{(2)}$, $\mathfrak{a}_{2n-1}^{(2)}$, $\mathfrak{b}_n^{(1)}$, $\mathfrak{c}_n^{(1)}$, $\mathfrak{d}_n^{(1)}$, $\mathfrak{d}_{n+1}^{(2)}$, there are $n + 1$ vertices.

$S(A) - \{\rho\}$ is connected. Thus we just have to consider the possibilities of adding a vertex ρ to a Dynkin diagram such that ρ is connected to only one vertex of this diagram (since we have ruled out cycles).

Consider for example $\mathfrak{f}_4$ in §3, no. 5. We cannot add a vertex by connecting it to an interior point (with the coefficients 3 and 4) without violating (a). Thus the only potential possibilities are $\mathfrak{f}_4^{(1)}$ and $\mathfrak{e}_6^{(2)}$ in the tables. Similar arguments with the other Dynkin diagrams leave us with Tables 1–3 as the only potential possibilities. That $S(A)$ determines A is clear since $a_{ij}a_{ji} = 0, 1, 2, 3, 4$ and the configurations

imply

$$a_{ij} = 0, \quad a_{ij} = -1, \quad a_{ij} = -2, \quad a_{ij} = -2, \quad a_{ij} = -3, \quad a_{ij} = -4,$$
$$a_{ji} = 0, \quad a_{ji} = -1, \quad a_{ji} = -1, \quad a_{ji} = -2, \quad a_{ji} = -1, \quad a_{ji} = -1.$$

It remains to verify that the possibilities in Tables 1–3 do occur and we must also verify the numerical labels.

Example 1. Let $\mathfrak{g}$ be simple and σ the identity automorphism e, so $m = 1$ and $L(\mathfrak{g}, e) = \sum_{j \in \mathbf{Z}} x^j \mathfrak{g}$. Let $\alpha_1, \ldots, \alpha_n$ be the simple roots in $\Delta(\mathfrak{g}, \mathfrak{h})$ with respect to some ordering and δ the highest root. Then each $(\alpha_j, 0)$ is a simple root of $L(\mathfrak{g}, e)$ with respect to $\mathfrak{h}$; $(-\delta, 1)$ is also a simple root (otherwise $(-\delta, 1) = (\beta, 1) + (\alpha, 0)$, $\alpha, \beta \in \Delta(\mathfrak{g}, \mathfrak{h})$, $\alpha > 0$ which is impossible). Thus

$$\tilde{\Pi} = \{(-\delta, 1), (\alpha_1, 0), \ldots, (\alpha_n, 0)\}.$$

We claim now that $\Pi = \{-\delta, \alpha_1, \ldots, \alpha_n\}$ realizes the diagrams $X_n^{(1)}$ in Table 1. For this we note the following consequence of Theorem 3.28:

$$\mathfrak{g} = \mathfrak{a}_1 \qquad a_{\alpha_1,-\delta}a_{-\delta,\alpha_1} = 4, \quad |a_{\alpha_1,-\delta}| = |a_{-\delta,\alpha_1}|;$$

$$\mathfrak{g} = \mathfrak{a}_n \quad (n > 1) \qquad a_{\alpha_i,-\delta}a_{-\delta,\alpha_i} = \begin{cases} 1 & \text{if } i = 1 \text{ or } i = n \\ 0 & \text{if } 1 < i < n; \end{cases}$$

$$\mathfrak{g} = \mathfrak{b}_n \quad (n \geqslant 2) \qquad a_{\alpha_i,-\delta}a_{-\delta,\alpha_i} = \delta_{i2};$$

$$\mathfrak{g} = \mathfrak{c}_n \quad (n \geqslant 3) \qquad a_{\alpha_i,-\delta}a_{-\delta,\alpha_i} = 2\delta_{i1};$$

$$\mathfrak{g} = \mathfrak{d}_n \quad (n \geqslant 4) \qquad a_{\alpha_i,-\delta}a_{-\delta,\alpha_i} = \delta_{i2}.$$

For $\mathfrak{g}$ exceptional we use (d) above, which guarantees that $a_{\alpha_i,\delta}$ is $\neq 0$

for exactly one i. Hence we find easily from the description of these root systems and Theorem 3.28:

$$\mathfrak{g} = \mathfrak{e}_6, \qquad a_{\alpha_i,-\delta} a_{-\delta,\alpha_i} = \delta_{i2};$$

$$\mathfrak{g} = \mathfrak{e}_7, \qquad a_{\alpha_i,-\delta} a_{-\delta,\alpha_i} = \delta_{i1};$$

$$\mathfrak{g} = \mathfrak{e}_8, \qquad a_{\alpha_i,-\delta} a_{-\delta,\alpha_i} = \delta_{i8};$$

$$\mathfrak{g} = \mathfrak{f}_4, \qquad a_{\alpha_i,-\delta} a_{-\delta,\alpha_i} = \delta_{i1};$$

$$\mathfrak{g} = \mathfrak{g}_2, \qquad a_{\alpha_i,-\delta} a_{-\delta,\alpha_i} = \delta_{i2}.$$

Thus we obtain precisely the diagrams in Table I; the numerical labels come from Theorem 3.28. The diagrams in Table I are known in the literature as the *extended Dynkin diagrams.*

Example 2. Consider for the simple Lie algebra $\mathfrak{g}$ the canonical generators X_i, Y_i, H_i $(1 \leqslant i \leqslant l)$ from Proposition 4.1. Let $\bar{\nu}$ be an automorphism of the corresponding Dynkin diagram of order $k > 1$ and let ν denote the automorphism of $\mathfrak{g}$ given by

$$\nu(X_i) = X_{\bar{\nu}(i)}, \qquad \nu(Y_i) = Y_{\bar{\nu}(i)}, \qquad \nu(H_i) = H_{\bar{\nu}(i)}$$

for $1 \leqslant i \leqslant l$. The existence and uniqueness of ν is clear from Theorem 4.15. Because of Theorem 3.29, $k = 2$ or 3. We shall now verify that by varying $\bar{\nu}$ the diagrams in Tables 2 and 3 are realized as $S(A)$ for the algebras $L(\mathfrak{g}, \nu)$. We refer to ν as an *automorphism of* $\mathfrak{g}$ *induced by an automorphism of the Dynkin diagram.* According to Theorem 3.29 only $\mathfrak{a}_l$, $\mathfrak{d}_l$, and $\mathfrak{e}_6$ have such automorphisms. We shall now list $\mathfrak{g}$ and $\mathfrak{g}_0$ for them.

Case 1 $\mathfrak{g} = \mathfrak{a}_{2n}, \qquad k = 2; \qquad \bar{\nu}(i) = 2n - i + 1.$

We put

$$\bar{H}_i = H_i + H_{2n-i+1} \quad (1 \leqslant i \leqslant n-1), \qquad \bar{H}_n = 2(H_n + H_{n+1}),$$
$$\bar{X}_i = X_i + X_{2n-i+1} \quad (1 \leqslant i \leqslant n-1), \qquad \bar{X}_n = X_n + X_{n+1},$$
$$\bar{Y}_i = Y_i + Y_{2n-i+1} \quad (1 \leqslant i \leqslant n-1), \qquad \bar{Y}_n = 2(Y_n + Y_{n+1}).$$

TABLE I

THE ORDER k OF ν AND THE ALGEBRA $\mathfrak{g}_0$

$\mathfrak{g}$	$\mathfrak{a}_{2n}$	$\mathfrak{a}_{2n-1}$	$\mathfrak{d}_{n+1}$	$\mathfrak{e}_6$	$\mathfrak{d}_4$
k	2	2	2	2	3
$\mathfrak{g}_0$	$\mathfrak{b}_n$	$\mathfrak{c}_n$	$\mathfrak{b}_n$	$\mathfrak{f}_4$	$\mathfrak{g}_2$

Since $\bar{\nu}$ interchanges the ith and the $(2n - i + 1)$th simple root, the elements $\bar{X}_i$, $\bar{Y}_i$, and $\bar{H}_i$ $(1 \leqslant i \leqslant n)$ belong to $\mathfrak{g}_0$. They satisfy the commutation relations

$$[\bar{X}_i, \bar{Y}_j] = \delta_{ij}\bar{H}_i, \qquad [\bar{H}_i, \bar{X}_j] = a_{ji}\bar{X}_j, \qquad [\bar{H}_i, \bar{Y}_j] = -a_{ji}\bar{Y}_j,$$

where (a_{ij}) is the Cartan matrix of $\mathfrak{b}_n$ (direct computation). The subalgebra $\mathfrak{h} = \sum_{i=1}^{n} \boldsymbol{C}\bar{H}_i$ is the set of fixed points of ν in the Cartan subalgebra $\sum_{i=1}^{2n} \boldsymbol{C}H_i$ of $\mathfrak{g}$. Since $\mathfrak{h}$ contains a regular element of this Cartan subalgebra (for example, the element H_ρ, which by Lemma 3.11 is regular), it is clear that $\sum_{i=1}^{2n} \boldsymbol{C}H_i$ is the centralizer $\mathfrak{z}(\mathfrak{h})$ of $\mathfrak{h}$ in $\mathfrak{g}$; in particular, $\mathfrak{h}$ is a maximal abelian subalgebra of $\mathfrak{g}_0$. The center $\mathfrak{c}_0$ of $\mathfrak{g}_0$ therefore belongs to $\mathfrak{h}$. But if $\sum c_i\bar{H}_i$ commutes with all $\bar{X}_j$, we deduce $\sum_i c_i a_{ji} = 0$ for all j; so since the matrix (a_{ji}) is nondegenerate, all $c_i = 0$. Thus $\mathfrak{c}_0 = 0$, $\mathfrak{g}_0$ is semisimple, and $\mathfrak{h} \subset \mathfrak{g}_0$ is a Cartan subalgebra. The space $\boldsymbol{C}\bar{X}_j$ is a corresponding root subspace; we denote by $\bar{\alpha}_j$ the corresponding root, so $\bar{\alpha}_j(\bar{H}_i) = a_{ji}$. The elements $\bar{\alpha}_1, ..., \bar{\alpha}_n$ are linearly independent (since (a_{ij}) is nonsingular), and we can order the dual of $\sum_{i=1}^{n} \boldsymbol{R}\bar{H}_i$ lexicographically with respect to the basis $\bar{\alpha}_1, ..., \bar{\alpha}_n$. We have the direct decomposition

$$\mathfrak{g} = \mathfrak{n}^- + \mathfrak{z}(\mathfrak{h}) + \mathfrak{n}^+,$$

where $\mathfrak{n}^+$ (resp. $\mathfrak{n}^-$) is the subalgebra spanned by the root spaces for the positive (resp. negative) members of $\Delta(\mathfrak{g}, \mathfrak{z}(\mathfrak{h}))$. According to Lemma 3.10, $\mathfrak{n}^+$ is spanned by the X_i and arbitrary r-fold commutators $[X_{i_1}, ..., X_{i_r}]$ of the X_i $(r > 1)$. Since ν permutes the positive roots, we derive the direct decomposition

$$\mathfrak{g}_0 = (\mathfrak{n}^- \cap \mathfrak{g}_0) + \mathfrak{h} + (\mathfrak{n}^+ \cap \mathfrak{g}_0),$$

and $\mathfrak{n}^+ \cap \mathfrak{g}_0$ is spanned by the $\bar{X}_i$ and arbitrary elements

$$\bar{X}_{(i)} = [X_{i_1}, ..., X_{i_r}] + \nu[X_{i_1}, ..., X_{i_r}], \qquad r > 1.$$

If $\alpha_1, ..., \alpha_{2n}$ are the simple roots in $\Delta(\mathfrak{g}, \mathfrak{z}(\mathfrak{h}))$ corresponding to $X_1, ..., X_{2n}$, the vector $\bar{X}_{(i)}$ is a joint eigenvector of the family $\text{ad}(\mathfrak{h})$ with eigenvalues $(\alpha_{i_1} + ... + \alpha_{i_r})(H)$, $H \in \mathfrak{h}$. But the restriction $\alpha_j \mid \mathfrak{h}$ coincides with $\bar{\alpha}_i$ $(i = j$ or $i = \bar{\nu}(j))$, so we deduce $\bar{X}_{(i)} \in \mathfrak{g}_0^\beta$ where β is the sum of r terms from $\bar{\alpha}_1, ..., \bar{\alpha}_n$. We conclude that the $\bar{\alpha}_i$ are simple roots (in $\Delta(\mathfrak{g}_0, \mathfrak{h})$); and since $n = \dim \mathfrak{h}$, they constitute all the simple roots. Hence $\mathfrak{g}_0 = \mathfrak{b}_n$.

Similar arguments apply to the other cases. We list below the necessary information.

Case 2 $\mathfrak{g} = \mathfrak{a}_{2n-1}$, $k = 2$; $\bar{\nu}(i) = 2n - i$.

We put

$$\bar{H}_i = H_i + H_{2n-i} \quad (1 \leqslant i \leqslant n-1), \qquad \bar{H}_n = H_n,$$
$$\bar{X}_i = X_i + X_{2n-i} \quad (1 \leqslant i \leqslant n-1), \qquad \bar{X}_n = X_n,$$
$$\bar{Y}_i = Y_i + Y_{2n-i} \quad (1 \leqslant i \leqslant n-1), \qquad \bar{Y}_n = Y_n.$$

Case 3 $\mathfrak{g} = \mathfrak{d}_{n+1}$ $(n > 1)$, $k = 2$;

$$\bar{\nu}(i) = i \quad (1 \leqslant i \leqslant n-1), \qquad \bar{\nu}(n) = n+1, \qquad \bar{\nu}(n+1) = n.$$

We put

$$\bar{H}_i = H_i \quad (1 \leqslant i \leqslant n-1), \qquad \bar{H}_n = H_n + H_{n+1},$$
$$\bar{X}_i = X_i \quad (1 \leqslant i \leqslant n-1), \qquad \bar{X}_n = X_n + X_{n+1},$$
$$\bar{Y}_i = \bar{Y}_i \quad (1 \leqslant i \leqslant n-1), \qquad \bar{Y}_n = Y_n + Y_{n+1}.$$

Case 4 $\mathfrak{g} = \mathfrak{e}_6$, $k = 2$; $\bar{\nu}$ interchanges 1 and 6, 3 and 5,

$$\bar{\nu}(4) = 4, \qquad \bar{\nu}(2) = 2.$$

We put

$$\bar{H}_1 = H_1 + H_6, \qquad \bar{H}_2 = H_3 + H_5, \qquad \bar{H}_3 = H_4, \qquad \bar{H}_4 = H_2,$$
$$\bar{X}_1 = X_1 + X_6, \qquad \bar{X}_2 = X_3 + X_5, \qquad \bar{X}_3 = X_4, \qquad \bar{X}_4 = X_2,$$
$$\bar{Y}_1 = Y_1 + Y_6, \qquad \bar{Y}_2 = Y_3 + Y_5, \qquad \bar{Y}_3 = Y_4, \qquad \bar{Y}_4 = Y_2.$$

Case 5 $\mathfrak{g} = \mathfrak{d}_4$, $k = 3$; $\bar{\nu}(1) = 4$, $\bar{\nu}(4) = 3$, $\bar{\nu}(3) = 1$.

We put

$$\bar{H}_1 = H_1 + H_3 + H_4, \qquad \bar{H}_2 = H_2,$$
$$\bar{X}_1 = X_1 + X_3 + X_4, \qquad \bar{X}_2 = X_2,$$
$$\bar{Y}_1 = Y_1 + Y_3 + Y_4, \qquad \bar{Y}_2 = Y_2.$$

We fix the Z_3-gradation $\mathfrak{g} = \bigoplus_{i \in Z_3} \mathfrak{g}_i$ corresponding to the primitive cube root $\epsilon_0 = e^{\frac{1}{3}2\pi i}$ of 1. Then

$$H_4 + \epsilon_0 H_1 + \epsilon_0^2 H_3 \in \mathfrak{g}_1, \qquad H_1 + \epsilon_0 H_4 + \epsilon_0^2 H_3 \in \mathfrak{g}_2.$$

Nothing new is added by considering the automorphism ν^2 instead of ν. In fact, ν and ν^2 are conjugate under the automorphism of $\mathfrak{d}_4$ defined under case (3).

Table I for $\mathfrak{g}_0$ is now readily verified for cases (1)–(5). Note that the fixed points of ν in the associated Cartan subalgebra of $\mathfrak{g}$ form a Cartan subalgebra of $\mathfrak{g}_0$ and that a basis of the roots for $\mathfrak{g}_0$ is obtained by restriction of a basis of the roots for $\mathfrak{g}$.

Now we compute the diagram $S(A)$ for these cases. Now that we know that $\mathfrak{g}_0$ is semisimple (even simple) in all these cases, we can state that if $\alpha \in \Delta(\mathfrak{g}_0, \mathfrak{h})$ is a simple root, then $(\alpha, 0)$ is a simple root of $L(\mathfrak{g}, \nu)$ with respect to $\mathfrak{h}$. Hence we conclude from Lemma 5.9 (i) that for a suitable $\rho_0 \in S(A)$ the complement $S(A) - \{\rho_0\}$ is the Dynkin diagram of $\mathfrak{g}_0$. (The Cartan matrix and the Dynkin diagram of $\mathfrak{g}_0$ can by Lemma 5.1 and the simplicity of $\mathfrak{g}_0$ be computed by means of B.) Since $S(A)$ determines A, Lemma 5.10 implies that if $\mathfrak{g}$ and $\mathfrak{g}'$ are simple and $L(\mathfrak{g}, \sigma)$ and $L(\mathfrak{g}', \sigma')$ have the same diagram $S(A)$, then $\mathfrak{g} \simeq \mathfrak{g}'$. We know also that the diagrams of Table 1 cannot occur for the present cases of $\mathfrak{g}$ because for them the diagram of $L(\mathfrak{g}, e)$ is simply linked, whereas the Dynkin diagrams of $\mathfrak{g}_0$ in Table I have multiple links.

Consider now case (1) above. As remarked, $(\bar{\alpha}_1, 0), ..., (\bar{\alpha}_n, 0)$ are simple roots of $L(\mathfrak{g}, \nu)$ with respect to $\mathfrak{h}$. Let $\delta \in \Delta(\mathfrak{g}, \mathfrak{z}(\mathfrak{h}))$ be the highest root and δ^* the restriction $\delta \mid \mathfrak{h}$. Then $(-\delta^*, 1)$ is a root of $L(\mathfrak{g}, \nu)$ with respect to $\mathfrak{h}$. In fact, since each segment $\alpha_i + ... + \alpha_j$ is a root in $\Delta(\mathfrak{g}, \mathfrak{z}(\mathfrak{h}))$ and $\delta = \alpha_1 + ... + \alpha_{2n}$, the vector

$$[\mathrm{ad}(Y_n)\,\mathrm{ad}(Y_{n-1}) \dots \mathrm{ad}(Y_2)(Y_1), \mathrm{ad}(Y_{n+1})\,\mathrm{ad}(Y_{n+2}) \dots \mathrm{ad}(Y_{2n-1})(Y_{2n})]$$

is a nonzero vector in $\mathfrak{g}^{-\delta}$ and it clearly belongs to $\mathfrak{g}_1$. But $x\mathfrak{g}^{-\delta} \cap \mathfrak{g}_1 \subset L(\mathfrak{g}, \nu)^{(-\delta^*, 1)}$, so by the dimensionality,

$$x\mathfrak{g}^{-\delta} \cap \mathfrak{g}_1 = L(\mathfrak{g}, \nu)^{(-\delta^*, 1)}.$$

The vectors $\bar{Y}_i$ $(1 \leqslant i \leqslant n)$ all commute with the left-hand side; and since the $\bar{\alpha}_i$ $(1 \leqslant i \leqslant n)$ constitute all the simple roots in $\Delta(\mathfrak{g}_0, \mathfrak{h})$ the $\bar{Y}_i$ generate $\mathfrak{g}_0 \cap \mathfrak{n}^-$. We conclude via Lemma 5.5′ (iii) that $(-\delta^*, 1) - (\gamma, 0) \notin \tilde{\Delta}^+$ for all $(\gamma, 0) \in \Delta_0^+$, so $(-\delta^*, 1)$ is simple.

Since the highest root $\bar{\delta}$ of $\mathfrak{b}_n$ is given by $\bar{\alpha}_1 + 2\bar{\alpha}_2 + ... + 2\bar{\alpha}_n$, we have because of the connection between $\alpha_j \mid \mathfrak{h}$ and $\bar{\alpha}_i$, that $\delta^* = \bar{\alpha}_1 + \bar{\delta}$. Also $\langle \bar{\alpha}_1, \bar{\delta} \rangle = 0$, so $\langle -\delta^*, \bar{\alpha}_1 \rangle \neq 0$ which implies that $S(A)$ is obtained by joining δ^* to $\bar{\alpha}_1$ of $\mathfrak{b}_n$; since, in addition, $\mid \delta^* \mid > \mid \bar{\alpha}_1 \mid$, $S(A)$ has to be $\mathfrak{a}_{2n}^{(2)}$. It follows that case (3) gives $S(A) = \mathfrak{d}_{n+1}^{(2)}$. Case (5) must give $S(A) = \mathfrak{d}_4^{(3)}$; case (4) must have $S(A) = \mathfrak{e}_6^{(2)}$; and now it is clear that case (2) gives $S(A) = \mathfrak{a}_{2n-1}^{(2)}$.

It remains to verify the numerical labels in Tables 2–3. For this we use the explicit determination of the matrix A from $S(A)$ explained above. For example, if $S(A) = \mathfrak{d}_4^{(3)}$, we get

$$A = \begin{pmatrix} 2 & -3 & 0 \\ -1 & 2 & -1 \\ 0 & -1 & 2 \end{pmatrix},$$

and the numerical labels are quickly verified.

We can now summarize the treatment of Examples 1 and 2 as follows.

Lemma 5.12. *Let $\mathfrak{g}$ be a simple Lie algebra over $\mathbf{C}$, ν an automorphism of $\mathfrak{g}$ of order k induced by an automorphism of the Dynkin diagram. Then $k = 1, 2,$ or 3 and the diagrams $S(A)$ of the various $L(\mathfrak{g}, \nu)$ exhaust all the diagrams in Tables* 1, 2, *and* 3.

Remark. In this lemma and in the rest of this section "an automorphism ν of $\mathfrak{g}$ induced by an automorphism of the Dynkin diagram" means the specific choice made in Examples 1 and 2.

Let $L(\mathfrak{g}, \nu)$ and k be as in Lemma 5.11. Since $k = 1, 2,$ or 3, the eigenspace $\mathfrak{g}_1$ is $\neq 0$. We recall that $\mathfrak{g}_0$ is simple. If $\beta \in \Delta(\mathfrak{g}_0, \mathfrak{h})$ is simple, then $\tilde{\beta} = (\beta, 0)$ is a simple root of $L(\mathfrak{g}, \nu)$ with respect to $\mathfrak{h}$. Let $\tilde{\alpha}_0$ be the lowest root of the form $(\alpha_0, 1)$ (such exist since the action of ad $\mathfrak{h}$ on $\mathfrak{g}_1$ can be diagonalized). Since no decomposition $(\alpha_0, 1) = (\beta, 1) + (\gamma, 0)$ into elements in $\tilde{\Delta}^+$ is possible, $\tilde{\alpha}_0$ is simple. Thus the simple elements in $\tilde{\Delta}^+$ are

$$\tilde{\alpha}_0, \ldots, \tilde{\alpha}_n, \qquad \text{where} \quad \tilde{\alpha}_0 = (\alpha_0, 1), \quad \tilde{\alpha}_i = (\alpha_i, 0) \quad (1 \leqslant i \leqslant n).$$

Now let $s_0, \ldots, s_n$ be a sequence of nonnegative integers, not all 0. We define a new $\mathbf{Z}$-gradation $L(\mathfrak{g}, \nu) = \bigoplus_{j \in \mathbf{Z}} L_j$ as follows. Writing a root $\tilde{\alpha} = \sum_0^n k_i \tilde{\alpha}_i$ (where by Lemma 5.9 (ii) the $k_i \in \mathbf{Z}$ are unique), let $\deg \tilde{\alpha} = \sum_i k_i s_i$ and put $L(\mathfrak{g}, \nu)_j = L_j = \sum_{\deg \tilde{\alpha} = j} L^{\tilde{\alpha}}$. This gradation is called a *gradation of type* $(s_0, \ldots, s_n)$.

Theorem 5.13. *Let $\mathfrak{g}$ be a simple Lie algebra over $\mathbf{C}$ and σ an automorphism of finite order. Then there exists an automorphism ν of $\mathfrak{g}$ induced by an automorphism of the Dynkin diagram and a $\mathbf{Z}$-gradation of $L(\mathfrak{g}, \nu)$ of type $(s_0, \ldots, s_n)$ in which $L(\mathfrak{g}, \nu)$ is isomorphic to the $\mathbf{Z}$-graded Lie algebra $L(\mathfrak{g}, \sigma)$ by an isomorphism under which the two $\mathbf{Z}$-gradations correspond.*

Proof. Because of Lemma 5.12, we can choose ν such that if $\tilde{\beta} = (\beta_0, s_0), \ldots, \tilde{\beta}_n = (\beta_n, s_n)$ are the simple roots of $L(\mathfrak{g}, \sigma)$ (in a suitable order), then under the bijection $\tilde{\beta}_i \to \tilde{\alpha}_i$ $(0 \leqslant i \leqslant n)$ the Cartan matrices of $L(\mathfrak{g}, \sigma)$ and of $L(\mathfrak{g}, \nu)$ coincide. We have by (7)

$$L(\mathfrak{g}, \sigma) = \bigoplus_{j \in \mathbf{Z}} L(\mathfrak{g}, \sigma)_j, \qquad L(\mathfrak{g}, \sigma)_j = \bigoplus_{\tilde{\beta}} L(\mathfrak{g}, \sigma)^{\tilde{\beta}}, \tag{10}$$

where $\tilde{\beta}$ runs over all the roots $\Sigma_0^n k_i\tilde{\beta}_i$ for which $\Sigma_0^n k_i s_i = j$. Also,

$$L(\mathfrak{g}, \nu) = \bigoplus_{j \in \boldsymbol{Z}} L(\mathfrak{g}, \nu)_j, \qquad L(\mathfrak{g}, \nu)_j = \bigoplus_{\deg \tilde{\alpha} = j} L(\mathfrak{g}, \nu)^{\tilde{\alpha}}. \tag{11}$$

The isomorphism $\tilde{\psi} : L(\mathfrak{g}, \sigma) \to L(\mathfrak{g}, \nu)$ from Lemma 5.10 (i) maps $L(\mathfrak{g}, \sigma)^{\tilde{\beta}}$ onto $L(\mathfrak{g}, \nu)^{\tilde{\alpha}}$, provided $\tilde{\beta} = \Sigma_0^n k_i\tilde{\beta}_i$, $\tilde{\alpha} = \Sigma_0^n k_i\tilde{\alpha}_i$, and the $\boldsymbol{Z}$-gradations (10) and (11) therefore correspond under $\tilde{\psi}$.

Lemma 5.14. *Let* $\mathfrak{g}$, ν, *and* k *be as in Lemma* 5.12, *and let* $a_0, ..., a_n$ *be the numerical labels in Table k corresponding to the simple roots* $\tilde{\alpha}_0, ..., \tilde{\alpha}_n$ *defined above. Let* $L(\mathfrak{g}, \nu)$ *be equipped with a gradation of type* $(s_0, ..., s_n)$. *Then*

$$x^k L_j \subset L_{j+m}, \qquad \text{where} \quad m = k \sum_0^n a_i s_i.$$

Proof. By definition of the labels a_i we have $\Sigma_0^n a_i\alpha_i = 0$. We note also that in all cases in the tables, $a_0 = 1$. For all roots $\tilde{\alpha} = (\alpha, j)$,

$$L^{\tilde{\alpha}} = \{x^j X : X \in \mathfrak{g}_{j \bmod k}, [H, X] = \alpha(H)X \text{ for } H \in \mathfrak{h}\},$$

so $x^k L^{\tilde{\alpha}} \subset L^{\tilde{\alpha}+(0,k)}$. Since the root $(0, k)$ can be written

$$(0, k) = k \sum_0^n a_i\tilde{\alpha}_i,$$

we deduce $\deg(0, k) = k\, \Sigma_0^n a_i s_i$ and the lemma follows.

Now we state the main theorem of this section.

Theorem 5.15. *Let* $\mathfrak{g}$ *be a simple Lie algebra over* $\boldsymbol{C}$, ν *a fixed automorphism of* $\mathfrak{g}$ *of order* k $(k = 1, 2, 3)$ *induced by an automorphism of the Dynkin diagram for a Cartan subalgebra* $\tilde{\mathfrak{h}}$ *of* $\mathfrak{g}$. *Let* $\mathfrak{g} = \bigoplus_i \mathfrak{g}_i^\nu$ *be the corresponding* $\boldsymbol{Z}_k$*-gradation. The fixed point set* $\mathfrak{h}^\nu$ *of* ν *in* $\tilde{\mathfrak{h}}$ *is a Cartan subalgebra of the (simple) Lie algebra* $\mathfrak{g}_0^\nu$. *Fix canonical generators* X_i, Y_i, H_i $(1 \leqslant i \leqslant n)$ *of* $\mathfrak{g}_0^\nu$ *corresponding to the simple roots* $\alpha_1, ..., \alpha_n$ *in* $\Delta(\mathfrak{g}_0^\nu, \mathfrak{h}^\nu)$. *Let* $\tilde{\alpha}_0$ *be the lowest root of* $L(\mathfrak{g}, \nu)$ *of the form* $(\alpha_0, 1)$ *and fix* $X_0 \neq 0$ *in* $\mathfrak{g}_1^\nu$ *such that* $xX_0 \in L(\mathfrak{g}_0, \nu)^{\tilde{\alpha}_0}$. *Let* $(s_0, ..., s_n)$ *be integers* $\geqslant 0$ *without nontrivial common factor, put* $m = k\, \Sigma_0^n a_i s_i$ *where the* a_i *are the labels from the diagram of* $L(\mathfrak{g}, \nu)$ *corresponding to the simple roots* $\tilde{\alpha}_0$, $\tilde{\alpha}_i = (\alpha_i, 0)$, $(1 \leqslant i \leqslant n)$. *Let* ϵ *be a fixed m-th root of unity. Then*:

(i) *The vectors* $X_0, X_1, ..., X_n$ *generate* $\mathfrak{g}$ *and the relations*

$$\sigma(X_i) = \epsilon^{s_i} X_i \qquad (0 \leqslant i \leqslant n)$$

define uniquely an automorphism of $\mathfrak{g}$ *of order m. It will be called an automorphism of type* $(s_0, ..., s_n; k)$.

(ii) *Let* $i_1, \ldots, i_t$ *be all the indices for which* $s_{i_1} = \ldots = s_{i_t} = 0$. *Then* $\mathfrak{g}_0$ (*that is,* $\mathfrak{g}_0^\sigma$) *is the direct sum of an* $(n - t)$-*dimensional center and a semisimple Lie algebra whose Dynkin diagram is the subdiagram of the diagram* $\mathfrak{g}^{(k)}$ *in Table k consisting of the vertices* $i_1, \ldots, i_t$.

(iii) *Except for conjugation, the automorphisms* σ *exhaust all m-th order automorphisms of* $\mathfrak{g}$.

Remark. Note that if $m > 1$, then $m > s_i$ $(0 \leqslant i \leqslant n)$ because if $s_i = m$ for some i, then $s_j = 0$ $(j \neq i)$, so m is a nontrivial common factor.

Proof. (i) To see that the X_i generate $\mathfrak{g}$ consider the covering mapping $\varphi : L(\mathfrak{g}, \nu) \to \mathfrak{g}$ and let $P = \sum_{j=1}^k x^j \mathfrak{g}_{j \bmod k}$. Then $\varphi(P) = \mathfrak{g}$. By the proof of Lemma 5.10 (i) the elements $e_0 = xX_0$, $e_1 = X_1, \ldots, e_n = X_n$ generate the subalgebra $L(\mathfrak{g}, \nu)^+ = \bigoplus_{\tilde{\alpha}>0} L(\mathfrak{g}, \nu)^{\tilde{\alpha}}$. But $L(\mathfrak{g}, \nu)^+ \supset P$ so $\varphi(L(\mathfrak{g}, \nu)^+) = \mathfrak{g}$ and $X_0, X_1, \ldots, X_n$ generate $\mathfrak{g}$.

Now let $\tilde{\sigma}$ be the well-defined and unique automorphism of $L(\mathfrak{g}, \nu)$ determined by

$$\tilde{\sigma}(e_{\tilde{\alpha}}) = \epsilon^{\sum k_i s_i} e_{\tilde{\alpha}} \qquad \text{if} \quad \tilde{\alpha} = \sum_0^n k_i \tilde{\alpha}_i, \quad e_{\tilde{\alpha}} \in L(\mathfrak{g}, \nu)^{\tilde{\alpha}}.$$

If $L(\mathfrak{g}, \nu) = \bigoplus_{j \in \boldsymbol{Z}} L_j$ is the $\boldsymbol{Z}$-gradation of type $(s_0, \ldots, s_n)$, then L_j belongs to the eigenspace of $\tilde{\sigma}$ with eigenvalue ϵ^j. But by Lemma 5.14, $x^k L_j = L_{j+m}$ which belongs to the same eigenspace of $\tilde{\sigma}$, so the ideal $(1 - x^k) L(\mathfrak{g}, \nu)$ is $\tilde{\sigma}$-invariant. Hence $\tilde{\sigma}$ induces an automorphisms σ of the quotient algebra $\mathfrak{g}$ with the stated property. Note that since $\sigma(Y_j) = \epsilon^{-s_j} Y_j$, the Cartan subalgebra $\mathfrak{h}^\nu$ is left pointwise fixed by σ.

For part (ii) we consider the $\boldsymbol{Z}_m$-gradation $\mathfrak{g} = \bigoplus_{i \in \boldsymbol{Z}_m} \mathfrak{g}_i$ defined by σ. By the above we have for each $r \in \boldsymbol{Z}$

$$\varphi(L_{j+rm}) = \varphi(x^{rk} L_j) = \varphi(L_j),$$

so $\varphi(L_j) = \mathfrak{g}_{j \bmod m}$. Also $L_j \cap (1 - x^k) L(\mathfrak{g}, \nu) = 0$, so φ gives an isomorphism of L_j onto $\mathfrak{g}_{j \bmod m}$. In particular,

$$\mathfrak{g}_0 \approx \bigoplus_{\deg \tilde{\alpha}=0} L(\mathfrak{g}, \nu)^{\tilde{\alpha}}.$$

But $\deg \tilde{\alpha} = 0$ if and only if $\tilde{\alpha} = \sum_{r=1}^t k_{i_r} \tilde{\alpha}_{i_r}$. As noted during the proof of Lemma 5.10, if $\tilde{\alpha} > 0$, then $L(\mathfrak{g}, \nu)^{\tilde{\alpha}}$ is spanned by the commutators $[e_{j_1}, \ldots, e_{j_s}]$ satisfying $\tilde{\alpha}_{j_1} + \ldots + \tilde{\alpha}_{j_s} = \tilde{\alpha}$. Hence the algebra

$$\bigoplus_{\deg \tilde{\alpha}=0} L(\mathfrak{g}, \nu)^{\tilde{\alpha}}$$

is generated by H_i, e_{i_r}, f_{i_r} $(1 \leqslant i \leqslant n,\ 1 \leqslant r \leqslant t)$. Putting $Y_j = \varphi(f_j)$, the vectors X_{i_r}, Y_{i_r} $(1 \leqslant r \leqslant t)$ generate the semisimple part of $\mathfrak{g}_0$; its Cartan matrix is the submatrix $(a_{i_k i_l})_{1 \leqslant k, l \leqslant t}$ of the Cartan matrix (a_{ij}) of $L(\mathfrak{g}, \nu)$. The center of $\mathfrak{g}_0$ is spanned by the vectors

$$H_j - \sum_{k=1}^{t} c_{jk} H_{i_k} \qquad (j \neq i_1, ..., i_t)$$

where the c_{jk} are (by Lemma 5.9(i)) determined by

$$\sum_{k=1}^{t} c_{jk} a_{i_l i_k} = a_{i_l j}.$$

This proves (ii).

Finally, for part (iii) let τ be an automorphism of $\mathfrak{g}$ of order m. With the given primitive mth root ϵ, τ gives a Z_m-gradation $\mathfrak{g} = \bigoplus_i \mathfrak{g}_i$. Let $\varphi' : L(\mathfrak{g}, \tau) \to \mathfrak{g}$ by the covering map, which of course satisfies $\varphi'(L(\mathfrak{g}, \tau)_j) = \mathfrak{g}_{j \bmod m}$. Consider now $L(\mathfrak{g}, \nu)$ with the Z-gradation of type $(s_0, ..., s_n)$ from Theorem 5.13 with the resulting isomorphism between $L(\mathfrak{g}, \tau)$ and $L(\mathfrak{g}, \nu)$. From Lemma 5.10 (ii) (noting that the automorphism $\mu_c : L(\mathfrak{g}, \nu) \to L(\mathfrak{g}, \nu)$ preserves each $L(\mathfrak{g}, \nu)^{\tilde{\alpha}}$), we then deduce that for a suitable automorphism ψ of $\mathfrak{g}$

$$\psi(\varphi'(L(\mathfrak{g}, \tau)_j)) = \varphi\Big(\bigoplus_{\deg \tilde{\alpha} = j} L(\mathfrak{g}, \nu)^{\tilde{\alpha}}\Big).$$

Thus ψ maps the ϵ^j-eigenspace of τ onto the ϵ^j-eigenspace of σ of type $(s_0, ..., s_n; k)$ constructed in (i). Hence $\psi\tau\psi^{-1} = \sigma$. Finally, the s_i have no common factor $a > 1$ because $(0, m)$ being a root of $L(\mathfrak{g}, \tau)$, m is an integral linear combination of the s_i, so a would divide m. This would contradict σ being of order m. This concludes the proof.

Theorem 5.16. *With the notation of Theorem* 5.15 *let σ be an automorphism of type $(s_0, ..., s_n; k)$. Then:*

(i) *σ is an inner automorphism if and only if $k = 1$.*

(ii) *If σ' is an automorphism of type $(s'_0, ..., s'_n; k')$, then σ and σ' are conjugate within* $\mathrm{Aut}(\mathfrak{g})$ *if and only if $k = k'$ and the sequence $(s_0, ..., s_n)$ can be transformed into the sequence $(s'_0, ..., s'_n)$ by an automorphism ψ_0 of the diagram $\mathfrak{g}^{(k)}$.*

Proof. (i) If $k = 1$, then $\nu = I$, so σ leaves pointwise fixed the Cartan subalgebra $\mathfrak{h}^\nu$ of $\mathfrak{g}$. By Exercise C.3, Chapter IX, σ is inner. On the other hand, if $k > 1$, then $\mathrm{rank}(\mathfrak{g}_0^\sigma) = \dim \mathfrak{h}^\nu < \mathrm{rank}(\mathfrak{g})$, whereas if $\sigma \in \mathrm{Int}(\mathfrak{g})$, then σ leaves elementwise fixed a Cartan subalgebra $\tilde{\mathfrak{h}}$ of $\mathfrak{g}$. (For example, by Theorem 2.1, Chapter VI, σ lies in a maximal compact subgroup U of Int($\mathfrak{g}$), so $\mathfrak{h}$ can be taken as the complexification of the Lie algebra of a maximal torus of U, containing σ.)

For part (ii) suppose first $k = k'$ and that $(s_0, ..., s_n)$ and $(s_0', ..., s_n')$ correspond under ψ_0. By Lemma 5.10, ψ_0 induces an automorphism $\tilde{\psi}$ of $L(\mathfrak{g}, \nu)$ satisfying $\tilde{\psi}\tilde{\sigma}\tilde{\psi}^{-1} = \tilde{\sigma}'$. But the automorphisms μ_c of $L(\mathfrak{g}, \nu)$ and ψ of $\mathfrak{g}$ from Lemma 5.10 (ii) satisfy $\varphi \circ \mu_c \circ \tilde{\psi} = \psi \circ \varphi$ and $\mu_c\tilde{\sigma}' = \tilde{\sigma}'\mu_c$. Since $\varphi \circ \tilde{\sigma} = \sigma \circ \varphi$, $\varphi \circ \tilde{\sigma}' = \sigma' \circ \varphi$, we deduce

$$\begin{aligned}\psi^{-1}\sigma'\psi\varphi &= \psi^{-1}\sigma'\varphi\mu_c\tilde{\psi} = \psi^{-1}\varphi\tilde{\sigma}'\mu_c\tilde{\psi} \\ &= \psi^{-1}\varphi\mu_c\tilde{\psi}\tilde{\sigma} = \varphi\tilde{\sigma} = \sigma\varphi,\end{aligned}$$

so $\sigma' = \psi\sigma\psi^{-1}$ as claimed.

For the converse we need a lemma.

Lemma 5.17. *Let $\bar{\nu}$ be the automorphism given by Theorem 5.13 for which $L(\mathfrak{g}, \sigma)$ and $L(\mathfrak{g}, \bar{\nu})$ have the same diagram $S(A)$. Then $\bar{\nu} = \nu$.*

Proof. We know that $\operatorname{rank}(\mathfrak{g}_0^\sigma) = \dim \mathfrak{h}^\nu$, so $\dim \mathfrak{h}^\nu = \dim \mathfrak{h}^{\bar{\nu}}$. But then Table I (for Example 2) shows that $\nu = \bar{\nu}$.

Turning to Theorem 5.16 suppose σ and σ' are conjugate. By Lemma 5.17 we have the isomorphisms $L(\mathfrak{g}, \sigma) \sim L(\mathfrak{g}, \nu)$, $L(\mathfrak{g}, \sigma') \sim L(\mathfrak{g}, \nu')$ (σ' having been defined by means of ν'). Hence $L(\mathfrak{g}, \nu)$ and $L(\mathfrak{g}, \nu')$ have the same diagram, so $k = k'$ and $\nu = \nu'$. Suppose $\tau\sigma\tau^{-1} = \sigma'$ ($\tau \in \operatorname{Aut}(\mathfrak{g})$) and let $\mathfrak{g} = \bigoplus_i \mathfrak{g}_i$, $\mathfrak{g} = \bigoplus_i \mathfrak{g}_i'$ be the $\boldsymbol{Z}_m$-gradations given by σ and σ', respectively. Then $\tau\mathfrak{g}_i = \mathfrak{g}_i'$. The algebra $\mathfrak{h} = \mathfrak{h}^\nu = \mathfrak{h}^{\nu'}$ is maximal abelian in $\mathfrak{g}_0$ and $\mathfrak{g}_0'$, and $\mathfrak{h}$ and $\tau(\mathfrak{h})$ are conjugate under a member of $\operatorname{Int}(\mathfrak{g}_0') \subset \operatorname{Int}(\mathfrak{g})$. Thus, replacing τ by $\tau_1\tau$ where τ_1 is suitably chosen in $\operatorname{Int}(\mathfrak{g}_0')$ (and thus commutes with σ'), we may assume that $\tau(\mathfrak{h}) = \mathfrak{h}$ and that the positive roots in $\Delta([\mathfrak{g}_0, \mathfrak{g}_0], \mathfrak{h} \cap [\mathfrak{g}_0, \mathfrak{g}_0])$ and in $\Delta([\mathfrak{g}_0', \mathfrak{g}_0'], \mathfrak{h} \cap [\mathfrak{g}_0', \mathfrak{g}_0'])$ correspond under τ. The extension $\tilde{\tau}$ of τ to $L(\mathfrak{g}, \sigma)$ given by $\tilde{\tau}(x^jY) = x^j\tau(Y)$ maps $L(\mathfrak{g}, \sigma)^{(\alpha, j)}$ onto $L(\mathfrak{g}, \sigma')^{(\alpha^\tau, j)}$ (with the same j since $\tau\mathfrak{g}_j = \mathfrak{g}_j'$; we have written α^τ for ${}^t\tau^{-1}\alpha$). Thus the simple roots $(\alpha_0, s_0), ..., (\alpha_n, s_n)$ of $L(\mathfrak{g}, \sigma)$ and the simple roots $(\alpha_0', s_0'), ..., (\alpha_n', s_n')$ of $L(\mathfrak{g}, \sigma')$ correspond under τ. Hence the sequences $(s_0, ..., s_n)$ and $(s_0', ..., s_n')$ correspond under an automorphism of the diagram $\mathfrak{g}^{(k)}$ of $L(\mathfrak{g}, \sigma) \sim L(\mathfrak{g}, \nu)$. This concludes the proof of Theorem 5.16.

We now illustrate Theorem 5.15 by a classification of all the automorphisms of order 2 up to conjugacy (problem B'' in §1).

The equation $m = k \sum_{i=0}^n a_i s_i = 2$ has the following solutions:

(A_1) $k = 1$, $a_{i_0} = 2$, $s_{i_0} = 1$, $s_i = 0$ for $i \neq i_0$.

(A_2) $k = 2$, $a_{i_0} = 1$, $s_{i_0} = 1$, $s_i = 0$ for $i \neq i_0$.

(B) $k = 1$, $a_{i_0} = a_{i_1} = 1$, $s_{i_0} = s_{i_1} = 1$, $s_i = 0$ for $i \neq i_0, i_1$.

In cases A_1, A_2, $\mathfrak{g}_0$ is semisimple, whereas in case B, $\mathfrak{g}_0$ has a one-dimensional center. Note that $(s_0, ..., s_n) = (2, 0, ..., 0)$ does not qualify as a solution because the s_i must not have any common factor.

Now we use Theorem 5.15 (ii) to write down $\mathfrak{g}_0$ for each case by means of the diagrams for $S(A)$.

Case A_1. We look for the label $a_{i_0} = 2$ in Table I. For $\mathfrak{a}_n^{(1)}$ there is no such label; for $\mathfrak{b}_n^{(1)}$ there are $n-1$ choices giving the possibilities $\mathfrak{d}_p \oplus \mathfrak{b}_{n-p}$ $(2 \leqslant p \leqslant n)$ for $\mathfrak{g}_0$. For $\mathfrak{c}_n^{(1)}$ there are again $n-1$ possibilities; however this diagram has an automorphism of order 2, so by Theorem 5.16 the possibilities $(s_0, ..., s_n) = (0, 1, ..., 0)$ and $(s_0, ..., s_n) = (0, 0, ..., 1, 0)$ give conjugate automorphisms, etc. Thus the possibilities for $\mathfrak{g}_0$ are restricted to $\mathfrak{c}_p \oplus \mathfrak{c}_{n-p}$ $(1 \leqslant p \leqslant [\frac{1}{2}n])$. Similar considerations apply to $\mathfrak{d}_n^{(1)}$; $\mathfrak{e}_6^{(1)}$ has three labels equal to 2, but by symmetry the corresponding automorphisms are all conjugate. Similarly for $\mathfrak{e}_7^{(1)}$.

Case A_2. For $\mathfrak{a}_{2n}^{(2)}$ there is only one possibility; for $\mathfrak{a}_{2n-1}^{(2)}$ there are three but the two for which $\mathfrak{g}_0 = \mathfrak{c}_n$ give conjugate automorphisms.

TABLE II

($\mathfrak{g}_0$ SEMISIMPLE)

$k = 1$		$k = 2$	
$\mathfrak{g}$	$\mathfrak{g}_0$	$\mathfrak{g}$	$\mathfrak{g}_0$
$\mathfrak{b}_n$ $(n > 2)$	$\mathfrak{d}_p \oplus \mathfrak{b}_{n-p}$ $(2 \leqslant p \leqslant n)$	$\mathfrak{a}_{2n}$ $(n \geqslant 1)$	$\mathfrak{b}_n$
$\mathfrak{c}_n$ $(n > 1)$	$\mathfrak{c}_p \oplus \mathfrak{c}_{n-p}$ $(1 \leqslant p \leqslant [\frac{1}{2}n])$	$\mathfrak{a}_{2n-1}$ $(n > 2)$	$\mathfrak{d}_n$
		$\mathfrak{a}_{2n-1}$ $(n > 2)$	$\mathfrak{c}_n$
$\mathfrak{d}_n$ $(n > 3)$	$\mathfrak{d}_p \oplus \mathfrak{d}_{n-p}$ $(2 \leqslant p \leqslant [\frac{1}{2}n])$	$\mathfrak{d}_{n+1}$ $(n > 1)$	$\mathfrak{b}_p \oplus \mathfrak{b}_{n-p}$ $(0 \leqslant p \leqslant [\frac{1}{2}n])$
$\mathfrak{g}_2$	$\mathfrak{a}_1 \oplus \mathfrak{a}_1$	$\mathfrak{e}_6$	$\mathfrak{c}_4$
$\mathfrak{f}_4$	$\mathfrak{b}_4$	$\mathfrak{e}_6$	$\mathfrak{f}_4$
$\mathfrak{f}_4$	$\mathfrak{a}_1 \oplus \mathfrak{c}_3$		
$\mathfrak{e}_6$	$\mathfrak{a}_1 \oplus \mathfrak{a}_5$		
$\mathfrak{e}_7$	$\mathfrak{a}_7$		
$\mathfrak{e}_7$	$\mathfrak{a}_1 \oplus \mathfrak{d}_6$		
$\mathfrak{e}_8$	$\mathfrak{a}_1 \oplus \mathfrak{e}_7$		
$\mathfrak{e}_8$	$\mathfrak{d}_8$		

Case B. For $\mathfrak{b}_n^{(1)}$, $\mathfrak{c}_n^{(1)}$, and $\mathfrak{e}_7$ the choice is unique. Because $\mathfrak{a}_n^{(1)}$ has rotational as well as axial symmetries only the listed possibilities for $\mathfrak{g}_0$ give nonconjugate automorphisms. All three possibilities for $\mathfrak{e}_6^{(1)}$ give conjugate automorphisms. For $\mathfrak{d}_n^{(1)}$ $(n > 4)$ there are two nonconjugate possibilities, but for $n = 4$ they coincide by virtue of Theorem 5.16.

We have now arrived at Tables II, III.

Remark. Our discussion showed that all the automorphisms giving rise to the same entry $(\mathfrak{g}, \mathfrak{g}_0)$ in Tables II, III are conjugate.

TABLE III

$(\dim(\text{center}(\mathfrak{g}_0)) = 1)$

$\mathfrak{g}$	$[\mathfrak{g}_0, \mathfrak{g}_0]$	$\mathfrak{g}$	$[\mathfrak{g}_0, \mathfrak{g}_0]$
$\mathfrak{a}_n$	$\mathfrak{a}_p \oplus \mathfrak{a}_{n-p-1}$	$\mathfrak{d}_n (n > 3)$	$\mathfrak{d}_{n-1}$
$(n > 1)$	$0 \leqslant p \leqslant [\frac{1}{2}(n-1)]$	$\mathfrak{d}_n (n > 4)$	$\mathfrak{a}_{n-1}$
$\mathfrak{b}_n (n > 2)$	$\mathfrak{b}_{n-1}$	$\mathfrak{e}_6$	$\mathfrak{d}_5$
$\mathfrak{c}_n (n > 1)$	$\mathfrak{a}_{n-1}$	$\mathfrak{e}_7$	$\mathfrak{e}_6$

§6. The Classifications

We can now give the solutions to the classification problems A, A′, B, B′, B″, and C stated in §1. We start with A, A′, and C. Table IV below follows from Theorem 3.23, 4.15, 3.32, Lemma 3.30 and (12), §3, as well the computation of the Lie algebras of $\boldsymbol{SU}(n)$, $\boldsymbol{SO}(n)$, and $\boldsymbol{Sp}(n)$ in §2.

1. The Simple Lie Algebras over C and Their Compact Real Forms. The Irreducible Riemannian Globally Symmetric Spaces of Type II and Type IV

In Table IV $\mathfrak{g}$ runs over all simple Lie algebras over $\boldsymbol{C}$; the subscript denotes the *rank* of $\mathfrak{g}$, that is, the dimension of a Cartan subalgebra. Moreover, G stands for a connected Lie group with Lie algebra $\mathfrak{g}^R$, U is an analytic subgroup of G whose Lie algebra is a compact real form of $\mathfrak{g}$. By Chapter VI, §2, U is a maximal compact subgroup of G. Let $\tilde{U}$ denote the universal covering group of U, $Z(\tilde{U})$ the center of $\tilde{U}$, and $\boldsymbol{Z}_p$ a cyclic group of order p. The dimension of U is also listed.

The five last are called the exceptional structures. The first four classes $\mathfrak{a}_n$, $\mathfrak{b}_n$, $\mathfrak{c}_n$, $\mathfrak{d}_n$ (the classical structures) are of course defined for all $n \geqslant 1$, but then the following isomorphisms occur (see Theorem 3.12) and $\mathfrak{d}_1$ is not semisimple.

$$\mathfrak{a}_1 = \mathfrak{b}_1 = \mathfrak{c}_1, \qquad \mathfrak{b}_2 = \mathfrak{c}_2, \qquad \mathfrak{a}_3 = \mathfrak{d}_3, \qquad \mathfrak{d}_2 = \mathfrak{a}_1 \times \mathfrak{a}_1. \tag{1}$$

With the restriction on the indices in the table each simple Lie algebra $\mathfrak{g}$ over $\boldsymbol{C}$ occurs exactly once.

TABLE IV

LIE GROUPS FOR THE SIMPLE LIE ALGEBRAS OVER $\boldsymbol{C}$ AND THEIR COMPACT REAL FORMS

$\mathfrak{g}$	G	U	$Z(\tilde{U})$	dim U
$\mathfrak{a}_n (n \geqslant 1)$	$\boldsymbol{SL}(n+1, \boldsymbol{C})$	$\boldsymbol{SU}(n+1)$	$\boldsymbol{Z}_{n+1}$	$n(n+2)$
$\mathfrak{b}_n (n \geqslant 2)$	$\boldsymbol{SO}(2n+1, \boldsymbol{C})$	$\boldsymbol{SO}(2n+1)$	$\boldsymbol{Z}_2$	$n(2n+1)$
$\mathfrak{c}_n (n \geqslant 3)$	$\boldsymbol{Sp}(n, \boldsymbol{C})$	$\boldsymbol{Sp}(n)$	$\boldsymbol{Z}_2$	$n(2n+1)$
$\mathfrak{d}_n (n \geqslant 4)$	$\boldsymbol{SO}(2n, \boldsymbol{C})$	$\boldsymbol{SO}(2n)$	$\boldsymbol{Z}_4$ if $n =$ odd	$n(2n-1)$
			$\boldsymbol{Z}_2 + \boldsymbol{Z}_2$ if $n =$ even	
$\mathfrak{e}_6$	$\boldsymbol{E}_6^{\boldsymbol{C}}$	$\boldsymbol{E}_6$	$\boldsymbol{Z}_3$	78
$\mathfrak{e}_7$	$\boldsymbol{E}_7^{\boldsymbol{C}}$	$\boldsymbol{E}_7$	$\boldsymbol{Z}_2$	133
$\mathfrak{e}_8$	$\boldsymbol{E}_8^{\boldsymbol{C}}$	$\boldsymbol{E}_8$	$\boldsymbol{Z}_1$	248
$\mathfrak{f}_4$	$\boldsymbol{F}_4^{\boldsymbol{C}}$	$\boldsymbol{F}_4$	$\boldsymbol{Z}_1$	52
$\mathfrak{g}_2$	$\boldsymbol{G}_2^{\boldsymbol{C}}$	$\boldsymbol{G}_2$	$\boldsymbol{Z}_1$	14

Using now Theorem 5.4, Chapter VIII, and Theorem 1.1, Chapter VI, we have:

The Riemannian globally symmetric spaces of type IV are the spaces G/U where G is a connected Lie group whose Lie algebra is $\mathfrak{g}^R$ where $\mathfrak{g}$ is a simple Lie algebra over $\boldsymbol{C}$, and U is a maximal compact subgroup of G. The metric on G/U is G-invariant and is uniquely determined (up to a factor) by this condition.

Secondly, in view of Prop. 1.2:

The Riemannian globally symmetric spaces of type II are the simple, compact, connected Lie groups U. The metric on U is two-sided invariant and is uniquely determined (up to a factor) by this condition.

2. The Real Forms of Simple Lie Algebras over $\boldsymbol{C}$. Irreducible Riemannian Globally Symmetric Spaces of Type I and Type IV

We now give the solutions to problems B, B′, and B″ in §1, the last of which is solved by Tables II and III at the end of §5. We state the result in Table V in terms of the classical groups discussed in §2.

Looking at Tables II and III we notice that in spite of the isomorphisms (1) the same pair $(\mathfrak{g}, \mathfrak{g}_0)$ does not occur twice. Taking the remark at the end of §5 into account we therefore deduce

Theorem 6.1. *Two involutive automorphisms of a simple Lie algebra* $\mathfrak{g}$ *over* $\boldsymbol{C}$ *are conjugate in* $\mathrm{Aut}(\mathfrak{g})$ *if and only if their fixed point algebras are isomorphic.*

Passing to the real forms, we can now prove the following result, already mentioned in Chapter IX, §5.

Theorem 6.2. *Suppose* $\mathfrak{g}_1$ *and* $\mathfrak{g}_2$ *are real forms of the same simple Lie algebra* $\mathfrak{g}$ *over* $\boldsymbol{C}$. *Let* $\mathfrak{g}_1 = \mathfrak{k}_1 + \mathfrak{p}_1$ *and* $\mathfrak{g}_2 = \mathfrak{k}_2 + \mathfrak{p}_2$ *be any Cartan decompositions. Then if* $\mathfrak{k}_1$ *and* $\mathfrak{k}_2$ *are isomorphic,* $\mathfrak{g}_1$ *and* $\mathfrak{g}_2$ *are isomorphic.*

Proof. Consider the compact real forms $\mathfrak{u}_1 = \mathfrak{k}_1 + i\mathfrak{p}_1$, $\mathfrak{u}_2 = \mathfrak{k}_2 + i\mathfrak{p}_2$. Let τ be an automorphism of $\mathfrak{g}$ such that $\tau\mathfrak{u}_2 = \mathfrak{u}_1$ and put $\mathfrak{g}_1' = \tau\mathfrak{g}_2$, $\mathfrak{k}_1' = \tau\mathfrak{k}_2$. Let s_1 and s_1' denote the involutions of $\mathfrak{u}_1$ with fixed points $\mathfrak{k}_1$ and $\mathfrak{k}_1'$, respectively. Because of Theorem 6.1 and Prop. 1.4, s_1 and s_1' are conjugate in $\mathrm{Aut}(\mathfrak{u}_1)$; hence by Prop. 2.2, Chapter V, $\mathfrak{g}_1$ and $\mathfrak{g}_1'$, and therefore $\mathfrak{g}_1$ and $\mathfrak{g}_2$, are isomorphic.

Let $\mathfrak{l}$ be a semisimple Lie algebra over $\boldsymbol{C}$, $\mathfrak{l}_0$ a real form of $\mathfrak{l}$. The *character* of $\mathfrak{l}_0$ is defined as $\delta = \dim \mathfrak{p}_0 - \dim \mathfrak{k}_0$, $\mathfrak{l}_0 = \mathfrak{k}_0 + \mathfrak{p}_0$ being a Cartan decomposition of $\mathfrak{l}_0$. The character reaches its minimum value $\delta = -\dim_C \mathfrak{l}$ if $\mathfrak{l}_0$ is a compact real form and its maximum value $\delta = \mathrm{rank}\, \mathfrak{l}$ if $\mathfrak{l}_0$ is a normal real form of $\mathfrak{l}$. For these extreme values of the character, the corresponding real forms are unique up to isomorphism (Cor. 7.3, Chapter III, and Theorem 5.10, Chapter IX). In contrast, the examples $\mathfrak{so}^*(18)$ and $\mathfrak{so}(12,6)$ which are real forms of $\mathfrak{so}(18, \boldsymbol{C})$ with character -9 show that two nonisomorphic real forms of a simple Lie algebra $\mathfrak{l}$ over $\boldsymbol{C}$ may have the same character. However, Tables II and III in §5 show that this cannot happen for the exceptional structures† so we label the real forms of the exceptional complex algebras by means of their character. Thus $\mathfrak{e}_{6(\delta)}$ denotes the real form of $\mathfrak{e}_6$ with character δ. The rank and dimension is also listed, although the rank was determined in §2 only for the classical spaces.

† In fact only for certain $\mathfrak{d}_n$ and certain $\mathfrak{a}_n$ (e.g., $\mathfrak{su}^*(14)$ and $\mathfrak{su}(9,5)$).

TABLE V

IRREDUCIBLE RIEMANNIAN GLOBALLY SYMMETRIC SPACES OF TYPE I AND TYPE III

	Noncompact	Compact	Rank	Dimension
A I	$SL(n, \boldsymbol{R})/SO(n)$	$SU(n)/SO(n)$	$n-1$	$\frac{1}{2}(n-1)(n+2)$
A II	$SU^*(2n)/Sp(n)$	$SU(2n)/Sp(n)$	$n-1$	$(n-1)(2n+1)$
A III	$SU(p, q)/S(U_p \times U_q)$	$SU(p+q)/S(U_p \times U_q)$	$\min(p, q)$	$2pq$
BD I	$SO_0(p, q)/SO(p) \times SO(q)$	$SO(p+q)/SO(p) \times SO(q)$	$\min(p, q)$	pq
D III	$SO^*(2n)/U(n)$	$SO(2n)/U(n)$	$[\frac{1}{2}n]$	$n(n-1)$
C I	$Sp(n, \boldsymbol{R})/U(n)$	$Sp(n)/U(n)$	n	$n(n+1)$
C II	$Sp(p, q)/Sp(p) \times Sp(q)$	$Sp(p+q)/Sp(p) + Sp(q)$	$\min(p, q)$	$4pq$
E I	$(\mathfrak{e}_{6(6)}, \mathfrak{sp}(4))$	$(\mathfrak{e}_{6(-78)}, \mathfrak{sp}(4))$	6	42
E II	$(\mathfrak{e}_{6(2)}, \mathfrak{su}(6) + \mathfrak{su}(2))$	$(\mathfrak{e}_{6(-78)}, \mathfrak{su}(6) + \mathfrak{su}(2))$	4	40
E III	$(\mathfrak{e}_{6(-14)}, \mathfrak{so}(10) + \boldsymbol{R})$	$(\mathfrak{e}_{6(-78)}, \mathfrak{so}(10) + \boldsymbol{R})$	2	32
E IV	$(\mathfrak{e}_{6(-26)}, \mathfrak{f}_4)$	$(\mathfrak{e}_{6(-78)}, \mathfrak{f}_4)$	2	26
E V	$(\mathfrak{e}_{7(7)}, \mathfrak{su}(8))$	$(\mathfrak{e}_{7(-133)}, \mathfrak{su}(8))$	7	70
E VI	$(\mathfrak{e}_{7(-5)}, \mathfrak{so}(12) + \mathfrak{su}(2))$	$(\mathfrak{e}_{7(-133)}, \mathfrak{so}(12) + \mathfrak{su}(2))$	4	64
E VII	$(\mathfrak{e}_{7(-25)}, \mathfrak{e}_6 + \boldsymbol{R})$	$(\mathfrak{e}_{7(-133)}, \mathfrak{e}_6 + \boldsymbol{R})$	3	54
E VIII	$(\mathfrak{e}_{8(8)}, \mathfrak{so}(16))$	$(\mathfrak{e}_{8(-248)}, \mathfrak{so}(16))$	8	128
E IX	$(\mathfrak{e}_{8(-24)}, \mathfrak{e}_7 + \mathfrak{su}(2))$	$(\mathfrak{e}_{8(-248)}, \mathfrak{e}_7 + \mathfrak{su}(2))$	4	112
F I	$(\mathfrak{f}_{4(4)}, \mathfrak{sp}(3) + \mathfrak{su}(2))$	$(\mathfrak{f}_{4(-52)}, \mathfrak{sp}(3) + \mathfrak{su}(2))$	4	28
F II	$(\mathfrak{f}_{4(-20)}, \mathfrak{so}(9))$	$(\mathfrak{f}_{4(-52)}, \mathfrak{so}(9))$	1	16
G	$(\mathfrak{g}_{2(2)}, \mathfrak{su}(2) + \mathfrak{su}(2))$	$(\mathfrak{g}_{2(-14)}, \mathfrak{su}(2) + \mathfrak{su}(2))$	2	8

3. Irreducible Hermitian Symmetric Spaces

In view of Theorem 6.1, Chapter VIII, it can be decided immediately which of the spaces in Table V are Hermitian symmetric. They are

$$\boldsymbol{A\ III},\ \boldsymbol{D\ III},\ \boldsymbol{BD\ I}(q=2),\ \boldsymbol{C\ I},\ \boldsymbol{E\ III},\ \boldsymbol{E\ VII}.$$

This exhausts the list of irreducible Hermitian symmetric spaces because the spaces of type II and IV cannot be Hermitian. According to Theorem 7.1, Chapter VIII, the noncompact spaces can be regarded as bounded domains in Euclidean space. In É. Cartan [19] such domains are constructed for the four large classes ***A III***, ***D III***, ***BD I*** $(q = 2)$, ***C I*** (cf. Exercise D.1 below).

4. Coincidences between Different Classes. Special Isomorphisms

Because of the isomorphism (1) there are some overlaps in Table V for small n. Using Theorem 6.2 we therefore derive the following isomorphisms (i)–(viii) already proved by Cartan in [2], pp. 352–355. His method was based on an analysis of the character of the real forms.

(i) $\boldsymbol{A\ I}\ (n = 2) = \boldsymbol{A\ III}\ (p = q = 1)$
$= \boldsymbol{BD\ I}\ (p = 2,\ q = 1) = \boldsymbol{C\ I}\ (n = 1)$

Corresponding isomorphisms:

$$\mathfrak{su}(2) \approx \mathfrak{so}(3) = \mathfrak{sp}(1)$$
$$\mathfrak{sl}(2, \boldsymbol{R}) \approx \mathfrak{su}(1, 1) \approx \mathfrak{so}(2, 1) \approx \mathfrak{sp}(1, \boldsymbol{R}).$$

(ii) $\boldsymbol{BD\ I}(p = 3,\ q = 2) = \boldsymbol{C\ I}(n = 2)$.
Corresponding isomorphisms:

$$\mathfrak{so}(5) \approx \mathfrak{sp}(2),$$
$$\mathfrak{so}(3, 2) \approx \mathfrak{sp}(2, \boldsymbol{R}).$$

(iii) $\boldsymbol{BD\ I}(p = 4,\ q = 1) = \boldsymbol{C\ II}(p = q = 1)$.
Corresponding isomorphisms:

$$\mathfrak{so}(5) \approx \mathfrak{sp}(2), \qquad \mathfrak{so}(4) \approx \mathfrak{sp}(1) \times \mathfrak{sp}(1),$$
$$\mathfrak{so}(4, 1) \approx \mathfrak{sp}(1, 1).$$

(iv) $\boldsymbol{A\ I}(n = 4) = \boldsymbol{BD\ I}(p = q = 3)$.
Corresponding isomorphisms:

$$\mathfrak{su}(4) \approx \mathfrak{so}(6), \qquad \mathfrak{so}(4) \approx \mathfrak{so}(3) \times \mathfrak{so}(3),$$
$$\mathfrak{sl}(4, \boldsymbol{R}) \approx \mathfrak{so}(3, 3).$$

(v) $\boldsymbol{A\ II}(n = 2) = \boldsymbol{BD\ I}(p = 5,\ q = 1)$.
Corresponding isomorphisms:

$$\mathfrak{su}(4) \approx \mathfrak{so}(6), \qquad \mathfrak{sp}(2) \approx \mathfrak{so}(5),$$
$$\mathfrak{su}^*(4) \approx \mathfrak{so}(5, 1).$$

(vi) $\boldsymbol{A\ III}(p = q = 2) = \boldsymbol{BD\ I}(p = 4,\ q = 2)$.
Corresponding isomorphisms:

$$\mathfrak{su}(4) \approx \mathfrak{so}(6),$$
$$\mathfrak{su}(2, 2) \approx \mathfrak{so}(4, 2).$$

(vii) $\boldsymbol{A\ III}(p = 3,\ q = 1) = \boldsymbol{D\ III}(n = 3)$.
Corresponding isomorphisms:

$$\mathfrak{su}(4) \approx \mathfrak{so}(6),$$
$$\mathfrak{su}(3, 1) \approx \mathfrak{so}^*(6).$$

(viii) $\boldsymbol{BD\ I}(p = 6,\ q = 2) = \boldsymbol{D\ III}(n = 4)$.
Corresponding isomorphisms:

$$\mathfrak{su}(4) \approx \mathfrak{so}(6),$$
$$\mathfrak{so}^*(8) \approx \mathfrak{so}(6, 2).$$

The last isomorphism does not occur in Cartan's paper cited above. However, the isometry of the space ***BD I***$(p = 6, q = 2)$ and ***D III***$(n = 4)$ is shown in Cartan [10], p. 459. The fact that they even coincide as Hermitian symmetric spaces (cf. Exercise D.2 below) is, however, not observed in Cartan [19], p. 152.

The spaces ***BD I***$(p + q = 4)$ and ***D III***$(n = 2)$ can of course be defined although $\mathfrak{so}(4)$ is not simple. The isomorphism $\mathfrak{d}_2 = \mathfrak{a}_1 \times \mathfrak{a}_1$ then yields additional isomorphisms corresponding to the three involutive automorphisms of $\mathfrak{a}_1 \times \mathfrak{a}_1$ given by: $(X, Y) \to (Y, X)$; $(X, Y) \to (\bar{X}, \bar{Y})$; $(X, Y) \to (X, \bar{Y})$.

(ix) ***BD I***$(p = 3, q = 1) = \mathfrak{a}_n(n = 1)$.
Corresponding isomorphisms:

$$\mathfrak{so}(4) \approx \mathfrak{su}(2) \times \mathfrak{su}(2),$$
$$\mathfrak{so}(3, 1) \approx \mathfrak{sl}(2, \boldsymbol{C}).$$

(x) ***BD I***$(p = 2, q = 2) =$ ***A I***$(n = 2) \times$ ***A I***$(n = 2)$.
Corresponding isomorphisms:

$$\mathfrak{so}(4) \approx \mathfrak{su}(2) \times \mathfrak{su}(2),$$
$$\mathfrak{so}(2, 2) \approx \mathfrak{sl}(2, \boldsymbol{R}) \times \mathfrak{sl}(2, \boldsymbol{R}).$$

(xi) ***D III***$(n = 2)$ and ***A I***$(n = 2)$.

$$\mathfrak{so}(4) \approx \mathfrak{su}(2) \times \mathfrak{su}(2),$$
$$\mathfrak{so}^*(4) \approx \mathfrak{su}(2) \times \mathfrak{sl}(2, \boldsymbol{R}).$$

EXERCISES AND FURTHER RESULTS

A. Special Isomorphisms

In Theorem 3.12 and §6, no. 4, we have general existence proofs of the isomorphisms which occur between different classes of simple Lie algebras. In Exercises 1–4 below it is indicated how some of these can be verified by *ad hoc* methods.

1. Exhibit the following local isomorphisms directly:

$$\boldsymbol{SL}(2, \boldsymbol{C}) \approx \boldsymbol{SO}(3, \boldsymbol{C}) \approx \boldsymbol{Sp}(1, \boldsymbol{C}) \tag{1}$$
$$\boldsymbol{SO}(5, \boldsymbol{C}) \approx \boldsymbol{Sp}(2, \boldsymbol{C}) \tag{2}$$
$$\boldsymbol{SL}(4, \boldsymbol{C}) \approx \boldsymbol{SO}(6, \boldsymbol{C}) \tag{3}$$
$$\boldsymbol{SO}(4, \boldsymbol{C}) \approx \boldsymbol{SL}(2, \boldsymbol{C}) \times \boldsymbol{SL}(2, \boldsymbol{C}) \tag{4}$$

(Hint: Identifying $\boldsymbol{Sp}(1)$ with the group of unit quaternions, the isomorphism $\boldsymbol{SU}(2) \approx \boldsymbol{Sp}(1)$ is immediate, so (1) and (4) follow from Example II in Chapter V, §2; for (3) consider the canonical bases $e_1, ..., e_4$ of $\boldsymbol{C}^4$, $f_1, ..., f_6$ of $\boldsymbol{C}^6$ and the linear bijection of $\Lambda^2\boldsymbol{C}^4$ onto $\boldsymbol{C}^6$ given by the map

$$(e_1 \wedge e_2, e_2 \wedge e_3, e_3 \wedge e_4, e_1 \wedge e_3, e_2 \wedge e_4, e_1 \wedge e_4) \to (f_1, ..., f_6). \qquad (5)$$

If the vectors $x, y \in \Lambda^2\boldsymbol{C}^4$ correspond to $\xi = \Sigma_1^6\, x_i f_i$, $\eta = \Sigma_1^6\, y_i f_i$, then

$$x \wedge y = (x_1 y_3 + x_3 y_1 + x_2 y_6 + x_6 y_2 - x_4 y_5 - x_5 y_4) e_1 \wedge e_2 \wedge e_3 \wedge e_4.$$

Each $g \in \boldsymbol{SL}(\boldsymbol{C}^4)$ extends to an automorphism $\tilde{g}$ of the Grassmann algebra of $\boldsymbol{C}^4$, whereby $\Lambda^4\boldsymbol{C}^4$ is left pointwise fixed; thus the linear transformation which $\tilde{g}$ (via (5)) induces on $\boldsymbol{C}^6$ belongs to $\boldsymbol{SO}(6, \boldsymbol{C})$. This gives (3); for (2) restrict the isomorphism (3) to a suitable subgroup).

2. The local isomorphism

$$\boldsymbol{SO}(3, 2) \approx \boldsymbol{Sp}(2, \boldsymbol{R}), \qquad (6)$$

observed in §6, no. 4, can also be proved more directly as follows (Siegel [1], §56): Consider the quadratic form

$$Q(W) = w_1 w_2 - w_3^2 - w_4 w_5 = {}^tWQW, \qquad w_j \in \boldsymbol{C}$$

and let $g = (v_{kl})$ be a *real* linear transformation leaving Q invariant. Let

$$X + iY = Z = w_5^{-1} W, \qquad \hat{W} = gW, \qquad \hat{Z} = \hat{w}_5^{-1}\hat{W}$$

so

$$\hat{Z} = \left(\sum_1^5 v_{sl} z_l\right)^{-1} gZ \qquad (\hat{z}_5 = z_5 = 1).$$

This transformation leaves invariant the surface $Q(Z) \equiv z_1 z_2 - z_3^2 - z_4 = 0$; since $Q(Y) = y_1 y_2 - y_3^2$ (by $z_5 = 1$), we deduce that the fractional linear transformation T_g given by

$$\hat{z}_k = \left(\sum_1^5 v_{5l} z_l\right)^{-1} \sum_1^5 v_{kl} z_l \qquad (k = 1, 2, 3)$$

where $z_5 = 1$, $z_4 = z_1 z_2 - z_3^2$, maps the set

$$\left\{\begin{pmatrix} z_1 & z_3 \\ z_3 & z_2 \end{pmatrix} : y_1 y_2 - y_3^2 > 0\right\} \qquad (7)$$

into itself. Since $y_1 \neq 0$, the set (7) decomposes into the parts with $y_1 > 0$ and $y_1 < 0$, which are either permuted or left invariant by T_g; the first part being the Siegel upper half-plane $\mathscr{S}_2$, the mapping $g \to T_g$ gives (6).

3. Exhibit local isomorphisms

$$\boldsymbol{SL}(2, \boldsymbol{C})^{\boldsymbol{R}} \approx \boldsymbol{SO}(3, 1) \tag{8}$$

$$\boldsymbol{SL}(4, \boldsymbol{R}) \approx \boldsymbol{SO}(3, 3) \tag{9}$$

$$\boldsymbol{SU}(2, 2) \approx \boldsymbol{SO}(2, 4) \tag{10}$$

(Hint: For (8) see Exercise 2, Chapter V; for (9) consider the bijection of $\Lambda^2\boldsymbol{R}^4$ onto $\boldsymbol{R}^6$ given by (5) and proceed as for (3); for (10) consider the bijection of $\boldsymbol{R}^6 = \{x_1, ..., x_6\}$ onto $\mathfrak{so}(4, \boldsymbol{C}) = \{A = (a_{ij}): {}^tA + A = 0\}$ given by

$$a_{12} = x_1 + ix_2, \qquad a_{13} = x_3 + ix_4, \qquad a_{14} = x_5 + ix_6$$

$$a_{23} = x_5 - ix_6, \qquad a_{24} = x_3 - ix_4, \qquad a_{34} = x_1 - ix_2$$

(compare D2 below) whereby

$$\operatorname{Tr}(AI_{2,2}{}^t\bar{A}I_{2,2}) = -x_1{}^2 - x_2{}^2 + x_3{}^2 + x_4{}^2 + x_5{}^2 + x_6{}^2. \tag{11}$$

If $g \in \boldsymbol{SU}(2, 2)$ the map $A \to gA^tg$ of $\mathfrak{so}(4, \boldsymbol{C})$ into itself leaves (11) invariant and so induces the desired element of $\boldsymbol{O}(2, 4)$.

4. Deduce a local isomorphism

$$\boldsymbol{SO}(2, 6) \approx \boldsymbol{SO}^*(8)$$

from Exercises D.1 and D.2 below.

5. The notation being as in §2, let σ and τ denote the conjugation of $\mathfrak{g}$ with respect to $\mathfrak{g}_0$ and $\mathfrak{u}$, respectively. Show that

For	***A I***	$\sigma(X) = \bar{X},$	$\tau(X) = -{}^t\bar{X}$
	A II	$\sigma(X) = J_n\bar{X}J_n^{-1},$	$\tau(X) = -{}^t\bar{X}$
	A III	$\sigma(X) = -I_{p,q}{}^t\bar{X}I_{p,q},$	$\tau(X) = -{}^t\bar{X}$
	B DI	$\sigma(X) = I_{p,q}\bar{X}I_{p,q},$	$\tau(X) = \bar{X}$
	D III	$\sigma(X) = J_n\bar{X}J_n^{-1},$	$\tau(X) = \bar{X}$
	C I	$\sigma(X) = \bar{X}$	$\tau(X) = J_n\bar{X}J_n^{-1}$
	C II	$\sigma(X) = -K_{p,q}{}^t\bar{X}K_{p,q},$	$\tau(X) = J_{p+q}\bar{X}J_{p+q}^{-1}.$

B. Root Systems and the Weyl Group

1. Let R be a reduced root system and suppose $\alpha, \beta, \alpha + \beta \in R$. Then the system of vectors

$$(\boldsymbol{Z}\alpha + \boldsymbol{Z}\beta) \cap R$$

is a root system of type $\mathfrak{a}_2$, $\mathfrak{b}_2$, or $\mathfrak{g}_2$.

2. Let R be an irreducible reduced root system, $(\alpha_1, ..., \alpha_l)$ a basis, $(\omega_1, ..., \omega_l)$ the "dual" basis given by $2\langle\omega_j, \alpha_i\rangle = \delta_{ij}\langle\alpha_i, \alpha_i\rangle$, and $\rho = \frac{1}{2}\sum_{\alpha>0}\alpha$. Then

$$\alpha_1 + ... + \alpha_l \in R, \qquad \omega_1 + ... + \omega_l = \rho.$$

3. Let R be irreducible and reduced. Then:

(i) $W(R)$ acts transitively on each subset of roots in R of the same length.

(ii) Only if $R = \mathfrak{b}_l, \mathfrak{c}_l, \mathfrak{f}_4, \mathfrak{g}_2$ does the closed positive Weyl chamber $\bar{C}^+$ contain a root different from δ. This root is unique and is given by

$$\begin{aligned} &\mathfrak{b}_l : \alpha_1 + ... + \alpha_l \\ &\mathfrak{c}_l : \alpha_1 + 2\alpha_2 + ... + 2\alpha_{l-1} + \alpha_l \\ &\mathfrak{f}_4 : \alpha_1 + 2\alpha_2 + 3\alpha_3 + 2\alpha_4 \\ &\mathfrak{g}_2 : 2\alpha_1 + \alpha_2. \end{aligned}$$

4. In the notation before Theorem 5.13 let $\alpha_1, ..., \alpha_n$ be the simple roots in $\Delta(\mathfrak{g}_0, \mathfrak{h})$ and $\tilde{\alpha}_0 = (\alpha_0, 1)$ the additional simple root in $\tilde{\Delta}^+$. Then $\alpha_0 \in \Delta(\mathfrak{g}_0, \mathfrak{h})$ except for the case $\mathfrak{a}_{2n}^{(2)}$ (in the $S(A)$ table) in which case $\frac{1}{2}\alpha_0 \in \Delta(\mathfrak{g}_0, \mathfrak{h})$.

5. Deduce from Cor. 6.6, Chapter VII, that for a simple compact Lie group the covering index equals 1 + the number of ones in the representation $\delta = d_1\alpha_1 + ... + d_l\alpha_l$. (Note that for $\mathfrak{u}$ simple, the polyhedron P_0 is given by $(2\pi i)^{-1}\alpha_j > 0$, $(2\pi i)^{-1}\delta < 1$.)

6. Let R be irreducible and reduced. Show that $-1 \notin W(R)$ if and only if $R = \mathfrak{a}_l\ (l > 1)$, $\mathfrak{d}_{2k+1}$ or $\mathfrak{e}_6$. (Hint: deduce from Prop. 5.3, Chapter IX, that $\mathrm{Aut}(\mathfrak{u})/\mathrm{Int}(\mathfrak{u})$ acts faithfully on $\tilde{Z}$ so, equivalently, $\mathrm{Aut}(R)/W(R)$ acts faithfully on $\tilde{T}(R)/T(R)$, so $-1 \in W(R)$ if and only if $T(R) \supset 2\tilde{T}(R)$.)

7. Let R be a reduced root sustem in V. Show that V has a unique positive definite inner product (,), invariant under $W(R)$, satisfying

$$(\lambda, \mu) = \sum_{\alpha\in R}(\lambda, \alpha)(\alpha, \mu), \qquad \lambda, \mu \in V.$$

Because of (9), Chapter III, §4, we call (,) the *Killing form*. The form $\langle\ ,\ \rangle$ used in the construction of the exceptional root systems in Theorem 3.23 is given by $\langle\lambda, \mu\rangle = \gamma(\lambda, \mu)$ $(\gamma \in \mathbf{R})$ and the formula

$$4(\beta, \beta)^{-1} = \sum_{\alpha\in R} a_{\alpha,\beta}^2, \qquad \beta \in R$$

implies

$$\gamma = 24, \quad 18, \quad 24, \quad 36, \quad 60$$

for the cases $\mathfrak{g}_2$, $\mathfrak{f}_4$, e_6, e_7, e_8, respectively.

8.* Let R be an irreducible and reduced root system in V, B a basis of R, δ the highest root. Let V_δ be the hyperplane in V perpendicular to δ and put $R' = V_\delta \cap R$, $B' = V_\delta \cap B$ (cf. Carles [1]).

(i) Show that R' is a root system with basis B' and determine its Dynkin diagram by means of the extended Dynkin diagram for R (Table 1 for the diagram $S(A)$ in §5).

(ii) Show that if L is the maximal length in R (for the Killing form), then

$$\text{card}\, R' = \text{card}\, R + 6 - 4L^{-2},$$

$$2\rho' = 2\rho - (L^{-2} - 1)\delta.$$

Here 2ρ denotes the sum of the positive roots.

9. Let $\mathfrak{u}$ be a simple compact Lie algebra and as in Chapter VII let Γ_Δ be the affine Weyl group, $\mathfrak{t}_\Delta \subset \Gamma_\Delta$ the subgroup of translations, P_0 the fundamental polyhedron, $\delta = \Sigma_1^l d_j \alpha_j$ the highest root. Put $\check{\alpha}_j = 2\alpha_j/\langle \alpha_j, \alpha_j \rangle$ and define ω_k by $\langle \omega_k, \check{\alpha}_j \rangle = \delta_{jk}$. Put

$$e_j = 4\pi i \langle \alpha_j, \alpha_j \rangle^{-1} H_{\omega_j}, \qquad f_j = 4\pi i \langle \alpha_j, \alpha_j \rangle^{-1} H_{\alpha_j} \qquad (1 \leqslant j \leqslant l).$$

Prove in steps (i), (ii), (iii) (cf. Weyl [2], Bourbaki [2], Chapter VI) that the order of W is given by

$$|W| = l! \cdot d_1 d_2 \ldots d_l \cdot a,$$

where a is the covering index:

(i) Show that P_0 is the polyhedron with vertices $0, d_1^{-1} e_1, \ldots, d_l^{-1} e_l$. Thus, by elementary geometry, the parallelotope

$$P = \left\{ \sum_{j=1}^{l} x_j e_j : 0 < x_1 < 1, \ldots, 0 < x_l < 1 \right\}$$

satisfies

$$\text{vol}(P)/\text{vol}(P_0) = l! \cdot d_1 \ldots d_l.$$

(ii) Let Q denote the parallelotope

$$Q = \left\{ \sum_{j=1}^{l} y_j f_j : 0 \leqslant y_1 < 1, \ldots, 0 \leqslant y_l < 1 \right\}.$$

Since $\bar{P}_0$ and Q are fundamental domains for Γ_Δ and $\mathfrak{t}_\Delta$, respectively, deduce

$$\text{vol}(Q)/\text{vol}(P_0) = (\Gamma_\Delta : \mathfrak{t}_\Delta) = |W|.$$

(iii) As in the proof of Prop. 3.31 show that

$$\mathfrak{t}(\mathfrak{u}) = \sum_{j=1}^{l} \boldsymbol{Z} e_j, \qquad \mathfrak{t}_\Delta = \sum_{j=1}^{l} \boldsymbol{Z} f_j$$

so, by Lemma 3.30 and Proposition 3.31,

$$\mathrm{vol}(Q)/\mathrm{vol}(P) = \text{covering index}.$$

10. Let $\mathfrak{g}$ be an arbitrary semisimple Lie algebra over $\boldsymbol{R}$, Σ its system of restricted roots, and Σ' and Σ'', respectively, the sets of indivisible and unmultipliable elements in Σ. Construct semisimple subalgebras $\mathfrak{g}', \mathfrak{g}'' \subset \mathfrak{g}$ which are normal real forms and have Σ' and Σ'', respectively, as their restricted root systems.

(Hint: if μ runs through a basis of Σ' (resp. Σ''), take $\mathfrak{g}'$ (resp. $\mathfrak{g}''$) as the subalgebra generated by the elements H, X, Y in the proof of Theorem 2.16, Chapter VII.)

Example from Exercise 5, Chapter VII: if $\mathfrak{g} = \mathfrak{su}(p, q)$ $(0 < p < q)$, then $\mathfrak{g}' = \mathfrak{so}(p, p+1)$, $\mathfrak{g}'' = \mathfrak{sp}(p, \boldsymbol{R})$.

C. The Classification of Compact, Locally Isometric, Globally Symmetric Spaces

1. Let $\tilde{U}$ be a simple, simply connected compact Lie group and $\tilde{Z}$ its center. Let S_1, S_2 be two subgroups of $\tilde{Z}$. Then $\tilde{U}/S_1$ and $\tilde{U}/S_2$ are isomorphic if and only if there is an automorphism σ of $\tilde{U}$ such that $\sigma S_1 = S_2$ (cf. Theorem 1.11, Chapter II).

Deduce from (12), §3, and Theorem 3.32 that the number of nonisomorphic connected Lie groups with Lie algebra $\mathfrak{u}$ is

1	if	$\mathfrak{u} = \mathfrak{g}_2, \mathfrak{f}_4, \mathfrak{e}_8$
2	if	$\mathfrak{u} = \mathfrak{b}_l, \mathfrak{c}_l, \mathfrak{e}_6, \mathfrak{e}_7$
3	if	$\mathfrak{u} = \mathfrak{d}_{2k+1}$ or $\mathfrak{d}_4$
4	if	$\mathfrak{u} = \mathfrak{d}_{2k}$ $(k \geqslant 3)$
$d(l+1)$	if	$\mathfrak{u} = \mathfrak{a}_l$,

where $d(n)$ is the number of divisors in n (cf. Cartan [9], §29, our Proposition 1.2, and Goto and Kobayashi [1]).

2*. Let θ be an involutive automorphism of the simple Lie algebra $\mathfrak{u}$. In the notation of Chapter VII, §9 the globally symmetric spaces associated with $(\mathfrak{u}, \theta)$ are $\tilde{U}/K^*$ and the maps

$$\tilde{U}/\tilde{K} \to \tilde{U}/K^* \to \tilde{U}/\tilde{K}_{\tilde{Z}} = \mathrm{Int}(\mathfrak{u})/\mathrm{Int}(\mathfrak{u})_\theta \qquad \text{(adjoint space)}$$

are coverings. Let Σ^+ denote the set of positive restricted roots, $\mu_1, \ldots, \mu_l$ the simple ones and $\delta = \sum_{i=1}^{l} d_i \mu_i$ the highest one (Lemma 11.8, Chapter VII). In analogy with the group case, we define:

Covering index of $(\mathfrak{u}, \theta)$ = order of $\pi_1(\mathrm{Int}(\mathfrak{u})/\mathrm{Int}(\mathfrak{u})_\theta)$. A proof analogous to that of Exercise B.5 gives

Covering index of $(\mathfrak{u}, \theta) = 1 + \mathrm{Card}\{i : d_i = 1\}$ (cf. Takeuchi [1]).

3. Using 2, deduce from Exercise 5 in Chapter VII that the space $\boldsymbol{SU}(p+q)/\boldsymbol{S}(\boldsymbol{U}_p \times \boldsymbol{U}_q)$ is for $p < q$ the only globally symmetric space in its local isometry class, while for $p = q$ it is the simply connected double covering of its adjoint space.

4. Using 2, Table VI below, and Theorem 3.28, prove the following formulas for the covering index a of $(\mathfrak{u}, \theta)$ (cf. Cartan [10], Ch. III):

(a) For ***A III*** ($2l \leqslant$ rank $\mathfrak{u}$), ***A IV***, ***C II*** ($2l <$ rank($\mathfrak{u}$)), ***D III*** (rank $\mathfrak{u}$ odd), ***E II***, ***E III***, ***E VI***, ***E VIII***, ***E IX***, ***F I***, ***F II***, ***G***,

$$a = 1.$$

(b) For ***A III*** ($2l =$ rank $\mathfrak{u} + 1$), ***B I***, ***B II***, ***C I***, ***C II*** ($2l =$ rank $\mathfrak{u}$), ***D I*** ($l <$ rank($\mathfrak{u}$)), ***D II***, ***D III*** (rank $\mathfrak{u}$ even), ***E V***, ***E VII***,

$$a = 2.$$

(c) For ***E I*** and ***E IV***,

$$a = 3.$$

(d) For ***D I*** ($l =$ rank $\mathfrak{u}$),

$$a = 4.$$

($\pi_1(\mathrm{Int}(\mathfrak{u})/\mathrm{Int}(\mathfrak{u})_\theta) \approx \boldsymbol{Z}_4$ if l is odd, $\approx \boldsymbol{Z}_2 \times \boldsymbol{Z}_2$ if l is even.

(e) For ***A I*** and ***A II***,

$$a = l + 1.$$

($\pi_1(\mathrm{Int}(\mathfrak{u})/\mathrm{Int}(\mathfrak{u})_\theta) \approx \boldsymbol{Z}_{l+1}$.)

For the case (a) all globally symmetric spaces associated with $(\mathfrak{u}, \theta)$ coincide; for (b), (c), and (e), $l + 1$ prime there are exactly two different globally symmetric spaces associated with $(\mathfrak{u}, \theta)$.

D. Bounded Symmetric Domains

1. Verify the following realizations of the classical noncompact Hermitian symmetric spaces as bounded domains. Let $M_{p,q}(K)$ denote the set of $p \times q$ matrices with coefficients in the field $K = \boldsymbol{R}$ or $\boldsymbol{C}$ and let $A < B$ signify that $B - A$ is positive definite, that is, ${}^tZ(B - A)\bar{Z} > 0$ for $Z \neq 0$.

A III: $\{Z \in M_{p,q}(C) : {}^tZ\bar{Z} < I_q\}$.

Action of $SU(p, q)$ by

$$T_g : Z \to (AZ + B)(CZ + D)^{-1} \qquad \text{if} \quad g = \begin{pmatrix} A & B \\ C & D \end{pmatrix}.$$

D III: $\{Z \in M_{n,n}(C) : {}^tZ = -Z,\ {}^tZ\bar{Z} < I_n\}$.

Action of $SO^*(2n)$ by

$$T_g : Z \to (AZ + B)(-\bar{B}Z + \bar{A})^{-1} \qquad \text{if} \quad g = \begin{pmatrix} A & B \\ -\bar{B} & \bar{A} \end{pmatrix}.$$

Here $SO^*(2n)$ is viewed as the invariance group of the forms

$$-w_1\bar{w}_1 - \dots - w_n\bar{w}_n + w_{n+1}\bar{w}_{n+1} + \dots + w_{2n}\bar{w}_{2n}$$
$$w_1w_{n+1} + \dots + w_nw_{2n}.$$

C I: $\{Z \in M_{n,n}(C) : {}^tZ = Z,\ Z\bar{Z} < I_n\}$.

Action of $Sp(n, C) \cap SU(n, n)$ by

$$T_g : Z \to (AZ + B)(\bar{B}Z + \bar{A})^{-1} \qquad \text{if} \quad g = \begin{pmatrix} A & B \\ \bar{B} & \bar{A} \end{pmatrix}.$$

BD I$_{(p=2)}$: $\{X \in M_{2,q}(R) : X{}^tX < I_2\}$.

Action of $SO_0(2, q)$ by

$$T_g : X \to (AX + B)(CX + D)^{-1} \qquad \text{if} \quad g = \begin{pmatrix} A & B \\ C & D \end{pmatrix}.$$

The complex structure is given by

$$J_0X = J_0 \begin{pmatrix} X_1 \\ X_2 \end{pmatrix} = \begin{pmatrix} X_2 \\ -X_1 \end{pmatrix}.$$

(See Cartan [19], Siegel [2], Hua [2], [3]; the exceptional domains ***E III*** and ***E VII*** are represented similarly by the 3×3 Hermitian matrices over the Cayley numbers by Hirzebruch [3]; together with D.2 below, these descriptions give another proof of Theorem 7.1, Chapter VIII.)

2*. (a) Show that the mapping

$$Z \to X = 2 \begin{pmatrix} Z{}^tZ + 1 & i(Z{}^tZ - 1) \\ \bar{Z}{}^t\bar{Z} + 1 & -i(\bar{Z}{}^t\bar{Z} - 1) \end{pmatrix}^{-1} \begin{pmatrix} Z \\ \bar{Z} \end{pmatrix}$$

is a holomorphic diffeomorphism of the domain

$$\boldsymbol{BD\ I}'_{(p=2)} = \{Z \in M_{p,q}(C) : |\, Z{}^tZ \,|^2 + 1 - 2\bar{Z}{}^tZ > 0,\ |\, Z{}^tZ \,| < 1\}$$

onto the bounded domain ***BD I***$_{(p=2)}$ in Exercise 1 (cf. Hua [3]).

(b) Show that the mapping

$$\begin{pmatrix} z_1 \\ z_2 \\ \vdots \\ z_6 \end{pmatrix} \to \begin{pmatrix} 0 & z_1 + iz_2 & z_3 + iz_4 & z_5 + iz_6 \\ -z_1 - iz_2 & 0 & z_5 - iz_6 & -z_3 + iz_4 \\ -z_3 - iz_4 & -z_5 + iz_6 & 0 & z_1 - iz_2 \\ -z_5 - iz_6 & z_3 - iz_4 & -z_1 + iz_2 & 0 \end{pmatrix}$$

is a holomorphic diffeomorphism of $\boldsymbol{BD\ I}_{(p=2,q=6)}$ onto $\boldsymbol{D\ III}$ $(n = 4)$ (Morita [1]).

3. In the notation of §7, Chapter VIII (Cor. 7.6) let ν denote the automorphism of $\mathfrak{g}$ defined by

$$\nu = \exp \frac{\pi}{4} \operatorname{ad} \Big(\sum_{\gamma \in \Gamma} (X_\gamma - X_{-\gamma}) \Big).$$

Prove that

$$\nu(X_\gamma + X_{-\gamma}) = \frac{2}{\langle \gamma, \gamma \rangle} H_\gamma \qquad \text{so} \qquad \nu(\mathfrak{a}_0) = \sum_{\gamma \in \Gamma} \boldsymbol{R} H_\gamma.$$

4*. In this exercise we summarize the determination in Harish-Chandra [5], pp. 585-588 and in Moore [2], pp. 359-363 of the restricted root system Σ for an irreducible noncompact Hermitian symmetric space G_0/K_0. Following the notation of Chapter VIII, §7, the idea is to investigate the root system $\Delta(\mathfrak{g}, \mathfrak{h})$ via its division into compact and noncompact roots and then use the "Cayley transform" ν in Exercise 3 above to relate $\Delta(\mathfrak{g}, \mathfrak{h})$ and Σ.

(a) Let $\lambda \to \bar{\lambda}$ denote the restriction from $\mathfrak{h}$ to the subspace $\Sigma_{i=1}^{s} \boldsymbol{C} H_{\gamma_i}$ (and write γ_i for $\bar{\gamma}_i$). Then there are two possibilities for $\Delta(\mathfrak{g}, \mathfrak{h})^-$:

(i) $\Delta(\mathfrak{g}, \mathfrak{h})^- = \{\pm\frac{1}{2}\gamma_i \pm \frac{1}{2}\gamma_j \ (1 \leqslant i < j \leqslant s), \pm\gamma_i \ (1 \leqslant i \leqslant s)\}$

(ii) $\Delta(\mathfrak{g}, \mathfrak{h})^- = \{\pm\frac{1}{2}\gamma_i \pm \frac{1}{2}\gamma_j \ (1 \leqslant i < j \leqslant s), \pm\frac{1}{2}\gamma_i, \pm\gamma_i \ (1 \leqslant i \leqslant s)\}$.

Moreover, all γ_i have the same length.

(b) If λ_i corresponds to $\frac{1}{2}\gamma_i$ via ν, then $(\lambda_1, \ldots, \lambda_s)$ is an orthogonal basis of the dual $\mathfrak{a}_0^*$ and the set $\Sigma = \Sigma(\mathfrak{g}_0, \mathfrak{a}_0)$ of restricted roots is

(i) $\Sigma = \{\pm\lambda_i \pm \lambda_j \ (1 \leqslant i < j \leqslant s), \pm 2\lambda_i \ (1 \leqslant i \leqslant s)\} = \mathfrak{c}_s$

(ii) $\Sigma = \{\pm\lambda_i \pm \lambda_j \ (1 \leqslant i < j \leqslant s), \pm\lambda_i, \pm 2\lambda_i \ (1 \leqslant i \leqslant s)\} = (\mathfrak{bc})_s$.

According to Korányi and Wolf [1], the first possibility happens if and only if G_0/K_0 is of tube type, that is, holomorphically equivalent to the tube over a self-dual cone. These cases are ***A III*** $(p = q)$, ***D III*** (n even), ***C I***, ***BD I*** $(p = 2)$, ***E VII***.

5. Consider the Hermitian form B_τ on $\mathfrak{g} \times \mathfrak{g}$ (§7, Chapter VIII) and let $\| \ \|$ denote the corresponding operator norm. Let ψ: $G_0/K_0 \to D \subset \mathfrak{p}_-$

be Harish-Chandra's holomorphic diffeomorphism of G_0/K_0 with a bounded domain D; under the differential $d\psi: \mathfrak{p}_0 \to \mathfrak{p}_-$ let $\mathfrak{D} \subset \mathfrak{p}_0$ be the domain which corresponds to $D \subset \mathfrak{p}_-$. Using the description of Σ in Exercise 4, prove that

$$\mathfrak{D} = \{X \in \mathfrak{p}_0 : \| \operatorname{ad} X \| < 2\}.$$

6. Let M be an irreducible bounded symmetric domain and in the notation of Chapter VIII, §7, let $\operatorname{Aut}_k(\Delta(\mathfrak{g}, \mathfrak{h})) \subset \operatorname{Aut}(\Delta(\mathfrak{g}, \mathfrak{h}))$, $W_k(\Delta(\mathfrak{g}, \mathfrak{h})) \subset W(\Delta(\mathfrak{g}, \mathfrak{h}))$ denote the subgroups leaving the set of compact roots invariant. Then

$$I(M)/I_0(M) = \operatorname{Aut}(\mathfrak{g}_0)/\operatorname{Int}(\mathfrak{g}_0) = \operatorname{Aut}_k(\Delta(\mathfrak{g}, \mathfrak{h}))/W_k(\Delta(\mathfrak{g}, \mathfrak{h}))$$

(cf. Exercise A.7, Chapter VI, and Theorem 5.4, Chapter IX).

7. (continuation) Recalling that $I_0(M) = H_0(M)$, a component of $I(M)$ corresponding to an element $t \in \operatorname{Aut}_k(\Delta(\mathfrak{g}, \mathfrak{h}))$ consists of holomorphic maps if and only if $t \mid \mathfrak{c} = 1$.

8*. (continuation) Using Exercise 7 it can be proved that if M is an irreducible bounded symmetric domain, then $H(M)$ is connected except for the cases ***A III*** ($p = q \geqslant 2$) and ***BD I*** ($p = 2$, q even) in which case it has two components (Cartan [19], Takeuchi [1]; in this statement we consider ***D III*** for $n > 4$ only, which, in view of the coincidences in §6, no. 4, is sufficient).

E. Automorphisms

1. Let an automorphism σ of finite order be called *regular* if its fixed points set $\mathfrak{g}_0$ is abelian. Show that the minimal order of a regular automorphism is $\sum_0^n a_i$ and that such automorphisms are inner and all conjugate.

2. Determine all automorphisms of order 3 of a simple Lie algebra over $\boldsymbol{C}$. Show that two such are conjugate if the fixed point algebras are isomorphic.

3*. For $\mathfrak{u}$ compact and simple the factor group $\operatorname{Aut}(\mathfrak{u})/\operatorname{Int}(\mathfrak{u})$ is determined by Theorem 5.4, Chapter IX, and Theorem 3.29, Chapter X. For $\mathfrak{g}_0$ simple and noncompact the factor group $\operatorname{Aut}(\mathfrak{g}_0)/\operatorname{Int}(\mathfrak{g}_0)$ (which by Exercise A.7, Chapter VI equals $I(M)$ $I_0(M)$) can also be determined in terms of the root structure (Cartan [10], Takeuchi [1], Murakami [1], [2], Matsumoto [2]), generalizing Exercise D.6 above.

4. Let σ be an automorphism of finite order of a complex semisimple Lie algebra $\mathfrak{g}$. Deduce from Lemma 5.3 and (4) §5 that:

(a) σ leaves invariant a Cartan subalgebra of $\mathfrak{g}$.

(b) σ leaves fixed a regular element $X \in \mathfrak{g}$.

(These results hold even if σ is just assumed semisimple, cf. Borel and Mostow [1]).

5. Construct two nonconjugate automorphisms of order 5 of the complex Lie algebra $\mathfrak{g} = \mathfrak{b}_4$ such that in both cases the fixed point algebra is isomorphic to $\mathfrak{a}_2 \oplus \boldsymbol{C}^2$.

6. Show that the complex algebra $\mathfrak{g} = \mathfrak{d}_4$ has automorphisms of order 2 and 3, respectively, such that in both cases the fixed point algebra is isomorphic to $\mathfrak{a}_3 \oplus \boldsymbol{C}$.

F. Restricted Roots and Multiplicities

Let $\mathfrak{g}$ be a simple noncompact Lie algebra over $\boldsymbol{R}$ and let θ, $\mathfrak{k}$, $\mathfrak{p}$, $\mathfrak{a}$, $\mathfrak{h}$, Σ, Δ, etc. be as in Chapter IX, §6. For $\psi \in \Sigma$ let m_ψ denote its multiplicity and put $\Delta_\psi = \{\alpha \in \Delta : \bar{\alpha} = \psi\}$. We now summarize some results from Araki's paper [1] which contains not only the classification of all simple Lie algebras over $\boldsymbol{R}$ but also determines Σ and the multiplicity function for each case.

1. (a) The space $\mathfrak{h}_{\boldsymbol{R}} = \sum_{\alpha \in \Delta} \boldsymbol{R} H_\alpha$ satisfies

$$\mathfrak{h}_{\boldsymbol{R}} = i(\mathfrak{h} \cap \mathfrak{k}) + \mathfrak{a}$$

and B is strictly positive definite on it.

(b) The restriction $\alpha \to \bar{\alpha}$ from $\mathfrak{h}_{\boldsymbol{R}}$ to $\mathfrak{a}$ coincides with the orthogonal projection $\alpha \to \frac{1}{2}(\alpha - \alpha^\theta)$ of the dual $\hat{\mathfrak{h}}_{\boldsymbol{R}}$ onto the dual $\hat{\mathfrak{a}}$. Thus we can consider Σ as a subset of $\hat{\mathfrak{h}}_{\boldsymbol{R}}$.

2. If $\alpha \in \Delta$, then $\alpha + \alpha^\theta \notin \Delta$, so $\langle \alpha, \alpha^\theta \rangle \geqslant 0$.

3. Let $\psi \in \Sigma$. Then $\psi \in \Delta_\psi$ if and only if m_ψ is odd. In this case, $\langle \alpha, \alpha^\theta \rangle = 0$ for all $\alpha \in \Delta_\psi - \{\psi\}$.

4. (a) If $\psi \in \Sigma$ and m_ψ is odd, then $2\psi \notin \Sigma$.

(b) If ψ, $2\psi \in \Sigma$ and m_ψ is even, then $m_{2\psi}$ is odd.

5. Let $\psi \in \Sigma$, m_ψ even. Then:

(a) $2\psi \in \Sigma$ if and only if there exists an $\alpha \in \Delta_\psi$ such that $\langle \alpha, \alpha^\theta \rangle > 0$; in this case

$$\langle \beta, \beta^\theta \rangle > 0 \qquad \text{for all} \quad \beta \in \Delta_\psi.$$

(b) $2\psi \notin \Sigma$ if an only if there exists an element $\alpha \in \Delta_\psi$ such that $\langle \alpha, \alpha^\theta \rangle = 0$; in this case $\langle \beta, \beta^\theta \rangle = 0$ for all $\beta \in \Delta_\psi$.

6. The normal form, split rank, and complex structure can be characterized as follows:

(a) $m_\psi = 1$ for all $\psi \in \Sigma$ if and only if $\mathfrak{g}$ is the normal real form of $\mathfrak{g}^C$.

(b) m_ψ is even for each $\psi \in \Sigma$ if and only if $\operatorname{rank}(G) = \operatorname{rank}(K) + \operatorname{rank}(G/K)$, or equivalently, if and only if all Cartan subalgebras of $\mathfrak{g}$ are conjugate.

(c) $m_\psi = 2$ for all $\psi \in \Sigma$ if and only if $\mathfrak{g}$ has a complex structure.

7. Let Σ and Δ have compatible orderings and put $\Delta_0 = \{\alpha \in \Delta : \bar{\alpha} \equiv 0\}$. Let $B = B(\Delta)$ be a basis of Δ, put $B_0 = B \cap \Delta_0$ and let $\bar{B} \subset \Sigma$ be the set of restrictions of $B - B_0$ to $\mathfrak{a}$. Then:

(a) Δ_0 is a root system with basis B_0.

(b) $\bar{B}$ is a basis of the root system Σ.

8. (continuation) Let

$$r = \operatorname{Card} B = \dim \mathfrak{h}_{\boldsymbol{R}}, \qquad r_0 = \operatorname{Card} B_0, \qquad l = \dim \mathfrak{a},$$

and define the *Satake diagram* of (B, θ) as follows. Every root of B_0 is denoted by a black circle • and every root of $B - B_0$ by a white circle ○. If $\alpha, \beta \in B - B_0$ are such that $\bar{\alpha} = \bar{\beta}$, then α and β are joined by a curved arrow ↷. In Table VI we list for each simple non-complex Lie algebra over $\boldsymbol{R}$:

(i) the Satake diagram of (B, θ);
(ii) the Dynkin diagram of $\bar{B}$;
(iii) the multiplicities m_λ, $m_{2\lambda}$ for $\lambda \in \bar{B}$.

The table is reproduced from Araki [1]; for the classical spaces the root system Σ and the multiplicities were given by Cartan [10]; for the exceptional spaces he stated the result without proof. In the table we follow the notation of Table V with the following refinements (Cartan [10]):

A IV	for	$\mathfrak{su}(p, 1)$,	$p > 1$.
B I	for	$\mathfrak{so}(p, q)$	$p + q$ odd, $p \geqslant q > 1$.
D I	for	$\mathfrak{so}(p, q)$	$p + q$ even, $p \geqslant q > 1$.
B II	for	$\mathfrak{so}(p, 1)$	p even.
D II	for	$\mathfrak{so}(p, 1)$	p odd.

The Dynkin diagrams for B with the same numbering as in §3, no. 5, occur in the second column; in the third column we list the restrictions $\lambda_i = \bar{\alpha}_i$ forming the basis $\bar{B}$.

TABLE VI

SATAKE DIAGRAMS, RESTRICTED ROOT SYSTEMS, AND MULTIPLICITIES

$\mathfrak{g}$	Satake diagram of (B, θ)	Dynkin diagram of $\bar{B}$	m_λ.	$m_{2\lambda}$.
A I	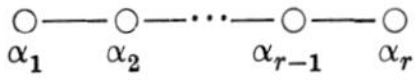	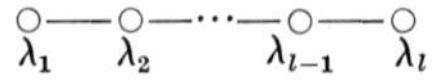	1	0
A II	 $r = 2l + 1$	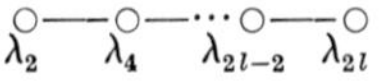	4	0
A III	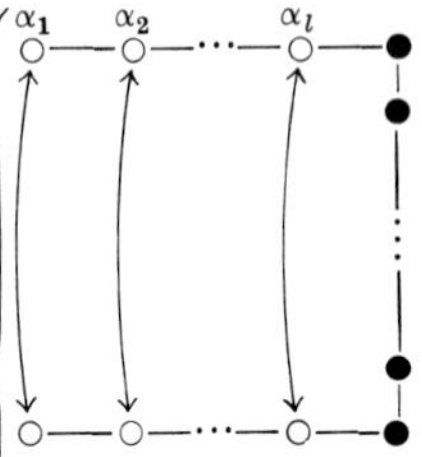	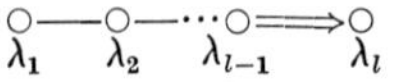 $(2 \leqslant l \leqslant \frac{1}{2}r)$	2 $(i < l)$	0
			$2(r - 2l + 1)$ $(i = l)$	1
	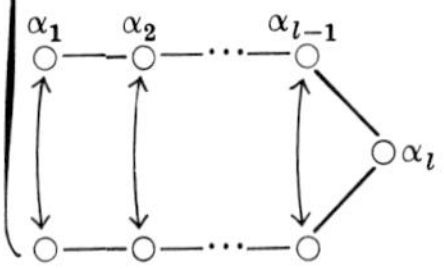	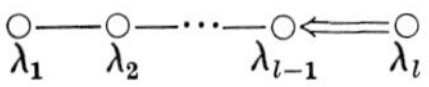 $r = 2l - 1$	2 $(i \leqslant l - 1)$	0
			1 $(i = l)$	0
A IV			$2(r - 1)$	1
B I	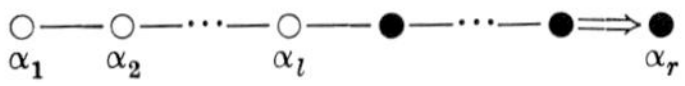	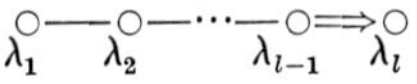 $(2 \leqslant l \leqslant r)$	1 $(i < l)$	0
			$2(r - l) + 1$ $(i = l)$	0
B II	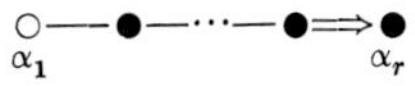		$2r - 1$	0
C I	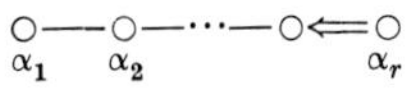	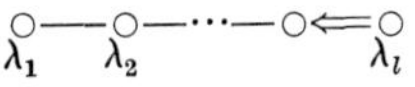 $(l = r)$	1	0
C II	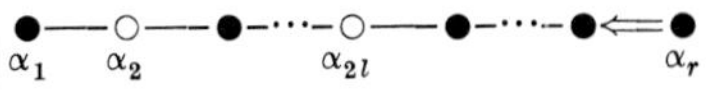	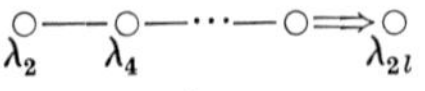 $1 \leqslant l \leqslant \frac{1}{2}(r - 1)$	4 $(i \neq 2l)$	0
			$4(r - 2l)$ $(i = 2l)$	3
	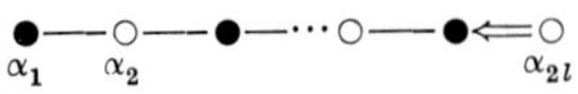	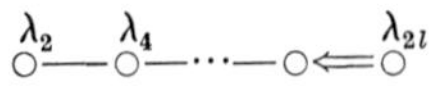 $(2 \leqslant l = \frac{1}{2}r)$	4 $(i \neq 2l)$	0
			3 $(i = 2l)$	0

Table continued

TABLE VI (*continued*)

$\mathfrak{g}$	Satake diagram of (B, θ)	Dynkin diagram of $\bar{B}$	m_{λ_i}	$m_{2\lambda_i}$
D I	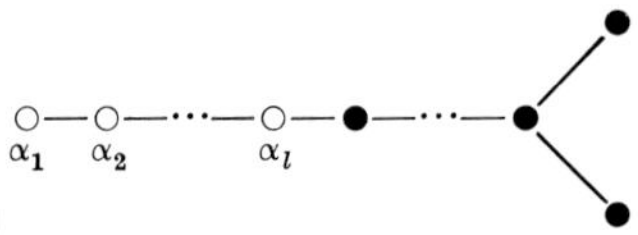	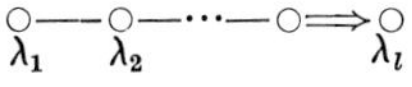 $(2 \leqslant l \leqslant r - 2)$	1 $(i < l)$ $2(r - l)$ $(i = l)$	0 0
	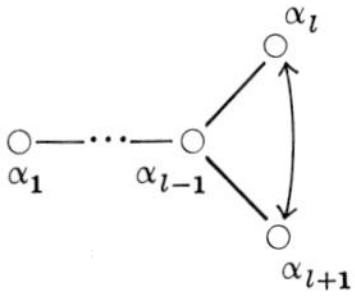	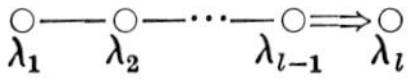	1 $(i < l)$ 2 $(i = l)$	0 0
	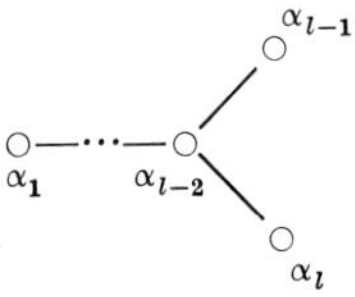	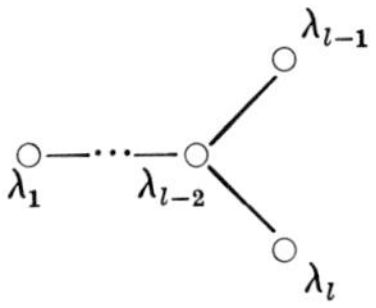	1	0
D II	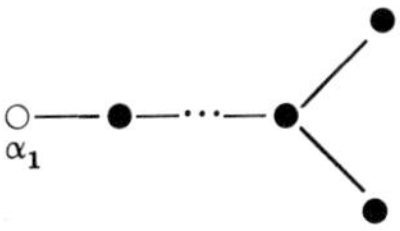		$2(r - 1)$	0
D III	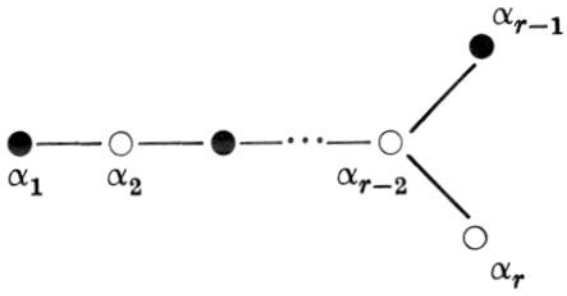	 $r = 2l$	4 $(i \neq 2l)$ 1 $(i = 2l)$	0 0
	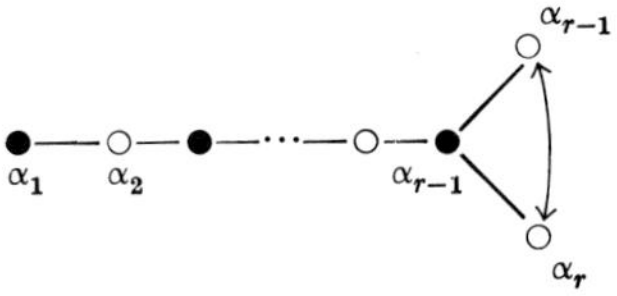	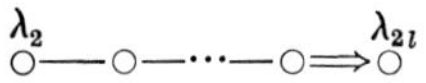 $r = 2l + 1$	4 $(i \neq 2l)$ 4 $(i = 2l)$	0 1

Table continued

TABLE VI (*continued*)

$\mathfrak{g}$	Satake diagram of (B, θ)	Dynkin diagram of $\bar{B}$	m_{λ}.	$m_{2\lambda}$.
E I	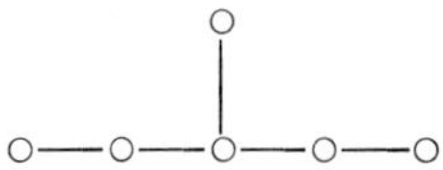	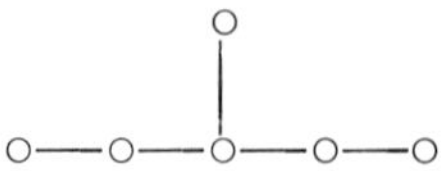	1	0
E II	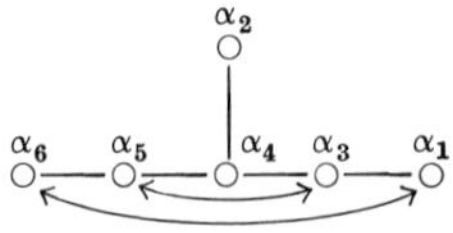	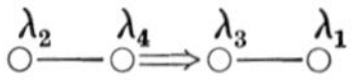	1 $(i = 2, 4)$ 2 $(i = 1, 3)$	0 0
E III	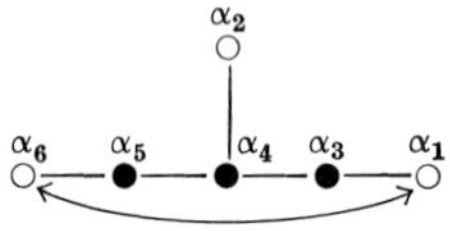	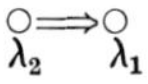	8 $(i = 1)$ 6 $(i = 2)$	1 0
E IV	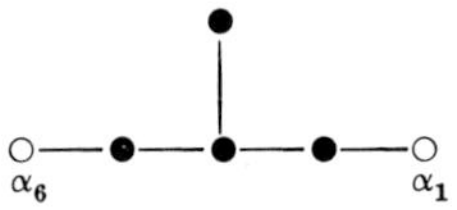	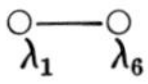	8	0
E V	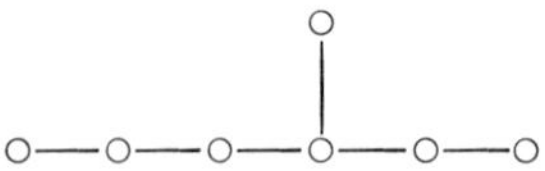	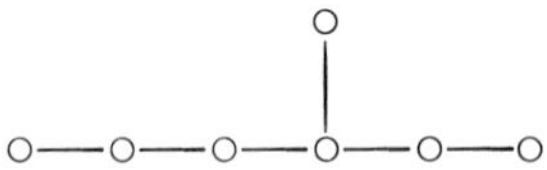	1	0
E VI	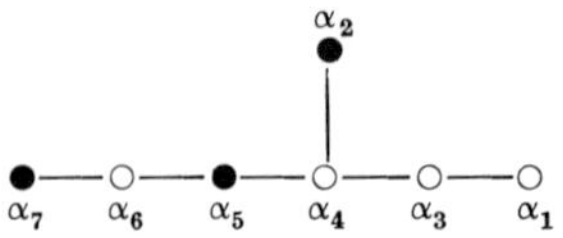	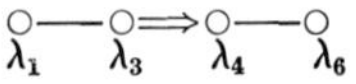	1 $(i = 1, 3)$ 4 $(i = 2, 4)$	0 0
E VII	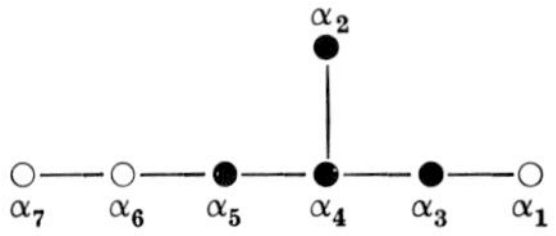	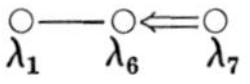	8 $(i = 1, 6)$ 1 $(i = 7)$	0 0
E VIII	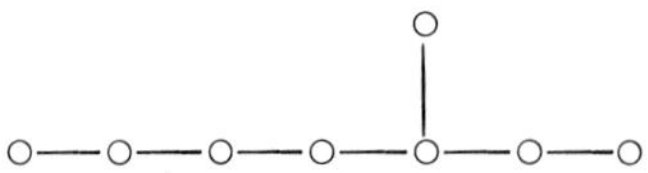	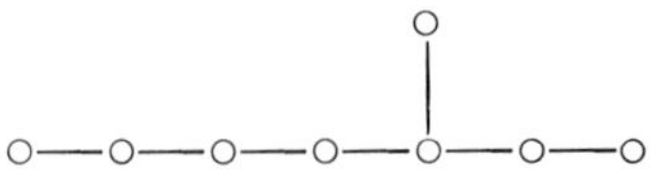	1	0
E IX	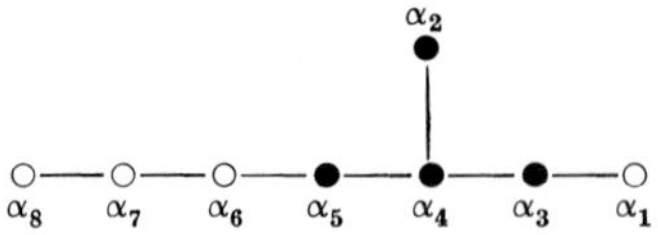	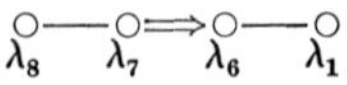	8 $(i = 6, 1)$ 1 $(i = 7, 8)$	
F I	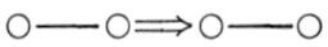		1	0
F II	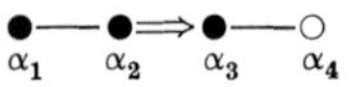		8	7
G		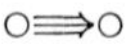	1	0

9. Verify the following statements on the basis of the table above:

(a) A simple Lie algebra $\mathfrak{g}$ over $\boldsymbol{R}$ is determined up to isomorphism by its Satake diagram.

(b) A simple Lie algebra $\mathfrak{g}$ over $\boldsymbol{R}$ is determined by the triple $(\mathfrak{a}, \Sigma, m)$ where Σ is the restricted root system and m is the multiplicity function.

G. Two-Point Homogeneous Spaces

A Riemannian manifold M with distance function d is called *two-point homogeneous* if for any two pairs $p, q \in M, p', q' \in M$, satisfying $d(p, q) = d(p', q')$ there exists an element $g \in I(M)$, the group of isometries of M, such that $g \cdot p = p'$, $g \cdot q = q'$. A Riemannian manifold M is called *isotropic* if for each $p \in M$ the linear isotropy group $d(I(M)_p)$ acts transitively on the unit sphere in the tangent space M_p.

1. A Riemannian manifold is two-point homogeneous if and only if it is isotropic.

2*. A Riemannian manifold is isotropic if and only if it is either a Euclidean space or a Riemannian globally symmetric space of rank one.

This result is a consequence of Wang's classification [1] of compact two-point homogeneous spaces and Tits' classification [1] of isotropic homogeneous manifolds (Riemannian or not). For the noncompact case an a priori proof is given in Nagano [1], Helgason [3], p. 252; for the compact case alternatives to Wang's proof can be found in Varma [1], Freudenthal [6], Wolf [4], Matsumoto [1], the last proof using no classification.

The two-point homogeneous spaces are therefore $\boldsymbol{R}^n$, the simply connected compact and noncompact spaces

$$\boldsymbol{A\,III}(q=1),\ \boldsymbol{BD\,I}(q=1),\ \boldsymbol{C\,II}(q=1),\ \boldsymbol{F\,II},$$

and, in accordance with Exercise C4, the real projective spaces $\boldsymbol{SO}(n+1)/\boldsymbol{O}(n)$.

NOTES

§1-§4. Lie's theorem about the differential equation $dy/dy = Y/X$ quoted in Chapter II, §8 and its various generalizations suggested to him the problem of classifying all local transformation groups of $\boldsymbol{R}^n$. Only for small n ($n \leqslant 6$) did this turn out to be a manageable problem (Lie [1], Page [1]). Killing realized that the problem could be split in two; first to classify the underlying abstract groups and then their representations as transformation groups. So Killing set himself the algebraic problem of finding all possible ways, up to isomorphism, in which an r-dimensional vector space can be turned into a Lie algebra. On 18 October 1887 he wrote to Engel that he had found a complete classification of the simple Lie algebras over $\boldsymbol{C}$. The exceptional Lie algebras $\mathfrak{e}_6$, $\mathfrak{e}_7$, $\mathfrak{e}_8$, $\mathfrak{f}_4$, $\mathfrak{g}_2$ are Killing's most remarkable discovery, although the indication in his second paper, p. 48, can hardly be considered a proof of their existence (except for $\mathfrak{g}_2$). The principal result of his papers is that these algebras, together with the classical ones $\mathfrak{a}_l$, $\mathfrak{b}_l$,$\mathfrak{c}_l$, and $\mathfrak{d}_l$, exhaust the class of all simple Lie algebras over $\boldsymbol{C}$. Killing's method was founded on a detailed analysis of the solutions of the characteristic equation $\det(\lambda I - \operatorname{ad} X) = 0$ and a classification of all possibilities for the matrix (a_{ij}) in §3, no. 2. (For $i \neq j$, $-a_{ij}$ is the largest integer q for which $\alpha_i + q\alpha_j$ is a root.)

In his thesis, Cartan [1] gives genuine proofs of the classification results stated by Killing, having pointed out a number of significant errors and gaps in Killing's papers. Generously he does not list the lack of existence proofs for the exceptional algebras among these gaps; however, in a paper in *Leipziger Berichte* 1893, published before his thesis, he indicates specific transformation groups in spaces of dimensions 16, 27, 57, 15 and 5, respectively, realizing the exceptional Lie algebras (for $\mathfrak{g}_2$ this was done simultaneously by Engel [1]) and states that such groups cannot exist in spaces of lower dimension. Some others were less charitable in their comments on Killing's papers: "C'était là un résultat d'une très haute importance; malheureusement toutes les démonstrations étaient fausses; il ne restait que des aperçus dénués de toute force probant" (Poincaré [2]); similar opinions are expressed by Lie [1], Vol. III.

Simplified treatments of the classification over $\boldsymbol{C}$ (not the construction of the algebras) based on Theorem 5.4, Chapter III were given by van der Waerden [1] and Dynkin [1]; see also Freudenthal [3]. While a basis for the root system was already used by Killing, the concept of a simple root seems to have originated with Dynkin [1]. Coxeter (in [1] and in Weyl [2]) and later Witt [2] classified all finite groups generated by reflections and applied the result to the classification problem.

The construction of the simple Lie algebras is a more difficult problem. One way of stating it is as follows: Let (a_{ij}) be a nonsingular $l \times l$ matrix such that (1) $a_{ii} = 2$, $a_{ij} \leqslant 0$ $(i \neq j)$ and $a_{ij} = 0$ whenever $a_{ji} = 0$; (2) the group W generated by the linear transformations $s_i : x_j \to x_j - a_{ji}x_i$ is finite. Show that there exists a semisimple Lie algebra $\mathfrak{g}$ over $\boldsymbol{C}$ whose Cartan matrix is (a_{ij}). A result of this type was proved geometrically by Witt [2], assuming it holds for $l \leqslant 4$. A general algebraic proof without this proviso is due to Chevalley [9] and Harish-Chandra [3]. Using similar tools, Serre [1], Chapter VI proved the existence with conditions (1) and (2) replaced by the axioms of a root system. This is a substantial simplification since (2) is cumbersome to verify for the exceptional Lie algebras. Tits [6] gives yet another method.

In §1 the classification problem is broken up into more specialized problems. Proposition 1.2 shows how the global symmetry of a space of type II forces on it a group structure. Prop. 1.5 goes back to Cartan [2], p. 267. In §2 we describe the customary group theoretic models of the classical symmetric spaces; a more unified description by means of positive involutions of semisimple associative algebras is given by Weil [3].

In §3 we give the now standard classification of root systems. We use the definition of Dynkin [1], Araki [1], Serre [1], and Bourbaki [2], Chapitre VI, and classify the possible Coxeter graphs by the method of Dynkin [1]. The next step is the construction of a reduced root system for each diagram (cf. Pontragin [1], §66,

Serre [1], Chapter V, Bourbaki [2], Chapter VI), giving Theorem 3.23. The factor group $\mathrm{Aut}(R)/W(R)$ (in Theorem 3.29) was determined by Cartan [4], and in [9] he determines $\tilde{Z}$ and the corresponding classification of compact simple Lie groups (cf. Exercise C.1). A detailed treatment was given by Dynkin and Oniščik[1]. In [10] he generalizes this method to give a global classification of the compact irreducible symmetric spaces (cf. Exercises C.2-4). The proof of Theorem 3.32 follows Bourbaki [2], Chapter VI. The method of presentation in §4 of the constructive part of the classification is due to Chevalley [9], Harish-Chandra [3], Part I and Serre [1], Chapter VI.

§5. In his paper [2] Cartan classifies the simple Lie algebras over $\boldsymbol{R}$. His method, which required formidable computations, used the signature of the Killing form although it often happens that two nonisomorphic real forms of the same complex algebra have the same signature. Cartan's statement ([2], p. 263): "Les groupes réels d'ordre r qui correspondent à une même type complexe d'ordre r se classent en général complètement d'après leur *caractère*," is therefore not to be taken literally; cf. Lardy ([1], p. 195). After noticing the equivalence of problems B and B′ (§1) Cartan (in [12]) simplified his original treatment (see also Lardy [1]). Following his general theory [1] of automorphism of complex simple Lie groups, Gantmacher [2] gave a simplified treatment of the real classification. For further developments of this method see Murakami [3], Wallach [2], and Freudenthal and de Vries [1]. While Gantmacher used a Cartan subalgebra $\mathfrak{h}$ of $\mathfrak{g}$ whose "toral part" $\mathfrak{h} \cap \mathfrak{k}$ is maximal abelian in $\mathfrak{k}$, Araki develops in [1] a new method using a Cartan subalgebra $\mathfrak{h} \subset \mathfrak{g}$ whose "vector part" $\mathfrak{h} \cap \mathfrak{p}$ is maximal abelian in $\mathfrak{p}$. In addition to a solution to problem B′ (§1) this method gives valuable information about the restricted roots and their multiplicities (cf. Exercises F). A modification is given by Sugiura [2]. In the present work we use the method of Kač [1] which at the same time gives a rather explicit description of the automorphism σ of finite order. The method amounts to a development of the weight theory for the natural representations of $\mathfrak{g}$ on the infinite-dimensional algebras $L(\mathfrak{g}, \sigma)$. In a much more general situation the algebras of this type ("Kač-Moody algebras") have been studied by Kač [4] and independently by Moody [1, 2]. The graphs in the diagrams $S(A)$ occur already in Coxeter [1].

SOLUTIONS TO EXERCISES

CHAPTER I

A. Manifolds

A.1. First take a covering $\{V_\alpha\}_{\alpha \in I}$ of A by open relatively compact sets V_α disjoint from B. Then take a covering $\{V_\beta\}_{\beta \in J}$ of the closed set $M - \bigcup_{\alpha \in I} V_\alpha$ by open relatively compact sets V_β disjoint from A. The covering $\{V_\alpha\}_{\alpha \in I}, \{V_\beta\}_{\beta \in J}$ of M has a locally finite refinement $\{W_\gamma\}_{\gamma \in \Gamma}$. If $\{\varphi_\gamma\}_{\gamma \in \Gamma}$ is a partition of unity subordinate to this covering, put $f = \sum_{W_\gamma \cap A \neq \emptyset} \varphi_\gamma$.

A.2. If p_1, $p_2 \in M$ are sufficiently close within a coordinate neighborhood U, there exists a diffeomorphism mapping p_1 to p_2 and leaving $M - U$ pointwise fixed. Now consider a curve segment $\gamma(t)$ $(0 \leqslant t \leqslant 1)$ in M joining p to q. Let t^* be the supremum of those t for which there exists a diffeomorphism of M mapping p on $\gamma(t)$. The initial remark shows first that $t^* > 0$, next that $t^* = 1$, and finally that t^* is reached as a maximum.

A.3. The "only if" is obvious and "if" follows from the uniqueness in Prop. 1.1. Now let $\mathfrak{F} = C^\infty(\boldsymbol{R})$ where $\boldsymbol{R}$ is given the ordinary differentiable structure. If n is an odd integer, let $\mathfrak{F}^n$ denote the set of functions $x \to f(x^n)$ on $\boldsymbol{R}$, $f \in \mathfrak{F}$ being arbitrary. Then $\mathfrak{F}^n$ satisfies $\mathfrak{F}_1$, $\mathfrak{F}_2$, $\mathfrak{F}_3$. Since $\mathfrak{F}^n \neq \mathfrak{F}^m$ for $n \neq m$, the corresponding δ^n are all different.

A.4. (i) If $d\Phi \cdot X = Y$ and $f \in C^\infty(N)$, then $X(f \circ \Phi) = (Yf) \circ \Phi \in \mathfrak{F}_0$. On the other hand, suppose $X\mathfrak{F}_0 \subset \mathfrak{F}_0$. If $F \in \mathfrak{F}_0$, then $F = g \circ \Phi$ where $g \in C^\infty(N)$ is unique. If $f \in C^\infty(N)$, then $X(f \circ \Phi) = g \circ \Phi$ ($g \in C^\infty(N)$ unique), and $f \to g$ is a derivation, giving Y.

(ii) If $d\Phi \cdot X = Y$, then $Y_{\Phi(p)} = d\Phi_p(X_p)$, so necessity follows. Suppose $d\Phi_p(M_p) = N_{\Phi(p)}$ for each $p \in M$. Define for $r \in N$, $Y_r = d\Phi_p(X_p)$ if $r = \Phi(p)$. In order to show that $Y : r \to Y_r$ is differentiable we use (by virtue of Theorem 15.5) coordinates around p and around $r = \Phi(p)$ such that Φ has the expression $(x_1, ..., x_m) \to (x_1, ..., x_n)$. Writing

$$X = \sum_1^m a_i(x_1, ..., x_m) \frac{\partial}{\partial x_i},$$

we have for q sufficiently near p

$$d\Phi_q(X_q) = \sum_1^n a_i(x_1(q), \ldots, x_m(q)) \left(\frac{\partial}{\partial x_i}\right)_{\Phi(q)},$$

so condition (1) implies that for $1 \leqslant i \leqslant n$, a_i is constant in the last $m - n$ arguments. Hence

$$Y = \sum_1^n a_i(x_1, \ldots, x_n, x_{n+1}(p), \ldots, x_m(p)) \frac{\partial}{\partial x_i}.$$

(iii) $f \in C^\infty(N)$ if and only if $f \circ \psi \in C^\infty(\boldsymbol{R})$. If $f(x) = x^3$, then $f \circ \psi(x) = x$, $(f' \circ \psi)(x) = 3x^{\frac{2}{3}}$, so $f \in C^\infty(N)$, $f' \notin C^\infty(N)$. Hence $f \circ \Phi \in \mathfrak{F}_0$, but $X(f \circ \Phi) \notin \mathfrak{F}_0$; so by (i), X is not projectable.

A.5. Obvious.

A.6. Use Props. 15.2 and 15.3 to shrink the given covering to a new one; then use the result of Exercise A.1 to imitate the proof of Theorem 1.3.

A.7. We can assume $M = \boldsymbol{R}^m$, $p = 0$, and that $X_0 = (\partial/\partial t_1)_0$ in terms of the standard coordinate system $\{t_1, \ldots, t_m\}$ on $\boldsymbol{R}^m$. Consider the integral curve $\varphi_t(0, c_2, \ldots, c_m)$ of X through $(0, c_2, \ldots, c_m)$. Then the mapping $\psi : (c_1, \ldots, c_m) \to \varphi_{c_1}(0, c_2, \ldots, c_m)$ is C^∞ for small c_i, $\psi(0, c_2, \ldots, c_m) = (0, c_2, \ldots, c_m)$, so

$$d\psi_0 \left(\frac{\partial}{\partial c_i}\right) = \left(\frac{\partial}{\partial t_i}\right)_0 \qquad (i > 1).$$

Also

$$d\psi_0 \left(\frac{\partial}{\partial c_1}\right)_0 = \left(\frac{\partial \varphi_{c_1}}{\partial c_1}\right)(0) = X_0 = \left(\frac{\partial}{\partial t_1}\right)_0.$$

Thus ψ can be inverted near 0, so $\{c_1, \ldots, c_m\}$ is a local coordinate system. Finally, if $c = (c_1, \ldots, c_m)$,

$$\begin{aligned} \left(\frac{\partial}{\partial c_1}\right)_{\psi(c)} f &= \left(\frac{\partial (f \circ \psi)}{\partial c_1}\right)_c \\ &= \lim_{h \to 0} \frac{1}{h} [f(\varphi_{c_1+h}(0, c_2, \ldots, c_m)) - f(\varphi_{c_1}(0, c_2, \ldots, c_m)] \\ &= (Xf)(\psi(c)) \end{aligned}$$

so $X = \partial/\partial c_1$.

A.8. Let $f \in C^\infty(M)$. Writing $\sim$ below when in an equality we omit terms of higher order in s or t, we have

$$
\begin{aligned}
& f(\psi_{-t}(\varphi_{-s}(\psi_t(\varphi_s(o))))) - f(o) \\
&\quad = f(\psi_{-t}(\varphi_{-s}(\psi_t(\varphi_s(o))))) - f(\varphi_{-s}(\psi_t(\varphi_s(o)))) \\
&\qquad + f(\varphi_{-s}(\psi_t(\varphi_s(o)))) - f(\psi_t(\varphi_s(o))) \\
&\qquad + f(\psi_t(\varphi_s(o))) - f(\varphi_s(o)) + f(\varphi_s(o)) - f(o) \\
&\quad \sim -t(Yf)(\varphi_{-s}(\psi_t(\varphi_s(o)))) + \tfrac{1}{2}t^2(Y^2f)(\varphi_{-s}(\psi_t(\varphi_s(o)))) \\
&\qquad - s(Xf)(\psi_t(\varphi_s(o))) + \tfrac{1}{2}s^2(X^2f)(\psi_t(\varphi_s(o))) \\
&\qquad + t(Yf)(\psi_t(\varphi_s(o))) - \tfrac{1}{2}t^2(Y^2f)(\psi_t(\varphi_s(o))) \\
&\qquad + s(Xf)(\varphi_s(o)) - \tfrac{1}{2}s^2(X^2f)(\varphi_s(o)) \\
&\quad \sim st(XYf)(\psi_t(\varphi_s(o))) - st(YXf)(\psi_t(\varphi_s(o))).
\end{aligned}
$$

This last expression is obtained by pairing off the 1st and 5th term, the 3rd and 7th, the 2nd and 6th, and the 4th and 8th. Hence

$$f(\gamma(t^2)) - f(o) = t^2([X, Y]f)(o) + O(t^3).$$

A similar proof is given in Faber [1].

B. The Lie Derivative and the Interior Product

B.1. If the desired extension of $\theta(X)$ exists and if $C : \mathfrak{D}_1^1(M) \to C^\infty(M)$ is the contraction, then (i), (ii), (iii) imply

$$(\theta(X)\omega)(Y) = X(\omega(Y)) - \omega([X, Y]), \qquad X, Y \in \mathfrak{D}^1(M).$$

Thus we define $\theta(X)$ on $\mathfrak{D}_1(M)$ by this relation and note that $(\theta(X)\omega)(fY) = f(\theta(X)(\omega))(Y)$ $(f \in C^\infty(M))$, so $\theta(X)$ $\mathfrak{D}_1(M) \subset \mathfrak{D}_1(M)$. If U is a coordinate neighborhood with coordinates $\{x_1, \ldots, x_m\}$, $\theta(X)$ induces an endomorphism of $C^\infty(U)$, $\mathfrak{D}^1(U)$, and $\mathfrak{D}_1(U)$. Putting $X_i = \partial/\partial x_i$, $\omega_j = dx_j$, each $T \in \mathfrak{D}_s^r(U)$ can be written

$$T = \sum T_{(i),(j)} X_{i_1} \otimes \ldots \otimes X_{i_r} \otimes \omega_{j_1} \otimes \ldots \otimes \omega_{j_s}$$

with unique coefficients $T_{(i),(j)} \in C^\infty(U)$. Now $\theta(X)$ is uniquely extended to $\mathfrak{D}(U)$ satisfying (i) and (ii). Property (iii) is then verified by induction on r and s. Finally, $\theta(X)$ is defined on $\mathfrak{D}(M)$ by the condition $\theta(X)T \mid U = \theta(X)(T \mid U)$ (vertical bar denoting restriction) because as in the proof of Theorem 2.5 this condition is forced by the requirement that $\theta(X)$ should be a derivation.

B.2. The first part being obvious, we just verify $\Phi \cdot \omega = (\Phi^{-1})^*\omega$. We may assume $\omega \in \mathfrak{D}_1(M)$. If $X \in \mathfrak{D}^1(M)$ and C is the contraction $X \otimes \omega \to \omega(X)$, then $\Phi \circ C = C \circ \Phi$ implies $(\Phi \cdot \omega)(X) = \Phi(\omega(X^{\Phi^{-1}})) = ((\Phi^{-1})^*\omega)(X)$.

B.3. The formula is obvious if $T = f \in C^\infty(M)$. Next let $T = Y \in \mathfrak{D}^1(M)$. If $f \in C^\infty(M)$ and $q \in M$, we put $F(t, q) = f(g_t \cdot q)$ and have

$$F(t, q) - F(0, q) = t \int_0^1 \left(\frac{\partial F}{\partial t}\right)(st, q)\, ds = t\, h(t, q),$$

where $h \in C^\infty(\boldsymbol{R} \times M)$ and $h(0, q) = (Xf)(q)$. Then

$$(g_t \cdot Y)_p f = (Y(f \circ g_t))(g_t^{-1} \cdot p) = (Yf)(g_t^{-1} \cdot p) + t(Yh)(t, g_t^{-1} \cdot p)$$

so

$$\lim_{t \to 0} \frac{1}{t} (Y - g_t \cdot Y)_p f = (XYf)(p) - (YXf)(p),$$

so the formula holds for $T \in \mathfrak{D}^1(M)$. But the endomorphism $T \to \lim_{t\to 0} t^{-1}(T - g_t \cdot T)$ has properties (i), (ii), and (iii) of Exercise B.1; it coincides with $\theta(X)$ on $C^\infty(M)$ and on $\mathfrak{D}^1(M)$, hence on all of $\mathfrak{D}(M)$ by the uniqueness in Exercise B.1.

B.4. For (i) we note that both sides are derivations of $\mathfrak{D}(M)$ commuting with contractions, preserving type, and having the same effect on $\mathfrak{D}^1(M)$ and on $C^\infty(M)$. The argument of Exercise B.1 shows that they coincide on $\mathfrak{D}(M)$.

(ii) If $\omega \in \mathfrak{D}_r(M)$, $Y_1, \ldots, Y_r \in \mathfrak{D}^1(M)$, then by B.1,

$$(\theta(X)\omega)(Y_1, \ldots, Y_r) = X(\omega(Y_1, \ldots, Y_r)) - \sum_i \omega(Y_1, \ldots, [X, Y_i], \ldots, Y_r)$$

so $\theta(X)$ commutes with A.

(iii) Since $\theta(X)$ is a derivation of $\mathfrak{A}(M)$ and d is a *skew-derivation* (that is, satisfies (iv) in Theorem 2.5), the commutator $\theta(X)d - d\theta(X)$ is also a skew-derivation. Since it vanishes on f and df ($f \in C^\infty(M)$), it vanishes identically (cf. Exercise B.1). For B.1–B.4, cf. Palais [3].

B.5. This is done by the same method as in Exercise B.1.

B.6. For (i) we note that by (iii) in Exercise B.5, $i(X)^2$ is a derivation. Since it vanishes on $C^\infty(M)$ and $\mathfrak{D}_1(M)$, it vanishes identically; (ii) follows by induction; (iii) follows since both sides are skew-derivations which coincide on $C^\infty(M)$ and on $\mathfrak{A}_1(M)$; (iv) follows because both sides are derivations which coincide on $C^\infty(M)$ and on $\mathfrak{A}_1(M)$.

C. Affine Connections

C.1. M has a locally finite covering $\{U_\alpha\}_{\alpha \in A}$ by coordinate neighborhoods U_α. On U_α we construct an arbitrary Riemannian structure g_α. If $1 = \sum_\alpha \varphi_\alpha$ is a partition of unity subordinate to the covering, then $\sum_\alpha \varphi_\alpha g_\alpha$ gives the desired Riemannian structure on M.

C.2. If Φ is an affine transformation and we write $d\Phi(\partial/\partial x_j) = \sum_i a_{ij}\, \partial/\partial x_i$, then conditions ∇_1 and ∇_2 imply that each a_{ij} is a constant. If A is the linear transformation (a_{ij}), then $\Phi \circ A^{-1}$ has differential I, hence is a translation B, so $\Phi(X) = AX + B$. The converse is obvious.

C.3. We have $\Phi^*\omega_j^i = \sum_k (\Gamma_{kj}^i \circ \Phi)\, \Phi^*\omega^k$, so by (5′), (6), (7) in §8

$$\Phi^*\omega_j^i = \sum_k (\Gamma_{kj}^i \circ \Phi)(a_k\, dt + t\, da_k) = 0.$$

This implies that $\Gamma_{kj}^i \equiv 0$ in normal coordinates, which is equivalent to the result stated in the exercise.

C.4. A direct verification shows that the mapping $\delta : \theta \to \sum_1^m \omega_i \wedge \nabla_{X_i}(\theta)$ is a skew-derivation of $\mathfrak{A}(M)$ and that it coincides with d on $C^\infty(M)$. Next let $\theta \in \mathfrak{A}_1(M)$, $X, Y \in \mathfrak{D}^1(M)$. Then, using (5), §7,

$$\begin{aligned}
2\,\delta\theta(X, Y) &= 2 \sum_i (\omega_i \wedge \nabla_{X_i}(\theta))(X, Y) \\
&= \sum_i \omega_i(X)\, \nabla_{X_i}(\theta)(Y) - \omega_i(Y)\, \nabla_{X_i}(\theta)(X) \\
&= \nabla_X(\theta)(Y) - \nabla_Y(\theta)(X) \\
&= X \cdot \theta(Y) - \theta(\nabla_X(Y)) - Y \cdot \theta(X) + \theta(\nabla_Y(X)),
\end{aligned}$$

which since the torsion is 0 equals

$$X\theta(Y) - Y \cdot \theta(X) - \theta([X, Y]) = 2\, d\theta(X, Y).$$

Thus $\delta = d$ on $\mathfrak{A}_1(M)$, hence by the above on all of $\mathfrak{A}(M)$.

C.5. No; an example is given by a circular cone with the vertex rounded off.

C.6. Using Props. 11.3 and 11.4 we obtain a mapping $\varphi : M \to N$ such that $d\varphi_p$ is an isometry for each $p \in M$. Thus $\varphi(M) \subset N$ is an open subset. Each geodesic in the manifold $\varphi(M)$ is indefinitely extendable, so $\varphi(M)$ is complete, whence φ maps M onto N. Now Lemma 13.4 implies that (M, φ) is a covering space of N, so M and N are isometric.

D. Submanifolds

D.1. Let $I : G_\Phi \to M \times N$ denote the identity mapping and $\pi : M \times N \to M$ the projection onto the first factor. Let $m \in M$ and $Z \in (G_\Phi)_{(m, \Phi(m))}$ such that $dI_m(Z) = 0$. Then $Z = (d\varphi)_m(X)$ where $X \in M_m$. Thus $d\pi \circ dI \circ d\varphi(X) = 0$. But since $\pi \circ I \circ \varphi$ is the identity mapping, this implies $X = 0$, so $Z = 0$ and I is regular.

D.2. Immediate from Lemma 14.1.

D.3. Consider the figure 8 given by the formula

$$\gamma(t) = (\sin 2t, \sin t) \qquad (0 \leqslant t \leqslant 2\pi).$$

Let $f(s)$ be an increasing function on $\boldsymbol{R}$ such that

$$\lim_{s \to -\infty} f(s) = 0, \qquad f(0) = \pi, \qquad \lim_{s \to +\infty} f(s) = 2\pi.$$

Then the map $s \to \gamma(f(s))$ is a bijection of $\boldsymbol{R}$ onto the figure 8. Carrying the manifold structure of $\boldsymbol{R}$ over, we get a submanifold of $\boldsymbol{R}^2$ which is closed, yet does not carry the induced topology. Replacing γ by δ given by $\delta(t) = (-\sin 2t, \text{sint } t)$, we get another manifold structure on the figure.

D.4. Suppose $\dim M < \dim N$. Using the notation of Prop. 3.2, let W be a compact neighborhood of p in M and $W \subset U$. By the countability assumption, countably many such W cover M. Thus by Lemma 3.1, Chapter II, for N, some such W contains an open set in N; contradiction.

D.5. For each $m \in M$ there exists by Prop. 3.2 an open neighborhood V_m of m in N and an extension of g from $V_m \cap M$ to a C^∞ function G_m on V_m. The covering $\{V_m\}_{m \in M}$, $N - M$ of N has a countable locally finite refinement $V_1, V_2, \ldots$. Let $\varphi_1, \varphi_2, \ldots$ be the corresponding partition of unity. Let $\varphi_{i_1}, \varphi_{i_2}, \ldots$ be the subsequence of the (φ_j) whose supports intersect M, and for each φ_{i_p} choose $m_p \in M$ such that $\operatorname{supp}(\varphi_{i_p}) \subset V_{m_p}$. Then $\Sigma_p \, G_{m_p} \varphi_{i_p}$ is the desired function G.

D.6. The "if" part is contained in Theorem 14.5 and the "only if" part is immediate from (2), Chapter V, §6.

E. Curvature

E.1. If (r, θ) are polar coordinates of a vector X in the tangent space M_p, the inverse of the map $(r, \theta) \to \operatorname{Exp}_p X$ gives the "geodesic polar coordinates" around p. Since the geodesics from p intersect sufficiently small circles around p orthogonally (Lemma 9.7), the Riemannian structure has the form $g = dr^2 + \varphi(r, \theta)^2 \, d\theta^2$. In these coordinates the Riemannian measure $f \to \int f \sqrt{\bar{g}} \, dx_1 \ldots dx_n$ and the Laplace-Beltrami operator are, respectively, given by

$$f \to \iint f(r, \theta) \, \varphi(r, \theta) \, dr \, d\theta,$$

and

$$\Delta f = \frac{\partial^2 f}{\partial r^2} + \varphi^{-1} \frac{\partial \varphi}{\partial r} \frac{\partial f}{\partial r} + \varphi^{-1} \frac{\partial}{\partial \theta} \left(\varphi^{-1} \frac{\partial f}{\partial \theta} \right).$$

In particular

$$\Delta(\log r) = -\frac{1}{r^2} + \frac{1}{r\varphi}\frac{\partial\varphi}{\partial r}.$$

On the other hand, if (x, y) are the normal coordinates of $\operatorname{Exp}_p X$ such that

$$r^2 = x^2 + y^2, \qquad \tan\theta = \frac{y}{x},$$

then, since $r\,dr = x\,dx + y\,dy$, $r^2\,d\theta = x\,dy - y\,dx$,

$$g = r^{-4}[(x^2r^2 + y^2\varphi^2)\,dx^2 + 2xy(r^2 - \varphi^2)\,dx\,dy + (y^2r^2 + x^2\varphi^2)\,dy^2]$$

so since the coefficients are smooth near $(x, y) = (0, 0)$ φ^2 has the form

$$\varphi^2 = r^2 + cr^4 + \ldots,$$

where $c = c(p)$ is a constant. But then

$$\lim_{r\to 0} \Delta(\log r) = c(p).$$

On the other hand,

$$A(r) = \int_0^r \int_0^{2\pi} \varphi(t, \theta)\,dt\,d\theta,$$

so using the definition in §12 we find $K = -3c(p)$ as stated.

This result is stated in Klein [1], p. 219, without proof (with opposite sign).

E.2. Let $X = \partial/\partial x_1$ and $Y = \partial/\partial x_2$ so γ_ϵ is formed by integral curves of X, Y, $-X$, $-Y$.

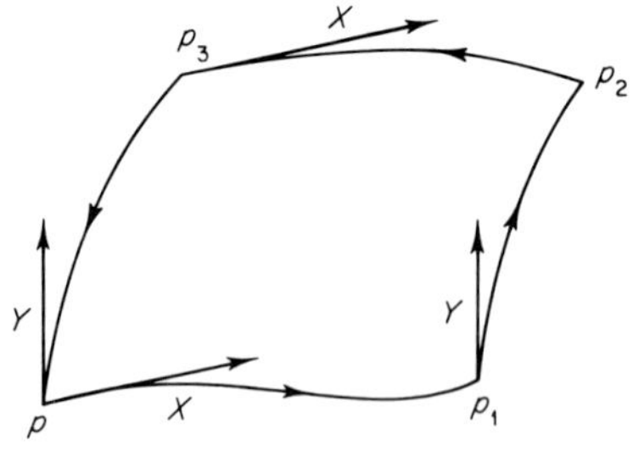

Let $p = p_0 = (0, 0, \ldots, 0)$

$p_1 = (\epsilon, 0, \ldots, 0)$

$p_2 = (\epsilon, \epsilon, \ldots, 0)$

$p_3 = (0, \epsilon, \ldots, 0)$

and τ_{ij} the parallel transport from p_j to p_i along γ_ϵ. Let T be any vector field on M, and write $T_i = T_{p_i}$. Then

$$\begin{aligned}
&\tau_{03}\tau_{32}\tau_{21}\tau_{10}T_0 - T_0 \\
&\quad = (\tau_{03}\tau_{32}\tau_{21}\tau_{10}T_0 - \tau_{03}\tau_{32}\tau_{21}T_1) + (\tau_{03}\tau_{32}\tau_{21}T_1 - \tau_{03}\tau_{32}T_2) \\
&\qquad + (\tau_{03}\tau_{32}T_2 - \tau_{03}T_3) + (\tau_{03}T_3 - T_0).
\end{aligned}$$

We use Theorem 7.1 and write $\sim$ when we omit terms of higher order in ϵ. Then our expression is

$$\sim \tau_{03}\tau_{32}\tau_{21}[-\epsilon(\nabla_X T)_1 + \tfrac{1}{2}\epsilon^2(\nabla_X^2 T)_1]$$

$$+ \tau_{03}\tau_{32}[-\epsilon(\nabla_Y T)_2 + \tfrac{1}{2}\epsilon^2(\nabla_Y^2 T)_2]$$

$$- \tau_{03}\tau_{32}[-\epsilon(\nabla_X T)_2 + \tfrac{1}{2}\epsilon^2(\nabla_X^2 T)_2]$$

$$- \tau_{03}[-\epsilon(\nabla_Y T)_3 + \tfrac{1}{2}\epsilon^2(\nabla_Y^2 T)_3].$$

Combining now the 1st and 5th term, 2nd and 6th term, etc., this expression reduces to

$$\sim \epsilon^2\tau_{03}\tau_{32}(\nabla_Y(\nabla_X(T)))_2 - \epsilon^2\tau_{03}(\nabla_X(\nabla_Y(T))_3$$

which, since $[X, Y] = 0$, reduces to

$$\sim \epsilon^2\tau_{03}(R(Y, X)T)_3 \sim \epsilon^2(R(Y, X)T)_0.$$

This proof is a simplification of that of Faber [1]. See Laugwitz [1], §10 for another version of the result. For curvature and holonomy groups, see e.g. Ambrose and Singer [2].

F. Surfaces

F.1. Let Z be a vector field on S and $\tilde{X}$, $\tilde{Y}$, $\tilde{Z}$ vector fields on a neighborhood of s in $\boldsymbol{R}^3$ extending X, Y, and Z, respectively. The inner product $\langle\ ,\ \rangle$ on $\boldsymbol{R}^3$ induces a Riemannian structure g on S. If $\tilde{\nabla}$ and ∇ denote the corresponding affine connections on $\boldsymbol{R}^3$ and S, respectively, we deduce from (2), §9

$$\langle \tilde{Z}_s, \tilde{\nabla}_{\tilde{X}}(\tilde{Y})_s\rangle = g(Z_s, \nabla_X(Y)_s).$$

But

$$\tilde{\nabla}_{\tilde{X}}(\tilde{Y})_s = \lim_{t\to 0} \frac{1}{t}(Y_{\gamma(t)} - Y_s),$$

so we obtain $\nabla = \nabla'$; in particular ∇' is an affine connection on S.

F.2. Let $s(u, v) \to (u, v)$ be local coordinates on S and if g denotes the Riemannian structure on S, put

$$E = g\left(\frac{\partial}{\partial u}, \frac{\partial}{\partial u}\right), \qquad F = g\left(\frac{\partial}{\partial u}, \frac{\partial}{\partial v}\right), \qquad G = g\left(\frac{\partial}{\partial v}, \frac{\partial}{\partial v}\right).$$

Let $r(u, v)$ denote the vector from 0 to the point $s(u, v)$. Subscripts denoting partial derivatives, r_u and r_v span the tangent space at $s(u, v)$, and we may take the orientation such that

$$\xi_{s(u,v)} = \frac{r_u \times r_v}{|r_u \times r_v|},$$

$\times$ denoting the cross product. We have

$$\dot{\gamma}_S = r_u \dot{u} + r_v \dot{v}$$
$$\ddot{\gamma}_S = r_{uu}\dot{u}^2 + 2r_{uv}\dot{u}\dot{v} + r_{vv}\dot{v}^2 + r_u\ddot{u} + r_v\ddot{v},$$

and

$$r_u \cdot r_u = E, \qquad r_u \cdot r_v = F, \qquad r_v \cdot r_v = G,$$

whence

$$r_{uu} \cdot r_u = \tfrac{1}{2}E_u, \qquad r_{uv} \cdot r_u = \tfrac{1}{2}E_v, \qquad r_{vv} \cdot r_v = \tfrac{1}{2}G_v,$$
$$r_{uv} \cdot r_v = \tfrac{1}{2}G_u, \qquad r_{uu} \cdot r_v = F_u - \tfrac{1}{2}E_v, \qquad r_{vv} \cdot r_u = F_v - \tfrac{1}{2}G_u.$$

From this it is clear that the geodesic curvature can be expressed in terms of $\dot{u}$, $\dot{v}$, $\ddot{u}$, $\ddot{v}$, E, F, G, and their derivatives, and therefore has the invariance property stated.

F.3. We first recall that under the orthogonal projection P of $\boldsymbol{R}^3$ on the tangent space $S_{\gamma_S(t)}$ the curve $P \circ \gamma_S$ has curvature in $\gamma_S(t)$ equal to the geodesic curvature of γ_S at $\gamma_S(t)$. So in order to avoid discussing developable surfaces we define the rolling in the problem as follows. Let $\pi = S_{\gamma_S(t_0)}$ and let $t \rightarrow \gamma_\pi(t)$ be the curve in π such that

$$\gamma_\pi(t_0) = \gamma_S(t_0), \qquad \dot{\gamma}_\pi(t_0) = \dot{\gamma}_S(t_0)$$

($t - t_0$ is the arc-parameter measured from $\gamma_\pi(t_0)$) and such that the curvature of γ_π at $\gamma_\pi(t)$ is the geodesic curvature of γ_S at $\gamma_S(t)$. The rolling is understood as the family of isometries $S_{\gamma_S(t)} \rightarrow \pi_{\gamma_\pi(t)}$ of the tangent planes such that the vector $\dot{\gamma}_s(t)$ is mapped onto $\dot{\gamma}_\pi(t)$. Under these maps a Euclidean parallel family of unit vectors along γ_π corresponds to a family $Y(t) \in S_{\gamma_S(t)}$. We must show that this family is parallel in the sense of (1), §5. Let τ denote the angle between $\dot{\gamma}_S(t)$ and $Y(t)$. Then

$$\begin{aligned}\dot{\tau}(t) &= -\text{curvature of } \gamma_\pi \text{ at } \gamma_\pi(t)\\ &= -\text{geodesic curvature of } \gamma_S \text{ at } \gamma_S(t)\\ &= -(\xi \times \dot{\gamma}_S \cdot \ddot{\gamma}_S)(t).\end{aligned}$$

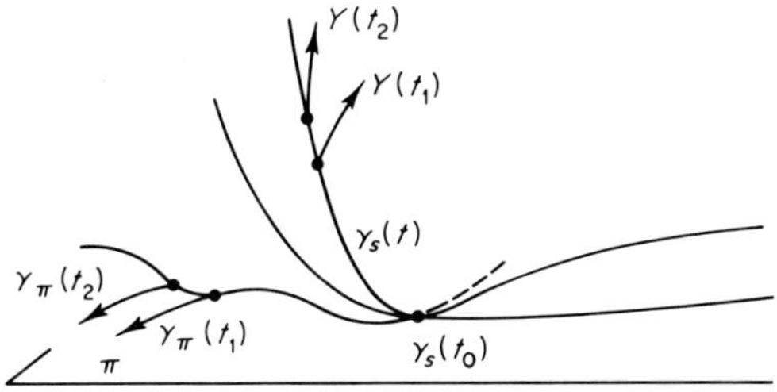

We can choose the coordinates (u, v) near $\gamma_S(t_0)$ such that for t close to t_0

$$u(\gamma_S(t)) = t, \qquad v(\gamma_S(t)) = \text{const.}, \qquad g_{\gamma_S(t)}\left(\frac{\partial}{\partial u}, \frac{\partial}{\partial v}\right) = 0.$$

(For example, let $r \to \delta_t(r)$ be a geodesic in S starting at $\gamma_S(t)$ perpendicular to γ_S; small pieces of these geodesics fill up (disjointly) a neigborhood of $\gamma_S(t_0)$; the mapping $\delta_t(r) \to (t, r)$ is a coordinate system with the desired properties.) Writing $Y(t) = Y^1(t)\, r_u + Y^2(t) r_v$ (using notation from previous exercise), we have

$$Y^1(t) = \cos \tau(t), \qquad Y^2(t) = G^{-1/2} \sin \tau(t) \tag{1}$$

and shall now verify (2), §5. By (2), §9 we have

$$2 \sum_l g_{lk} \Gamma^l_{ij} = \frac{\partial}{\partial x_i} g_{jk} + \frac{\partial}{\partial x_j} g_{ik} - \frac{\partial}{\partial x_k} g_{ij}.$$

On the curve γ_S we have $E \equiv 1$, $F \equiv 0$, so

$$\Gamma^1_{11} = 0, \qquad \Gamma^2_{11} = -\frac{E_v}{2G}, \qquad \Gamma^1_{12} = \frac{E_v}{2},$$

$$\Gamma^1_{22} = F_v - \frac{G_u}{2}, \qquad \Gamma^2_{22} = \frac{G_v}{2G}, \qquad \Gamma^2_{12} = \frac{G_u}{2G}.$$

Thus we must verify

$$\dot{Y}^1 + \frac{1}{2} E_v Y^2 = 0, \qquad \dot{Y}^2 - \frac{E_v}{2G} Y^1 + \frac{G_u}{2G} Y^2 = 0. \tag{2}$$

But using formulas from Exercise F.2 we find

$$\dot{\tau}(t) = -(\xi \times \dot{\gamma}_S \cdot \ddot{\gamma}_S)(t) = \tfrac{1}{2}(G^{-1/2}E_v)(\gamma_S(t)))$$

and now equations (2) follow directly from (1).

G. The Hyperbolic Plane

1. (i) and (ii) are obvious. (iii) is clear since

$$\frac{x'(t)^2}{(1-x(t)^2)^2} \leqslant \frac{x'(t)^2 + y'(t)^2}{(1-x(t)^2-y(t)^2)^2}$$

where $\gamma(t) = (x(t), y(t))$. For (iv) let $z \in D$, $u \in D_z$, and let $z(t)$ be a curve with $z(0) = z$, $z'(t) = u$. Then

$$d\varphi_z(u) = \left\{\frac{d}{dt}\varphi(z(t))\right\}_{t=0} = \frac{z'(0)}{(\bar{b}z+\bar{a})^2} \qquad \text{at} \quad \varphi \cdot z,$$

and $g(d\varphi(u), d\varphi(u)) = g(u, u)$ now follows by direct computation. Now (v) follows since φ is conformal and maps lines into circles. The first relation in (vi) is immediate; and writing the expression for $d(0, x)$ as a cross ratio of the points -1, 0, x, 1, the expression for $d(z_1, z_2)$ follows since φ in (iv) preserves cross ratio. For (vii) let τ be any isometry of D. Then there exists a φ as in (iv) such that $\varphi\tau^{-1}$ leaves the x-axis pointwise fixed. But then $\varphi\tau^{-1}$ is either the identity or the complex conjugation $z \to \bar{z}$. For (viii) we note that if $r = d(0, z)$, then $|z| = \tanh r$; so the formula for g follows from (ii). Part (ix) follows from

$$v = \frac{1-|z|^2}{|z-i|^2}, \qquad dw = -2\frac{dz}{(z-i)^2}, \qquad d\bar{w} = -2\frac{d\bar{z}}{(\bar{z}+i)^2}.$$

CHAPTER II

A. On the Geometry of Lie Groups

A.1. (i) follows from $\exp \operatorname{Ad}(x)tX = x \exp tXx^{-1} = L(x)\, R(x^{-1}) \exp tX$ for $X \in \mathfrak{g}$, $t \in \boldsymbol{R}$. For (ii) we note $J(x \exp tX) = \exp(-tX)\, x^{-1}$, so $dJ_x(dL(x)_e X) = -dR(x^{-1})_e X$. For (iii) we observe for $X_0, Y_0 \in \mathfrak{g}$

$$\begin{aligned}\Phi(g \exp tX_0, h \exp sY_0) &= g \exp tX_0 h \exp sY_0\\ &= gh \exp t \operatorname{Ad}(h^{-1})\, X_0 \exp sY_0,\end{aligned}$$

so

$$d\Phi(dL(g)X_0, dL(h)Y_0) = dL(gh)(\mathrm{Ad}(h^{-1})X_0 + Y_0).$$

Putting $X = dL(g)X_0$, $Y = dL(h)Y_0$, the result follows from (i).

A.2. Suppose $\gamma(t_1) = \gamma(t_2)$ so $\gamma(t_2 - t_1) = e$. Let $L > 0$ be the smallest number such that $\gamma(L) = e$. Then $\gamma(t + L) = \gamma(t)\,\gamma(L) = \gamma(t)$. If τ_L denotes the translation $t \to t + L$, we have $\gamma \circ \tau_L = \gamma$, so

$$\dot{\gamma}(0) = d\gamma\left(\frac{d}{dt}\right)_0 = d\gamma\left(\frac{d}{dt}\right)_L = \dot{\gamma}(L).$$

A.3. The curve σ satisfies $\sigma(t + L) = \sigma(t)$, so as in A.2, $\dot{\sigma}(0) = \dot{\sigma}(L)$.

A.4. Let (p_n) be a Cauchy sequence in G/H. Then if d denotes the distance, $d(p_n, p_m) \to 0$ if $m, n \to \infty$. Let $B_\epsilon(o)$ be a relatively compact ball of radius $\epsilon > 0$ around the origin $o = \{H\}$ in G/H. Select N such that $d(p_N, p_m) < \frac{1}{2}\epsilon$ for $m \geqslant N$ and select $g \in G$ such that $g \cdot p_N = o$. Then $(g \cdot p_m)$ is a Cauchy sequence inside the compact ball $B_\epsilon(o)^-$, hence it, together with the original sequence, is convergent.

A.5. For $X \in \mathfrak{g}$ let $\tilde{X}$ denote the corresponding left invariant vector field on G. From Prop. 1.4 we know that (i) is equivalent to $\nabla_{\tilde{Z}}(\tilde{Z}) = 0$ for all $Z \in \mathfrak{g}$. But by (2), §9 in Chapter I this condition reduces to

$$g(\tilde{Z}, [\tilde{X}, \tilde{Z}]) = 0 \qquad (X, Z \in \mathfrak{g})$$

which is clearly equivalent to (ii). Next (iii) follows from (ii) by replacing X by $X + Z$. But (iii) is equivalent to $\mathrm{Ad}(G)$-invariance of B so Q is right invariant. Finally, the map $J : x \to x^{-1}$ satisfies $J = R(g^{-1}) \circ J \circ L(g^{-1})$, so $dJ_g = dR(g^{-1})_e \circ dJ_e \circ dL(g^{-1})_g$. Since dJ_e is automatically an isometry, (v) follows.

A.6. Assuming first the existence of ∇, consider the affine transformation $\sigma : g \to \exp \frac{1}{2}Yg^{-1} \exp \frac{1}{2}Y$ of G which fixes the point $\exp \frac{1}{2}Y$ and maps γ_1, the first half of γ, onto the second half, γ_2. Since

$$\sigma = L(\exp \tfrac{1}{2}Y) \circ J \circ L(\exp -\tfrac{1}{2}Y),$$

we have $d\sigma_{\exp \frac{1}{2}Y} = -I$. Let $X^*(t) \in G_{\exp tY}$ $(0 \leqslant t \leqslant 1)$ be the family of vectors parallel with respect to γ such that $X^*(0) = X$. Then σ maps $X^*(s)$ along γ_1 into a parallel field along γ_2 which must be the field $-X^*(t)$ because $d\sigma(X^*(\frac{1}{2})) = -X^*(\frac{1}{2})$. Thus the map $\sigma \circ J = L(\exp \frac{1}{2}Y)\, R(\exp \frac{1}{2}Y)$ sends X into $X^*(1)$, as stated in part (i). Part (ii) now follows from Theorem 7.1, Chapter I, and part (iii) from Prop. 1.4. Now (iv) follows from (ii) and the definition of T and R.

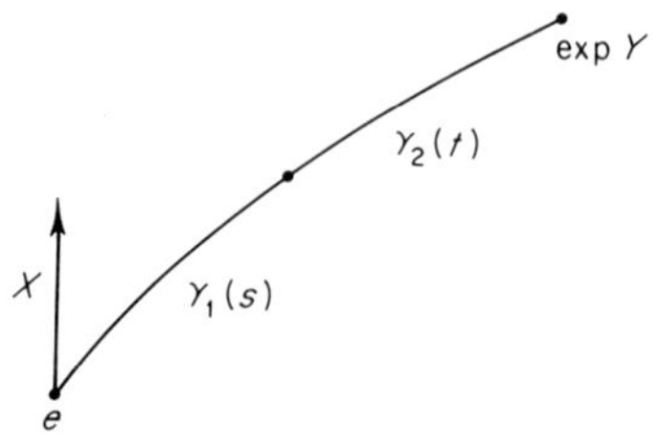

Finally, we prove the existence of ∇. As remarked before Prop. 1.4, the equation $\nabla_{\tilde{X}}(\tilde{Y}) = \frac{1}{2}[\tilde{X}, \tilde{Y}]$ $(X, Y \in \mathfrak{g})$ defines uniquely a left invariant affine connection ∇ on G. Since $\tilde{X}^{R(g)} = (\mathrm{Ad}(g^{-1})X)^{\sim}$, we get

$$\nabla_{\tilde{X}^{R(g)}}(\tilde{Y}^{R(g)}) = \tfrac{1}{2}\{\mathrm{Ad}(g^{-1})[X, Y]\}^{\sim} = (\nabla_{\tilde{X}}(\tilde{Y}))^{R(g)};$$

this we generalize to any vector fields Z, Z' by writing them in terms of $\tilde{X}_i$ $(1 \leqslant i \leqslant n)$. Next

$$\nabla_{J\tilde{X}}(J\tilde{Y}) = J(\nabla_{\tilde{X}}(\tilde{Y})).$$

Since both sides are right invariant vector fields, it suffices to verify the equation at e. Now $J\tilde{X} = -\bar{X}$ where $\bar{X}$ is right invariant, so the problem is to prove

$$(\nabla_{\bar{X}}(\bar{Y}))_e = -\tfrac{1}{2}[X, Y].$$

For a basis $X_1, \ldots, X_n$ of $\mathfrak{g}$ we write $\mathrm{Ad}(g^{-1})Y = \sum_i f_i(g)X_i$. Since $\bar{Y}_g = dR(g)Y = dL(g)\,\mathrm{Ad}(g^{-1})Y$, it follows that $\bar{Y} = \sum_i f_i\tilde{X}_i$, so using ∇_2 and Lemma 4.2 from Chapter I, §4,

$$(\nabla_{\bar{X}}(\bar{Y}))_e = (\nabla_{\tilde{X}}(\bar{Y}))_e = \sum_i (Xf_i)_e\, X_i + \tfrac{1}{2}\sum_i f_i(e)[\tilde{X}, \tilde{X}_i]_e$$

Since $(Xf_i)(e) = \{(d/dt)\, f_i(\exp tX)\}_{t=0}$ and since

$$\left\{\frac{d}{dt}\,\mathrm{Ad}(\exp(-tX))(Y)\right\}_{t=0} = -[X, Y],$$

the expression on the right reduces to $-[X, Y] + \frac{1}{2}[X, Y]$, so (1) follows. As before, (1) generalizes to any vector fields Z, Z'.

The connection ∇ is the 0-connection of Cartan-Schouten [1].

B. The Exponential Mapping

B.1. At the end of §1 it was shown that $GL(2, \boldsymbol{R})$ has Lie algebra $\mathfrak{gl}(2, \boldsymbol{R})$, the Lie algebra of all 2×2 real matrices. Since $\det(e^{tX}) =$

$e^{t\,\mathrm{Tr}(X)}$, Prop. 2.7 shows that $\mathfrak{sl}(2, \boldsymbol{R})$ consists of all 2×2 real matrices of trace 0. Writing

$$X = a\begin{pmatrix}1 & 0\\0 & -1\end{pmatrix} + b\begin{pmatrix}0 & 1\\0 & 0\end{pmatrix} + c\begin{pmatrix}0 & 0\\1 & 0\end{pmatrix} = \begin{pmatrix}a & b\\c & -a\end{pmatrix}$$

a direct computation gives for the Killing form

$$B(X, X) = 8(a^2 + bc) = 4\,\mathrm{Tr}(XX),$$

whence $B(X, Y) = 4\,\mathrm{Tr}(XY)$, and semisimplicity follows quickly. Part (i) is obtained by direct computation. For (ii) we consider the equation

$$e^X = \begin{pmatrix}\lambda & 0\\0 & \lambda^{-1}\end{pmatrix} \qquad (\lambda \in \boldsymbol{R},\quad \lambda \neq 1).$$

Case 1: $\lambda > 0$. Then $\det X < 0$. In fact $\det X = 0$ implies

$$I + X = \begin{pmatrix}\lambda & 0\\0 & \lambda^{-1}\end{pmatrix},$$

so $b = c = 0$, so $a = 0$, contradicting $\lambda \neq 1$. If $\det X > 0$, we deduce quickly from (i) that $b = c = 0$, so $\det X = -a^2$, which is a contradiction. Thus $\det X < 0$ and using (i) again we find the only solution

$$X = \begin{pmatrix}\log\lambda & 0\\0 & -\log\lambda\end{pmatrix}.$$

Case 2: $\lambda = -1$. For $\det X > 0$ put $\mu = (\det X)^{1/2}$. Then using (i) the equation amounts to

$$\cos\mu + (\mu^{-1}\sin\mu)a = -1, \qquad (\mu^{-1}\sin\mu)b = 0,$$
$$\cos\mu - (\mu^{-1}\sin\mu)a = -1, \qquad (\mu^{-1}\sin\mu)c = 0.$$

These equations are satisfied for

$$\mu = (2n + 1)\pi \quad (n \in \boldsymbol{Z}), \qquad \det X = -a^2 - bc = (2n + 1)^2\,\pi^2.$$

This gives infinitely many choices for X as claimed.

Case 3: $\lambda < 0$, $\lambda \neq -1$. If $\det X = 0$, then (i) shows $b = c = 0$, so $a = 0$; impossible. If $\det X > 0$ and we put $\mu = (\det X)^{1/2}$, (i) implies

$$\cos\mu + (\mu^{-1}\sin\mu)a = \lambda, \qquad (\mu^{-1}\sin\mu)b = 0,$$
$$\cos\mu - (\mu^{-1}\sin\mu)a = \lambda^{-1}, \qquad (\mu^{-1}\sin\mu)c = 0.$$

Since $\lambda \neq \lambda^{-1}$, we have $\sin \mu \neq 0$. Thus $b = c = 0$, so $\det X = -a^2$, which is impossible. If $\det X < 0$ and we put $\mu = (-\det X)^{1/2}$, we get from (i) the equations above with sin and cos replaced by sinh and cosh. Again $b = c = 0$, so $\det X = -a^2 = -\mu^2$; thus $a = \pm\mu$, so

$$\cosh \mu \pm \sinh \mu = \lambda, \qquad \cosh \mu \mp \sinh \mu = \lambda^{-1},$$

contradicting $\lambda < 0$. Thus there is no solution in this case, as stated.

B.2. The Killing form on $\mathfrak{sl}(2, \boldsymbol{R})$ provides a bi-invariant pseudo-Riemannian structure with the properties of Exercise A.5. Thus (i) follows from Exercise B.1. Each $g \in \boldsymbol{SL}(2, \boldsymbol{R})$ can be written $g = kp$ where $k \in \boldsymbol{SO}(2)$ and p is positive definite. Clearly $k = \exp T$ where $T \in \mathfrak{sl}(2, \boldsymbol{R})$; and using diagonalization, $p = \exp X$ where $X \in \mathfrak{sl}(2, \boldsymbol{R})$. The formula $g = \exp T \exp X$ proves (ii).

B.3. Follow the hint.

B.4. Considering one-parameter subgroups it is clear that $\mathfrak{g}$ consists of the matrices

$$X(a, b, c) = \begin{pmatrix} 0 & c & 0 & a \\ -c & 0 & 0 & b \\ 0 & 0 & 0 & c \\ 0 & 0 & 0 & 0 \end{pmatrix} \qquad (a, b, c \in \boldsymbol{R}).$$

Then $[X(a, b, c), \ X(a_1, b_1, c_1)] = X(cb_1 - c_1 b, c_1 a - c a_1, 0)$, so $\mathfrak{g}$ is readily seen to be solvable. A direct computation gives

$$\exp X(a, b, c) = \begin{pmatrix} \cos c & \sin c & 0 & c^{-1}(a \sin c - b \cos c + b) \\ -\sin c & \cos c & 0 & c^{-1}(b \sin c + a \cos c - a) \\ 0 & 0 & 1 & c \\ 0 & 0 & 0 & 1 \end{pmatrix}.$$

Thus $\exp X(a, b, 2\pi)$ is the same point in G for all $a, b \in \boldsymbol{R}$, so exp is not injective. Similarly, the points in G with $\gamma = n2\pi$ $(n \in \boldsymbol{Z})$ $\alpha^2 + \beta^2 > 0$ are not in the range of exp. This example occurs in Auslander and MacKenzie [1]; the exponential mapping for a solvable group is systematically investigated in Dixmier [2].

B.5. Let N_0 be a bounded star-shaped open neighborhood of $0 \in \mathfrak{g}$ which exp maps diffeomorphically onto an open neighborhood N_e of e in G. Let $N^* = \exp(\frac{1}{2}N_0)$. Suppose S is a subgroup of G contained in N^*, and let $s \neq e$ in S. Then $s = \exp X$ $(X \in \frac{1}{2}N_0)$. Let $k \in \boldsymbol{Z}^+$ be such that $X, 2X, \ldots, kX \in \frac{1}{2}N_0$ but $(k+1)X \notin \frac{1}{2}N_0$. Since N_0 is star-shaped, $(k+1)X \in N_0$; but since $s^{k+1} \in N^*$, we have $s^{k+1} = \exp Y$, $Y \in \frac{1}{2}N_0$. Since exp is one-to-one on N_0, $(k+1)X = Y \in \frac{1}{2}N_0$, which is a contradiction.

C. Subgroups and Transformation Groups

C.1. The proofs given in Chapter X for $SU^*(2n)$ and $Sp(n, C)$ generalize easily to the other subgroups.

C.2. Let G be a commutative connected Lie group, $(\tilde{G}, \pi)$ its universal covering group. By facts stated during the proof of Theorem 1.11, $\tilde{G}$ is topologically isomorphic to a Euclidean group R^p. Thus G is topologically isomorphic to a factor group of R^p and by a well-known theorem on topological groups (e.g. Bourbaki [1], Chap. VII) this factor group is topologically isomorphic to $R^n \times T^m$. Thus by Theorem 2.6, G is analytically isomorphic to $R^n \times T^m$.

For the last statement let $\bar{\gamma}$ be the closure of γ in H. By the first statement and Theorem 2.3, $\bar{\gamma} = R^n \times T^m$ for some n, $m \in Z^+$. But γ is dense in $\bar{\gamma}$, so either $n = 1$ and $m = 0$ (γ closed) or $n = 0$ ($\bar{\gamma}$ compact).

C.3. By Theorem 2.6, I is analytic and by Lemma 1.12, dI is injective. Q.E.D.

C.4. The mapping ψ_g turns $g \cdot N_0$ into a manifold which we denote by $(g \cdot N_0)_x$. Similarly, $\psi_{g'}$ turns $g' \cdot N_0$ into a manifold $(g' \cdot N_0)_y$. Thus we have two manifolds $(g \cdot N_0 \cap g' \cdot N_0)_x$ and $(g \cdot N_0 \cap g' \cdot N_0)_y$ and must show that the identity map from one to the other is analytic. Consider the analytic section maps

$$\sigma_g : (g \cdot N_0)_x \to G, \qquad \sigma_{g'} : (g' \cdot N_0)_y \to G$$

defined by

$$\sigma_g(g \exp(x_1X_1 + \dots + x_rX_r) \cdot p_0) = g \exp(x_1X_1 + \dots + x_rX_r),$$
$$\sigma_{g'}(g' \exp(y_1X_1 + \dots + y_rX_r) \cdot p_0) = g' \exp(y_1X_1 + \dots + y_rX_r),$$

and the analytic map

$$J_g : \pi^{-1}(g \cdot N_0) \to (g \cdot N_0)_x \times H$$

given by

$$J_g(z) = (\pi(z), [\sigma_g(\pi(z))]^{-1}z).$$

Furthermore, let $P : (g \cdot N_0)_x \times H \to (g \cdot N_0)_x$ denote the projection on the first component. Then the identity mapping

$$I : (g \cdot N_0 \cap g' \cdot N_0)_y \to (g \cdot N_0 \cap g' \cdot N_0)_x$$

can be factored:

$$(g \cdot N_0 \cap g' \cdot N_0)_y \xrightarrow{\sigma_{g'}} \pi^{-1}(g \cdot N_0) \xrightarrow{J_g} (g \cdot N_0)_x \times H \xrightarrow{P} (g \cdot N_0)_x.$$

In fact, if $p \in g \cdot N_0 \cap g' \cdot N_0$, we have

$$p = g \exp(x_1 X_1 + \dots + x_r X_r) \cdot p_0 = g' \exp(y_1 X_1 + \dots + y_r X_r) \cdot p_0,$$

so for some $h \in H$,

$$\begin{aligned} P(J_g(\sigma_{g'}(p))) &= P(J_g(g' \exp(y_1 X_1 + \dots + y_r X_r))) \\ &= P(\pi(g' \exp(y_1 X_1 + \dots + y_r X_r)), h) \\ &= P(\pi(g \exp(x_1 X_1 + \dots + x_r X_r)), h) \\ &= g \exp(x_1 X_1 + \dots + x_r X_r)) \cdot p_0. \end{aligned}$$

Thus I is composed of analytic maps so is analytic, as desired.

C.5. The subgroup $H = G_p$ of G leaving p fixed is closed, so G/H is a manifold. The map $I : G/H \to M$ given by $I(gH) = g \cdot p$ gives a bijection of G/H onto the orbit $G \cdot p$. Carrying the differentiable structure over on $G \cdot p$ by means of I, it remains to prove that $I : G/H \to M$ is everywhere regular. Consider the maps on the diagram

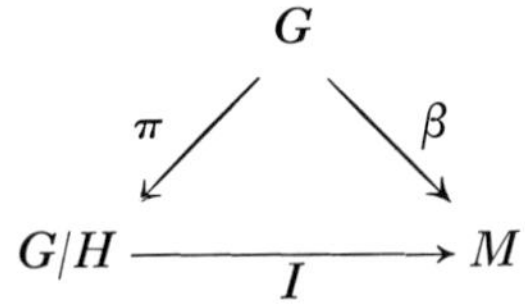

where $\pi(g) = gH$, $\beta(g) = g \cdot p$ so $\beta = I \circ \pi$. If we restrict π to a local cross section, we can write $I = \beta \circ \pi^{-1}$ on a neighborhood of the origin in G/H. Thus I is C^∞ near the origin, hence everywhere. Moreover, the map $d\beta_e : \mathfrak{g} \to M_p$ has kernel $\mathfrak{h}$, the Lie algebra of H (cf. proof of Prop. 4.3). Since $d\pi_e$ maps $\mathfrak{g}$ onto $(G/H)_H$ with kernel $\mathfrak{h}$ and since $d\beta_e = dI_H \circ d\pi_e$, wee that dI_H is one-to-one. Finally, if $T(g)$ denotes the diffeomorphism $m \to g \cdot m$ of M, we have $I = T(g) \circ I \circ \tau(g^{-1})$, whence

$$dI_{gH} = dT(g)_p \circ dI_H \circ d\tau(g^{-1})_{gH},$$

so I is everywhere regular.

C.6. By local connectedness each component of G is open. It acquires an analytic structure from that of G_0 by left translation. In order to show the map $\varphi : (x, y) \to xy^{-1}$ analytic at a point $(x_0, y_0) \in G \times G$ let G_1 and G_2 denote the components of G containing x_0 and y_0, respectively. If $\varphi_0 = \varphi \mid G_0 \times G_0$ and $\psi = \varphi \mid G_1 \times G_2$, then

$$\psi = L(x_0 y_0^{-1}) \circ I(y_0) \circ \varphi_0 \circ L(x_0^{-1}, y_0^{-1}),$$

where $I(y_0)(x) = y_0 x y_0^{-1}$ $(x \in G_0)$. Now $I(y_0)$ is a continuous automorphism of the Lie group G_0, hence by Theorem 2.6, analytic; so the expression for ψ shows that it is analytic.

C.8. If N with the indicated properties exists we may, by translation, assume it passes through the origin $o = \{H\}$ in M. Let L be the subgroup $\{g \in G : g \cdot N = N\}$. If $g \in G$ maps o into N, then $gN \cap N \neq \emptyset$; so by assumption, $gN = N$. Thus $L = \pi^{-1}(N)$ where $\pi : G \to G/H$ is the natural map. Using Theorem 15.5, Chapter I we see that L can be given the structure of a submanifold of G with a countable basis and by the transitivity of G on M, $L \cdot o = N$. By C.7, L has the desired property. For the converse, define $N = L \cdot o$ and use Prop. 4.4 or Exercise C.5. Clearly, if $gN \cap N \neq \emptyset$, then $g \in L$, so $gN = N$.

For more information on the primitivity notion which goes back to Lie see e.g. Golubitsky [1].

D. Closed Subgroups

D.1. $\boldsymbol{R}^2/\Gamma$ is a torus (Exercise C.2), so it suffices to take a line through 0 in $\boldsymbol{R}^2$ whose image in the torus is dense.

D.2. $\mathfrak{g}$ has an Int($\mathfrak{g}$)-invariant positive definite quadratic form Q. The proof of Prop. 6.6 now shows $\mathfrak{g} = \mathfrak{z} + \mathfrak{g}'$ ($\mathfrak{z}$ = center of $\mathfrak{g}$, $\mathfrak{g}' = [\mathfrak{g}, \mathfrak{g}]$ compact and semisimple). The groups Int($\mathfrak{g}$) and Int($\mathfrak{g}'$) are analytic subgroups of $\boldsymbol{GL}(\mathfrak{g})$ with the same Lie algebra so coincide.

D.3. We have

$$\alpha_{0,\frac{1}{3}}(c_1, c_2, s) = (c_1, e^{2\pi i/3}c_2, s)$$

$$\begin{aligned}(a_1, a_2, r)&(c_1, c_2, s)(a_1, a_2, r)^{-1}\\ &= (a_1(1 - e^{2\pi is}) + c_1e^{2\pi ir}, a_2(1 - e^{2\pi ihs}) + c_2e^{2\pi ihr}, s)\end{aligned}$$

so $\alpha_{0,\frac{1}{3}}$ is not an inner automorphism, and $A_{0,\frac{1}{3}} \notin$ Int($\mathfrak{g}$). Now let $s_n \to 0$ and let $t_n = hs_n + hn$. Select a sequence $(n_k) \subset \boldsymbol{Z}$ such that $hn_k \to \frac{1}{3}$ (mod 1) (Kronecker's theorem), and let τ_k be the unique point in $[0, 1)$ such that $t_{n_k} - \tau_k \in \boldsymbol{Z}$. Putting $s_k = s_{n_k}$, $t_k = t_{n_k}$, we have

$$\alpha_{s_k,t_k} = \alpha_{s_k,\tau_k} \to \alpha_{0,\frac{1}{3}}.$$

Note: G is a subgroup of $H \times H$ where $H = \begin{pmatrix} 1 & 0 \\ c & \alpha \end{pmatrix}$, $c \in C_m$, $|\alpha| = 1$.

E. Invariant Differential Forms

E.1. The affine connection on G given by $\nabla_{\tilde{X}}(\tilde{Y}) = \frac{1}{2}[\tilde{X}, \tilde{Y}]$ is torsion free; and by (5), §7, Chapter I, if ω is a left invariant 1-form,

$$\nabla_{\tilde{X}}(\omega)(\tilde{Y}) = -\omega(\nabla_{\tilde{X}}(\tilde{Y})) = -\tfrac{1}{2}\omega(\theta(\tilde{X})(\tilde{Y})) = \tfrac{1}{2}(\theta(\tilde{X})\omega)(\tilde{Y}),$$

so $\nabla_{\tilde{X}}(\omega) = \frac{1}{2}\theta(\tilde{X})(\omega)$ for all left invariant forms ω. Now use Exercise C.4 in Chapter I.

E.2. The first relation is proved as (4), §7. For the other we have $g{}^tg = I$, so $(dg){}^tg + g{}^t(dg) = 0$. Hence $(g^{-1}\,dg) + {}^t(dg)({}^tg)^{-1} = 0$ and $\Omega + {}^t\Omega = 0$.

For $\boldsymbol{U}(n)$ we find similarly for $\Omega = g^{-1}\,dg$,

$$d\Omega + \Omega \wedge \Omega = 0, \qquad \Omega + {}^t\bar{\Omega} = 0.$$

For $\boldsymbol{Sp}(n) \subset \boldsymbol{U}(2n)$ we recall that $g \in \boldsymbol{Sp}(n)$ if and only if

$$g{}^t\bar{g} = I_{2n}, \qquad gJ_n{}^tg = J_n$$

(cf. Chapter X). Then the form $\Omega = g^{-1}\,dg$ satisfies

$$d\Omega + \Omega \wedge \Omega = 0, \qquad \Omega + {}^t\bar{\Omega} = 0, \qquad \Omega J_n + J_n{}^t\Omega = 0.$$

E.3. A direct computation gives

$$g^{-1}\,dg = \begin{pmatrix} 0 & dx & dz - x\,dy \\ 0 & 0 & dy \\ 0 & 0 & 0 \end{pmatrix}$$

and the result follows.

CHAPTER III

A. Solvable and Nilpotent Lie Algebras

A.1. Consider the derived algebra and observe, as a consequence of Cor. 6.3, Chapter II that a semisimple Lie algebra equals its derived algebra.

A.2. A direct computation shows $[\mathfrak{t}(n), \mathfrak{t}(n)] = \mathfrak{n}(n)$ and also that center $(\mathfrak{t}(n)) = \boldsymbol{R}(E_{11} + \ldots + E_{nn})$ and center $(\mathfrak{n}(n)) = \boldsymbol{R}E_{1n}$. By Theorem 2.4 (i), $\mathfrak{n}(n)$ is nilpotent, thus by Cor. 2.6 solvable, whence $\mathfrak{t}(n)$ is solvable. Thus (i) and (ii) are proved. For (iii) we have

$$B(X, [Y, Z]) = \mathrm{Tr}(\mathrm{ad}\,X\,\mathrm{ad}\,Y\,\mathrm{ad}\,Z - \mathrm{ad}\,X\,\mathrm{ad}\,Z\,\mathrm{ad}\,Y) = 0$$

because ad X, ad Y, and ad Z can, on the complexification, be expressed by upper triangular matrices and thereby the two matrix products on the right have the same diagonal elements.

A.3. We indicate a proof of this except for the second implication $\Leftarrow$, for which see e.g. Bourbaki [2], I, §5. If $\mathfrak{D}\mathfrak{g}$ is nilpotent, then it is solvable (Cor. 2.6); so, by definition, $\mathfrak{g}$ is solvable. Conversely, if $\mathfrak{g}$ is solvable, we extend the action of ad X ($X \in \mathfrak{g}$) to the algebraic closure extension of $\mathfrak{g}$, whereby they can be viewed in upper triangular form. Hence, if $X \in \mathfrak{D}\mathfrak{g}$, then by Exercise A.2, ad X is nilpotent, whence $\mathfrak{D}\mathfrak{g}$ is nilpotent (Theorem 2.4). That $\mathfrak{g}$ solvable $\Rightarrow B(\mathfrak{g}, [\mathfrak{g}, \mathfrak{g}]) = 0$ is seen just as in Exercise A.2.

A.4. G is analytically isomorphic to the closed subgroup of $\boldsymbol{GL}(2, \boldsymbol{R})$ consisting of the matrices $\left(\begin{smallmatrix} a & b \\ 0 & 1 \end{smallmatrix}\right)$, $a > 0$, $b \in \boldsymbol{R}$, whose Lie algebra $\mathfrak{g}$ is identified with the subalgebra $\{\left(\begin{smallmatrix} x & y \\ 0 & 0 \end{smallmatrix}\right) : x, y \in \boldsymbol{R}\}$ of $\mathfrak{gl}(2, \boldsymbol{R})$, which by A.2 is solvable. If $\mathfrak{s}$ is a noncommutative Lie algebra over $\boldsymbol{R}$ of dimension 2, then there exists a basis $U, V \in \mathfrak{s}$ such that $[U, V] = V$. The map $\alpha U + \beta V \to \alpha E_{11} + \beta E_{12}$ is the desired isomorphism.

A.5. The map $\rho : X \in \mathfrak{g} \to \text{ad } X \mid \mathfrak{a}$ ($\mid$ denoting restriction) is a representation of $\mathfrak{g}$ on $\mathfrak{a}$. An invariant subspace is an ideal in $\mathfrak{g}$ contained in $\mathfrak{a}$. Thus by the minimality ρ is irreducible, so by Lie's theorem $\dim \mathfrak{a} = 1$.

The analog for a solvable Lie algebra $\mathfrak{g}_0$ over $\boldsymbol{R}$ is as follows: Let $\mathfrak{a}_0 \subset \mathfrak{g}_0$ be a minimal proper ideal. Then $\dim \mathfrak{a}_0 \leqslant 2$. In fact, let $\mathfrak{a} \subset \mathfrak{g}$ denote the complexifications. Applying Lie's theorem to the representation $X \to \text{ad } X \mid \mathfrak{a}$, we find a vector $H \neq 0$ in $\mathfrak{a}$ and a linear form γ on $\mathfrak{g}_0$ such that $[X, H] = \gamma(X)H$ for $X \in \mathfrak{g}_0$. Decompose $H = H_1 + iH_2$ and $\gamma = \alpha + i\beta$ where H_1, H_2, α, β are real, giving the relations $[X, H_1] = \alpha(X)H_1 - \beta(X)H_2$, $[X, H_2] = \alpha(X)H_2 + \beta(X)H_1$ ($X \in \mathfrak{g}_0$), which show that $\boldsymbol{R}H_1 + \boldsymbol{R}H_2$ is an ideal in $\mathfrak{g}_0$, contained in $\mathfrak{a}_0$. Thus $\dim \mathfrak{a}_0 \leqslant 2$ (cf. Mostow [5]).

B. Semisimple Lie Algebras

B.1. (i) Let B and B_Γ denote the Killing forms of $\mathfrak{g}$ and $\mathfrak{g}_\Gamma$, respectively. Following the notation of Theorem 5.5 for $\mathfrak{g}$ let $X_i = H_i + \sum_{\alpha \in \Gamma} a_\alpha^i X_\alpha$ ($i = 1, 2$) where $H_i = \sum_{\alpha \in \Gamma} b_\alpha^i H_\alpha$ ($i = 1, 2$). Then since $B_\Gamma(X_\alpha, X_\beta) = \text{Tr}_{\mathfrak{g}_\Gamma}(\text{ad } X_\alpha \text{ ad } X_\beta) = 0$ if $\alpha, \beta \in \Gamma$, $\alpha + \beta \neq 0$, we have

$$B_\Gamma(X_1, X_2) = B_\Gamma(H_1, H_2) + \sum_{\alpha \in \Gamma} (a_\alpha^1 a_{-\alpha}^2 + a_{-\alpha}^1 a_\alpha^2)\, B_\Gamma(X_\alpha, X_{-\alpha}),$$

But by Lemma 5.1 and Theorem 5.5,

$$\text{ad } X_{-\alpha} \text{ ad } X_\alpha(X_\beta) = N_{-\alpha,\alpha+\beta} N_{\alpha,\beta} X_\beta = N_{\alpha,\beta}^2 X_\beta,$$
$$\text{ad } X_{-\alpha} \text{ ad } X_\alpha(H_\beta) = \alpha(H_\beta)\, H_\alpha,$$

so

$$B_\Gamma(X_\alpha, X_{-\alpha}) = \alpha(H_\alpha) + \sum_{\beta \in \Gamma} N_{\alpha,\beta}^2 > 0,$$

$$B_\Gamma(H_1, H_2) = \sum_{\alpha \in \Gamma} \alpha(H_1)\, \alpha(H_2).$$

Suppose $B_\Gamma(X_1, X_2) = 0$ for all $X_2 \in \mathfrak{g}_\Gamma$. Then $B_\Gamma(X_1, H_2) = 0$, so $B_\Gamma(H_1, H_2) = 0$ for all $H_2 \in \sum_{\alpha\in\Gamma} \boldsymbol{R}H_\alpha$. Hence $\alpha(H_1) = 0$ for each $\alpha \in \Gamma$, that is, $B(H_1, H_\alpha) = 0$ for all $\alpha \in \Gamma$, so $H_1 = 0$ (Theorem 4.4 (i)). Now $X_1 = 0$ follows using $B_\Gamma(X_\alpha, X_{-\alpha}) > 0$. (This proof occurs in Bruhat [1], p. 200.)

(ii) We have $\mathfrak{z}_H = \mathfrak{h} + \sum_\Gamma \mathfrak{g}_\gamma$ where $\Gamma = \{\alpha \in \Delta : \alpha(H) = 0\}$. By Theorem 4.2 (v) $\mathfrak{h} \cap \mathfrak{g}_\Gamma = \sum_{\gamma\in\Gamma} \boldsymbol{C}H_\gamma$; so if $\mathfrak{c}$ is the joint nullspace of the Γ in $\mathfrak{h}$, $\mathfrak{z}_H = \mathfrak{c} + \mathfrak{g}_\Gamma$ is the desired decomposition.

B.2. (i) and (iii) are obvious; and if ad $X = (\alpha_{ij})$, then it is clear from the determinant expansion that $a_{n-2}(X) = \sum_{i<j} (\alpha_{ii}\alpha_{jj} - \alpha_{ij}\alpha_{ji})$, so (ii) follows; (iv) is clear from $\mathfrak{g} = [\mathfrak{g}, \mathfrak{g}]$. For (v) consider for $X \in \mathfrak{g}$ the characteristic polynomial

$$\det(\lambda I - \operatorname{ad} X) = \lambda^{d_1(X)}(\lambda^{n-d_1(X)} + \cdots + b(X))$$

where $d_1(X) = \dim \mathfrak{g}(X, 0)$, $b(X) \neq 0$ (cf. Prop. 1.1). We have also

$$\det(\lambda I - \operatorname{ad} X) = \lambda^l(\lambda^{n-l} + \cdots + a_l(X))$$

so $d_1(X) \geqslant l$ with equality holding if and only if $a_l(X) \neq 0$. This proves (v). Finally (vii) (and therefore (vi)) are contained in the discussion of $\mathfrak{a}_n$ in §8.

B.3. (i) Suppose first V is real. Since a compact group of linear transformations of V leaves invariant a positive definite quadratic form, this part follows (as Prop. 6.6 in Chapter II) by orthogonal complementation. If V is complex, we use a positive definite Hermitian form instead.

For (ii) we suppose first V is complex. Then π extends to a representation of the complexification $\mathfrak{g}^C$ on V. Let $\mathfrak{u}$ be a compact real form of $\mathfrak{g}^C$, U the (compact) simply connected Lie group with Lie algebra $\mathfrak{u}$, and extend π to a representation of U on V, also denoted π. If $W \subset V$ is $\pi(\mathfrak{g})$-invariant, it is also $\pi(\mathfrak{g}^C)$- and $\pi(U)$-invariant and a $\pi(U)$-invariant complementary subspace will also be $\pi(\mathfrak{g}^C)$-invariant. Finally, we consider the case when V is real using a trick from Freudenthal and de Vries [1], §35. We view π as a representation of $\mathfrak{g}$ on the complexification V^c of V and then each member of $\pi(\mathfrak{g})$ commutes with the conjugation σ of V^c with respect to V. Let $W \subset V$ be a $\pi(\mathfrak{g})$-invariant subspace. Then the complexification $W^c = W + iW$ is a $\pi(\mathfrak{g})$-invariant subspace of V^c, so by the first case W^c has a $\pi(\mathfrak{g})$-invariant complement $Z' \subset V^c$. Let $Z = (1 + \sigma)(Z' \cap (1 - \sigma)^{-1}(iW))$. Since $\sigma(1 + \sigma) = \sigma + 1$ and $\pi(X)\sigma = \sigma\pi(X)$ $(X \in \mathfrak{g})$, we have $Z \subset V$, $\pi(\mathfrak{g})Z \subset Z$. Also $Z \cap W = \{0\}$. In fact, if $z \in Z \cap W$, there exists a $z' \in Z'$ such that $(1 - \sigma)z' \in iW$,

$(1+\sigma)z' = z$. Hence $z' = \frac{1}{2}(1-\sigma)z' + \frac{1}{2}(1+\sigma)z' \in W^c$, so $z' = 0$ and $z = 0$. Finally, $W + Z = V$. In fact, if $v \in V$, then $v = w' + z'$ ($w' \in W^c$, $z' \in Z'$). Then $w' + z' = v = \sigma v = \sigma w' + \sigma z'$, so $(1-\sigma)z' = (1-\sigma)(-w') \in iW$, so $z' \in Z' \cap (1-\sigma)^{-1}(iW)$ and $(1+\sigma)z' \in Z$. Hence $v = \frac{1}{2}(1+\sigma)w' + \frac{1}{2}(1+\sigma)z' \in W + Z$.

(This "theorem of complete reducibility" was first proved by H. Weyl [1], I, §5 by a similar method; algebraic proofs were later found by Casimir and van der Waerden [1] and by Whitehead [4].)

B.4. The automorphism $N = \sigma\theta$ of $\mathfrak{g}$ is symmetric with respect to the positive definite bilinear form B_θ. Put $P = N^2$. Then $\varphi = P^{\frac{1}{4}}$ has the desired property. (The result, with a different proof, occurs in Berger [2], p. 100, based on results of Murakami [1].)

B.5. Similar to Exercise B.6.

B.6. The quadratic form is given by the matrix

$$s = \begin{pmatrix} I_1 & 0 & 0 \\ 0 & 0 & I_n \\ 0 & I_n & 0 \end{pmatrix}$$

and $X \in \mathfrak{b}_n$ if and only if

$$^t(\exp tX)s \exp tX = s \qquad \text{for all} \quad t \in \boldsymbol{R}.$$

This condition is equivalent to $^tX = -sXs^{-1}$; and writing out $sX + {}^tXs = 0$ with X in block form corresponding to that of s, we derive (i). We already know $\mathfrak{b}_n$ is semisimple, so (ii) amounts to verifying the bracket relations which is straightforward.

B.7. Because of Exercise B.3 (ii), V is a direct sum $V = \sum_i V_i$ where each V_i is G-invariant and irreducible under G. Let $\pi_i(g)$ denote the restriction $g \mid V_i$. If z is in the center of G, $\pi_i(z)$ is in the center of $\pi_i(G)$ and by Schur's lemma (cf. Chevalley [2], p. 183) a scalar. The closed subgroup G' of G generated by the commutators $xyx^{-1}y^{-1}$ ($x, y \in G$) has Lie algebra containing $[\mathfrak{g}, \mathfrak{g}]$ (cf. Lemma 1.8, Chapter II) which however equals $\mathfrak{g}$ (by Cor. 6.3, Chapter II), so $G' = G$. Consequently, $\det \pi_i(g) = 1$ for $g \in G$. Thus there are only finitely many possibilities for z. This proves (i).

For (ii) consider the closure $\bar{G}$ of G in $\boldsymbol{GL}(V)$ and its Lie algebra $\bar{\mathfrak{g}} \subset \mathfrak{gl}(V)$. Since $\mathrm{Ad}(G)\mathfrak{g} \subset \mathfrak{g}$, we have $\mathrm{Ad}(\bar{G})\mathfrak{g} \subset \mathfrak{g}$, so $\mathfrak{g}$ is an ideal in $\bar{\mathfrak{g}}$. Applying Exercise B.3 (ii) to $\mathfrak{g}$ acting on $\bar{\mathfrak{g}}$, we get a subspace $\mathfrak{g}' \subset \bar{\mathfrak{g}}$ such that $\bar{\mathfrak{g}} = \mathfrak{g} + \mathfrak{g}'$ (direct sum), $[\mathfrak{g}, \mathfrak{g}'] \subset \mathfrak{g}'$, whence $[\mathfrak{g}, \mathfrak{g}'] = 0$. By continuity, each $g \in \bar{G}$ leaves each V_i invariant and the restriction $\pi_i(g) = g \mid V_i$ satisfies $\det \pi_i(g) = 1$. Thus each $X \in \mathfrak{g}'$ has trace 0 on each V_i.

But (by $[\mathfrak{g}, \mathfrak{g}'] = 0$) this action commutes with the irreducible action of G on V_i; so again by Schur's lemma, $X \mid V_i$ is a scalar. Hence $X = 0$, so $\mathfrak{g} = \bar{\mathfrak{g}}$ and G is closed.

For (iii) consider the composite $\mathrm{Ad} \circ \rho$. Because of Theorem 2.6, Lemma 1.12, and Cor. 6.2, Chapter II, ρ is analytic and the differential $d(\mathrm{Ad} \circ \rho) = \mathrm{ad} \circ d\rho$ is an isomorphism. Thus the kernel of $\mathrm{Ad} \circ \rho$ is a discrete normal subgroup of G and is therefore contained in the center Z of G; in other words, the inverse image $\rho^{-1}(Z)$ satisfies $\rho^{-1}(Z) \subset Z$. Since ρ is surjective, this implies $Z \subset \rho(Z)$ and, ρ being in addition a homomorphism, $\rho(Z) \subset Z$. Since by (i) Z is finite, $\rho(Z) = Z$ implies ρ injective on Z; so by $\rho^{-1}(e) \subset Z$, ρ is injective on G. For (iv) consider $\rho : x \to x^2$ on the circle group as an example concerning (i) and (iii).

B.8. Let $\mathfrak{n} = \{Z_0 \in \mathfrak{g} : B(Z_0, \mathfrak{g}) = 0\}$. Suppose $\mathfrak{g}$ were not semisimple. Then $\mathfrak{n}$ is a nonzero ideal in $\mathfrak{g}$, so $\mathfrak{n} = \mathfrak{g}$ and $B(\mathfrak{g}, \mathfrak{g}) = 0$. Let $X \in \mathfrak{g}$. Since $[\mathfrak{g}, \mathfrak{g}]$ is an ideal in $\mathfrak{g}$ and since $\mathfrak{g}$ is non-abelian, we have $[\mathfrak{g}, \mathfrak{g}] = \mathfrak{g}$, so we write $X = \sum_{i=1}^{k} [S_i, T_i]$ for $S_i, T_i \in \mathfrak{g}$. If $D : \mathfrak{g} \to \mathfrak{g}$ is any derivation, we have $\mathrm{ad}(DX) = [D, \mathrm{ad}\, X]$ and by a direct computation

$$\mathrm{Tr}(\mathrm{ad}\, XD) = \sum_{i=1}^{k} B(T_i, DS_i) = 0,$$

so by the criterion, $\mathrm{ad}\, X$ is nilpotent. This shows $\mathfrak{g}$ nilpotent, and Cor. 2.8 gives a contradiction.

C. Geometric Properties of the Root Pattern

C.1. If $\beta \in \Delta$, then $s_\alpha(H_\beta) = H_\beta + (p + q)H_\alpha$ and $p \leqslant p + q \leqslant q$. Thus $\beta + (p + q)\alpha$ is in the α-series containing β, hence in Δ.

C.2. Immediate from Theorem 4.3 (i).

C.3. By the Schwartz' inequality $| a_{\beta,\alpha} a_{\alpha,\beta} | \leqslant 4$, and the equality sign holds only if β and α are proportional. In that case $\beta = \pm\alpha$ or 0 and the inequality obvious. Since the α-series containing β can be written $\beta + p\alpha + m\alpha$ $(0 \leqslant m \leqslant q - p)$, we can conclude from the first part and from Theorem 4.3 (i) that $0 \leqslant q - p \leqslant 3$, so the α-series has at most four members.

C.4. Since M is a projection, Trace M = rank M.

C.5. Since $| a_{\alpha,\beta} a_{\beta,\alpha} | \leqslant 4$, the assumptions imply $a_{\alpha,\beta} a_{\beta,\alpha} = 4 \cos^2 \theta$ is an integer $m = 1, 2, 3$, and the smallest of the integers $| a_{\alpha,\beta} |$, $| a_{\beta,\alpha} |$ is 1. But then if $\langle \beta, \beta \rangle \geqslant \langle \alpha, \alpha \rangle$,

$$1 = a_{\alpha,\beta}^2 = 4 \cos^2 \theta \langle \alpha, \alpha \rangle \langle \beta, \beta \rangle^{-1}$$

so $\langle \beta, \beta \rangle = m \langle \alpha, \alpha \rangle$.

C.6. Cases: (a) $\langle\alpha, \beta\rangle = 0$; (b) $\langle\alpha, \beta\rangle \neq 0$, $\langle\alpha, \alpha\rangle \geqslant \langle\beta, \beta\rangle$; (c) $\langle\alpha, \beta\rangle \neq 0$, $\langle\beta, \beta\rangle > \langle\alpha, \alpha\rangle$.

(a): Here $p + q = 0$, so by $\alpha + \beta \in \Delta$ and Exercise C.3 the α-series containing β is $\beta - \alpha, \beta, \beta + \alpha$. Now α and $\beta - \alpha$ are nonproportional, so

$$0 < a_{\alpha,\beta-\alpha}a_{\beta-\alpha,\alpha} < 4.$$

Since $a_{\beta-\alpha,\alpha} = -2$, we conclude $a_{\alpha,\beta-\alpha} = -1$, whence $\langle\beta, \beta\rangle = \langle\alpha, \alpha\rangle$. Thus $\langle\beta + \alpha, \beta + \alpha\rangle/\langle\beta, \beta\rangle = 2$ which equals $(-p + 1)/q$.
For (b) and (c) we use the identity (Steinberg [2])

$$\frac{-p+1}{q} - \frac{\langle\alpha+\beta, \alpha+\beta\rangle}{\langle\beta, \beta\rangle} = (a_{\beta,\alpha} + 1)\left(\frac{1}{q} - \frac{\langle\alpha, \alpha\rangle}{\langle\beta, \beta\rangle}\right),$$

which is a direct consequence of $-a_{\beta,\alpha} = p + q$. We write P and Q for the two factors on the right and have to prove that P or Q is 0.

(b): We know $a_{\alpha,\beta}a_{\beta,\alpha} = 1, 2$, or 3, so in the present case $a_{\beta,\alpha} = -1$ or 1. In the first case $P = 0$. In the second case, $\beta - \alpha$ is a root (Exercise C.2), and so is $\beta + \alpha$. Since $p + q < 0$, we conclude from Exercise C.3 that $p = -2$, $q = 1$. Thus $a_{\beta-2\alpha,\alpha} = a_{\beta,\alpha} - 4 = -3$, whence $a_{\alpha,\beta-2\alpha} = -1$. This implies $2\langle\alpha, \beta\rangle = \langle\beta, \beta\rangle$, which together with $a_{\beta,\alpha} = 1$ gives $\langle\alpha, \alpha\rangle = \langle\beta, \beta\rangle$, so $Q = 0$.

(c): As in case (b) we have $a_{\alpha,\beta} = -1$ or $+1$. If $a_{\alpha,\beta} = 1$, then $\beta - \alpha$ is a root (Exercise C.2) and since $a_{\beta,\alpha} > 0$ we have $p + q < 0$. Thus by Exercise C.3 the α-series containing β is $\beta - 2\alpha$, $\beta - \alpha$, β, $\beta + \alpha$. But then $a_{\beta,\alpha} = 1$, so $m = a_{\alpha,\beta}a_{\beta,\alpha} = 1$, contradicting $\langle\beta, \beta\rangle > \langle\alpha, \alpha\rangle$. Hence $a_{\alpha,\beta} = -1$. Thus $a_{\beta,\alpha} < 0$ and $p + q > 0$. If $a_{\beta,\alpha} = -1$, then $P = 0$. If $a_{\beta,\alpha} = -2$, then $\beta + 2\alpha \in \Delta$, $\beta + 3\alpha \notin \Delta$, so $q = 2$, $p = 0$. By Exercise C.3 $a_{\alpha,\beta}a_{\beta,\alpha} = 2$, so $\langle\beta, \beta\rangle = 2\langle\alpha, \alpha\rangle$ and $Q = 0$. Finally, if $a_{\beta,\alpha} = -3$, then by Exercise C.5 $\langle\beta, \beta\rangle = 3\langle\alpha, \alpha\rangle$, whereas by Exercise C.3, $q = 3$ so $Q = 0$.

C.7. It is clear that

$$(N'_{\alpha,\beta})^2 = N^2_{\alpha,\beta}\frac{2\langle\alpha+\beta, \alpha+\beta\rangle}{\langle\alpha, \beta\rangle\langle\beta, \beta\rangle},$$

which by Lemma 5.2 and Exercise C.6 equals $(1 - p)^2$.

C.8. Use Theorem 4.3 (i) and (ii) and Theorem 5.5.

CHAPTER IV

A. Geometry of Homogeneous Spaces

A.1. Assuming first the existence of ∇ consider for $X \in \mathfrak{m}$ the affine transformation $\tau(\exp X) \circ s_o$ of G/H, which fixes the point $(\exp \frac{1}{2}X)H$, and has differential $-I$ at this point because $\tau(\exp X)s_o = \tau(\exp \frac{1}{2}X) \circ s_o \circ \tau(\exp(-\frac{1}{2}X))$. Then (iii) follows as (i) in Exercise A.6, Chapter II. Since $d\tau(\exp tX)_0(X) = d\gamma_X((d/dt)_t)$, (ii) follows. For the existence of ∇ let $U \subset \mathfrak{m}$, $N_0 \subset G/H$ be as in §4, Chapter II. For $X \in \mathfrak{m}$, let $X^* \in \mathfrak{D}^1(N_o)$ be defined by $X^*_{\exp Y \cdot 0} = d\tau(\exp Y)_0(X)$ if $Y \in U$. If $X_1, \ldots, X_r$ is a basis of $\mathfrak{m}$, then $(X_i^*)_{1 \leqslant i \leqslant r}$ is a basis of the module $\mathfrak{D}^1(N_o)$ and we can define a unique affine connection ∇^0 on N_o by the requirements

(a) $\nabla^0_{X_i^*}(X_j^*)_0 = 0$;

(b) $\nabla^0_{X_i^*}(X_j^*)_{x \cdot 0} = x \cdot (\nabla^0_{x^{-1} \cdot X_i^*}(x^{-1} \cdot X_j^*))_0 , \qquad x = \exp X, \quad X \in U.$

Since for $h \in H$, $d\tau(h)(X^*) = (\mathrm{Ad}(h)X)^*$ on a neighborhood of o, it is clear that

$$d\tau(h)_0\, (\nabla^0_Z(Y))_0 = (\nabla^0_{d\tau(h)Z}(d\tau(h)Y))_0$$

for $Z = X_i^*$, $Y = X_j^*$ and therefore, by Prop. 3.3 (∇_1) and (∇_2), §4 in Chapter I, for all Z, $Y \in \mathfrak{D}^1(N_o)$. Thus we can define for Z, $Y \in \mathfrak{D}^1(G/H)$, $g \in G$,

$$(\nabla_Z(Y))_{g \cdot o} = d\tau(g)_o \{(\nabla^0_{d\tau(g^{-1})Z}(d\tau(g^{-1})Y))_o\},$$

and then ∇ is a G-invariant affine connection. Replacing G by $\tau(G) \cup s_0$ we see that ∇ is invariant under s_o. The uniqueness in (i) is clear since condition (a) above is forced by the invariance under s_0 and since (b) and the definition of ∇ are forced by the G-invariance. Now (ii) shows that s_o is the geodesic symmetry at o, so the geodesic symmetry s_{gH} equals $\tau(g) \circ s_o \circ \tau(g^{-1})$. This being an affine transformation, $T = 0$ and $\nabla_V R = 0$ (Theorem 1.1). For the formula for R_0 let X^+, Y^+, Z^+ be the vector fields on G/H defined in §3, Chapter I. Then $X_0^+ = X$, etc., and by Theorem 3.4, Chapter II, and the relation $[\mathfrak{m}, \mathfrak{m}] \subset \mathfrak{h}$ we have $[X^+, Y^+]_o = 0$. Also by (iii) above, Theorem 7.1 and Exercise B.3, Chapter I, we have $(\nabla_{X^+}(V))_0 = (\theta(X^+)V)_o$ for $V \in \mathfrak{D}'(G/H)$. Also the relation

$$\nabla_{d\tau(\exp tX)Y^+}(d\tau(\exp tX)V) = d\tau(\exp tX)(\nabla_{Y^+}(V))$$

gives by differentiation.

$$\nabla_{[X^+, Y^+]} + \nabla_{Y^+}\theta(X^+) = \theta(X^+)\nabla_{Y^+}.$$

Thus

$$(\nabla_{X^+}\nabla_{Y^+}(Z^+))_0 = (\theta(Y^+)\,\theta(X^+)\,Z^+)_0 + (\nabla_{[X^+,Y^+]}(Z^+))_0,$$

so, since $[X^+, Y^+]_0 = 0$, we obtain

$$(R(X^+, Y^+)(Z^+))_0 = -\{\theta([X^+, Y^+])(Z^+)\}_0.$$

Thus by Theorem 3.4, Chapter II,

$$R_o(X, Y)(Z) = -[[X, Y], Z].$$

A.2. (i) The tangent spaces G_g and $(G/H)_{gH}$ are given by $G_g = dL(g)\mathfrak{g}$, $(G/H)_{gH} = d\tau(g)\mathfrak{m}$; and if $X \in \mathfrak{g}$, the natural map $\pi: G \to G/H$ satisfies $d\pi\, dL(g)X = d\tau(g)\, d\pi X$. Thus the vector $X_0 = dL(g)X$ satisfies $\| X_0 \| \geqslant \| d\pi(X_0)\|$, the equality sign holding only if $X \in \mathfrak{m}$. Hence the lengths of a curve $\gamma^* \subset G$ and of its projection $\gamma = \pi \circ \gamma^*$ satisfy $L(\gamma^*) \geqslant L(\gamma)$, with equality holding only if $\dot{\gamma}^*(t) \perp (\gamma^*(t)H)_{\gamma^*(t)}$ for all t. Such a curve is said to be *transversal*. Given a curve γ in G/H, there exists a transversal curve $\tilde{\gamma}$ in G such that $\pi\tilde{\gamma} = \gamma$. By the local minimizing property of geodesics $\tilde{\gamma}$ is a transversal geodesic in G (and therefore of the form $\tilde{\gamma}(t) = \exp tX$ ($X \in \mathfrak{m}$)) if and only if γ is a geodesic in G/H. This proves (i).

For (ii) let $S \subset \mathfrak{m}$ be the subspace spanned by X and Y and $K(S)$ and $K(\tilde{S})$ the corresponding sectional curvatures of G/H and G, respectively. Let f and F denote the Radon-Nikodym derivatives of the restrictions of exp and Exp, respectively, to S. Then if $Z \in S$ and we put $B_Z = \sum_0^\infty (-\text{ad}\, Z)^m/(m+1)!$, we have with the notation of §4

$$f(Z) = | B_Z(X) \vee B_Z(Y)|, \qquad F(Z) = | d\pi B_Z(X) \vee d\pi B_Z(Y)|.$$

Let (X_k) be an orthonormal basis of $\mathfrak{m}$ such that $X_1 = X$, $X_2 = Y$, and (X_ρ) an orthonormal basis of $\mathfrak{h}$. We write $Z = z_1X_1 + z_2X_2$, $[X_i, X_j] = \sum_k c^k{}_{ij}X_k + \sum_\rho c^\rho{}_{ij}X_\rho$. When computing $(\Delta f)(0)$ and $(\Delta F)(0)$, we can cancel terms of order $\geqslant 2$ in z_1 and z_2. Writing $(+)$ for such terms we have

$$B_Z(X_1) = (1 + p_1)X_1 + p_2X_2 + \dots + \sum_\rho p_\rho X_\rho\, (+),$$

$$B_Z(X_2) = q_1X_1 + (1 + q_2)X_2 + \dots + \sum_\rho q_\rho X_\rho\, (+),$$

where the p_i, q_j are linear in z_1 and z_2 and $p_\rho = -\tfrac{1}{2}c^\rho{}_{21}z_2$, $q_\rho = -\tfrac{1}{2}c^\rho{}_{12}z_1$. The expressions for $d\pi B_Z(X_1)$ and $d\pi B_Z(X_2)$ are obtained by cancelling the X_ρ from the expressions above. Comparing with the proof of Theorem 4.2, we find

$$(\Delta f)(0) = (\Delta F)(0) + \tfrac{1}{2} \sum_{\rho} (c^{\rho}{}_{12})^2.$$

But $K(S) = -\frac{3}{2}(\Delta F)(0)$, $K(\tilde{S}) = -\frac{3}{2}(\Delta f)(0)$, so the result follows since we know $K(\tilde{S}) = \frac{1}{4}\|[X, Y]\|^2$ both from Exercise A.6 (iv), Chapter II, and from Theorem 4.2 in the present chapter (cf. §6).

A.3. In the notation of §6 the map $d\pi$ maps the (-1)-eigenspace of $d\sigma$ onto $\mathfrak{g}$ as follows: $d\pi(X, -X) = 2X$. But it is easy to verify

$$2B_{\mathfrak{g}\times\mathfrak{g}}((X, -X), (X, -X)) = B_{\mathfrak{g}}(2X, 2X),$$

and this is equivalent to $2Q^* = Q$.

A.4. Lemma 1.2 gives an isometry between two normal neighborhoods. The method of proof of Theorem 5.6 gives a local isometry of M_1 onto M_2. Now use Lemma 13.4, Chapter I.

A.5. We know from Theorem 15.4, Chapter I that M is separable. The results of §2 remain valid for M because the Riemannian structure only enters in the proofs through the associated metric. In particular, Theorem 2.2 holds, so the orbit $H \cdot p$ is obviously closed.

A.6. Immediate from Lemma 1.2 and Exercise C.6, Chapter I.

B. Cohomology of Symmetric Spaces

B.1. If ω is invariant, so is $s_m\omega$ for each $m \in M$ (Prop. 3.4). But if ω is a p-form $s_m\omega = (-1)^p\omega$ ((6), §1). Thus

$$d\omega = (-1)^p\, d(s_m\omega) = (-1)^p\, s_m(d\omega) = (-1)^{2p+1}\, d\omega, \quad \text{so} \quad d\omega = 0.$$

Another proof comes from Exercise C.4, Chapter I (cf. Cartan [14], p. 191).

B.2. The $*$ operator commutes with the action of G on $\mathfrak{U}(G/K)$, so by Exercise B.1 a G-invariant form on G/K is harmonic. The converse follows as Theorem 7.8, Chapter II, except that we replace the left invariant vector fields $\tilde{X}$ by the induced vector field X^+ on G/K ($X \in \mathfrak{g}$).

CHAPTER V

1. By Theorem 6.4 for the group $G = \boldsymbol{SO}(3)$ we may take $X \in \mathfrak{t}$ where $\mathfrak{t}$ by §8, Chapter III can be chosen as $\boldsymbol{R}(E_{12} - E_{21})$. Then the formula follows trivially.

2. Let $\boldsymbol{H}$ denote the space of all 2×2 skew-Hermitian matrices which we write

$$h = \begin{pmatrix} i(x_1 - x_3) & -x_2 + ix_4 \\ x_2 + ix_4 & i(x_1 + x_3) \end{pmatrix} \qquad (x_j \in \boldsymbol{R})$$

Then $\boldsymbol{SL}(2, \boldsymbol{C})$ acts on $\boldsymbol{H}$ as follows. If $g \in \boldsymbol{SL}(2, \boldsymbol{C})$, the transformation $\sigma_g : h \to gh^t\bar{g}$ maps $\boldsymbol{H}$ into itself preserving $\det h = -x_1^2 + x_2^2 + x_3^2 + x_4^2$. Thus, since $\boldsymbol{SL}(2, \boldsymbol{C})$ is connected, $g \to \sigma_g$ is a continuous homomorphism of $\boldsymbol{SL}(2, \boldsymbol{C})$ into $\boldsymbol{SO}_0(1, 3)$. The kernel consists of $\{-I, I\}$; and using Lemma 5.1 Chapter II, we get the analytic isomorphism $\boldsymbol{SL}(2, \boldsymbol{C})/\boldsymbol{Z}_2 \cong \boldsymbol{SO}_0(1, 3)$.

A more geometric realization is as follows: Consider the cone $x_1^2 - x_2^2 - x_3^2 - x_4^2 = 0$ $(x_1 > 0)$ and intersect it with the hyperplane $x_1 = \frac{1}{2}$, obtaining the sphere $x_2^2 + x_3^2 + x_4^2 = \frac{1}{4}$ in this hyperplane. Each $g \in \boldsymbol{SO}_0(1, 3)$ permutes the rays from the origin lying on the cone, thereby permuting their intersections with the hyperplane $x_1 = \frac{1}{2}$ and hence g induces a transformation τ_g of the sphere, which via a stereographic projection corresponds to a member of $\boldsymbol{SL}(2, \boldsymbol{C})/\boldsymbol{Z}_2$ (since g maps planes into planes, τ_g maps circles into circles).

3. If $X \in \mathfrak{e}$ is in the normalizer of $\mathfrak{u}$, then $[X, \mathfrak{u}] \subset \mathfrak{u} \cap \mathfrak{e} = 0$, so by Cor. 1.7, $X = 0$. For (ii) note that $\mathfrak{e}_0 = 0$, so $\mathfrak{u}_0$ is an ideal in $\mathfrak{l}$, hence $\mathfrak{u}_0 = 0$. Thus $[\mathfrak{e}, \mathfrak{e}] = [\mathfrak{e}_-, \mathfrak{e}_-] + [\mathfrak{e}_+, \mathfrak{e}_+] = \mathfrak{u}_- + \mathfrak{u}_+ = \mathfrak{u}$.

4. A compact subgroup of dimension $\geqslant 1$ contains a torus (compact connected abelian subgroup $\neq e$) hence an element $x \neq e$ such that $x^2 = e$. Let H be the fixed point set of the involution $g \to xgx^{-1}$.

5. If M has compact type, then $I_0(M)$ is semisimple and compact. In particular M is compact and so is $I(M)$. Conversely, if $I(M)$ is semisimple and compact and $p \in M$, the map $\theta : g \to s_p g s_p$ is an automorphism of $I_0(M)$ whose fixed point set has the same Lie algebra as the isotropy subgroup at p.

If M is of the noncompact type, let $\mathfrak{g}_0 = \mathfrak{k}_0 + \mathfrak{p}_0$ be a Cartan decomposition of the Lie algebra $\mathfrak{g}_0$ of $I(M)$. If $\mathfrak{u} \neq 0$ is a compact ideal in $\mathfrak{g}_0$ it would be semisimple (Prop. 6.1, Chapter II) hence compactly imbedded in $\mathfrak{g}_0$. Decomposing $\mathfrak{g}_0 = \mathfrak{u} + \mathfrak{g}_1 + \ldots + \mathfrak{g}_s$ where the $\mathfrak{g}_i$ are simple ideals, we get a Cartan decomposition of $\mathfrak{g}_0$ where $\mathfrak{u}$ is contained in the compact part. By conjugacy of all Cartan decompositions $\mathfrak{k}_0$ would contain a compact ideal $\neq 0$ in $\mathfrak{g}_0$. This is impossible since $I_0(M)$ is effective. The converse is clear from Theorem 1.1

6. The first statement is easy since the restriction $g_p \mid S$ has the form $x_1^2 \pm x_2^2$ in a suitable basis. The remaining statements follow from Lemma 12.5 and Theorem 12.2 in Chapter I.

7. (sketch)(i). The group $O(p, q+1)$ acts transitively on Q_{-1} and the subgroup fixing $p_0 = (0, ..., 0, 1)$ is isomorphic to $O(p, q)$. If s_0 denotes the linear transformation

$$(x_1, ..., x_{p+q}, x_{p+q+1}) \to (-x_1, ..., -x_{p+q}, x_{p+q+1}),$$

then the mapping $\sigma : g \to s_0 g s_0$ is an involutive automorphism of $O(p, q+1)$ whose differential $d\sigma$ has fixed point set $\mathfrak{o}(p, q)$. The (-1)-eigenspace of $d\sigma$, say $\mathfrak{m}$, is spanned by

$$Z_i = E_{i,p+q+1} + E_{p+q+1,i} \qquad (1 \leqslant i \leqslant p),$$

$$Z_j = E_{j,p+q+1} - E_{p+q+1,j} \qquad (p+1 \leqslant j \leqslant p+q).$$

The Killing form B on $\mathfrak{o}(p, q+1)$ is by (16), §8 and Lemma 6.1 in Chapter III given by $B(X, Y) = (p+q-1)\,\mathrm{Tr}(XY)$, so we find

$$B(Z_i, Z_i) = -B(Z_j, Z_j) = 2(p+q-1).$$

Using Exercise A.1 (iv) in Chapter IV, we find by a direct computation that $O(p, q+1)/O(p, q)$, with the invariant pseudo-Riemannian structure induced by the restriction of $B/2(p+q-1)$ to $\mathfrak{m}$, has constant sectional curvature -1.

On the other hand, the mapping $\psi : gO(p, q) \to g \cdot p_0$ has a differential $d\psi$ which maps $\mathfrak{m}$ onto the tangent plane $x_{p+q+1} = 1$ to Q_{-1} at p_0, and $d\psi(X) = X \cdot p_0$ $(X \in \mathfrak{m})$. Thus

$$d\psi(Z_n) = (\delta_{1,n}, ..., \delta_{p+q+1,n}) \qquad (1 \leqslant n \leqslant p+q),$$

so $B_{-1}(d\psi(X)) = B(X, X)/2(p+q-1)$ for $X \in \mathfrak{m}$; hence ψ is an isometry. This proves (i) and (ii). Part (iii) follows from Lemmas 1.2 and 1.4 in Chapter IV.

For (iv) we note the following decompositions into the eigenspaces of $d\sigma$

$$\mathfrak{o}(p, q+1) = \mathfrak{o}(p, q) + \mathfrak{m}, \qquad \mathfrak{o}(p+1, q) = \mathfrak{o}(p, q) + \mathfrak{n}$$

where $\mathfrak{n}$ is spanned by the matrices

$$Y_i = E_{i,p+q+1} - E_{p+q+1,i} \qquad (1 \leqslant i \leqslant p),$$

$$Y_j = E_{j,p+q+1} + E_{p+q+1,j} \qquad (p+1 \leqslant j \leqslant p+q).$$

As in Example I in the text we see that the isomorphism

$$X \to \begin{pmatrix} -iI_{p+q} & 0 \\ 0 & I_1 \end{pmatrix} X \begin{pmatrix} iI_{p+q} & 0 \\ 0 & I_1 \end{pmatrix}$$

maps the dual algebra $\mathfrak{o}(p, q) + i\mathfrak{m}$ onto $\mathfrak{o}(p+1, q)$ fixing $\mathfrak{o}(p, q)$, sending $i\mathfrak{m}$ onto $\mathfrak{n}$. This proves (iv).

CHAPTER VI

A. Geometric Features of the Cartan Decomposition

A.1. Obvious from Theorem 6.9, Chapter II, and Theorem 1.1.

A.2. Clear from Theorem 1.1.

A.3. Let $N(K)$ be the normalizer of K in G. If $(g_n) \subset N(K)$ is a sequence converging to $g \in G$, then $\mathrm{Ad}_G(g_n)\mathfrak{k} \subset \mathfrak{k}$, so $\mathrm{Ad}_G(g)\mathfrak{k} \subset \mathfrak{k}$, so $g \in N(K)$ and $N(K)$ is closed. The Lie algebra of $N(K)$ is the normalizer of $\mathfrak{k}_0$ in $\mathfrak{g}_0$, which by Chapter V, Exercise 3, equals $\mathfrak{k}_0$. Now (i) follows from Theorem 1.1. Next, Lemma 1.2 implies (ii). Also (iii) follows from Theorem 2.3, Chapter II, Prop. 5.1, 5.2, Chapter VIII, and Theorem 1.1. For (iv) suppose $H \subset G$ a subgroup, properly containing K. By (iii) $\bar{H} = G$. The vector subspace $\mathfrak{k}'$ generated by $\mathrm{Ad}_G(H)\mathfrak{k}$ is invariant under $\mathrm{Ad}_G(\bar{H})$, so $\mathfrak{k}' = \mathfrak{g}$. Thus $\mathfrak{g} = \Sigma_1^m \mathrm{Ad}(h_i)\mathfrak{k}$ for a suitable subset $h_1, \ldots, h_m$ of H. But then the mapping $(k_1, \ldots, k_m) \to \Pi_{i=1}^m h_i k_i h_i^{-1}$ of K^m into G has differential at e given by $(T_1, \ldots, T_m) \to \Sigma_i \mathrm{Ad}(h_i)T_i$, where $T_i \in \mathfrak{k}$, and therefore is a submersion at e. Hence H contains a neighborhood of e in G, so H is open, hence closed, so we have a contradiction.

A.4. (i) If p and q are fixed points, the unique geodesic joining them is left pointwise fixed. Let $S \subset M$ denote the set of fixed points. Fix $p \in S$ and let

$$\mathfrak{s} = \{X \in M_p : daX = X \text{ for all } a \in A\}.$$

Then $\mathfrak{s}$ is a linear subspace of M_p, and by Theorem 13.3, Chapter I, $T = \mathrm{Exp}_p\, \mathfrak{s}$ is a submanifold of M, and $T \subset S$. On the other hand, if $s \in S$, then the unique geodesic δ joining p to s lies in S and $\dot{\delta}_p \in \mathfrak{s}$. Thus $s \in T$, so $S = T$. If an M-geodesic γ is tangent to S at p consider the S-geodesic Γ tangent to γ at p. Then Γ is left pointwise fixed by A so $\dot{\gamma}_p \in \mathfrak{s}$ and $\gamma \in S$ as desired. (For a generalization of (i) see Kobayashi [3].)

For (ii) let $X \in \mathfrak{p}_0$. Then $a(\exp X)K = (\exp X)K$ for all $a \in A$ if and only if $\exp(-X)A \exp X \in K$ which by connectedness is equivalent to

$\mathrm{Ad}_G(\exp(-X))\mathfrak{a} \subset \mathfrak{k}_0$. For (iii) take $T \in \mathfrak{a}$, $X \in \mathfrak{b}$. Taking the $\mathfrak{p}_0$-component in the relation $e^{\mathrm{ad}\, X}(T) \in \mathfrak{k}_0$, we obtain

$$\sinh(\mathrm{ad}(X))T = 0.$$

Now ad X can be put in real diagonal form, so this implies easily ad $X(T) = 0$, proving (iii).

A.5. In terms of the notation of Theorem 1.1 the mapping in question is $\mathrm{Exp}\, X \to \exp 2X$ $(X \in \mathfrak{p}_0)$. Because of Theorem 1.1 it just remains to prove that S is totally geodesic. Since the one-parameter subgroups are geodesics in $I_0(M)$ (Exercise A.5, Chapter II), the submanifold S is geodesic at e. Since the maps

$$T_x : g \to xg\sigma(x^{-1})$$

are isometries of $I_0(M)$, map $S = \{g^{-1}\sigma(g) : g \in G\}$ into S, form a transitive group of isometries of S, S is geodesic at each of its points, that is, totally geodesic.

A.6. The first part is clear from Theorems 1.1 and 2.1, and Exercise A.3 (i) (which shows that the mapping is one-to-one). Next, note that if $t = \exp X$ is a transvection along γ_{pq} and $k \in K_p \cap K_q$, then $ktk^{-1} = \exp \mathrm{Ad}(k)X = \exp X = t$.

A.7. Part (i) is already proved in Chapter II, §5. In order to prove (ii) invoke the group $\mathrm{Aut}(\mathfrak{g}_0)$ which by conjugation permutes the maximal compact subgroups of $\mathrm{Int}(\mathfrak{g}_0)$, hence acts transitively on $\mathrm{Int}(\mathfrak{g}_0)/K$ and we have the identification

$$\mathrm{Aut}(\mathfrak{g}_0)/\tilde{K} = \mathrm{Int}(\mathfrak{g}_0)/K, \tag{1}$$

where $\tilde{K}$ is the normalizer of K in $\mathrm{Aut}(\mathfrak{g}_0)$. In terms of this identification the action of Σ on $\mathrm{Int}(\mathfrak{g}_0)/K$ is (cf. Exercise A.6) given by

$$I_\sigma : g\tilde{K} \to \sigma g\tilde{K}. \tag{2}$$

But $\mathrm{Ad}_{\mathrm{Aut}(\mathfrak{g}_0)}(\tilde{K})$ maps $\mathfrak{k}_0$, the Lie algebra of K (and $\tilde{K}$) into itself, hence maps the orthogonal complement $\mathfrak{p}_0$ into itself, and leaves the restriction of the Killing form to $\mathfrak{p}_0$ invariant. This gives rise to an invariant metric on $\mathrm{Aut}(\mathfrak{g}_0)/\tilde{K}$, and (1) holds as an isometry. The mapping (2) being an isometry, (ii) is proved.

(iii) Assuming $\mathfrak{g}_0$ without compact ideals, the mapping $\sigma \to I_\sigma$ is one-to-one. In fact, $I_\sigma = e$ implies $\sigma g\tilde{K} = g\tilde{K}$ for all g so $\sigma \in \tilde{K}$ and $e^{\mathrm{ad} X}\tilde{K} = \sigma e^{\mathrm{ad} X}\sigma^{-1}\tilde{K} = e^{\mathrm{ad}(\sigma X)}\tilde{K}$, so $\sigma X = X$ and $\sigma = e$. On the other hand, let $\tau \in I(M)$. The Lie algebra $\mathfrak{L}(I(M))$ is isomorphic to $\mathfrak{g}_0$, so $\sigma = \mathrm{Ad}(\tau)$ belongs to $\mathrm{Aut}(\mathfrak{g}_0)$. Also $\mathrm{Ad}(I_\sigma) = \sigma$ so the isometry $\rho = I_\sigma\tau^{-1}$ satisfies $\mathrm{Ad}(\rho) = e$. Then ρ commutes with each $g \in I_0(M)$, the identity component of $I(M)$. If 0 is the fixed point of K then $k \cdot \rho \cdot 0 = \rho k \cdot 0 = \rho \cdot 0$ so $\rho \cdot 0 = 0$. Hence $\rho g \cdot 0 = g\rho \cdot 0 = g \cdot 0$ so $\rho = e$. Thus $I_\sigma = \tau$, so (iii) is proved.

A.8. (i) According to Exercise A.7, M induces a compact group of isometries of $\mathrm{Int}(\mathfrak{g}_0)/U$, U being any maximal compact subgroup. Let $S \subset \mathrm{Int}(\mathfrak{g}_0)$ be the analytic subgroup with Lie algebra $\mathfrak{s}_0$. Then S fixes a point p in the symmetric space if and only if S normalizes the corresponding isotropy group K_p, that is, if and only if $S \subset K_p$ (Exercise A.3). The fixed points of S form a simply connected totally geodesic submanifold (Exercise A.4), which M leaves invariant. Thus M has a fixed point in this manifold (Theorem 13.5, Chapter I), proving (i).

(ii) By (i) the compact real form $\mathfrak{u}' = \mathfrak{k}_0' + i\mathfrak{p}_0'$ of $\mathfrak{g}'$ extends to a σ-invariant compact real form $\mathfrak{u}$ of $\mathfrak{g}$. Then $\mathfrak{g}_0 = \mathfrak{u} \cap \mathfrak{g}_0 + (i\mathfrak{u}) \cap \mathfrak{g}_0$, and this is the desired decomposition $\mathfrak{g}_0 = \mathfrak{k}_0 + \mathfrak{p}_0$.

A.9. Part (i) follows from (2), §9, Chapter I, and Lemma 1.2. For (ii) we note that by (ii′), §9, Chapter I,

$$\frac{df_Y(s)}{ds} = B_\theta(\nabla_{\dot\gamma(s)}\dot\gamma(s), \tilde{Y}) + B_\theta(\dot\gamma(s), \nabla_{\dot\gamma(s)}\tilde{Y}).$$

The first term vanishes (γ being a geodesic) so the expression is (by (i))

$$\begin{aligned}
&\tfrac{1}{2}B_\theta(T(s) + X(s), [T(s) + X(s), Y] + [T(s) - X(s), Y] + [\theta Y, T(s) + X(s)])\\
&\quad = -B_\theta([T(s), X(s)], Y) - \tfrac{1}{2}B(T(s) + X(s), [Y, T(s) - X(s)])\\
&\quad = -B_\theta([T(s), X(s)], Y) - B([T, X], Y)\\
&\quad = -B_\theta([T(s), X(s)], Y - \theta Y).
\end{aligned}$$

(iii) By definition, $f_Y(s) = B_\theta(T(s) + X(s), Y)$ so

$$\dot{f}_Y(s) = B_\theta(\dot{T}(s) + \dot{X}(s), Y).$$

Combining this with (ii), we get $\dot{T}(s) = 0$, $\dot{X}(s) = -2[T(s), X(s)]$ so $T(s) = T_0$, $X(s) = e^{-2\mathrm{ad}(sT_0)}X_0$.

(iv) Let $\alpha(s) = \exp s(X_0 - T_0)$, $\beta(s) = \exp 2sT_0$ and $\delta(s) = \alpha(s)\,\beta(s)$. Then, if $I(x)g = xgx^{-1}$,

$$\begin{aligned}
\dot\delta(s) &= dL(\alpha(s))\,\dot\beta(s) + dR(\beta(s))\,\dot\alpha(s)\\
&= dL(\alpha(s))\,\dot\beta(s) + dL(\beta(s))\,dI(\beta(s)^{-1})\,\dot\alpha(s).
\end{aligned}$$

But $I(x)(g \exp Z) = xgx^{-1} \exp \mathrm{Ad}(x)Z$ so

$$dI(x)_g(\tilde{Z}_g) = (\mathrm{Ad}(x)Z)^{\sim}_{xgx^{-1}}.$$

Thus

$$\dot{\delta}(s) = 2(\tilde{T}_0)_{\delta(s)} + (e^{-2s \operatorname{ad} T_0}(X_0 - T_0))^{\sim}_{\delta(s)},$$

which by (iii) shows $\gamma(s) \equiv \delta(s)$.

B. The Iwasawa Decomposition

B.1. Let $e_1, \ldots, e_n$ be the standard orthonormal basis of $\boldsymbol{R}^n$ and let $g \in \boldsymbol{SL}(n, \boldsymbol{R})$. We orthonormalize the vectors $u_i = g \cdot e_i$ $(1 \leqslant i \leqslant n)$ by $v_1 = u_1/|u_1|$, $v_2 = (u_2 - cu_1)/|u_2 - cu_1|$ $(c = u_1 \cdot u_2/|u_1|^2)$, etc., so $v_j = \sum_{i \leqslant j} s_{ij} u_i$. Then determine $k \in \boldsymbol{O}(n)$ by $ke_j = v_j$. If g and k have the expression $g = (g_{kl})$, $k = (k_{pq})$, we have $\sum_i g_{li} s_{ij} = k_{lj}$. This shows $g = kan$ where a has positive diagonal and n is supertriangular with 1 on the diagonal. But then we conclude $\det k \det a = 1$, so $\det k = \det a = 1$. Thus we have $K = \boldsymbol{SO}(n)$, $A_{\mathfrak{p}}$ = set of diagonal matrices with determinant one and positive diagonal and N consists of all unipotent supertriangular matrices.

B.2. (i). Suppose $n_1\theta(n_2) = k$ for some $n_1, n_2 \in N$, $k \in K$. In terms of the basis (X_i) from Lemma 3.5 $\operatorname{Ad}(n_1\theta(n_2))_{ij} = k_{ij}$. The matrix $\operatorname{Ad}(n_1)$ (resp. $\operatorname{Ad}(\theta n_2)$) is lower (resp. upper) triangular with 1 in diagonal. Hence $k_{11} = 1$ so $k_{1j} = k_{j1} = 0$ for $j > 1$, (k_{ij}) being unitary. This in turn implies $\operatorname{Ad}(\theta n_2)_{1j} = 0$ $(j > 1)$ whence $k_{22} = 1$ and $k_{2j} = 0$ for $j \neq 2$. By induction $\operatorname{Ad}(k) = e$ so $\operatorname{Ad}(\theta n_2) = \operatorname{Ad}(n_1^{-1})$, whence $\operatorname{Ad}(n_1) = e$. Thus $n_1 = n_2 = e$.

For (ii) suppose $n_1\theta(n_2) = ma$. Applying θ and using $ma = am$, we get

$$\theta(n_1)\, n_2 n_1 \theta(n_2) = m^2.$$

Since $m\theta(N)m^{-1} = \theta(N)$, we conclude from (i) that $m^2 = e$ and $n_1 n_2 = e$. Then $ma = n_1\theta(n_1^{-1}) \in P$ so $m = e$. Thus $\operatorname{Ad}(n_1)\operatorname{Ad}(\theta n_1^{-1})$ is a diagonal matrix; so using the above matrix representations of $\operatorname{Ad}(n_1)$ and $\operatorname{Ad}(\theta n_1^{-1})$, we conclude $n_1 = e$.

For (v) let N_H denote the right-hand side of the relation. Then $N \subset N_H$. Suppose $ka \in N_H$ $(k \in K,\ a \in A_{\mathfrak{p}})$. Then by assumption

$$\operatorname{Ad}(\exp(-tH)k\exp(tH)) \to \operatorname{Ad}(a^{-1}),$$

whereas

$$\operatorname{Ad}(\exp(-tH)k\exp(tH))_{ij} = e^{t(\lambda_j - \lambda_i)(H)} k_{ij},$$

where $\lambda_1 < \lambda_2 < \ldots$ are the roots of $\mathfrak{g}$ with respect to $\mathfrak{h}$ in increasing order, the root 0 counted with multiplicity $\dim \mathfrak{h}$. These relations imply $k_{ij} = 0$ for $i < j$ and $k_{ii} > 0$ for all i. But (k_{ij}) is unitary, so we deduce $\operatorname{Ad}(k) = e$ so $k \in Z$ and $a = e$. But $N_H \cap Z = \{e\}$ so (v) follows.

For (iii) let $\mathfrak{e}_0$ denote the subspace of $\mathfrak{p}_0$ which $d\pi$ maps onto the tangent space $(N \cdot o)_0$ which of course equals $d\pi(\mathfrak{n}_0)$. Let $Y \in \mathfrak{e}_0$. Then there exists an $X \in \mathfrak{n}_0$ such that $d\pi(Y) = d\pi(X)$, that is, $Y - X \in \mathfrak{k}_0$. But then if $H \in \mathfrak{h}_{\mathfrak{p}_0} = (A \cdot o)_o$, we have $B(Y, H) = B(X, H) = 0$, proving (iii).

For (iv) let n^* be the element in N such that the point $an^* \cdot o$ minimizes the distance from o to the manifold $aN \cdot o$. Then the geodesic in G/K joining o to $an^* \cdot o$ is perpendicular to this manifold at $an^* \cdot o$ (Lemma 13.6, Chapter I). The manifold $A \cdot o$ contains all the geodesics intersecting $N \cdot o$ orthogonally at o (by (iii)). Hence the manifold $an^*A \cdot o$ contains all the geodesics intersecting the horocycle $aN \cdot o$ orthogonally at $an^* \cdot o$. In particular, there exists a $b \in A$ such that $an^*b \cdot o = o$. Hence $n^* = e$, proving (iv).

B.3. Here we use §8, Chapter III, with the same notation.

(i) **The algebra** $\mathfrak{a}_n = \mathfrak{sl}(n+1, C)$. From the formula for the Killing form it is immediately seen that

$$\mathfrak{u} = \mathfrak{su}(n+1)$$

is a compact real form. Then $\mathfrak{h} \cap \mathfrak{u}$ consists of the purely imaginary diagonal matrices of trace 0. Thus we can take

$$\mathfrak{h}_{\mathfrak{p}_0} = \{\text{real diagonal matrices of trace } 0\}.$$

Taking lexicographic ordering with respect to the basis $e_i - e_{i+1}$ $(1 \leqslant i \leqslant n)$ of the dual of $\mathfrak{h}_{\mathfrak{p}_0}$, we see that

$$\mathfrak{n}_+ = \sum_{i<j} CE_{ij},$$

so consists of the upper triangular matrices with 0 in the diagonal.

(iii) **The algebra** $\mathfrak{c}_n = \mathfrak{sp}(n, C)$. As already explained the algebra

$$\mathfrak{u} = \mathfrak{sp}(n) = \left\{\begin{pmatrix} U & V \\ -\bar{V} & \bar{U} \end{pmatrix} : \begin{matrix} U \text{ skew Hermitian,} \\ V \text{ symmetric} \end{matrix}\right\}$$

is a compact real form, and as a maximal abelian subspace of $\mathfrak{p}_0$ we can take the set of matrices

$$\mathfrak{h}_{\mathfrak{p}_0} = \left\{\begin{pmatrix} X & 0 \\ 0 & -X \end{pmatrix} : X \text{ real diagonal matrix}\right\}.$$

Taking lexicographic ordering with respect to the basis $e_i - e_{i+1}$ $(1 \leqslant i \leqslant n-1)$, e_n of the dual of $\mathfrak{h}_R$ we see that Δ^+ consists of the roots $e_i + e_j$ $(i \leqslant j)$ and $e_i - e_j$ $(i < j)$. Consequently

$$\mathfrak{n}_+ = \sum_{i \leqslant j} C(E_{i\,n+j} + E_{j\,n+i}) + \sum_{i<j} C(E_{ij} - E_{n+j\,n+i}).$$

Thus

$$\mathfrak{n}_+ = \left\{ \begin{pmatrix} Z_1 & Z_2 \\ 0 & -{}^tZ_1 \end{pmatrix} : \begin{matrix} Z_2 \text{ symmetric, } Z_1 \text{ upper triangular,} \\ 0 \text{ in diagonal} \end{matrix} \right\}.$$

(iii) **The algebra** $\mathfrak{d}_n \approx \mathfrak{so}(2n, C)$. Here we use the model $\mathfrak{d}_n$ from Exercise B.5, Chapter III. The transformation

$$S: \begin{pmatrix} z_1 \\ \vdots \\ z_{2n} \end{pmatrix} \to \left(\begin{array}{c|c} I_n & I_n \\ \hline iI_n & -iI_n \end{array} \right) \begin{pmatrix} z_1 \\ \vdots \\ z_{2n} \end{pmatrix} = \begin{pmatrix} u_1 \\ \vdots \\ u_{2n} \end{pmatrix}$$

satisfies

$$\sum_1^{2n} u_i^2 = 4 \sum_{i=1}^{n} z_i z_{n+i},$$

so the mapping $A \to S^{-1}AS$ maps $\mathfrak{so}(2n, C)$ onto our present model $\mathfrak{d}_n$. The compact real form $\mathfrak{so}(2n) \subset \mathfrak{so}(2n, C)$ (consisting of the real skew symmetric matrices) is mapped onto the space of matrices

$$\mathfrak{u} = \left\{ \begin{pmatrix} X + iY & Z \\ \bar{Z} & X - iY \end{pmatrix} : \begin{matrix} X \text{ and } Z \text{ skew, } Y \text{ symmetric,} \\ X \text{ and } Y \text{ real} \end{matrix} \right\},$$

which therefore is a compact real form of $\mathfrak{d}_n$. Clearly the purely imaginary diagonal matrices in $\mathfrak{u}$ form a maximal abelian subalgebra. Thus we can take

$$\mathfrak{h}_{\mathfrak{p}_0} = \left\{ \begin{pmatrix} X & 0 \\ 0 & -X \end{pmatrix} : X \text{ real diagonal matrix} \right\}.$$

As in (ii) we can take

$$\mathfrak{n}_+ = \sum_{i<j} C \begin{pmatrix} E_{ij} & 0 \\ 0 & -E_{ji} \end{pmatrix} + \sum_{i<j} C \begin{pmatrix} 0 & F_{ij} \\ 0 & 0 \end{pmatrix}.$$

Thus

$$\mathfrak{n}_+ = \left\{ \begin{pmatrix} Z_1 & Z_2 \\ 0 & -{}^tZ_1 \end{pmatrix} : \begin{matrix} Z_2 \text{ skew symmetric, } Z_1 \text{ upper triangular,} \\ 0 \text{ in diagonal.} \end{matrix} \right\}.$$

(iv) **The algebra** $\mathfrak{b}_n \approx \mathfrak{so}(2n+1, C)$. We use the model $\mathfrak{b}_n$ from Exercise B.6, Chapter III, and can proceed as in (iii). Then we find:

$$\mathfrak{u} = \left\{ \begin{pmatrix} 0 & Z_0 & \bar{Z}_0 \\ -{}^t\bar{Z}_0 & X+iY & Z \\ -{}^tZ_0 & \bar{Z} & X-iY \end{pmatrix} : \begin{matrix} X, Z \text{ skew, } Y \text{ symmetric,} \\ X \text{ and } Y \text{ real} \\ Z_0 \text{ any complex, } 1 \times n \text{ matrix} \end{matrix} \right\}$$

$$\mathfrak{h}_{\mathfrak{p}_0} = \left\{ \begin{pmatrix} 0 & 0 & 0 \\ 0 & X & 0 \\ 0 & 0 & -X \end{pmatrix} : \begin{matrix} X \text{ real diagonal,} \\ n \times n \text{ matrix} \end{matrix} \right\},$$

$$\mathfrak{n}_0 = \left\{ \begin{pmatrix} 0 & Z_0 & 0 \\ 0 & Z_1 & Z_2 \\ -{}^tZ_0 & 0 & -{}^tZ_1 \end{pmatrix} : \begin{matrix} Z_0\ 1 \times n \text{ matrix, } Z_2 \text{ skew symmetric,} \\ Z_1 \text{ upper triangular,} \\ 0 \text{ in diagonal} \end{matrix} \right\}.$$

4. Immediate from the proof of Lemma 6.2.

CHAPTER VII

1. Here $\mathfrak{p}_*$ consists of all symmetric purely imaginary $n \times n$ matrices of trace 0, and $\mathfrak{h}_{\mathfrak{p}_*}$ is the diagonal in $\mathfrak{p}_*$. Thus $\mathfrak{m}_0 = \{0\}$. The group M is given by the diagonal matrices with diagonal $\{\epsilon_1, \ldots, \epsilon_n\}$ such that each $\epsilon_i = \pm 1$ and $\epsilon_1 \ldots \epsilon_n = 1$. The group M' consists of the signed permutation matrices, that is, the matrices where each row and each column has all entries 0 except one which is ± 1. The signs are subject only to the condition that the determinant is 1. Thus $W(U, K) = M'/M$ consists of the group of all permutations on n letters.

2. The mapping Φ is clearly one-to-one and differentiable. Now let $\mathfrak{u} = \mathfrak{h} + \mathfrak{m}$ be the decomposition of the Lie algebra $\mathfrak{u}$ of U into the eigenspaces of $d\sigma$, $\mathfrak{h}$ denoting the Lie algebra of H. Let π: $U \to U/H$ be the natural map; and if $u \in U$, let $\tau(u)$ denote the translation $xH \to uxH$. If $X, Y \in \mathfrak{m}$, then

$$\Phi(\exp X \exp tY) = \exp 2X \exp(-X) \exp 2tY \exp X,$$

so

$$d\Phi_{\text{Exp } X}(d\tau(\exp X)\, d\pi(Y)) = dL(\exp 2X) \operatorname{Ad}(\exp(-X))(2Y),$$

proving the regularity of Φ. Let M denote the image $\Phi(U/H)$ and consider the subset

$$\mathfrak{m}^* = \{(X, -X) \in \mathfrak{u} \times \mathfrak{u} : X \in \mathfrak{m}\}.$$

Then $\mathfrak{m}^*$ is a Lie triple system contained in $\mathfrak{u} \times \mathfrak{u}$ so by Theorem 7.2, Chapter IV, Exp $\mathfrak{m}^*$ is a totally geodesic submanifold of $(U \times U)/U^*$.

Under the identification of $(U \times U)/U^*$ with U, Exp $\mathfrak{m}^*$ becomes exp $\mathfrak{m}$ which clearly equals M.

3. Immediate from (13), §11 for the curvature.

4. Modify the proof of Theorem 2.12 replacing W' by the group W'' generated by the reflections in the walls of a fixed Weyl chamber C, and showing, as before, that W'' is transitive.

5. (i) is immediate and so is the maximality of $\mathfrak{a}_*$. The Lie algebra $\mathfrak{h}$ is the image of the algebra of diagonal matrices under the automorphism $X \to AXA^{-1}$ of $\mathfrak{sl}(p+q, \boldsymbol{C})$ where $A = I_{p+q} - \Sigma_1^p(E_{ip+i} - E_{p+i\,i})$. Under this automorphism the diagonal elements correspond to the forms e_i. Thus $\mathfrak{h}$ is a Cartan subalgebra and Δ is as described (cf. §8, Chapter III). The description of Σ is now immediate; for the multiplicities note for example that $\pm f_i$ is the restriction $\pm(\bar{e}_i - \bar{e}_j)$ $(2p+1 \leqslant j \leqslant p+q)$ and also the restriction $\mp(\bar{e}_{i+p} - \bar{e}_j)$ $(2p+1 \leqslant j \leqslant p+q)$, so $\pm f_i$ has multiplicity $2(q-p)$. (iv) is immediate. For (v) note that $2f_1 = e_1 - e_{p+1} \in \Delta$, so by (3), §6, $\| 2f_1 \| = 2(p+q)((2(p+q))^{-2} \cdot 2) = (p+q)^{-1}$.

6. The points H' and H'' in §6 exponentiate to the identity element in the adjoint group (Lemma 6.5); and since they are not Weyl group equivalent (Theorem 2.22), the one-parameter subgroups exp tH' and exp tH'' are not conjugate (Prop. 2.2).

8. We have $A_o = \boldsymbol{SU}(n)/S$ where S is the centralizer in $\boldsymbol{SU}(n)$ of the element exp $H(\delta)$. But $\delta = e_1 - e_n$, so we find

$$\exp H(\delta) = \begin{pmatrix} -1 & & & & \\ & 0 & & & \\ & & \ddots & & \\ & & & 0 & \\ & & & & -1 \end{pmatrix},$$

so the description of S follows easily.

9. Clear from Prop. 2.2 and Cor. 8.9.

10. In the notation of Theorem 9.1 we have a subgroup $S_0 \subset \tilde{Z}$ (perhaps not $\tilde{\theta}$-invariant) such that $U = \tilde{U}/S_0$. Let φ: $\tilde{U} \to U$ be the natural projection and put $N = \varphi(\tilde{K} \cap \tilde{Z})$, $U' = U/N$, $K' = K/N$. We shall construct an involutive automorphism θ' of U' such that $d\theta' = \theta$ and θ' leaves K' pointwise fixed. Since $U/K = U'/K'$, this will prove that U/K is globally symmetric. Let ψ: $U \to U/N$ be the natural projection.

Let $u' \in U'$; choose $\tilde{u} \in \tilde{U}$ such that $\psi\varphi\tilde{u} = u'$ and put $\theta'u' = \psi\varphi\tilde{\theta}\tilde{u}$. To show that this is a valid definition suppose $\tilde{u}_1$, $\tilde{u}_2 \in \tilde{U}$ such that

$\psi\varphi\tilde{u}_1 = \psi\varphi\tilde{u}_2 = u'$. Then $\varphi(\tilde{u}_1) = \varphi(\tilde{u}_2)\,\varphi(k)$ where $k \in \tilde{K} \cap \tilde{Z}$ so $\tilde{u}_1 = \tilde{u}_2 ks$, where $s \in S_0$. It follows that $\psi\varphi\tilde{\theta}\tilde{u}_1 = (\psi\varphi\tilde{\theta}\tilde{u}_2)(\psi\varphi\tilde{\theta}s)$ so we just have to prove

$$\varphi\tilde{\theta}s \in \varphi(\tilde{K} \cap \tilde{Z}).$$

But $\varphi(s) = e$ and $s\tilde{\theta}(s) \in \tilde{K} \cap \tilde{Z}$ so $\varphi\tilde{\theta}s = \varphi(s\tilde{\theta}s) \in \varphi(\tilde{K} \cap \tilde{Z})$. Thus θ' is an involutive automorphism and $d\theta' = d\tilde{\theta} = \theta$. Finally let $k \in K'$. Then there exists an element $\tilde{k} \in \tilde{K}$ such that $k' = \psi\varphi\tilde{k}$; also $\theta'(k') = \psi\varphi\tilde{\theta}\tilde{k} = \psi\varphi\tilde{k} = k'$, as claimed.

CHAPTER VIII

A. Complex Structures

A.1. ω is bilinear and $\omega(X, X) = g(X, JX) = g(JX, J^2X) = -g(JX, X) = -\omega(X, X)$, so ω is alternate, hence a 2-form. By (7), §7, Chapter I, the Kähler condition means $\nabla_X(JY) = J\nabla_X(Y)$ for all $X, Y \in \mathfrak{D}^1(M)$.

(i) ⇒ (ii): We have by Prop. 7.2, Chapter I,

$$\begin{aligned}(\nabla_X\omega)(Y, Z) &= X(\omega(Y, Z)) - \omega(\nabla_X Y, Z) - \omega(Y, \nabla_X(Z)) \\ &= X(g(Y, JZ)) - g(\nabla_X(Y), JZ) - g(Y, J\nabla_X(Z))\end{aligned}$$

which by $J(\nabla_X(Z)) = \nabla_X(JX)$ equals $(\nabla_X g)(Y, JZ) = 0$.

(ii) ⇒ (iii): By Exercise C.4, Chapter I.

(iii) ⇒ (i): We have by (2), §9, Chapter I,

$$\begin{aligned}&2g(\nabla_X(JY) - J\nabla_X(Y), Z) \\ &\quad = 2g(\nabla_X(JY), Z) + 2g(\nabla_X(Y), JZ) \\ &\quad = Xg(Z, JY) + g(X, [Z, JY]) + (JY)g(Z, X) + g(JY, [Z, X]) \\ &\qquad - Zg(JY, X) - g(Z, [JY, X] + Xg(JZ, Y) + g(X, [JZ, Y]) \\ &\qquad + Yg(JZ, X) + g(Y, [JZ, X]) - (JZ)g(Y, X) - g(JZ, [Y, X]).\end{aligned}$$

Using (1), (3), §1 on the 2nd and 8th term, this becomes

$$\begin{aligned}&X\omega(Z, Y) + Y\omega(X, Z) - Z\omega(X, Y) \\ &\qquad + \omega([Z, X], Y) - \omega([Y, X], Z) + \omega(X, [Z, Y]) \\ &\qquad + X\omega(JY, JZ) - (JY)\omega(X, JZ) + (JZ)\omega(X, JY) \\ &\qquad + \omega([JY, X], JZ) - \omega([JZ, X], JY) + \omega([JZ, JY], X) \\ &\quad = -3\, d\omega(X, Y, Z) + 3\, d\omega(X, JY, JZ) = 0.\end{aligned}$$

Hence $\nabla_X(JY) = J\nabla_X(Y)$ proving (i).

A.2. The complex structure J of $\mathfrak{g}$ commutes with each ad X ($X \in \mathfrak{g}$); so, as indicated, exp is an almost complex mapping of $\mathfrak{g}$ into G. Hence the left invariant complex structure on G is integrable and exp a holomorphic mapping.

A.3. The Riemannian structure will be invariant under $\boldsymbol{SU}(1, 1)$ so will be a constant multiple of the standard one (Exercise G.1, Chapter I).

B. Bounded Symmetric Domains

B.1. In order to verify the relation $B(X, X) = 2 \operatorname{Tr}(T_X)$ we may assume $X = H \in \mathfrak{a}_0$. But then by Lemma 3.6, Chapter VI, both sides are equal to $2 \sum_{\alpha \in P_+} \alpha(H)^2$.

B.2. Clearly $G(\gamma) \cap KP_+$ has Lie algebra $\boldsymbol{C}H_\gamma + \mathfrak{g}^\gamma$, so

$$\dim_{\boldsymbol{C}} G(\gamma) \cdot o_c = 1.$$

Since $U \cap KP_+ = K_0$, the group $U(\gamma) \cap KP_+$ equals $U(\gamma) \cap K_0$ and has Lie algebra $\boldsymbol{R}(iH_\gamma)$ (each root is real on $i\mathfrak{h}_0$). Thus $U(\gamma) \cdot o_c = U(\gamma)/U(\gamma) \cap K_0$ is a compact open submanifold of $G(\gamma) \cdot o_c$, so

$$U(\gamma)/(U(\gamma) \cap K_0) = U(\gamma) \cdot o_c = G(\gamma) \cdot o_c = G(\gamma)/G(\gamma) \cap KP_+.$$

But $U(\gamma)$ is θ-invariant, so $(U(\gamma), U(\gamma) \cap K_0)$ is a Riemannian symmetric pair; the relation above shows that $U(\gamma)/(U(\gamma) \cap K_0)$ has a $U(\gamma)$-invariant complex structure, so by Prop. 4.2 and Theorem 4.6 it is simply connected. The group $G^* = \boldsymbol{SL}(2, \boldsymbol{C})$ being also simply connected, σ induces a holomorphic diffeomorphism of the Riemann sphere $G^*/K^*P_+^*$ onto $G(\gamma)/G(\gamma) \cap KP_+$ which maps the orbits of 0 under P_-^* and G_0^* into $\exp(\mathfrak{g}^{-\gamma}) \cdot o_c$ and $G_0(\gamma) \cdot o_c$, respectively.

For a generalization, see "Polydisk theorem" in Wolf [3].

B.3. Putting $Y_\gamma = X_\gamma + X_{-\gamma}$, the matrix identity implies (cf. Lemma 7.7),

$$\exp(yY_\gamma) \exp(xX_{-\gamma}) = \exp(x_1 X_{-\gamma}) \exp(s[X_\gamma, X_{-\gamma}]) \exp(y_1 X_\gamma).$$

Thus

$$\exp(yY_\gamma) \cdot \xi(xX_{-\gamma}) = \xi(x_1 X_{-\gamma}),$$

and the formula for $a \cdot \xi(X)$ follows by taking products over all $\gamma \in \Gamma$ and using the strong orthogonality.

B.4. From Cor. 7.13 we have $J_0 = \operatorname{ad} H_0$. Also

$$\operatorname{ad} H_0 \circ \operatorname{Ad}_{I(M)}(h) = \operatorname{Ad}_{I(M)}(h) \circ \operatorname{ad}(\operatorname{Ad}_{I(M)}(h^{-1})H_0);$$

and since H_0 spans $\mathfrak{c}_0$, $\mathrm{Ad}_{I(M)}(h^{-1})H_0 = \pm H_0$. Each $g \in I(M)$ can be written $g = \exp Xh$ ($X \in \mathfrak{p}_0$, $h \in H$) and since $\exp X$ is holomorphic (Lemma 4.3) the result follows.

C. Siegel's Generalized Upper Half-Plane

C.1-C.2. $\mathfrak{g}_0$ is a real form of $\mathfrak{sp}(n, \boldsymbol{C})$, hence semisimple (§8, Chapter III); the map $\theta : X \to JXJ^{-1}$ is a Cartan involution. In fact writing (cf. Chapter X)

$$X = \begin{pmatrix} X_1 & X_2 \\ X_3 & -{}^tX_1 \end{pmatrix} : \begin{array}{l} X_1, X_2, X_3 \text{ real } n \times n \text{ matrices,} \\ X_2, X_3 \text{ symmetric,} \end{array}$$

we find

$$-B(X, \theta X) = (2n + 2)\,\mathrm{Tr}(2X_1^t X_1 + X_2^2 + X_3^2).$$

The fixed points under θ are given by $X_1 = -{}^tX_1$, $X_2 = -X_3$, so $\mathfrak{k}_0 = \mathfrak{sp}(n, \boldsymbol{R}) \cap \mathfrak{so}(2n)$.

C.3. Obvious.

C.4. $\mathfrak{p}_0$ consists of the matrices X above for which $X_1 = {}^tX_1$, $X_2 = X_3$, so $\mathfrak{p}_0$ is the set of symmetric matrices in $\mathfrak{sp}(n, \boldsymbol{R})$. Now use Prop. 5.3. Chapter VI.

C.5. $\{g \in G : T_g(iI) = iI\}$ is given by $A = D$, $B = -C$, so this group is $\boldsymbol{Sp}(n, \boldsymbol{R}) \cap \boldsymbol{SO}(2n)$. For the surjectivity, see Exercise 6.

C.6. The matrix $I - iZ$ ($Z \in \mathscr{S}_n$) is clearly nonsingular, so $W = (I + iZ)(I - iZ)^{-1} = (I - iZ)^{-1}(I + iZ)$ exists and

$$\begin{aligned} I - \bar{W}W &= (I + i\bar{Z})^{-1}[(I + i\bar{Z})(I - iZ) - (I - i\bar{Z})(I + iZ)](I - iZ)^{-1} \\ &= 4AY\bar{A} \qquad \text{where} \quad A = (I + i\bar{Z})^{-1}. \end{aligned}$$

Thus $I - \bar{W}W$ is positive definite. That the map $Z \to W$ is surjective is seen in the same way, using the inverse map $Z \to i(I - W)(I + W)^{-1}$.

The group $\boldsymbol{Sp}(n, \boldsymbol{C}) \cap \boldsymbol{SU}(n, n)$ acts on $\{W\}$ by

$$T_g : W \to (AW + B)(\bar{B}W + \bar{A})^{-1} \qquad \text{if } g = \begin{pmatrix} A & B \\ \bar{B} & \bar{A} \end{pmatrix}.$$

Given W_0, let C be any matrix such that $\bar{C}(I - \bar{W}_0 W_0)^t C = I_n$ and then the transformation $W \to C(W - W_0)(I - \bar{W}_0 W)^{-1}\,\bar{C}^{-1}$ maps W_0 to 0; thus the domain is homogeneous. (For more details, see Siegel [1], II.)

D. An Alternative Proof of Prop. 4.2

D.1. We may assume G acting effectively on M. Thus the center $\mathfrak{z}_0$ of $\mathfrak{g}_0$ satisfies $\mathfrak{z}_0 \cap \mathfrak{k}_0 = 0$. But $\mathfrak{z}_0 \cap \mathfrak{p}_0 = 0$ by the assumed absence of

K_0-fixed vectors in M_0. Thus $\mathfrak{g}$ is centerless, so we can take G^c as the universal covering group of its adjoint group.

D.2. Using our assumption we see that $\mathfrak{k} + \mathfrak{p}_-$ is its own normalizer in $\mathfrak{g}$, so L is the identity component of the normalizer of $\mathfrak{k} + \mathfrak{p}_-$ in G^c, hence closed.

D.3-D.5. Clear.

CHAPTER IX

A. The Decompositions

A.1. Let $B = \{g = (g_{ij}) \in G : g_{21} = 0\}$, $m^* = \left(\begin{smallmatrix} 0 & 1 \\ -1 & 0 \end{smallmatrix}\right)$. Then $G = B \cup Bm^*B$.

A.2. (i) Let t_1, t_2 be upper triangular matrices. The operation $g \to t_1 g$ amounts to multiplying the kth row in g by a number, replacing the $(k-1)$st row in g by a linear combination of the kth and $(k-1)$th row, etc. The operation $g \to gt_2$ has the same effect on the columns. Thus the desired form of s can be reached.

(ii) Given $g = (g_{pq}) \in G$ we want $n = (n_{ij}) \in N$ such that $gn \in \bar{N}MA$, that is,

$$\sum_p g_{ip} n_{pj} = 0 \qquad \text{for} \quad i < j.$$

We write this as

$$g_{ij} + \sum_{p<j} g_{ip} n_{pj} = 0 \qquad \text{for} \quad i < j.$$

If all $\Delta_i(g)$ are $\neq 0$, n certainly exists. On the other hand, if $g = \bar{n}d(g)n$, then a direct computation shows $\Delta_i(g) = \Delta_i(d(g)) = \Pi_{j=1}^{i} d(g)_{jj}$.

(iii) Immediate since $\theta(g) = ({}^t g)^{-1}$.

A.3. Given $m \in M$ there exists a maximal torus $T \subset U$ containing m and $\exp(i\mathfrak{a})$. Its Lie algebra $\mathfrak{t}$ is θ-invariant because if $Z \in \mathfrak{t}$, then $Z - \theta Z \in i\mathfrak{p}$ and commutes with $i\mathfrak{a}$; so $Z - \theta Z \in i\mathfrak{a}$ whence $\theta Z \in \mathfrak{t}$. Thus $\mathfrak{t} = \mathfrak{t} \cap \mathfrak{k} + i\mathfrak{a}$ and $m \in \exp \mathfrak{t} \subset \exp(\mathfrak{t} \cap \mathfrak{k}) \exp(i\mathfrak{a}) \subset M^0 \exp(i\mathfrak{a})$.

A.4. Suppose $\mathrm{Ad}(g)H_1 = H_2$. Writing $g = kan$, we have $\mathrm{Ad}(an)H_1 = H_1 + X$ $(X \in \mathfrak{n})$, so $\mathrm{Ad}(k^{-1})H_2 - H_1 = X$. Since $\mathfrak{p} \cap \mathfrak{n} = 0$, this implies $X = 0$, whence $\mathrm{Ad}(k)H_1 = H_2$. Now use Prop. 2.2, Chapter V.

B. The Rank-One Case

B.1. If $\mathfrak{g}$ is not isomorphic to $\mathfrak{sl}(2, \boldsymbol{R})$, then $\dim \mathfrak{p} > 2$ (if $\dim \mathfrak{p} = 2$, G/K has constant curvature) K/M is sphere of dimension $\geqslant 2$, hence simply connected.

B.2. (i) In view of Theorem 5.4, Chapter VIII, it suffices to prove that $\mathfrak{k}^\alpha$ contains no nonzero ideal $\mathfrak{c}$ in $\mathfrak{g}^\alpha$. This $\mathfrak{c}$ would satisfy $[\mathfrak{c}, \mathfrak{p}^\alpha] = 0$, so in particular $\mathfrak{c} \subset \mathfrak{m}^\alpha$. But then $[\mathfrak{c}, \mathfrak{g}_\alpha] \subset [\mathfrak{m}^\alpha, \mathfrak{g}_\alpha] \subset \mathfrak{g}_\alpha$. Hence if $X \in \mathfrak{g}_\alpha$ so $X - \theta X \in \mathfrak{p}^\alpha$, the relation $[\mathfrak{c}, X - \theta X] = 0$ implies $[\mathfrak{c}, X] = 0$. Thus $\mathfrak{c}$ lies in the center of $\mathfrak{g}^\alpha$, so $\mathfrak{c} = 0$.

(ii) If J denotes the complex structure of $\mathfrak{g}$, then if $H \in \mathfrak{a}$, $X \in \mathfrak{g}_\alpha$, $[H, JX] = J[H, X] = \alpha(H)JX$ so $J\mathfrak{g}_\alpha \subset \mathfrak{g}_\alpha$. Hence $\mathfrak{g}^\alpha$ has a complex structure and by Theorem 6.3, Chapter VI, $\mathfrak{a}^\alpha + J\mathfrak{a}^\alpha$ is a Cartan subalgebra of $\mathfrak{g}^\alpha$ considered as a Lie algebra over $\boldsymbol{C}$. Now (ii) follows, for example from Theorem 5.4, Chapter III.

B.3. (i) Immediate from Theorem 3.8. For (ii) we imbed $\bar{n}$ in a group locally isomorphic to $\boldsymbol{SU}(2, 1)$ and note that it suffices to prove it for $\boldsymbol{SU}(2, 1)$ itself. As in Lemma 3.9 we take $m^* = I_{2,1}$. Then if $\bar{n} = k(\bar{n}) \exp H(\bar{n})\, n(\bar{n})$, we have, since $\theta(g) = m^*gm^*$,

$$\begin{aligned} k(\bar{n})^{-1}m^* &= m^*m^*k(\bar{n})^{-1}m^* = m^* \exp(-H(\bar{n}))\, \theta(n(\bar{n}))\, \theta(\bar{n})^{-1} \\ &= m^*k[\exp(-H(\bar{n}))\, \theta(n(\bar{n}))\, \exp(H(\bar{n}))], \end{aligned}$$

so

$$\bar{n}_0 = \exp(-H(\bar{n}))\, \theta(n(\bar{n}))\, \exp(H(\bar{n})).$$

Writing

$$\bar{n}_0 = \exp \begin{pmatrix} it_0 & z_0 & it_0 \\ -\bar{z}_0 & 0 & -\bar{z}_0 \\ -it_0 & -z_0 & -it_0 \end{pmatrix},$$

we obtain from Lemma 3.9

$$\begin{aligned} t_0 &= -t, \\ z_0 &= -z(1 + |z|^2 - 2it)^{-1}((1 + |z|^2)^2 + 4t^2)^{1/2}. \end{aligned}$$

In particular, $|t_0| = |t|$, $|z_0| = |z|$, so (ii) follows.

B.4. Obvious by computing isotropy group of 0 (cf. Chapter X, §2, no. 3)

C. Cartan Subalgebras

C.1. Denote the algebras by $\boldsymbol{R}X$, $\boldsymbol{R}H$, and $\boldsymbol{R}T$, respectively. They are easily seen to be maximal abelian subalgebras. Also ad X has all eigenvalues 0, ad H is real semisimple (with eigenvalues $\pm\frac{1}{2}$), and ad T is semisimple (with eigenvalues $\pm\frac{1}{2}i$), so the statements made follow.

C.2. We know fromTheorem 6.1 that if $\mathfrak{h}$ is a θ-invariant Cartan subalgebra such that $\mathfrak{h} = \mathfrak{h} \cap \mathfrak{k} + \mathfrak{a}$ ($\mathfrak{a} \subset \mathfrak{p}$ maximal abelian), then $\mathfrak{h} \cap \mathfrak{k}$ is maximal abelian in $\mathfrak{k}$. In particular it contains the center $\mathfrak{c}$ of $\mathfrak{k}$. But then Lemma 3.6, Chapter VI, shows that all the roots of $\mathfrak{g}^C$ with respect to $\mathfrak{h}^C$ vanish identically on $\mathfrak{c}$, so $\mathfrak{c} = \{0\}$, $\mathfrak{k}$ is semisimple, and K compact (Wallach [1], Ch. 7, §9).

C.3. The result follows from the proofs of Theorem 5.1 and Proposition 5.3.

CHAPTER X

B. Root Systems and the Weyl Group

B.1. The system defined is in fact an irreducible root system.

B.2. The second relation follows from Lemma 3.11; the first is immediate by classification.

B.3. (i) Assume $\alpha, \beta \in R$ such that $|\alpha| = |\beta|$. By the irreducibility $\langle \alpha, s\beta \rangle < 0$ for some $s \in W(R)$. Thus we can assume $\langle \alpha, \beta \rangle < 0$. It follows that $a_{\alpha,\beta} = a_{\beta,\alpha} = -1$, so $s_\alpha s_\beta \alpha = \beta$.

(ii) The first statement is clear from (i) and the fact that except for the R listed all $\alpha \in R$ have the same length. The uniqueness is clear from Theorem 2.22, Chapter VII, and the formulas are readily verified by means of the root description in §3, no. 3.

B.4. For Table 1 for $S(A)$ this is obvious because $-\alpha_0 =$ the highest root. For Tables 2 and 3 we see that $-\alpha_0$ (resp. $-\frac{1}{2}\alpha_0$) is precisely the root given in Exercise B.3.

B.7. By decomposing V and R into irreducible components we may assume R irreducible. Let $\langle\ ,\ \rangle$ be any $W(R)$-invariant inner product on V and consider the new inner product

$$\{\lambda, \mu\} = \sum_{\alpha \in R} \langle \lambda, \alpha \rangle \langle \alpha, \mu \rangle$$

which is $W(R)$-invariant and therefore by irreducibility $\{\lambda, \mu\} = c\langle \lambda, \mu \rangle$, where c is a constant > 0. Put $(\lambda, \mu) = c^{-1}\langle \lambda, \mu \rangle$.

To compute γ, say for $\mathfrak{g}_2$, take $\beta = \alpha_1$, so

$$(\alpha_1, \alpha_1)^{-1} = \langle \alpha_1, \alpha_1 \rangle^{-2} \sum_{\alpha} \langle \alpha, \alpha_1 \rangle^2$$

$$= 4^{-1} \cdot 2(1 + 1 + 4 + 9 + 9 + 0) = 12 = 24\langle \alpha_1, \alpha_1 \rangle^{-1}.$$

For the others see, e.g., Bourbaki [2], Chapter VI.

B.10. Let $\alpha_1, \ldots, \alpha_l$ be the simple roots in Σ and for $\mu = \alpha_i$ $(1 \leqslant i \leqslant l)$ denote the vectors H, X, Y from Theorem 2.16, Chapter VII by H_i, X_i, Y_i. By the quoted proof we have, if $a_{ji} = 2\langle\alpha_j, \alpha_i\rangle\langle\alpha_i, \alpha_i\rangle^{-1}$,

$$[H_i, H_j] = 0, \qquad [X_i, Y_j] = \delta_{ij} H_i$$

$$[H_i, X_j] = a_{ji} X_j, \qquad [H_i, Y_j] = -a_{ji} Y_j$$

$$(\text{ad } X_i)^{1-a_{ji}}(X_j) = (\text{ad } Y_i)^{1-a_{ji}}(Y_i) = 0.$$

Let $\mathfrak{g}'$ be the subalgebra of $\mathfrak{g}$ generated by H_i, X_i, Y_i $(1 \leqslant i \leqslant l)$. Using Theorem 4.15, we see that $\mathfrak{g}'$ has the desired property (cf. Borel and Tits [1], p. 117; Kostant and Rallis [1], p. 792).

D. Bounded Symmetric Domains

D.1. The descriptions can be verified by means of the following steps: (i) determine the dimension of the matrix space in question; (ii) determine the isotropy group at the origin (point iI for $\boldsymbol{C}\ \boldsymbol{I}$) and verify that dim G_0/K_0 equals the dimension of the matrix space. The transitivity can be proved by the method of Exercises C.5 and C.6 in Chapter VIII.

D.3. As in Lemma 7.11, it suffices to verify this for $\mathfrak{g} = \mathfrak{sl}(2, \boldsymbol{C})$, in which case it follows by straightforward computation. (cf. Harish-Chandra [5], p. 584).

D.5. Since $d\psi$ commutes with Ad(k) $(k \in K_0)$, since $\| \ \|$ is Ad(K_0)-invariant, and since $\mathfrak{p}_0 = \text{Ad}(K_0)\mathfrak{a}_0$, $\mathfrak{D} = \text{Ad}(K_0)(\mathfrak{D} \cap \mathfrak{a}_0)$, it suffices to show that

$$\mathfrak{D} \cap \mathfrak{a}_0 = \{H \in \mathfrak{a}_0 : \| \text{ad } H \| < 2\}.$$

But by Cor. 7.13 and Cor. 7.17, Chapter VIII,

$$\psi(\mathfrak{D} \cap \mathfrak{a}_0) = D \cap \mathfrak{a}_- = \Big\{\sum_{\gamma \in \Gamma} x_\gamma \psi(X_\gamma + X_{-\gamma}) : | x_\gamma | < 1\Big\},$$

so

$$\mathfrak{D} \cap \mathfrak{a}_0 = \Big\{\sum_{\gamma} x_\gamma (X_\gamma + X_{-\gamma}) : | x_\gamma | < 1\Big\}.$$

But by Exercise 3 the eigenvalues of ad$(X_\gamma + X_{-\gamma})$ are those of $2\langle\gamma, \gamma\rangle^{-1}$ ad H_γ, which by Exercise 4b are $0, \pm 1, \pm 2$. Since $\|$ ad $H \|$ equals the maximum of the absolute value of the eigenvalues, the description of $\mathfrak{D} \cap \mathfrak{a}_0$ follows. (The result was proved by R. Hermann (unpublished); cf. Moore [2], p. 371; also Langlands [1], Lemma 2.)

D.6. The proof is analogous to that of Theorem 5.4, Chapter IX, using in addition the conjugacy of the Cartan decompositions of $\mathfrak{g}_0$.

D.7. Let k be an element in said component leaving the origin fixed. It is holomorphic if and only if $\mathrm{Ad}_{I(M)}(k)$ commutes with J_0. But by solution to Exercise B.4 in Chapter VIII this is equivalent to $\mathrm{Ad}_{I(M)}(k) \mid \mathfrak{c} = 1$. Since k could be taken such that $\mathrm{Ad}_{I(M)}(k)$ maps $\mathfrak{h}$ into itself, we obtain the condition $t \mid \mathfrak{c} = 1$ (cf. Takeuchi [1]).

E. Automorphisms

E.1. In the notation of Theorem 5.15, $\mathfrak{h}^\nu$ is maximal abelian in $\mathfrak{g}_0$, so each s_i is >0 for σ regular. The minimal order is reached for $s_0 = \dots = s_n = k = 1$, so by Theorem 5.16 such automorphisms are inner and all conjugate. (This corollary of Theorems 5.15 and 5.16, observed by V. Kač, had been proved by Kostant [4], p. 1027 for the case when σ is assumed inner.)

E.2. We must discuss the solutions of the equation $3 = k \sum_0^n a_i s_i$; we use part (iii) of Theorem 5.15 to determine $\mathfrak{g}_0$. The solutions are:

(a) $k = 3, \quad a_{i_0} = s_{i_0} = 1, \quad s_i = 0 \quad \text{for} \quad i \neq i_0.$

(b) $k = 1, \quad a_{i_0} = 1, \quad s_{i_0} = 2, \quad s_{i_1} = 1, \quad a_{i_1} = 1, \quad s_i = 0$ for $i \neq i_0, i_1$.

(c) $k = 1, \quad a_{i_0} = a_{i_1} = a_{i_2} = s_{i_0} = s_{i_1} = s_{i_2} = 1, \quad s_i = 0$ for $i \neq i_0, i_1, i_2$.

(d) $k = 1, \quad a_{i_0} = 2, \quad a_{i_1} = 1, \quad s_{i_0} = s_{i_1} = 1, \quad s_i = 0$ for $i \neq i_0, i_1$.

(e) $k = 1, \quad a_{i_0} = 3, \quad s_{i_0} = 1, \quad s_i = 0 \quad \text{for} \quad i \neq i_0.$

For case (a) we get using $S(A)$ — Table 3,

$$\mathfrak{g} = \mathfrak{d}_4, \qquad \mathfrak{g}_0 = \mathfrak{g}_2 \qquad \text{or} \qquad \mathfrak{g}_0 = \mathfrak{a}_2.$$

For case (e) we get using $S(A)$ — Table 1,

$$\mathfrak{g} = \mathfrak{e}_6, \qquad \mathfrak{g}_0 = \mathfrak{a}_2 \oplus \mathfrak{a}_2 \oplus \mathfrak{a}_2$$

$$\mathfrak{g} = \mathfrak{e}_7, \qquad \mathfrak{g}_0 = \mathfrak{a}_2 \oplus \mathfrak{a}_5$$

$$\mathfrak{g} = \mathfrak{e}_8, \qquad \mathfrak{g}_0 = \mathfrak{a}_2 \oplus \mathfrak{e}_6 \quad \text{or} \quad \mathfrak{g}_0 = \mathfrak{a}_8$$

$$\mathfrak{g} = \mathfrak{f}_4 \qquad \mathfrak{g}_0 = \mathfrak{a}_2 \oplus \mathfrak{a}_2$$

$$\mathfrak{g} = \mathfrak{g}_2 \qquad \mathfrak{g}_0 = \mathfrak{a}_2.$$

For the other cases the possibilities for $\mathfrak{g}_0$ are determined by the same method. We combine all cases in the following table. Here $\mathfrak{a}_0 = \mathfrak{b}_0 = \mathfrak{c}_0 = \mathfrak{d}_0$ means 0. The table confirms the classification in Wolf and Gray [1], obtained by different methods.

That $\mathfrak{g}_0$ determines σ up to conjugacy can be verified by the argument of Theorem 6.1.

TABLE VII

AUTOMORPHISMS OF ORDER 3 OF SIMPLE LIE ALGEBRAS

$\mathfrak{g}$	$\mathfrak{g}_0$
$\mathfrak{a}_n, n \geqslant 1$	$C \oplus \mathfrak{a}_{n-p} \oplus \mathfrak{a}_{p-1}\ (1 \leqslant p \leqslant n,\ C^2 \oplus \mathfrak{a}_{n-p} \oplus \mathfrak{a}_{p-q-1} \oplus \mathfrak{a}_{q-1}\ (1 \leqslant q < p \leqslant n)$
$\mathfrak{b}_n, n \geqslant 2$	$C \oplus \mathfrak{a}_{n-p} \oplus \mathfrak{b}_{p-1}\ (1 \leqslant p \leqslant n)$
$\mathfrak{c}_n, n \geqslant 2$	$C \oplus \mathfrak{a}_{n-p} \oplus \mathfrak{c}_{p-1}\ (1 \leqslant p \leqslant n)$
$\mathfrak{d}_n, n \geqslant 5$	$C \oplus \mathfrak{a}_{n-p} \oplus \mathfrak{d}_{p-1}\ (1 \leqslant p \leqslant n),\ C^2 \oplus \mathfrak{a}_{n-2}$
$\mathfrak{d}_4$	$C \oplus \mathfrak{a}_3,\ C^2 \oplus \mathfrak{a}_2,\ C \oplus \mathfrak{a}_1 \oplus \mathfrak{a}_1 \oplus \mathfrak{a}_1,\ \mathfrak{a}_2,\ \mathfrak{g}_2$
$\mathfrak{e}_6$	$C \oplus \mathfrak{a}_5,\ C \oplus \mathfrak{d}_5,\ C \oplus \mathfrak{a}_1 \oplus \mathfrak{a}_4,\ C^2 \oplus \mathfrak{d}_4,\ \mathfrak{a}_2 \oplus \mathfrak{a}_2 \oplus \mathfrak{a}_2$
$\mathfrak{e}_7$	$C \oplus \mathfrak{e}_6,\ C \oplus \mathfrak{d}_6,\ C \oplus \mathfrak{a}_6,\ C \oplus \mathfrak{a}_1 \oplus \mathfrak{d}_5,\ \mathfrak{a}_2 \oplus \mathfrak{a}_5$
$\mathfrak{e}_8$	$C \oplus \mathfrak{e}_7,\ C \oplus \mathfrak{d}_7,\ \mathfrak{a}_2 \oplus \mathfrak{e}_6,\ \mathfrak{a}_8$
$\mathfrak{f}_4$	$C \oplus \mathfrak{c}_3,\ C \oplus \mathfrak{b}_3,\ \mathfrak{a}_2 \oplus \mathfrak{a}_2$
$\mathfrak{g}_2$	$C \oplus \mathfrak{a}_1,\ \mathfrak{a}_2$

E.4. With the notation of Lemma 5.3 and (4), §5, let $H_0 \in \mathfrak{h}$ be an element such that if a root $(\alpha, i) \in \bar{\Delta}$ satisfies $\alpha(H_0) = 0$, then $\alpha \equiv 0$. Writing an element $Z \in \mathfrak{g}$ according to (4), $Z = Z_{\mathfrak{h}} + \Sigma_{\bar{\alpha}} Z_{\bar{\alpha}}$, we see that $[Z, H_0] = 0 \Leftrightarrow [Z, \mathfrak{h}] = 0$. Hence the centralizer of H_0 in $\mathfrak{g}$ coincides with the Cartan subalgebra $\mathfrak{z}(\mathfrak{h})$, so H_0 is regular.

E.5. Consider the diagram $\mathfrak{b}_4^{(1)}$ with the vertices enumerated,

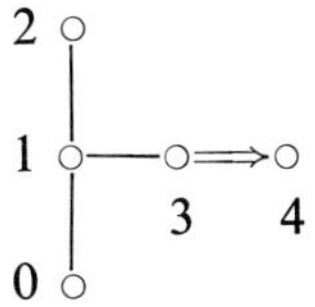

$$(a_0, a_1, a_2, a_3, a_4) = (1, 2, 1, 2, 2).$$

Consider the automorphisms by types:

$$\sigma = (1, 0, 0, 1, 1; 1), \qquad \sigma' = (2, 0, 1, 0, 1; 1).$$

They have the stated properties.

E.6. Clear from Table III and the solution to Exercise E.2.

F. Restricted Roots and Multiplicities

F.1. See Chapter VI, §3.

F.2. If $\alpha + \alpha^\theta \in \Delta$, then $(\mathfrak{g}^C)^{(\alpha+\alpha^\theta)} \in \mathfrak{k}^C$ (Lemma 3.3, Chapter VI) while

$$(\mathfrak{g}^C)^{(\alpha+\alpha^\theta)} = C[X_\alpha, X_{\alpha\theta}] \subset \mathfrak{p}^C.$$

F.3. The map $\alpha \to -\alpha^\theta$ permutes Δ_ψ and has a fixed point if and only if $\psi \in \Delta_\psi$. This proves the first statement. If $\alpha \in \Delta_\psi - \{\psi\}$ such that $\langle\alpha, \alpha^\theta\rangle \neq 0$, then by Exercise 2 and Lemma 3.4 $\beta = \alpha - \alpha^\theta \in \Delta$, so $2\psi \in \Sigma$ and $\langle\beta, \beta\rangle \geqslant \langle 2\psi, 2\psi\rangle$. On the other hand, $\langle\beta, \psi\rangle = \langle 2\psi, \psi\rangle > 0$, so since $\psi \in \Delta_\psi$, we conclude from Exercise C.5, Chapter III, that $\langle\beta, \beta\rangle/\langle\psi, \psi\rangle = 1, 2,$ or 3, a contradiction.

F.4. (a) This was proved (taking $\beta \in \Delta_{2\psi}$) in the solution to Exercise 3.

(b) If $m_{2\psi}$ were even, select (by Exercise 3) $\beta \in \Delta_{2\psi}$ $\beta \neq -\beta^\theta$, and $\alpha \in \Delta_\psi$, $\alpha \neq -\alpha^\theta$. By Exercise 2, $\langle\beta, \beta^\theta\rangle \geqslant 0$. If $\langle\beta, \beta^\theta\rangle > 0$, then $\Delta \ni \beta - \beta^\theta = 4\psi$, which is impossible. Thus $\langle\beta, \beta^\theta\rangle = 0$, whence $\langle\beta, \beta\rangle = 8\langle\psi, \psi\rangle$. On the other hand, $m_{2\psi}$ even implies by Exercises 3 that $2\psi = \alpha - \alpha^\theta \notin \Delta$. Hence $\langle\alpha, \alpha^\theta\rangle = 0$ and $\langle\alpha, \alpha\rangle = 2\langle\psi, \psi\rangle$. Thus $\langle\beta, \beta\rangle = 4\langle\alpha, \alpha\rangle$, which is a contradiction.

F.5. (a) If $\psi, 2\psi \in \Sigma$, m_ψ even, then for any $\alpha \in \Delta_\psi$, $2\psi = \alpha - \alpha^\theta \in \Delta$. If $\langle\alpha, \alpha^\theta\rangle = 0$, then $\Delta \ni \alpha - \alpha^\theta - s_\alpha(\alpha - \alpha^\theta) = 2\alpha$, which is impossible. Thus $\langle\alpha, \alpha^\theta\rangle > 0$. Conversely, this relation implies $2\psi = \alpha - \alpha^\theta \in \Delta_\psi$, so $2\psi \in \Sigma$. This proves (a) and (b) as well.

F.6. Parts (a) and (b) are proved in Chapter IX, §6. For (c) we see from Exercises 4b and 5b that for each $\psi \in \Sigma$, Δ_ψ consists of two elements α, β, $\beta = -\alpha^\theta$, $\langle\alpha, \beta\rangle = 0$. Let $\gamma \in \Delta$ such that $\bar{\gamma} = 0$. Then if $\langle\alpha, \gamma\rangle < 0$, we have $\alpha + \gamma \in \Delta$ and $\bar{\alpha} + \bar{\gamma} = \psi$, so $\alpha + \gamma = \beta$. Thus $\beta - \alpha \in \Delta$, so by $\langle\alpha, \beta\rangle = 0$, $\Delta \ni \beta + \alpha = 2\psi$, which is a contradiction. Similarly $\langle\alpha, \gamma\rangle > 0$ is impossible, so $\langle\alpha, \gamma\rangle = 0$. This shows that $\Delta = (\Delta - \Delta_\mathfrak{p}) \cup \Delta_\mathfrak{p}$ is an orthogonal decomposition of Δ. The ideal in $\mathfrak{g}^C$ generated by the root spaces from $\Delta - \Delta_\mathfrak{p}$ is contained in $\mathfrak{k}^C$ (Lemma 3.3, Chapter VI), so is 0; hence $\Delta = \Delta_\mathfrak{p}$.

Now introduce compatible orderings in Δ and Σ and let B be a basis of Δ of simple roots. Then $\alpha \in B$ implies $-\alpha^\theta \in B$. We must show $\mathfrak{g}^C$ not simple, that is, B decomposable. Suppose B is indecomposable. Let $\alpha \in B$; then $-\alpha^\theta \in B$ and we can find a chain (§3) $\{\alpha_1, \ldots, \alpha_k\}$ in B such that $\alpha = \alpha_1$, $-\alpha^\theta = \alpha_k$. Take the smallest $i \geqslant 2$ such that $-\alpha_i^\theta = \alpha_j \in \{\alpha_1, \ldots, \alpha_k\}$. Then $\{\alpha_{j+1}, \ldots, \alpha_k, -\alpha_2^\theta, \ldots, -\alpha_i^\theta\}$ is a cycle in B, so by Corollary 3.14, $i = 2$, $j = k - 1$; that is, $-\alpha_2^\theta = \alpha_{k-1}$. Iterating, we end up with a root β such that $-\beta^\theta = \beta$ or $-\beta^\theta = \delta$ with $\langle\beta, \delta\rangle \neq 0$. This is a contradiction. (This proof was given by Loos [2], Chapter VI.)

F.7. (a) Obvious. (b) Following Satake [3], we show the members of $\bar{B}$ are linearly independent by proving that if $\lambda \in \bar{B}$, then there exists an $H \in \mathfrak{a}$ such that $\lambda(H) = 1$, $\mu(H) = 0$ if $\mu \neq \lambda$, $\mu \in \bar{B}$. Let $r = \dim \mathfrak{h}$, $r_0 = \text{Card } B_0$, $l = \dim \mathfrak{a}$, and let

$$B = \{\alpha_1, \ldots, \alpha_{r-r_0}, \alpha_{r-r_0+1}, \ldots, \alpha_r\},$$

where $B_0 = \{\alpha_{r-r_0+1}, \ldots, \alpha_r\}$, Let

$$-\alpha_i^\theta = \sum_{j=1}^{r} c_j^{(i)} \alpha_j, \qquad 1 \leqslant i \leqslant r - r_0,$$

where $c_j^{(i)} \in \boldsymbol{Z}^+$. Applying $\alpha \to -\alpha^\theta$ to this identity and substituting, we find $c_j^{(i)} = 0$ for all $j \in [1, r - r_0]$ except one, say $j = i'$. Thus

$$-\alpha_i^\theta = \alpha_{i'} + \sum_{j=r-r_0+1}^{r} c_j^{(i)} \alpha_j,$$

and $i \to i'$ is a permutation of $[1, r - r_0]$ of order 2. We can therefore assume the α_i ordered such that $i' = i$ $(1 \leqslant i \leqslant p_1)$ $i' = i + p_2$ $(p_1 + 1 \leqslant i \leqslant p_1 + p_2)$ $i' = i - p_2$ $(p_1 + p_2 + 1 \leqslant i \leqslant p_1 + 2p_2 = r - r_0)$. Now fix $i \in [1, r - r_0]$ such that $\bar{\alpha}_i = \lambda$ and select $H \in \mathfrak{h}_{\boldsymbol{R}}$ such that $\alpha_i(H) = \alpha_{i'}(H) = 1$, $\alpha_k(H) = 0$ for $k \in [1, r]$, $k \neq i, i'$. Then $\theta H = -H$, so $H \in \mathfrak{a}$ as desired. That $\bar{B}$ is a basis of Σ is now obvious since the members of Σ are the restrictions of Δ to $\mathfrak{a}$.

G. Two-Point Homogeneous Spaces

G.1. Let M be two-point homogeneous, let $p \in M$ and N_0 and N_p be spherical normal neighborhoods of 0 in M_p and $p \in M$. Using Lemma 9.3 Chapter I, we see that M is isotropic.

Conversely, suppose M is isotropic. Given any points $p, q \in M$, we can join them by a broken geodesic each part of which lies in a spherical, convex, normal neighborhood. Combining the isometries which reverse each of these segments, we find a $g \in M$ such that $gp = q$. Thus $I(M)$ is transitive on M, and by Theorem 3.2, Chapter II and Theorem 2.5, Chapter IV, $I(M)\,I(M)_p$ is homeomorphic to M. Hence by Exercise A.4 in Chapter II, M is complete, so any $q \in M$ can be joined to a fixed $p \in M$ by a geodesic of length $d(p, q)$. Now the isotropy of M implies that the isotropy group $I(M)_p$ is transitive on each sphere $S_p(p)$, whence the two-point homogeneity.

BIBLIOGRAPHY

ADAMS, J. F.

1. "Lectures on Lie Groups." Benjamin, New York, 1969.

ADLER, A.

1. Characteristic classes of homogeneous spaces. *Trans. Amer. Math. Soc.* **86** (1957), 348-365.

ALEKSSEEVSKI, D. V.

1. Riemannian spaces with exceptional holonomy group. *Funct. Anal. Appl.* **2** (1968), 97-105.
2. Compact quaternionic spaces. *Functional Anal. Appl.* **2** (1968), 106-114.

ALEXANDROFF, P.

1. Über die Metrisation der im Kleinen kompakten topologischen Räume. *Math. Ann.* **92** (1924).

ALLAMIGEON, A. C.

1. Espaces homogènes symétriques harmoniques à groupe semisimple. *C. R. Acad. Sci. Paris* **243** (1956), 121-123.
2. Propriétés globales des espaces de Riemann harmoniques. *Ann. Inst. Fourier* **15** (1965), 91-132.

AMBROSE, W.

1. Parallel translation of Riemannian curvature. *Ann. of Math.* **64** (1956), 337-363.
2. The Cartan structural equations in classical Riemannian geometry. *J. Indian Math. Soc.* **24** (1960), 23-76.

AMBROSE, W., and SINGER, I. M.

1. On homogeneous Riemannian manifolds. *Duke Math. J.* **25** (1958), 647-669.
2. A theorem on holonomy. *Trans. Amer. Math. Soc.* **75** (1953), 428-443.

AOMOTO, K.

1. On some double coset decompositions of complex semisimple Lie groups. *J. Math. Soc. Japan* **18** (1966), 1-44.

ARAKI, S. I.

1. On root systems and an infinitesimal classification of irreducible symmetric spaces. *J. Math. Osaka City Univ.* **13** (1962), 1-34.
2. On Bott-Samelson K-cycles associated with symmetric spaces. *J. Math. Osaka City Univ.* **13** (1963), 87-133.
3. On the Brouwer degree of some maps of compact symmetric spaces. *Topology* **3** (1965), 281-290.

ARENS, R.

1. Topologies for homeomorphism groups. *Amer. J. Math.* **68** (1946), 593-610.

ASTRAHANCEV, V. V.

1. Symmetric spaces of corank 1. *Mat. Sb. (N.S.)* **96** (138) (1975), 135-151, 168.

AUSLANDER, L., and MACKENZIE, R. E.
1. "Introduction to Differentiable Manifolds." McGraw-Hill, New York, 1963.

AVEZ, A.
1. Espaces harmoniques compacts. *C. R. Acad. Sci. Paris* **258** (1964), 2727-2729.

AZENCOTT, R., and WILSON, E. N.
1. Homogeneous manifolds with negative curvature I, II. *Trans. Amer. Math. Soc.* **215** (1976), 323-362; *Mem. Amer. Math. Soc.* **8** (1976).

BAILY, W. L.
1. "Introductory Lectures on Automorphic Forms." I. Shoten and Princeton Univ. Press, Princeton, New Jersey, 1973.

BAILY, W. L., and BOREL, A.
1. Compactifications of arithmetic quotients of bounded symmetric domains. *Ann. of Math.* **84** (1966), 442-528.

BARRETT, W.
1. Sur la structure des groupes semi-simples réels. *Bull. Sci. École Polytechn. Timisoara* **9** (1939), 22-90.

BARUT, A. O., and RACZKA, R.
1. "Theory of Group Representations and Applications." Polish Scientific Publishers, Warsaw, 1977.

BECK, R. E.
1. Connections on semisimple Lie groups. *Trans. Amer. Math. Soc.* **164** (1972), 453-460.

BERGER, M.
1. Sur les groupes d'holonomie homogène des variétés a connexions affine et des variétés riemanniennes. *Bull. Soc. Math. France* **83** (1955), 279-330.
2. Les espaces symétriques non compacts. *Ann. Sci. École Norm. Sup.* **74** (1957), 85-177.
3. Sur quelques variétés riemanniennes suffisamment pincées. *Bull. Soc. Math. France* **88** (1960), 57-71.
4. Les variétés riemanniennes homogènes normales simplement connexes à courbure strictement positive. *Ann. Scuola Norm. Sup. Pisa* **15** (1961), 179-246.
5. "Lectures on Geodesics in Riemannian Geometry." Tata Institute of Fundamental Research, Bombay, 1965.

BERGMAN, S.
1. Über die Entwicklung der harmonischen Funktionen der Ebene und des Raumes nach Orthogonalfunktionen. *Math. Ann.* **96** (1922), 237-271.
2. Über die Kernfunktion eines Bereiches und ihr Verhalten am Rande. *J. Reine Angew. Math.* **169** (1933), 1-42.

BIRKHOFF, G.
1. Representability of Lie algebras and Lie groups by matrices. *Ann. of Math.* **38** (1937), 526-533.
2. Analytical groups. *Trans. Amer. Math. Soc.* **43** (1938), 61-101.

BISHOP, R. L., and CRITTENDEN, R. J.
1. "Geometry of Manifolds." Academic Press, New York, 1964.

BOCHNER, S.
1. Vector fields and Ricci curvature. *Bull. Amer. Math. Soc.* **52** (1946), 776-797.
2. Curvature in Hermitian metric. *Bull. Amer. Math. Soc.* **53** (1947), 179-195.
3. Über orthogonale Systeme analytischer Funktionen, *Math. Z.* **14** (1922), 180-207.

BOCHNER, S., and MARTIN, W. T.
1. "Several Complex Variables." Princeton Univ. Press, Princeton, New Jersey, 1948.

BOOTHBY, W. M.
1. "An Introduction to Differentiable Manifolds and Riemannien Geometry." Academic Press, New York, 1975.

BOOTHBY, W. M., and WANG, H.-C.
1. On the finite subgroups of connected Lie groups. *Comment. Math. Helv.* (1965), 281-294.

BOOTHBY, W. M., and WEISS, G. L., (eds.)
1. "Symmetric Spaces," Short Courses at Washington Univ. Dekker, New York, 1972.

BOREL, A.
1. Exposé 12-13 in Séminaire H. Cartan, Paris, 1949-1950.
2. Les fonctions automorphes de plusieurs variables complexes. *Bull. Soc. Math. France* **80** (1952), 167-182.
3. Les espaces hermitiens symétriques. Séminaire Bourbaki, 1952.
4. Sous-groupes compact maximaux des groupes de Lie. Séminaire Bourbaki, 1952.
5. Kählerian coset spaces of semi-simple Lie groups. *Proc. Nat. Acad. Sci., U.S.A.* **40** (1954), 1147-1151.
6. Topology of Lie groups and characteristic classes. *Bull. Amer. Math. Soc.* **61** (1955), 398-432.
7. On the curvature tensor of the Hermitian symmetric manifolds. *Ann. of Math.* **71** (1960), 508-521.
8. Sous-groupes commutatifs et torsion des groupes de Lie compacts connexes. *Tôhoku Math. J.* **13** (1961), 216-240.
9. Travaux de Mostow sur les espaces homogènes, Séminaire Bourbaki, 1957.
10. Compact Clifford-Klein forms of symmetric spaces. *Topology* **2** (1963), 111-122.
11. Some metric properties of arithmetic quotients of symmetric spaces and an extension theorem. *J. Differential Geometry* **6** (1972), 543-560.
12. "Introduction aux groupes arithmétiques." Hermann, Paris, 1969.

BOREL, A., and HARISH-CHANDRA
1. Arithmetic subgroups of algebraic groups. *Ann. of Math.* **75** (1967), 485-535.

BOREL, A., and HIRZEBRUCH, F.
1. Characteristic classes and homogeneous spaces, Part I. *Amer. J. Math.* **80** (1958), 459-538.

BOREL, A., and LICHNEROWICZ, A.
1. Espaces riemanniens et hermitiens symétriques. *C. R. Acad. Sci. Paris* **234** (1952), 2332-2334.
2. Groupes d'holonomie des variétés riemanniennes. *C. R. Acad. Sci. Paris* **234** (1952), 1835-1837.

BOREL, A., and MOSTOW, G. D.
1. On semisimple automorphisms of Lie algebras. *Ann. of Math.* **61** (1955), 389-404.

BOREL, A., and SERRE, J.-P.
1. Corners and arithmetic groups. *Comment. Math. Helv.* **48** (1973), 439-491.

BOREL, A., and DE SIEBENTHAL, J.
1. Les sous-groupes fermés de rang maximum des groupes de Lie clos. *Comment. Math. Helv.* **23** (1949), 200-221.

BOREL, A., and TITS, J.
1. Groupes réductifs. *Publ. Math.* IHES No. 27 (1965), 659-756.

BOTT, R.
1. An application of the Morse theory to the topology of Lie groups. *Bull. Soc. Math. France* **84** (1956), 251-282.
2. The stable homotopy of the classical groups. *Ann. of Math.* **70** (1959), 313-337.

BOTT, R., and SAMELSON, H.
1. Applications of the theory of Morse to symmetric spaces. *Amer. J. Math.* **78** (1958) 964-1028; correction: *Amer. J. Math.* **83** (1961), 207-208.

BOURBAKI, N.
1. "Éléments de mathématiques," Vol. III, "Topologie générale." Hermann, Paris, 1949.
2. "Éléments de mathématique." Chapter I-Chapter VIII, "Groupes et algèbres de Lie." Hermann, Paris, 1960-1975.

BRAUER, R.
1. On the relation between the orthogonal group and the unimodular group. *Arch. Rat. Mech.* **18** (1965), 97-99.

BREMERMANN, H. J.
1. Holomorphic continuation of the kernel function and the Bergman metric in several complex variables. *In* "Lectures on functions of a complex variable," pp. 349-383. Univ. of Mich. Press, Ann Arbor, Michigan, 1955.

BROWN, G.
1. A remark on semisimple Lie algebras, *Proc. Amer. Math. Soc.* **15** (1964), 518.

BRUHAT, F.
1. Sur les représentations induites des groupes de Lie. *Bull. Soc. Math. France* **84** (1956), 97-205.
2. "Lectures on Lie Groups and Representations of Locally Compact Groups." Tata Institute of Fundamental Research, Bombay, 1958.

BRUN, J.
1. Sur la simplification par les variétes homogènes. *Math. Ann.* **235** (1977), 175-183.

BUSEMAN, H.
1. "The Geometry of Geodesics." Academic Press, New York, 1955.

CAHEN, M., LEMAIRE, L., and PARKER, M.
1. Relèvements d'une structure symétrique dans des fibrés associés à un espace symétrique. *Bull. Soc. Math. Belg.* **24** (1972), 227-237.

CAHEN, M., and PARKER, M.
1. Sur les classes d'espaces pseudo-riemanniens symétriques. *Bull. Soc. Math. Belg.* **22** (1970), 339-354.

CAHEN, M., and WALLACH, N.
1. Lorentzian symmetric spaces. *Bull. Amer. Math. Soc.* **76** (1970), 585-591.

CALABI, E., and VESENTINI, E.
1. On compact locally symmetric Kähler manifolds. *Ann. of Math.* **71** (1960), 472-507.

CARLES, R.
1. Méthode récurrente pour la classification des systèmes de racines réduits et irréductibles. *C. R. Acad. Sci. Paris* **276** (1973), 355-358.

Cartan, É.
1. "Sur la structure des groupes de transformations finis et continus." Thèse, Paris, Nony, 1894; 2nd éd. Vuibert, 1933.
2. Les groupes réels simples finis et continus. *Ann. Sci. École Norm. Sup.* **31** (1914), 263–355.
3. Sur la réduction a sa forme canonique de la structure d'un groupe de transformations fini et continu. *Amer. J. Math.* **18** (1896), 1-61.
4. Le principe de dualité et la théorie des groupes simples et semisimples. *Bull. Sci. Math.* **49** (1925), 361-374.
5. Sur les espaces de Riemann dans lesquels le transport par parallélism conserve la courbure. *Rend. Acc. Lincei* 3i (1926), 544-547.
6. Sur une classe remarquablz d'espaces de Riemann. *Bull. Soc. Math. France* **54** (1926), 214-264.
7. Sur une classe remarquable d'espaces de Riemann. *Bull. Soc. Math. France* **55** (1927), 114-134.
8. La géométrique des groupes de transformations. *J. Math. Pures Appl.* **6** (1927), 1-119.
9. La géométrie des groupes simples. *Ann. Mat. Pura Appl.* **4** (1927), 209-256. Compléments, *Ann. Mat. Pura Appl.* **5** (1928), 253-260.
10. Sur certaines formes riemanniennes remarquables des géométries a groupe fondamental simple. *Ann. Sci. École Norm. Sup.* **44** (1927), 345-467.
11. Complément au mémoire "Sur la géométrie des groupes simples." *Ann. Math. Pura Appl.* **5** (1928), 253-260.
12. Groupes simples clos et ouverts et géométrie riemannienne. *J. Math. Pures Appl.* **8** (1929), 1-33.
13. Sur la détermination d'un système orthogonal complet dans un espace de Riemann symétrique clos. *Rend. Circ. Mat. Palermo* **53** (1929), 217-252.
14. Sur les invariants intégraux de certains espaces homogènes clos et les propriétés topologiques de ces espaces. *Ann. Soc. Pol. Math.* **8** (1929), 181-225.
15. Sur les représentations linéaires des groupes clos. *Comm. Math. Helv.* **2** (1930), 269-283.
16. La théorie des groupes finis et continus et l'Analysis situs. *Mém. Sci. Math. Fasc.* **XLII** (1930).
17. "Leçons sur la géométrie projective complexe." Gauthier-Villars, Paris, 1931.
18. Les espaces riemanniens symétriques. *Verh. Int. Math. Kongr. Zürich* **1** (1932), 152-161.
19. Sur les domaines bornés homogènes de l'espace de n variables complexes. *Abh. Math. Sem. Univ. Hamburg* **11** (1935), 116-162.
20. "La topologie des espaces représentatifs des groupes de Lie." *Actual. Scient. Ind.* No. 358. Hermann, Paris, 1936.
21. La topologie des espaces homogènes clos. *Mém. Sémin. Anal. Vect., Moscow* **4** (1937), 388-394.
22. "Leçons sur la géométrie des espaces de Riemann." Gauthier-Villars, Paris, 1928, 2nd ed., 1946.
23. "Leçons sur la géométrie projective complexe," 2nd ed. Gauthier-Villars, 1950.
24. "Oeuvres complètes," Partie I, Vols. 1 and 2. Gauthier-Villars, Paris, 1952.
25. Le troisième théorème fondamental de Lie, *C. R. Acad. Sci. Paris* **190** (1930), 914-916, 1005-1007.
26. "Théorie des groupes finis et continus et la géométrie différentielle traitées par la méthode de repère mobile." Gauthier-Villars, Paris, 1951.

Cartan, É., and Scouten, J. A.
1. On the geometry of the group-manifold of simple and semi-simple groups. *Proc. Akad. Wetensch., Amsterdam* **29** (1926), 803-815.

CARTAN, H.

1. Notions d'algèbre différentielle; applications aux groupes de Lie et aux variétés où opère un groupe de Lie. Colloque de topologie (espaces fibrés), Bruxelles, 1950, 15-27.
2. La transgression dans un groupe de Lie et dans un espace fibré principal. Colloque de topologie (espaces fibrés), Bruxelles, 1950, 57-71.
3. Fonctions automorphes. Séminaire 1957-1958, Paris.

CARTER, R. W.

1. "Simple Groups of Lie Type." Wiley, New York, 1972.
2. Conjugacy classes in the Weyl group. *Compositio Math.* **25** (1972), 1-59.

CASIMIR, H., and VAN DER WAERDEN, B. L.

1. Algebraischer Beweis der vollständigen Reducibilität der Darstellungen halbeinfachen Liescher Gruppen. *Math. Ann.* **111** (1935), 1-12.

CHAE, Y.

1. Symmetric spaces which are mapped conformally on each other. *Kyungpook Math. J.* **2** (1959), 65-72.

CHAVEL, I.

1. "Riemannian Symmetric Spaces of Rank One." Dekker, New York, 1972.

CHEEGER, J., and EBIN, D. G.

1. "Comparison Theorems in Riemannian Geometry." North Holland, Amsterdam, 1975.

CHEN, B. Y., and LUE, H. S.

1. Differential Geometry of $\mathbf{SO}(n+2)/\mathbf{SO}(2) \times \mathbf{SO}(n)$, I. *Geometriae Dedicata* **4** (1975), 253-261.

CHEN, B.-Y., and NAGANO, T.

1. Totally geodesic submanifolds of symmetric spaces, I. *Duke Math. J.* **44** (1977), 745-755; II (preprint).

CHEN, K. T.

1. On the composition functions of nilpotent Lie groups. *Proc. Amer. Math. Soc.* **8** (1957), 1158-1159.

CHEN, S. S., and GREENBERG, L.

1. Hyperbolic spaces. *In* "Contributions to Analysis" (a collection of papers dedicated to Lipman Bers), pp. 49-87. Academic Press, New York, 1974.

CHENG, C.-H.

1. On the Cartan subalgebra of the real semisimple Lie algebra I, II. *Sci. Record* **3** (1959), 214-219, 220-224.
2. On the Weyl group of a real semisimple Lie algebra, *Sci. Record* **3** (1959), 385-389.
3. On the maximal solvable subalgebra of a real semisimple Lie algebra. *Sci. Record* **3** (1959), 390-392.

CHERN, S. S.

1. On integral geometry in Klein spaces. *Ann. of Math.* **43** (1942), 178-189.
2. On a generalization of Kähler geometry. *In* "Algebraic Geometry and Topology" (S. Lefschetz symposium), pp. 103-121. Princeton Univ. Press, Princeton, New Jersey, 1957.
3. Topics in differential geometry. Inst. Adv. Study, Princeton, New Jersey. Notes. 1951.
4. The geometry of G-structures. *Bull. Amer. Math. Soc.* **72** (1966), 167-219.

CHERN, S. S., and CHEVALLEY, C.
1. Élie Cartan and his mathematical work. *Bull. Amer. Math. Soc.* **58** (1952), 217-250.

CHEVALLEY, C.
1. An algebraic proof of a property of Lie groups. *Amer. J. Math.* **63** (1941), 785-793.
2. "Theory of Lie Groups," Vol. I. Princeton Univ. Press, Princeton, New Jersey, 1946.
3. Algebraic Lie algebras. *Ann. of Math.* **48** (1947), 91-100.
4. The Betti numbers of the exceptional simple Lie groups. *Proc. Int. Congr. of Math. Harvard* **II** (1950), 21-24.
5. Sur certains groupes simples. *Tôhoku Math. J.* **7** (1955), 14-66.
6. "Théorie des Groupes de Lie," Vols. II and III. Hermann, Paris, 1951 and 1955.
7. Invariants of finite groups generated by reflections. *Amer. J. Math.* **77** (1955), 778-782.
8. On the topological structure of solvable groups. *Ann. of Math.* **42** (1941), 668-675.
9. Sur la classification des algèbres de Lie simples et de les représentations. *C. R. Acad. Sci. Paris* **227** (1948), 1136-1138.

CHEVALLEY, C., and EILENBERG, S.
1. Cohomology theory of Lie groups and Lie algebras. *Trans. Amer. Math. Soc.* **63** (1948), 85-124.

CHOW, W. L.
1. On the geometry of algebraic homogeneous spaces. *Ann. of Math.* **50** (1949), 32-67.

CHU, H., and KOBAYASHI, S.
1. The automorphism group of a geometric structure. *Trans. Amer. Math. Soc.* **113** (1964), 141-150.

COHN, P. M.
1. "Lie Groups." Cambridge Univ. Press, London and New York, 1957.

CONLON, L.
1. The topology of certain spaces of paths on a compact symmetric space. *Trans. Amer. Math. Soc.* **112** (1964), 228-248.
2. Variational completeness and K-transversal domains. *J. Differential Geometry* **5** (1971), 135-147.
3. Applications of affine root systems to the theory of symmetric spaces. *Bull. Amer. Math. Soc.* **75** (1969), 610-613.

CORNWELL, J. F.
1. Semisimple subgroups of linear semisimple Lie groups. *J. Math. Phys.* **16** (1975), 394-395.
2. Direct determination of the Iwasawa decomposition for noncompact semisimple Lie algebras. *J. Math. Phys.* **16** (1975), 1992-1999.

COXETER, H. S. M.
1. Discrete groups generated by reflections. *Ann. of Math.* **35** (1934), 588-621.

CRITTENDEN, R.
1. Minimum and conjugate points in symmetric spaces. *Canad. J. Math.* **14** (1962), 320-328.

CURTIS, M., and WIEDERHOLD, A.
1. A note on the Cartan integers. *Houston J. Math.* **1** (1975), 29-33.

DALLA VOLTA, V.
1. Geodesic manifolds in the spaces of Siegel-Hua. *Ann. Mat. Pura Appl.* **46** (1958), 19-42.

DAO, C. T.
1. Algebraic questions on the realization of cycles in symmetric spaces. *Vestnik Moscov Univ. Ser. I, Mat. Mec.* **31** (1976), 62-66.

D'ATRI, J. E.
1. Geodesic conformal transformations and symmetric spaces. *Kodai Math. Sem. Rep.* **26** (1974/1975), 201-203.

DE RHAM, G.
1. Sur la reductibilité d'un espace de Riemann. *Comment. Math. Helv.* **26** (1952), 328-344.
2. "Variétés différentiables." Hermann, Paris, 1955.

DIEUDONNÉ, J.
1. Une généralisation des espaces compacts. *J. Math. Pures Appl.* **23** (1944), 65-76.
2. Éléments d'Analyse, Vol. I-V, Hermann, Paris, 1971-1975.
3. A simplified method for the study of complex semisimple Lie algebras. *Tensor* **24** (1972), 239-242.

DIXMIER, J.
1. "Algèbres enveloppants." Gauthier-Villars, Paris 1974.
2. L'application exponentielle dans les groupes de Lie résolubles. *Bull. Soc. Math. France* **85** (1957), 113-121.

DONNELY, H.
1. Symmetric Einstein spaces and spectral geometry. *Indiana Math. J.* **24** (1974), 603-606.

DUFLO, M., and VERGNE, M.
1. Une propriété de la représentation coadjointe d'une algèbre de Lie. *C. R. Acad. Sci. Paris* **268** (1969), 583-585.

DYNKIN, E. B.
1. The structure of semi-simple Lie algebras. *Uspekhi Mat. Nauk.* **2** (1947), 59-127.
2. Topological characteristics of homomorphisms of compact Lie groups. *Mat. Sb.* **35** (1954), 129-177.
3. Semisimple subalgebras of semisimple Lie algebras. *Mat. Sb.* **30** (1952), 349-462.
4. Maximal subgroups of the classical groups. *Trudy Moscov. Mat. Obsc.* **1** (1952), 39-166.

DYNKIN, E. B., and ONIŠČIK, A. L.
1. Compact global Lie groups. *Uspekhi Mat. Nauk.* **10** (1955), 3-74; *Amer. Math. Soc. Transl.* **21** (1962).

EHRESMANN, C.
1. Sur certains espaces homogènes de groupes de Lie. *Enseignement Math.* **35** (1936), 317-333.
2. Les connexions infinitésimales dans un espace fibré differentiable. Colloque de Topologie, Bruxelles, 1950, 29-55.

EISENHART, L. P.
1. "Continuous Groups of Transformations." Princeton Univ. Press, Princeton, New Jersey, 1933.

ELÍASSON, H.
1. Die Krümmung des Raumes $\mathbf{Sp}(2)/\mathbf{SU}(2)$ von Berger. *Math. Ann.* **164** (1966), 317-323.
2. Über die Anzahl geschlossener Geodätischen in gewissen Riemannschen Mannigfaltigkeiten. *Math. Ann.* **166** (1966), 119-147.

ENGEL, F.
1. Sur un groupe simple à quatorze paramètres. *C. R. Acad. Sci. Paris* **116** (1893), 786-788.
2. Sophus Lie (obituary). *Jber. Deutsch. Math.-Verein.* **8** (1900), 30-46.
3. Wilhelm Killing (obituary). *Jber. Deutsch. Math.-Verein.* **39** (1930), 140-154.

FABER, R. J.
1. The Lie bracket and the curvature tensor. *Enseignement Math.* (2), **22** (1976), 29-34.

FEDENKO, A. S.
1. Symmetric spaces with simple non-compact fundamental groups. *Doklady Akad. Nauk. S.S.S.R.* **108** (1956), 1026-1028.
2. Periodic automorphisms of the classical compact groups. *Rev. Roumaine Math. Pures Appl.* **15** (1970), 1375-1378.

FLANDERS, H.
1. "Differential Forms." Academic Press, New York, 1963.

FLENSTED-JENSEN, M.
1. Spherical functions on rank-one symmetric spaces and generalizations. Proc. of Symposia in Pure Math. Vol. 26, 339-342, Amer. Math. Soc., 1973.
2. Spherical functions on a real semisimple Lie group. A method of reduction to the complex case. *J. Functional Analysis* (1978) (to appear).

FRANKEL, T. T.
1. Critical submanifolds of the classical groups and Stiefel manifolds. *In* "Differential and Combinatorial Topology" (S. S. Cairns, ed.), pp. 37-53. Princeton Univ. Press, Princeton, New Jersey, 1965.

FREUDENTHAL, H.
1. Sur les invariants caractéristiques des groupes semi-simple. *Nederl. Akad. Wetensch. Proc. Ser. A* **56** (1953), 90-94.
2. Zur Berechnung der Charaktere der halbeinfachen Lieschen Gruppen, I. II. *Nederl. Akad. Wetensch. Proc. Ser. A* **57** (1954), 369-376, 487-491.
3. Zur Klassifikation der einfachen Lie-Gruppen. *Nederl. Akad. Wetensch. Proc. Ser. A* **60** (1958), 379-383.
4. Die Topologie der Lieschen gruppen als algebraisches Phänomen, I. *Ann. of Math.* **42** (1941), 1051-1074.
5. Clifford-Wolf-Isometrien symmetrischer Räume. *Math. Ann.* **150** (1963), 136-149.
6. Zweifach Homogenität und Symmetrie. *Nederl. Akad. Wetensch. Proc. Ser. A* **70** (1966), 18-22.
7. Lie groups in the foundations of geometry. *Advan. Math.* **1** (1964), 145-190.
8. L'algèbre topologique en particulier les groupes topologiques et de Lie. *Revue de Synthèse* (1968), 223-243.

FREUDENTHAL, H., and DEVRIES, H.
1. "Linear Lie Groups." Academic Press, New York, 1969.

FRÖLICHER, A.
1. Zur Differentialgeometrie der komplexen Strukturen. *Math. Ann.* **129** (1955), 50-95.

FUJIMOTO, A.
1. On decomposable symmetric affine spaces. *J. Math. Soc. Japan* **9** (1957), 158-170.

FURSTENBERG, H.
1. A Poisson formula for semisimple Lie groups. *Ann. of Math.* **77** (1963), 335-386.

GANTMACHER, F.

1. Canonical representation of automorphisms of a complex semi-simple Lie group. *Mat. Sb.* **5** (1939), 101-144.
2. On the classification of real simple Lie groups. *Mat. Sb.* **5** (1939), 217-249.

GARLAND, H., and RAGUNATHAN, M. S.

1. Fundamental domains for lattices in R-rank 1 semisimple Lie groups. *Ann. of Math.* **92** (1970), 279-326.

GELFAND, I. M., and FOMIN, S. V.

1. Geodesic flows on manifolds of constant negative curvature. *Uspekhi Mat. Nauk.* **7** (1952), 118-137.

GELFAND, I. M., and GRAEV, M. I.

1. The geometry of homogeneous spaces, group representations in homogeneous spaces and questions in integral geometry related to them, I. *Trudy Moskov. Mat. Obšč.* **8** (1959), 321-390.

GELFAND, I. M., and KIRILLOV, A.

1. Structure of the sfield connected with a semisimple decomposable Lie algebra. *Funkcional. Anal. i Prilozen.* **3** (1969), 7-26.

GELFAND, I. M., and NAIMARK, M. A.

1. Unitary representations of the classical groups. *Trudy Mat. Inst. Steklov.* **36**). Moskov-Leningrad, (1950); German transl.: Akademie-Verlag, Berlin, 1957.

GILKEY, P. B.

1. The spectral geometry of symmetric spaces. *Trans. Amer. Math. Soc.* **500** (1977), 341-353.

GILMORE, R.

1. "Lie Groups, Lie Algebras and Some of Their Applications." Wiley, New York, 1974.

GINDIKIN, S., and KARPELEVIČ, F.

1. Plancherel measure of Riemannian symmetric spaces of non-positive curvature. *Soviet Math. Dokl.* **3** (1962), 962-865.

GINDIKIN, S., PYATETZKI-SHARIRO, I. I., and VINBERG, E. B.

1. Homogeneous Kähler manifolds. Engl. transl. by A. Korányi. C.I.M.E. Summer Course on Geometry of Bounded Homogeneous Domains. Edizioni Cremonese, Roma, 1968, 3-87.

GLEASON, A.

1. Groups without small subgroups. *Ann. of Math.* **56** (1952), 193-212.

GODEMENT, R.

1. A theory of spherical functions, I. *Trans. Amer. Math. Soc.* **73** (1952), 496-556.
2. Articles in H. Cartan [3].

GOLUBITSKY, M.

1. Primitive actions and maximal subgroups of Lie groups. *J. Differential Geometry* **7** (1972), 175-192.

GOLUBITSKY, M., and ROTHSCHILD, B.

1. Primitive subalgebras of exceptional Lie algebras. *Bull. Amer. Math. Soc.* **77** (1971), 983-986.

Goto, M.
1. Faithful representations of Lie groups, I. *Math. Japon.* **1** (1948), 107-119.
2. On algebraic homogeneous spaces. *Amer. J. Math.* **76** (1954), 811-818.
3. On an arcwise connected subgroup of a Lie group. *Proc. Amer. Math. Soc.* **20** (1960), 157-162.

Goto, M., and Jakobsen, H. P.
1. On self-intersecting geodesics, Aarhus Universitet, 1973, (preprint).

Goto, M., and Kobayashi, E. T.
1. On the subgroups of the centers of simply connected simple Lie groups—Classification of simple Lie groups in the large. *Osaka J. Math.* **6** (1969), 251-281.

Gray, A.
1. Riemannian manifolds with geodesic symmetries of order 3. *J. Differential Geometry* (1972), 343-369.
2. Isometric immersions in symmetric spaces. *J. Differential Geometry* **3** (1969), 237-244.

Greub, W., Halperin, S., and Vanstone, R.
1. "Connections, Curvature and Cohomology," 3 Vols. Academic Press, New York, 1972, 1973, 1976.

Griffiths, P., and Schmid, W.
1. Locally homogeneous complex manifolds. *Acta Math.* **123** (1969), 253-302.

Gromoll, D., Klingenberg, W., and Meyer, W.
1. "Riemannsche Geometrie im Grossen." Lecture Notes in Math. 55, Springer, New York, 1968.

Guillemin, V. W., and Sternberg, S.
1. An algebraic model of transitive differential geometry. *Bull. Amer. Math. Soc.* **70** (1964), 16-47.

Günther, P.
1. Einige Sätze über das Volumenelement eines Riemannschen Raumes. *Publ. Math. Debrechen* **7** (1960), 78-93.
2. Sphärische Mittelwerte in kompakten harmonischen Riemannschen Mannigfaltigkeiten. *Math. Ann.* **165** (1966), 281-296.

Hadamard, J.
1. Les surfaces à courbures opposées. *J. Math. Pures Appl.* **4** (1898), 27-73.
2. Sur les éléments linéaires a plusieurs dimensions. *Bull. Soc. Math. France* **25** (1901), 37-40.

Hano, J.
1. On Kählerian homogeneous spaces of unimodular groups. *Amer. J. Math.* **69** (1957), 885-900.

Hano, J., and Matsushima, Y.
1. Some studies on Kählerian homogeneous spaces. *Nagoya Math. J.* **11** (1957), 77-92.

Hano, J., and Morimoto, A.
1. Note on the group of affine transformations of an affinely connected manifold. *Nagoya Math. J.* **8** (1955), 71-81.

Harish-Chandra
1. On representations of Lie algebras. *Ann. of Math.* **50** (1949), 900-915.
2. Lie algebras and the Tannaka duality theorem. *Ann. of Math.* **51** (1950), 299-330.

3. On some applications of the universal enveloping algebra of a semi-simple Lie algebra. *Trans. Amer. Math. Soc.* **70** (1951), 28-96.
4. Representations of semi-simple Lie groups, I, II, III. *Trans. Amer. Math. Soc.* **75** (1953), 185-243; **76** (1954), 26-65, 234-253.
5. Representations of semi-simple Lie groups, IV, V, VI. *Amer. J. Math.* **77** (1955), 743-777; **78** (1956), 1-41, 564-628.
6. On a lemma of F. Bruhat. *J. Math. Pures Appl.* **35** (1956), 203-210.
7. The characters of semi-simple Lie groups. *Trans. Amer. Math. Soc.* **83** (1956), 98-163.
8. Fourier transforms on a semi-simple Lie algebra, I, II. *Amer. J. Math.* **79** (1957), 193-257, 653-686.
9. Spherical functions on a semi-simple Lie group, I, II. *Amer. J. Math.* **80** (1958), 241-310, 553-613.

HAUSNER, M., and SCHWARTZ, J. T.
1. "Lie groups; Lie algebras." Gordon and Breach, New York, 1968.

HELGASON, S.
1. The surjectivity of invariant differential operators on symmetric spaces I. *Ann. of Math.* **98** (1973), 451-479.
2. On Riemannian curvature of homogeneous spaces. *Proc. Amer. Math. Soc.* **9** (1958), 831-838.
3. Differential operators on homogeneous spaces. *Acta Math.* **102** (1959), 239-299.
4. Some remarks on the exponential mapping for an affine connection. *Math. Scand.* **9** (1961), 129-146.
5. Some results on invariant theory. *Bull. Amer. Math. Soc.* **68** (1962).
6. Duality and Radon transform for symmetric spaces. *Amer. J. Math.* **85** (1963), 667-692.
7. An analogue of the Paley-Wiener theorem for the Fourier transform on certain symmetric spaces. *Math. Ann.* **165** (1966), 297-308.
8. Totally geodesic spheres in compact symmetric spaces. *Math. Ann.* **165** (1966), 309-317.
9. A duality for symmetric spaces with applications to group representations. *Advan. Math.* **5** (1970), 1-154.
10. Invariant differential equations on homogeneous manifolds. *Bull. Amer. Math. Soc.* **83** (1977), 751-774.
11. Fundamental solutions of invariant differential operators on symmetric spaces. *Amer. J. Math.* **86** (1964), 565-601.
12. Lie groups and symmetric spaces. Battelle Rencontres, 1-71, Benjamin, New York, 1968.
13. "Differential Geometry and Symmetric Spaces." Academic Press, New York, 1962.

HELWIG, K.-H.
1. Jordan-Algebren und symmetrische Räume I. *Math. Z.* **115** (1970), 315-349.

HENRICH, C. J.
1. How to find the top root form of a complex simple Lie algebra. *J. Reine Angew. Math.* **265** (1974), 138-141.

HERMANN, R.
1. Geodesics of bounded symmetric domains. *Comment. Math. Helv.* **35** (1961), 1-8.
2. Geometric aspects of potential theory in symmetric bounded domains, I, II, III. *Math. Ann.* **148** (1962), 349-366; **151** (1963), 143-149; **153** (1964), 384-394.

3. Variational completeness for compact symmetric spaces. *Proc. Amer. Math. Soc.* **11** (1960), 544-546.

HICKS, N.
1. A theorem on affine connexions. *Illinois J. Math.* **3** (1959), 242-254.
2. An example concerning affine connexions. *Proc. Amer. Math. Soc.* **11** (1960), 952-956.
3. "Notes on Differential Geometry." Van Nostrand Reinhold, Princeton, New Jersey, 1965.
4. On the curvature of an invariant connexion. *Boll. Un. Mat. Ital.* **3** (1970), 768-772.

HIRAI, T.
1. A note on automorphisms of real semi-simple Lie algebras. *J. Math. Soc. Japan* **28** (1976), 250-256.

HIRZEBRUCH, F.
1. Automorphe Formen und der Satz von Riemann-Roch. Int. Symposium on Algebraic Topology, Mexico, 1958.

HIRZEBRUCH, U.
1. Halbräume und ihre holomorphen Automorphismen. *Math. Ann.* **153** (1964), 395-417.
2. Über Jordan-Algebren und kompakte Riemannsche symmetrische Räume vom Rang 1. *Math. Z.* **90** (1965), 339-354.
3. Über Jordan-Algebren und beschränkte symmetrische Gebiete. *Math. Z.* **94** (1966), 387-390.
4. Über eine Realisierung der Hermiteschen symmetrischen Räume. *Math. Z.* **115** (1970), 371-382.

HOCHSCHILD, G.
1. The automorphism group of a Lie group. *Trans. Amer. Math. Soc.* **72** (1952), 209-216.
2. "The Structure of Lie Groups." Holden-Day, San Francisco, 1965.

HODGE, W. V. D.
1. "Theory and Applications of Harmonic Integrals." Cambridge Univ. Press, London and New York, 1952.

HOPF, H.
1. Über den Rang geschlossenen Liescher Gruppen. *Comment. Math. Helv.* **13** (1940), 119-143.
2. Maximale Toroide und singuläre Elemente in geschlossenen Lieschen Gruppen. *Comment. Math. Helv.* **15** (1943), 59-70.

HOPF, H., and RINOW, W.
1. Über den Begriff der vollständigen differentialgeometrischen Fläche. *Comment. Math. Helv.* **3** (1931), 209-225.

HOTTA, R., and WALLACH, N.
1. On Matsushima's formula for the Betti numbers of a locally symmetric space. *Osaka J. Math.* **12** (1975), 419-431.

HUA, L. K.
1. On the Riemannian curvature of the non-Euclidean space of several complex variables. *Acta Math. Sinica* **4** (1954), 141-168.
2. "Harmonic Analysis of Functions of Several Complex Variables in the Classical Domains." Peking, 1958; Russian transl. Moscow, 1959.
3. On the theory of Fuchsian functions of several variables. *Ann. of Math.* **47** (1946), 167-191.

HUMPHREYS, J. E.
1. "Introduction to Lie Algebras and Representation Theory." Springer-Verlag, Berlin, 1972.

HUNT, G. A.
1. A theorem of Élie Cartan. *Proc. Amer. Math. Soc.* **7** (1956), 307-308.

HUREWICZ, W., and WALLMAN, H.
1. "Dimension Theory." Princeton Univ. Press, Princeton, New Jersey, 1948.

HURWITZ, A.
1. Über die Erzeugung der Invarianten durch Integration. *Gött. Nachr.* (1897), 71-90.

ISE, M.
1. Realizations of bounded symmetric domains as matrix spaces. *Nagoya Math. J.* **42** (1971), 115-133.

IWAHORI, N.
1. On discrete reflection groups on symmetric Riemannian manifolds. Proc. U.S.-Japan Seminar on Differential Geometry, Kyoto, 1965.
2. On the structure of a Hecke ring of a Chevalley group over a finite field. *J. Fac. Sci. Univ. Tokyo* **10** (1964), 215-236.

IWAHORI, N., and MATSUMOTO, H.
1. On some Bruhat decompositions and the structure of the Hecke ring of p-adic Chevalley group. *Publ. Math. I.H.E.S.* no. 25, (1965), 5-48.

IWAMOTO, H.
1. On integral invariants and Betti numbers of symmetric Riemannian spaces. I, II. *J. Math. Soc. Japan* **1** (1949), 91-110, 235-243.
2. Sur les espaces riemanniens symmétriques, I. *Japan J. Math.* **19** (1948), 513-523.

IWASAWA, K.
1. On some types of topological groups. *Ann. of Math.* **50** (1949), 507-558.

JACOBSON, N.
1. "Lie Algebras." Wiley (Interscience), New York, 1962.
2. Completely reducible Lie algebras of linear transformations. *Proc. Amer. Math. Soc.* **2** (1951), 105-133.
3. "Exceptional Lie Algebras." Dekker, New York, 1971.

JAFFEE, H. A.
1. Real forms in Hermitian symmetric spaces and real algebraic varieties. Dissertation, State Univ. of New York, Stony Brook, 1974.
2. Antiholomorphic automorphisms of the exceptional symmetric domains. (preprint).

KAČ, V.
1. Automorphisms of finite order of semisimple Lie algebras. *Funkcional. Anal. i Prilozen.* **3** (1969), 94-96.
2. Graded Lie algebras and symmetric spaces. *Funkcional. Anal. i Prilozen.* **2** (1968), 93-94.
3. Infinite-dimensional algebras, Dedekind η-function, classical Möbius function and the very strange formula. *Advances in Math.* 1978.
4. Simple irreducible graded Lie algebras of finite growth. *Izv. Akad. Nauk SSSR Ser. Mat.* **32** (1968), 1323-1367.

KÄHLER, E.
1. Über eine bemerkenswerte Hermitische Metrik. *Abh. Math. Sem. Univ. Hamburg* **9** (1933), 173-186.

KAJDAN, D. A.
1. On arithmetic varieties. *In* "Lie Groups and their Representations" (I. M. Gelfand, ed.), pp. 151-217. Halsted Press (Wiley), New York, 1975.

KANEYUKI, S.
1. "Homogeneous bounded Domains and Siegel Domains." Lecture Notes in Math. Springer, New York, 1971.

KANEYUKI, S., and NAGANO, T.
1. On certain quadratic forms related to symmetric Riemannian spaces. *Osaka Math. J.* **14** (1962), 241-252.

KANTOR, I. L., SIROTA, A. I., and SOLODOVNIKOV, A. S.
1. A class of symmetric spaces with an extensible group of motions and a generalization of the Poincaré model. *Dokl. Akad. Nauk. SSSR* **173** (1967), 511-514.

KAPLANSKY, I.
1. "Lie Algebras and Locally Compact Groups." Univ. of Chicago Press, Chicago, Illinois, 1971.

KARPELEVIČ, F. I.
1. Surfaces of transitivity of a semi-simple subgroup of the group of motions of a symmetric space. *Dokl. Akad. Nauk. SSSR* **93** (1953), 401-404.
2. The simple subalgebras of the real Lie algebras. *Trudy Moskov. Mat. Obšč.* **4** (1955), 3-112.
3. Geodesics and harmonic functions on symmetric spaces. *Dokl. Akad. Nauk. SSSR* **124** (1959), 1199-1202.
4. The geometry of geodesics and the eigenfunctions of the Laplace-Beltrami operator on symmetric spaces. *Trans. Moscow Math. Soc.* **14** (1965), 48-185.

KELLEY, J. L.
1. "General Topology." Van Nostrand Reinhold, Princeton, New Jersey, 1955.

KELLY, E. F.
1. Tight equivariant imbeddings of symmetric spaces. *J. Differential Geometry* **7** (1972), 535-548.

KILLING, W.
1. Die Zusammensetzung der stetigen endlichen Transformationsgruppen, I, II, III, IV. *Math. Ann.* **31** (1888), 252-290; **33** (1889), 1-48; **34** (1889), 57-122; **36** (1890), 161-189.

KIRILLOV, A.
1. "Éléments de la théorie des représentations." Édition MIK, Moscow, 1974.

KLEIN, F.
1. "Höhere Geometrie." Chelsea, New York, 1957.

KLINGEN, H.
1. Diskontinuierliche Gruppen in symmetrischen Räumen, I, II. *Math. Ann.* **129** (1955), 345-369; **132** (1956), 134-144.
2. Analytic automorphisms of bounded symmetric complex domains. *Pacific J. Math.* **10** (1960), 1327-1332.

KLINGENBERG, W.
1. Über Riemannsche Mannigfaltigkeiten mit nach oben beschränkter Krümmung. *Ann. Mat. Pura Appl.* **60** (1963), 49-59.
2. Manifolds with restricted conjugate locus I. *Ann. of Math.* **78** (1963), 527-547.

KNAPP, A. W.
1. Weyl group of a cuspidal parabolic. *Ann. Sci. École Norm. Sup.* **8** (1975), 275-294.

KOBAYASHI, S.
1. Une remarque sur la connexion affine symétrique. *Proc. Japan Acad.* **31** (1955), 13-14.
2. Espaces a connexions affines et riemanniens symétriques. *Nagoya Math. J.* **9** (1955), 25-37.
3. Fixed points of isometries. *Nagoya Math. J.* **13** (1958), 63-68.
4. Geometry of bounded domains. *Trans. Amer. Math. Soc.* **92** (1959), 267-290.
5. Isometric imbeddings of compact symmetric spaces. *Tôhoku Math. J.* **20** (1968), 21-25.
6. "Transformation Groups in Differential Geometry." Springer, New York, 1972.

KOBAYASHI, S., and NAGANO, T.
1. On filtered Lie algebras and geometric structures, I, II. *J. Math. Mech.* **13** (1964), 875-908; **14** (1965), 513-522.

KOBAYASHI, S., and Nomizu, K.
1. "Foundations of Differential Geometry I, II." Wiley (Interscience), New York, 1963, and 1969.

KOECHER, M.
1. Positivitätsbereiche im $\mathbf{R}^n$. *Amer. J. Math.* **79** (1957), 575-596.
2. Die Geodätischen von Positivitätsbereichen. *Math. Ann.* **135** (1958), 192-202.
3. An elementary approach to bounded symmetric domains, Rice University, Houston, Texas. Notes. 1969.

KOH, S.
1. Affine symmetric spaces. *Trans. Amer. Math. Soc.* **119** (1965), 291-301.

KOLCHIN, E.
1. Algebraic matric groups and the Picard-Vessiot theory of homogeneous linear ordinary differential equations. *Ann. of Math.* **49** (1948), 1-42.

KORÁNYI, A.
1. Holomorphic and harmonic functions on bounded symmetric domains. C.I.M.E. Summer Course on Geometry of Bounded Homogeneous Domains. Edizioni Cremonese, Roma 1968, 125-197.
2. Analytic invariants of bounded symmetric domains. *Proc. Amer. Math. Soc.* **19** (1968), 279-284.

KORÁNYI, A., and WOLF, J. A.
1. Realization of Hermitian symmetric spaces as generalized half planes. *Ann. of Math.* **81** (1965), 265-288.

KOSTANT, B.
1. On the conjugacy of real Cartan subalgebras. I. *Proc. Nat. Acad. Sci. U.S.A.* **41** (1955), 967-970.
2. On differential geometry and homogeneous spaces, I, II. *Proc. Nat. Acad. Sci. U.S.A.* **42** (1956), 258-261, 354-357.
3. A characterization of the classical groups. *Duke Math. J.* **25** (1958), 107-124.
4. The principal three-dimensional subgroup and the Betti-numbers of a complex simple Lie group. *Amer. J. Math.* **81** (1959), 973-1032.
5. A characterization of invariant affine connections. *Nagoya Math. J.* **16** (1960), 35-50.
6. On invariant skew-tensors. *Proc. Nat. Acad. Sci. U.S.A.* **40** (1955), 148-151.

7. Lie algebra cohomology and the generalized Borel-Weil theorem. *Ann. of Math.* **74** (1961), 329-387.
8. On convexity, the Weyl group and the Iwasawa decomposition. *Ann. Sci. École Norm. Sup.* **6** (1973), 413-455.

Kostant, B., and Rallis, S.
1. Orbits and representations associated with symmetric spaces. *Amer. J. Math.* **93** (1971), 753-809.

Koszul, J. L.
1. Sur la forme hermitienne canonique des espaces homogènes complexes. *Canad. J. Math.* **7** (1955), 562-576.
2. Variétés Kähleriennes. Universidade de São Paulo, 1957, 100 p.
3. Exposés sur les espaces homogènes symétriques. Soc. Math. São Paulo, 1959, 71 p.
4. Homologie et cohomologie des algèbres de Lie. *Bull. Soc. Math. France* **78** (1950), 66-127.
5. Domaines bornés homogènes et orbites de groupes de transformations affines. *Bull. Soc. Math. France* **89** (1961), 515-533.
6. Ouverts convexes homogènes des espaces affines. *Math. Z.* **79** (1962), 254-259.

Kowalsky, O.
1. Riemannian manifolds with general symmetries. *Math. Z.* **136** (1974), 137-150.

Kuga, M.
1. Fibred variety over symmetric spaces whose fibers are abelian varieties. Proc. U.S.-Japan Seminar on Differential Geometry, Kyoto, 1965, 72-81.

Kuga, M., and Sampson, J.
1. A coincidence formula for locally symmetric spaces. *Amer. J. Math.* **94** (1972), 486-500.

Kulkarni, R. S.
1. Curvature and metric. *Ann. of Math.* **91** (1970), 211-231.

Lang, S.
1. "$SL_2(R)$." Addison-Wesley, Reading, Massachusetts, 1975.

Langlands, R. P.
1. The dimension of the space of automorphic forms. *Amer. J. Math.* **85** (1963), 99-125.

Lardy, P.
1. Sur la détermination des structures réelles de groupes simples, finis et continus, au moyen des isomorphies involutives. *Comment. Math. Helv.* **8** (1935-1936), 189-234.

Laugwitz, D.
1. "Differential and Riemannian Geometry." Academic Press, New York, 1965.

Ledger, A. J.
1. Symmetric harmonic spaces. *J. London Math. Soc.* **32** (1957), 53-56.

Ledger, A. J., and Obata, M.
1. Affine and Riemannian s-manifolds. *J. Differential Geometry* **2** (1968), 451-459.

Ledger, A. J., and Pettet, B. R.
1. Compact quadratic s-manifolds. *Comment. Math. Helv.* **51** (1976), 105-131.

Ledger, A. J., and Yano, K.
1. The tangent bundle of a locally symmetric space. *J. London Math. Soc.* **40** (1965), 487-492.

LEICHTWEISS, K.
1. Zur Riemannschen Geometrie in Grassmannschen Mannigfaltigkeiten. *Math. Z.* **76** (1961), 334-366.

LEPOWSKY, J.
1. A generalization of H. Weyl's "unitary trick." *Trans. Amer. Math. Soc.* **216** (1976), 229-236.

LEPOWSKY, J., and MCCOLLUM, G. W.
1. Cartan subspaces of symmetric Lie algebras. *Trans. Amer. Math. Soc.* **216** (1976), 217-228.

LEUNG, D. S. P.
1. Reflection principle for minimal submanifolds of Riemannian symmetric spaces. *J. Differential Geometry* **8** (1973), 153-161.

LEVI, E. E.
1. Sulla struttura dei Gruppi e continui. *Atti Accad. Sci. Torino* **60** (1905), 551-565.

LEVI-CIVITA, T.
1. Nozione di parallelismo in una varieta qualunque. *Rend. Palermo* **42** (1917), 73-205.

LEVY, H.
1. Forma canonica dei ds^2 per i quali si annullano i simboli di Riemann a cinque indici. *Rend. Acc. Lincei* **3** (1926), 65–69.

LIBERMAN, P.
1. Sur les structures presque complexes et autres structures infinitésimales régulières. *Bull. Soc. Math. France* **83** (1955), 195-224.

LICHNEROWICZ, A.
1. Courbure, nombres de Betti, et espaces symétriques. Proc. Int. Congr. of Math. Harvard II (1950), 216-223.
2. Espaces homogènes kähleriens. Coll. Int. Geom. Diff., Strasbourg, (1953), 171-184.
3. Équations de Laplace et espace harmoniques. Premier Colloq. sur les équations aux dérivées partielles, Louvain, 1953, 9-23.
4. Un théorème sur les espaces homogènes complexes. *Arch. Math.* **5** (1954), 207-215.
5. "Théorie globale des connexions et des groupes d'holonomie." Éd. Cremonese, Rome, 1955.
6. "Géométrie des groupes de transformations." Dunod, Paris, 1958.

LIE, S.
1. "Theorie der Transformationsgruppen I, II, III." Unter Mitwirkung von F. Engel. Leipzig, 1888, 1890, 1893.
2. Zur Theorie des Integrabilitetsfaktors. Christiania Forh. 1874, 242-254; reprint Gesamm. Abh. Vol. III, no. XIII, 176-187.
3. Beiträge zur allgemeinen Transformationstheorie. Leipziger Ber. 1888, 14-21. Gesamm. Abh. VI, 230-236.

LISTER, W. G.
1. A structure theory of Lie triple systems. *Trans. Amer. Math. Soc.* **72** (1952), 217-242.

LOOS, O.
1. Spiegelungsräume und homogene symmetrische Räume. *Math. Z.* **99** (1967), 141-170.
2. "Symmetric Spaces I, II," Benjamin, New York, 1969.
3. Kompakte Unterräume symmetrischer Räume. *Math. Z.* **125** (1972), 264-270.

4. Jordan triple systems, R-spaces, and bounded symmetric domains. *Bull. Amer. Math. Soc.* **77** (1971), 558-561.
5. An intrinsic characterization of the fibre bundles associated with homogeneous spaces defined by Lie group automorphisms. *Abh. Math. Sem. Univ. Hamburg* **37** (1972), 160-179.

MAKAREVIČ, B. O.
1. Jordan algebras and orbits in symmetric R-spaces. *Funkcional. Anal. i Priložen.* **8** (1974), 89-90.

MALCEV, A.
1. On the theory of Lie groups in the large. *Mat. Sb.* **16** (58) (1945), 163-190.

MARCUS, L.
1. Exponentials in algebraic matrix groups. *Advan. Math.* **11** (1973), 351-367.

MARGULIS, G. A.
1. Non uniform lattices in semisimple algebraic groups. *In* "Lie Groups and Their Representations" (I. M. Gelfand, ed.), pp. 371-553. Halsted Press (Wiley), New York, 1975.
2. Discrete groups of motions of manifolds of nonpositive curvature. Proc. Int. Congr. Math., Vancouver, 1974, Vol. 2, 21-34.

MATSUKI, T.
1. The orbits of affine symmetric spaces under the action of minimal parabolic subgroups (preprint).

MATSUMOTO, H.
1. Quelques remarques sur les espaces riemanniens isotropes. *C. R. Acad. Sci. Paris* **272** (1971), 316-319.
2. Quelques remarques sur les groupes de Lie algébriques réels. *J. Math. Soc. Japan* **16** (1964), 419-446.
3. Sur un théorème de point fixe de É. Cartan. *C. R. Acad. Sci. Paris* **274** (1972), 955-958.

MATSUSHIMA, Y.
1. Un théorème sur les espaces homogènes complexes. *C. R. Acad. Sci. Paris* **241** (1955), 785-787.
2. Espaces homogènes de Stein des groupes de Lie complexes. *Nagoya Math. J.* **16** (1960), 205-218.
3. On the first Betti number of compact quotient spaces of higher dimensional symmetric spaces. *Ann. of Math.* **75** (1962), 312-330.
4. A formula for the Betti numbers of compact locally symmetric Riemannian manifolds. *J. Differential Geometry* **1** (1967), 99-109.
5. "Differentiable Manifolds," Dekker, New York, 1972.

MATSUSHIMA, Y., and MURAKAMI, S.
1. On certain cohomology groups attached to Hermitian symmetric spaces. *Osaka J. Math.* **2** (1965), 1-35.
2. On vector bundle valued harmonic forms and automorphic forms on symmetric Riemannian manifolds. *Ann. of Math.* **78** (1963), 365-416.

MAURER, L.
1. Über allgemeine Invarianten-Systeme. *München Sitzungber.* **18** (1888), 103-150.

MAUTNER, F.
1. Geodesic flows on symmetric Riemann spaces. *Ann. of Math.* **65** (1957), 416-431.

Mayer, W., and Thomas, T. Y.
1. Foundations of the theory of Lie groups. *Ann. of Math.* **30** (1935), 770-822.

Meschiari, M.
1. On the reflections in bounded symmetric domains. *Ann. Scuola Norm. Sup. Pisa* **26** (1972), 403-435.

Michel, R.
1. Les tenseurs symétriques dont l'énergie est nulle sur toutes les geodesiques des espaces symétriques de rang 1. *J. Math. Pures Appl.* **53** (1974), 271-278.

Miller, W.
1. "Symmetry Groups and Their Applications." Academic Press, New York, 1973.

Miller, K. S., and Murray, F. J.
1. "Existence Theorems for Ordinary Equations." New York Univ. Press. New York, 1954.

Milnor, J.
1. "Morse Theory." Ann. of Math. Studies Vol. 51, Princeton Univ. Press, Princeton, New Jersey, 1963.
2. Curvatures of left invariant metrics on Lie groups. *Advances in Math.* **21** (1976), 293-329.

Montgomery, D., and Zippin, L.
1. "Topological Transformation Groups." Wiley (Interscience), New York, 1955.

Moody, R. V.
1. A new class of Lie algebras. *J. Algebra* **10** (1968), 211-230.
2. Euclidean Lie algebras. *Canad. J. Math.* **21** (1969), 1432-1454.

Moore, C. C.
1. Compactifications of symmetric spaces. *Amer. J. Math.* **86** (1964), 201-218.
2. Compactification of symmetric spaces, II. The Cartan domains. *Amer. J. Math.* **86** (1964), 358-378.

Morita, K.
1. On the kernel functions of symmetric domains. Sci. Rep. Tokyo Kyoiku Daigaku, Sect. A 5 (1956), 190-212.

Morozov, V. V.
1. On a nilpotent element of a semisimple Lie Algebra. *Dokl. Akad. Nauk. SSSR* **36** (1942), 83-86.

Mostow, G. D.
1. A new proof of É. Cartan's theorem on the topology of semi-simple groups. *Bull. Amer. Math. Soc.* **55** (1949), 969-980.
2. The extensibility of local Lie groups of transformations and groups on surfaces. *Ann. of Math.* **52** (1950), 606-636.
3. Some new decomposition theorems for semi-simple Lie groups. *Mem. Amer. Math. Soc.* **14** (1955), 31-54.
4. Equivariant embeddings in Euclidean space. *Ann. of Math.* **65** (1957), 432-446.
5. Factor spaces of solvable groups. *Ann. of Math.* **60** (1954), 1-27.
6. "Strong Rigidity of Locally Symmetric Spaces." Ann. Math. Studies, 78, Princeton Univ. Press, 1973.

7. Continuous groups. Encyclopaedia Britannica, 1967.
8. On maximal subgroups of real Lie groups. *Ann. of Math.* **74** (1961), 503-517.
9. Self-adjoint groups. *Ann. of Math.* **62** (1955), 44-55.
10. Covariant fiberings of Klein spaces, I, II. *Amer. J. Math.* **77** (1955), 247-277; **84** (1962), 466-474.

MURAKAMI, S.
1. On the automorphisms of a real semi-simple Lie algebra. *J. Math. Soc. Japan* **4** (1952), 103-133.
2. Supplements and corrections to [1]. *J. Math. Soc. Japan* **5** (1953), 105-112.
3. Sur la classification des algèbres de Lie réelles et simples. *Osaka J. Math.* **2** (1965), 291-307.
4. Plongements holomorphes de domains symétriques. C.I.M.E. Summer Course on Geometry of Bounded Homogeneous Domains. Edizioni Cremonese, Roma, 1968, 281-287.

MYERS, S. B.
1. Riemannian manifolds in the large. *Duke Math. J.* **1** (1935), 39-49.

MYERS, S. B., and STEENROD, N.
1. The group of isometrics of a Riemannian manifold. *Ann. of Math.* **40** (1939), 400-416.

NAKAJIMA, K.
1. Symmetric spaces associated with Siegel domains. *Proc. Japan Acad.* **50** (1974), 188-191.

NAGANO, T.
1. Homogeneous sphere bundles and the isotropic Riemannian manifolds. *Nagoya Math. J.* **15** (1959), 29-55.
2. Transformation groups on compact symmetric spaces. *Trans. Amer. Math. Soc.* **118** (1965), 428-453.

NAGANO, T., and YANO, K.
1. Les champs des vecteurs géodésiques sur les espaces symétriques. *C. R. Acad. Sci. Paris* **252** (1961), 504-505.

NEWLANDER, A., and NIRENBERG, L.
1. Complex analytic coordinates in almost complex manifolds. *Ann. of Math.* **65** (1957), 391-404.

NOMIZU, K.
1. On the group of affine transformation of an affinely-connected manifold. *Proc. Amer. Math. Soc.* **4** (1953), 816-823.
2. Invariant affine connections on homogeneous spaces. *Amer. J. Math.* **76** (1954), 33-65.
3. Reduction theorem for connections and its application to the problem of isotropy and holonomy groups of a Riemannian manifold. *Nagoya Math. J.* **9** (1955), 57-66.
4. "Lie Groups and Differential Geometry." Math. Soc. Japan Publ., Vol. 2, 1956.
5. On infinitesimal holonomy and isotropy groups. *Nagoya Math. J.* **11** (1957), 111-114.
6. On local and global existence of Killing vector fields. *Ann. of Math.* **72** (1960), 105-120.

OCHIAI, T.
1. Transformation groups on Riemannian symmetric spaces. *J. Differential Geometry* **3** (1969), 231-236.

OCHIAI, T., and TAKAHASHI, T.
1. The group of isometries of a left invariant Riemannian metric on a Lie group. *Math. Ann.* **223** (1976), 91-96.

ONIŠČIK, A. L.
1. Lie groups transitive on compact manifolds, I, II, III. *Amer. Math. Soc. Translations* **73** (1968), 59-72; *Mat. Sb.* **116** (1967), 373-388; *Mat. Sb.* **117** (1968), 255-263.

OSHIMA, T.
1. A realization of Riemannian symmetric spaces. *J. Math. Soc. Japan* **30** (1978), 117–132.

OSHIMA, T., and SEKIGUCHI, J.
1. Boundary value problem on symmetric homogeneous spaces. *Proc. Japan Acad.* **53** (1977), 81-83.

OZOLS, V.
1. Critical points of the displacement function of an isometry. *J. Differential Geometry* **3** (1969), 411-432.
2. Clifford translations of symmetric spaces. *Proc. Amer. Math. Soc.* **44** (1974), 169-175.

PAGE, J. M.
1. On the primitive groups of transformations in space of four dimensions. *Amer. J. Math.* **10** (1888), 293-346.

PALAIS, R. S.
1. On the differentiability of isometries. *Proc. Amer. Math. Soc.* **8** (1957), 805-807.
2. Imbeddings of compact differentiable transformation groups in orthogonal representations. *J. Math. Mech.* **6** (1957), 673-678.
3. A definition of the exterior derivative in terms of Lie derivatives. *Proc. Amer. Math. Soc.* **5** (1954), 902-908.
4. Natural operations on differential forms. *Trans. Amer. Math. Soc.* **92** (1959), 125-141.
5. A global formulation of the Lie theory of transformation groups. *Memoirs Amer. Math. Soc.* No. 22m, 1957.

PASIENCIER, S., and WANG, H.-C.
1. Commutators in a semisimple Lie group. *Proc. Amer. Math. Soc.* **13** (1962), 907-913.

PATERA, J., WINTERNITZ, P., and ZASSENHAUS, H.
1. The maximal solvable subgroups of the $\mathbf{SU}(p, q)$ groups and all subgroups of $\mathbf{SU}(2, 1)$. *J. Math. Phys.* **15** (1974), 1378-1393.
2. The maximal solvable subgroups of $\mathbf{SO}(p, q)$ groups. *J. Math. Phys.* **15** (1974), 1932-1938.

PERROUD, M.
1. The maximal solvable subalgebras of the real classical Lie algebras. *J. Math. Phys.* **17** (1976), 1028-1033.

PETER, F., and WEYL, H.
1. Die Vollständigkeit der primitiven Darstellungen einer geschlossenen kontinuierlichen Gruppe. *Math. Ann.* **97** (1927), 737-755.

PFLUGER, A.
1. "Theorie der Riemannschen Flächen." Springer, Berlin, 1959.

POINCARÉ, H.
1. Sur les groupes continus. *Cambr. Phil. Trans.* **18** (1899), 220-255.
2. Rapport sur les travaux de M. Cartan. *Acta Math.* **38** (1921), 137-145.

PONTRJAGIN, L. S.
1. "Topological Groups," 2nd ed. Moscow, 1954; German transl. Teubner, Leipzig, 1957, 1958.

PRASAD, G.
1. Strong rigidity of Q-rank 1 lattices. *Invent. Math.* **21** (1973), 255-286.

PYATETZKI-SHAPIRO, I.
1. On a problem proposed by É. Cartan. *Dokl. Akad. Nauk. SSSR* **124** (1959), 272-273.
2. Geometry and classification of homogeneous bounded domains. *Russian Math. Surv.* **20** (1966), 1-48.
3. Arithmetic groups in complex domains. *Uspehi Mat. Nauk.* **19** (1964), 93-121.

RAGUNATHAN, M. S.
1. "Discrete Subgroups of Lie Groups." Springer, New York, 1972.

RAMANUJAM, S.
1. Morse theory of certain symmetric spaces. *J. Differential Geometry* **3** (1969), 213-230.
2. Topology of classical groups. *Osaka J. Math.* **6** (1969), 243-249.

RASEVSKII, P. K.
1. On the geometry of homogeneous spaces. *Trudy Sem. Vektor. Tenzor. Anal.* **9** (1952), 49-74.
2. On real cohomologies of homogeneous spaces. *Uspehi Mat. Nauk.* **24** (1969), 23-90.
3. A theorem on the connectedness of a subgroup of a simply connected Lie group commuting with any of its automorphisms. *Trans. Moscow Math. Soc.* **30** (1974), 1-24.

RAUCH, H. E.
1. Geodesics, symmetric spaces and differential geometry in the large. *Comment. Math. Helv.* **27** (1953), 294-320.
2. The global study of geodesics in symmetric and nearly symmetric Riemannian manifolds. *Comment. Math. Helv.* **35** (1961), 111-125.
3. "Geodesics and Curvature in Differential Geometry in the Large." Yeshiva Univ., New York, 1959.
4. A contribution to differential geometry in the large. *Ann. of Math.* **54** (1951), 38-55.

RICHARDSON, R.
1. Compact real forms of a complex semisimple Lie algebra. *J. Differential Geometry* **2** (1968), 411-420.

RIEMANN, B.
1. "Über die Hypothesen, welche der Geometrie zu Grunde liegen." Habilitationvorlesung, 1854; Gesamm. Werke, 2nd ed., Leipzig, 1892.

RINOW, W.
1. Über die Zusammenhänge zwischen der Differentialgeometrie im grossen und im kleinen. *Math. Z.* **35** (1932), 512-538.

ROSSMANN, W.
1. The structure of semisimple symmetric spaces. Queens University, 1977, (preprint).

ROTHAUS, O. S.
1. Domains of positivity. *Abh. Math. Sem. Univ. Hamburg* **24** (1960), 189-235.
2. The construction of homogeneous convex cones. *Ann. of Math.* **83** (1966), 358-376; correction, *Ann. of Math.* **87** (1968), 399.

ROTHSCHILD, L. P.
1. Invariant polynomials and conjugacy classes of real Cartan subalgebras. *Indiana Univ. Math. J.* **21** (1971), 115-120.
2. On the uniqueness of quasi-split real semisimple Lie algebras. *Proc. Amer. Math. Soc.* **24** (1970), 6-8.

ROZENFELD, B. A.
1. "Non-Euclidean Geometries." Moscow, 1955.
2. A geometric interpretation of symmetric spaces with simple fundamental group. *Dokl. Akad. Nauk. SSSR* **110** (1956), 23-26.
3. On the theory of symmetric spaces of rank one. *Mat. Sb.* **41** (83) (1957), 373-380.

SABININ, L. V.
1. Trisymmetric spaces with simple compact groups of motions. *Trudy Sem. Vektor. Tenzor. Anal.* **16** (1972), 202-226.

SAGLE, A., and WALDE, R.
1. "Introduction of Lie Groups and Lie Algebras." Academic Press, New York, 1973.

SAMELSON, H.
1. A note on Lie groups. *Bull. Amer. Math. Soc.* **52** (1946), 870-873.
2. Topology of Lie groups. *Bull. Amer. Math. Soc.* **58** (1952), 2-37.
3. On curvature and characteristic of homogeneous spaces. *Michigan Math. J.* **5** (1958), 13-18.
4. "Notes on Lie Algebras." Van Nostrand Reinhold, Princeton, New Jersey, 1969.

SATAKE, I.
1. On a theorem of É. Cartan. *J. Math. Soc. Japan* **2** (1951), 284-305.
2. A remark on bounded symmetric domains. *Sci. Papers College Gen. Ed. Univ. Tokyo* **3** (1953), 131-144.
3. On representations and compactifications of symmetric Riemann spaces. *Ann. of Math.* **71** (1960), 77-110.
4. On compactifications of the quotient space for arithmetically defined discontinuous groups. *Ann. of Math.* **72** (1960), 555-580.
5. "Classification Theory of Semisimple Algebraic Groups." Dekker, New York, 1971.
6. Holomorphic imbeddings of symmetric domains into a Siegel space. *Amer. J. Math.* **87** (1965), 425-461.
7. A note on holomorphic imbeddings and compactification of symmetric domains. *Amer. J. Math.* **90** (1968), 231-247.

SCHOUTEN, J. A.
1. "Ricci Calculus." 2nd ed. Springer, Berlin, 1954.

SCHREIER, O.
1. Abstrakte kontinuierliche Gruppen. *Abh. Math. Sem. Univ. Hamburg* **4** (1925), 15-32.

SCHUR, F.
1. Neue Begründung der Theorie der endlichen Transformationsgruppen. *Math. Ann.* **35** (1890), 161-197.
2. Zur Theorie der endlicher Transformationsgruppen, *Math. Ann.* **38** (1891), 273-286.
3. Über den analytischen Character der eine endliche continuierliche Transformationsgruppe darstellende Funktionen. *Math. Ann.* **41** (1893), 509-538.

SCHUR, I.
1. Neue Anwendungen der Integralrechnung auf Probleme der Invariantentheorie. *Sitzungsber. Preuss. Akad. Wiss.* (1927), 189, 297, 346.

SEIFERT, H.
1. Zum Satz von O. Bonnet über den Durchmesser einer Eifläche. *Math. Z.* **77** (1961), 125-130.

SEIFERT, H., and THRELFALL, W.
1. "Lehrbuch der Topologie." Teubner, Leipzig, 1934.

SELBERG, A.
1. On discontinuous groups in higher-dimensional symmetric spaces. Internat. Colloq. Function Theory, Bombay, 1960, 147-164; Tata Institute of Fundamental Research, Bombay, 1960.
2. Recent developments in the theory of discontinuous groups of motions of symmetric spaces. *Proc. 15th Scand. Congr. Oslo*, 1968; Lecture notes 118, Springer, New York, 1970, 99-120.

SELIGMAN, G. B.
1. "Modular Lie Algebras." Springer, New York, 1967.

SÉMINAIRE "SOPHUS LIE"
1. "Théorie des algèbres de Lie, Topologie des groupes de Lie." École Norm. Sup. Paris, 1955.

SERRE, J. P.
1. "Algèbres de Lie complexes." Benjamin, New York, 1966.
2. "Lie Algebras and Lie Groups." Benjamin, New York, 1965.

SHAPIRO, R. A.
1. Pseudo-Hermitian symmetric spaces. *Comment. Math. Helv.* **46** (1971), 529-548.

SIEGEL, C. L.
1. Symplectic geometry. *Amer. J. Math.* **65** (1943), 1-86.
2. "Analytic functions of several complex variables." Inst. Adv. Study, Princeton, New Jersey, 1949.

SIMONS, J.
1. On the transitivity of holonomy systems. *Ann. of Math.* **76** (1962), 213-234.

SINGER, I. M.
1. Infinitesimally homogeneous spaces. *Comm. Pure Appl. Math.* **13** (1960), 685-697.

SINYUKOV, N. S.
1. On geodesic mappings of Riemannian spaces onto symmetric Riemannian spaces. *Dokl. Akad. Nauk. SSSR* **98** (1954), 21-23.

SIROKOV, P. A.
1. On a certain type of symmetric spaces. *Mat. Sb.* **41** (83) (1957), 361-372.

SIROTA, A. I., and SOLODOVNIKOV, A. S.
1. Non-compact semisimple Lie groups. *Uspehi Mat. Nauk.* **18** (1963), 87-144.

SOLOMON, L.
1. Invariants of finite reflection groups. *Nagoya Math. J.* **22** (1963), 57-64.

SPIVAK, M.
1. "A Comprehensive Introduction to Differential Geometry I." Publish or Perish, Boston, 1970.

STEENROD, N.
1. "Topology of Fibre Bundles." Princeton Univ. Press, Princeton, New Jersey, 1951.

STEINBERG, R.
1. Finite reflection groups. *Trans. Amer. Math. Soc.* **91** (1959), 493-504.

2. "Lectures on Chevalley Groups." Yale University, New Haven, Connecticut, 1967.
3. Automorphisms of classical Lie algebras. *Pacific J. Math.* **11** (1961), 1119-1129.

STERNBERG, S.
1. "Lectures on Differential Geometry." Prentice-Hall, Englewood Cliffs, New Jersey, 1964.

STIEFEL, E.
1. Kristallographische Bestimmung der Charaktere der geschlossenen Lie'schen Gruppen. *Comment. Math. Helv.* **17** (1944-1945), 165-200.
2. Über eine Beziehung zwischen geschossenen Lie'schen Gruppen und diskontinuierlichen Bewegungsgruppen euklidischer Räume und ihre Anwendung auf die Aufzählung der einfachen Lie'schen Gruppen. *Comment. Math. Helv.* **14** (1941-1942), 350-380.

STRUIK, D. J.
1. "Grundzüge der Mehrdimensionalen Differentialgeometrie in direkter Darstellung." Springer, Berlin, 1922.

SUGIURA, M.
1. Conjugate classes of Cartan subalgebras in real semi-simple Lie algebras. *J. Math. Soc. Japan* **11** (1959), 374-434; correction, *J. Math. Soc. Japan* **23** (1971), 374-383.
2. Classification of root systems with involutions and real simple Lie algebras. Appendix in Satake [5].
3. "Unitary Representations and Harmonic Analysis—An Introduction." Kodansha, Tokyo and Wiley, New York, 1975.

SULANKE, R., and WINGEN, P.
1. "Differential geometrie und Faserbündel." VEB Deutscher Verlag der Wissenschaften, Berlin, 1972.

SUMITOMO, T.
1. Projective and conformal transformations in compact Riemannian manifolds. *Tensor (N.S.)* **9** (1959), 113-135.
2. On a certain class of Riemannian homogeneous spaces. *Colloq. Math.* **26** (1972), 129-133.

TAI, S. S.
1. Minimum imbeddings of compact symmetric spaces of rank one. *J. Differential Geometry* **2** (1968), 55-66.

TAKAHASHI, R.
1. Sur les représentations unitaires des groupes de Lorentz généralisés. *Bull. Soc. Math. France* **91** (1963), 289-433.

TAKEUCHI, M.
1. On the fundamental group and the group of isometries of a symmetric space. *J. Fac. Sci. Univ. Tokyo* **X** (1964), 88-123.
2. On Pontrjagin classes of compact symmetric spaces. *J. Fac. Sci. Univ. Tokyo* (1962), 313-328.
3. Cell decompositions and Morse equalities on certain symmetric spaces. *J. Fac. Sci. Univ. Tokyo* (1965), 81-192.
4. On the fundamental group of a simple Lie group. *Nagoya Math. J.* **40** (1970), 147-159.
5. On orbits in a compact Hermitian symmetric space. *Amer. J. Math.* **90** (1968), 657-680.
6. Nice functions on symmetric spaces. *Osaka J. Math.* **6** (1969), 283-289.

TAKEUCHI, M., and KOBAYASHI, S.
1. Minimal imbeddings of *R*-spaces. *J. Differential Geometry* **2** (1968), 203-215.

TELEMAN, C.
1. Sur une classe d'espaces riemanniens symétriques. *Rev. Math. Pures Appl.* **2** (1957), 445-470.

TILGNER, H.
1. On the use of globally symmetric pseudo-Riemannian spaces in cosmology. *Rep. Mathematical Phys.* **5** (1974), 51-64.

TITS, J.
1. Sur certains classes d'espaces homogènes de groupes de Lie. *Acad. Roy. Belg. Cl. Sci. Mem. Coll.* **29** (1955), no. 3.
2. Automorphismes à déplacement borné des groupes de Lie. *Topology* **3** (1964), 97-107.
3. Espaces homogènes complexes compacts. *Comment. Math. Helv.* **37** (1962), 111-120.
4. "Tabellen zu den einfachen Lie Gruppen und ihre Darstellungen." Lecture Notes no. 40, Springer-Verlag, Heidelberg, 1967.
5. "Buildings of spherical type and finite *BN*-pairs." Lecture Notes in Math. 386, Springer, New York, 1974.
6. Sur les constantes de structure et le théorème d'existence des algèbres de Lie semi-simples. *I.H.E.S. Publ. Math.* **31** (1966).

TOPOGONOV, V. A.
1. A certain characteristic property of a four dimensional symmetric space of rank 1. *Sibirsk. Mat. Ž.* **13** (1972), 884-902.

VAN DANTZIG, D., and VAN DER WAERDEN, B. L.
1. Über metrische homogene Räume. *Abh. Math. Sem. Univ. Hamburg* **6** (1928), 367-376.

VAN DER WAERDEN, B. L.
1. Die Klassifizierung der einfachen Lie'schen Gruppen. *Math. Z.* **37** (1933), 446-462.

VAN EST, W. T.
1. Dense imbeddings of Lie groups. *Nederl. Akad. Wetensch., Proc. Ser. A* **54** (1951), 321-328.

VARADARAJAN, V. S.
1. "Lie Groups, Lie Algebras and their Representations." Prentice Hall, Englewood Cliffs, New Jersey, 1974.
2. "Harmonic Analysis on Real Reductive Groups." Lecture Notes in Math. No. 576, Springer Verlag, New York, 1977.

VARMA, H. O. SINGH
1. Two-point homogeneous manifolds. *Proc. Akad. Wetenschappen, Amsterdam* **68** (1965), 746-753.

VINBERG, E. B.
1. Theory of homogeneous convex cones. *Trudy Moscov. Mat. Obšč.* **12** (1963), 303-358.
2. Invariant linear connections in a homogeneous space. *Trudy Moscov. Mat. Obšč.* **9** (1960), 191-210.
3. Morozov-Borel theorem for real Lie groups. *Dokl. Akad. Nauk. SSSR* **141** (1961), 270-273.
4. Discrete groups generated by reflection in Lobatchevski spaces. *Mat. Sb.* **114** (1967), 471-488; correction, *Mat. Sb.* **115** (1967), 303.

5. The Weyl group of a graded Lie algebra. *Izv. Akad. Nauk. SSSR Ser. Mat.* **40** (1976), 488-526.
6. Discrete groups generated by reflections. *Izv. Akad. Nauk. SSSR Ser. Mat.* **35** (1971), 1072-1112.

Von Neumann, J.
1. Über die analytischen Eigenschaften von Gruppen linearen Transformationen und ihrer Darstellungen. *Math. Z.* **30** (1929), 3-42.

Vrànceanu, G.
1. Sur une classe d'espaces symétriques. "Bericht von der Riemann-Tagung des Forschungsinstituts fur Mathematik," pp. 112-123. Akademie-Verlag, Berlin, 1957.

Wallach, N.
1. "Harmonic Analysis on Homogeneous Spaces." Dekker, New York, 1973.
2. On maximal subsystems of root systems. *Canad. J. Math.* **20** (1968), 555-574.
3. Compact homogeneous Riemannian manifolds with strictly positive curvature. *Ann. of Math.* **96** (1972), 277-295.
4. Minimal immersions of symmetric spaces into spheres. Symmetric Spaces. Short courses at Washington Univ., Dekker, New York, 1972, 1-40.

Wan, Z. X.
1. "Lie algebras." Pergammon Press, New York, 1975.

Wang, H. C.
1. Two-point homogeneous spaces. *Ann. of Math.* **55** (1952), 177-191.
2. Discrete nilpotent subgroups of Lie groups. *J. Differential Geometry* **3** (1969), 481-492.
3. Root systems and Abelian subgroups of compact Lie groups. *Proc. U.S.-Japan Seminar on Differential Geometry, Kyoto*, 1965, 151-152.
4. Closed manifolds with a homogeneous complex structure. *Amer. J. Math.* **76** (1954), 1-32.

Warner, F.
1. "Foundations of Differentiable Manifolds and Lie Groups." Scott Foresman, Glenview, Illinois, 1970.

Warner, G.
1. "Harmonic Analysis on Semisimple Lie Groups I, II." Springer, Berlin, 1972.

Weil, A.
1. "L'intégration dans les groupes topologiques et ses applications." Hermann, Paris, 1940.
2. "Variétés Kähleriennes." Hermann, Paris, 1958.
3. Algebras with involutions and the classical groups. *J. Indian Math. Soc.* **24** (1960), 589-623.
4. On discrete subgroupes of Lie groups, I, II. *Ann. of Math.* **72** (1960), 369-384; **75** (1962), 578-602.

Weyl, H.
1. Theorie der Darstellung kontinuierlicher halbeinfacher Gruppen durch lineare Transformationen, I, II, III, und Nachtrag. *Math. Z.* **23** (1925), 271-309; **24** (1926), 328-376, 377-395, 789-791.
2. The structure and representations of continuous groups. Inst. Adv. Study, Princeton, New Jersey. Notes. 1935.
3. "The Classical Groups." Princeton Univ. Press, Princeton, New Jersey, 1939.

WHITEHEAD, J. H. C.
1. Affine spaces of paths which are symmetric about each point. *Math. Z.* **35** (1932), 644-659.
2. Convex regions in the geometry of paths. *Quart. J. Math.* **3** (1932), 33-42.
3. On the covering of a complete space by the geodesics through a point. *Ann. of Math.* **36** (1935), 679-704.
4. On the decomposition of an infinitesimal group. *Proc. Camb. Phil. Soc.* **32** (1936), 229-237.
5. Obituary: Élie Joseph Cartan 1869-1951. Obit. Notices Roy. Soc. London 8, 71-95, 1952.

WITT, E.
1. Treue Darstellung Lie'scher Ringe. *J. Reine Angew. Math.* **177** (1937), 152-166.
2. Spiegelungsgruppen und Aufzählung halb-einfacher Lie'scher Ringe. *Abh. Math. Sem. Univ. Hamburg* **14** (1941), 289-322.

WOLF, J. A.
1. Homogeneous manifolds of constant curvature. *Comment. Math. Helv.* **36** (1961), 112-147.
2. Locally symmetric homogeneous spaces. *Comment. Math. Helv.* **37** (1962-1963), 65-101.
3. Fine structure of Hermitian symmetric spaces. *In* "Symmetric Spaces," pp. 271-357. Dekker, New York, 1972.
4. "Spaces of Constant Curvature." McGraw-Hill, New York, 1967.
5. Discrete groups, symmetric spaces, and global holonomy. *Amer. J. Math.* **84** (1962), 527-542.
6. Geodesic spheres in Grassmann manifolds. *Illinois J. Math.* **7** (1963), 425-446.
7. Elliptic spaces in Grassmann manifolds. *Illinois J. Math.* **7** (1963), 447-462.
8. On the classification of Hermitian symmetric spaces. *J. of Math. Mech.* **13** (1964), 489-496.
9. The action of a real semisimple Lie group on a compact flag manifold I: Orbit structure and holomorphic arc components. *Bull. Amer. Math. Soc.* **75** (1969), 1121-1137.
10. Symmetric spaces which are real cohomology spheres. *J. Differential Geemetry* **3** (1969), 59-68.
11. Curvature in nilpotent Lie groups. *Proc. Amer. Math. Soc.* **15** (1964), 271-274.

WOLF, J., and GRAY, A.
1. Homogeneous spaces defined by Lie group automorphisms I; II. *J. Differential Geometry* **2** (1968), 77-159.

WOLF, J. A., and KORÁNYI, A.
1. Generalized Cayley transformations of bounded symmetric domains. *Amer. J. Math.* **87** (1965), 899-939.

YAMABE, H.
1. On an arcwise connected subgroup of a Lie group. *Osaka Math. J.* **2** (1950), 13-14.
2. Generalization of a theorem of Gleason. *Ann. of Math.* **58** (1953), 351-365.

YAN, ZHI-DA
1. Sur les espaces symétriques non compacts. *Sci. Sinica* **14** (1965), 31-38.
2. On the quasi-inner automorphisms of a semisimple real Lie algebra. *Acta Math. Sinica* **14** (1964), 387-391.

YANO, K.
1. "The Theory of Lie Derivatives and Its Applications." North-Holland, Amsterdam, and Wiley (Interscience), New York, 1957.
2. "Differential Geometry on Complex and Almost Complex Spaces." Pergamon Press, New York, 1965.

YEN, C. T.
1. Sur les représentations linéaires de certains groupes et les nombres de Betti des espaces homogènes symétriques. *C. R. Acad. Sci. Paris* **228** (1949), 1367-1369.

YOSIDA, K.
1. A theorem concerning the semisimple Lie groups. *Tohoku Math. J.* **43** (1937), 81-84.

ŽELOBENKO, D. P.
1. "Compact Lie Groups and their Representations." Amer. Math. Soc., Providence, Rhode Island, 1973.

ZILLER, W.
1. The free loop space of globally symmetric spaces. *Invent. Math.* **41** (1977), 1-22.

LIST OF NOTATIONAL CONVENTIONS

I. Set theory. Let A and B be sets. The symbol $A \subset B$ means that A is a subset of B. If $A \subset B$ and $A \neq B$, then A is called a *proper* subset of B. The empty set is denoted by $\emptyset$. The set $A - B$ is the set of elements in A not in B. The symbols $\cap, \cup$, respectively, denote intersection and union of sets. The symbol $x \in A$ ($x \notin A$) means that x is (x is not) an element of the set A. The subset of A consisting of x alone is denoted $\{x\}$. If M and N are sets the symbol $f\colon M \to N$ means a mapping of M into N. If $M \subset N$ and $f(m) = m$ for all $m \in M$, f is called the *identity mapping* of M into N and is denoted by I or 1. If $f\colon M \to N$ and $g\colon N \to P$, then the mapping which assigns to every $m \in M$ the element $g(f(m)) \in P$ is denoted $g \circ f$. If $f\colon M \to N$ and $A \subset N$, then $f^{-1}(A)$ denotes the set of elements $m \in M$ for which $f(m) \in A$. If $\mathscr{P}$ is a property and M a set then $\{x \in M : x \text{ has property } \mathscr{P}\}$ denotes the set of $x \in M$ with property $\mathscr{P}$. Thus $f^{-1}(A) = \{m \in M : f(m) \in A\}$. The sign $\Rightarrow$ means "implies." In order to save parentheses, the image $f(m)$ of m under a mapping f will sometimes be denoted $f \cdot m$. A mapping $f\colon M \to N$ is said to be *one-to-one* if $m_1 \neq m_2 \Rightarrow f(m_1) \neq f(m_2)$. If $f(M) = N$, f is said to map M *onto* N ("f is onto").

II. Algebra. The identity element of a group will usually be denoted by e. If K is a subgroup of a group G, the symbol G/K denotes the set of left cosets gK, $g \in G$. When K is considered as an element in G/K it will sometimes be denoted by $\{K\}$. If $x \in G$, the mapping $gK \to xgK$ of G/K onto itself will be denoted by $\tau(x)$.

By a *field* we shall always mean a commutative field of characteristic 0. Let V be a vector space over a field K. The *dual space*, consisting of all linear mappings of V into K, is denoted by V^* or $V^\wedge$. The dimension of the vector space V is denoted by $\dim V$ or $\dim_K V$. If $\dim V < \infty$, then $(V^*)^*$ can be identified with V. If $e_1, \ldots, e_n$ is a basis of V and $f_1, \ldots, f_n$ are linear mappings of V into K such that $f_i(e_j) = \delta_{ij}$ (Kronecker symbol), then $f_1, \ldots, f_n$ is called the *basis* of V^* *dual* to $e_1, \ldots, e_n$. Let W be a subspace of V. A *basis of* V (mod W) is a set of elements in V which together with a basis of W constitute a basis of V. The number of elements in a basis of V (mod W) is called the *codimension* of W. If A and B are subspaces of a vector space such that each $v \in V$ can be written $v = a + b$ where $a \in A$, $b \in B$, then we write $V = A + B$. If, in addition $A \cap B = \{0\}$, V is called the *direct sum* of A and B and the subspace B is said to be *complementary* to A. If V and W are vector spaces over the same field K, then the *product* of V and W, denoted $V \times W$, is the set of all pairs (v, w) where $v \in V$, $w \in W$, turned into a vector space over K by the rules

$$(v, w) + (v', w') = (v + v', w + w'), \quad \alpha(v, w) = (\alpha v, \alpha w), \quad \alpha \in K.$$

The subsets $\{(v, 0): v \in V\}$ and $\{(0, w): w \in W\}$ are subspaces of $V \times W$, isomorphic to V and W, respectively, and $V \times W$ is the direct sum of those subspaces.

If $v \in V$, $v^* \in V^*$, then the value $v^*(v)$ will sometimes be denoted by $\langle v, v^* \rangle$. Let A be a linear mapping of a vector space V into a vector space W over the same field. The *transpose* of A (the *dual* of A), denoted ${}^tA$, is the linear map $W^* \to V^*$ determined by $\langle Av, w^* \rangle = \langle v, {}^tAw^* \rangle$. A linear map $A: V \to V$ will often be called *endomorphism* of V. If V has finite dimension, the *determinant* and *trace* of A will be denoted by $\det(A)$ and $\mathrm{Tr}\,(A)$, respectively.

Let V and W be vector spaces over the same field K. A *bilinear form* on $V \times W$ is a mapping $B: V \times W \to K$ such that for each $v \in V$, the mapping $B_v: w \to B(v, w)$ belongs to W^* and such that for each $w \in W$, the mapping $B^w: v \to B(v, w)$ belongs to V^*. Thus a bilinear form on $V \times W$ gives rise to linear mappings $V \to W^*$ and $W \to V^*$. The bilinear form B is called *nondegenerate* if $v \neq 0$ implies $B_v \not\equiv 0$ and if $w \neq 0$ implies $B^w \not\equiv 0$. The set of all bilinear forms on $V \times W$ is a vector space over K whose dual is denoted $V \otimes W$ and called the *tensor product* of V and W. Each element $(v, w) \in V \times W$ gives rise to an element $v \otimes w$ in $V \otimes W$ determined by $(v \otimes w)(B) = B(v, w)$. The direct sum $K + V + V \otimes V + V \otimes V \otimes V + \ldots$ is an associative algebra, the *tensor algebra* $T(V)$ over V, the multiplication being $\otimes$.

Let $\boldsymbol{R}$ and $\boldsymbol{C}$, respectively, denote the fields of real and complex numbers. Let $\boldsymbol{Z}$ denote the ring of integers. Let V be a vector space over $\boldsymbol{R}$. A bilinear form B on $V \times V$ is called *symmetric* if $B(v, v') = B(v', v)$ for $v, v' \in V$, *positive definite* if $B(v, v) \geqslant 0$ for $v \in V$, *strictly positive definite* if $B(v, v) > 0$ for $v \neq 0$ in V. Let W be a vector space over $\boldsymbol{C}$. A mapping $B: W \times W \to \boldsymbol{C}$ is called a *Hermitian form* if for each $w_o \in W$ the mapping $w \to B(w, w_o)$ is linear and if for each pair $(w', w'') \in W \times W$ the numbers $B(w', w'')$ and $B(w'', w')$ are conjugate complex numbers.

By a *ring* we shall always mean a commutative ring with an identity element. Let A be a ring. A commutative group M is called a *module* over A (or an A-module) if for each $a \in A$ and $m \in M$ an element am is defined such that

$$a(m_1 + m_2) = am_1 + am_2, \quad (a_1 + a_2)m = a_1m + a_2m,$$
$$(a_1a_2)m = a_1(a_2m), \quad 1m = m.$$

A subset $N \subset M$ such that $n_1, n_2 \in N$ implies $n_1 + n_2 \in N$, $an_1 \in N$ for each $a \in A$ is called a *submodule* of M.

A vector space V over a field K is called an (associative) *algebra* (over K) if there exists a multiplication in V with the properties: $\alpha(v_1v_2) = (\alpha v_1)v_2 = v_1(\alpha v_2)$, $1v = v$, $(v_1v_2)v_3 = v_1(v_2v_3)$, $v_1(v_2 + v_3) = v_1v_2 + v_1v_3$, $(v_1 + v_2)v_3 = v_1v_3 + v_2v_3$ for $\alpha \in K$, $v, v_1, v_2, v_3 \in V$. Let A be in algebra (with identity 1) over a field K and V a vector space over K. A *representation* of A on V is a homomorphism ρ of A into the algebra of all endomorphisms of V such that $\rho(1) = I$.

III. Topology. A topological space shall always mean a topological space in which the Hausdorf separation axiom holds. Let M be a topological space. A collection $\{U_\alpha\}$, $(\alpha \in A)$ of open subsets of M is called a *basis for the open sets* if each open set can be written as a union of some U_α. If $p \in M$, a *neighborhood* of p is a subset of M containing an open subset of M containing p. A *fundamental system of neighborhoods* of p is a system $\{N_\alpha\}_{\alpha \in A}$ of neighborhoods of p such that each neighborhood of p contains some N_α. A topological space is called *separable* if it has a countable dense subset. For metric spaces, separability is equivalent to the existence of a countable basis for the open sets. A topological space is called *compact* if each open covering has a finite subcovering. A subset of a topological space is called *relatively compact* if its closure is compact. A mapping $f: M \to N$ of a topological space M onto a topological space N is called a *local homeomorphism* of M onto N if each point $m \in M$ has an open neighborhood which f maps homeomorphically onto an open neighborhood of $f(m)$ in N. A *domain* in a topological space is an open connected subset. A *path* (or a *continuous curve*) in a topological space is a continuous mapping of a closed interval $[a, b]$ into the space. A space is called *pathwise connected* if any two points in the space can be joined by means of a path. A topological space is said to be *locally connected* (*locally pathwise connected*) if each neighborhood of any point p in the space contains a connected (pathwise connected) neighborhood of p.

SYMBOLS FREQUENTLY USED

In addition to the preceding conventions the list below contains many of the symbols whose meaning is usually fixed throughout the book. The symbols from Chapter V, §5 have not been relisted.

$\mathfrak{a}$, $\mathfrak{a}_*$: maximal abelian subspaces, 385, 401, 319
$\mathfrak{a}_e$: unit lattice, 319
$\mathfrak{a}_K$, $\mathfrak{a}_\Sigma$: lattices for K and for Σ: 322, 321
$\mathfrak{a}_\lambda$: line $\boldsymbol{R}iA_\lambda$, 336
$\mathfrak{a}_r$: complement $\mathfrak{a}_* - D(U, K)$, 319
$a_{\beta,\alpha}$: Killing-Cartan integer, 456
$\mathfrak{a}^+$: positive Weyl chamber, 402
$A^+(g)$: component in Cartan decomposition, 402
$\mathfrak{a}_l$: classical Lie algebra, 186
ad: adjoint representation of a Lie algebra, 99
Ad: adjoint representation of a Lie group, 127
$A(M)$, $A_0(M)$: group of holomorphic isometries, and its identity component, 372
Aut($\mathfrak{a}$): group of automorphisms of $\mathfrak{a}$, 126
Aut(Δ): group of automorphisms of a root system Δ, 421
$\mathfrak{A}(M)$: Grassmann algebra, 17
$\mathfrak{A}_s(M)$: set of s-forms on M, 17
$B_r(p)$: open ball with center p, radius r, 51
$B(X, Y)$: Killing form, 131
B_θ: modified Killing form, 253
$B(g)$: component in Bruhat decomposition, 407
$\mathfrak{b}_l$: classical Lie algebra, 186
$\boldsymbol{C}^n$: complex n-space, 4
C^∞: indefinitely differentiable, 4, 6
$C^\infty(M)$, $C_c^\infty(M)$: set of differentiable functions, set of differentiable functions of compact support, 6
$\mathscr{C}^0\mathfrak{g} \supset \mathscr{C}^1\mathfrak{g} \supset \mathscr{C}^2\mathfrak{g} \supset \cdots$: central descending series, 161
c_{jk}^i: structural constants, 137
$\mathfrak{c}_l$: classical Lie algebra, 186
Card: cardinality, 531
γ_X: maximal geodesic determined by X, 31
d, δ: exterior differentiation and its adjoint, 20, 143
$\boldsymbol{D}(G)$: algebra generated by $\tilde{\mathfrak{g}}$, 107
$d\Phi_p$: differential of Φ at p, 22
$D(U, K)$, $D(\mathfrak{u}, \theta)$: diagram of (U, K), 295
$D(U)$, $D(\mathfrak{u})$: diagram of U, 299, 300
$\mathfrak{D}\mathfrak{g}$: derived algebra of $\mathfrak{g}$, 158
$\mathfrak{D}^1(M)$, $\mathfrak{D}_1(M)$: set of vector fields (1-forms), 9, 11
$\mathfrak{D}(M)$, $\mathfrak{D}(p)$, $\mathfrak{D}^*(M)$, $\mathfrak{D}^*(p)$, $\mathfrak{D}_*(M)$, $\mathfrak{D}_*(p)$: tensor algebras, 15, 16
$\mathfrak{D}_s^r(M)$: set of tensor fields of type (r, s), 13
$\Delta = \Delta(\mathfrak{g}, \mathfrak{h})$: set of nonzero roots of $\mathfrak{g}$ with respect to $\mathfrak{h}$, 166
Δ^+, $\Delta_\mathfrak{p}$: set of positive roots; set of roots nonzero on $\mathfrak{h}_\mathfrak{p}$, 260
$\bar{\Delta}$, $\bar{\Delta}^0$, $\tilde{\Delta}$, $\tilde{\Delta}^0$, $\tilde{\Delta}^+$: sets of roots, 492, 493

Φ^*T: transform of a covariant tensor field T, 25
R: curvature tensor, 43
$\mathbf{R}^n$: Euclidean n-space, 2
R_ρ, $R(\rho)$: right translation by ρ, 122, 135
$S_r(p)$: sphere with center p, radius r, 51
s_p: geodesic symmetry with respect to p, 198
s_α: the reflection in the plane $\alpha = 0$, 194
$\Sigma(\Sigma^+)$: the set of (positive) restricted roots, 263
$*$: the star operator, 142
Tr: trace, 131
$\mathfrak{t}_0$, $\mathfrak{t}$: maximal abelian subalgebras, 298
$\mathfrak{t}_\alpha$, T_α: the set $\alpha^{-1}(2\pi i \boldsymbol{Z})$ and its centralizer, 301
$\mathfrak{t}_\Delta \subset \mathfrak{t}_e \subset \mathfrak{t}(\mathfrak{u})$: the Δ-lattice, the unit lattice and the central lattice, 314, 305, 311
$\mathfrak{t}_r$: the complement $\mathfrak{t}_0 - D(U)$, 307
T_X: restriction of $(\text{ad } X)^2$, 214
$T(V)$: tensor algebra over V, 100
$\tau(x)$: mapping $gH \to xgH$, 120
$U(\mathfrak{g})$: universal enveloping algebra of $\mathfrak{g}$, 100
$\tilde{U}$, $\tilde{Z}$: compact, simply connected semisimple Lie group and its center, 326
$W(U)$, $W(\mathfrak{u})$, $W(U, K)$, $W(\mathfrak{u}, \theta)$, $W(\mathfrak{g}, \theta)$: Weyl groups, 284, 289, 299, 300, 402
W^c: complexification of W, 179
$\tilde{X}$, $\bar{X}$: left and right invariant vector fields induced by X, 99, 122
X^+, X^*: vector field induced by X, vector field adapted to X, 122, 36
$\mathbf{Z}$, $\mathbf{Z}_\rho$, $\mathbf{Z}^+$: the integers, the integers mod ρ, the nonnegative integers
Z_x: centralizer of x, 301
[,]: bracket, 9, 99
$\wedge$: exterior multiplication, 19
$\langle\ ,\ \rangle$: scalar (inner) product, 166
$\|\ \|$: length of a vector, norm, 48, 338
${}^tA$: transpose of a matrix A, 444
$\oplus$: direct sum
$f \mid S$: restriction of f to S
$\succ$: special ordering, 337

INDEX

G

H

I

J

K

L

R

S

T

Pure and Applied Mathematics

A Series of Monographs and Textbooks

RECENT TITLES

ROBERT A. ADAMS. Sobolev Spaces

JOHN J. BENEDETTO. Spectral Synthesis

D. V. WIDDER. The Heat Equation

IRVING EZRA SEGAL. Mathematical Cosmology and Extragalactic Astronomy

J. DIEUDONNÉ. Treatise on Analysis: Volume II, enlarged and corrected printing; Volume IV; Volume V; Volume VI

WERNER GREUB, STEPHEN HALPERIN, AND RAY VANSTONE. Connections, Curvature, and Cohomology: Volume III, Cohomology of Principal Bundles and Homogeneous Spaces

I. MARTIN ISAACS. Character Theory of Finite Groups

JAMES R. BROWN. Ergodic Theory and Topological Dynamics

C. TRUESDELL. A First Course in Rational Continuum Mechanics: Volume 1, General Concepts

GEORGE GRATZER. General Lattice Theory

K. D. STROYAN AND W. A. J. LUXEMBURG. Introduction to the Theory of Infinitesimals

B. M. PUTTASWAMAIAH AND JOHN D. DIXON. Modular Representations of Finite Groups

MELVYN BERGER. Nonlinearity and Functional Analysis: Lectures on Nonlinear Problems in Mathematical Analysis

CHARALAMBOS D. ALIPRANTIS AND OWEN BURKINSHAW. Locally Solid Riesz Spaces

JAN MIKUSINSKI. The Bochner Integral

THOMAS JECH. Set Theory

CARL L. DEVITO. Functional Analysis

MICHIEL HAZEWINKEL. Formal Groups and Applications

SIGURDUR HELGASON. Differential Geometry, Lie Groups, and Symmetric Spaces

ROBERT B. BURCKEL. An Introduction To Classical Complex Analysis: Volume 1

In preparation

C. TRUESDELL AND R. G. MUNCASTER. Fundamentals of Maxwell's Kinetic Theory of a Simple Monatomic Gas: Treated as a Branch of Rational Mechanics

LOUIS HALLE ROWEN. Polynominal Identities in Ring Theory

JOSEPH J. ROTMAN. An Introduction to Homological Algebra

ROBERT B. BURCKEL. An Introduction To Classical Complex Analysis: Volume 2

A B C 8 D 9 E 0 F 1 G 2 H 3 I 4 J 5